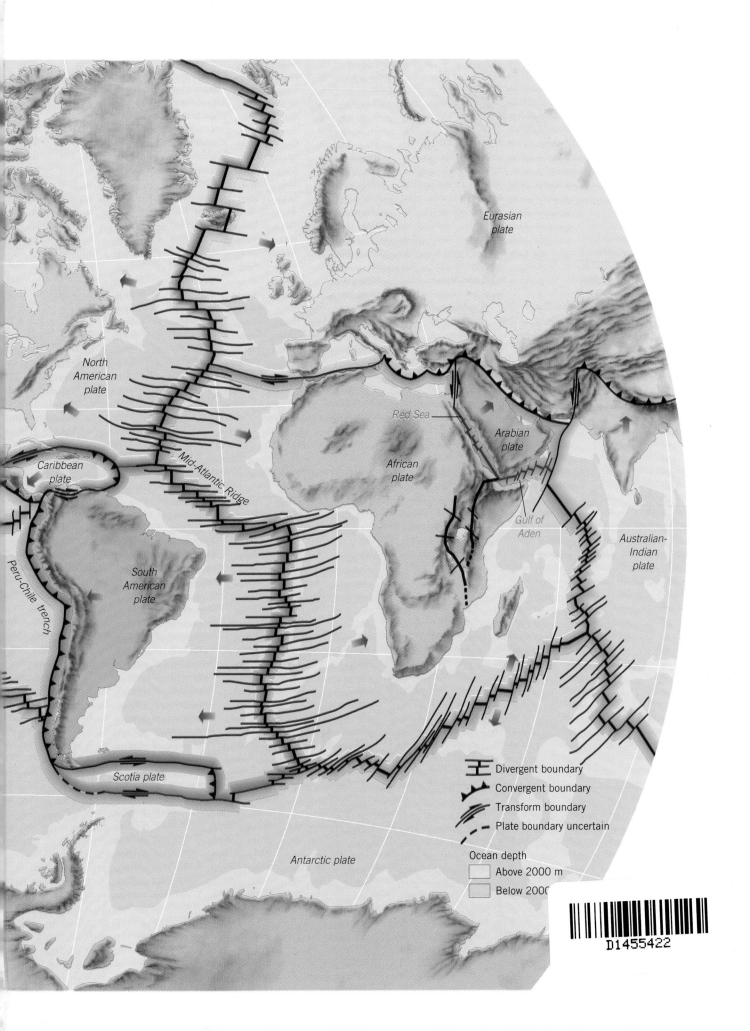

Eurasian
plate

North
American
plate

Red Sea

Arabian
plate

African
plate

Gulf of
Aden

Australian-
Indian
plate

Caribbean
plate

Mid-Atlantic Ridge

Peru-Chile trench

South
American
plate

Scotia plate

Antarctic plate

⊞ Divergent boundary

◄◄◄ Convergent boundary

≡ Transform boundary

╌╌ Plate boundary uncertain

Ocean depth

☐ Above 2000 m

☐ Below 2000

D1455422

PHYSICAL

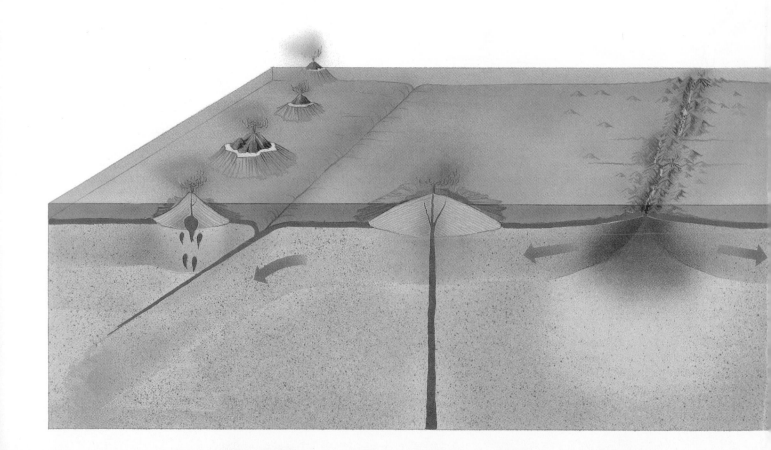

GEOLOGY
UPDATED VERSION

ANATOLE DOLGOFF

City University of New York–New York City Technical College

with

MARY FALCON

HOUGHTON MIFFLIN COMPANY
Boston   New York

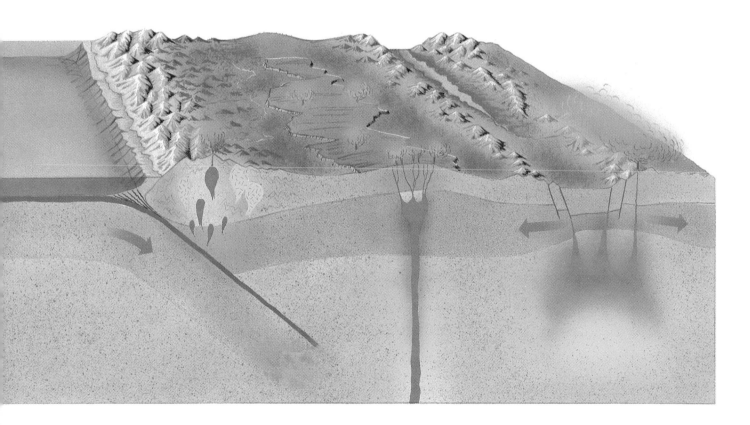

Editor-in-Chief: Kathi Prancan
Project Editor: Elizabeth Gale Napolitano
Senior Production/Design Coordinator: Jill Haber
Manufacturing Coordinator: Sally Culler
Marketing Manager: Karen Natale

Acknowledgments for reprinted material appear beginning on page 619.

Printed in the U.S.A.

Library of Congress Catalog Number: 97-72461

ISBN: 0-669-33923-7

3456789-DS-02

*This book is dedicated to the memory
of my parents,
Esther and Samuel Dolgoff.*

Preface

This textbook is designed for students enrolled in the introductory physical geology course offered at most two- and four-year colleges. This first course typically attracts a variety of students, from geology and other science majors to business, education, and humanities majors fulfilling a science requirement. My purpose in writing this book has been to present geological principles in a manner that conveys the delight that geologists take in their subject. I have been guided by several major goals:

- *To provide basic information about geology in a way that is inviting for students who have probably taken little interest in the subject before this course.* Many students have grown up in an environment in which the indoor shopping mall has almost completely supplanted the natural world. The sense of connection they should feel with the Earth has largely been stifled, and a good geology textbook must work toward motivating students to reestablish this link. To this end, I have woven the dramatic interactions between geological processes and human activities into the very fabric of the book. I have highlighted such timely issues as successful and failed efforts to design earthquake-resistant buildings; the advantages and disadvantages of flood-control measures on the Mississippi River; the impact of human activity on fragile beaches and hillsides; and the threat of human-induced global warming. I have explored such serious environmental problems as famine and land degradation in the Sahel; the undesirable effects of water diversion in arid climates; groundwater pollution and overuse; and subsurface nuclear- and toxic-waste disposal. Throughout the book I have discussed geologists' practical contributions toward solving these problems.

    I have not hesitated to stress that many problems geologists, engineers, and regional planners thought they had overcome or were on the verge of overcoming have turned out to be thornier than they originally believed—for example, predicting earthquakes in light of such recent events as the Northridge and Kobe disasters; constructing artificial levee systems strong enough to withstand the likes of the Mississippi and Sacramento floods; and finding a safe place to bury nuclear waste in view of recent discoveries about the long-term explosive potential of plutonium. Probing these matters emphasizes the challenge of geology and reinforces the concept that science is largely about unfinished business. Indeed, there is much that we leave for the next generation of geologists to accomplish.

- *To maintain the traditional core of the subject while giving proper weight to the advances of recent years.* The list of recent geological advances is virtually endless, and for the author of an introductory textbook, the main problem that arises is determining what to include and what to leave out. The material I have chosen to include strikes a reasonable balance between the theoretical and the practical, between classical ideas and topics now of wide interest. For example, the chapter on earthquakes treats plate tectonics and elastic rebound theory on the one hand, and engineers' new findings as a result of the Mexico City, Northridge, and Kobe disasters on the other.

• *To involve students actively in the scientific process.* For some students, physical geology is only the first of numerous science courses they will take as undergraduates, but for many, it may be the only science studied in college. For this reason, an introductory textbook must demonstrate how scientists look at problems and attempt to solve them.

I have set the tone of science as inquiry in the introductory chapter, where I have posed the intriguing geological question, What is the meaning of the thick salt deposits beneath the floor of the Mediterranean Sea? For more than twenty years, the dominant view has been that 6 million years ago, the Mediterranean dried up and left behind a huge sunken desert. But the emphasis of the discussion is not so much on the answer—which, in any case, is in dispute—but on how geologists have gone about *seeking* the answer.

I have maintained the same inquiry-based approach throughout the book. No topic, no matter how familiar, is presented as a finished piece of business. Rather, I have described explanations and theories as working hypotheses, some more firmly established than others. I have discussed some theories as much for where they can lead the student as for the facts they explain. For example, I have presented the probable Cretaceous meteoroid impact as much for its influence on the traditional concept of uniformitarianism as for its role in the extinction of the dinosaurs.

• *To demonstrate geology's core role in advancing human knowledge.* Finally, I have tried to convey the pivotal knowledge that geology has contributed to human intellectual development. This knowledge includes the concepts of the Earth's vast age and of slow evolutionary change; geologists' explanations of the natural causes of catastrophic events such as earthquakes and volcanoes; and the application of rational methods in evaluating competing ideas.

Organization

Some authors offer a skeletal description of plate tectonics in the first chapter and reserve major discussion of the theory for the end of the book. Others treat it somewhere in the middle. While these traditional approaches have merit, I believe that the introductory physical geology course is most solid when it is built upon a foundation of plate tectonics. Traditional organizations force an author to play catch-up, dragging in various aspects of plate tectonics theory by the hair. Not only does the student get a disjointed view, but by the time the theory is presented, it has lost much of its impact.

In this book I have laid the cornerstone of plate tectonics early. After a brief introductory chapter on the process of scientific inquiry and career opportunities in the geosciences, Chapter 1 takes readers from the basics of classical geology—uniformitarianism, the age of the Earth, and the rock cycle—to continental drift, the precursor of plate tectonics. In Chapter 2, I have employed an inductive, quasi-historical approach in developing the case for plate tectonics, describing the features of the ocean floor and explaining the significance of the supporting evidence as I go along. The chapter concludes with a preview of the impact of plate tectonics on continental evolution. Thus Chapters 1 and 2 give students the solid foundation and context needed to understand subsequent concepts.

Chapters 3 and 4 focus on earthquakes and the Earth's interior. I have included such generally neglected subjects as paleoseismology, mantle convection, and the relationship between theories of the Earth's interior and theories of the Earth's origin. Then, after discussing minerals, rocks, volcanism, geologic time, and rock deformation (Chapters 5–11), I have concluded this section—which is mainly concerned with internal processes—with plate tectonics and continental structure (Chapter 12). I have treated Chapter 12 as a synthesis of Chapters 1 through 11. Students are then in a position to appreciate the surficial processes (Chapters 13–19) that act upon the continental crust.

Chapter 20, on mineral resources, and Chapter 21, on planetary geology, reprise the major themes of the textbook from markedly different perspectives. Chapter 20 describes how natural processes directly affect our material well-being. Although I have examined problems of overconsumption and pollution, I have avoided apocalyptic statements, which I believe have hindered public appreciation of very real environmental issues. Chapter 21 seeks to place the Earth in the context of the geological evolution of the terrestrial planets, emphasizing the reasons why the Earth, the Moon, Mars, and Venus differ from one another.

Students like this sequence of topics—which I have used with minor variations for many years—because it presents the evidence in context. Moreover, earthquakes, continental drift, and plate tectonics intrigue them, so introducing these topics early in the book affords the opportunity to exploit a keen interest. As I have learned in the classroom, the earlier you capture students' interest, the more likely they will engage actively with the rest of the course. Furthermore, three decades have passed since the plate tectonics revolution transformed geology. Scientists almost universally accept the theory, and although they have modified it, no one has seriously challenged its basic premises. The time has come to put plate tectonics up front where it belongs.

I am sensitive, however, to the fact that instructors have personal preferences and practical needs that determine the sequence of topics in their courses. I have therefore made the chapters as self-contained as possible, so as to ensure maximum flexibility of use. For example, there are two traditional ways of presenting surficial processes. One is to place the chapter on weathering before sedimentary rocks; the other is to present it later in the book, as the first chapter in a series presenting gradational processes. I have taken the latter approach because there is more to weathering and soils than simply supplying the raw materials for sediments, and I feel that it is important for students to see the entire gradational sequence. The concepts that students need to know about weathering and sediments are summarized at the beginning of the sedimentary chapter, but they are covered in depth in the weathering chapter. Yet instructors who would prefer to assign the weathering chapter before sedimentary rocks can do so easily and without redundancy.

For those who prefer a traditional approach, Chapter 5, on minerals, can be assigned either immediately before or after Chapter 2, on plate tectonics; and Chapters 3 and 4, on earthquakes and the Earth's interior, can be taught after the mineral and rock chapters (Chapters 5–9) without any loss of coherence. Instructors who opt to cover planetary geology as the introductory topic in the course can assign Chapter 21 early, either before Chapter 1 or after Chapter 2.

Pedagogy

Some introductory physical geology textbooks divide information into short, easily digestible bits, presumably to make the material accessible to students. Without the context that supports the facts, however, students have no choice but to memorize the material; moreover, educational studies have shown that they retain very little of such information after the final exam. In contrast, the chapters in this book are narratives built around themes and concepts; the details and definitions evolve naturally from the flow of the discussion. For example, in Chapter 8, on sedimentary rocks, the characteristics of coarse detrital sediments emerge logically from a description of their weathering, transport, and depositional history as we travel hypothetically from mountaintop to sea basin. Of course, geology is a very visual science, and I have illustrated and reinforced the major themes of this textbook with a rich and varied program of photographs, diagrams, graphs, and maps.

Two special features in each chapter allow students the freedom to read the material for understanding, without the need for underlining (or highlighting) passages and without the annoyance of breaking their train of thought to look up a word elsewhere.

- **Study Outline.** This detailed outline at the end of each chapter covers all the important concepts and uses every key term. The outline provides a more functional cognitive map than the ordinary paragraph-style summary because it visually conveys the relationships between topics and subtopics. Furthermore, in the paragraph format, many extra words are needed to complete full sentences and provide transitions from one sentence to the next. Because the outline format is economical, more information can be packed into the same number of words.
- **Marginal Definitions.** Each study term is defined in the margin next to the place in the text where it is introduced.

Also, to help students master material and study for exams, the textbook includes the following pedagogical devices:

- **Chapter Outline**. Each chapter begins with a preview outline of the headings within the chapter.
- **Study Terms.** This end-of-chapter list of key terms features references to the page where each term is first used and defined.
- **Critical Thinking Questions.** Five to ten questions at the end of each chapter test students' recall *and understanding* of the chapter material.

And in the back of the book:

- **Appendixes.** The appendixes feature a conversion table of metric and English units of measurement, the periodic table of the elements, mineral identification tables, and a basic guide to geologic maps.
- **Glossary.** The glossary provides complete definitions of terms and identifies the text pages where they are introduced.
- **Selected Bibliography.** This brief supplementary reading list, organized by chapter, includes both classic books and current articles.

New to this Update

The major changes in this Updated Version of the book may be divided into two categories: (1) an expanded and more accessible treatment of issues stemming from human interaction with the natural environment and (2) a significant recasting of certain key topics in light of recent discoveries.

Humans and the Natural Environment

For purposes of this book, an environmental "issue" or "problem" is defined as *any human intervention in the natural environment that may cause actual or potential harm to either humans or the natural environment.* The criterion is broad enough to include the material most of us want to teach, but places limits on what is characterized as "environmental." Thus hurricanes and earthquakes themselves are not environmental issues, but groins and retaining walls that strip beaches of their ability to withstand hurricanes—and dangerous construction and zoning practices in earthquake-prone regions—are.

As in the earlier version of this book, environmental discussions appear throughout the text where relevant. However, in this Update, topics relating to the environment are bracketed with icons. For those instructors who wish to emphasize the practical application of geology to environmental issues, the icons point to where such topics begin and where they end.

An innovation I personally am excited about is the Apply and Decide feature interspersed throughout the text. Each Apply and Decide describes how geological information in the chapter can be applied to a better understanding of—and possibly the solution to—an important environmental problem. The purpose is to encourage students to think critically about connections between scientific and social issues by asking: What inferences or conclusions can be drawn from the geological data? What options exist toward solving or ameliorating the problem? What are the social, economic, and political consequences of choosing a given course of action?

For example, Apply and Decide 8.1 describes studies of Guatemalan lake sediments suggesting that severe environmental degradation was a major contributing factor in the sudden collapse of the Mayan civilization. Students are then asked: Are there any parallels concerning our own civilization that can be drawn from the Guatemalan lake studies? How students arrive at the answer to this question is more important than their conclusions, which are always open to debate. At the very least, students will have explored an interesting linkage between sedimentology, human ecology, anthropology, and population dynamics.

Similarly, Apply and Decide 3.1 discusses how seismologists work with engineers to design seismically "safe" buildings. Students are then asked to consider such questions as: How much weight should be given to "expert" opinion? Who decides what is "acceptable" risk? In Apply and Decide 15.1 on the impact of large dams, students are asked to consider such issues as: Who pays? Who benefits? What are the tradeoffs? In 21.1 they are asked whether taxpayers' money should be spent to identify potential meteoroid hazards, and so on.

Updating of Key Topics

This Updated Version also reflects the need to keep pace with fast-breaking developments that have occurred in a number of fields since initial publication in 1995. To this end, for example, I have:

- added a discussion of recent discoveries that the inner core is anisotropic and rotates at a faster rate than the Earth as a whole (Chapter 4, Interior of the Earth).

- extended and revised the treatment of meteoroid impacts. The main innovation is that impacts are treated as a normal geologic process and are referred to as necessary throughout the text. Thus you will find a new Aside feature devoted to impact craters as landforms in Chapter 11, Rock Deformation, and an Apply and Decide discussion of their potential future menace in Chapter 21, Planetary Geology. This is in addition to previous discussions of the role of impacts in mass extinctions (Chapter 10, Geologic Time) and of the role of impact cratering in dating planetary surfaces.

- extensively revised the section describing the geology of Venus (Chapter 21, Planetary Geology). The surface of the planet everywhere bears the imprint of volcanism, but impact cratering data tell us that, unlike the Earth, the entire surface of Venus is 500 million years old. The gradualistic uniformitarian model no longer seems to apply to Venus.

An Essential Alternative

There may be instructors who are interested in the organization and approach of this book but prefer to use an "essentials" text. To better meet their needs, I have prepared a new text entitled *Essentials of Physical Geology*. It is a condensed version of this Updated edition that has been recast for use by nonscience majors in a one-semester or one-quarter course. Designed as well for instructors who include a unit on historical geology in their course, *Essentials of Physical Geology* closes with a four-chapter sequence, which includes Chapters 16 (Geologic Time) and 17 (Formation of the Continental Crust) plus two entirely new chapters: 18 (Earth History: From the Origin of the Planet Through the Proterozoic Eon) and 19 (Earth History: From the Paleozoic Era to the Present).

For those instructors who prefer the Updated Version of this book but also want to include historical geology in their course, the two new Earth history chapters may be packaged with the Updated Version. Contact your local Houghton Mifflin sales representative for further information. To create an expanded unit on historical geology, the instructor can assign Update chapters 10 (Geologic Time) and 12 (Plate Tectonics and the Continental Crust) and then assign the two supplementary historical chapters. Note, however, that because the historical geology chapters begin with a complete discussion of the origin of the solar system, you may want to delete the equivalent portion of Chapter 21 (Planetary Geology)—or the entire chapter for that matter—from your reading assignments.

Supplements

This textbook is supplemented by a number of useful learning and teaching aids:

- **Student Study Guide.** For students who benefit from pencil-and-paper exercises, drills, and practice tests, the study guide offers an additional avenue of review. Dozens of interactive drills, along with practice multiple-choice exams with answers and self-evaluation charts, allow students to identify areas of weakness and pinpoint the text pages that they should review again.

- **Laboratory Manual for Students.** This manual offers lab studies and activities on topics closely tied to *Physical Geology.* Included are labs on plate tectonics, mineral identification, rock deformation, streams, groundwater, and glacial landscapes. Other units help to develop and hone students' skills in working with and interpreting geologic and topographic maps. Physical Geology Interactive is a CD-ROM version of the lab manual that extends all activities, using web resources for information and data to be used in the exercises.

- **Test Item File and Computerized Testing.** The printed test bank, offering over 1200 multiple-choice questions and approximately 200 additional essay and illustration-based questions, is conveniently organized by chapter. Also featured are 40 questions, covering the material in the first eleven chapters, that are suitable for use in a midterm exam, plus 40 questions, covering the material in the last eleven chapters, that are designed for use in a final exam. The computerized testing program offers the same questions that appear in the printed test bank but in handy electronic format for IBM-compatible and Macintosh computers.

- **Instructor's Manual.** An invaluable aid for the instructor, this comprehensive manual includes brief chapter summaries, detailed chapter outlines, lecture suggestions, ideas for student activities, teaching tips, and up-to-date information on media resources.

- **Overhead Transparencies.** This set of some 100 color acetates of useful illustrations is grouped by topic for ease of presentation. Most illustrations are taken from the textbook, but some supplementary visuals are included to expand the range of options for the instructor.

- **The Earth Sciences Videodisc Set.** This state-of-the-art videodisc set features full-motion video with narration and numerous animations, as well as more than

3000 still images. All the core course topics are covered, including plate tectonics, minerals, volcanoes, faulting and folding, and the rock cycle, among many
others.

• **The Earth Sciences Geology Slide Set.** This generous collection of full-color
images features original photographs of rocks, minerals, landforms, and important geologic sites and phenomena from around the world.

Acknowledgments

Writing this book was a major undertaking, and its completion would have been impossible without the unselfish contributions of numerous individuals. Their interest, support, and advice have sustained me throughout the process. It is only proper that I
acknowledge them here, with deep thanks.

I am very grateful to Luigia and Herbert Miller, my oldest and dearest friends, who
encouraged me every step of the way and kept me focused on my writing. Isolda, my
wife, put up with my long, work-related absences and occasional ill humor with a good
grace I did not deserve. Maria Chen, my colleague at New York City Technical College,
logged hour upon hour helping me with manuscript preparation and research for no
other reasons than friendship and belief in the book. My brother, Abraham Dolgoff, an
engineering geologist, was a patient and expert consultant. Jinny Joyner, a skilled editor, became involved in the book during the early stages of development. She sharp-
ened my prose and made many astute suggestions for fine-tuning the chapter organization. My special thanks go to her, as well as to Kent Porter Hamann, the editor who
launched this project at D. C. Heath, and to Jim Porter Hamann, my first marketing
manager at Heath. Their hard work and enthusiastic support will not be forgotten.

Sylvia Mallory, geology editor at Heath, joined the team midstream but quickly
proved her ability to sponsor the book with editorial acuity and market savvy. Behind
the scenes, production editor Anne Starr guided the manuscript, photos, and illustrations through every phase of the production process, keeping me organized and holding me to a tight schedule, all with gracious (albeit relentless) persistence. Billie Porter
and Judy Mason worked together to acquire the text's many beautiful photographs, and
Robert E. Morency, Jr., assisted me in reviewing these photos for content. The elegant
look of the book reflects the talents of our designer, Alwyn Velásquez, and the artistry
of the page make-up was Irene Cinelli's contribution. Art editor Jim Roberts creatively
and capably coordinated the rendering of the line illustrations. Many of the more complex line illustrations were the work of a wonderful artist, Elizabeth Morales, an anthropologist by training, but with a good deal of formal background in geology. The Heath
staff calls her work reflective art, but I call it suitable for framing.

There follows a long list of geologists who reviewed the manuscript and offered
detailed suggestions and criticisms, often line by line; most were points well taken. Several of these reviewers deserve special mention. Craig Manning at the University of California, Los Angeles, urged me to trust my judgment concerning the decision to place
plate tectonics at the beginning of the book and supplied expert advice on a number of
chapters. Wang-Ping Chen at the University of Illinois, Urbana-Champaign, reviewed
every manuscript chapter and most of the illustrations for accuracy. No concept or
detail was too insignificant to escape his scrutiny, and the book is infinitely better for
his careful attention. John Nicholas at the University of Bridgeport; Pamela Martin at
the University of California, Santa Barbara; and John Geissman at the University of New
Mexico also devoted many tedious hours to checking the illustrations for accuracy. I
thank them and all of the other excellent reviewers for their time and wise counsel:

Charles Alpers, *United States Geological Survey*
Robert Behling, *West Virginia University*

immersive8...

Mary Lou Bevier, *University of British Columbia*
Marcia Bjornerud, *Miami University*
Thomas Broadhead, *University of Tennessee, Knoxville*
Donald Burt, *Arizona State University*
Karl Chauffe, *St. Louis University*
Chu-Yung Chen, *University of Illinois, Urbana-Champaign*
Kevin Cole, *Grand Valley State University*
Lorence Collins, *California State University, Northridge*
Brian Cooper, *Sam Houston State University*
Spencer Cotkin, *University of Arkansas*
Robert Corbett, *Illinois State University*
Larry Davis, *Washington State University*
Joseph DiPietro, *University of Southern Indiana*
David Dockstader, *Jefferson Community College*
William Dupré, *University of Houston*
Goran A. Ekström, *Harvard University*
Terry Engelder, *Pennsylvania State University*
G. Lang Farmer, *University of Colorado, Boulder*
Michael Gibson, *University of Tennessee, Martin*
Bryce Hand, *Syracuse University*
John Howe, *Bowling Green State University*
Larry Knox, *Tennessee Technological University*
Lawrence G. Kodosky, *Oakland Community College*
David Lea, *University of California, Santa Barbara*
Jonathan Lincoln, *Montclair State University*
Steve Loftouse, *Pace University*
Constantine Manos, *State University of New York, New Paltz*
Sandra McBride, *Queen's University*
Elizabeth McClellan, *Western Kentucky University*
James McClurg, *University of Wyoming*
Eileen McLellan, *University of Maryland, College Park*
James I. Mead, *Northern Arizona University*
David Mogk, *Montana State University*
Anne Pasch, *University of Alaska*
Lincoln Pratson, *University of Colorado*
Fredrick Rich, *Georgia Southern University*
Mary Jo Richardson, *Texas A & M University*
Jeanette Sablock, *Salem State College*
Robert Schoch, *Boston University*
Karl Seifert, *Iowa State University*
Lynn Shelby, *Murray State University*
William Smith, *Western Michigan University*
Donald Spano, *University of Southern Colorado*
Neptune Srimal, *Florida Atlantic University*
George Stephens, *The George Washington University*
Monte Wilson, *Boise State University*

Finally, I wish to express my deepest gratitude to Mary and Ricardo Falcon. For untold hours over the past two years, Ricardo has devoted his computer skills, keen intelligence, and dedication to the project. Mary tirelessly applied her extensive publishing experience, professional know-how, and editorial talent to the book's development. Every page bears the imprint of her sound judgment. No one could ask for a better editor and adviser.

A. D.

Brief Contents

Contents

"Aside" Feature Essays

"Apply and Decide" Feature Essays

PHYSICAL GEOLOGY

Introduction

Geology is the science of the Earth, the study of its composition, structure, and history. Above all, it is the study of those processes that shaped the Earth of the past and those that continue to mold the Earth of the present. It is a science concerned with everything from the migration of sand dunes across the desert to the migration of continents across the globe; from the flow of molten rock down the sides of volcanoes to the flow of rivers to the sea. In this age of space exploration, geology has even exceeded its earthly boundaries. Planetary geology has become a thriving branch of the science.

The scope of geological investigations ranges from the atomic to the global, from events that occur in seconds to those that take billions of years to unfold. Much of what **geologists** do involves solving problems that affect millions of people, such as earthquake prediction, estimations of groundwater reserves, or petroleum exploration. Geologists also address questions of theoretical interest just for the joy of discovery, such as whether asteroids, volcanic eruptions, or neither caused the extinction of the dinosaurs.

Geology is an integrative science. Its goal of improved understanding of the Earth depends heavily upon the application of physics, mathematics, chemistry, and biology to geological problems. However, geologists work with the natural world in all its complexity, and for this reason, their research requires more than the mechanical application of physics and chemistry. The successful geologist is usually a generalist with the ability to interpret information from many sources.

The wide-ranging nature of the research is one of the reasons geologists find the subject fascinating. You may never know at the start of an investigation where a seemingly insignificant fact may lead. Nevertheless, when the paths suggested by many facts lead toward a significant discovery or a deeper understanding of what is already known, your sense of achievement is real, even if your contribution is but a part of the whole.

geology
The study of the materials, processes, products, and history of the Earth.

geologist
One who investigates the materials, processes, products, and history of the Earth. Geologists conduct basic scientific research in order to increase our understanding of the Earth, and they apply their knowledge to improve our lives in many ways.

◄ View from the top of the drill tower on an ocean-going research vessel.

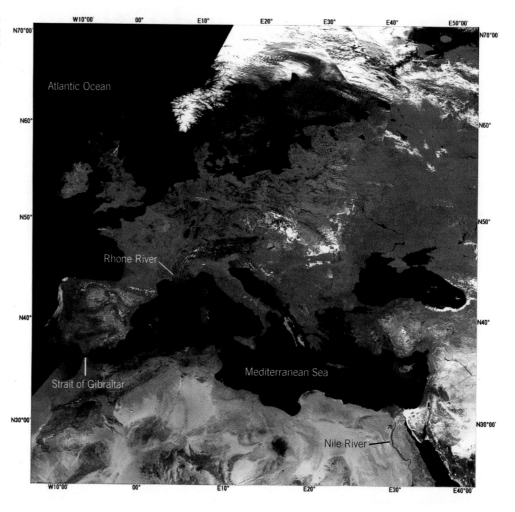

A Geological Puzzle

hypothesis
A tentative explanation of a phenomenon or process that is tested for validity by repeated observation or experimentation.

We will begin this book by describing the development of a **hypothesis** that attempts to explain a key chapter in the geological history of the Mediterranean Sea (Figure I.1). We will present the hypothesis in full, as well as the intense debate and rival explanations it has fostered. In the course of exploring this geological puzzle, we shall demonstrate some of the methods, observations, and reasoning of geology and perhaps transmit a bit of the special enjoyment geologists derive from the study of the Earth.

At the core of the puzzle lies the discovery of two seemingly unrelated facts: the first, a series of buried canyons in the lands that surround the Mediterranean Sea; the second, a thick layer of salt that lies about 150 meters (492 feet) beneath the floor of the Mediterranean Sea.

Buried River Canyons

The Aswan High Dam of Egypt is one of the world's largest dams. An engineering wonder of the modern world, it dwarfs the nearby pyramids, those magnificent achievements of an ancient technology. The dam's function is to regulate the flow of the Nile River northward to the Mediterranean, thereby producing hydroelectric power and a controlled supply of water for irrigation. The Nile flows on a riverbed of gravel, sand, and silt, but in the 1950s, Russian geologists working on the design of the dam determined that the huge foundation should rest on solid rock. In order to locate rock

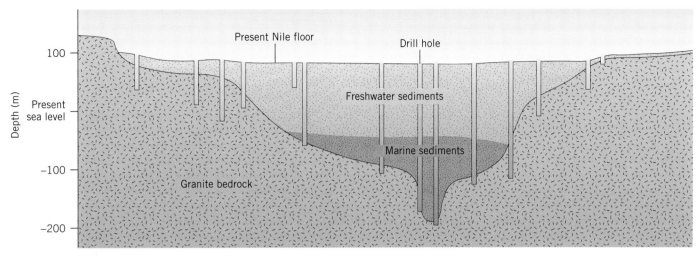

Figure I.2 Cross section of the buried Nile canyon at Aswan. Canyons like these are carved by streams above sea level, but most of the Nile canyon is well below present sea level. Note the V-shaped profile cut in granite and the sequence of marine sediments overlain by river sediments. (Adapted from Kenneth J. Hsü, *The Mediterranean Was a Desert,* Princeton, N.J.: Princeton University Press, 1990, Fig. 38.)

beneath the loose river sediments, they took samples of the subsurface by drilling 15 holes across the Nile Valley at the dam site. What they discovered amazed them.

Although the present floor of the Nile at Aswan stands about 100 meters (328 feet) above sea level, a narrow canyon, cut in granite, lies beneath it (Figure I.2). The classic V-shaped cross section of the canyon is typical of those carved by swift rivers flowing from high elevations. Yet the canyon floor is 183 meters (600 feet) *below* present sea level—that is, 183 meters lower than the surface of the Mediterranean Sea of today. The ancient canyon is filled with marine sediments, muds, and other materials that settled out of seawater. These sediments in turn are covered by freshwater river sediments, including gravel, sand, and silt. The placement of the marine sediments indicates that after the canyons were cut, the sea rose and invaded the river valley.

The Aswan Dam site is about 1200 kilometers (744 miles) from the Mediterranean coast. Intrigued by their discovery of a below-sea-level canyon so far from the coast, the geologists attempted to determine its extent. Downstream of the dam, on the Nile Delta near the coast, they drilled as deep as their equipment would allow but were unable to reach bedrock. The depth of the canyon floor at that point was greater than 1500 meters (nearly 1 mile). The geologists concluded that the deep V-shaped canyon slopes toward the Mediterranean Sea. It is difficult to imagine that this gorge was carved by any agent other than water running downhill. But what hill? The canyon is buried below sea level.

What the Russian geologists discovered beneath the Nile is not unique. Similar buried canyons have since been found to rim the entire Mediterranean. Because all of them were carved below present sea level, they are impossible to explain unless we assume that the Mediterranean Sea stood thousands of meters lower at the time the canyons were carved. Furthermore, like the Nile, all these canyons are filled with marine sediments that contain fossils—traces of ancient life preserved in the sediment. The fossils in the various canyons match, indicating close correspondence in the ages of the marine deposits—strong evidence that the sea invaded these canyons at about the same time.

In sum, the first piece of our Mediterranean puzzle is the presence of buried canyons. These canyons were cut by rivers that flowed into a Mediterranean Sea that stood much lower than it does today. Marine sediments and fossils were deposited in the canyons by a rising sea.

Salt Layers Beneath the Sea Floor

The 1950s and 1960s marked a period of rapid technological advance in marine geology. Especially important were two developments. The first was the perfection of the continuous seismic profiler, a device that allows a ship on the surface to take an

Figure I.3 A seismic reflection profile revealing the M reflector beneath the floor of the western Mediterranean Sea. The protruding, fingerlike structures are salt domes that pierce the overlying strata. (Courtesy of the Lamont-Doherty Earth Observatory.)

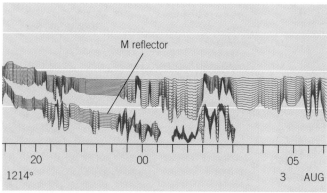

(b)

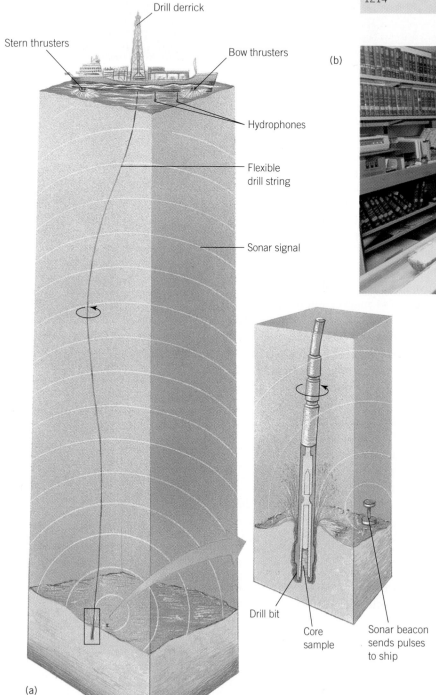

(a)

Figure I.4 (a) A research vessel taking a deep-sea core. Sonar is used to keep the ship in place above the drill site. (b) Core samples of Mediterranean salt layers. The cores are removed from the drill casings, sliced lengthwise, catalogued, and stored for later study.

"acoustical X ray" of the layers of sedimentary deposits and rocks beneath the sea floor. The second was construction of deep-sea drilling vessels that are capable of obtaining rock and sediment samples from one thousand meters or more beneath the ocean bottom. Geologists finally had at their disposal the means to explore the 70.8 percent of the Earth's surface that is covered by water. Thus at about the same time that the Russian geologists were probing the subsurface beneath the Nile River, a team of American and European geologists were probing the subsurface beneath the Mediterranean Sea—which leads us to the second piece of the Mediterranean puzzle.

Figure I.3 is a continuous seismic profile of what lies beneath the Mediterranean

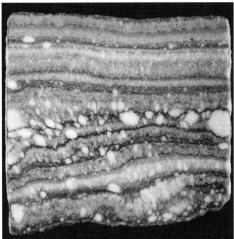

(a)

(b)

Figure I.5 (a) This deep-sea core of calcium sulfate salts, split to reveal its layering, was taken from the M reflector after drilling through soft marine sediments. The vertical crack in the lower part of the core is believed to have resulted from drying of the sediment under the hot sun. (b) Close-up of another Mediterranean salt core. The dark layers were formed by mats of blue-green algae; the light layers are calcium sulfate salts that were trapped between the mats. (c) A typical coastal desert environment in Baja California. The present-day colonial algae grow in mats that are similar in form to the algal structures in the salt core seen in (b).

(c)

Sea basin. To produce this profile, powerful sound waves were transmitted from the research ship *Robert Conrad* to the ocean floor. The waves penetrated the ocean bottom, reflected off the layers beneath, and returned to receivers towed behind the ship. The extent to which a wave is reflected depends in large part upon the difference in density between the adjacent layers; the greater the contrast, the stronger the reflections. Therefore, the reflecting surface, labeled "M reflector" in Figure I.3, is a boundary that separates soft marine muds from something denser. Wherever research vessels probed, that reflecting surface was found at depths of around 150 meters (492 feet) beneath the sea floor. What was the composition of this material that caused such sharp wave reflections?

In the summer of 1970, geologists Kenneth Hsü and William Ryan led a team of scientists onboard the drilling ship *Glomar Challenger* on an expedition to drill into this hard layer and extract drill-core samples, one of which is shown in Figure I.4. All the cores from this layer are composed of calcium sulfate salts. These salts bear striking resemblance to those we observe today forming from the evaporation of seawater in shallow seas (Figure I.5). The matlike structures in some of the cores were particularly interesting. They appeared to be similar to the mats of certain types of algae that grow in the tidal flats of modern coastal deserts. Hsü and his colleagues thus hypothesized that the salt layers were, at some time in the past, deposited at or near sea level under scorching desert conditions. The geologists also noted that the salt layers in the drill cores were sealed between layers of deep ocean sediments—that is, with older marine deposits directly below and younger marine deposits directly above.

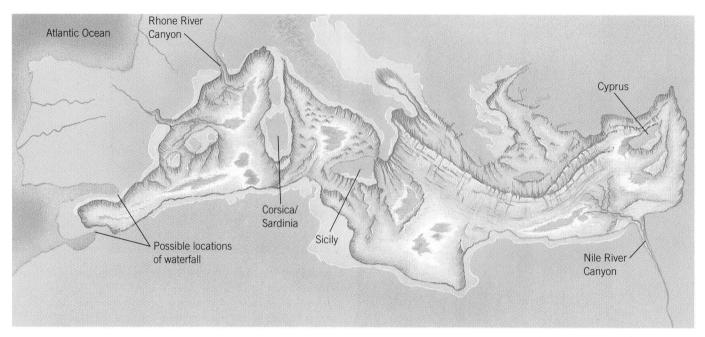

Figure I.6 This artist's interpretation of what the hypothesized Mediterranean desert might have looked like is based on the submarine topography of the present-day sea floor.

Did the Mediterranean Sea Dry Up?

The combination of all these facts led Hsü and his colleagues to propose an intriguing explanation: that the blue Mediterranean, cradle of ancient civilizations, the "lake" of a dozen nations, had in the past evaporated and left a parched, sunken desert 2 to 3 kilometers (nearly 2 miles) beneath the surface of the surrounding continents (Figure I.6). By analyzing fossils in strata above and below the salt layer, they estimated that the evaporation occurred a geologically recent 6 million years ago and took a rapid 200,000 years to complete.

Can you imagine the Mediterranean Sea turned into an enormous plain of white salt? In Hsü's vision, Sicily, Corsica, Cyprus, and other islands stood as massive mountains above the plain. What are now the margins of the African and Eurasian continents were towering highlands sculpted with magnificent canyons; scattered briny lakes, fed by rivers, dotted the landscape.

There was one other factor that Hsü's hypothesis had to explain. The buried salt layer is 1 to 2 kilometers thick; yet if all the water of the modern Mediterranean were to evaporate, it would yield a layer of less than 30 meters (98 feet). It was clear to Hsü that the salt lakes and streams of the basin floor had to have been constantly nourished with additional seawater in order to deposit the salts. The question was, Where could the water have come from? For all its great size and depth, the Mediterranean is a secluded sea with a single connection to the Atlantic Ocean—the shallow Strait of Gibraltar. Hsü proposed that the Mediterranean dried up when that connection was briefly severed, either by mountain building that uplifted the Gibraltar Strait or by a lowering of the worldwide sea level that exposed the strait. In either case, a formidable dam existed between the Atlantic Ocean and the Mediterranean Sea.

Water spilled over the dam whenever the level of the Atlantic rose, thereby periodically replenishing the briny lakes. Then about 6 million years ago, the dam broke, adding to Hsü's picture of a desolate Mediterranean plain an immense cascade of seawater at its western border that dwarfed any waterfall now on Earth. Although the volume of water roaring over the fall was greater than a thousand Niagaras, it took more than one hundred years to fill the Mediterranean basin and reestablish the two-way circulation pattern with the Atlantic.

Hsü's hypothesis also accounts for the buried river canyons that surround the basin. As the Mediterranean dried up, the rivers cut deeper into the land to keep pace with the falling level of the sea. Later, the rising sea flooded the canyons, and marine sediments filled the former stream valleys. The rivers then deposited freshwater sands and gravels over the marine sediments. Some rivers, like the Nile and Rhone, laid down enormous wedges of sediment where they entered the sea. These river deltas are spread over the marine muds that overlie the buried salts of the basin.

The Debate

Many geologists take issue with Hsü's model of a vast, sunken Mediterranean desert. Before discussing their objections, however, we will note the many points of agreement. No one disputes that the salt layers exist or that they were deposited when the sea was isolated in an arid climate. No one disputes that the buried river canyons exist or that they were cut well below present sea level. These are facts, the realities that any theory regarding the history of the Mediterranean must explain. What *is* in contention is Hsü's interpretation of the facts—that the entire Mediterranean Sea evaporated and left behind a vast salt-covered desert, thousands of meters below sea level, at the bottom of what had been the deep Mediterranean Sea basin (Figure I.7a).

Critics of Hsü's "deep-basin desert" model have proposed an alternative explanation, which can be characterized as a "shallow-basin, shallow-sea" model. They contend that at the time the salts were deposited, the Mediterranean Sea was a series of interconnected shallow basins with a constricted connection to the Atlantic (Figure I.7b). The arid climate caused evaporation at the sea surface, leaving the water in the basins more and more salty. When the briny water became so saturated that it could hold no more salt, excess salt precipitated from the water and settled to the shallow bottom. Meanwhile, seawater continually seeped in from the Atlantic to replenish the basins. It too evaporated, and over time, thick salt deposits were laid down in this shallow sea that never went dry. Hsü's critics contend that only after the salts were deposited did the floor of the Mediterranean sink to its present depth. Accordingly, the deep river canyons were cut as a consequence of the Mediterranean basin being lowered. They were not cut because the sea dried up.

Hsü's critics point out that the Earth's outer shell, the crust, is very unstable in the Mediterranean region. It is a zone of intense earthquake and volcanic activity. The earthquakes are associated with faults, or fractures, along which blocks of the crust have moved. Geologists can date these movements by determining the ages of the rocks cut by the faults. The critics cite evidence of fault movements *younger* than the salt deposits that have brought the basin to its present depth. In their eyes, therefore, the salts were deposited in a Mediterranean basin much shallower than the present one.

The alternative model takes away the desert, the deep basin, and the miles-high waterfall. However, it is more consistent with observations of how salt deposits form in the modern world—that is, in the shallow waters of shallow basins. Hsü's critics feel that the deep-basin desert theory is inconsistent with the fundamental tenet of geology: that the present is the key to the past. It leaves too great a disparity between what is happening on the Earth today and what supposedly occurred in the past, for there is no past or present record of an ocean as deep and extensive as the Mediterranean Sea ever completely evaporating.

Defenders of the deep-basin desert hypothesis counter their critics by pointing out that the sedimentary record is on their side. Because the salt layer is sealed between layers of deep-ocean sediments, they say that the Mediterranean Sea floor would have had to bob up and down like an elevator to accommodate the shallow-basin model. Did the floor of the Mediterranean rise 3000 meters, sit there for 200,000 years while the salt collected on its bottom, and then sink again?

Which camp is correct—those who propose that the Mediterranean Sea evaporated, leaving behind a spectacular desert, or those who argue that it never happened? If you are looking for a definitive answer, you will not receive it here. As you will learn, the

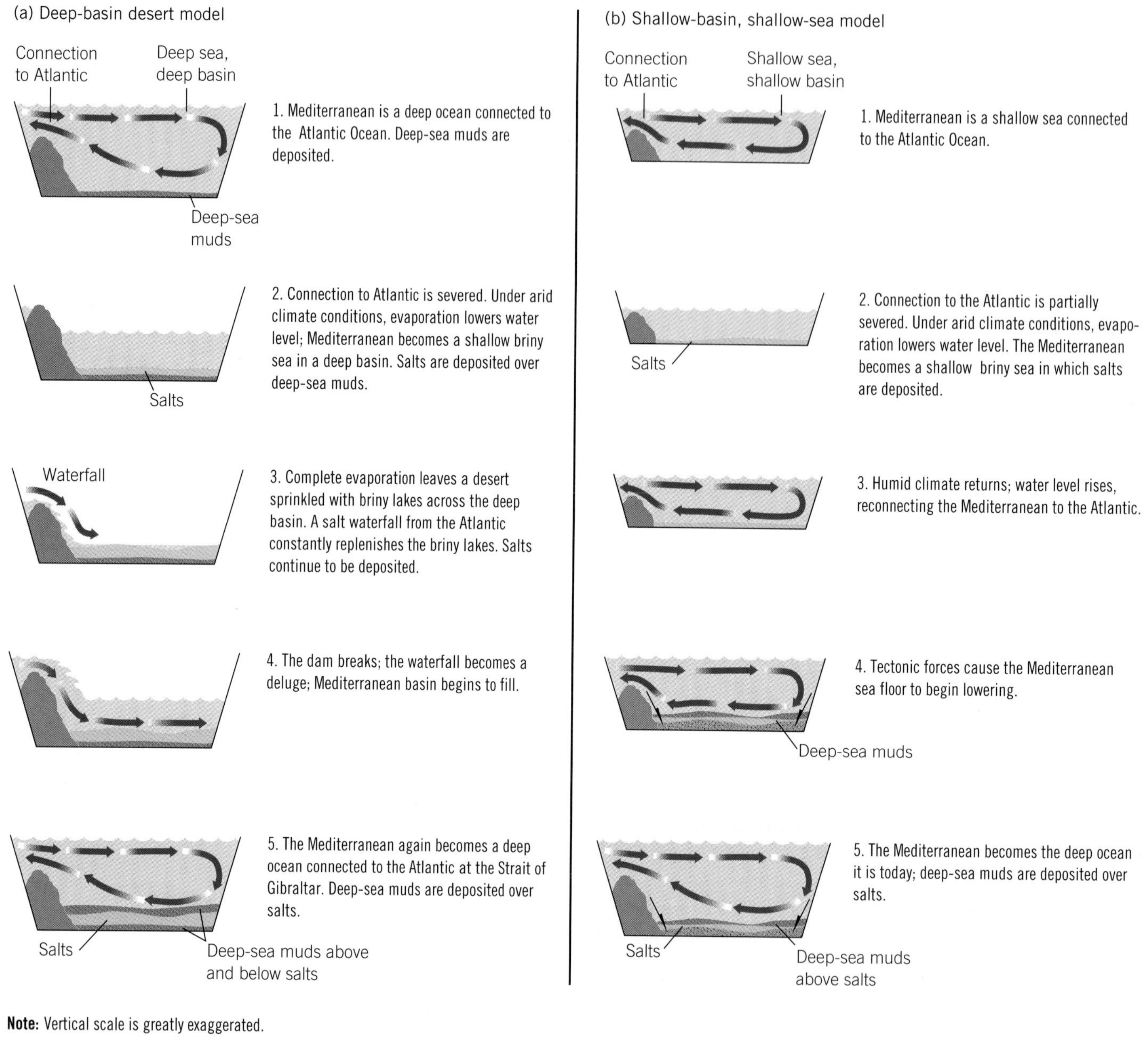

(a) Deep-basin desert model
Connection to Atlantic
Deep sea, deep basin
1. Mediterranean is a deep ocean connected to the Atlantic Ocean. Deep-sea muds are deposited.
Deep-sea muds
2. Connection to Atlantic is severed. Under arid climate conditions, evaporation lowers water level; Mediterranean becomes a shallow briny sea in a deep basin. Salts are deposited over deep-sea muds.
Salts
Waterfall
3. Complete evaporation leaves a desert sprinkled with briny lakes across the deep basin. A salt waterfall from the Atlantic constantly replenishes the briny lakes. Salts continue to be deposited.
4. The dam breaks; the waterfall becomes a deluge; Mediterranean basin begins to fill.
5. The Mediterranean again becomes a deep ocean connected to the Atlantic at the Strait of Gibraltar. Deep-sea muds are deposited over salts.
Salts
Deep-sea muds above and below salts
(b) Shallow-basin, shallow-sea model
Connection to Atlantic
Shallow sea, shallow basin
1. Mediterranean is a shallow sea connected to the Atlantic Ocean.
Salts
2. Connection to the Atlantic is partially severed. Under arid climate conditions, evaporation lowers water level. The Mediterranean becomes a shallow briny sea in which salts are deposited.
3. Humid climate returns; water level rises, reconnecting the Mediterranean to the Atlantic.
4. Tectonic forces cause the Mediterranean sea floor to begin lowering.
Deep-sea muds
5. The Mediterranean becomes the deep ocean it is today; deep-sea muds are deposited over salts.
Salts
Deep-sea muds above salts
Note: Vertical scale is greatly exaggerated.

Figure I.7 Two different hypotheses provide possible explanations of the origin of the Mediterranean salt deposits. (a) The salts accumulated in a deep basin similar to what exists today. (b) The salts accumulated in a shallow basin that was subsequently lowered to present depths.

Mediterranean basin is trapped between two colliding landmasses, Africa and Europe. The effects of this collision make it one of the most geologically complex regions on Earth. Reconstructing its history is an ongoing process involving the energies of many talented people. The larger point is that even those scientists who favor what may ultimately prove to be the incorrect model have as useful a role to play as those who advocate the correct model. They sharpen the correct model by challenging inconsistencies and weaknesses, and their arguments open new avenues of research. In the long view, both "correct" and "incorrect" advocates are collaborators in the scientific search for the truth.

Research and Theory in the Geosciences

Science is the knowledge of the natural or material world gained through systematic observation and experimentation. A key word in this definition is *observation*. Science differs from other human endeavors in its insistence that observations about the world

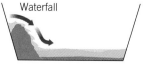

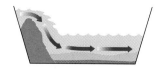

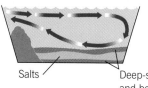

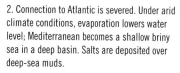

take precedence over assertions that are based on how the world should be or how we would like it to be. Another key word in the definition is *systematic;* it means "having a plan or method." Within the story of any geological inquiry, including the story of the Mediterranean puzzle, we can discern the various components of the systematic process of science, or the **scientific method.**

The Scientific Method

The scientific method can be narrowly defined as a series of steps that begins with a question or problem to be solved and ends with an answer or solution (Figure I.8). The solution is based on a solid body of meticulously collected evidence that has withstood rigorous testing and evaluation by many scientists over time. In broader terms, however, the scientific method is less a procedure than a way of thinking. Its purpose is to ensure objectivity and the dogged determination to get at the truth—no matter what it is! What cannot be described in textbook fashion is that scientific progress often leaps ahead on bursts of insight and key discoveries that change perspectives overnight.

Questions and Tentative Answers

A scientist conducting a scientific investigation can be compared to a detective in a criminal investigation. For the geologist, the "crime" is a question about a geological phenomenon, based on tangible and measurable facts, such as the 6-million-year-old salt deposits beneath the floor of the Mediterranean Sea. Hsü and his colleagues asked, "What happened to the Mediterranean Sea in the past to cause the salt to be deposited?" In fact, all scientific inquiries begin with a question or series of questions about the natural world. The subject of inquiry may never have been addressed previously, or it may have been addressed by others but not satisfactorily answered. For this reason, researchers usually pay close attention to what other scientists have written on the subject before proceeding with their own investigation.

Like a detective who uses the evidence gathered at the scene of a crime to identify a suspect, geologists use evidence gathered from the field to develop a hypothesis or **model**—that is, a tentative answer to the question. Exactly when the geologists feel ready to make an educated guess about the answer to the question depends upon the extent of previous knowledge about the phenomenon being researched; the speed with which they are able to obtain evidence; and, perhaps most important, the knowledge, experience, and intuition of the researchers themselves. In his book *The Mediterranean Was a Desert,* Kenneth Hsü recalls that the deep-basin desert hypothesis occurred to him one sleepless night at sea aboard the *Glomar Challenger,* following the day in which they had extracted the first salt core from the sea bottom.

Agatha Christie's detective hero Hercule Poirot proclaimed, "Until the real culprit is found, everyone is a suspect!" Geologists may approach the solution to a problem in a similar manner using the method of **multiple working hypotheses.** Take, for example, the question of the Mediterranean salt layer. Two competing hypotheses have been proposed: (1) The salts were deposited in the briny lakes of a desert at the bottom of a deep-ocean basin, and (2) the salts were deposited in the shallow waters of

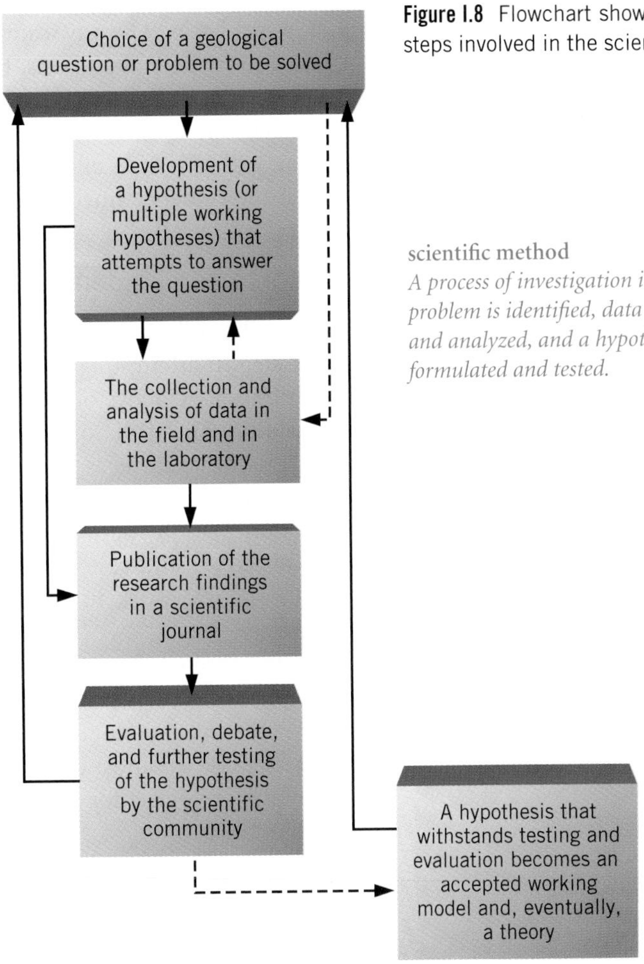

Figure I.8 Flowchart showing the major steps involved in the scientific method.

Choice of a geological question or problem to be solved

Development of a hypothesis (or multiple working hypotheses) that attempts to answer the question

The collection and analysis of data in the field and in the laboratory

Publication of the research findings in a scientific journal

Evaluation, debate, and further testing of the hypothesis by the scientific community

A hypothesis that withstands testing and evaluation becomes an accepted working model and, eventually, a theory

scientific method
A process of investigation in which a problem is identified, data are collected and analyzed, and a hypothesis is formulated and tested.

model
A hypothesis expressed as a visual or statistical simulation, or as a description by analogy of phenomena or processes that are difficult to observe and describe directly.

multiple working hypotheses
An approach to geological research in which several possible explanations of a phenomenon are developed and evaluated simultaneously and impartially.

a shallow-sea basin. The advocates of the first model have gathered an impressive array of evidence for their hypothesis, but the evidence in support of the second hypothesis cannot be ignored. For this reason, other researchers pursuing an answer to the origin of the salts would do best to use both models as their multiple working hypotheses, gathering data and following all leads as impartially as possible, ruling out neither explanation until the evidence unequivocally contradicts it.

The Search for Evidence

data
Items of factual or statistical information.

The systematic collection of **data**—that is, information about the subject of inquiry—is a crucial step in attempting to prove or disprove a hypothesis. The data may be in the form of physical evidence obtained directly from a location in question (in the Mediterranean inquiry, the drill cores), or the evidence may be measurements obtained from instruments (for example, seismic profiles). Geological researchers today may also access data from a host of other sources, such as the satellite images, sea-floor maps, and other types of information routinely collected by government agencies like the United States Geological Survey and research institutions such as the Lamont-Doherty Earth Observatory. Geological work in the field and in the laboratory is discussed later in this chapter.

Having collected the evidence, geologists next need to organize and analyze it. The primary goal of analysis is to highlight relevant similarities and relationships among the data. Only then can geologists reach conclusions regarding what they have found. This stage may well be the most creative and challenging aspect of the endeavor. Some of the tools used by geologists for analysis of data are also discussed later in this chapter.

Evaluation by the Scientific Community and Further Research

When they are satisfied that the evidence supports their hypothesis and that their findings are significant enough to share with the scientific community, the researchers prepare a report on their findings, commonly called a "paper," and submit it for publication to a scientific journal. The paper is then sent out for peer review to experts on the subject, who transmit their confidential opinions to the journal editors. Their role is to determine whether the paper is of a quality worthy of wider circulation, not to determine whether the hypothesis is correct or to argue why they do or do not agree with it. Upon publication, the paper is then informally evaluated by the geological profession at large, especially by those working on the same or similar problems. The authors may also be invited to present their paper at academic meetings of professional associations (Figure I.9). The largest of these include the Geological Society of America and the American Geophysical Union, and there are myriad specialized groups affiliated with these organizations.

Now that the researchers' work has been presented to the scientific community for evaluation, the next step is taken by other researchers who have been intrigued by the paper and begin their own investigations. Indeed, hypotheses are the lifeblood of geology, and of science in general, because they stimulate debate and further research. A new hypothesis may have little impact; or it may be debated and tested immediately, acting as a springboard to fresh investigations—the ultimate compliment, as many geologists would argue.

It also is entirely possible that the hypothesis will be ignored or even scorned at first, only to be revived decades later when new data and fresh insights prove its worth. In the next chapter you will learn about the birth, death, and rebirth

Figure I.9 The frank exchange of ideas is an essential element of geological research. These geologists are participating in a seminar at a conference on oil and energy resources.

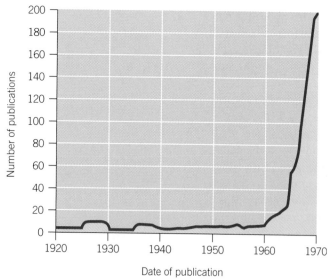

Figure I.10 The rapid increase in the number of professional papers devoted to continental drift reflects a scientific breakthrough. Prior to the 1960s, the subject was ignored by most American geologists. (John A. Stewart, *Colliding Continents and Colliding Paradigms*, Bloomington, Ind.: Indiana U. Press, 1990, p. 46.)

of the famous theory of *continental drift,* first proposed by Alfred Wegener in 1915. Long reviled by North American geologists, it was revived in the 1960s when new evidence was found to support it. Figure I.10 shows that the sudden revival of interest in the hypothesis triggered a steep rise in the number of papers published on the subject. These papers were the results of further investigations stimulated by the hypothesis. They reported new data, offered variations on the hypothesis and counterproposals, and suggested new directions. With each new investigation, the steps of the scientific method were followed once again.

Development of a Theory

All hypotheses contain the germ of a theory. Hypotheses become theories when, over time, they meet the challenges presented by newly discovered data and competing hypotheses. Scientists define a **theory** as a set of propositions that is widely accepted by the scientific community as the explanation of a broad group of natural phenomena. A good theory is a work of creative imagination, similar in some respects to an oil painting or sculpture. Painters and sculptors, however, are free to bend their materials to their will. They are governed only by the restrictions they choose to place on the forms they create, whereas the restrictions placed on scientists are far more stringent. A scientific theory is grounded in physical reality, and it is valid only insofar as it is able to explain and predict that reality as accurately as possible. A scientific theory must proceed from cause to effect in a logical manner and account for *all* the relevant facts, not just those that support the theory. Also, it must be consistent with the laws of nature—that is, observations proven true time and again. A theory should not violate the law of gravity, for example.

theory
A widely accepted explanation for a group of known facts. A theory is a hypothesis that has been elevated to a high level of confidence by repeated confirmation through testing and experimentation.

The theory of *plate tectonics* is the backbone of modern geology. It plays a role analogous to the theory of evolution in biology in that it provides a unifying set of explanations for what had previously been considered unrelated geological phenomena. Plate tectonics explains such things as the origin of oceans and mountain belts, the drift of continents, the occurrence of earthquakes and volcanoes, and much more. Beginning with the next chapter, we will see how plate tectonics amply fulfills the criteria of a good theory.

Finally, keep in mind that the establishment of a theory, no matter how widely accepted, does not place it beyond debate. In science, nothing is beyond debate, although the validity of an argument against an established theory will be judged by the strength of the supporting evidence. The revision of established theories is considered to be part of the scientific process.

Geology in the Field and in the Laboratory

It is in the pursuit of evidence that geologists drill through thousands of meters of Antarctic ice to collect data on ancient climates, camp on the rim of a crater to monitor the pulse of a stirring volcano, trek across the Mongolian Steppes to find dinosaur bones, or cast a worldwide network of seismographs to more accurately predict earthquakes. Gathering hard evidence from the direct study of the Earth is the foundation of all geological research, and it is what most geologists spend most of their time doing (Figure I.11a).

The special nature of geology, a small portion of which was illustrated in our Mediterranean example, is based on the kinds of questions geologists attempt to answer and upon the technological tools at their disposal. All sciences benefit from advances in technology and none more so than geology, for much of the Earth is inaccessible to direct observation. Geology draws not only from the basic sciences, such as physics and chemistry, but also from the applied sciences, such as computer science, acoustical engineering, and electronics. We have seen, for example, that improved coring, navigational,

(a)

(b)

Figure I.11 (a) A team of geologists on the rim of a volcano measures an erupting lava fountain. (b) Taking rock samples is a fundamental activity of geological fieldwork. (c) A U.S. Geological Survey geologist uses a mass spectrometer to analyze the chemistry of a rock sample.

(c)

and acoustical-profiling techniques gave geologists the ability to explore the bottom of the Mediterranean Sea in a manner not previously possible.

We also saw in the Mediterranean example that, on occasion, there is an unplanned aspect to scientific data collection called *serendipity,* the good fortune of finding something of value purely by chance. The Russian geologists discovered an important piece of the Mediterranean puzzle (the buried Nile canyon) while they were working on something entirely different (the design of the Aswan Dam). They quickly realized that they had found something valuable and decided to pursue it—and their reaction was no accident, because scientists are trained to take advantage of a lucky find. As baseball legend Branch Rickey once said, "Luck is the residue of design." More important than how a geologist comes upon a piece of evidence is his or her ability to recognize its value.

Geologists gather most of the evidence concerning Earth processes and history from rocks, which are the principal components of the Earth's surface. Much of this evidence comes from studying the rocks in their natural settings (Figure I.11b). For example, you can determine whether a rock is older or younger than neighboring rocks, whether it has been disturbed, and whether the disturbance was some local event, such as a landslide, or the result of regional uplift. You can also examine it for visible features, such as ripple marks, mud cracks, or fossils, or for the alignment of minerals—all of which, as you will learn in this course, help the geologist reconstruct the environment in which the rock was formed.

Back in the lab, you can grind a piece of the rock down to a translucent sliver 30 micrometers thick and place it under a polarizing microscope in order to determine its texture and mineral content. You can examine the minerals on an even finer scale by using X-ray devices that will enable you to better understand their atomic structure. You can extract radioactive isotopes from the rock and pass them through a mass spectrometer in order to determine the rock's age (Figure I.11c). You can use sophisticated chemical techniques to ferret out extremely rare elements, and you can conduct all manner of experiments in an attempt to duplicate the physical and chemical conditions under which the rock was formed.

Laboratory and field work are combined in a variety of ways to solve geological problems. For example, in Chapter 10, Geologic Time, you will learn about a particular 1-centimeter-thick layer of clay in the limestone cliffs of northern Italy. Field mapping has located the clay layer at the precise boundary between rocks of the Cretaceous and Tertiary periods—that is, between the Age of the Dinosaurs and the Age of the Mammals. Chemical analysis tells us the clay layer is proportionately richer than ordinary clay in the extremely rare element iridium. Optical analysis of glass beads within the clay indicates that they were once molten droplets that could have been formed by the high-velocity impact of an extraterrestrial object with the Earth. Finally, thin deposits of clay having similar properties have been found at Cretaceous-Tertiary boundaries all over the world. This evidence, obtained from field and laboratory work, has led a number of geologists to hypothesize that an asteroid collided with the Earth 65 million years ago and caused the extinction of the dinosaurs.

We witnessed a vivid example of a large-scale collision in our solar system in July 1994, when remains of the comet Shoemaker-Levy 9 collided with Jupiter.

Geologic Maps

For most regional studies, data are best presented visually. The classic method of visual presentation is the **geologic map,** which displays the distribution, ages, and structural features of the rocks present (Figure I.12). In this way, the geologist can interpret the sequence of events that have affected the region and possibly reach some conclusions about the causes of these events. A century ago, the map was constructed by field parties who physically trekked across the region, meticulously recording what they saw. Classic mapping techniques are still vital, but they are supplemented with satellite imaging and a variety of devices that record information about the nonvisible parts of the Earth: hidden rock formations, heat flow, magnetic patterns, variations in gravity, and earthquakes.

geologic map
A visual display of the rock types, distribution, ages, and structural features of a given portion of the Earth's surface.

Mathematical Analysis

Sometimes it takes many months or even years in the field to collect the data for a geological study. For example, it may take several seasons of drilling in Greenland and Antarctica to obtain the ice cores that provide information on long-term climatic change. Other data may take only seconds to collect; modern digital seismometers, for instance, can sample earthquake vibrations 1000 times per second. The measurements from networks of recording stations are relayed to central receiving stations over computer links, so that data analysis can begin while the earthquake is still in progress. However, sorting and organizing these mountains of information—whether it is from hundreds of cores or thousands of seismic recordings—would be impossible without computerized mathematical analysis, which allows researchers to detect areas of similarity and contrast among the data.

The Use of Computer Models

Often, geological theories are presented through computer models designed to simulate the behavior of natural phenomena too complex to predict simply through direct observation. A computer model is generally constructed from a series of equations, derived

(a)

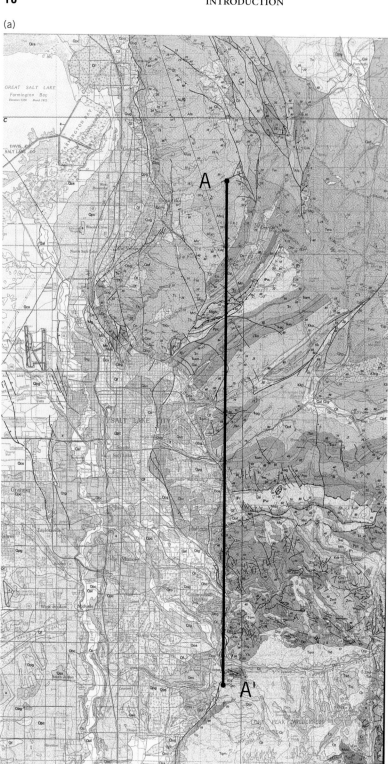

SCALE 1:100 000

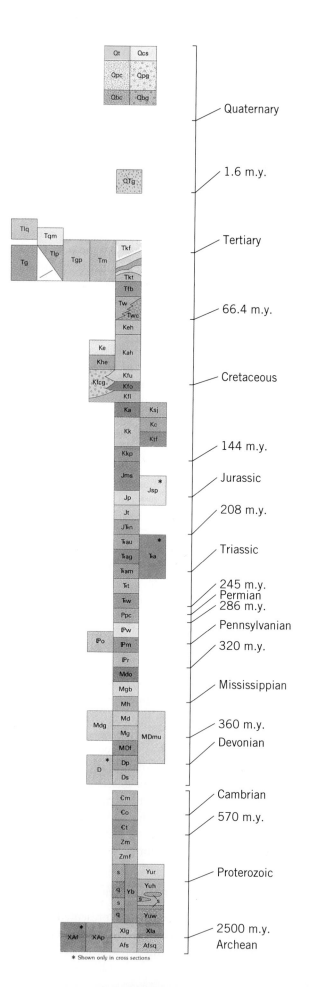

Figure I.12 (a) This geologic map, showing Salt Lake City, is one of many kinds used to analyze the geologic structure and history of a region. The various rock units are represented by colors and symbols that are keyed to their geologic ages in the legend. (b) The cross section (next page) illustrates what would be seen if the Earth were sliced open along line A–A'. A sequence of folded strata (green, blue, and brown) is prominent in its central portion. The dark line intersecting the sequence is a fault—a fracture along which one side of the Earth moved with respect to the other. See Appendix D for more on how geologic maps are made. (Adapted from Bruce A. Bryant, 1990. Courtesy of the U.S. Geological Survey.)

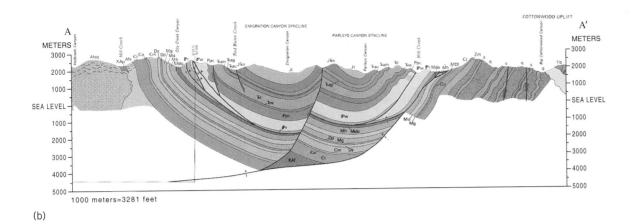

(b)

1000 meters=3281 feet

from the laws of physics and chemistry, that govern the behavior of the phenomena being studied. Data collected from direct observation of the Earth and from historical records are fed into the equations, and the computer calculates the many complex interactions between them. The goal is for the model to mimic the behavior of the phenomena so closely that it can be used with confidence as a predictive tool. For example, we will see in Chapter 17, Glaciers and Climate, that researchers concerned with the possibility of global warming are busily constructing computer models of Earth's climate system in order to predict the effects of carbon dioxide buildup in the atmosphere.

Geology as an Applied Science

Like doctors and lawyers, most geologists specialize. Table I.1 summarizes specialist research interests as compiled by the American Geological Institute (AGI). We present it here as a preview of the scope of the geosciences.

Nearly every specialty in geology also has an applied aspect. However, what the list does not convey is how interrelated the specialties are. As an example, consider the variety of skills petroleum geologists must employ to locate subsurface oil and natural gas deposits (Figure I.13). Oil and gas form, migrate, and accumulate within sedimentary rock strata, so petroleum geologists must have a firm grounding in sedimentology—the study of sediment formation, transport, and deposition. In addition, the valuable fluids are found in specific reservoir rocks within these thick accumulations of strata. To be able to identify these formations, they must apply their knowledge of stratigraphy—that is, the study of the time and space relationships of sedimentary strata.

The time relationships of strata are frequently determined through fossils, so the petroleum geologist must also have a grounding in paleontology. However, not only are oil and gas found in specific strata, but they also migrate under pressure to traps *within* the strata, where they accumulate in economically extractable quantities. The most productive traps are those associated with folds, faults, or rising salt domes. Analysis of these features is within the province of structural geology, another area of knowledge useful to the petroleum geologist. Finally, most traps are hidden deep in the Earth's crust, and a host of gravitational, electrical, and seismic techniques are required to locate them. Thus the petroleum geologist must know something about applied geophysics. Our point is that one could randomly choose any specialty on the AGI list and find similar connections to other specialties.

Geology is a practical science, and nearly everything learned by geological research has its human applications. Seismologists who attempt to predict earthquakes are concerned with the safety of entire cities (Figure I.14). Geomorphologists study landforms to identify landslide hazards, to minimize flood damage, or to prevent the loss of precious farmland to soil erosion. Hydrologists track the movement of dangerous carcinogens through groundwater to prevent contamination of wells, streams, and lakes. Glacial

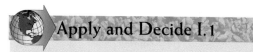

Apply and Decide I.1

Geology, Society, and the Environment

Sixty-one centuries ago (about 4100 B.C.), one of the earliest and most successful of agricultural civilizations flourished on the flood plain between the Tigris and Euphrates rivers in Mesopotamia, the region that is now Iraq. The Mesopotamians dealt with the demands of their burgeoning population by building an elaborate system of irrigation, which brought more land under cultivation. Although this solution worked for many centuries, irrigation in the dry climate that prevailed brought with it the scourge of salinization—a problem associated with the cultivation of arid lands that vexes geoscientists to this day. Contemporary accounts described how, as the salt content of soil gradually increased, food production declined. By 1700 B.C., the Euphrates delta had been abandoned. Agricultural success had resulted in population growth, which in turn placed an intolerable burden on the environment. It was a burden that the technology of the period (irrigation) alleviated in the short run but intensified in the long run. Impoverishment of the land and its people were the ultimate result.

The environmental effects of ancient civilizations were local and limited by relatively small populations. For example, the population of Mesopotamia was never greater than about half a million people. However, the growth of the human population has been relentless. At the time Christ lived, there were some 200 million people in the world. By the year 1900, the world population had climbed to a billion; by 1950 it rose to 2 billion. Today, scarcely 50 years later and 2000 years since the time of Christ, there are 6 billion people on the Earth. This increase has had an even greater environmental impact than the numbers might suggest, because it has been accompanied by unprecedented per capita demands on the Earth's resources and on the capacity of the Earth to cleanse itself of pollution.

Effects of Modern Society on the Natural Environment

Each year the world's rivers transport to the ocean 22 billion tons of sediment—rock fragments and soil eroded from the land surface. This amount appears trivial, however, when compared with the 2700 billion tons moved each year by human hands and machines in activities such as mining, quarrying, land clearing, and construction (see the figure). In these and other ways, the human species, through its command of technology and social organization, has come to rival the impact of natural systems on the Earth's surface environment. A summary of the major environmental changes wrought by the human species over the past few thousand years would include the following developments:

- We have modified at least half of the land habitats of the Earth and made a significant impact on the other half. (No region is remote enough to escape the human imprint. Traces of DDT and 19 other pesticides, for example, have been found in the tree bark of remote Brazilian rain forests.)

- We have converted to farmland an area equal to the entire continent of South America. If you add modern cities, suburbs, roads, golf courses, and theme parks to that area, it more than doubles the sprawl of human civilization across the Earth's surface.

- Not surprisingly, we have drastically reduced those species in competition with us. Mass extinctions of animal and plant species have been accomplished through pollution, hunting, and most importantly, through habitat destruction—especially the clearing of the world's forests, prairies, and wetlands. Biologist E. O. Wilson of Harvard University estimates that at least 25 percent of the species still alive today will perish in the next 50 years—an extinction rate not approached since the dinosaurs died out at the end of the Cretaceous period, 65 million years ago.

- We are rapidly depleting the water stored in underground aquifers, our primary source of fresh water. We use 35 times the amount our ancestors consumed three centuries ago. Most of it goes to farming and approximately 25 percent is used by industry. Excessive groundwater pumping leads to the sinking of land surfaces (subsidence), and saltwater encroachment in coastal regions.

- To supplement the underground supply of water and create hydroelectric power as a byproduct, nearly every major river on the Earth's surface has been dammed or leveed, or will be in the near future. Reservoirs alone submerge a land area the size of France. The trapping of sediments behind dams leads to the starvation and erosion of downstream flood plains, deltas, and beaches.

- The gases released by factories and automobiles turn atmospheric moisture into acid rain that destroys forests and kills off the aquatic life in lakes.

- By adding carbon dioxide, methane, sulfur dioxide, and other "greenhouse gases" to the atmosphere, we have altered its composition in ways that may profoundly affect the Earth's climate. These gases trap the heat rising off the Earth's surface and are strongly suspected to

wood, cow dung, or agricultural wastes are used as fuel for heating and cooking.

Most of Africa and the Indian subcontinent have no wastewater treatment facilities; raw and human sewage are discharged directly into the same bodies of water used for drinking. In India, for example, ritual bathers are exposed to a wide range of toxicants and pathogens, both solid and suspended, in the waters of the polluted Ganges River. In China, billions of tons of unfiltered industrial pollutants are dumped directly into the waterways; in fact, that one country produces more toxic water pollution than the whole of the western world.

Construction of this access highway in Orange County, California depicts the human capacity to disrupt the natural environment on a monumental scale.

Geoscience and the Environment

Our purpose here is not to compile a list of depressing facts about humanity's impact on the environment, but to emphasize that the fate of the Earth and the human race are inextricably intertwined. Modern science and industry have been remarkably successful at meeting human needs, but at a cost. Still, energy and mineral resources are indispensable if we are to maintain current living standards. In the nineteenth and early twentieth centuries, society's goal was to exploit the Earth's resources as quickly and cost-efficiently as technology would allow. Today we know that those resources are limited and that exploitation must be balanced by careful ecological restoration and preservation. As recognition of our responsibilities expands, the role of the geoscientist grows ever more vital. Throughout this book, you will find examples of how geoscientists:

cause global warming. Although some distinguished scientists disagree with the whole concept of human-induced global warming, most appear satisfied that the case has been made. For them, the major questions are: How much global warming can we expect in the foreseeable future? Will the glaciers melt significantly and cause a corresponding rise in sea level? How might the world's ecosystems adjust to the climatic change? Will shifting rainfall patterns turn the fertile farmlands of North America into deserts? Will diseases now confined to the tropics spread to regions at higher latitudes as those regions become warmer?

- The products associated with our daily consumption— the hulks of cars and trucks, tires, TV sets and computers, plastic food wrappers and containers, ash from coal furnaces, transmission fluid, battery acid, and nuclear waste—end up in junkyards, holes in the ground, landfills, and the oceans. Can the Earth's normal systems of erosion and deposition process these wastes in a time frame meaningful to humans?

Though you would hardly guess the fact from the terms within which most environmental debates are framed, the effects of land, water, and air degradation and pollution tend to fall most harshly on the poorest people of the Third World nations. Each year those nations lose four million children under the age of five to acute respiratory diseases. According to United Nations Development reports, most of these deaths stem from children living in poorly ventilated huts where

- discover the mineral and energy resources necessary for the survival of the human race while searching for alternative energy sources that are less harmful to the environment;

- solve problems of environmental pollution and alteration;

- assess the impact of human-induced change on the natural environment and deduce long-term trends;

- work to protect people and property from natural disasters; and

- recommend changes in land and water use, energy consumption, and resource development to preserve the natural environment while raising living standards.

Table I.1 Selected Geoscience Specialties

- **Geophysicists** decipher the Earth's interior and magnetic, electric, and gravitational fields.
- **Geochemists** investigate the nature and distribution of chemical elements in rocks and minerals.
- **Petroleum geologists** are involved in exploration and production of oil and natural gas.
- **Economic geologists** explore for and develop geologic materials that have profitable uses.
- **Hydrologists** study the circulation and distribution of global water and ice on and below the surface and in the atmosphere.
- **Hydrogeologists** study the abundance, distribution, and quality of groundwater and related geological aspects of surface water.
- **Engineering geologists** investigate geological factors that affect engineering structures such as bridges, buildings, airports, and dams.
- **Environmental geologists** work to solve problems of pollution, waste disposal, urban development, and hazards such as flooding and erosion.
- **Seismologists** determine the location, magnitude, and origin of earthquakes and trace the behavior of earthquake waves to interpret the structure of the Earth.
- **Geochronologists** determine the age of rocks by calculating the rates of decay of certain radioactive elements and thus help reconstruct the geologic history of the Earth.

- **Planetary geologists** study the Moon and other planets to understand the evolution of the Solar System.
- **Geomorphologists** study the effects of Earth processes and investigate the nature, origin, and development of present landforms and their relationship to underlying structures.
- **Glaciologists** study the physical properties and movement of glaciers and ice sheets.
- **Marine geologists** investigate the oceans and continental shelves.
- **Mineralogists** study the occurrence, formation, composition, structure, and properties of minerals.
- **Paleontologists** study fossils to understand past life-forms and their changes through time and to reconstruct past environments.
- **Petrologists** determine the origin and genesis of rocks by analyzing mineral or grain relationships.
- **Sedimentologists** study sedimentary rocks and the processes of sediment formation, transportation, and deposition.
- **Stratigraphers** investigate the time and space relationships of layered rocks and their fossil and mineral content.
- **Structural geologists** study deformation, fracturing, and folding that has occurred in the Earth's crust.
- **Volcanologists** investigate volcanoes and volcanic phenomena.

geologists study ice cores to discern climatic cycles and to test various hypotheses of global warming or cooling. Structural geologists search for concealed faults and folds in order to aid engineers in designing tunnels, dams, and roads.

There are 22 chapters in this book, and in nearly every one of them, you will find examples of geologists doing practical work—from discovering deposits of oil, coal, metals, and gems to preserving beaches, harbors, and wetlands. You will see how geologists attempt to solve problems that stem from society's uses and misuses of the natural world, and you will learn about how you can be part of the solution, not part of the problem.

Figure I.13 An offshore drilling platform. Petroleum geologists are employed by oil companies not only to find oil deposits, but also to determine the dimensions of an oil pool and to monitor the quality of the oil extracted.

Figure I.14 This geologist is using a laser measuring device as part of an intensive research program to predict earthquakes along California's San Andreas fault.

STUDY OUTLINE

Geology is the study of the composition, structure, and history of the Earth and the processes that shape it. **Geologists** conduct geological research studies and apply their knowledge to benefit human society.

I. A GEOLOGICAL PUZZLE

A. A deep river-cut canyon filled with marine sediments and fossils was discovered buried below sea level under the Nile River. At about the same time, a thick layer of salt was located, first by seismic waves and then by drilling, beneath the floor of the Mediterranean. These possibly related findings led some geologists to propose an unusual **hypothesis** about the history of the Mediterranean Sea: that it had in the past evaporated and left a sunken desert, thickly coated with salt deposits, thousands of meters lower than the surface of the surrounding continents.

B. Other geologists argued that the Mediterranean began as a shallow briny sea in which the salts were deposited in the same way they are deposited on the Earth today. The sea floor was lowered and the Mediterranean became a deep ocean after the salts were deposited.

C. More information is needed before we can determine which **model** is correct. In the meantime, both correct and incorrect models play important roles in the scientific process, serving as springboards for debate and future research.

II. RESEARCH AND THEORY IN THE GEOSCIENCES

A. Science is the study of the natural or material world through careful observation and experimentation. Geologists employ the **scientific method,** a systematic approach to gaining knowledge of the natural world.

    1. Geologists begin an investigation by asking a question about a geological phenomenon. Before proceeding with their own investigation, they learn what other scientists have written on the subject.

    2. Development of a hypothesis or model, a tentative answer to the question, may be the next step in the investigation, or it may come later during the collection of data. Sometimes, in geological research, it is best to employ

   multiple working hypotheses, ruling out no proposed explanation until the evidence unequivocally contradicts it.

   3. The evidence, or **data,** may be physical samples, instrument measurements, or statistics collected by the researchers in the field or obtained from government agencies and research institutions.

   4. The evidence is then organized and analyzed to identify relationships among the data.

   5. When they are ready to share what they have found with the scientific community, the researchers prepare a paper describing their hypothesis and supporting evidence, and submit it to a scientific journal for publication.

   6. The hypothesis is subjected to evaluation, debate, and further testing by the scientific community. This assessment leads to new research, and the process begins again.

   7. A hypothesis becomes a **theory** when it has been elevated to a higher level of confidence through extensive testing and debate. In science, however, no theory is ever elevated beyond debate.

B. Much of geology involves studying Earth materials and processes in the field before samples and other information are taken to the laboratory for further analysis.

   1. The **geological map,** which displays the distribution, ages, and structural features of the rocks present, is an important tool in field study.

   2. Seismic recordings and chemical and optical analyses are among the techniques used in laboratory studies. Computerized mathematical analysis is necessary for most studies involving vast and complex sets of data. Computer models that simulate the behavior of natural phenomena are also used.

III. GEOLOGY AS AN APPLIED SCIENCE. There are many areas of specialization in the geosciences (see Table I.1), and nearly all have both basic research and applied aspects. Geological specialties are also interrelated. For example, a petroleum geologist must have a firm grounding in such areas as sedimentology, stratigraphy, paleontology, structural geology, and geophysics.

STUDY TERMS

data (p. 12)   model (p. 11)
geologic map (p. 15)   multiple working hypotheses (p. 11)
geologist (p. 3)   scientific method (p. 11)
geology (p. 3)   theory (p. 13)
hypothesis (p. 4)

METRICS REVIEW

In this introductory chapter, English conversions appear in parentheses after metric units; in all other chapters, measurements will be in metric units alone. See Appendix A for a complete chart of metric/English unit conversions. If you are comfortable with the metric system, skip this section. If you are not, here are a few tips that will help you develop the same feel for metric units that you have for English units. Take, for example, a layer of volcanic ash 2.3 inches thick. You don't know *exactly* how thick 2.3 inches is, but you can imagine it. You may have learned to think of an inch as the distance from the tip to the first joint of your thumb, and you can envision the thickness of the ash as a little more than twice that. In the same way, you can learn to estimate centimeters with your fingernail. One centimeter is about the width of an adult female's index nail, or a male's pinky nail.

You might also want to memorize a few key, approximate conversions from metric to English units. For an exact conversion from centimeters to inches, multiply the number of centimeters by 0.39. For approximate conversions, remember that

- 1 centimeter is a little less than half an inch
- 8 centimeters is a little more than 3 inches
- 30 centimeters is about 12 inches

For an exact conversion from meters to feet, multiply the number of meters by 3.28. For approximate conversions, remember that

- 1 meter is a little more than 3 feet
- 20 meters is about 66 feet
- 300 meters is a little less than 1000 feet

For an exact conversion from kilometers to miles, multiply the number of kilometers by 0.62. For approximate conversions, remember that

- 1 kilometer is about two-thirds of a mile
- 40 kilometers is about 25 miles
- 100 kilometers is 62 miles

CRITICAL THINKING QUESTIONS

1. Identify the steps of the scientific method in the story of the Mediterranean puzzle.
2. A geologist is often compared to a detective visiting the scene of an event millions of years after it occurred. In what ways is this comparison reasonable? In what ways is it not? How does a scientist's work compare to an artist's, a minister's, or a politician's work?
3. Explain the statement, "Luck is the residue of design." What does it have to do with science?
4. How might a broad academic background in the sciences and liberal arts add to a geologist's competence ? In what ways might a geology course benefit the liberal arts or business major?

Basic Geological Concepts

Geology is a relatively young sci-ence, having begun to evolve into something like its modern form in the late eighteenth century. Little was known in those early days of the composition and structure of the Earth's crust. The interior of the Earth was a complete mystery, the vast expanses of the ocean floor a matter of conjecture. If you wanted to visit a region, you had to travel by boat, horse, or foot. Nevertheless, many concepts established during that period remain at the core of the science.

We begin this chapter with a discussion of the concepts of classical geology, which, as the science developed, were applied to an ever-expanding base of knowledge of the rocks and structures of the Earth. In the second half, we discuss a triumph of the appli-cation of these concepts, the theory of continental drift. Drift was the forerunner of plate tectonics, the unifying theory of modern geology. The development of plate tec-tonics is discussed in full in Chapter 2, and its applications will be evident throughout this text. Of the many threads that connect the pioneering geologists to their modern spiritual descendants, the most important is the widening recognition that the Earth is an ever-changing, dynamic planet.

geology
The study of the materials, processes, products, and history of the Earth.

The Huttonian Revolution

The founding of modern geology is called the Huttonian revolution, in honor of the contributions of the Scottish physician and farmer, James Hutton (1726–1797). His ideas may be expressed in three closely related principles.

1. The processes that presently shape the Earth have acted in much the same man-ner throughout geologic time, so direct observation of the Earth of today enables us to interpret the past.

◄ Much of what geologists know about the composition of the Earth's surface and the processes that shape it comes from the study of rocks in their natural settings.

Figure 1.1 The rock outcrop at Siccar Point, Scotland, from which Hutton drew profound inferences concerning the Earth's age and ability to cycle materials.

2. Observation of the Earth reveals that these processes cause change in very small increments, and yet the sum of the changes we observe is enormous. Thus we conclude that the Earth's age is also enormous—far greater than all of human history.
3. Earth is a dynamic planet whose surface is in a constant state of change and whose materials are ceaselessly cycled and recycled.

Hutton's first principle—which in effect established that the Earth is subject to direct scientific inquiry—is called **uniformitarianism.** It is often summarized by the simple statement, "The present is the key to the past."

Uniformitarianism was an idea not widely accepted in Hutton's day. Some held that the Earth is quite young, having reached its present form through a series of sudden violent events. This position was called **catastrophism.** Its advocates struggled to find a place in science for such biblical events as Noah's Flood, and they sought to justify through science the biblical interpretation that the Earth is only a few thousand years old. In contrast, the uniformitarians pointed to numerous examples of how gradual change could cumulatively account for most of what was known about the Earth. Hence, uniformitarianism became the accepted view of most geologists. Sudden, violent episodes such as landslides, earthquakes, and volcanoes were seen as events that occur within an overall context of gradual change (see Aside 1.1).

Hutton reached his conclusions regarding the Earth's great age and dynamism by applying the principle of uniformitarianism to rocks he observed in the field. In his trips around the British Isles, Hutton observed the everyday effects of rain and wind as they beat down upon rocks and soil. He saw rivulets of water carrying rock fragments, sand, and soil to the streams. He saw muddy waters flowing downstream and, in rainy seasons, sand and gravel moving with the swift currents along the stream bottoms. He thus realized that running water ceaselessly wears down the surface of the land and carries it, particle by particle, from the high mountains down to the oceans, where these particles are deposited as **sediment** on the sea floor.

Hutton also saw rocks in the mountains that were made of cemented sand particles and rock fragments, and he wondered how these sediments had gotten back up the mountains from the sea. Then one day, as he was exploring the low cliffs of Siccar Point along the east coast of Scotland, the answer hit him so suddenly that he was reported to have jumped up and down in delight. Hutton visited Siccar Point by boat in 1788; we will visit it using Figure 1.1.

Hutton observed that the layers, or **strata,** at Siccar Point were **sedimentary rocks,** consisting of fragments derived from the wearing away of older rocks. He also noted that there were two distinct sequences of strata present: a nearly horizontal sequence

uniformitarianism
The principle that is based on the concept that past geological events can be explained by forces occurring today. "The present is the key to the past."

catastrophism
The doctrine that a series of sudden, violent, and short-lived worldwide events are responsible for the state of the Earth's crust and for the variety of life-forms that live on it.

sediment
Solid particles transported by water, wind, or ice and deposited in loose layers on the Earth's surface.

strata
Visually distinguishable layers of sedimentary rock.

sedimentary rock
A layered rock formed from the consolidation of sediment.

Figure 1.2 The sequence of events deduced by Hutton from the outcrop at Siccar Point. The surface between the lower and upper strata is erosional—an unconformity that separates rocks of markedly different ages.

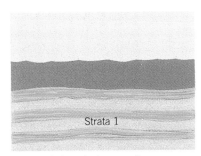

1. Strata 1 deposited on sea floor

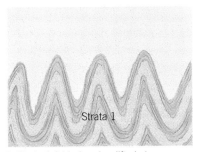

2. Strata 1 folded and uplifted above sea level

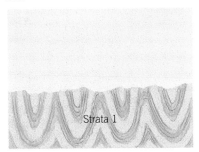

3. Strata 1 eroded

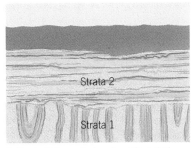

4. Strata 2 deposited on top of eroded surface

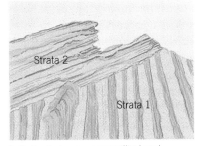

5. Both strata sequences tilted and uplifted above sea level. Strata 2 eroded

resting atop one that had been compressed into a series of wavelike folds. So tight were the folds in the lower sequence that the strata appeared to stand upright, or vertically—especially so because the folds were truncated. The boundary between the two sequences is what geologists today call an *unconformity*.

Hutton knew that sediments are deposited on the sea floor in horizontal layers, parallel to the Earth's surface, so that the oldest layer is on the bottom with successively younger layers resting on top. This simple principle, called **superposition,** was established long before Hutton. He had seen vertical layers similar to the lower strata at Siccar Point many times before in the folded strata of the highlands of the British Isles. Therefore, it was immediately clear to him how the two sequences of vertical and horizontal rocks at Siccar Point had been formed (Figure 1.2). The vertical sequence had originally been deposited horizontally on the sea floor and was the remnant of a much thicker accumulation of sedimentary rock. Subsequently, the rocks were folded and uplifted by powerful forces from within the Earth to form a mountain range. With uplift came the renewed attack by running water and all the other things that wear away rocks exposed to the Earth's surface. Eventually, the folded rocks were worn down to sea level, and the younger, nearly horizontal strata were spread over the truncated remnants of the once proud mountains. Thus clearly visible at Siccar Point is the story of how each sequence of strata formed, but missing at the unconformity is the entire record of the time it took to wear the mountain range to a nub.

Having observed the slow rate at which streams erode the land, Hutton knew this gap in the record amounted to an enormous time span compared with that of a human life. He also realized that it represented but a short chapter in the history of the Earth, because the Siccar Point rocks had been derived from still more ancient rocks. Furthermore, Hutton saw that the processes that had created the unconformity remained at work at Siccar Point, for the very cliffs he stood upon were being worn low and were supplying sediment to the adjacent beach and ocean floor. He knew this sediment was the raw material of future mountains, because much of the British highlands were composed of identical stuff—uplifted marine strata made of sediment washed in from former landmasses.

superposition
The principle that in a sequence of sedimentary strata, the oldest layer is located on the bottom and followed in turn by successively younger layers, up to the top of the sequence.

ASIDE 1.1 *Modern Catastrophism*

Throughout the nineteenth century, geologists struggled to establish the study of the Earth as a science and to banish the superstitious and sensationalistic interpretations to which the subject was prone. For this reason, *gradualism*—incremental change based only on directly observable processes—came to be viewed by many as the sole embodiment of rational investigation of the Earth.

In recent years, however, geologists have uncovered evidence that sudden, catastrophic events may indeed have fundamentally changed the Earth in ways that cannot be explained within the classic uniformitarian framework. For example, consider the well-regarded hypothesis that the 5000-square-kilometer scablands of Washington State were carved in *a few days* by a gargantuan flood of glacial meltwaters. First advanced in the early 1920s, this idea was ridiculed and debated for nearly fifty years. Finally, in the late 1960s, the overwhelming evidence in support of the theory broke down its orthodox gradualistic opposition.

Because most geologic change is observed to be gradual, many modern geologists look upon catastrophic interpretations with some skepticism. For example, in Chapter 10, Geologic Time, we will discuss the hypothe-

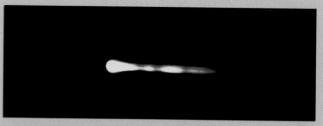

A meteor breaks up as it enters the Earth's atmosphere. One factor that gradualism does not account for is the effect of occasional collisions of extraterrestial objects with the Earth.

sis that an asteroid crashing into the Earth was responsible for the sudden extinction of the dinosaurs—currently one of the hottest topics of debate in the geosciences. Whereas strong evidence has been amassed in favor of the asteroid hypothesis, much evidence also supports other explanations of the dinosaurs' demise.

The important point is that the modern concept of catastrophism is based on physical evidence that can be evaluated and debated; it agrees with the age of the Earth as established by modern geology and is unrelated to the religion-driven version of Hutton's day.

For these reasons, Hutton interpreted the events recorded in the strata at Siccar Point as linked to an apparently endless cycle of uplift, erosion, deposition, and renewal—the sum of which constitutes the Earth's past, present, and future. Two statements by Hutton sum up his ideas:

> "The Earth, like the body of an animal, is wasted at the same time it is repaired . . . it is thus destroyed in one part but is renewed in another."

> "We see no vestige of a beginning, no prospect of an end,"—a statement known today to be not precisely correct, but close to the mark in spirit.

Establishing Geologic Time

relative age
The age of an object or event expressed relative to the age of other objects or events but not in relation to time units such as years.

fossil
The remains, trace, or imprint of a plant or animal preserved in rock.

Hutton was able to determine the sequence in which the rocks at Siccar Point formed—that is, their **relative ages**—by tracing the physical relationships among the rocks. However, such methods cannot be applied to rocks that are in separate locations. In order to compare the ages of the rocks at Siccar Point to those in the Grand Canyon or in China, for example, we require the existence of objects that are preserved in strata throughout the world and are the same relative age wherever they are found. Fortunately, such objects do exist. They are called **fossils,** the physical evidence of past plant and animal life preserved in rocks (Figure 1.3).

We can interpret the time significance of fossils through the application of uniformitarian principles. We begin by noting that many animal and plant species are distributed over wide regions of the Earth's surface. Some, like pollen or plankton, are swept around the Earth by winds and ocean currents. Others, like certain species of clams, are spread over wide expanses of the shallow ocean floor. We can observe the

Figure 1.3 A typical fossil assemblage in ancient limestone strata. We observe that shells of contemporary organisms are incorporated in present-day sediments at the time of deposition. Thus we may confidently assume that the shells, or fossils, of ancient organisms are the same age as the strata within which they are found.

shells of these and other organisms being preserved today in the sedimentary layers deposited on lake and sea bottoms and on river flood plains—proof that the shells are the same age as the layers. If the present is the key to the past, we conclude that a plant or animal fossil must be the same age as the sedimentary layer in which it is found. We further reason that if the principle of superposition applies to rock layers, it must also apply to the fossils in the layers. The oldest fossils are found in the oldest layer, which is generally at the bottom of a sedimentary sequence, and the fossils in each succeeding layer are younger than those in the layer directly beneath them.

None of these observations would matter if we found identical fossils in all rock layers, but such is decidedly *not* the case. Wherever geologists have studied thick sequences of strata, they have found that the fossils in them change from bottom to top. Some are found only in the lower levels; others at intermediate levels; still others at the highest levels. In other words, organisms exist over a certain range, or time span, and then become extinct. What is more, study of the entire fossil record proves that fossil organisms succeed one another in definite and recognizable order, so that rocks containing similar fossils are similar in age. This is a statement of an important principle, **faunal succession,** and it holds true no matter how widely separated are the rock strata that contain the fossils. When you climb the cliffs of North Dakota and Montana, you discover the bones of dinosaurs in layers below those that contain the bones of ancestral horses. At no place on Earth is it the other way around.

Faunal succession is, of course, a manifestation of organic evolution—the fact that living things have changed through time, giving rise to new species; that the life on Earth today, including human life, is derived from previous life-forms. The biology of evolution is a fascinating subject in its own right, but geologists are also interested in faunal succession because it allows them to arrange rocks in chronological order and to *correlate* them (that is, to prove their age equivalence). Once the rocks are arranged and correlated, so too are the events they record.

By comparing and correlating rock strata around the world, geologists were able to construct a chronological ordering of geologic events called the relative **geologic time scale**—*relative* because it denotes the sequence of events and not the actual time that they occurred. Largely the work of nineteenth-century geologists, it is one of the great achievements of science. As shown in Figure 1.4, the earliest division of the time scale is listed on the bottom, with successively younger ones on top, just the way the ages of rock strata are deduced in the field. We will have much more to say about the time scale in Chapter 10, Geologic Time. For the present, it is sufficient to acquaint yourself with the general vocabulary and subdivisions of the scale.

faunal succession
The principle that fossil organisms succeed one another in definite and recognizable order, so that rocks containing identical fossils are identical in age.

geologic time scale
A chronological ordering of geologic events and time units listed in sequence from the oldest on the bottom to the youngest on the top.

Figure 1.4 The units of the relative geologic time scale are arranged with the oldest divisions at the bottom and successively younger ones on top. The development of radiometric dating techniques enabled geologists to transform the relative time scale into an absolute time scale that shows the age of each unit in years.

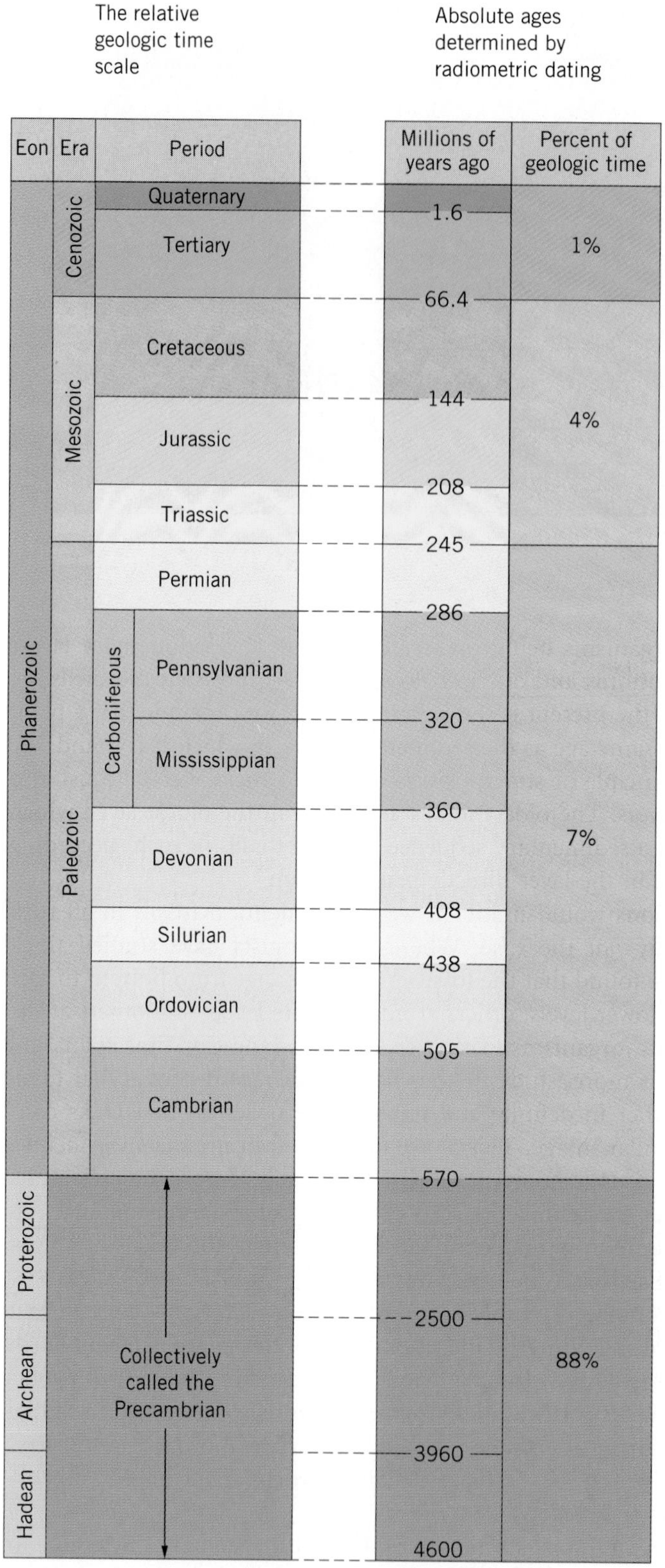

The relative geologic time scale

Absolute ages determined by radiometric dating

Eon	Era	Period	Millions of years ago	Percent of geologic time
Phanerozoic	Cenozoic	Quaternary	1.6	1%
		Tertiary	66.4	
	Mesozoic	Cretaceous	144	4%
		Jurassic	208	
		Triassic	245	
	Paleozoic	Permian	286	7%
		Pennsylvanian (Carboniferous)	320	
		Mississippian (Carboniferous)	360	
		Devonian	408	
		Silurian	438	
		Ordovician	505	
		Cambrian	570	
Proterozoic		Collectively called the Precambrian	2500	88%
Archean			3960	
Hadean			4600	

Absolute Time

absolute age
The age of an object or event in years as determined radiometrically or by other quantifiable means.

Notice that Figure 1.4 also lists the **absolute ages,** in millions of years, of the various subdivisions of the scale. This was a later addition that came with the development of *radiometric-dating* techniques early in the twentieth century. Until then, geologists had no actual way of determining the age of rocks or, indeed, the age of the Earth. They could offer only the informed opinion that these things must be very old.

Radiometric age determinations are based on the decay of unstable (that is, radioactive) atoms into stable (or nonradioactive) atoms within rocks and other materials. Because the rates of decay are constant, we know that the radioactive atoms are, in effect, nuclear clocks that tick away within the rocks over vast time periods. The age of a rock can thus be calculated from knowledge of the decay rates and the proportion of radioactive to stable products present. We will also postpone discussion of radiometric dating until Chapter 10. For the present, let us just note that geologists have dated a wide variety of rocks, and from these dates, they have been able to determine the ages of the various divisions of the geologic time scale, as well as the age of the Earth.

When radiometric dating revealed the true scope of geologic time, geologists, like other people, were quite surprised. Who would have guessed that analysis of meteorites would prove the Earth to be 4.6 billion years old, or that nearly the entire fossil record upon which the relative time scale is based would span only the past 570 million years? The 4.6-billion-year age of the Earth is a time span incomprehensible in terms of human experience and so is perhaps best treated by analogy. An interesting one follows.

Meteorites that have fallen to the Earth's surface from the Solar System are believed to be similar in composition to the matter from which the Earth was formed.

> Compress, for example, the entire 4.6 billion years of geologic time into a single year. On that scale, each second represents about 150 years. The oldest Earth rocks we know date from early February and living things first appeared in the sea in the last week of March. Land plants and animals emerged in late November, and the widespread swamps that formed the Pennsylvania coal deposits flourished for about four days in early December. Dinosaurs became dominant in mid-December but disappeared on December 26, at about the time the Rocky Mountains were first uplifted. Humanlike creatures appeared sometime during the evening of December 31, and the most recent continental ice sheets began to recede from the Great Lakes area and from northern Europe about 1 minute and 15 seconds before midnight on that day. Rome ruled the Western world for 5 seconds, from 11:59:45 to 11:59:50. Columbus arrived in America 3 seconds before midnight, and the science of geology was born with the writings of James Hutton just slightly more than 1 second before the end of our eventful year of years.*

The Earth Machine

Though Hutton's portrayal of the Earth as an animal wasted at the same time it is repaired makes a striking metaphor, modern geologists find that thinking of the Earth as if it were a gigantic machine is a more useful approach. The energy sources that drive the machine are both external and internal. Solar and gravitational energy power those processes that wear down rocks and redistribute fragmental and dissolved materials over the Earth's surface. Internal heat, much of it generated by the slow decay of radioactive atoms, is ultimately responsible for volcanoes, mountains, earthquakes, the creation and destruction of oceanic crust, continental drift, and related phenomena. The Earth machine constitutes a closed system, one whose mass has remained essentially constant since the time the Earth solidified as a planet. It is only energy that enters or leaves the system (aside from meteorites, a few astronauts and space probes, and negligible quantities of cosmic dust).

External Processes

James Hutton knew that each stream is linked to the continuous circulation pattern of the world's water, called the **hydrologic cycle** (Figure 1.5). The cycle begins when ocean

hydrologic cycle
The continuous circulation pattern of the world's water, from ocean to atmosphere to Earth's land surface to ocean again.

---

*Adapted from the analogy created by Don L. Eicher for his book *Geologic Time,* 2nd ed. (Englewood Cliffs, N.J.: Prentice Hall, 1978). The calendar dates reflecting the age of the Earth, the oldest rock, and the first living things have been updated in light of recent findings.

Figure 1.5 Solar energy powers the Earth's hydrologic cycle. The circulation of water among land, atmosphere, and ocean is unique to our planet and is largely responsible for the redistribution of matter over the Earth's surface.

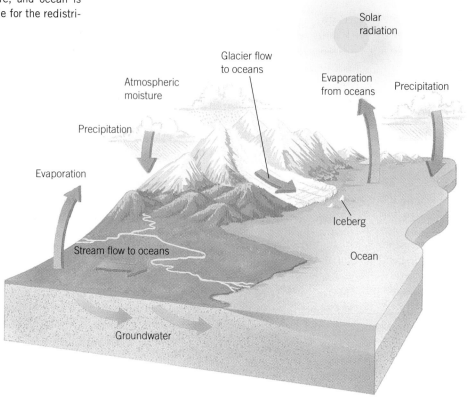

water, having absorbed solar heat, evaporates and enters the atmosphere, leaving its dissolved salts behind. Some of the water vapor gets caught up in the wind systems, condenses, and falls on the land as rain or snow. A portion of the water striking the land surface evaporates back to the atmosphere. Some of it is also taken up in the roots of plants and then passed back to the atmosphere. About a third of it, however, either returns directly to the sea as stream runoff, seeps beneath the surface as underground water, or freezes and becomes incorporated in glaciers. Glaciers melt at their edges, however, and most groundwater eventually seeps back to the surface; so in time, both join the general stream runoff back to the sea.

Water plays the key role in a vast system of **erosion** and **deposition** in which the rocks of the continents are broken down and the fragments are transported to the lowlands and margins of the continents. The system begins with **weathering,** the physical and chemical alteration of rocks exposed on the Earth's surface. Though many processes are involved, naturally occurring acids play the key role of converting the hard minerals of the rock to clay and other products. Bacteria, plants, and animals, themselves dependent on water, also aid in the breakdown; and the weathering by-products and organic residues mix to produce soil, the prerequisite of plant and animal life on the land.

Water that falls on the land as rain initially flows downhill in thin sheets. The sheets seek out weaknesses and irregularities in the land surface, and the water soon becomes funneled into stream channels (Figure 1.6a). These channels concentrate the energy of the flowing water as it runs downhill. For this reason, streams are efficient implements for carving valleys. As they do so, hillside slopes are made unstable, and the loose weathering products slide or slump downhill to the streams under the influence of gravity— a process called **mass wasting**. Thus valleys are widened, and the material carved out of the land is eventually brought to the sea. Along the way, the sediments are spread on the flood plains, deltas, marshes, and beaches and on the margins of the continents beneath the sea (Figure 1.6b). Some of the fine clays and silts are carried in suspension far from the coasts and settle to the bottom, forming deep-sea muds.

erosion
The wearing down of rocks or soil, by weathering, mass wasting, running water, wind, and ice.

deposition
The gravitational settling of rock-forming materials out of such natural agents as water, wind, or ice.

weathering
The physical and chemical alteration of rocks exposed to atmospheric influences on the Earth's surface.

mass wasting
The downslope movement of soil and rock material under direct influence of gravity.

(a)

(b)

The geologic work of water on the Earth's surface is both destructive and constructive. The destructive work—erosion—involves weathering, mass wasting, and transport of sediment by streams, and it reduces the elevation of the continents. The constructive work—deposition—redistributes the continental real estate. Glacial ice, wind, ocean currents, and waves also contribute to these constructive and destructive processes. This ceaseless movement of Earth materials downhill, from high places to low, should reduce the surface of the Earth to a featureless sea-level plain. However, such leveling has not occurred on a worldwide basis because forces from within the Earth uplift the land and form mountains and plateaus. Thus internal and external forces work together to disperse matter over the surface of the Earth.

Figure 1.6 (a) Over time, a small stream may erode large quantities of rock and soil. Many such small streams feed into large rivers. (b) The Ganges River Delta, shown here, illustrates the constructive work of streams. Built of sediment that was eroded and transported from the Himalaya Mountains, the delta underlies the fertile farmlands of Bangladesh.

The Rock Cycle

Mountain building requires energy. Hutton believed an abundance of it was available in the form of internal heat. If this heat were sufficient to uplift and deform rocks, he said, it was also capable of melting them deep inside the Earth. The molten masses could then rise and pierce the rocks of the Earth's thin outer layer, the crust. Eventually, some of them would spill onto the surface as lava. Although this Scottish physician and farmer had never seen a live volcano, he pioneered the idea that **magma** is derived from the melting of solid rock and that much of the Earth's surface is composed of **igneous rocks**—rocks that have cooled from the molten state (Figure 1.7).

In our modern age, when television provides spectacular footage of lava pouring down the flanks of erupting volcanoes, it may seem unlikely that intelligent people would have ever doubted the existence of igneous rocks. However, in Hutton's day, many people regarded the few eruptions that they had heard about or witnessed as purely local events, and they did not think that lava had anything to do with fundamental cycles of nature. Debate at the time centered on the possible igneous origin of granite and

magma
Molten (hot-liquid) rock material.

igneous rock
Rock that has cooled from the molten, or magmatic, state.

(a)

PINK GRANITE

GRAPHIC GRANITE

PEGMATITE

OBSIDIAN

LAVA

(b)

Figure 1.7 (a) This geologist is collecting a magma sample of a recent Hawaiian eruption. The magma will eventually cool and solidify to form igneous rock. (b) Four types of igneous rocks and the volcanic glass, obsidian.

lithification

The process by which sediment is converted to sedimentary rock through compaction, cementation, or crystallization.

basalt—rocks common to Scotland and other parts of the British Isles. Hutton's view would prevail if he could prove that the granites and basalts were younger than surrounding rocks and had pierced them while still molten. He was able to do so by closely observing the kinds of boundaries that exist between granite or basalt and neighboring rocks. He noticed that in many locations, basalt cut across ancient sedimentary rocks, which appeared baked (as if in an oven) at the contact with the basalt. He also observed sedimentary rock fragments incorporated within the basalt. Clearly, the basalt was extremely hot when it pierced the sedimentary rock. Just as clearly, the only way the rock fragments could have found their way into the basalt was if the basalt had been molten at the time.

Granite masses also showed clear evidence of having been molten. In places, the granite was intermingled with the ancient rock of the countryside, so that it was difficult to tell where one ended and the other began. In other localities, the country rock close to the granite was contorted and deformed, as if softened by the intense heat of the granitic magma. Hutton had proved his case in the field. Years later, the pioneering experimental geologist James Hall clinched matters by successfully melting granite and basalt in an oven at eight to ten times the temperature of boiling water.

If the Earth machine continuously *cycles* materials, then somehow the sediment deposited on the continents and on the ocean floor must later be transformed into magma. Hutton strongly suspected this to be the case, but it was left to later geologists to prove that, under certain conditions, sediments are indeed transformed in stages into magma. No one can actually see the process as it occurs, but we can trace the change from sediment to magma in the field. For example, we can see that the major difference between loose sediment and sedimentary rocks is that in rocks, particles are bound more or less tightly together (Figure 1.8). Binding occurs after burial, when the sediment is compacted and the grains are held together either by chemical cements or clay. The process by which sediment is converted to sedimentary rock is **lithification**.

Most sedimentary rocks are simply uplifted and eroded and the particles deposited once again, as at Siccar Point—but not all. We can trace some rock formations from the flanks into the deeply eroded cores of mountain chains and observe the changes in them that occur along the way. We see that the original horizontal layering of the strata becomes increasingly crumpled, or folded, toward the center of the mountain range. Pebbles in the strata, rounded by stream transport and deposition, are now found to be elongated in the deformed strata (Figure 1.9a). Farther into the mountain belt, we see that the original sedimentary features have become fainter, almost ghostlike in appearance, while traces of new minerals and structures have grown to dominance. We can see clearly that the sedimentary rock has been transformed into another rock. But what is

(a)

(b)

Figure 1.8 (a) Unconsolidated sediments typical of a changing coastal environment. The dark layer of bay mud, rich in organic matter, lies between windblown beach sand deposits. (b) The textures, colors, and mineral contents of these sedimentary rocks are largely determined by their erosional and depositional histories.

the nature of the transformation? Had the rock melted, the sedimentary structures would not have faded; they would have been obliterated. Therefore, we must conclude that the transformation took place while the original materials remained solid. The term for this process is **metamorphism,** and rocks transformed in this manner are called **metamorphic rocks** (Figure 1.9b–d). The agents of metamorphism are heat, pressure, and, in many cases, chemically active fluids that bathe the rock.

Still farther into the mountain belt, we find that the structures of the metamorphic rock break down, and the rock blends with granite or similar igneous rock. Evidently,

metamorphism
The structural and mineralogical changes that occur in solid rock through the action of heat, pressure, and chemically active fluids.

metamorphic rock
A rock that has been altered from its original state through metamorphism.

Figure 1.9 (a) Distorted sedimentary structures showing early-stage metamorphism of a sedimentary rock from Death Valley, California. Compression has elongated and aligned the pebbles. (b) Slate is derived from the metamorphism of shale; the stripes across the surface of this sample are ghosts of sedimentary layering. (c) Gneiss represents a higher degree of metamorphism in which new minerals have grown in size and are segregated into dark and light bands. (d) This metamorphic rock, coarse-grained marble, was originally a limestone deposited in a shallow sea.

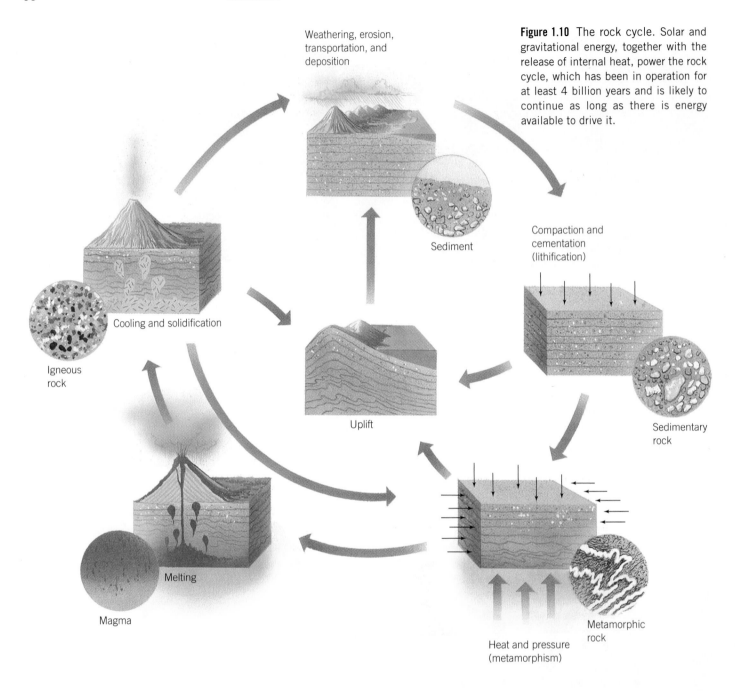

Weathering, erosion, transportation, and deposition

Sediment

Cooling and solidification

Igneous rock

Uplift

Compaction and cementation (lithification)

Sedimentary rock

Melting

Magma

Heat and pressure (metamorphism)

Metamorphic rock

Figure 1.10 The rock cycle. Solar and gravitational energy, together with the release of internal heat, power the rock cycle, which has been in operation for at least 4 billion years and is likely to continue as long as there is energy available to drive it.

rock cycle
A model that describes the formation, breakdown, and re-formation of a rock as a result of sedimentary, igneous, and metamorphic processes.

the heat and pressure generated during mountain building were so great that the metamorphic rock finally began to melt, producing magma. Thus igneous, sedimentary, and metamorphic rocks form a closed loop, the **rock cycle,** reflective of the manner in which internal heat, solar energy, water, and gravity act upon and transform the materials of the Earth's crust (Figure 1.10).

The rock cycle is not a smooth-functioning loop. Sediment is transformed to metamorphic rock only if it is buried deeply and in a region undergoing mountain building or igneous activity. We will soon learn, too, that most magma is not derived from recycled crustal rocks in the manner we have just described, but rather from the melting of solid rock far beneath the crust. In addition, not all materials are cycled at an equal rate, because external and internal processes do not act with equal intensity throughout the Earth. Nevertheless, the rock cycle is a powerful concept. Together with the principle of uniformitarianism and the recognition of Earth's great age, it remains a central theme of geology.

The Earth's Interior

Throughout the nineteenth and early twentieth centuries, the concepts of the Huttonian revolution were refined and applied to an expanding base of factual knowledge concerning the Earth. It is a process that continues to this day. One area critical to modern geology is our evolving understanding of the physical and chemical properties of the Earth's interior—the source of volcanism, earthquakes, mountain building, and metamorphism. In fact, Hutton rightly considered the Earth's surface to be largely the expression of its internal activity.

Of course, the major problem in studying the interior has always been our inability to drill far beneath the surface. (The deepest hole drilled so far is about 15 kilometers.) Geologists have had to infer the properties of the interior by measuring the Earth's gravitational and magnetic fields, by sampling the heat that escapes through the surface, and, most important, by detecting the vibrations set off by earthquakes, or *seismic waves,* that travel through the Earth's interior.

By analyzing the wave patterns detected at recording stations located around the surface of the Earth, geologists were able to reconstruct the paths of the waves through the interior. They discovered that at certain depths, seismic waves bend abruptly or are reflected back to the surface, while at other depths, certain waves die out altogether. It had been proven experimentally that the velocity and the path of seismic waves depend upon the density and other physical properties of the substances through which they pass. Therefore, geologists realized that the abrupt changes at various depths mark the boundaries between materials that differ from one another either in composition or in their physical state. By the early twentieth century, geologists had determined that the Earth is largely solid and consists of four major concentric layers: crust, mantle, outer core, and inner core, which differ in thickness, density, composition, and other physical properties (Figure 1.11).

The outermost layer, the **crust,** is the thinnest of the four. It is divided into high-standing *continental crust,* from 20 to 40 kilometers thick, and low-standing *oceanic crust,* from 2 to 10 kilometers thick. Their contrasting thicknesses reflect a fundamental difference in their compositions and densities. The continents are composed largely of coarse-grained, light-colored *granitic* rocks rich in silicon, aluminum, sodium, and potassium—all light **elements.** Oceanic crust consists of dark, fine-textured *basaltic* rocks, which contain some of these same elements, but in lesser proportions, and, in addition, are composed of a large percentage of the denser elements iron and magnesium.

Both continental and oceanic crust rest on the denser **mantle.** Extending from the base of the crust to the Earth's outer core at depths of 2900 kilometers, this vast layer comprises most of the Earth's volume and mass. Judging from sea-floor volcanic eruptions and from seismic wave speeds, the mantle is composed of silicate rock rich in iron and magnesium. No doubt the mantle is the ultimate source of all crustal materials. What is more, like an iceberg in water, the lighter crust floats on the denser mantle—a state of balance called **isostasy** (Figure 1.12). Because continental crust is thicker and lighter than oceanic crust, it sinks deeper into the mantle at the same time that it rises higher above it. Thus the mantle exhibits dual properties. In response to short-term stresses, such as the vibrations triggered by earthquakes, it behaves more rigidly than cold steel. However, in response to long-term stresses, such as the weight of the continents, it yields like a viscous fluid.

The boundary between the mantle and the **outer core** is marked by abrupt changes in seismic wave patterns. Most strikingly, waves capable of traveling only through solids die out altogether, indicating that the Earth is molten at that depth. Judging from the velocity of the waves that do pass through and from the inferred pressures and temperatures at that depth, the outer core is composed mostly of iron mixed with lighter elements, possibly carbon, sulfur, or silicon. Later in this chapter, we will discuss an important aspect of the Earth—its magnetic field. Geologists believe that motions

crust
The outermost zone, or layer, of the Earth, consisting of continental and oceanic components of distinctly different density and composition.

element
A substance that cannot be broken down into other substances by ordinary chemical or physical means. Water, for example, is not an element because it can be broken down into two elements, oxygen and hydrogen.

mantle
The zone of the Earth's interior between the crust and the outer core.

isostasy
The condition of balance, or equilibrium, in which the crust floats on the mantle.

outer core
The molten outermost zone of the Earth's core, between the mantle and the inner core.

Figure 1.11 The structure of the interior of the Earth as revealed by analysis of seismic waves and other means. The crust is so thin it cannot be drawn to scale.

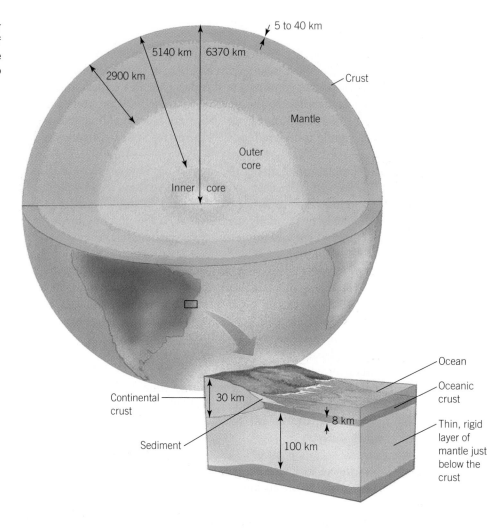

within the molten outer core generate this field. The outer core reaches to a depth of 5140 kilometers. Seismic waves are reflected sharply at that depth, and those waves that do pass through speed up markedly. From these observations and the estimated pressures and temperatures at that depth, seismologists infer that the **inner core** is a solid metallic ball about 2460 kilometers in diameter. The inner core is probably composed of iron alloyed with other heavy metals, such as nickel.

Such was the approximate state of knowledge concerning the interior up until the middle of this century. It was essentially a simple picture of four cleanly defined, independent layers. Since then, our knowledge of the interior has grown more sophisticated, spurred by a more sophisticated technology. For example, the mantle has turned out to be far more complex than previously imagined and, in fact, consists of at least three subdivisions exhibiting varying degrees of fluidity and rigidity. The thin, uppermost layer of the mantle is very rigid and is welded to the rigid crust as a structural unit. This rigid zone of crust and upper mantle, known today as the *lithosphere,* rests on the *asthenosphere,* a warm plastic zone that stays close to its melting point. The properties of the lower mantle are still a matter of debate. Most geologists agree that although it is extremely rigid, over millions of years, it is capable of slowly flowing and deforming.

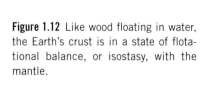

Figure 1.12 Like wood floating in water, the Earth's crust is in a state of flotational balance, or isostasy, with the mantle.

Continental Drift

It is ironic that for nearly two centuries, mainstream geologists who championed Hutton's concept of a restless, mobile Earth had, in fact, grossly underestimated the restlessness and mobility of our planet. They thought that it was fundamental to the idea of uniformitarianism that the continents and ocean basins be permanently frozen in place. Because both were composed mainly of igneous rocks, geologists assumed that they had solidified in their present positions early in the history of the Earth. This rigidity of thinking—more fixed than the continents, as it turned out—prevented many geologists from considering rival theories dispassionately. The prominent American geologist Bailey Willis in 1907 summed up the prevailing opinion: "The vast depressions have been oceanic basins throughout known history. We may state this unequivocally." This view held until the 1960s.

Nevertheless, there were contrary views. The rival hypothesis to permanence was **continental drift,** which envisioned a far more restless, mobile Earth. Its central concept was that the continents have not been locked permanently in their present positions on the Earth's surface but have moved relative to one another, and to the ocean floor, in the past. The twentieth-century scientist deservedly given the most credit for marshaling the facts that refute the idea of permanently fixed continents and ocean basins is the German astronomer, meteorologist, and geophysicist Alfred Wegener (1880–1930). He presented his case for continental drift in *The Origin of the Continents and Oceans,* first published in 1915. Wegener theorized that all the world's continental crust had existed as a single landmass, which he called **Pangaea** (whole Earth), and that the continents we see today are the result of its fragmentation. Figure 1.13a shows Pangaea according to the interpretation of Wegener's colleague Alexander du Toit. He divided Pangaea into two subcontinents: *Laurasia,* consisting of North America, Greenland, and Eurasia; and *Gondwanaland,* consisting of South America, Africa, India, Australia, and Antarctica. Oceans covered the rest of the world, and a triangular sea, the Tethys, partially divided the two subcontinents. Figures 1.13b–e show a later interpretation of how the supercontinent Pangaea divided and the continental blocks that we know today drifted to their present positions.

The key weakness in the hypothesis of continental drift was the apparent lack of an adequate mechanism by which the continents can move across the globe in the manner that Wegener had envisioned. This led some early advocates to search for catastrophic or single-event explanations for drift. Wegener, however, viewed drift as an ongoing, continuous process, explainable according to sound uniformitarian principles. He suggested that as a result of both the Earth's rotation about its axis and the tidal attraction between the Moon and the Earth, the continents are subjected to forces that impel them westward and toward the equator. Like ships plowing through water, the continents push their way through the yielding oceanic crust.

Opponents were quick to point out several flaws in Wegener's proposed mechanism for drift. First, ships are obviously more rigid than the water they plow through; in contrast, ocean floor rock is quite rigid. In fact, experimental studies and the analysis of seismic waves reveal oceanic rock to be more rigid than continental rock. How, then, could the continents push their way through oceanic crust without being torn apart or completely distorted in the process? Second, the Earth's rotation and tidal forces are about a million times too small to accomplish the task of propelling the continents.

Another proposed drift mechanism was internal heat convection that drives the continents from below. However, prominent physicists at the time mistakenly ruled out that possibility. With external forces too small and internal forces considered nonexistent, the hypothesis of continental drift seemed to be impaled on a self-generated contradiction.

Still, the concept of continental drift would not go away. The crux of the debate can be summed up in the following analogy: If the claim is made that there is an elephant in the next room, the only action that will positively prove or disprove the assertion is

continental drift
A hypothesis suggesting that the continents move over the Earth's surface.

Pangaea
The name of the supercontinent that existed between 200 and 300 million years ago, which has since fragmented into the present continents.

Figure 1.13 Stages in the fragmentation of Pangaea leading to the modern configuration of the continents. Note the break-up of Gondwanaland and the collision of Africa and India with Eurasia, closing out the ancient Tethys Sea. Note also that the Atlantic Ocean did not form all at once. The equatorial and southern segments opened first, followed by the northern Atlantic. (Adapted from Robert S. Dietz and John C. Holden, *Journal of Geophysical Research* 75: pp. 943–956.)

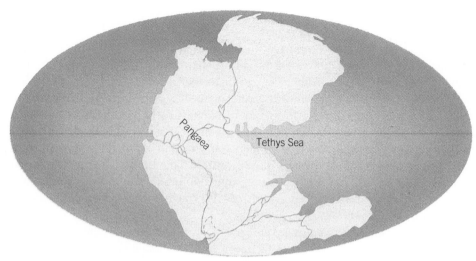

(a) 200 million years ago

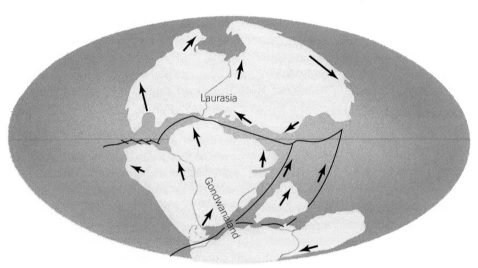

(b) 180 million years ago

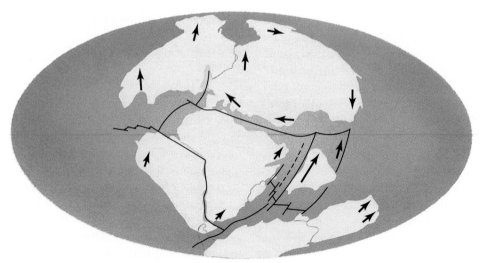

(c) 135 million years ago

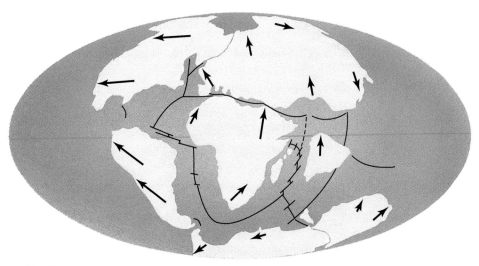

(d) 65 million years ago

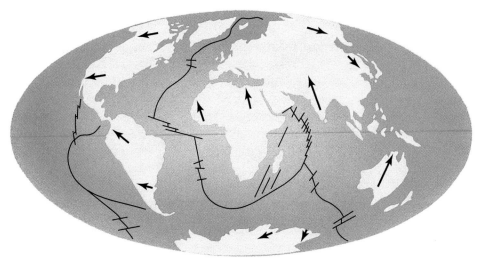

(e) Present

to check. If the elephant turns out to be there after all, you cannot explain it away by saying that you do not know how it got there. In science, facts are the final arbiter of theory. Drift advocates attempted to provide incontrovertible evidence that the "elephant" was indeed in the next room—that the continents had indeed moved. For drift advocates, precisely *how* the continents had moved was an important problem to be addressed *after* the fact that they had moved was firmly established. In the following sections, we will examine some of the evidence for drift.

Physical Evidence

Consider the matching coastlines of South America and Africa in Figure 1.14. Their complementary shapes were noted as far back as 1620 by the natural philosopher Francis Bacon, using the crude maps of a largely unexplored Earth. Bacon wrote that each continent "has similar isthmuses and similar capes, which is no mere accidental occurrence. So, too, the New and Old World are conformable in this, that both worlds are broad and extend toward the north, but narrow and pointed toward the south." Every curious schoolchild since then has noticed that the coastlines of the continents display a jigsaw puzzle fit.

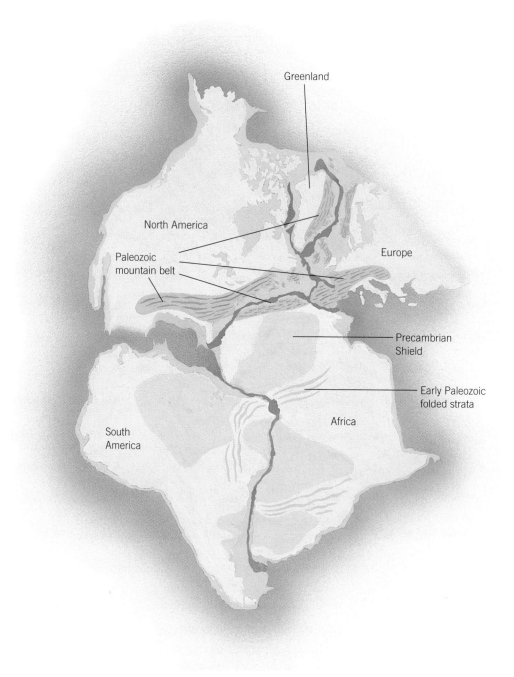

Figure 1.14 Strong evidence in support of continental drift includes the jigsaw puzzle fit of the continents (about halfway down the continental margins). Note the positions of Paleozoic mountain belts of the Northern Hemisphere and the complex structural trends of the Southern Hemisphere. All of these geological patterns match on either side of the junction. (Adapted from E. C. Bullard, J. E. Everett, and A. G. Smith, "The Fit of Continents Around the Atlantic," in P. M. S. Blacket, E. C. Bullard, and S. K. Runcorn, eds., *A Symposium on Continental Drift,* London: Royal Society of London, 1965, vol. 1088, p. 41.)

Modern geologists do not consider the coastlines the true edges of the continents, because they know that even slight sea-level fluctuations can cause large-scale changes in the configuration of the coast, especially in low areas. They take the true edge of the continents to be considerably deeper, about halfway down the continental margin. Figure 1.14 shows a computer-generated jigsaw puzzle fit of the continents at the 900-meter depth. Notice that the fit is far more exact than is implied from the shape of the coastlines. The fit holds not only for South America and Africa but also for North America and Eurasia. Putting these landmasses together produces a world that closely matches the Pangaea supercontinent of Figure 1.13.

The jigsaw puzzle fit of the continents itself does not constitute proof of drift. A random collection of odd shapes, with creative arrangement, might also be made to fit quite well. Skeptics had great fun manipulating the continental blocks into positions never suggested by Wegener and his supporters. For example, Australia and New Guinea can be placed rather snugly between Africa, India, and Arabia. Why not this fit, critics of drift asked.

Their criticisms were relevant. It was not enough to show that the continents *could* fit together—it had to be demonstrated that they had indeed been together. The way to do this was to establish that various patterns and events are explainable only when the continents are assembled in Wegener's jigsaw puzzle configuration. After the jigsaw puzzle is assembled, the pieces must make a coherent picture.

Structural Evidence

If a continent divides, rock structures located where the division occurred should be interrupted, like a picture torn in two. Unless the rocks have been worn away or buried beneath younger deposits of the newly formed continental margins, the structures should align again when the pieces are reassembled properly.

And they do! The most remarkable case is between Africa and South America. In Ghana, Nigeria, and Guinea, metamorphic rock formations of the Sahara shield, formed some 2 billion years ago, are overlain by younger, less deformed ones, 600 million years of age. The unconformity between the two formations can be traced right out to the Atlantic Ocean. If South America is nestled against Africa, in accordance with the computer fit, the same rocks, bearing the same relationship to one another, continue through the coastal city of São Luis, Brazil, and wind the same snakelike path that they do in Africa.

Less immediately striking, but no less convincing, is the correspondence of the Appalachian, European, and African mountain belts. The Appalachians had an extremely complicated history during the Paleozoic era, 245 to 570 million years ago. Four distinct periods of mountain building can be deciphered from their intricate structures. Similar structures of the same age are found in the British Isles, Scandinavia, southern Europe, and Africa. Reassembled as Pangaea, they form a continuous chain of identical rocks and structures with the Appalachians (see Figure 1.14).

Evidence in the Distribution of Ancient Glaciers

What was an enormous glacier doing astride the African equator about 300 million years ago? This was no mere high-altitude mountain glacier but a continental ice sheet similar in extent to the one atop present-day Antarctica. Markings made by the ice as it moved over underlying rock and distinctive deposits left upon melting leave no doubt that there was indeed such a glacier centered in Africa. On the other hand, North America and Europe show no evidence of glaciation during that time. Even in northern Canada, the climate apparently was warm, as recorded by coal deposits, rocks lithified from desert sands, and reptile fossils.

Field investigations leave little doubt that glacial ice also moved over parts of South America, India, and Australia at the same time that the African ice sheet was advancing. If these continents were in their present locations at that time, where did the ice come from? Tracing the direction of ice movement on each continent places the source of the ice in each case in the deep ocean, an impossibility (Figure 1.15a). (As we will see in Chapter 17, Glaciers and Climate, glaciers require the support of solid land beneath them in order to grow and spread.) However, if one assumes the jigsaw puzzle fit of Figure 1.15b, the continents cluster around Africa and Antarctica at the South Pole, and the spread of glaciers from there is now more easily understood.

Figure 1.15 (a) The distribution of ancient glacial features suggests that equatorial Africa was covered by a continental ice sheet 300 million years ago. Also, glacial features of the same age located on the other southern continents appear to have sources in the deep ocean. (b) When the continents are reassembled as Pangaea, these same patterns point to a huge southern hemisphere glacier centered in Africa, which was close to the South Pole at the time. (Adapted from C. K. Seyfert and L. A. Sirkin, *Earth History and Plate Tectonics,* New York: Harper and Row, 1973, Fig. 7.6).

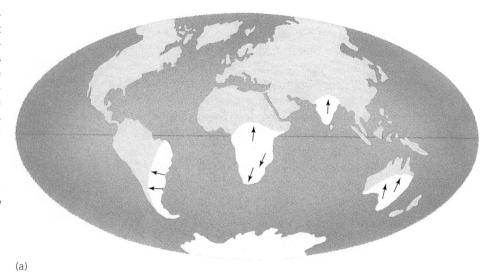

(a)

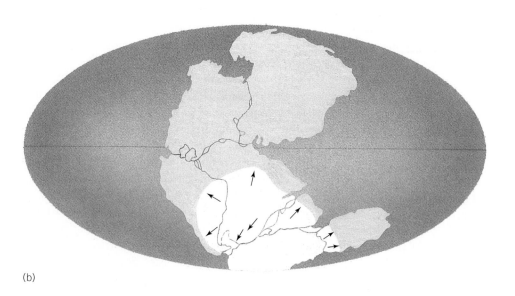

(b)

Evidence in the Distribution of Ancient Life-Forms

The facts of the distribution of ancient life as seen in fossils are equally hard to ignore. For example, present-day shallow-water corals attach to the sea bottom and often grow in reef-forming colonies. The reefs occur in the warm waters from 30 degrees north to 30 degrees south of the equator. Yet extensive fossil deposits of extinct corals and associated organisms are found as far north as the Arctic Circle, and they seem to have lived in shallow seas that cut sharply across today's climatic zones. When the continents are reassembled as Pangaea, the ancient reefs are restored to latitudes that match the distribution of modern coral reefs.

Consider, too, the curious case of *Lystrosaurus,* a prime example of the puzzling distribution of the fossil record of certain land organisms. *Lystrosaurus* was a reptile, about 1.5 meters long, whose habitat was the swamps and river deltas of the Triassic period, some 240 million years ago (Figure 1.16). *Lystrosaurus* bones are found in India and South Africa. They have also been hacked out of rocks of the same age in Antarctica. How could *Lystrosaurus* have made it to these far-flung locations if the continents were locked permanently in their present positions? Did it swim? Its skeleton indicates that it was a land animal incapable of swimming thousands of kilometers. Assuming that it was a cold-blooded reptile, unable to regulate its body temperature the way mammals can, how could it have survived Antarctica's climate, even if that climate was warmer than today's?

A similar paradox is presented by the distribution of *Glossopteris* (Figure 1.17), an ancient seed fern found in all continents of the Southern Hemisphere but only in India in the Northern Hemisphere. Drift opponents suggested that in the past, there were land bridges, which were similar to today's Isthmus of Panama or to island chains like the Aleutians, that would have provided an avenue for the spread of life into these far-flung continents. Yet surveys of the ocean bottom around Antarctica and other southern continents reveal no evidence of such features.

Lystrosaurus and *Glossopteris* are not oddball fossils. They are but two of the many groups of animals and plants that occupied vast regions of the present Southern Hemisphere and India. When the continents are reconstructed to resemble Wegener's Pangaea, these flora and fauna make sense as occupants of a broad habitat, probably a vast swampy plain built of the sediment washed in by streams issuing from mountain chains to the west and south.

Figure 1.16 Because *Lystrosaurus* could not have swum across oceans, it is difficult to imagine how it migrated to all the southern continents. However, if the continents were once a single unit, this large land reptile could have migrated across the huge landmass. (Illustration by Juan Barbera, *New York Times*, July 19, 1988.)

Similarities in the Sedimentary Record

This broad expanse of the Earth's surface remained a coherent geologic setting for a far longer span of time than that represented by the 100-million-year life span of *Lystrosaurus*. This setting is preserved in a sequence of strata, the *Gondwana succession,* named for the region in India where they were first described in detail (Figure 1.18). The Gondwana strata are a record of the environmental changes that occurred over hundreds of millions of years. From coal deposits that indicate lush swamps, to rocks made of windblown sands that tell of deserts, to coarse debris that tells of melting glaciers, this record is mirrored in striking detail on all the southern continents (South America, Africa, Antarctica, and Australia).

In South America, Africa, and Antarctica, the sequence is capped by marine deposits that alternate with basaltic lava. Geologists have interpreted this as marking the initial rifting, or splitting, of Gondwanaland that led eventually to the separate continents. They believe that huge cracks developed in the crust as Gondwanaland began to split. A sea—actually the beginnings of the South Atlantic—invaded the narrow space between the separating continental blocks. Marine deposits accumulated in the space, as well as basaltic lava that rose through the cracks in the crust, and some of these features were preserved on the separating continents. The Red Sea and Gulf of California regions mark similar, but more recent, early-stage continental rifting (Figure 1.19). Notice the close fit of the coastlines on either side of these narrow seaways; they fit as well as the coastlines of South America and Africa.

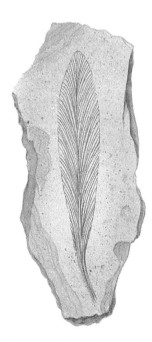

Figure 1.17 *Glossopteris* is a fossil seed fern widely distributed throughout the Gondwanaland continents.

Paleomagnetic Evidence

Though impressive, much of the evidence of continental drift presented so far is circumstantial and subject to differences of interpretation. It is the study of the Earth's past magnetism, or **paleomagnetism,** that compellingly proves continental drift. It also proves that continental drift has been an ongoing phenomenon that began long before the breakup of Pangaea. To understand this evidence, we must first discuss briefly the Earth's magnetism and how it is recorded in the rock record.

At any given location, a compass needle will rotate horizontally and point in the direction of the Earth's magnetic North and South poles, which are close to, but not the same as, the geographic poles. This is how a compass is generally used. A compass needle designed to rotate vertically forms an angle with the horizontal plane (or the

paleomagnetism
The study of the Earth's past magnetism as recorded in rocks at the time of their formation.

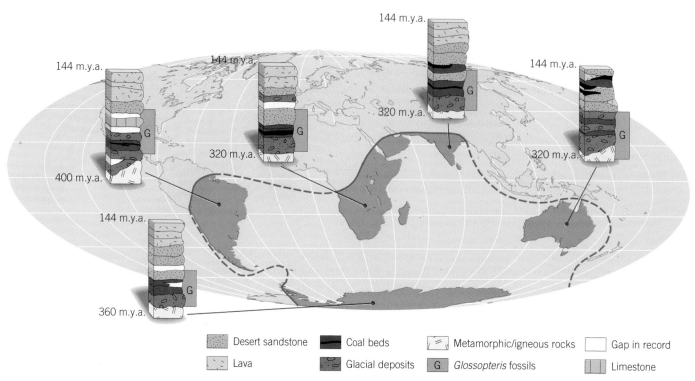

Desert sandstone	Coal beds	Metamorphic/igneous rocks	Gap in record
Lava	Glacial deposits	G *Glossopteris* fossils	Limestone

Figure 1.18 The now-separated southern continents all contain a sequence of similar sedimentary deposits of the same age called the Gondwana succession. It is unlikely that identical sequences would form simultaneously on five widely separated continents.

horizon) called its *magnetic inclination*. Figure 1.20 shows that the magnetic inclination varies between perfectly horizontal (zero) at the magnetic equator and perfectly vertical (pointing straight into the ground) at the magnetic North Pole. Note again that the magnetic poles and equator are also close to their geographical counterparts, so that the inclination of the compass needle closely measures latitude. Now, how can we use this information to measure the past positions of the continents?

Magnetite is an iron oxide mineral found in the basaltic lavas that spill onto the surface of the continents or ocean floor. As soon as the lava begins to cool, the tiny minerals form as solid crystals. They become magnetized when their temperature falls below 580 °C (the *Curie point*), and they acquire the magnetic orientation characteristic of the latitude at which they are located. As the lava solidifies, the magnetites are locked in place and thus preserve a record of their relation to north at the time the rock formed (Figure 1.21). Similarly, magnetites eroded from igneous rocks become oriented to magnetic north as they sink to the sea floor and are cemented in place within the sur-

Figure 1.19 (a) The Red Sea and (b) the Gulf of California are present-day examples of early-stage continental separation. The matching coastlines on either side of these narrow seaways are obvious, but no more so than the matching continental margins on either side of the Atlantic Ocean.

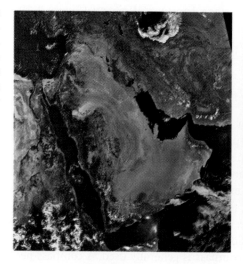

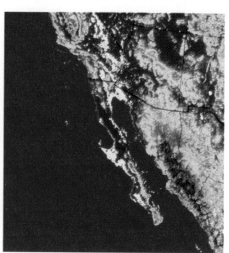

(a)   (b)

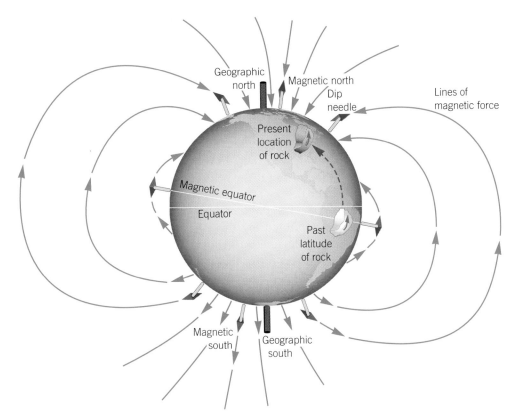

Figure 1.20 Lines of force delineate the Earth's magnetic field and show the inclination of a compass needle at any given latitude. The horizontal inclination of the rock sample presently found in North America indicates that it was formed at the equator.

rounding sediments. Unless disturbed by erosion, metamorphism, or remelting, these magnetites are reliable records of the past position of the magnetic poles.

The magnetite crystals of a lava or of a sedimentary rock are in effect tiny compass needles that preserve the inclination of the field at the latitude where the rock formed. For example, the magnetic inclination of the rock shown in Figure 1.20 is horizontal. This orientation indicates that the landmass in which the rock is embedded once straddled the equator, as South America does today.

Because the rock's present location is North America, we may conclude that the landmass within which it is embedded has been moved from the equator to its present position. By examining the changes of inclination in a sequence of strata in a given location, such as from the bottom to the top of the Grand Canyon, we can determine the path of North America through time with respect to magnetic north. These changes, when plotted on a present-day globe, define an *apparent polar-wandering curve* (Figure 1.22a).

Figure 1.21 (a) Like tiny compass needles, magnetite crystals rotate in the liquid lava until they are aligned parallel to the Earth's magnetic field. (b) This direction is preserved as the lava cools to solid rock.

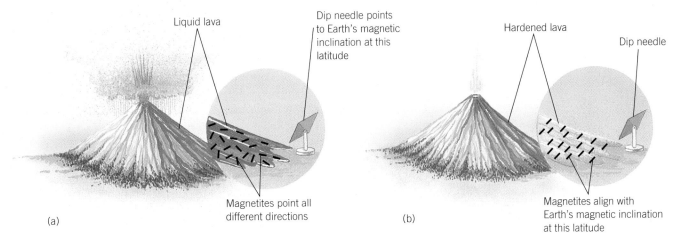

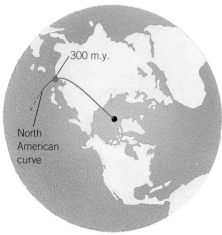

(a) North American polar-wandering
    curve

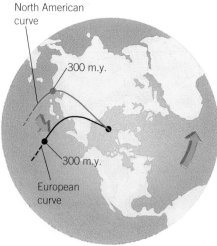

(b) North American and European
    curves

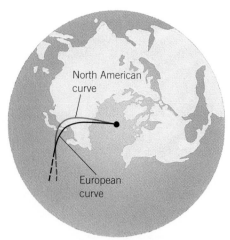

(c) Match of polar-wandering curves

Figure 1.22 Paleomagnetic evidence strongly supports continental drift. When the polar-wandering curves of Europe and North America are aligned so that they match, the Atlantic Ocean is closed out and the continents fit together in a manner that closely resembles Pangaea.

Things really get interesting when we plot the European and North American polar-wandering curves on the same globe (Figure 1.22b). The curves of the two continents, Europe and North America, are nearly identical but are separated by about 40 degrees longitude. But wait! How can there be *two* curves if there is only *one* North Pole at any given time, whether the rocks are located in Europe, North America, or Tasmania?

The answer is clear if we consider that 40 degrees is the width of the Atlantic Ocean. When the curves are fitted together, the Atlantic Ocean closes and the margins of Europe and North America fit together as well (Figure 1.22c). This demonstrates that it was the opening of the Atlantic Ocean and the resulting drift of the continents that created the two curves. In fact, we can rearrange all the continents using paleomagnetic evidence. The resulting configuration confirms the jigsaw puzzle fit of the continental margins, the matching geologic structures, and the climatic patterns—in sum, a world closely resembling Wegener's Pangaea.

The paleomagnetic data indicate further that the clustering of the continental blocks into the single landmass, Pangaea, was a relatively short-term episode in the long-running history of continental drift. Continental crust, in one configuration or another, drifted over the globe prior to the breakup of Pangaea and probably has done so throughout geologic time. For example, the paleomagnetic evidence shows that 400 million years ago, ancestral North America straddled the equator (Figure 1.23). Now the northward distribution of extinct corals mentioned earlier becomes less exotic; the corals had not spread north to the Arctic Circle but had once occupied roughly the same latitude zones in which present-day corals are living.

The paleomagnetic confirmation of continental drift was largely the work of the British geophysicist S. Keith Runcorn and his colleagues during the 1950s and early 1960s. Unfortunately, it came too late to comfort Alfred Wegener, who was routinely ridiculed for his beliefs; he died in 1930 while on an expedition to the Greenland ice cap in search of evidence to support his ideas.

Continental Drift and Plate Tectonics

Continental drift is now viewed as one aspect of a more comprehensive process, *plate tectonics*. Instead of being thought of as rafts plowing *through* crust, the continents are now considered "cargo" embedded in larger rafts, called *lithospheric plates*. These plates consist of oceanic as well as continental crust and include a thin part of the upper mantle down to the soft, ductile zone mentioned previously, the *asthenosphere*. The plates move as coherent slabs over the asthenosphere; and as they wander, so do the embedded continents. In their nomadic wanderings over the Earth, the continents have split and reassembled in different combinations many times in the geologic past.

The evidence for plate tectonics was gathered in the decades following World War II, when geologists put to sea with a technology Wegener would probably have considered science fiction. This technology enabled them to detect an ocean floor topography as varied as that on land, to study the record of the soft muds that lay on the hard-rock oceanic crust, to probe the secrets of that crust, and to get a clearer picture of the Earth beneath the crust itself. In the next chapter, we will describe plate tectonics fully, beginning with the evidence revealed by studies of the ocean floor. In later chapters, we will explore the close connection

Equator

Figure 1.23 Paleomagnetic evidence indicates that 400 million years ago, North America was positioned 60° with respect to its present orientation and straddled the equator.

between plate tectonics, mountain building, earthquakes, and volcanic activity. In Chapter 12, Plate Tectonics and Continental Crust, we will discuss the influence of plate tectonics on continental evolution.

STUDY OUTLINE

Geology began to evolve as a science in the late eighteenth century.

I. **THE HUTTONIAN REVOLUTION. Uniformitarianism,** the basic principle proposed by James Hutton, says that we can interpret the past through observation of the Earth today because geologic processes have remained the same throughout geologic time.

    A. Hutton saw the Earth's surface in a constant state of change, its materials ceaselessly cycled and recycled. He observed that the **strata** at Siccar Point were **sedimentary,** consisting of fragments of older rocks. He knew the simple principle of **superposition,** which states that sediments are deposited in horizontal layers, with the oldest on the bottom and successively younger layers on top. Thus he realized that enormous geologic changes occur in very small increments, and he concluded that the Earth's age is also enormous.

    B. The opposing view, **catastrophism,** held that the Earth is quite young, having reached its present form through a series of sudden violent events.

C. Establishing geologic time
1. The principle of **faunal succession** holds that, because **fossils** succeed one another in order, rocks containing similar fossils are similar in age. This principle allowed geologists to create the **geologic time scale,** by which the **relative ages** of rocks can be measured.
2. Later development of *radiometric dating* enabled geologists to determine the absolute ages of rocks. The age of the Earth is 4.6 billion years.

D. The Earth machine
1. In the **hydrologic cycle,** seawater evaporates and enters the atmosphere; it condenses and falls to the land, where it eventually flows back to the sea. Water is the key agent of **erosion** and **deposition.** It carries sediment from high to low places and in this manner levels the land surface. The leveling process begins with **weathering,** the physical and chemical alteration of rocks. As running water forms streams and carves valleys, loose weathering products slump downhill toward the streams in **mass wasting.** As external processes work to level the land, internal forces work to uplift it.
2. In the **rock cycle,** sediment is cemented into sedimentary rock (**lithification**), which is buried beneath other layers; heat and pressure transform it into **metamorphic rock,** which is melted and turned into magma. **Igneous rock** results when **magma** is cooled. The igneous rock is weathered and converted to sediment, and the rock cycle begins again.

E. The Earth's interior. Geologists infer the properties of the interior primarily by analysis of seismic waves.
1. The Earth is largely solid and consists of four major concentric zones: **crust, mantle, outer core,** and **inner core,** which differ in thickness, density, composition, and other physical properties.
2. The crust floats on the mantle in a state of balance called **isostasy.** Because continental crust is thicker and is made of lighter **elements** than oceanic crust, the continents sink deeper but also rise higher than the ocean basins.

II. **CONTINENTAL DRIFT.** Early in this century, Alfred Wegener proposed that all the world's continental crust had existed as a single landmass, called **Pangaea,** and that the continents we see today are the result of its fragmentation.
A. Physical evidence. The jigsaw puzzle fit of the continental margins suggests that the continents were a single landmass that was later torn apart.
1. When the continents are recombined as Pangaea, the trends of rock structures make a continuous line, as though they had been torn in two.
2. Ancient sedimentary rocks currently located near the equator contain evidence of past glaciation. However, if the continents are moved into their former positions as Pangaea, the contradiction is cleared up.
3. The locations of many animal and plant fossils also make sense only if the continents were a single landmass in the past.
4. Sequences of strata found on all the southern continents display remarkable similarities of rock types and fossils. They also contain evidence of the initial rifting that led to the separation of the continents.

B. Paleomagnetic evidence. Above all other evidence, it was **paleomagnetism,** the study of Earth's past magnetism, that proved continental drift.
1. An unconstrained compass needle forms an angle with the horizontal plane of the Earth, called its *magnetic inclination.* Magnetized crystals within sedimentary rocks are "fossil magnets" that point to the position of the magnetic poles of the past.
2. Geologists used the magnetic inclination of ancient rocks to trace the *apparent polar-wandering curves* for each continent. Because the magnetic

poles remain essentially stationary, the curves strongly suggest that the continents moved. The polar-wandering curves are aligned only when the continents are in their Pangaea positions.

III. **CONTINENTAL DRIFT AND PLATE TECTONICS.**   Continental drift is one aspect of *plate tectonics.* The continents are imbedded in *lithospheric plates,* which move as coherent slabs over the *asthenosphere.* The continents have split and reassembled in various combinations many times in the geologic past.

STUDY TERMS

absolute age (p. 30)
catastrophism (p. 26)
continental drift (p. 39)
crust (p. 37)
deposition (p. 32)
element (p. 37)
erosion (p. 32)
faunal succession (p. 29)
fossil (p. 28)
geologic time scale (p. 29)
geology (p. 25)
hydrologic cycle (p. 31)
igneous rock (p. 33)
inner core (p. 38)
isostasy (p. 37)
lithification (p. 34)

magma (p. 33)
mantle (p. 37)
mass wasting (p. 32)
metamorphic rock (p. 35)
metamorphism (p. 35)
outer core (p. 37)
paleomagnetism (p. 45)
Pangaea (p. 39)
relative age (p. 28)
rock cycle (p. 36)
sediment (p. 26)
sedimentary rock (p. 26)
strata (p. 26)
superposition (p. 27)
uniformitarianism (p. 26)
weathering (p. 32)

CRITICAL THINKING QUESTIONS

1. How might the concept of uniformitarianism, if taken to an extreme, inhibit the advance of geological knowledge? Give two examples cited in this chapter.
2. Why does an unconformity represent a gap in the geologic record?
3. Explain the relationship between the principles of superposition and faunal succession.
4. Explain the relationship between gravity and the hydrologic cycle.
5. What conditions must exist for a planet to exhibit a rock cycle?
6. Can you think of materials other than the Earth's mantle that exhibit the dual properties of rigidity and plasticity?
7. Why is the representation of the paths of the continents over the globe through geologic time referred to as an *apparent polar-wandering curve?*
8. Were Wegener's opponents merely short-sighted and intolerant, or did they have sound reasons for disagreeing with his hypothesis of continental drift? Explain.

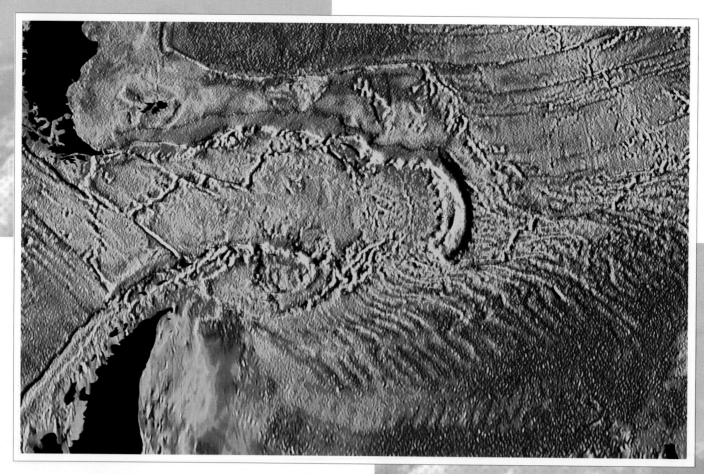

Plate Tectonics and the Ocean Floor

Save for a few scattered expeditions employing crude equipment, the ocean floor was uncharted territory through the first half of the twentieth century. World War II changed all that. By the early 1950s, the United States government, oil companies, universities, and foundations were pouring millions of dollars into marine geology. It was the beginning of the age of "big science" and team research, modeled in some ways after wartime projects. It was also an age of great technological advances, as instruments designed for the war effort were adapted to probe the sea floor and new ones were invented. The ocean floor was the new frontier of the geosciences, offering many opportunities for creative people who were willing to work hard.

Bruce Heezen and Marie Tharp of the Lamont Geological Observatory* were among the many scientists seeking to forge careers in marine geology during the 1950s and 1960s. Although both were in their twenties, they were entrusted with the important task of making sense of an immense amount of data collected by Lamont research ships in the North Atlantic. Tharp had the difficult assignment of turning thousands of depth-sounding numbers into a series of highly detailed profiles, or cross sections, of the North Atlantic floor. As shown in Figure 2.1, the submerged mountainous region towering above the general level of the ocean floor in the center of each profile is the Mid-Atlantic Ridge. From the form of the mountains and from samples dredged from the ocean floor, it was apparent that the Mid-Atlantic Ridge consists of a series of submarine volcanoes, many of them active. Tharp noted, "I was struck by the fact that

*Now called the Lamont-Doherty Earth Observatory.

◀ Topographic image of the South Atlantic Ocean floor, based on data gathered by the U.S. Navy's Geosat satellite. To the right of the southern tip of South America (black, top-left) and the northern tip of Antarctica (black, bottom-left) lies the Scotia plate. Just right of the plate is the curved South Sandwich trench. At the upper right of the image is the southern end of the Mid-Atlantic Ridge, an undersea mountain chain that extends down the entire length of the Atlantic Ocean. The theory of plate tectonics offers a compelling explanation of these and other large-scale features of the Earth's crust.

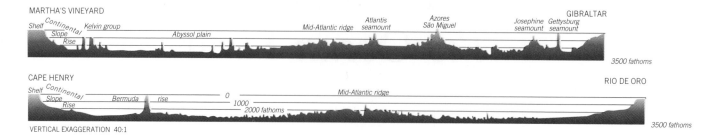

MARTHA'S VINEYARD

Shelf *Continental* *Kelvin group*   *Abyssol plain*   *Mid-Atlantic ridge*   *Atlantis seamount*   *Azores São Miguel*   *Josephine seamount* *Gettysburg seamount*   GIBRALTAR

Slope
Rise

3500 *fathoms*

CAPE HENRY   RIO DE ORO

Shelf *Continental*   *Bermuda*   *rise*   0   *Mid-Atlantic ridge*

Slope
Rise   1000
2000 *fathoms*

VERTICAL EXAGGERATION 40:1   3500 *fathoms*

Figure 2.1 These profiles of the Mid-Atlantic Ridge were constructed by Bruce Heezen and Marie Tharp from data collected by Lamont Geological Observatory expeditions. Near the center of each profile is a prominent cleft, which Tharp identified as a central rift valley. (Marie Tharp)

fault
A fracture in bedrock along which rocks on one side have moved relative to the other side.

mid-ocean ridge
A seismically and volcanically active mountain range, marked by a central rift valley, that extends continuously through the major ocean basins.

the only consistent match-up when I compared the profiles was a V-shaped indentation located in the center of each profile. The individual mountains didn't match up, but the cleft did."* Tharp believed she had discovered a *central rift valley* in the Mid-Atlantic Ridge.

Tharp's belief was based on the cleft's form, identical to regions of continents where a central valley has dropped down between huge parallel **faults,** or fractures in the crust. The valleys are tensional features, formed by stretching or extension of the crust. Many continental rifts are still active and marked by numerous shallow earthquakes that are triggered when the crustal blocks, hung up along the faults by friction, suddenly slip. So Heezen proposed that they plot the locations of several thousand Atlantic earthquakes to see whether they matched the central rift valley of the ridge. It was a perfect fit.

The next step was to refer to the earthquake records and plot the locations of earthquakes in the world's oceans. The researchers found that a line of earthquakes, continuous with the Mid-Atlantic Ridge, extended around the globe. What is more, the earthquake belt could be traced into the continental East African Rift Valley, whose profile was identical to the central rift valley of the Atlantic (Figure 2.2). Heezen and Tharp had discovered "a huge tension crack, caused by a splitting apart of the Earth's crust." Actually, as Tharp pointed out, it is a branching network of cracks splitting all the ocean basins of the world and, as Figure 2.3 shows, parts of the continents as well. Stretching more than 64,000 kilometers, the **mid-ocean ridge** is the longest continuous feature of the Earth's crust.

Heezen was somewhat ambivalent about their discovery in the beginning. According to Tharp, "He groaned and said, 'It cannot be. It looks too much like continental drift.' He and almost everyone else at Lamont, and in the United States, thought continental drift was impossible. North American geologists considered it to be almost a form of scientific heresy, and to suggest that someone believed in it was comparable to saying there must be something wrong with him or her."†

Heezen knew that the discovery of the mid-ocean ridge system was a serious blow to the theory of permanently fixed continents and ocean basins. He saw the striking symmetry between the Mid-Atlantic Ridge and the margins of the continents on either side; and he realized that this huge tension crack was proof that the crust was being pulled apart in the same directions predicted by the theory of continental drift.

In this chapter, we will see that Heezen and Tharp's discovery was one of a series that spurred a revolution in geology not seen since James Hutton's day. That revolution

---

*Marie Tharp, "Discovery of the Mid-Ocean Rift System," in *Yearbook* (New York: Lamont-Doherty Geological Observatory of Columbia University, 1989).

†*Ibid.*

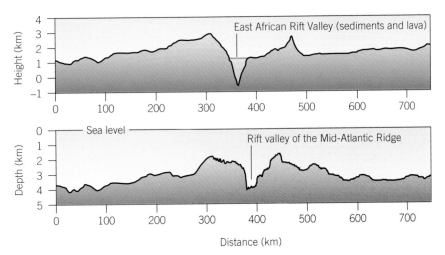

is plate tectonics, the unifying theory of the Earth's crust. Plate tectonics provides the mechanism that validates Wegener's hypothesis of continental drift; it explains the origins of the ocean basins, continents, and mountains. Its impact is felt in all branches of the geosciences. From forecasting violent volcanoes and earthquakes to searching for oil and mineral deposits, the theory of plate tectonics is also a practical theory.

The story of the plate tectonics revolution begins on the ocean floor, and Marie Tharp's detailed depiction of ocean floor topography is a useful guide to the following discussion (see Figure 2.3).

An Emerging Picture of the Sea Floor

Once the mid-ocean ridge was discovered, marine geologists proceeded to map its topography in detail. They found that many seemingly isolated volcanic islands, such as Iceland, are located on the ridge system. Also, the mid-ocean ridge is crossed by narrow canyons dividing it into segments that do not connect; in other words, the segments are offset (see Figure 2.3). The narrowness of the canyons, plus the occurrence of numerous earthquakes between the offset rift valley segments, indicate that the canyons are in fact long cracks, or **fracture zones**. A number of them have been traced hundreds, even thousands, of kilometers until they pass under ocean sediments. Fracture zones, although strikingly displayed on the ocean floor, are not exclusively marine features; the San Andreas fault of California is a prominent fracture zone in the continental crust.

Marine geologists also employed a variety of techniques to determine what lies beneath the mid-ocean ridge. Heat probes across the ridge recorded abnormally high quantities of heat escaping from the central rift valley. Tests using sensitive pendulums and weights attached to springs also recorded relatively weak gravitational attraction across the rift valley, pointing to lighter-than-normal materials beneath the surface. Natural and artificially induced seismic waves slowed beneath the rift, indicating that they had passed through softer, lighter, and presumably warmer substances (Figure 2.4). The combined data made a strong case for the existence of a thick wedge of semimolten rock beneath the ridge crest, the likely source of the erupting lava. This inference was later confirmed by geologists who descended in submersibles through the dark, deep water to the rift valleys of the North Atlantic and the East Pacific Rise. There they observed, firsthand, lava oozing from cracks in the ocean floor and jets of hot fluids issuing from chimneylike vents (Figure 2.5).

A striking, counterintuitive fact is that the shallowest features of the ocean basins—the mid-ocean ridges—are located toward the basin centers, whereas the deepest features are found along the basin margins. Long known from the laying of transoceanic telephone cables, the **deep-sea trenches** are a series of narrow, curving depressions that

fracture zone
A region of closely spaced cracks or faults in rocks. In the context of this chapter, a prominent crack or fault that runs perpendicular to, and offsets, the mid-ocean ridge.

deep-sea trench
A deep, narrow, elongated depression in the sea floor.

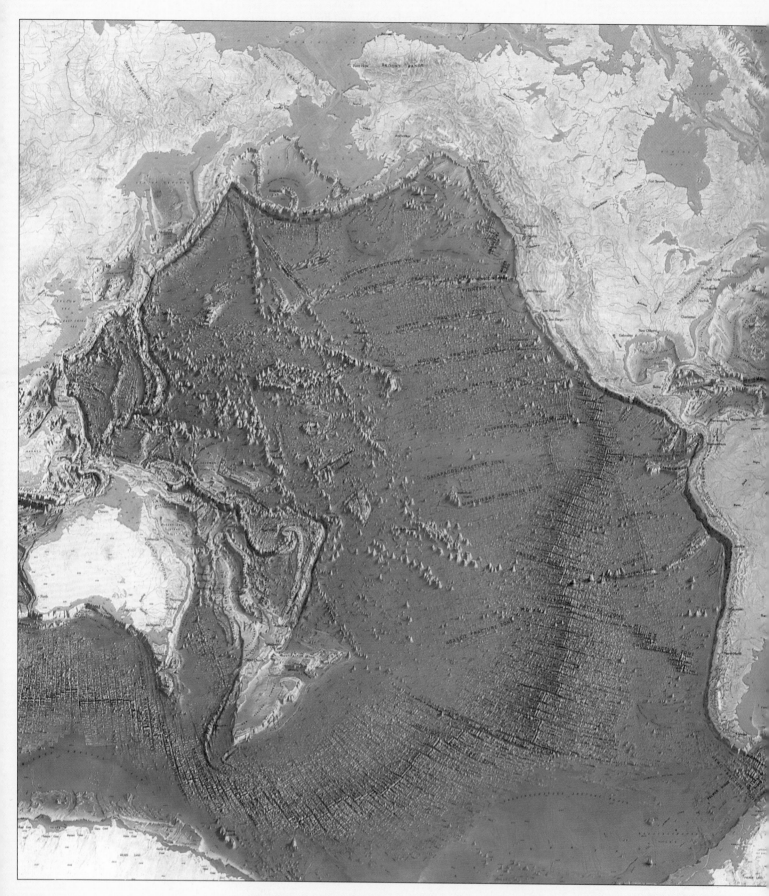

Figure 2.3 The main features of the ocean floor are shown in this model, which was created from Marie Tharp and Bruce Heezen's research. The mid-ocean ridge system divides the world's ocean basins and intersects the continents (note, for example, the Red Sea and the Gulf of California). Deep-sea trenches rim the Pacific Ocean. (Marie Tharp)

WORLD OCEAN FLOOR

BY BRUCE C. HEEZEN AND MARIE THARP

Lamont-Doherty Geological Observatory, Palisades, New York 10964

Based on Research and Exploration Initiated and Supported by the

UNITED STATES NAVY   OFFICE OF NAVAL RESEARCH

1977

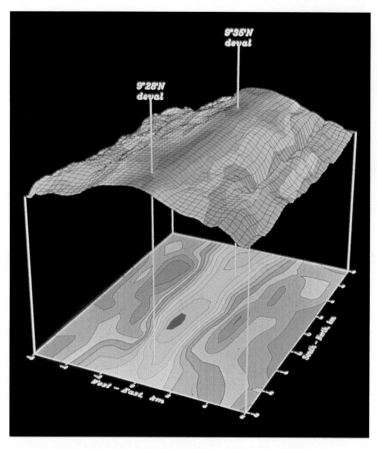

Figure 2.4 The yellow and red colors (lower image) depict the decrease of seismic wave velocities two kilometers beneath the crest of the East Pacific Rise (upper image). A decrease of this sort indicates the presence of molten or semimolten rock beneath the ridge crest. (Woods Hole Oceanographic Institute)

Figure 2.5 Jets of hot fluid escaping from submarine vents along the East Pacific Rise. These features, plus lava erupting onto the sea floor, are evidence that molten rock is present beneath the mid-ocean rift valleys.

island arc
A curved chain of islands with its convex side to the ocean, emerging from the deep-sea floor. Generally located close to a continent.

together form the lowest portions of the Earth's surface (see Figure 2.3). The Marianas trench, for example, is at a greater depth beneath sea level than the Himalayas are above it. The trenches mainly encircle the Pacific and parts of the Indian Ocean, a notable Atlantic exception being the Puerto Rico trough.

The great archipelagos, or volcanic **island arcs,** of the world—the Aleutians, Japan, the Philippines, Indonesia, and the West Indies—are closely associated with the trenches. The island chains are convex toward the ocean, and the trenches run parallel to them along their seaward side (see Figure 2.3). Heat escapes from the trenches at a lower rate than it does from the mid-ocean ridge and the ocean floor in general—evidence that the oceanic crust beneath the trenches is cold and, therefore, dense.

In the eastern Pacific, deep-sea trenches border continental landmasses rather than island arcs. There too, however, the trenches are closely associated with volcanic activity. The Peru-Chile trench is virtually offshore of the west coast of South America, and not surprisingly, it is close by a volcanic chain, the western Andes Mountains (see Figure 2.3). Central America, itself largely of volcanic origin, is bordered by a similar trench. Farther north, the mid-American trench is directly off the west coast of Mexico. This pattern, interrupted in southern California, is resumed again off the Oregon-Washington coast, seaward of the volcanic Cascade mountain range.

The volcanoes that rim the Pacific are active, flickering on and off like fireflies in the dark. Indeed, the Pacific Rim is often referred to as the *Ring of Fire.* Most of the world's volcanoes are confined to this ring, a few other regions of active mountain building, and the mid-ocean ridge system. Except for isolated hot spots such as the volcanic Hawaiian Islands, the vast areas in between are calm by comparison.

Over 99 percent of the world's earthquakes follow essentially the same narrow bands as the volcanoes (Figure 2.6). Earthquakes are triggered by the sudden slippage of rock along faults. At the central rift valley of the mid-ocean ridge, they occur in the uppermost crust. Those occurring landward of the deep-sea trenches are of a different nature, however, as Kiyoo Wadati, a Japanese seismologist, found in 1933. Wadati traced a narrow, slanting zone of earthquake activity beneath the Kurile trench north of Japan to depths of 720 kilometers into the mantle (Figure 2.7). Twenty years later, surveys by American seismologist Hugo Benioff confirmed Wadati's discovery, which is now known as the *Wadati-Benioff zone*. It is only beneath the world's trenches that such a pattern of earthquakes is found.

The evidence we have presented thus far (with much detail excluded) is the approximate state of knowledge that geologists of the late 1950s had of the major features of the ocean floor. They reasoned roughly as follows: First, we have the mid-ocean ridge, a worldwide system of high mountains rising from the ocean floor, with magma erupting from rift valleys that run down the center of the ridge crest. We also have high heat flow—thus, warm and light rocks—and shallow earthquakes at the ridge crest. Second, we have the deep-sea trenches—long, deep depressions that run along the margins of the Pacific Ocean floor. There we have low heat flow—thus, cold and dense rocks—and zones of intense earthquake activity that extend deep into the mantle. Hot magma rising at the ridge crests and cold oceanic crust sinking at the trenches—geologists saw a certain complementary symmetry in this arrangement.

The Sea-Floor Spreading Hypothesis

In the early 1960s, Harry Hess of Princeton University offered an elegant explanation for the apparent symmetry between mid-ocean ridges and deep-sea trenches. He proposed that, like a wound refusing to heal, oceanic crust separates along the rift valley in the mid-ocean ridge, allowing magma to bleed to the surface and form fresh oceanic crust. The crust then spreads laterally, cooling and contracting as it moves toward the

Figure 2.6 Most earthquakes occur along deep-sea trenches, mid-ocean ridges, and regions of active mountain building. The mid-ocean ridges are sites of shallow-focus earthquakes exclusively. Intermediate- and deep-focus earthquakes are confined to zones beneath and landward of the trenches. Also, note the close association between the location of active volcanoes and earthquakes in the vicinity of the trenches.

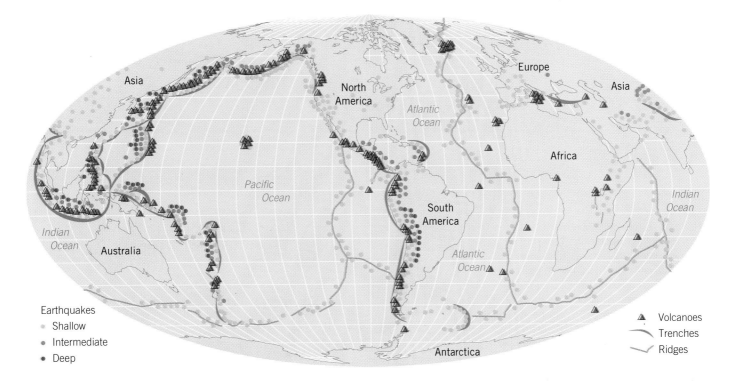

Figure 2.7 Shallow-, intermediate-, and deep-focus earthquakes occur on a deep-slanting plane in the Wadati-Benioff zone beneath the Kurile trench. The Kurile island arc is of volcanic origin.

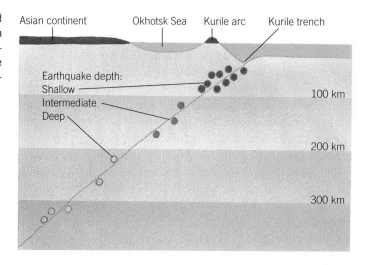

trenches. There it descends back into the mantle and triggers the deep line of earthquakes beneath the trenches. Thus the ocean floor is in a state of slow motion—a vast conveyor belt. American geologist Robert Dietz, an early advocate and contributor to this concept, named the process **sea-floor spreading** (Figure 2.8).

Hess's idea was exciting, but it also seemed unreasonable to geologists schooled in the idea that the continents are locked permanently in place. After all, the mid-ocean ridge and the deep-sea trenches are typically separated by thousands of kilometers; in

sea-floor spreading
The theory that new oceanic crust is created at the mid-ocean ridges, spreads laterally, and descends back into the mantle at the deep-sea trenches.

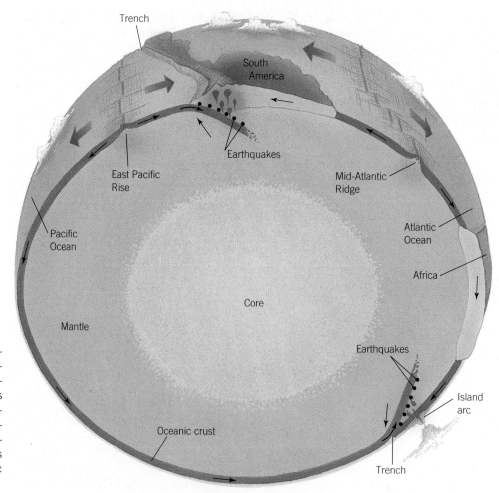

Figure 2.8 A representation of the sea-floor spreading hypothesis. Newly created oceanic crust spreads from mid-ocean rift valleys as fresh magma wells up to replace it. Old oceanic crust descends into the mantle beneath deep-sea trenches. There it triggers deep-focus earthquakes and generates magma that feeds bordering volcanic arcs.

many parts of the world, entire continents intervene. More evidence was needed of the ocean floor between these features and of the mantle beneath the crust in order to test the spreading hypothesis.

What constitutes a valid test of this hypothesis? One approach is to reason as follows: If the sea floor is created at the mid-ocean ridges and spreads toward the trenches, then we should expect the age of the sea floor to increase steadily from ridge crest to deep-sea trench. On the other hand, if the ocean basins are locked permanently in place, then we should expect the sea floor to be the same age throughout. Because the sea floor presumably formed early in the history of the Earth, if it is locked in place, it should be billions of years old.

Evidence in Magnetic Reversals

Clearly, in order to test sea-floor spreading, geologists had to find a way to date the age of the ocean floor—no easy task considering its vast extent and depth. However, there was a solution to the problem, and it relates to a seemingly strange bit of evidence that British geologist Ronald Mason uncovered in the mid-1950s, when he had a magnetometer towed from a naval vessel off the California-Oregon coast. The magnetometer detected parallel bands of strong and weak magnetism in the basaltic ocean floor that run parallel to the axis of a mid-ocean ridge segment, the Juan de Fuca Ridge (Figure 2.9). The bands extend very far on both sides of the ridge until they are lost under sediment.

As we discussed in the previous chapter, lava preserves the direction of the Earth's magnetism at the time of eruption because tiny magnetic crystals act like compass needles and are frozen in place when the basaltic lava solidifies. The magnetism that Mason measured is the sum of the Earth's present magnetism and the magnetism frozen into the oceanic crust at the time the lava erupted (see Figure 2.9). When the magnetic field of each is pointing in the same direction, the total magnetic field is stronger than normal; when the magnetic field of the oceanic crust is opposite to the polarity of the Earth's magnetic field, the total intensity is weaker than normal. In other words, the alternating bands of sea-floor magnetism are the record of past reversals of the Earth's magnetic field. If the sea-floor spreading hypothesis is correct, then the ages of the magnetic bands should increase steadily away from the mid-ocean ridge. However, to prove this hypothesis, geologists had to establish the ages of the **magnetic reversals.**

With this purpose in mind, Alan Cox and his colleagues at the United States Geological Survey set out to measure the magnetism in lava layers of volcanoes along the Hawaiian coast. The volcanoes are constructed of thick layers of lava that record almost continuous eruptions going back millions of years; landslides have left a thick cross section of these layers exposed in the cliffs facing the ocean. The geologists climbed the cliffs and carefully sampled each layer, bottom to top. Back in the laboratory, the age of each sample was determined from the decay rates of radioactive atoms present, and its magnetization was identified by using a magnetometer. Combining the data, researchers were able to date the age of the magnetic reversals in the layers reaching back 4.5 million years. Later, other geologists would extend their data several hundred million years into the past by measuring the magnetism of older lavas in all parts of the world. The results are displayed as the *polarity reversal time scale* of Figure 2.9. Notice that the upper 10-million-year part of the scale matches perfectly the reversal pattern of the Juan de Fuca Ridge.

Now, refer to Figure 2.10 and notice that the irregular, distinctive pattern of the polarity reversal time scale is reproduced in uncanny detail by the magnetic bands on either side of the worldwide mid-ocean ridge. The match is perfect; the most recent reversals occur at the mid-ocean ridge, and progressively older ones are found progressively farther away toward the flanks of the ocean basins. Reversals of short duration are matched by relatively narrow bands, and those of long duration by broad bands.

magnetic reversal
A switch in the direction of the Earth's magnetic field such that a compass needle that today points north would, at the time of reversal, have pointed south.

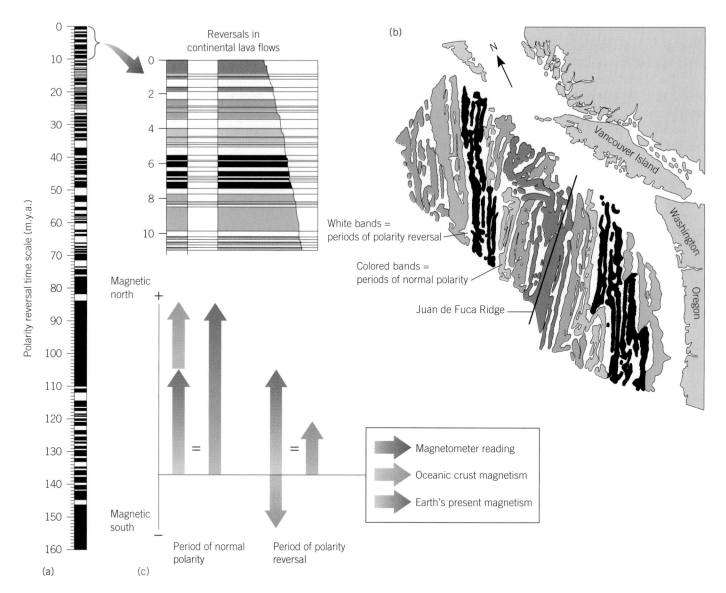

Figure 2.9 Magnetic polarity reversals are used to determine the age of oceanic crust. (a) Scientists detected a pattern of magnetic reversals in successive lava flows on land and dated them by radiometric methods, thus establishing the polarity reversal time scale. (b) In the Juan de Fuca Ridge, an identical pattern occurs on both sides of the central rift valleys of the mid-ocean ridges, with the youngest reversals closest to the ridge crests. (c) Determination of a polarity reversal. A magnetometer records a stronger-than-average reading where the magnetism of the oceanic crust adds to the polarity of Earth's magnetic field. A weaker-than-average reading occurs where sea-floor magnetism is reversed—that is, where it is opposite to the polarity of the Earth's field. (Adapted from H. W. Menard, *Marine Geology of the Pacific,* New York: McGraw-Hill, 1964.)

When they made their discovery, Fred Vine was a young graduate student at Cambridge University and Drummond Matthews was his advisor.

In the early 1960s, British geologists Fred Vine and Drummond Matthews realized that there is only one way to explain this correspondence. As lava bleeds out of the rift valley of the mid-ocean ridge, it acquires the polarity of the Earth's magnetic field at the time of eruption (Figure 2.11). Upon solidification, the lava forms oceanic crust, which is moved aside as younger lava wells up from beneath the rift valley. In this manner, subsequent magnetic reversals are then recorded in fresh lava. Therefore, each segment of the oceanic crust was originally formed at the rift valley and has moved gradually away from the mid-ocean ridge, bearing the magnetic polarity of its time of solidification.

Referring again to Figure 2.10, we see from the magnetic reversal pattern that the age of the Pacific Ocean floor increases steadily from the mid-ocean ridge to the deep-

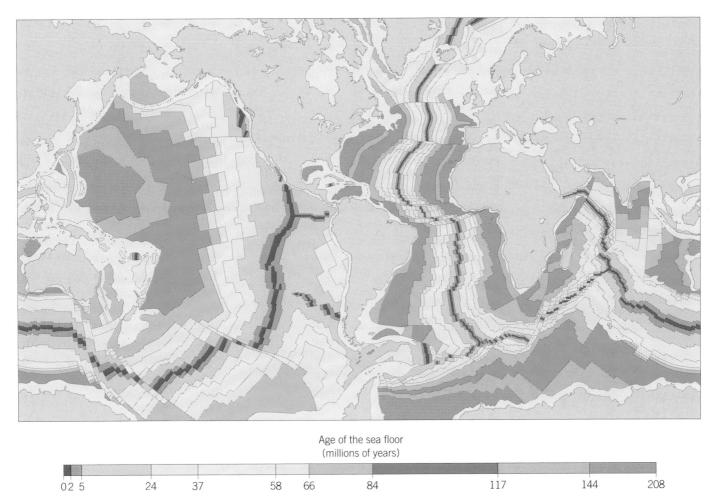

Age of the sea floor
(millions of years)

0 2 5   24   37   58 66   84   117   144   208

Figure 2.10 The age of the sea floor as determined primarily by polarity reversal data. Notice the similarity of the pattern in all the ocean basins. The Pacific basin is spreading at a faster rate than the Atlantic, as indicated by the broader pattern. (Adapted from R. L. Larson and W. C. Pitman, *The Bedrock Geology of the World,* New York: W. H. Freeman, 1985.)

sea trenches on either side of the ocean basin, where the pattern ends abruptly. This evidence lends strong support to the idea that the spreading ocean floor descends beneath the trenches. Thus geologists came to visualize the oceanic crust spreading from the ridge crests as a very wide, very slow-motion conveyor belt—or better yet, because polarity reversals are imprinted on it, as a gigantic tape recording.

We can calculate the velocity of sea-floor spreading by measuring the distance from the ridge crest to a given magnetic band and then dividing it by the age of that band as determined by its position in the polarity reversal time scale (Figure 2.12). Notice in Figure 2.10 that spreading rates vary considerably. The Atlantic Ocean floor is spreading from the Mid-Atlantic Ridge about as fast as your fingernails grow. The Pacific Ocean floor is spreading from the East Pacific Rise at up to ten times that speed. The more rapid Pacific spreading rate is reflected in the wider magnetic bands across the Pacific floor.

Evidence in Sea-Floor Sediment

Studies of sea-floor sediments provide an independent check of the sea-floor spreading hypothesis. To collect these samples, a consortium of research institutions launched the Joint Oceanographic Institution's Deep Earth Sampling project (JOIDES). At the heart of the project was the *Glomar Challenger,* the specially designed research ship mentioned in the Introduction (see Figure 1.4). This vessel was designed specifically to recover cores of sediment and basaltic crust from the deep ocean floor. Its six independent engines, coordinated by computer, kept the ship on target over kilometers of ocean while the crew drilled through thousands of meters of marine mud, sand, gravel, and rock. The

64

CHAPTER 2

geologists, rotating 24-hour shifts aboard the *Glomar Challenger,* sampled the world's ocean floors and discovered a remarkably consistent pattern.

Toward the margins of the continents, the oceanic crust is covered by a wedge of sediment thick enough to bury the irregularities of the ocean floor and produce a flat, nearly featureless **abyssal plain.** However, the wedge gradually tapers off toward the mid-ocean ridge, where the rocks of the crust are bare—save for thin patches of recent plankton shells that have settled out of the ocean above the ridge. Furthermore, as we approach the mid-ocean ridge, progressively younger sediment is found in contact with basaltic crust of the same age (Figure 2.13). Both the thinning of the sediment and the age pattern are to be expected if the sea floor is spreading. As new crust forms along the mid-ocean ridge, older crust and sediment are moved aside, and the younger sediment blankets not only the older sea floor but also the newly exposed crust. One of the triumphs of the *Glomar* expeditions was the discovery that the ages of the oldest parts of the Atlantic Ocean basin—that is, its flanks—match the time of fragmentation of Wegener's Pangaea.

Another significant point concerns the sediment found on the floor of the western Pacific. At 200 million years of age, it is the oldest known sediment of the ocean and rests upon the oldest known oceanic crust, also 200 million years old—which represents a mere 5 percent of the Earth's 4.6-billion-year age. Thus, surprisingly, the present ocean basins—which compose 60 percent of the Earth's surface—have been in existence for less than 5 percent of geologic time. The ocean basins are young features that have been cycled and recycled many times in the past.

Evidence in Sea-Floor Topography

Sea-floor spreading offers a plausible explanation for many formerly puzzling aspects of ocean topography, including the fact that depths increase with distance from the mid-ocean ridge crests. The mid-ocean ridge stands high above the general level of the ocean floor because it is warmer, which makes it less dense and more buoyant. As oceanic crust

abyssal plain
A flat, level area of the ocean floor that begins at the foot of the continental rise.

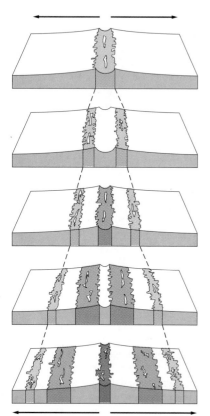

Figure 2.11 The explanation of the polarity reversal pattern supports the theory of sea-floor spreading. As oceanic crust spreads away from the ridge crest, younger oceanic crust is created by the upwelling of fresh magma from beneath the rift valley. Thus the age of the sea floor increases steadily away from the ridge crest.

Figure 2.12 This example shows how the polarity reversal pattern may be used to calculate the velocity of sea-floor spreading relative to the position of the ridge crests.

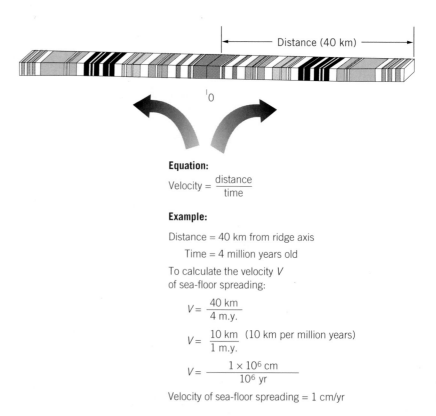

Equation:

Velocity = distance / time

Example:

Distance = 40 km from ridge axis

Time = 4 million years old

To calculate the velocity *V* of sea-floor spreading:

$V = \dfrac{40 \text{ km}}{4 \text{ m.y.}}$

$V = \dfrac{10 \text{ km}}{1 \text{ m.y.}}$ (10 km per million years)

$V = \dfrac{1 \times 10^6 \text{ cm}}{10^6 \text{ yr}}$

Velocity of sea-floor spreading = 1 cm/yr

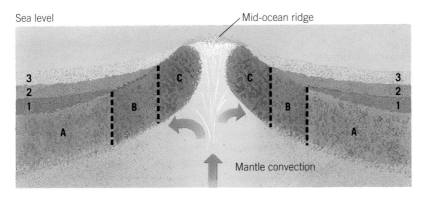

Figure 2.13 As the ocean basins widen, sediment is deposited on the newly created oceanic crust in a pattern that is identical in all the oceans. Younger sediment overlaps older sediment in the direction of the ridge crest. Wherever sediment rests directly on the oceanic crust, it matches the age of that segment of crust. Thus sediment layer 1 matches the age of crust segment A, layer 2 matches the age of segment B, and layer 3 matches the age of segment C.

spreads from the central rift valley, it cools, grows denser, and sinks lower into the mantle. For this reason, the mountainous topography of the mid-ocean ridge does not extend across the entire ocean floor. However, like a hot skillet just removed from the stove, cooling is most rapid in the portion of the crust that has most recently moved away from the heat source below the rift valley. The cooling rate then levels off with time—that is, with distance from the ridge crest. These differences in cooling rates are reflected in the profile of the ocean floor: steep at the mid-ocean ridge and gentle at the margins of the basins. In general, the depth of each location on the ocean floor is determined by its age (Figure 2.14). Differences in cooling rates are also responsible for the subtle contrast between the topography of the Atlantic and Pacific floors across their respective mid-ocean ridges. The slow-spreading Atlantic basin has had more time to deepen; hence, the profile of the Mid-Atlantic Ridge is generally steep and rugged. On the other hand, the Pacific crust, spreading more rapidly from the East Pacific Rise, has had less time to cool; its broader, gentler profile is the reason it is called a "rise."

The ocean floor is pockmarked with thousands of volcanoes, and their distribution and depth confirm the effects of sea-floor spreading on sea-floor topography. Most are **seamounts**—recognized by their cone-shaped cross sections—that grew to their maximum height without ever reaching the surface. Other undersea volcanoes were once islands, as shown by cores containing ancient coral reefs, beach gravel, and shallow-water fossils. Wave erosion has planed the tops of these **guyots** so thoroughly that their older name *tablemount* is descriptive. Many are currently thousands of meters below sea level; their submergence offers strong evidence that the sea floor sinks as it spreads.

In the 1880s, American geologist James Dana surveyed another group of Pacific islands, the Hawaiian chain. There he observed that active or recent volcanism was confined to the islands of Hawaii and Maui. Also, the more northwestward the islands that he visited, the more eroded their volcanoes were. He therefore concluded that the ages of the islands increase from southeast to northwest; radiometric dating of the island lavas in the mid-twentieth century bore out his conclusion. At about the same time, oceanographic surveys discovered that the Hawaiian Islands are the most visible of a 7000-kilometer-long line of sea-floor volcanoes (Figure 2.15a). The line extends from the active Loihi Seamount off the southeast coast of Hawaii and northwest to Midway Island; then it bends northward as the Emperor Seamount chain toward the Aleutian trench. The ages of the volcanoes increase steadily from Hawaii to the trench. The oldest of them last erupted 65 million years ago.

seamount
A sea-floor volcano that has never risen above sea level.

guyot
A sea-floor volcano that at one time rose above sea level, where its top was planed by wave erosion.

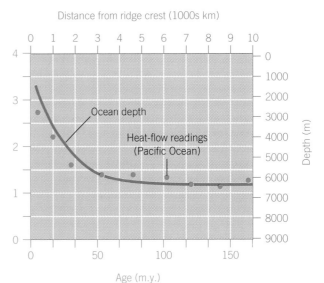

Figure 2.14 Crustal heat flow diminishes with distance from the mid-ocean ridge crest as the depth and age of the crust increase. This suggests that the cooling and spreading rates control the general contour of the ocean bottom.

ASIDE 2.1 *Darwin's Subsidence Theory*

Charles Darwin is best known for his theory of natural selection as the fundamental mechanism of organic evolution. Few people know, however, that in addition to being a great biologist, he was also an excellent geologist. He made shrewd observations concerning the origin of the Pacific Islands, which he visited while employed as a naturalist aboard the *HMS Beagle* in the 1830s (Figure 2A). He noted that a newly emergent volcano will acquire a *fringing reef* that grows directly offshore. But as the volcanic island sinks, the reef—needing sunlight—continues to grow up to sea level. In time, it evolves into a barrier reef, with a lagoon between the reef and the island. The last stage is a roughly circular-shaped *atoll* encircling the submerged volcano, which twentieth-century surveys would show to be a guyot (Figure 2B).

Figure 2A The Kagangel Atoll, Belau, is a coral reef surrounding a lagoon. The reef is anchored to a submerged volcanic island.

From the distribution of these islands, Darwin concluded that subsidence occurred on a grand scale and took millions of years to accomplish. His theory was an early demonstration of mobility and change in the Earth's crust. Darwin had also observed relatively recent marine fossils preserved in the strata of the Andes Mountains and concluded that the western South American coast had been uplifted. He inferred that the Pacific sea-floor subsidence was in some way related to the uplift. Today, we believe that subsidence of the sea floor is the inevitable result of the cooling of the oceanic crust as it spreads from the mid-ocean ridge, and that uplift of the Andes is caused by Pacific Ocean crust descending beneath South America.

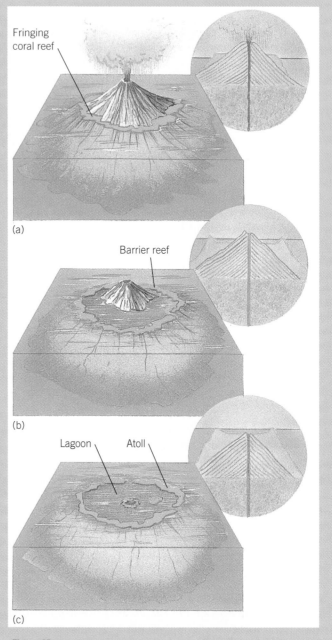

Fringing coral reef

(a)

Barrier reef

(b)

Lagoon   Atoll

(c)

Figure 2B Darwin's subsidence theory of Pacific island coral reefs. As a volcanic island sinks, the attached coral continues to grow upward toward the sunlight at sea level, and, in the process, evolves from a fringing reef to a barrier reef to an atoll.

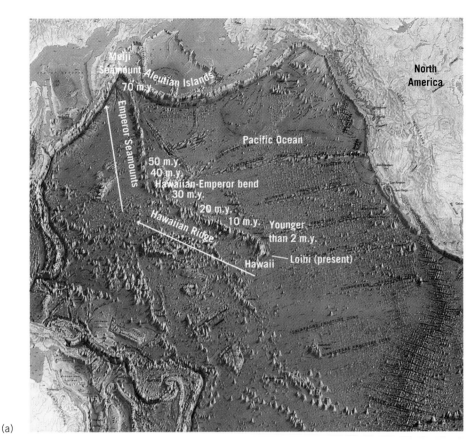

(a)

Figure 2.15 (a) Scientists observed that the ages of the volcanic Hawaiian Islands and the Emperor Seamount chain increase steadily as they approach the Aleutian trench. (Marie Tharp) (b) The probable explanation is that each volcano formed over the stationary magma source, or hot spot, over which Hawaii is currently situated and then moved away from the hot spot as the sea floor spread to the north and then northwest. As the oceanic crust cooled, the sea floor deepened, and former volcanic islands were submerged.

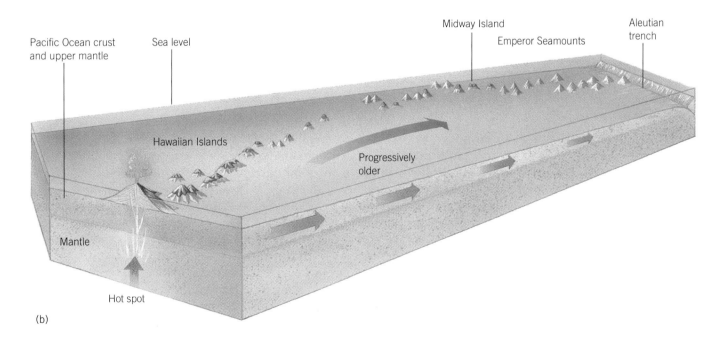

(b)

Significantly, of this long chain, only the southeasternmost Hawaiian volcanoes are active. They are situated over a magma reservoir 150 kilometers within the mantle. If the sea-floor spreading hypothesis is correct, then the oceanic crust now at the Aleutian trench was created at a mid-ocean ridge far to the southeast and passed over the spot now occupied by Hawaii. For this reason, Canadian geologist J. Tuzo Wilson proposed in 1963 that the volcanoes of the Hawaiian-Emperor Seamount chain were punched out as the Pacific oceanic crust passed over the same magma source that spews out the lava currently building Hawaii (Figure 2.15b). Wilson called this long-active

J. Tuzo Wilson of the University of Toronto was an especially insightful geoscientist who contributed significantly to many aspects of the theories of sea-floor spreading and plate tectonics.

Figure 2.16 The distribution of some prominent hot spots.

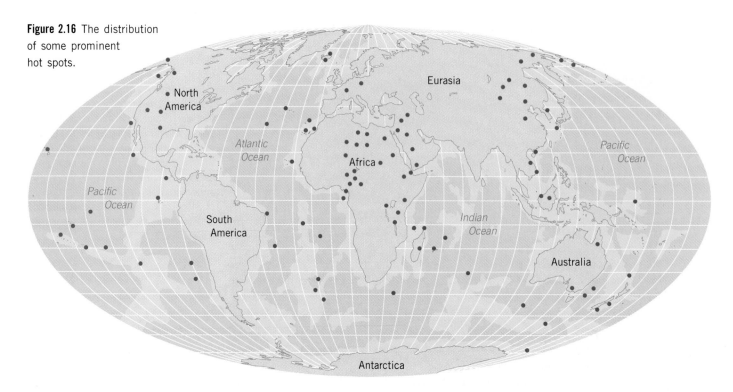

hot spot
An area of volcanic activity produced by a plume of magma rising from the mantle.

plate tectonics
The theory that proposes that the lithosphere is divided into plates that interact with one another at their boundaries, producing tectonic activity.

seismic belt
A long, narrow zone of earthquake activity associated with lithospheric plate boundaries.

plate
A rigid segment of the lithosphere that moves as a unit over the asthenosphere.

region of volcanic activity a **hot spot.** Figure 2.16 shows that many hot spots have since been identified; they are distributed throughout the continents and ocean basins, although they are heavily concentrated in Africa. Because of their number and wide occurrence, hot spots are fundamental to our understanding of crustal processes; we shall return to them shortly.

Plate Tectonics

By the early 1960s, geologists working primarily on land had assembled overwhelming evidence of the essential correctness of Alfred Wegener's theory that the continents had for a time existed as a single landmass and have since fragmented and drifted to their present positions. At about the same time, marine geologists had assembled overwhelming evidence of sea-floor spreading. What is more, it was obvious from the timing and geometry of drift and spreading that the two processes are related. The pursuit of an explanation for this relationship led to a new synthesis, **plate tectonics,** which, as we will see, goes far beyond merely combining drift and spreading to offer a comprehensive theory of Earth's crustal structure.

Confirmation of sea-floor spreading established that new oceanic crust is created at the mid-ocean ridge by magma erupting from beneath the central rift valley floor, and that the new crust is then moved as if it were on a conveyor belt to the deep-sea trenches, where it descends back into the mantle. Moreover, these same features—trenches and mid-ocean ridge—coincide with the major earthquake or **seismic belts;** that is, the earthquakes trace a network of stress lines around the globe. Furthermore, geologists realized that the crust within the seismic belts must behave rigidly; if the sea floor behaved like soft clay, then the parallel bands of alternating magnetism would deform beyond recognition on their journey from ridge crest to trench.

In 1968, a handful of geologists, independently of one another, proposed the theory of plate tectonics: that the regions encircled by seismic belts are rigid slabs, or **plates,** and that the earthquakes are generated when these plates grind against one another as they move. According to the theory, the seismic belts—that is, the blizzard of dots confined to the narrow strips shown in Figure 2.6—delineate plate boundaries. These boundaries segment the Earth's surface into a mosaic of about seven or eight large plates and many smaller ones (Figure 2.17). Some, such as the Nazca and Pacific plates, contain oceanic crust exclusively. Most of the others include continental as well as oceanic crust; as these plates spread away from the mid-ocean ridge and toward the trenches, the continents move along with them. So the continents really do drift—but passively, embedded within the moving plates, and not like ships plowing through oceanic crust, as Alfred Wegener had envisioned.

How thick are the plates? The answer to this question must be found indirectly through analysis of seismic waves, for we lack the technology to drill deep into the Earth. The velocity of a seismic wave is determined by the properties of the substance through which it travels; the more rigid the substance, the higher is its velocity. Figure 2.18 shows that wave velocities increase to a depth of about 100 kilometers, confirming that this outer zone of the Earth is quite rigid. This depth marks the base of the **lithosphere,** which is composed of the crust and a thin layer of the upper mantle. Its thickness amounts to a sixty-fourth of the radius of the Earth, so on the scale of Earth's dimensions, it is a mere eggshell. The lithosphere is divided into the mosaic of lithospheric plates that constitute Earth's outer surface; these plates are what geologists refer to in discussing plate tectonics. Seismic-wave velocities decrease markedly beneath the lithosphere and do not recover until they reach 250 kilometers, which suggests a zone in the mantle that is close to melting and has lost some of its rigidity. The plates glide over, or are carried along on, this soft yielding zone in the mantle, the **asthenosphere.**

Up to this point, we have said a great deal about the *plate* in plate tectonics but have neglected the latter term. *Tectonics* is derived from the same Greek root as architecture, *tekton,* meaning "to construct." It refers to those processes responsible for the origin of the large-scale structural elements of the crust. Thus plate tectonics is a theory stating that the Earth's surface is divided into lithospheric plates and that the interactions occurring at plate boundaries are responsible for the production and destruction of oceanic crust; the rifting, drifting, and collision of continents; the formation of mountains, volcanoes, and earthquakes; and more.

Because the plates are rigid slabs packed tightly together on the surface of a sphere, there are only three ways they can interact (Figure 2.19): they may diverge, converge, or slide by one another. The narrow zones where these interactions occur constitute the three principal types of plate boundaries, each displaying characteristic landform features, earthquake patterns, and igneous activity. **Divergent boundaries,** where the plates separate and grow, are characterized by mid-ocean ridge rift valleys and their continental extensions. **Convergent boundaries,** where the plates collide and cycle lithosphere back to the mantle, are characterized by deep-sea trenches, volcanic arcs, and fold-mountain belts. **Transform boundaries,** where the plates slide by one another, are characterized by narrow fracture zones that connect offset segments of the mid-ocean ridge or deep-sea trenches.

Divergent Boundary Processes

The narrow band of earthquakes that trace the elevated central rift valley of the mid-ocean ridge and its continental extensions defines the divergent boundaries of the world.

The scientists credited with establishing the plate tectonics model include Dan McKenzie of Cambridge University and Robert Parker of the Scripps Institute of Oceanography; Jason Morgan of Princeton University; Xavier LePichon of Lamont-Doherty Geological Observatory; and Don Isacks, Jack Oliver, and Lynn Sykes, also of Lamont. At the time, they were all in their twenties or early thirties and were either graduate students or recent Ph.D.s.

lithosphere
The rigid outermost layer of the Earth, which includes the crust and a sliver of the upper mantle.

asthenosphere
A semimolten zone of the mantle just below the lithosphere.

divergent boundary
The border between two plates that are moving away from one another as new crust is formed.

convergent boundary
The border between two plates that are colliding with one another. Crustal material is subducted back into the mantle at these boundaries.

transform boundary
The border between two plates that are sliding by one another horizontally without either creating or destroying oceanic crust. It connects offset ridge segments, offset trenches, or ridge segments to trenches.

Figure 2.17 (on pages 70-71) According to the theory of plate tectonics, the Earth's surface is divided into a mosaic of lithospheric plates whose boundaries are delineated by narrow seismic belts. Some of the plates consist solely of oceanic crust; others include continental crust as well. The arrows show that the plates move away from the ridge crests and toward the trenches. (Adapted from "This Dynamic Planet: World Map of Volcanoes, Earthquakes, and Plate Tectonics," by Tom Simkin and Robert I. Tilling of the Smithsonian Institution and James N. Taggart, William J. Jones, and Henry Spall of the U.S. Geological Survey, 1989.)

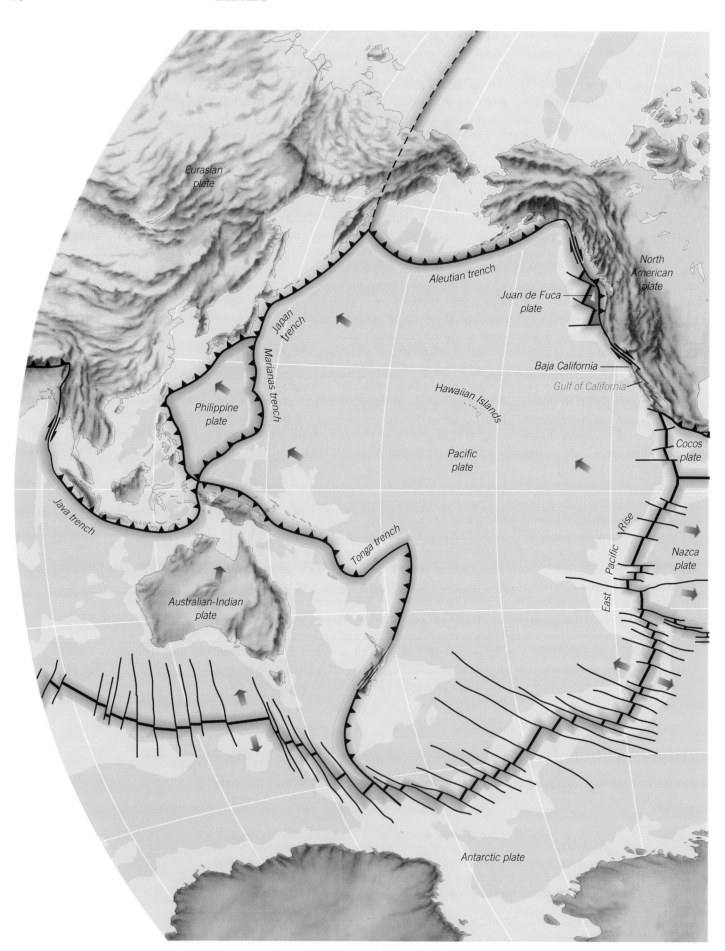

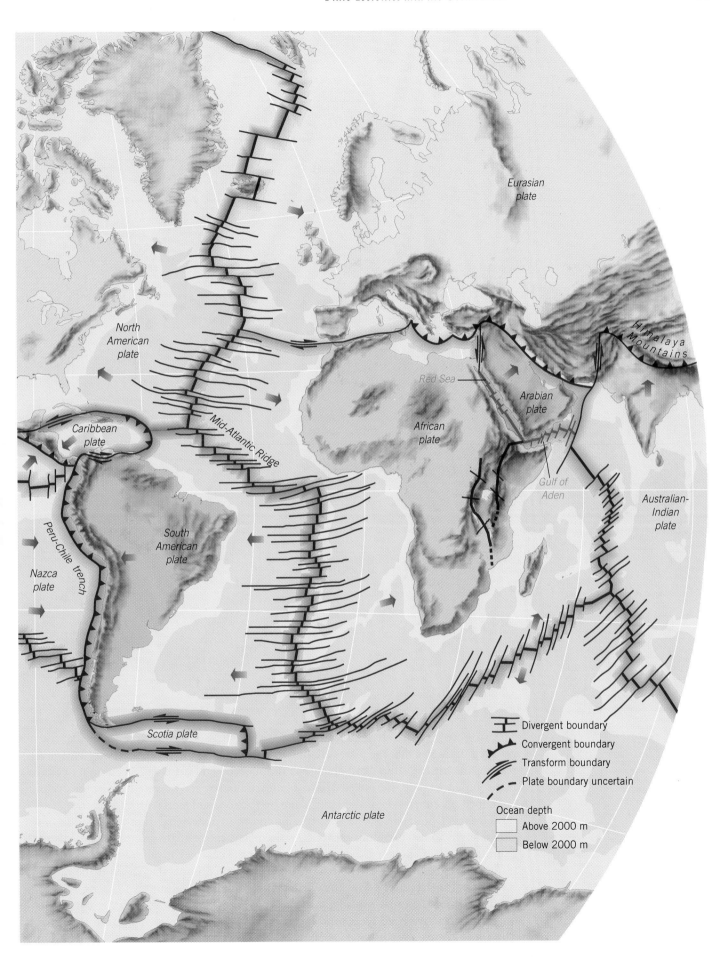

Eurasian plate

North American plate

Caribbean plate

Mid-Atlantic Ridge

Peru-Chile trench

Nazca plate

South American plate

African plate

Red Sea

Arabian plate

Gulf of Aden

Himalaya Mountains

Australian-Indian plate

Scotia plate

Antarctic plate

Divergent boundary
Convergent boundary
Transform boundary
Plate boundary uncertain

Ocean depth
Above 2000 m
Below 2000 m

Figure 2.18 The abrupt change in seismic velocities at depths of about 100 kilometers marks the boundary between the lithosphere and the asthenosphere, over which the plates slide.

Figure 2.19 There are three types of plate boundaries: (a) divergent boundaries, where the plates are created and spread away from a mid-ocean ridge; (b) transform boundaries, where the plates slide by one another horizontally; and (c) convergent boundaries, where the plates collide, causing one to descend into the mantle beneath the deep-sea trench.

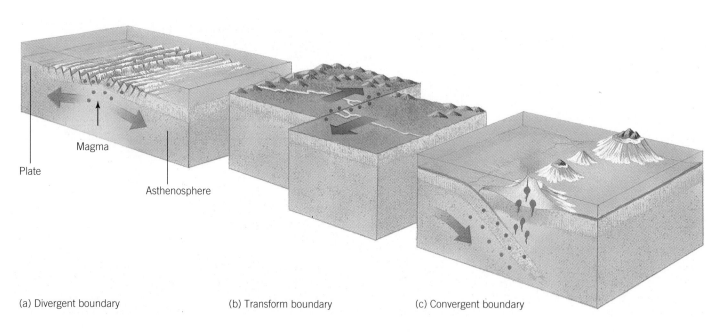

(a) Divergent boundary   (b) Transform boundary   (c) Convergent boundary

spreading center

The segment of the mid-ocean ridge that is the site of active rifting and sea-floor spreading.

The earthquakes there are shallow, and analysis of the associated fault motions that trigger them indicates that the plates are under tension along the boundary. This tension is responsible for the rift valley, which has dropped downward along steep faults as the plates on either side are pulled apart. The earthquakes are triggered when the rocks, restrained on either side of the fault plane by friction, suddenly break free.

Divergent boundaries, the sites of simultaneous plate separation and growth, are also referred to as **spreading centers.** As the plates move apart, magma that wells up from the partially molten asthenosphere beneath the central rift valley is accreted to the trailing edges of the plates. In this way, the plates widen in parallel strips at the rate of 2 to 10 centimeters per year as they diverge from the ridge crest.

If new oceanic crust is being created at divergent boundaries, then the continents on either side have been separating for as long as these basins have been in existence. Thus, for example, North America, South America, Europe, and Africa were joined prior

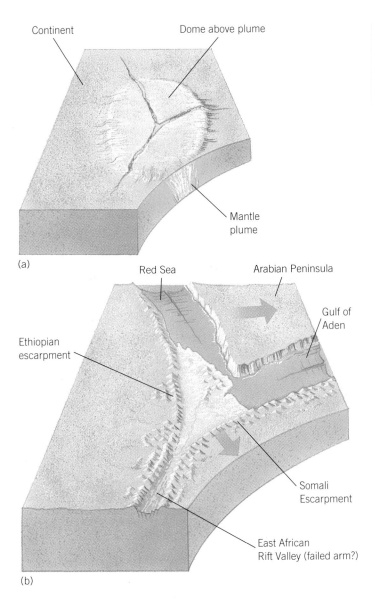

(a)

(b)

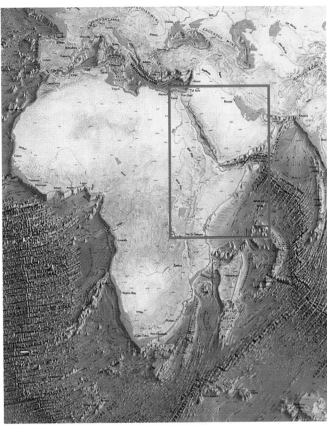

(c)

Figure 2.20 Hypothesis of how a divergent plate boundary develops. (a) A rising mantle plume beneath the continental crust forms a dome with three branching fractures (a triple junction). (b) The fractures spread apart, forming narrow rift valleys that widen and deepen far enough for magma to well up from the mantle and form oceanic crust. The third rift does not completely separate the continental crust; instead, it remains a failed arm. (c) This detail of Marie Tharp's model shows the triple-junction relationships among the Red Sea, the Gulf of Aden, and the East African Rift Valley. Notice the connection of the triple junction to the Carlsberg Ridge, the segment of the mid-ocean ridge directly to the east. (Marie Tharp)

to the opening of the Atlantic Ocean. Hence, the hypothesis of continental drift, so long the subject of intense debate, fits easily within the framework of plate tectonics.

It follows, therefore, that the divergent boundary that today is the Mid-Atlantic Ridge began as a fracture in continental crust. Indeed, we can observe what may be a very early stage of the same process at the Afar Triangle, where northeast Africa and the Arabian Peninsula fit hand-in-glove. Notice that three rift zones meet there and form a *triple junction* (Figure 2.20). Two of them, the Red Sea and Gulf of Aden, are continuous with the volcanic Carlsberg Ridge that splits the Indian Ocean; the third, the East African Rift Valley, intersects continental crust. The triple junction is in fact located over a hot spot, a point of intense volcanic activity formed by a plume of magma and hot rock rising from the mantle. Earlier, we described how these plumes of magma punch out volcanoes on the thin, moving oceanic crust; but they may act differently on thicker, stationary continental crust.

Geologists believe that about 15 million years ago, a plume that rose beneath what is now the Afar Triangle heated and stretched the overlying continental crust until it fractured in three directions. The continents separated cleanly along two of the fractures, and basaltic magma welled up from the mantle between the continental blocks, forming denser oceanic crust. Eventually, the Indian Ocean flooded these canyons to create what are now the Red Sea and Gulf of Aden. Today, these narrow seaways are volcanically active spreading centers, and they continue to widen with the passage of time.

Volcanic and seismic activity are less frequent along the third branch of the Afar Triangle, the East African Rift Valley, which is still underlain by continental crust. Evidently, the East African rift is not as fully developed as the Red Sea and Gulf of Aden rifts. It may well remain as a deep, linear trough or *failed arm* in the African continental crust.

The Mid-Atlantic Ridge was probably formed as a series of connecting triple junctions. If we mentally close out the Atlantic Ocean by joining the continents so that they resemble Pangaea, many continental rift zones, which today trend toward the Atlantic Ocean, meet at junctions that resemble the Afar Triangle (Figure 2.21). These rift zones may be considered failed arms similar to the East African Rift Valley. However, because they are millions of years older than the East African rift, the surrounding cliffs have long since been eroded. The ancient valleys are mostly filled with sediment, and some determine the courses of major rivers, such as the Mississippi, the Amazon, and the Niger.

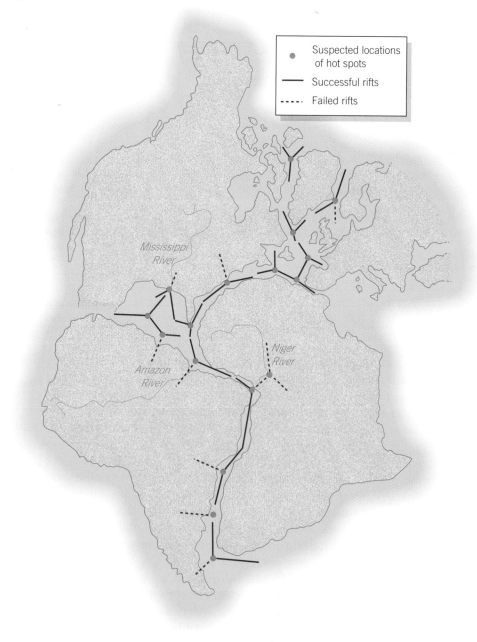

Figure 2.21 By locating ancient and active hot spots along the present Mid-Atlantic Ridge, we may infer how the Atlantic Ocean basin formed. When the continents existed as Pangaea, the Mid-Atlantic Ridge may have developed where the active rifts of the triple junctions joined. Today, many of the failed arms of these triple junctions are sediment-filled troughs occupied by major rivers on the separate continents.

More important are the arms of these junctions that have *not* failed. Notice in Figure 2.21 that they join up and form something that closely resembles the Mid-Atlantic Ridge. Thus a reasonable assumption is that continental rifting and ocean basin formation are initiated by a number of mantle plumes or hot spots whose lengthening arms meet and "unzip" the continental lithosphere. This action releases pressure on the hot mantle and causes the shallow asthenosphere to melt along the entire length of the fracture. The magma then rises into the rift valley—and thus begins the processes we associate with sea-floor spreading.

The Atlantic Ocean of today is in a mature stage of development. It is characterized by a wide ocean basin between separated continents and by **passive continental margins** that are nearly free of earthquakes and igneous activity because they are far from plate boundaries (Figure 2.22). The old oceanic crust near the continental margins has cooled and deepened, and the adjacent continental crust has itself subsided well below sea level. The continent-ocean boundary, now stable for many millions of years, has accumulated thousands of meters of land-derived sediment, which thins toward the mid-ocean ridge because of sea-floor spreading.

passive continental margin
A margin between an ocean and a continent that does not include a plate boundary. It is also free of volcanic or seismic activity.

The wide passive continental margin, built of this sediment brought down from the continents, includes an extensive, shallow **continental shelf** and a broad **continental slope** that leads down to the deep water (see Figure 2.22). An apronlike **continental rise** fringes the base of the margin and trails off into the flat abyssal plains that smother the irregularities of the basaltic oceanic crust. These margins, located on the trailing edges of the separating continents, retain vestiges of the rift valleys, lavas, and continental sediments associated with the ancient separation of the continents.

continental shelf
A very gently sloping surface that extends from the shoreline to the continental slope.

continental slope
A more steeply sloping surface that extends from the continental shelf to the continental rise or oceanic trench.

continental rise
That part of the continental margin that extends gently downward from the continental slope to the abyssal plain.

You may have noticed that the discussion of plate divergence has been confined mainly to the Atlantic Ocean, which opened up as Pangaea was rifted apart. However, not all plates are created by the separation of continents. For example, those spreading from the East Pacific Rise may have formed independently of continental crust.

Convergent Boundary Processes

The narrow bands of earthquakes that occur along deep-sea trenches delineate most of the world's convergent boundaries. These boundaries are the sites where two plates, spreading toward one another from the mid-ocean ridges, collide. Where they meet, the denser, colder plate bends and descends into the mantle in a process called **subduction,** meaning "to pass under." The descent of the subducting plate is traced by the curving Wadati-Benioff zone of earthquakes deep into the mantle. **Subduction zones** are areas where the plates that are created at the mid-ocean ridge are eventually destroyed or, more precisely, recycled into the mantle.

subduction
The downward plunging of one lithospheric plate under another.

subduction zone
The region where one lithospheric plate thrusts downward under another.

The most numerous and varied examples of convergence are located along the margins of the Pacific and Indian oceans. There we are able to distinguish three convergence modes based on the nature of the lithosphere involved: *ocean-ocean, ocean-continent,* and *continent-continent.*

Ocean-Ocean Convergence

Ocean-ocean convergence produces island arc and deep-sea trench systems, of which there are about twenty, mainly in the western Pacific. No two are identical, but the Japan arc displays features and processes common to most of them (Figure 2.23).

We can see that as the cold and dense Pacific plate plunges beneath the Eurasian plate, it forms a deep-sea trench. Scraped off the descending plate is much of the sediment deposited on the sea floor during its long journey from mid-ocean ridge to subduction zone. This sediment is then plastered to the edge of the bordering plate as an **accretionary wedge,** which forms a low ridge that rims the trench.

At about 100 kilometers beneath the surface, the subducting plate generates magma that rises through the denser mantle to the surface and fuels the island arc volcanoes

accretionary wedge
A large mass of sediment that accumulates in a deep-sea trench and is scraped onto the neighboring plate by the subducting slab.

Figure 2.22 The development of an ocean basin from triple junction to full maturity. Note the gradual deepening of the sea floor as the ocean basin widens and the building of elaborate passive margins by the deposition of thick sediment wedges on the faulted oceanic crust.

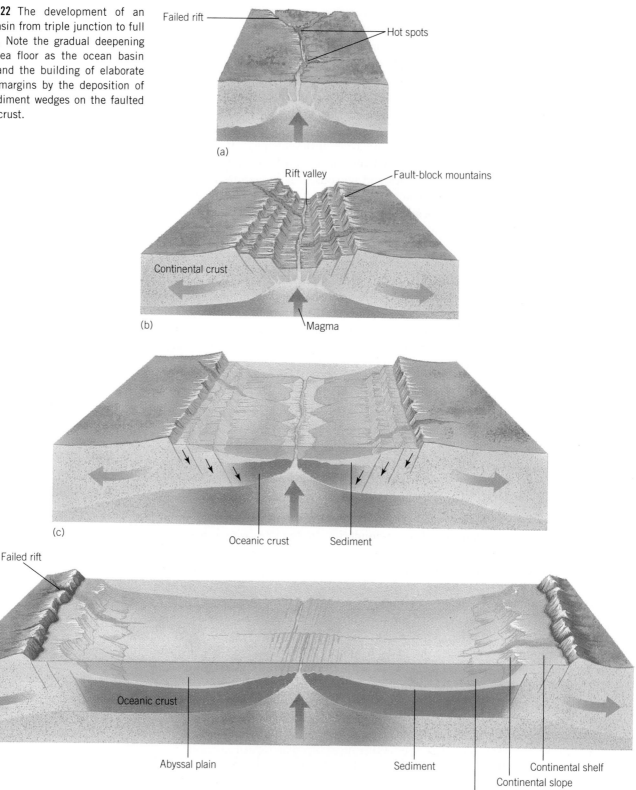

just landward of the trench. Precisely how the magma is generated remains unknown. One hypothesis is that the descending plate absorbs enough heat from the warm mantle to partially melt. Another hypothesis suggests that the subducting plate triggers melting in the wedge of upper mantle above it because of water rising from the plate—water that was trapped within the subducting plate and was expelled because of the increas-

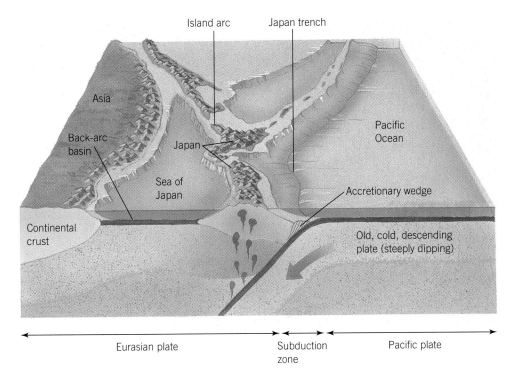

Figure 2.23 Ocean-ocean convergence. Features of this type of plate interaction include: (1) a deep-sea trench where the denser plate is subducted; (2) an accretionary sediment wedge scraped onto the adjacent plate; (3) a volcanic island arc fed by the magma generated by plate subduction; and (4) a back-arc basin between the island arc and the continental mainland.

ing pressure. The water in turn lowers the melting point of the hot mantle, generating the magma.

A back-arc basin separates the Japan arc from the mainland of Asia; similar basins are a consistent feature of most island arc–trench systems. At first glance, these basins seem unexceptional, merely regions where the oceanic crust is separated from the main ocean basin by the island arcs. But detailed oceangoing surveys have revealed strong evidence of *back-arc spreading* in some of them—that is, sea-floor spreading *within* the back-arc basins. Rift fractures divide the basins, and they display magnetic reversal patterns similar to those that run parallel to the mid-ocean ridges. Apparently, magma erupting at the rift replaces older crust that has moved aside toward the flanks of the back-arc basin.

The more we study island arcs, the more evidence we find of the power of plate tectonics to account for the subtle differences we observe. For example, the reason the trenches curve is that they are slanted depressions on a sphere, and the geometry of a sphere dictates that all but vertical depressions must curve. In fact, the shallower the angle of subduction, the tighter is the curve traced by the trench. Because magma is generated from a subducting plate at depths beginning at about 100 kilometers, the island arcs of a gently dipping plate will be farther from the trench than the islands of a steeply dipping one.

But what governs the angle of descent of the subducting plate? One critical factor is its density, which in turn is controlled by the temperature of the plate: the colder the plate, the steeper the angle. However, the temperature of a plate is determined by how long it has been spreading from the mid-ocean ridge—in other words, by its age. Therefore, ocean-ocean convergences, most of them far from the spreading centers of the mid-ocean ridge, probably occur because an old plate has cooled to the point that it grows dense enough to descend into the mantle, thus initiating subduction.

Ocean-Continent Convergence

In contrast to the aged crust of the western Pacific, the oceanic crust of the eastern Pacific is quite young, as shown by the sea-floor age patterns on the map in Figure 2.10. The reason for the age difference is also shown on the map. The North and South

Figure 2.24 Ocean-continent convergence. Buoyant oceanic crust subducts into the mantle at a relatively low angle. Volcanoes erupt on the bordering continental crust and build mountain chains such as the Andes and the Cascades.

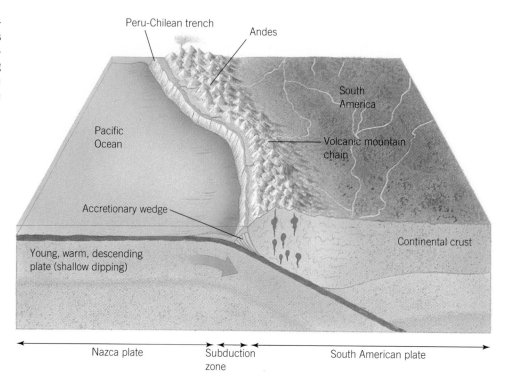

American plates, spreading westward as a consequence of the breakup of Pangaea and the opening of the Atlantic, are in the process of overriding the plates that are spreading eastward from the Pacific mid-ocean ridge. All that remains of a presumably huge expanse of plate that lay west of North America is the tiny Juan de Fuca plate. The Cocos and Nazca plates to the south are much larger than the Juan de Fuca plate but are small in comparison to the huge Pacific plate that is spreading from the East Pacific Rise.

Subduction zones that border nearly the entire length of South, Central, and North America mark the boundaries where plates spreading from the east and from the west meet and clash. This vast stretch of the Earth's surface is undergoing ocean-continent convergence wherein a subducting plate descends beneath a deep-sea trench directly off the shore of continental crust (Figure 2.24). Consequently, the magma generated by the subducting plate erupts on land to build the great volcanic chains that parallel the west coasts of these continents—the Andes and the Cascades, for example.

Ocean-continent convergence is responsible for the rugged coast of western South America and the Andes. In addition, the region is shaken by frequent and powerful earthquakes set in motion by the subducting Nazca plate. The continental margin is narrow and steep; sediments brought to the coast are washed quickly into the Peru-Chile trench directly offshore. Western South America is an **active continental margin,** because it is on the leading edge of a continental plate astride a subduction zone. Contrast it with the South American Atlantic coast. Far from plate boundaries, the eastern coast is nearly earthquake-free, and its only rugged topography consists of the resistant remnants of very ancient mountains that predate Pangaea. Its passive margin consists of a wide continental shelf and slope built of sediment deposited on a stable crust (see also Figure 2.22). Refer again to Figure 2.17, and you will see that the margins of North America also have features and tectonic settings that are broadly similar to what we have just described: a rugged, active western margin at the intersection of plate boundaries, and a broad, passive eastern margin far from plate boundaries.

active continental margin
A boundary between an ocean and a continent marked by plate interaction and thus by frequent volcanic or seismic activity.

Continent-Continent Convergence

If rifting separates continental blocks, subduction brings them together, and the resulting collision builds mountains. The Himalayas are the classic example of continent-

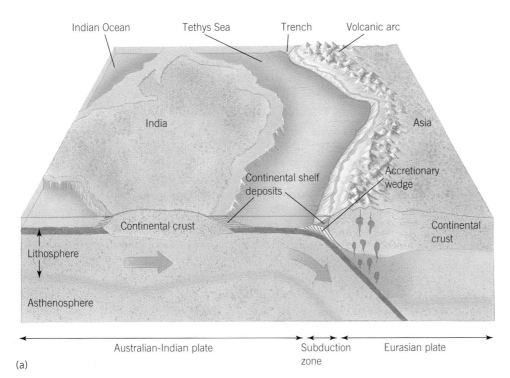

(a)

Figure 2.25 Continent-continent convergence. The Himalayas were created by this type of plate interaction. (a) Sediment accumulated in the now-vanished Tethys Sea as the Australian-Indian plate subducted beneath the Eurasian plate. (b) As the Indian and Eurasian landmasses collided, the Himalayas developed from the uplifted and deformed Tethys sediment.

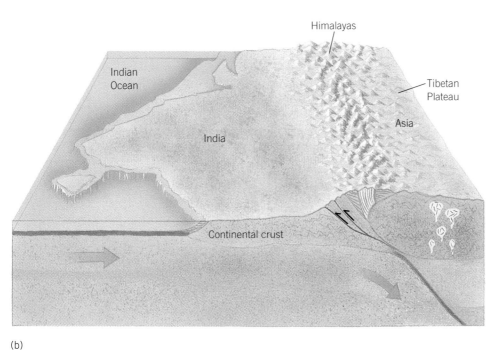

(b)

continent convergence. They were formed as a result of the breakup of Pangaea when India separated from what is now Antarctica and rode northward on a spreading plate toward the stationary Eurasian plate (Figure 2.25). The leading edge of the Australian-Indian plate was subducted beneath Asia in ocean-continent convergence until the intervening ocean basin, the *Tethys,* closed up. Too light to sink, the Indian landmass had no alternative but to collide with Asia. Trapped between the closing landmasses, as in a vise, was all the sediment that had accumulated in the Tethys Sea before and during India's long journey and that had been scraped off the descending plate. The Himalayas are constructed of this highly deformed sedimentary wedge. They rest on the broad, thickened foundation of continental crust formed by the collision, as does the high Tibetan plateau directly to the north. The thickened foundation inhibits further subduction.

The Tethys Sea no longer exists. The evidence of its former existence is, however, preserved in the sediment and slivers of oceanic crust that compose the Himalayas. Indeed, the places to search for any former oceans, and all the history of subduction and collision they represent, are in the mountain belts of continents. As you have learned, the ocean basins are recycled every 200 million years or so, while continental crust, which is too light to sink into the mantle, remains on the Earth's surface to absorb the impact of this recycling.

The continents are a collage of fragments thrown together by collision, welded by metamorphism and igneous activity, torn apart by rifting, and reshuffled by huge and prolonged horizontal displacements. Their origin and history are influenced by plate tectonics as much as the ocean basins are. Just as processes associated with plate divergence play the key role in creating oceanic crust, those associated with plate convergence are mainly responsible for the evolution of continental crust. It is at convergent margins that new continental crust is generated. In Chapter 12, we will explore fully the role of plate tectonics in building the continents.

The eastern margin of the Indian Ocean is a virtual panorama of the convergence processes we have discussed (Figure 2.26). There we see the same Australian-Indian plate that collided with Asia to form the Himalayas is subducting beneath the Java trench at the rate of 6.5 centimeters per year. The volcanoes that occur in an island arc along the length of the trench provide a good example of ocean-ocean convergence. Next, take a look at Australia, just south of the trench. It is headed for collision with southeastern Asia in a manner similar to the earlier collision involving India. The intervening sediment and island arc volcanoes will be bulldozed onto the continent.

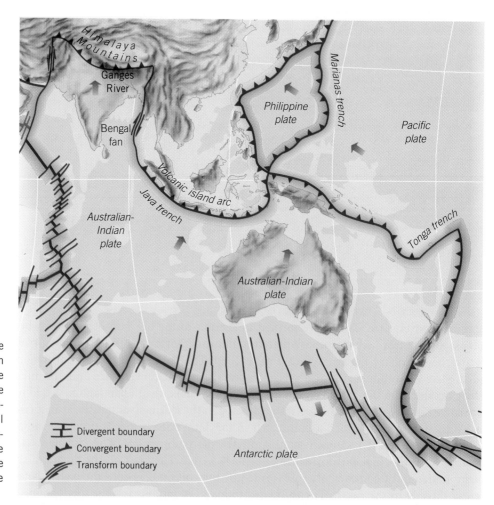

Figure 2.26 Many plate convergence processes are visible along the eastern margin of the Indian Ocean. If the present ocean-ocean convergence along the Java trench continues, Australia and the Asian mainland will eventually collide. As the Australian-Indian plate subducts beneath the Eurasian plate, the sediments of the Bengal fan are being scraped onto the mainland as an accretionary wedge.

Notice also the enormous Bengal fan that spreads across the abyssal plain between India and the Malay Peninsula. It is composed of the sediment the Ganges River has eroded from the Himalayas. The fan is in the process of being accreted to the margin of southeastern Asia as the Australian-Indian plate upon which it rides descends beneath the Java trench. In sum, the Himalayan chain, itself derived from the sediment of a vanished sea, is providing the raw materials for continental growth. This is a large-scale example of what James Hutton had in mind over two centuries ago when he concluded that the Earth's crust is in a simultaneous state of destruction and renewal, although he did not realize that plate tectonics is responsible for the process.

Transform Boundaries

Refer again to Figure 2.17; notice that plates not only separate and collide but also slide by one another horizontally. The transform boundaries along which sliding occurs are named after the kinds of faults that create them, **transform faults,** first described by J. Tuzo Wilson. Inspection will reveal the reason for the name: they connect either offset mid-ocean ridge segments, offset deep-sea trench segments, or ridges to trenches (Figure 2.27). In each case, the relative motion between plates is altered or transformed along the connection. For example, divergence at a mid-ocean ridge spreading center is converted to convergence at a trench subduction zone.

The prominent fracture zones that cross offset segments of the mid-ocean ridge are good places to study transform fault motion (Figure 2.28a). According to the theory of plate tectonics, as new crust spreads away from the central rift valleys, it is only in the narrow zone between ridge crests that the plates move in opposite directions. This motion triggers earthquakes within the zone. Outside that zone, the lithosphere on either side of the fracture moves in the same direction at the same speed; hence, we should expect to find no earthquakes beyond the ridge crests.

All these findings are consistent with plate tectonics but are anti-intuitive: the reasonable assumption would be that the motion on either side of the fault is in the direction of the offset rift valleys and that earthquakes are triggered throughout the length of the fracture (Figure 2.28b). It was considered a triumph of plate tectonics in 1968 when seismologist Lynn Sykes and his colleagues at the Lamont-Doherty Geological Observatory proved that these "reasonable assumptions" were untrue and that the earthquakes at transform fault boundaries conform to the expectations of the theory of plate tectonics.

The plate tectonics explanation of transform faults, however, does not explain what causes the offset of the ridge segments. At this point, it is fair to say that there is no agreement on what causes them. Perhaps the simplest theory is that the crust fractures and offsets as the cooling plates contract. Another suggestion is that magma may rise from the asthenosphere in the form of isolated plumes that shatter and segment the plates. This is a variation on the idea that the segments form in the process of rifting above triple-junction hot spots that do not line up; the transform faults serve as the connecting arms between these junctions. A third, older theory is that the ridge segments and transforms occur along ancient zones

transform fault
Type of fault where the plates slide by one another horizontally without either creating or destroying oceanic crust.

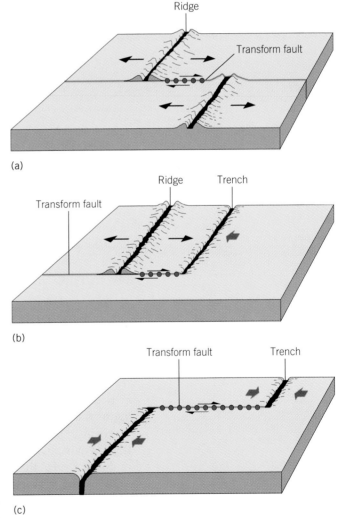

Figure 2.27 A transform fault may connect (a) two offset ridge segments, (b) a ridge segment to a trench segment, or (c) two offset trench segments. Notice that earthquakes (indicated by dots) occur only within the offset segments of the fault.

Figure 2.28 Two explanations of the movement along a fracture zone that intersects the mid-ocean ridge. In the transform fault explanation, shown in part (a), the plates move in opposite directions only between the offset rift valleys, and that is where we should expect earthquakes. In the non-transform explanation, shown in part (b), earthquakes should occur along the entire length of the fracture. The seismic data support the transform fault explanation.

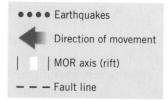

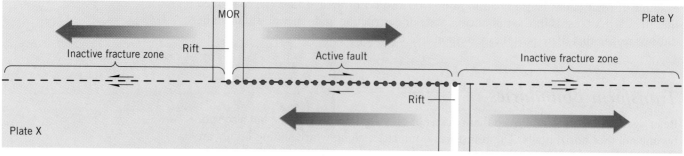

(a)

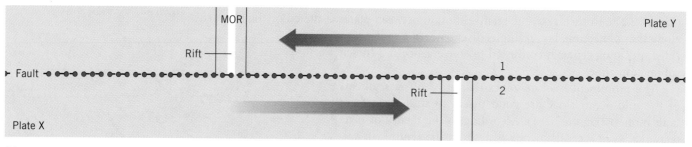

(b)

of weakness in the continental crust that were reactivated by the episode of rifting that created the current ocean basins. All these theories agree that the position of the offset ridge segments predates movement along transform faults.

Transform faults may displace continental as well as oceanic crust. The dangerous San Andreas fault of California is a transform fault boundary connecting two offset mid-ocean ridge segments: the East Pacific Rise in the Gulf of California and the Juan de Fuca Ridge off the northwest Pacific coast. Between these segments, southern California is moving northward relative to the rest of North America, and earthquakes result. (More on this in the next chapter.)

Absolute Plate Motion

Keep in mind that when we discussed spreading rates earlier, we referred to the velocity of the ocean floor *relative to the ridge*. The magnetic bands that reflect the polarity reversal time scale give no information about which feature is moving: sea floor, ridge, or both! The situation is similar to peering out the window of a train that is just starting up and, for a split second, not being sure whether it is the train or the station that is moving. Of course, our intuition—or prior knowledge—quickly reminds us that the train is moving; in the same manner, we *assume* that it is the oceanic crust and not the mid-ocean ridge that migrates.

However, this assumption is not always a correct one. For example, Antarctica is surrounded by mid-ocean ridges, which means that the sea floor is expanding on all sides of the continent. Clearly, this expansion can occur only if the ridges are migrating away from the continent. Assuming that the Earth has maintained a constant size, this sea-floor expansion must be balanced by an equivalent loss of oceanic crust elsewhere. The likely places are the Pacific and Indian oceans, which, as you remember, are surrounded by subduction zones.

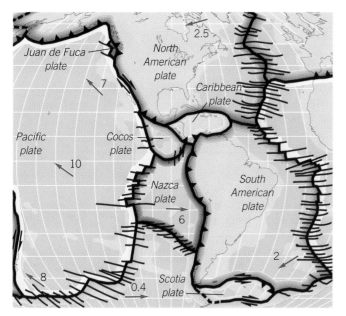

Figure 2.29 Absolute plate motion is measured (in centimeters per year) relative to hot spots, which are presumed to be stationary.

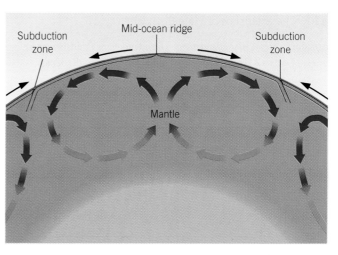

Figure 2.30 A simple model of a mantle convection current: the plates separate above the rising limb of the current and sink into the mantle above the descending limb.

Problems like these motivate geologists to find ways to measure plate motion relative to a frame of reference independent of the plates, which they call *absolute plate motion.* Fortunately, we have a convenient reference. Earlier, we described how the Hawaiian hot spot was used to measure the motion of the Pacific plate relative to the mantle as it moved toward the Aleutian trench. Because hot spots originate as plumes deep in the mantle and persist for millions of years, they are believed to be stationary. Furthermore, if they are in fact stationary, they can provide an objective measure of the rate at which lithospheric plates spread over the mantle. Figure 2.29 gives absolute plate velocities using hot spots as a reference. Notice that the Pacific plate, which consists entirely of oceanic crust, moves much faster than plates that have continents embedded in them. Evidently, the continents act as anchors.

The Causes of Plate Motion

One major unresolved problem posed by the theory of plate tectonics is that no one has been able to pin down precisely what causes the plates to move. Most theories emphasize the role of mantle **convection currents,** in which warm rock rises from the interior, surrenders its heat as it spreads, and sinks back into the mantle where it is reheated, much like spaghetti in a pot of boiling water. In a simple model, mid-ocean ridge spreading centers overlie the rising portions of the convection current, and subduction zones overlie the cooler sinking portions (Figure 2.30). The plates created by the upwelling magma at the spreading centers are dragged along by the spreading currents toward the subduction zones. However, the depth of the convection currents and the manner in which they rise and force the plates to move remain as open questions. Mantle convection is a major concept, and we will discuss it at length in Chapter 4, which deals with the interior of the Earth.

There, painted with a wide brush, is the theory of plate tectonics. Many of the features and processes that it explains are summarized in Table 2.1, and we will be adding to and refining that list in subsequent chapters. Nevertheless, you should not think that all geologists agree with every aspect or interpretation of the theory. There is no such thing as a closed subject in science—and certainly not in geology, where so much theory must, of necessity, rest on fragmentary information.

convection current
The flow resulting from the rise of warmer, less dense material and the descent of cooler, denser material.

Table 2.1   Characteristics of Plate Boundaries

Boundary Type	Lithosphere Involved	Relative Motion	Predominant Process	Features	Examples
Divergent	Continent	Away from one another	Continental rifting	Narrow rift valleys, some traceable to triple junctions; shallow earthquakes; lava flows; volcanoes	Afar Triangle, Red Sea, East African Rift Valley; Gulf of California
	Ocean		Sea-floor spreading	Mid-ocean ridges and rises; central rift valleys; shallow earthquakes; widening and deepening oceanic crust; submarine volcanic peaks	Atlantic Ocean
Convergent	Ocean-ocean	Toward one another	Subduction	Deep-sea trenches, island arcs, back-arc basins; active volcanism; huge quantities of magma added to crust. Earthquakes trace descending plates deep into mantle	Marianas, Aleutians, Japan, Philippines
	Ocean-continent			Deep-sea trenches bordering on continents; volcanoes erupting on land, forming coastal chains; numerous powerful earthquakes triggered by downward action of subducting oceanic lithosphere	Western South America, Washington, Oregon coast
	Continent-continent		Continental collision	High continental mountain chains formed of sediment deposited in closed-out ocean basins. Continental crust folded and thickened	Himalayas, Alps
Transform	Ocean or continent	Past one another	Horizontal slippage	Long, narrow fracture zones that run perpendicular to mid-ocean ridges and/or deep-sea trenches; powerful shallow earthquakes between offset ridge segments	San Andreas fault; Alpine fault, New Zealand; Mendocino fracture zone

STUDY OUTLINE

Discovered in the 1950s, the high-standing **mid-ocean ridge** is a string of active submarine volcanoes cut down the middle by a central rift valley. Similar tensional features caused by **faults** had been found previously on the continents. The mid-ocean ridge traces a seismic (earthquake) belt that extends through the ocean basins of the world.

   I. **SEA-FLOOR TOPOGRAPHY.**   The mid-ocean ridge is the shallowest feature of the ocean basins. Segments of the rift valley are offset and marked by **fracture zones.** Tests and observations found high heat flow beneath the ridge crest and seismic activity confined to the fracture zones between offset ridge segments.

     A. Deep-sea trenches, found around the margins of the Pacific Ocean, are the deepest features of the ocean basins. They are characterized by seismic activity and low heat flow. Volcanically active continental mountain chains, or **island arcs,** are located parallel to the trenches.

II. THE SEA-FLOOR SPREADING HYPOTHESIS.   In the early 1960s, geologists proposed that, in a process called **sea-floor spreading,** oceanic crust separates along the rift valley in the mid-ocean ridge, allowing magma to bleed to the surface and form fresh oceanic crust. The crust then spreads laterally, cooling and contracting as it moves toward the trenches. There it descends back into the mantle and triggers the deep line of earthquakes beneath the trenches.

A. Evidence in magnetic reversals
   1. Parallel bands of strong and weak magnetism in the ocean floor run parallel to the axis of the mid-ocean ridge. Identical alternating bands of magnetism, found in continental lava layers, record **magnetic reversals**—periods in the Earth's history when a compass needle that would today point north pointed south, alternating with periods when the needle pointed north again.
   2. The most recent reversals occur at the mid-ocean ridge, and progressively older ones are found toward the flanks of the ocean basins, evidence that each segment of the oceanic crust was originally formed at the rift valley and has been moved gradually away from the ridge.

B. Evidence in sea-floor sediment.   Further evidence is found in the flat **abyssal plain** along the continental margins, a wedge of sediment that becomes thinner and younger toward the ridge crest. The oldest known sediment and oceanic crust, found in the western Pacific, is 200 million years old.

C. Evidence in sea-floor topography
   1. Differences in cooling rates are reflected in the topography of the ocean floor: steep at the ridge crest and gentle at the margins of the basins. In general, each portion of the ocean floor is at a depth determined by its age.
   2. **Seamounts** and **guyots** are submerged islands—strong evidence that the sea floor sinks as it spreads. Scientists proposed that the volcanoes of the Hawaiian-Emperor Seamount chain were punched out as the Pacific oceanic crust passed over the same magma source that is currently beneath Hawaii. Many such **hot spots** are distributed throughout the continents and ocean basins.

III. PLATE TECTONICS.   Proposed in the mid-1960s, **plate tectonics** went beyond continental drift and sea-floor spreading to offer a comprehensive theory of the Earth's crustal structure. The **lithosphere,** consisting of oceanic and continental crust and upper mantle, is divided into rigid slabs, or **plates.** Plate boundaries occur along **seismic belts** and are marked by earthquakes generated when the plates grind against one another as they move over the **asthenosphere.** There are only three ways in which the plates can interact: they may diverge, converge, or slide by one another.

A. Divergent boundaries, the sites of simultaneous plate separation and growth, are also referred to as **spreading centers.** As the plates move apart, magma that wells up through the central rift valley is accreted to the trailing edge of the plates.
   1. The Mid-Atlantic Ridge is a divergent boundary that began as a fracture in the continental crust of Pangaea. In a similar rifting process happening today in the Afar Triangle, a hot spot stretched and fractured the crust, forming a triple junction. The Mid-Atlantic Ridge was probably formed as a series of connecting triple junctions.
   2. The Atlantic Ocean of today is in a mature stage of development. It is characterized by a wide ocean basin between separated continents and by **passive continental margins**—divided into a shallow **continental shelf,** a

continental slope, and an apronlike **continental rise**—that are nearly free of earthquakes and igneous activity because they are far from plate boundaries.

B. Where plates meet at **convergent boundaries,** or **subduction zones,** the denser, colder plate bends and descends into the mantle in a process called **subduction.** There are three types of convergence.

   1. Ocean-ocean convergence.   Where one plate plunges beneath another, it forms a deep-sea trench. Sea-floor sediment is scraped off the descending plate and plastered to the edge of the bordering plate as an **accretionary wedge.** Magma generated by the subducting plate rises to the surface and fuels the island arc volcanoes just landward of the trench. Two forces are at work: compression at the trench and arc, and tension and back-arc spreading in the ocean basin behind the arc.

   2. Ocean-continent convergence.   Where a plate descends beneath a deep-sea trench directly off the shore of continental crust, the magma generated by the subducting plate erupts on land to build the great volcanic mountain chains that parallel the coasts. Characterized by frequent and powerful earthquakes, the **active continental margin** is narrow and steep.

   3. Continent-continent convergence.   When continental crust on both sides of a subduction zone converges, ocean sediment trapped between is deformed and pushed up into a mountain belt as the continents collide.

C. **Transform boundaries** occur where two plates slide by one another horizontally. As new crust spreads away from the central rift valleys, the plates move in opposite directions in the narrow section of the fracture zone between the offset ridge crests, called a **transform fault.** The motion triggers the earthquakes that are confined only to that zone. Beyond the ridge crests, the crust on either side of the fracture moves in the same direction at the same speed.

D. Plate motion relative to a frame of reference independent of the plates is called absolute plate motion. If hot spots are determined to be stationary, they can be used to measure absolute plate motion.

E. We do not as yet know precisely what causes the plates to move. Most theories emphasize the role of mantle **convection currents,** in which warm rock rises from the interior, surrenders its heat as it spreads, and sinks back into the mantle, where it is reheated.

STUDY TERMS

abyssal plain (p. 64)

accretionary wedge (p. 75)

active continental margin (p. 78)

asthenosphere (p. 69)

continental rise (p. 75)

continental shelf (p. 75)

continental slope (p. 75)

convection current (p. 83)

convergent boundary (p. 69)

deep-sea trench (p. 55)

divergent boundary (p. 69)

fault (p. 54)

fracture zone (p. 55)

guyot (p. 65)

hot spot (p. 68)

island arc (p. 58)

lithosphere (p. 69)

magnetic reversal (p. 61)

mid-ocean ridge (p. 54)

passive continental margin (p. 75)

plate (p. 68)

plate tectonics (p. 68)

sea-floor spreading (p. 60)

seamount (p. 65)

seismic belt (p. 68)

spreading center (p. 72)

subduction (p. 75)

subduction zone (p. 75)

transform boundary (p. 69)

transform fault (p. 81)

CRITICAL THINKING QUESTIONS

1. How did the discovery of magnetic reversals confirm the sea-floor spreading hypothesis?
2. Sketch and describe what the sediment pattern of the ocean floor might look like if sea-floor spreading did not occur.
3. Use the sea-floor spreading concept to explain why the shallowest part of an ocean basin is toward the middle and the deepest part of the basin is along its margins.
4. In what ways does the worldwide pattern of earthquakes support the theory of plate tectonics?
5. What is the importance to plate tectonics theory of the decrease in seismic wave velocities at depths of about 100 kilometers? What are some possible explanations for the fact that deep-focus earthquakes are untraceable to depths greater than 700 kilometers?
6. In Chapter 1, we learned that 200 million years ago, the continents existed as a single landmass called Pangaea. Use the concepts of plate tectonics to explain how the continents moved to their present locations on the Earth.
7. In what ways does back-arc spreading resemble sea-floor spreading?
8. Compare and contrast the features of active and passive continental margins. How does plate tectonics account for the differences?
9. Explain the possible role of hot spots in the initial stages of ocean-floor creation. What is the evidence that connecting hot spots may have "unzipped" the Atlantic?
10. Assuming that the processes initiated by the hot spot at the Afar Triangle continue, sketch and describe the geologic future of the Red Sea and the Gulf of Aden.
11. As its name implies, the mid-ocean ridge runs down the center of the Atlantic Ocean. Explain why the ridge does not run down the center of the Pacific Ocean but is located along its eastern margin.

CHAPTER 3

The Causes of Earthquakes

Seismic Waves

The Seismograph

Locating the Earthquake

Determining the Sense of Motion Along the Fault

Measuring Earthquake Magnitudes and Energy

Earthquakes and Plate Tectonics

The San Andreas Fault

The Los Angeles Faults

Intraplate Earthquakes

Other Causes of Earthquakes

Forecasting Earthquakes

Seismic Gaps

Recurrence Studies

Precursor Studies

Minimizing Earthquake Damage

Construction Problems

Earthquakes

Earthquakes are shock waves, vibrations. Most are triggered by the sudden slippage of rock along fault planes in the crust and in the subduction zones of the upper mantle. The only one I have personally experienced occurred in 1965. I was at the University of Washington in Seattle when it struck in the early morning of April 29. The frame hut, a rabbit warren of desks and partitions that I shared with other graduate students, shook violently. Pen in hand, I found my body transformed into a crude seismograph as I lurched forward and involuntarily recorded a jagged spike across the map on my desk. The ground shaking and rattling continued for perhaps 10 seconds. It was an unpleasant sensation, this violation of one's faith in *terra firma.*

In the midst of my confusion, I fell in line with the other earth scientists, and together we staggered to the safety of the open parking lot. We couldn't help laughing once our fright wore off; after all, it is not often you get 10 geologists in one room blind to the fact that they are sitting on an earthquake. Once we were outside, the jellolike ground swayed again—but then again, the jello could have been in my own legs. The entire show lasted only a minute or two and nothing seemed to have changed, except for prominent ripples in the rain puddles. However, there was scattered damage around the university campus, which, I learned later, was typical of the kind inflicted on the rest of the city. Seven people lost their lives in the Seattle earthquake, and damage was estimated in the tens of millions of dollars.

The Seattle earthquake was more powerful than over 99 percent of the hundreds of thousands of earthquakes and slight tremors that instruments around the world record each year. It was the most powerful earthquake to strike the United States that year. Nevertheless, it was a popgun compared with the one that had leveled Anchorage,

earthquake
Vibrations within the Earth set in motion by the sudden release of accumulated strain energy.

◀ This photograph was taken in the wake of the earthquake in Kobe, Japan, on January 16, 1995.

Alaska, 13 months earlier, killing 131 people. The Seattle event was the kind of earthquake that strikes somewhere on the globe over a hundred times a year, whereas earthquakes of the size that struck Anchorage occur only once or twice a decade.

Earthquakes of similar strength can result in different levels of human misery. The Seattle earthquake was about as powerful as the Northridge (Los Angeles) earthquake of January 17, 1994. However, the Northridge event and the aftershocks that followed knocked down freeway overpasses, destroyed or severely damaged hundreds of buildings (including a large shopping mall), started numerous fires, burst water mains and gas lines, necessitated the evacuation of hospitals, and rendered unsafe the enormous Los Angeles Coliseum. In all, 55 people were killed, and there were many billions of dollars in damage. Exactly one year after the Northridge event, Kobe, Japan, was struck by an earthquake of similar size. Deaths attributable to that event totaled in the thousands and damage was in the hundreds of billions. As we will see in this chapter, many factors contribute to the destructiveness of an earthquake.

Human history is replete with examples of catastrophic earthquakes (Table 3.1). In this chapter, we examine the causes of earthquakes, the close relationship between earthquakes and plate tectonics, the detection and analysis of earthquakes through instruments, and, finally, the efforts of geologists to predict them and minimize their impact.

The Causes of Earthquakes

In 1884, geologist G. K. Gilbert was engaged in constructing a geological map of a part of the rugged Basin and Range province of Nevada and Utah. Following the occurrence

Table 3.1 Selected Historic Earthquakes

Year	Location	Magnitude	Number of Deaths
365	Crete, Knossos		50,000
1201	Egypt, Syria		1,000,000
1755	Portugal, Lisbon		70,000
1811	Missouri, New Madrid		Several
1886	South Carolina, Charleston		60
1906	California, San Francisco	8.2	700
1906	Ecuador	8.9	1,000
1920	China, Kansu	8.5	180,000
1923	Japan, Kwanto	8.2	143,000
1939	Turkey, Erzincan	8.0	23,000
1960	Southern Chile	8.5	5,700
1964	Alaska	8.6	131
1970	Peru	7.8	66,000
1971	California, San Fernando	6.5	65
1975	China, Haicheng	7.4	300
1976	China, Tangshan	7.6	650,000
1985	Mexico, Mexico City	8.1	5,600
1988	Armenia	6.9	25,000
1989	California, Loma Prieta	7.1	62
1990	Iran	7.3	50,000
1992	California, Landers	7.5	1
1994	California, Northridge	6.7	55
1995	Japan, Kobe	6.9	5,160

Source: U.S. National Oceanic and Atmospheric Administration.

Figure 3.1 A freshly exposed fault scarp at the base of Borah Peak, Idaho. G. K. Gilbert hypothesized that earthquakes are triggered by the release of strain energy that accompanies sudden movement along faults. Fault movement creates scarps such as this one.

of a swarm of earthquakes in the vicinity of Salt Lake City, he observed the appearance of freshly exposed steep cliffs, or *scarps,* at the base of the mountains nearby (Figure 3.1). The fresh scarps were proof that fault movement was uplifting the mountains, and Gilbert strongly suspected that the earthquakes had been generated by the fault motion. He concluded:

> The upthrust [of the mountains] produces a strain in the crust, involving a certain amount of distortion, and this strain increases until it is sufficient to overcome the starting friction along the fractured surface. Suddenly, and almost instantaneously, there is an amount of motion sufficient to relieve the strain, and this is followed by a long period of quiet, during which the strain is gradually reimposed.*

Gilbert's insight gained important quantitative support from geologist H. Fielding Reid's classic investigation into the cause of the catastrophic San Francisco earthquake of 1906. The earthquake occurred near Point Reyes, a few kilometers north of the city along a segment of the San Andreas fault. Reid and his colleagues mapped such disrupted features as stream courses, fence posts, roads, and railroad tracks (Figure 3.2). Then they compared the position of these features with their position before the earthquake, as located by surveys taken prior to 1906. They also measured the orientation of ground breaks in the fault zone. Their observations enabled Reid to determine that, in the San Francisco region at the time of the earthquake, the west side of the San Andreas fault moved as much as 6.5 meters north with respect to the east side. This movement represented the sudden release of huge energy, enough to trigger a great earthquake.

Exert an external force on a body and you are subjecting it to *stress.* The deformation (change of shape and/or volume) that the body undergoes in response to the stress is called *strain;* it is a way of storing the energy that the force has transmitted to the body. The Reid-Gilbert explanation of the cause of earthquakes, now known as the **elastic-rebound theory,** is based on this essentially simple relationship (Figure 3.3). The crux of the theory is that earthquakes result from the sudden release of strain energy that occurs when rocks rupture and lurch past one another along fault planes. The motion is always driven locally by **shear stress** that forces the rocks on opposite sides

elastic-rebound theory
The concept that earthquakes are generated by the sudden slippage of rocks on either side of a fault plane. In the process, the rocks release gradually accumulated strain energy and are returned to an unstrained condition.

shear stress
Stress (force per unit area) that acts parallel to a (fault) plane and tends to cause the rocks on either side of the plane to slide by one another.

*G. K. Gilbert, "A Theory of Earthquakes of the Great Basin with a Practical Application," *American Journal of Science,* 1884.

(a)

Figure 3.2 (a) This classic photograph of a fence offset was taken by G. K. Gilbert following the 1906 San Francisco earthquake. (b) The San Andreas fault where it passes under Tomales Bay on Point Reyes Peninsula, California. The fog bank in the distance hides the entrance to San Francisco Bay. (c) Modern geological map of the San Francisco Bay area, showing major fault lines. Point Reyes and Tomales Bay are on the upper left. The map also illustrates the unstable nature of the heavily urbanized terrain and the resulting vulnerability to earthquakes.

(b)

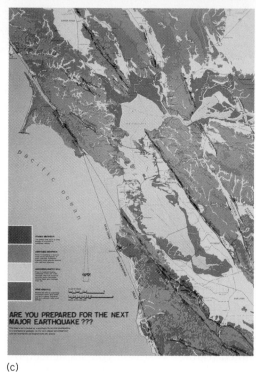

ARE YOU PREPARED FOR THE NEXT
MAJOR EARTHQUAKE ???

(c)

seismic wave
A wave generated within the Earth by sudden fault slippage or an explosion.

focus
The initial point of rupture within the Earth from which seismic waves are generated.

epicenter
The point on the Earth's surface that is directly above the focus of an earthquake.

of the fault plane to slide by one another. The local shear, in turn, is generated by regional stresses in the crust. The regional stress may be a large-scale shear that causes lithospheric plates to slide by one another horizontally. Other regional stresses are *tensile;* that is, they tend to pull the crust apart. Still others are *compressive,* tending to drive the rocks together (Figure 3.4).

Most of the huge amount of energy released during an earthquake is expended in pulverizing rock and generating heat. Only a relatively minor portion of the energy causes vibrations, or **seismic waves.** The waves fan out in all directions from the earthquake **focus,** which is the point of initial rupture beneath the surface of the Earth. The vertical projection of the focus onto the surface of the Earth is the earthquake **epicenter** (Figure 3.5). For example, the focus of the 1906 San Francisco earthquake was several kilometers below the Point Reyes epicenter.

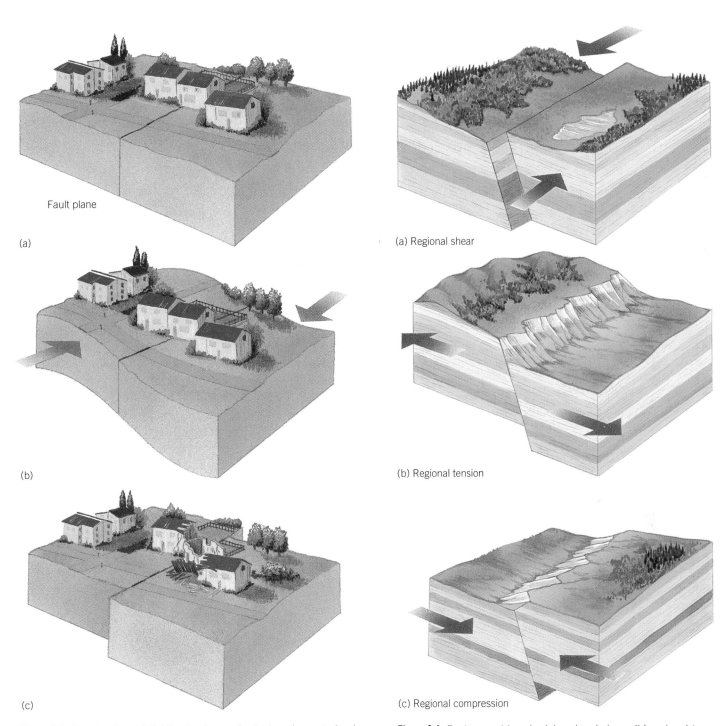

Figure 3.3 Elastic rebound. (a) Rocks along a fault plane in unstrained condition. (b) Strain builds as stress is applied and the blocks on either side of the fault plane are held together by friction. (c) The blocks suddenly lurch past one another to unstrained positions, releasing strain energy in the form of an earthquake.

Figure 3.4 Faults are driven by (a) regional shear, (b) regional tension, and (c) regional compression. Each generates shear stress along the fault plane as rocks slide by one another.

Figure 3.5 The focus of an earthquake is the point of initial rupture, where seismic waves originate. The epicenter is the surface projection of the focus.

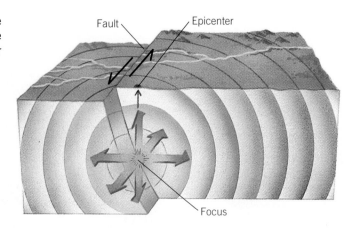

Seismic Waves

body wave
A seismic wave that travels through the Earth's interior.

surface wave
A seismic wave that travels along the Earth's surface and affects the interior to depths that are dependent upon its wavelength.

primary (P) wave
A seismic wave that propagates through the Earth as a series of compressions and expansions.

secondary (S) wave
A seismic wave that causes the components of a rock to vibrate perpendicularly to the direction of wave propagation. Also called a shear wave.

An earthquake releases two classes of seismic waves: **body waves** that travel through the interior and **surface waves** that travel along the surface. Body waves are in turn classified into two types: the faster-traveling **primary (P) waves** and the slower **secondary (S) waves.** P waves alternately compress and expand the rocks they pass through by shortening and lengthening atomic bonds. As shown in Figure 3.6a, the rock vibrates back and forth along the wave in an accordionlike motion. P waves are virtually identical to common sound waves; the major difference is that they generally have vibration rates (or frequencies) lower than the human ear can detect. An S wave causes the rock to vibrate at right angles to the direction of the wave path, much like what happens when you shake the free end of a rope tied to a pole (Figure 3.6b). Because the rock particles slide by one another as they vibrate, S waves are also called *shear* waves.

The velocities of seismic waves depend on the elastic properties and density of the material through which they pass. Thus we can learn a great deal about the composition and physical state of the Earth's interior by studying these waves. For example, P waves travel more rapidly through rigid, relatively dense crustal rocks than through softer rocks or liquids. S waves cannot travel through liquids at all, because liquids lack the internal strength to support shearing motions. The fact that S waves do not penetrate beyond 2900 kilometers is strong evidence that the Earth is molten at that depth.

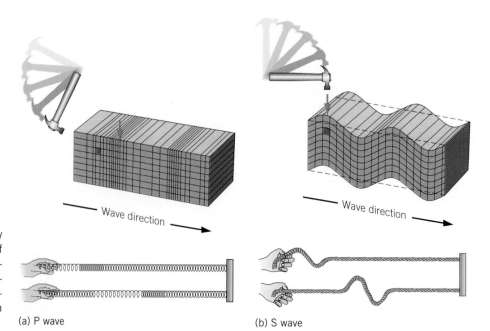

Figure 3.6 The principal seismic body waves. (a) P waves generate a series of alternating compressions and expansions in the direction of wave propagation. (b) S waves generate a series of vibrations perpendicular to the direction of wave propagation.

(a) P wave   (b) S wave

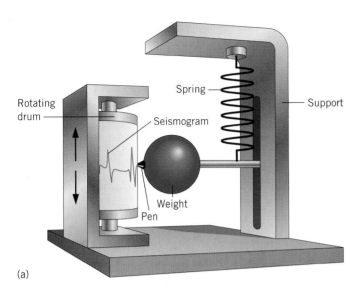

(a)

(b)

Figure 3.7 (a) A simple model of a seismograph designed to detect vertical ground motion. (b) A modern station employs a variety of instruments to detect and analyze seismic waves. The map shows the region that was subjected to the most intense ground shaking during the earthquake in Loma Prieta, California, in 1989.

Surface waves are a combination of two complex vibrations. One sets the Earth's surface swaying in a horizontal, snakelike motion at right angles to the wave path. They are similar to S waves and, like them, cannot travel through a lake or ocean—although they can shake the basins holding these water bodies. The other vibration causes the surface to rise and fall like ocean waves. During a powerful earthquake, the ground rises and falls at the same time it sways from side to side, which is one reason for damage to foundations, sewers, and pipelines.

The Seismograph

The waves generated during an earthquake are detected by an instrument called a **seismograph.** Modern versions are sensitive enough to record the Earth's every hiccup, and they never sleep. Several large-scale seismograph networks cover the globe. Each consists of hundreds of stations that constantly monitor for earthquakes and routinely exchange vast amounts of data.

The seismograph is based on the well-known principle of inertia, which states that the greater the mass of a body, the greater its resistance to any change of motion (Figure 3.7). A simple seismograph can be constructed by suspending a large mass (that is, a heavy weight) by thin springs from a rigid frame that is attached firmly to a platform embedded in bedrock. Also attached to the platform is a drum that rotates at a fixed rate. When the seismic waves of an earthquake arrive, they shake the platform, frame, and drum, but the suspended weight remains nearly stationary because of its inertia. A sensor—in simplest form, a pen attached to the weight—records on the rotating drum the difference in motion between the vibrating platform and the stationary weight. The resulting wave pattern, the record of the earthquake, is a **seismogram.** At least three seismographs are necessary to fully record the ground motion of an earthquake: two detect horizontal north-south and east-west vibrations, and the third detects vibrations in the vertical plane. Modern seismographs use microelectronics to record and transmit ground-motion data directly to digital computers.

Seismograms yield valuable information. Reading them properly enables us to locate the epicenter and focal depth of an earthquake, gauge its magnitude and energy, and determine the fault motions responsible for it in the first place. Seismograms are also our main sources of information concerning the forces that drive the Earth's lithospheric plates, and they are essential in determining the structure and composition of the interior. This use of seismograms will be discussed in the next chapter.

seismograph
A device that detects and records seismic waves.

seismogram
A record of seismic waves detected by a seismograph.

Figure 3.8 A typical seismogram, which records the time of arrival of P, S, and surface waves.

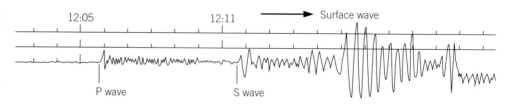

Locating the Earthquake

Figure 3.8 is a typical seismogram of a distant earthquake. Note first the series of 1-minute markers across the top of the seismogram that enable us to precisely date arrival times of waves at the station. Reading the trace from left to right, we first encounter a line that is nearly straight. The slight wiggles are the equivalent of background radio static. These *microseisms* have many causes: local traffic, landslides, waves breaking on a nearby shore. They prove that, like the ocean, the solid Earth is never at rest.

The abrupt change at 12:06 in the nearly straight-line pattern of Figure 3.8 marks the arrival of the P waves spreading from an earthquake. To the right (which, on the seismogram, means a later time), we see other changes in the pattern that record in sequence the arrival of the S waves and surface waves. The spacing between P and S wave arrivals is caused by velocity differences; because P waves outrun the S waves in their dash through the interior, they reach the station first. The last to arrive are the surface waves, because they are the slowest and travel the longer route of the Earth's circumference.

Since all seismic waves originate at the same point and then spread from that point at different speeds, the farther they travel, the greater the interval is between their arrival times. In Figure 3.9, we see that an interval of 6 minutes between P and S arrivals corresponds to an epicenter that is 4100 kilometers from the station. This information enables us to determine the distance to the epicenter but not its location, because the waves could have come from any direction. That is, as far as we know, the epicenter may lie anywhere on a circle of a 4100-kilometer radius. To pinpoint the epicenter, we need to pool information with at least two other stations that have followed our methods. Then we can construct three circles on a globe; their point of intersection marks the epicenter (Figure 3.10).

Pinpointing the focus of an earthquake is a trickier problem than finding the epicenter. However, the principle used is the same: the time gap between wave arrivals depends upon the distance traveled. One method compares the time interval between the P waves that take the direct route to the station and those that reach the station after having first reflected off the Earth's surface. The greater the difference in arrival times, the greater the focal depth of the earthquake.

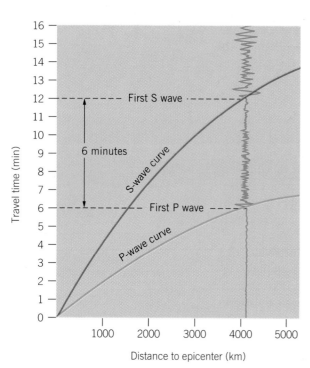

Figure 3.9 The interval between the arrival times of P and S waves depends upon the distance from the epicenter to the station. Therefore, a specific interval is associated with a specific distance. For example, the 6-minute interval shown in this graph corresponds to a distance of 4100 kilometers.

Determining the Sense of Motion Along the Fault

Because P waves are simply a series of compressions and expansions in bedrock, it is these motions that are transmitted to the seismograph. Compressions are recorded as upticks (∧) and expansions as downticks (∨) on the seismogram. Of special interest is the distribution of the *first* P-wave impulses recorded by seismographs around the globe, because their patterns of compressions and expansions reveal the *direction* of the fault motion that caused the earthquake.

Figure 3.11 shows a hypothetical fault divided into quadrants; the epicenter is at point A, and the movement along the fault is indicated by the arrows. The first signals recorded by seismographs in quadrants 1 and 3 will be compressional, because the

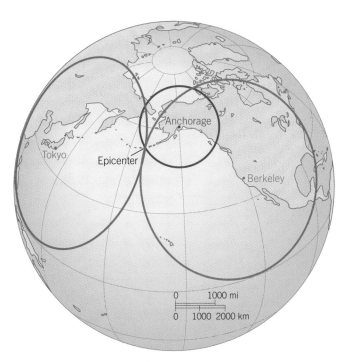

Figure 3.10 The intersection of circles drawn from three recording stations locates the epicenter. The radius of each circle is the distance from the epicenter to the station.

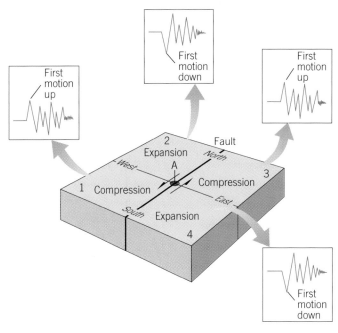

Figure 3.11 The first motion signal along a fault is received by seismographs in alternating quadrants as either a compression (quadrants 1 and 3) or an expansion (quadrants 2 and 4). Thus we may use seismograms to deduce the direction of fault movement by analyzing the signals received by different recording stations surrounding the fault.

ground is being driven toward the instruments; the first signals recorded by seismographs in quadrants 2 and 4 will be expansional, because ground motion is directed away from them. We can reconstruct the direction of motion along a real fault by using the data of many stations to plot a quadrant diagram similar to the one in Figure 3.11.

Measuring Earthquake Magnitudes and Energy

The more energy you impart to a rope you shake or a tuning fork you strike, the greater will be the *amplitude* of their vibrations—that is, the greater their displacement from the undisturbed position. The same is true of seismic wave amplitudes: the greater the energy released during an earthquake, the larger the amplitudes recorded on the seismogram. Therefore, wave amplitude can serve as an objective measure of an earthquake's size or **magnitude** (Figure 3.12). In 1935, California Institute of Technology seismologist Charles Richter developed a practical application of this concept, which has since become known as the **Richter magnitude scale.**

A Richter magnitude is a measurement of the amplitude of the highest wave recorded on a standard seismograph divided by its *period*, which is the time required for that wave to complete the vibration. Because wave amplitudes diminish with distance from the epicenter, Richter realized that a seismograph cannot distinguish between a weak quake nearby and a strong one far away. (Astronomers have the same trouble judging the distance of a star from its light output: Is it the light of a distant, bright star or of a near, faint star?). Therefore, he added a factor that corrects the measurement for distancing effects and adjusts it to the amplitude the seismograph would record if it were located 100 kilometers from the epicenter. Thus an earthquake should register the same magnitude at close and distant stations (see Figure 3.12).

The Richter scale is logarithmic, which means that adjacent numbers on the scale signify wave amplitudes that differ by a factor of 10. For example, the amplitude of a magnitude 5 earthquake is 10 times that of a magnitude 4 earthquake. Earthquakes of

magnitude
A measure of earthquake strength as interpreted from the maximum wave amplitude recorded by a seismograph.

Richter magnitude scale
A logarithmic scale of earthquake magnitudes based on the maximum-amplitude wave recorded by a standard seismograph and corrected for distance of the seismograph from the epicenter.

Figure 3.12 Calculating the magnitude of an earthquake from a seismogram. (B. Bolt, *Earthquakes,* San Francisco: W. H. Freeman & Co., 1978, p. 105.)

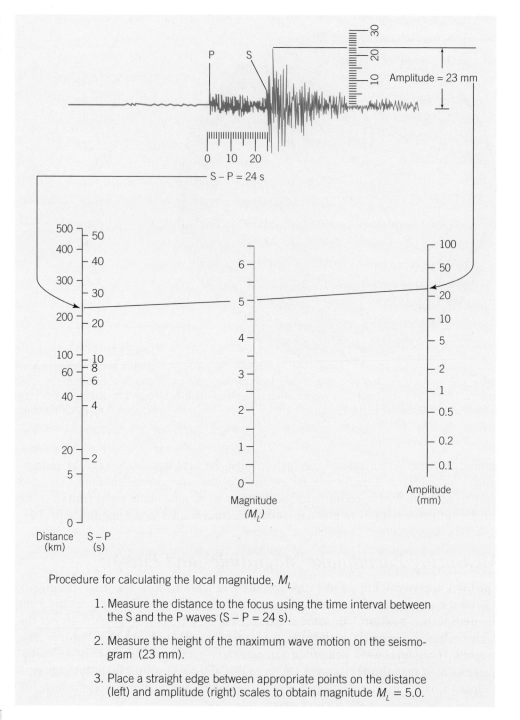

Procedure for calculating the local magnitude, M_L

1. Measure the distance to the focus using the time interval between the S and the P waves (S – P = 24 s).

2. Measure the height of the maximum wave motion on the seismogram (23 mm).

3. Place a straight edge between appropriate points on the distance (left) and amplitude (right) scales to obtain magnitude M_L = 5.0.

Table 3.2 Frequency of Earthquakes Worldwide

Magnitude	Average Number per Year
Over 8.0	1 to 2
7.0 to 7.9	20
6.0 to 6.9	120
5.0 to 5.9	800
4.0 to 4.9	6,200
3.0 to 3.9	49,000
2.0 to 2.9	300,000

Source: Adapted from *Earthquake Information Bulletin.*

magnitude 8 and above, which occur on average once every few years, are classified as *great earthquakes* (Table 3.2). Though the scale has no upper limit, no earthquake greater than 8.9 has ever been recorded. The magnitude of the 1906 San Francisco earthquake is in debate owing to the use of unstandardized instruments at that time, but it is estimated to have been 8.2. The 1964 Anchorage earthquake registered 8.6, and the 1985 Mexico City earthquake 8.1.

Keep in mind that the difference in effect between two magnitude points on the low end of the scale is not the same as the difference in effect between two points on the high end. For example, a magnitude 2 quake, although 10 times the amplitude of a magnitude 1, is still a very mild quiver, undetectable by all but the most sensitive instruments. By contrast, a magnitude 8 earthquake is an enormous jolt 10 times the amplitude of a very powerful magnitude 7 earthquake.

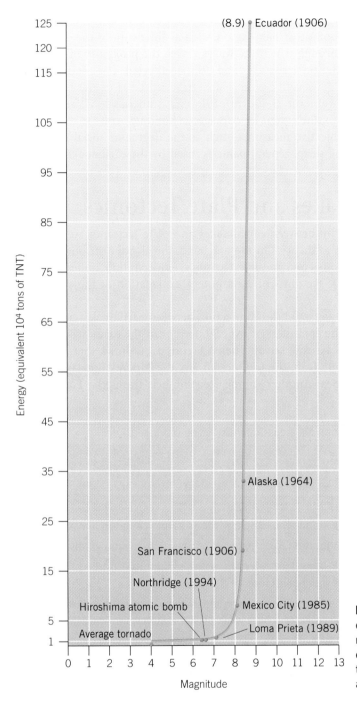

Figure 3.13 The relationship between magnitude and energy release of an earthquake. A single great earthquake releases far more energy than many thousands of small earthquakes combined. Earthquakes of similar magnitude on the Richter scale differ enormously in the amount of energy that they release.

Researchers have established that energy increases roughly 30 times for each magnitude number on the Richter scale (Figure 3.13). Thus a magnitude 7 earthquake releases 900 (30 × 30) times the energy of a magnitude 5 earthquake. Also, two earthquakes that register closely on the Richter scale nevertheless differ significantly in their energy release. For example, the magnitude 8.2 San Francisco earthquake of 1906 released three times the energy of the magnitude 8.1 Mexico City earthquake of 1985. Because of the magnitude-energy relationship, the energy released during a great earthquake far exceeds the combined energy of the hundreds of thousands of small earthquakes that occur each year.

In recent years, seismologists have been using a magnitude scale that differs slightly from the Richter scale because it is based on more refined instrumentation and greater knowledge of seismic waves. Neither the old scale nor the new, however, is a measure of the destructiveness of an earthquake. Certainly, the greater the magnitude of an earthquake, the greater is its potential for destruction. The extent of the damage, however, depends upon other factors, too, such as the depth of the earthquake, the types of rocks and soil through which the waves travel, the proximity of the epicenter to population centers, and the structures and utilities affected. The modified **Mercalli intensity scale** measures the severity of an earthquake in terms of the level of destruction and panic it causes. Values in the form of Roman numerals I–XII are assigned to objective measures, such as the extent of damage to buildings constructed of brick, mortar, or steel, and to subjective measures, such as the degree of social disruption (hardly felt, awakened during sleep, general panic) caused by the earthquake (Table 3.3).

Earthquakes and Plate Tectonics

In the relationship between earthquakes and plate tectonics, we have a near-perfect fit of observation and theory. On the one hand, the global distribution of earthquakes and their fault motions confirm plate tectonics—indeed, they underpin the theory. On the other hand, plate tectonics offers a potent explanation of the stresses that cause most earthquakes.

To better understand the connection between earthquakes and plate tectonics, consider this fact: the epicenters of over 99 percent of the earthquakes that occur each year are confined to narrow seismic belts. Figure 3.14 shows that these belts are located around the Pacific Rim, the Indonesian arc, the Alpine-Himalayan mountain belt, and the Caribbean Sea. Another thinner seismic belt traces the Earth-girdling mid-ocean

Table 3.3   The Modified Mercalli Intensity Scale (Abridged)

Intensity	Effects
I	Not felt.
II	Felt by persons at rest.
III	Felt indoors; hanging objects swing; vibration like passing of light trucks.
IV	Vibration like passing of heavy trucks; windows, dishes rattle.
V	Felt outdoors; sleepers wakened; liquids disturbed, some spilled; doors swing open, closed.
VI	Felt by all; many are frightened and run outdoors; windows, dishes, glassware broken; weak plaster and masonry cracked; furniture moved or overturned.
VII	Difficult to stand; hanging objects quiver; fall of plaster, loose bricks, stones, tiles, and architectural ornaments; waves on ponds.
VIII	Steering of cars affected; fall of stucco and some masonry walls; twisting, fall of chimneys, factory stacks, monuments, elevated tanks; branches broken from trees.
IX	General panic; frame structures, if not bolted, shifted off foundations; underground pipes broken; conspicuous cracks in ground; in alluviated areas, sand and mud ejected; earthquake fountains, sand craters.
X	Most masonry and frame structures destroyed; some well-built wooden structures and buildings destroyed; serious damage to dams and embankments; large landslides.
XI	Rails bent greatly; underground pipelines completely out of service.
XII	Damage nearly total; lines of sight and level distorted; objects thrown into the air.

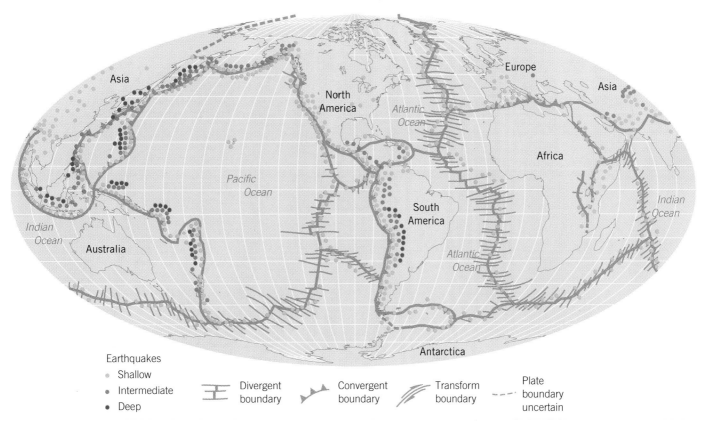

Earthquakes
- Shallow
- Intermediate
- Deep

Divergent boundary

Convergent boundary

Transform boundary

Plate boundary uncertain

Figure 3.14 Most earthquakes occur in narrow belts that define the plate boundaries. Intermediate- and deep-focus earthquakes follow the path of descending plates within subduction zones, while rift valleys and fracture zones are the sites of shallow earthquakes.

ridge system. There is a striking contrast in activity between these seismic belts and the vast intervening regions. According to plate tectonics theory, the seismic belts mark the boundaries of the Earth's lithospheric plates. Only a few great earthquakes occur far from plate boundaries, and many of them may also have connections to plate movements, as we shall soon see.

The distribution of earthquake focal depths proves as informative as the distribution of epicenters. Those within the narrow mid-ocean ridge seismic belts are shallow; they occur high up on the ridge crest in the rigid basaltic crust of the central rift valley. This observation fits well with plate tectonics theory, which indicates that the rift valleys are the sites of plate separation. Just beneath the basaltic crust are reservoirs of upwelling magma and the semimolten asthenosphere. Minimal seismic activity occurs there because neither region is capable of storing the strain energy necessary to generate earthquakes.

More striking is the pattern of focal depths along subduction zones (Figure 3.15). There are shallow earthquakes near the trenches, but there are also intermediate and deep earthquakes to depths of 700 kilometers. In fact, virtually all the Earth's deep-focus earthquakes are associated with subduction zones. The patterns of deep-earthquake focal points form continuous lines within the oceanic crust that trace the paths of subducting plates as they descend into the mantle.

Not only are the locations and depths of earthquakes consistent with plate tectonics theory, but also the fault movements observed at plate boundaries match the motions predicted by the theory (see Figure 3.15). First-motion studies confirm that earthquakes are generated by compression at convergent boundaries, by tension at divergent boundaries, and by horizontal shear at transform boundaries. Earthquakes at transform boundaries also satisfy the theory's requirement that they occur between offset mid-ocean ridge segments. Finally, the direction of motion along the transform fault is opposite to the apparent displacement of the ridge crests.

Thus there is compelling evidence that plate boundary interactions—divergence, convergence, and transform—are responsible for over 99 percent of earthquakes. But

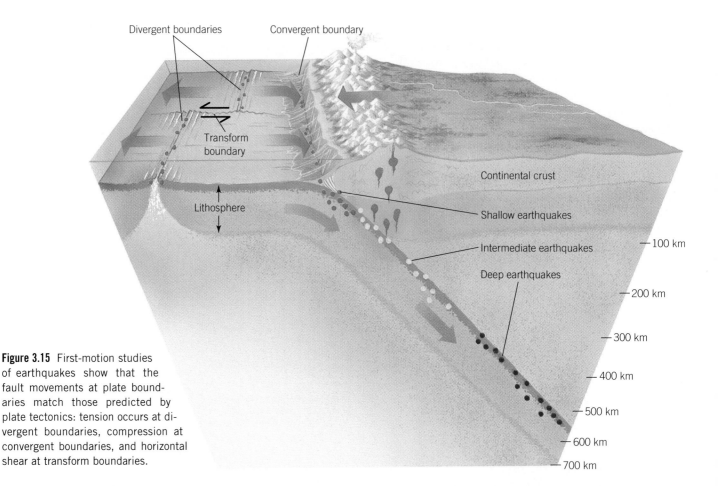

Figure 3.15 First-motion studies of earthquakes show that the fault movements at plate boundaries match those predicted by plate tectonics: tension occurs at divergent boundaries, compression at convergent boundaries, and horizontal shear at transform boundaries.

precisely how do these interactions trigger elastic rebound? A big clue lies in the distribution of earthquake focal depths: With the exception of deep-focus earthquakes of subduction zones, where the cold lithosphere is jammed far into the mantle, the vast majority of earthquakes occur within 30 to 40 kilometers of the surface; that is, within the upper parts of the plates. There, the rocks are rigid enough to store strain energy and brittle enough to rupture suddenly and generate an earthquake. Toward the base of the plates, however, higher temperatures and pressures soften the rocks somewhat, which causes them to flow plastically, rather than rupture, in response to stress.

The difference in response to stress between the upper and lower parts of the plate may be responsible for most earthquakes. According to one hypothesis, as two plates move past one another, the warm, soft lower layers of the plates move freely and smoothly, dragging the rigid upper layers of the plates with them. The progress of the upper layers is halted wherever they get hung up at an *asperity,* a rough point along the fault plane. Strain builds as the lower layers continue to move and the upper layers remain locked. Finally, the shear stress overcomes the hang-up, and the rocks lurch to unstrained positions, releasing energy in the form of seismic waves—the elastic-rebound effect. The greater the accumulation of strain at the asperity, the more powerful the earthquake will be. **Foreshocks** usually precede the main event as the rocks in the vicinity of the asperity begin to crack. **Aftershocks** follow as the rocks along the fault readjust and strain is transferred to the next asperity. Then there is a period of quiet while strain accumulates again.

foreshocks
A series of lower-magnitude earthquakes that may directly precede a higher-magnitude earthquake.

aftershocks
A series of lower-magnitude earthquakes that may directly follow a higher-magnitude earthquake.

The San Andreas Fault

The San Andreas fault is a transform boundary connecting two offset segments of the mid-ocean ridge, one in the Gulf of California and the other off the California-Oregon coast (Figure 3.16). Spreading eastward from these segments are the Cocos and Juan de Fuca plates, which are subducting beneath North America. Current theory is based largely on the pioneering work of Tanya Atwater and her colleagues at the University of California, Santa Barbara. As early as 1970, she proposed that these small plates are remnants of a much larger Farallon plate, which has been overridden by the westward-spreading North American plate—a process that began at least 80 million years ago. However, plate motions changed about 30 million years ago when the North American plate finally reached the East Pacific Rise, a segment of the mid-ocean ridge that lay to the west of the southern California–Baja coast. This change brought the North American plate in contact with the Pacific plate, which, rather than subducting, was spreading to the northwest. Today, the Pacific plate continues to spread northwest relative to the North American plate, and the San Andreas fault is the boundary between these moving plates.

Currently, a slice of California that includes Los Angeles and the Baja Peninsula is attached to the Pacific plate and is gliding northwestward relative to the North American plate along the San Andreas fault. The overall rate of movement is about 5 centimeters per year. Judging from the location of identical rock formations that are now separated by the fault, the total displacement of the west side of the San Andreas fault relative to the east side has amounted to at least 350 kilometers since the fault has been in existence (Figure 3.17). Some segments of the fault appear to be moving smoothly; they are peaceful and earthquake-free. Great earthquakes occur along fault segments that remain locked and release their strain energy sporadically.

The Los Angeles Faults

California's problems are not confined to the San Andreas fault; the state is sliced by active faults of all kinds. The region is subject to complicated stresses generated as the San Andreas fault grinds northward and shears, compresses, and extends the crust at many locations along the way (Figure 3.18). Southern California in general and the Los Angeles area in particular are especially vulnerable.

Figure 3.16 Parts (a) and (b) show that the Juan de Fuca and Cocos plates are remnants of the Farallon plate, which was subducted beneath the North American plate. The San Andreas is a transform fault that formed when the North American plate intersected the East Pacific Rise, as seen in part (c).

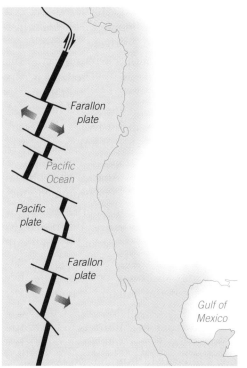

(a) 60 m.y.a.

(c) 0 m.y.a.

(b) 30 m.y.a.

Figure 3.17 Offset stream beds show the direction of movement along the San Andreas fault. The San Andreas is a right lateral fault, meaning that when you face the fault, the movement is to the right.

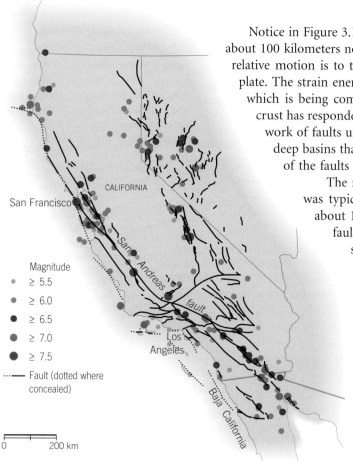

Magnitude
● ≥ 5.5
● ≥ 6.0
● ≥ 6.5
● ≥ 7.0
● ≥ 7.5

·····── Fault (dotted where concealed)

0   200 km

Figure 3.18 Some branches of California's complex network of faults. The stress that drives the Los Angeles faults is caused by the sharp bend in the San Andreas fault about 100 kilometers northeast of the city.

Notice in Figure 3.18 that the San Andreas fault makes a sharp westward bend about 100 kilometers northeast of Los Angeles. As a result, the Pacific plate, whose relative motion is to the northwest, abuts obliquely against the North American plate. The strain energy thus generated is transferred to the Los Angeles region, which is being compressed at the average rate of 1 centimeter per year. The crust has responded by buckling and fracturing. A 9400-square-kilometer network of faults underlies Los Angeles and the San Fernando Valley. Both are deep basins that hold thick accumulations of sedimentary rock, and many of the faults are in these rocks.

The magnitude 6.7 Northridge earthquake of January 17, 1994, was typical of those that plague this region. Its focus was located about 14 kilometers deep along a gently dipping *thrust fault.* In a fault of this type, the block of the crust overlying the fault plane slides upward toward the Earth's surface with respect to the block beneath the plane (Figure 3.19). The fault occurs in response to compression, and the result shortens the crust. Strain energy may remain locked within a Los Angeles or San Fernando Valley thrust fault for a decade or a century, until the fault suddenly ruptures and moves a meter or two, setting off an earthquake.

Many thrust faults around Los Angeles have broken the surface and have been mapped. Some prominent ones run parallel to the foothills of the San Gabriel Mountains that border the east side of the valley. Whenever the mountains are suddenly driven upward and to the south along these faults, earthquakes occur. Sudden movements along distant thrust faults also cause uplift because they compress the valley, and the mountains respond by rising. The fault movement associated with the Northridge earthquake thrust the Santa Susana Mountains about 33 centimeters vertically and about 4 centimeters to the north. Although the waves set off by the initial fault movement lasted only 10 seconds, they reverberated within the basin and reinforced one another, which prolonged the ground shaking and multiplied the damage.

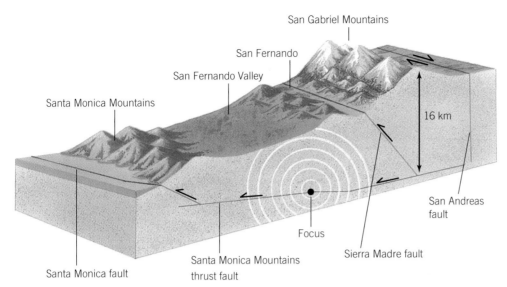

San Gabriel Mountains

San Fernando

San Fernando Valley

Santa Monica Mountains

16 km

San Andreas fault

Focus

Sierra Madre fault

Santa Monica Mountains thrust fault

Santa Monica fault

Figure 3.19 Regional compression is shortening the San Fernando Valley and uplifting the San Gabriel and Santa Monica mountains along thrust faults. Many of the faults in the valley are blind thrusts, meaning they lack surface expression.

The Northridge earthquake occurred along a previously unmapped fault, which was located after the fact by study of the distribution of aftershocks in the vicinity of the focus. These shocks clustered in a plane, delineating the fault. Some days later, a series of ground breaks defined the trace of the fault on the Earth's surface. Many of the faults beneath Los Angeles are *blind thrusts,* meaning that they occur at depth and leave no surface trace. As we learn about them after an earthquake, we can only guess how many blind thrusts remain to be discovered.

Intraplate Earthquakes

Fewer than 1 percent of all earthquakes occur away from plate boundaries, but some of the more powerful of these **intraplate earthquakes** have caused rampant destruction to populated regions. The magnitude 7.6 earthquake that hit the heavily populated industrial city of Tangshan, China, in 1976 killed at least 250,000 people.* In fact, China, far from plate boundaries, has a long history of devastating earthquakes.

Closer to home, the most powerful series of earthquakes in the history of the United States took place not in San Francisco or Anchorage but in New Madrid, Missouri, in 1811 and 1812 (magnitudes were later estimated to have been as high as 8.5). The shifting crust rerouted the Mississippi River, and significant damage was recorded as far away as Cincinnati, Ohio. From time to time, a number of moderate-sized earthquakes have shaken other supposedly stable regions of the interior of North America. The 1886 Charleston, South Carolina, earthquake not only wrecked the city but also sent out vibrations powerful enough to rock Boston and Chicago. No location in the United States—or the world, for that matter—should be considered risk-free (Figure 3.20).

We do not fully understand the mechanics of intraplate earthquakes. One plausible theory relates them to the reactivation of ancient faults, many deeply buried in the continental crust, that formed during episodes of rifting and subduction that predate present plate boundary interactions. As the present plates slide slowly over the asthenosphere, stresses are transferred to the old faults. Like the weak links in a chain, these faults then release their strain energy as earthquakes. The reactivation theory may well explain the New Madrid earthquakes. Recently, detailed seismic studies have located an ancient fault zone thousands of meters below the lower Mississippi Valley (Figure 3.21). Swarms of low-magnitude earthquakes have been traced to that zone, and the 1811–1812 New Madrid epicenters also fall within it.

intraplate earthquake
An earthquake whose epicenter is far from a lithospheric plate boundary.

---

*The official death toll was 250,000 people, but other estimates have placed it at 650,000.

Figure 3.20 This seismic risk map of the United States shows that there are regions far away from plate boundaries where we can expect significant earthquake activity. (Courtesy of National Oceanic and Atmospheric Administration.)

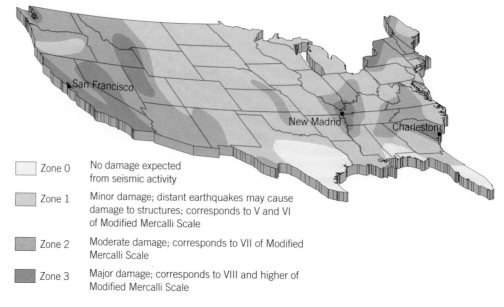

☐ Zone 0   No damage expected from seismic activity

☐ Zone 1   Minor damage; distant earthquakes may cause damage to structures; corresponds to V and VI of Modified Mercalli Scale

☐ Zone 2   Moderate damage; corresponds to VII of Modified Mercalli Scale

☐ Zone 3   Major damage; corresponds to VIII and higher of Modified Mercalli Scale

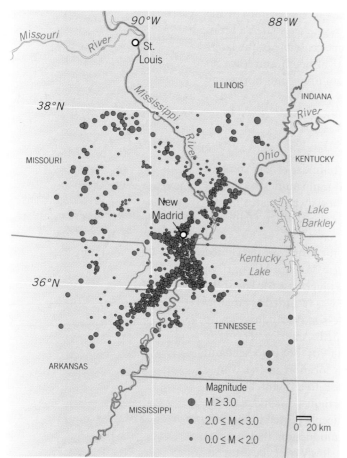

Figure 3.21 The epicenters of earthquakes recorded in the New Madrid region mark the trend of an active subsurface rift. Note the close connection of this rift with the course of the Mississippi River. (Adapted from A. Johnston, "New Madrid: The Rift, the River, and the Earthquake," *Earth* 1, no. 1, 1992.)

Another hypothesis that seeks to explain intraplate earthquakes suggests that in some regions, continental collisions are transmitting huge stresses deep into the interior of the plates. One such collision is probably the cause of many earthquakes in China, including the Tangshan disaster. The epicenters of the Chinese earthquakes are located along active faults that were set in motion by the ongoing convergence of India and Asia. Still another hypothesis proposes that some intraplate earthquakes are triggered by the doming effects of hot spots. As we discussed in Chapter 2, these plumes of magma rising from the mantle may cause rifts in the continent that will lead eventually to the formation of new plates.

Other Causes of Earthquakes

Some geologists caution that we should resist the temptation to link every geological event to plate tectonics and suggest that some earthquakes may have causes other than plate tectonics. One possible earthquake-producing mechanism involves the rebound effect accompanying the melting of the great ice sheets that covered much of North America, Europe, and Asia thousands of years ago. Melting relieved the load that depressed the crust, and the hypothesis is that stress has been transferred to old faults as the crust slowly rises.

Subsurface water injection is another suggested cause. The series of small earthquakes that struck seismically quiet Denver, Colorado, in 1964 offer striking evidence in favor of this idea (Figure 3.22). They were found to correlate precisely with the injection of fluid waste into highly fractured rocks, thousands of meters beneath the outskirts of the city, from a plant that manufactured nerve gas for the army. As the water seeped into ancient faults, it lowered the confining pressure between the rocks and caused them to slip suddenly, triggering the earthquakes. When the pumping stopped, so did the earthquakes. This incident offers the intriguing possibility that some shallow earthquakes may be prevented or controlled by pumping water in and out of faults.

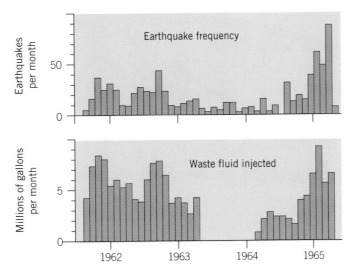

Figure 3.22 These graphs show the correlation between earthquake frequency and deep-well fluid injections in the vicinity of Denver, Colorado, during the 1960s. The number of earthquakes increased as the amount of fluid increased and decreased when injections ceased. (Adapted from D. Evans, "Manmade Earthquakes in Denver," *Geotimes* 10, no. 9, 1966, pp. 11–18.)

Forecasting Earthquakes

In 1975, a magnitude 7.4 earthquake that struck Haicheng, China, set off 90 seconds of intense ground shaking that reduced 90 percent of the city's buildings to rubble. With 3 million people living near the epicenter, this earthquake could have been another mind-numbing tragedy, but it was not. Chinese seismologists were aware that Haicheng was situated astride a major fault zone They had also noted that over a period of years, the earthquake epicenters along the fault were drawing ever closer to the city and that the earthquakes were increasing in magnitude and frequency. Analysis of this trend enabled them to predict the earthquake with enough lead time to allow for evacuation of the city and surrounding villages, thus saving tens of thousands of lives.

Haicheng 1975 marked the first fully documented, successful earthquake prediction. Unfortunately, the Chinese were unable to prepare for the magnitude 7.6 earthquake that struck Tangshan without warning one year later. Because Haicheng is closer to Tangshan than Los Angeles is to San Francisco, no two events so closely matched in time, space, and magnitude better illustrate the potential benefit of earthquake prediction research.

Given the imperfect state of our knowledge regarding earthquakes, many geologists prefer to distinguish between a *prediction* and a *forecast*. A prediction specifies the time, location, and magnitude of an event within narrow limits of uncertainty, whereas a forecast projects trends over a broad time frame. Each has its purpose. Forecasts are used for long-range planning: legislating zoning and building codes, setting up evacuation routes and emergency procedures. Predictions are used for immediate preparations: evacuating people, turning off gas lines, stopping trains, closing off bridges. Unfortunately, methods of prediction have so far proved to be largely unreliable, although we have made progress in forecasting earthquakes. In the following section, we describe some of the strategies geologists employ to forecast and predict earthquakes.

Seismic Gaps

Close examination of the seismic belt rimming the Pacific shows that segments that have recently experienced a great earthquake alternate with those that have long been quiescent—so-called **seismic gaps** (Figure 3.23). An interesting hypothesis uses this observation to forecast great earthquakes. The model is based on the mechanics of fault motion along plate boundaries. Earlier, we mentioned that while the lower layers of two lithospheric plates slide by one another smoothly at the plate boundary, the rigid upper layers of the plates are repeatedly hung up by rough spots (asperities). As an earthquake

seismic gap
A seismically inactive segment of a fault within which strain is accumulating; these gaps are bracketed by the epicenters of relatively recent great earthquakes.

Figure 3.23 This map shows the probability of great earthquakes occurring between 1989 and 1999 along segments of the seismic belts as determined by the gap theory. The segments that are currently quiescent and are bracketed by epicenters of relatively recent great earthquakes have the highest probability of future earthquakes. The dates and magnitudes refer to the most recent large earthquakes in those regions where the historic record is incomplete. (Adapted from "Circum-Pacific Seismic Potential, 1989–1999," by Stuart P. Nishenko, U.S. Geological Survey, 1989.)

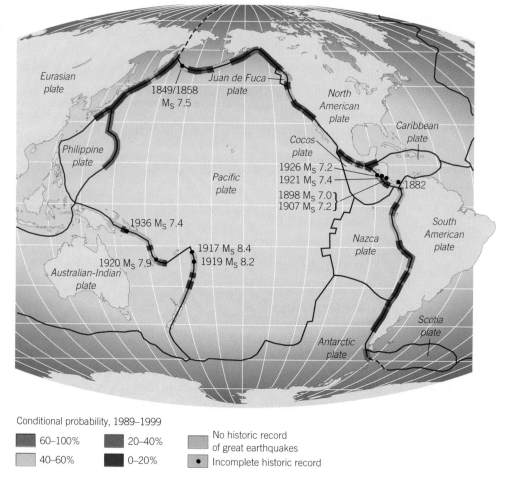

Conditional probability, 1989–1999

■ 60–100%	■ 20–40%	No historic record of great earthquakes
■ 40–60%	■ 0–20%	• Incomplete historic record

relieves the strain at an asperity, the strain shifts to neighboring asperities that are currently quiescent—the seismic gaps. According to the seismic gap hypothesis, quiet segments of the fault bordered by the epicenters of recent great earthquakes are the places to look for the next great earthquake. With each lurching movement of the plates, all segments along the fault will generate great earthquakes in sequence—except for those portions of the plate boundary that are smooth and free of asperites.

The gap hypothesis has had some successes; for example, the magnitude 7.1 Loma Prieta earthquake of 1989, which devastated downtown Santa Cruz and rocked San Francisco 95 kilometers to the north, occurred within a segment of the San Andreas fault that USGS geologists had identified as a seismic gap. However, in other large earthquakes, the epicenters did *not* fall within seismic gaps; that is, they occurred in places not predicted by the model. Some geologists thus conclude that the model is either mistaken or in need of refinement.

Recurrence Studies

One of the problems in testing forecasting tools such as seismic gaps is the insufficiency of historical records. They do not extend back far enough in time to establish reliably how often earthquakes have occurred in a given locale—that is, to establish their *recurrence intervals.*

Paleoseismology, as applied by geologist Kerry Sieh and his colleagues at the California Institute of Technology, is probably the most accurate method of extending the record back in time; their work in southern California has added enormously to our knowledge of the history of the San Andreas fault. They have found that Pallet Creek, which crosses the fault about 55 kilometers north of Los Angeles, has been repeatedly

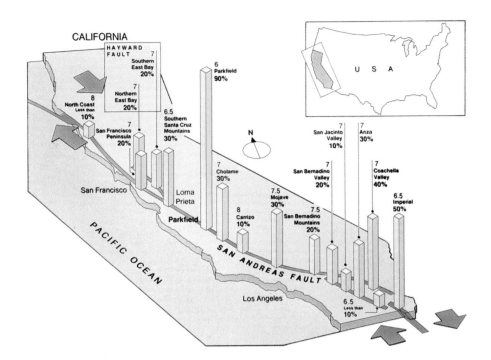

Figure 3.24 Probabilities of major earthquakes occurring along various segments of the San Andreas and Hayward faults over the next 30 years. There is a 90 percent probability of an earthquake of magnitude 6 or greater occurring at the tiny village of Parkfield, California, population 34. In the late 1980s, seismologists made the firm prediction that this earthquake would occur before 1993. That the event still had not occurred by late 1997 illustrates the uncertainties of earthquake prediction.

disrupted by sudden movements along the fault. Each episode accomplished two things: it cut the layers of riverbed sediments previously deposited, and it temporarily dammed the river, allowing a layer of organically rich peat to be deposited directly on top of the faulted sediment. Sieh obtained the age of each peat layer by the carbon-14 method, and, using these dates, he was able to closely date each episode of fault movement.

Sieh has been able to discern 12 major earthquakes during the past 1400 years in the Pallet Creek sediment. Recurrence intervals vary between 50 and 300 years, the average being 140 to 150 years. In fact, it has been almost 140 years since the last great earthquake struck the region in 1857, an interval uncomfortably close to the average determined by Sieh. For each year that an earthquake does not occur, the chance that one will occur increases by about 2 to 5 percent, and each year, there is a greater probability that it will carry a more wicked punch.

The recurrence studies suggest that Los Angeles is at greater risk than previously thought. San Francisco, on the other hand, may have more breathing room (Figure 3.24). The recurrence interval for the northern segment of the San Andreas fault is about 150–300 years, and the last great earthquake struck in 1906, not quite a century ago. However, we must remember that recurrence studies offer only one line of evidence. According to seismic gap evidence, the rocks of the San Francisco region remain locked, and estimates of the strain accumulating in the region are similar to 1906 levels. The 1989 Loma Prieta earthquake that rocked San Francisco was not the "big one" forecast for that great city; its epicenter was 95 kilometers to the south. More important, the "big one" is expected to release 50 times the energy and to be accompanied by four times the fault displacement of the Loma Prieta rupture.

See Chapter 10, Geologic Time, for a discussion of the carbon-14 and other radiometric dating methods.

Precursor Studies

Subtle changes that warn us that the rocks are close to failure and an earthquake is imminent may take place in the vicinity of an active fault. These changes, deviations from normal or expected patterns, are called *precursors*. Figure 3.25 shows the idealized patterns of some of the more promising precursors. Among them are accelerated land uplift, increased emissions of radon gas from wells, a precipitous rise in groundwater

Figure 3.25 Earthquake precursors. An idealized depiction of how measuring small changes that precede large earthquakes may be used as a predictive tool.

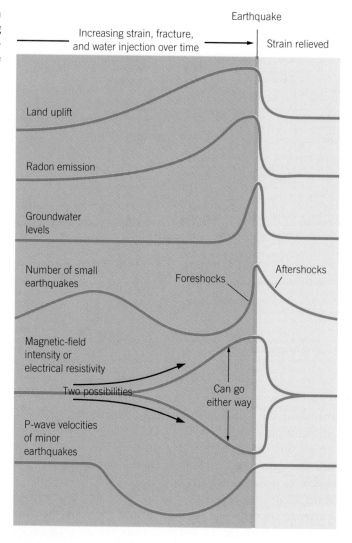

levels, and fluctuations in the frequency of small earthquakes and in the P-wave velocity of the waves generated. All of these changes are related to the response of the rocks to accumulation of strain energy. Ideally, changes would build to a crescendo just before the earthquake so that the geologist tracking the trends could issue warnings to the appropriate authorities.

Though all the precursors listed in Figure 3.25 have been detected prior to one earthquake or another, not all of them have been observed prior to the *same* earthquake, nor do they always follow the simple patterns depicted in the figure. In some instances, precursors have pointed to earthquakes that did not, in fact, occur; by the same token, earthquakes have struck without these warning signs. For these reasons, geologists have yet to determine if any precursors are reliable enough to form the basis of a prediction or a forecast. At our present level of knowledge, predicting earthquakes is as much an art as it is a science.

Recently, however, there has been some success in identifying those submarine earthquakes that are likely to trigger a **tsunami,** a giant ocean wave (see Aside 3.1).

tsunami
The Japanese word for a large, often deadly sea wave set in motion by an undersea earthquake.

Minimizing Earthquake Damage

Why is it that in 1988, a magnitude 6.9 earthquake killed 25,000 people but in 1992 a 7.5 earthquake—six times stronger—killed only one unfortunate individual? The answer

ASIDE 3.1 *Tsunamis*

The sea rose boiling in the harbor and broke up all the craft harbored there; the city burst into flames, and ashes covered the streets and squares; the houses came crashing down, roofs piling upon foundations, and even the foundations were smashed to pieces. Thirty thousand inhabitants of both sexes and all ages were crushed to death in the ruins.*

When Voltaire wrote of the sea boiling up in the harbor, he was describing a deadly phenomenon: a tsunami. Tsunamis are giant ocean waves generated by submarine earthquakes. The vibrations set off by a fault in the oceanic crust are transferred to the water surface and spread across the ocean in a series of low waves (see the figure). Most tsunamis originate above the subduction zones that rim the Pacific basin. There, cold and brittle oceanic crust fractures as it descends into the mantle at the deep-sea trenches.

With amplitudes of less than a meter and wave-lengths of many kilometers, a tsunami is difficult to detect in the open ocean. Nevertheless, it travels at jet plane speeds (800 to 1000 kilometers per hour). Then when it reaches the shallow depths of the coast, friction slows the wave down and, at the same time, causes it to roll up into a wall of water as high as 30 meters. These effects are intensified in the confined spaces of bays where most harbors are located—thus the name *tsunami*, Japanese for "harbor wave." The wall of water descending upon a harbor is preceded by an emptying of the harbor, and many deaths are attributed to the curiosity of people who run out to examine the exposed harbor bottom.

Tsunamis have destroyed entire cities and almost always cause extreme damage. A 1960 earthquake off the Chilean coast, for instance, set off a tsunami that traveled 14,500 kilometers across the Pacific Ocean to Japan, where it caused 180 deaths. The Japanese coastline is especially vulnerable; history records at least 15 major disasters there, some killing hundreds of thousands of people. For these reasons, submarine earthquakes are monitored closely, and an early-alert system warns of the possibility of a tsunami striking coastal regions. However, this system is not the complete solution, for a large-magnitude earthquake may not trigger a tsunami, whereas a relatively minor one may send huge waves crashing into an unsuspecting harbor.

Recently, though, geophysicists have developed a promising method of identifying those earthquakes likely to lead to a tsunami. They have found a correlation between the period (vibration rates) of tsunamis and certain very-long-period seismic waves. Earthquakes that have an appreciable amount of their energy in this long-period range are likely to trigger a sizable tsunami. Because seismic waves travel at speeds on the order of kilometers per second, rather than kilometers per hour, they outrun the tsunamis to recording stations. Their arrival may provide seismologists with sufficient lead time to issue warning bulletins.

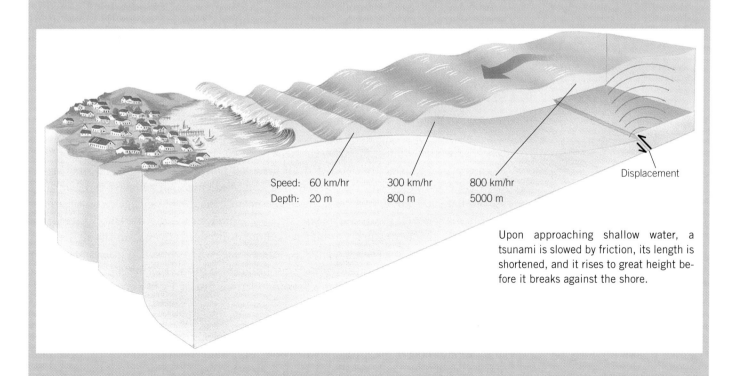

| Speed: | 60 km/hr | 300 km/hr | 800 km/hr |
| Depth: | 20 m | 800 m | 5000 m |

Displacement

Upon approaching shallow water, a tsunami is slowed by friction, its length is shortened, and it rises to great height before it breaks against the shore.

---

*Account of the 1755 Lisbon, Portugal, earthquake in Voltaire's *Candide*. (Modern estimates place the death toll closer to 70,000.)

Figure 3.26 Rescue workers toil atop a poorly constructed building in Spitak, Armenia, that was pancaked by the magnitude 6.9 earthquake on December 7, 1988.

Figure 3.27 Life on the San Andreas fault. Urbanization and its relation to earthquake hazards is dramatized by these housing tracts on a segment of the fault just south of San Francisco.

is that the 6.9 earthquake struck a heavily populated region of Armenia and caused thousands of old and poorly constructed buildings to collapse on their inhabitants (Figure 3.26). The 7.5 quake was centered in Landers, California, in the sparsely populated Mojave Desert region. Buildings there are few and far between, and the newer ones have been constructed according to California's earthquake building codes. These two earthquakes are testimony to the adage that earthquakes don't kill people, buildings do. The major problems of earthquakes are the result of civilization—namely, the types of structures built and the patterns of land use brought on by population growth, custom, and urban-suburban development (Figure 3.27).

After an earthquake, surveys are taken and the various Mercalli intensities are mapped so that patterns of damage may be established. (For more on the Mercalli intensity scale, see Table 3.3.) We have learned much from these surveys. For example, we can observe the connection between the type of subsurface in the region and the levels of earthquake damage that occurred. In the 1989 Loma Prieta earthquake, the Marina District of San Francisco was the site of severe residential damage. Figure 3.28a shows that in the 1906 San Francisco earthquake, most of the direct damage also occurred in the Marina District. Figure 3.28b explains why. Homes in the Marina District were built on landfill made of unconsolidated sediment dredged from San Francisco Bay. Saturated sediments have a tendency to lose contact when shaken, a process called **liquefaction.** This is what transformed the solid ground of the Marina District into loose mud incapable of supporting loads; thus foundations buckled and buildings pitched into the street.

Plans to minimize the damage of earthquakes should include surveys of the secondary effects of earthquakes as well as the more obvious damage. Investigators must make careful note of the proximity of population centers to potential landslide hazards and unsafe dams; of the possibility of fires and the disruption of communications and utilities; and of the possibility that damage to water mains and sewage pipelines can lead to contamination and disease.

liquefaction
The transformation of saturated sediment or soil to liquid when ground shaking causes the particles to lose contact.

Figure 3.28 The relationship between (a) earthquake intensity and (b) subsurface conditions. Houses built on a landfill made of unconsolidated sediment dredged from San Francisco Bay are especially vulnerable, while those built on firm igneous and metamorphic bedrock are better protected. (Adapted from B. Bolt, *Earthquakes,* San Francisco: W. H. Freeman & Co., 1978, p. 103. After H. O. Wood, 1907.)

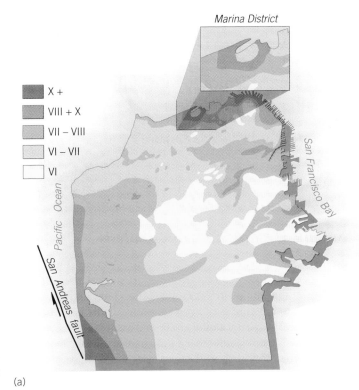

(a)

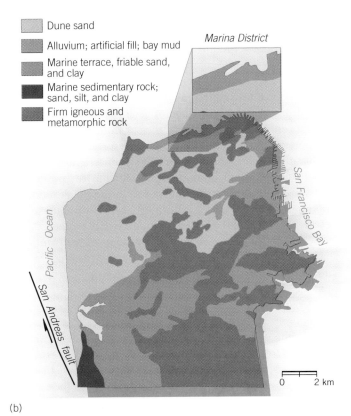

(b)

Construction Problems

Engineers should follow two important design principles to make a building earthquake-resistant. The first is to ensure that all beams and columns supporting the structure are firmly connected to one another and to the ground. The second principle is to make sure that the columns or beams that support the building's walls and floors transmit load directly to the ground. This design will minimize the effects of the strong shearing stresses that develop when the building is shaken from side to side.

Buildings, dams, and bridges are basically inverted pendulums, with their lower ends anchored to the ground and their upper ends free to vibrate. Tall structures are likely to magnify ground shaking, creating a tough problem for engineers working in downtown areas of modern cities. Conditions on the ground may not match those on, say, the thirtieth floor of an improperly designed building, which may sway like wheat in the wind. The engineer must design the building to flex during an earthquake and in this way dissipate energy within the steel, concrete, brick, and mortar of the building. Of course, the building must not flex so much as to rip the floors and roof from the support columns or to shear off the support columns at their base. As you may guess, designing a skyscraper to withstand a great earthquake can be expensive; the cost of damage prevention may increase tenfold for every order of earthquake magnitude required.

Faulty design and defects in construction are the reasons most buildings collapse during an earthquake and the reasons most people are killed. Figure 3.29 traces the damage to various buildings during the 1985 Mexico City earthquake. Traditional adobe houses crumbled; modern buildings with odd shapes were twisted apart. Adjacent short and tall buildings, vibrating at different rates, bumped against one another, which caused the short buildings to collapse. On the other hand, certain earthquake-resistant buildings survived; those built on specially designed rollers were effectively decoupled from ground vibrations and survived with little or no damage.

Earthquakes are not new to Mexico City, and many of its buildings were constructed according to strict codes. California's engineers thus took careful notes of the earthquake's effects. Attracting special attention was the poor performance of modern buildings with

Designing Seismically Safe Buildings— A Matter of Probabilities

With vast cities and highway systems being built in the vicinity of major fault systems, seismologists are being pressured to develop technology that can predict earthquakes. Until that technology is available, the best they can do is to identify high-risk areas and to aid engineers and planners in minimizing the damage.

Seismic Hazards and Risk Analysis

Seismologists concerned with earthquake damage control work closely with engineers in the effort to design earth-quake-safe buildings, bridges, and dams. They have two functions: to determine the *seismic hazard,* or the potential for damage, at a specific building site, and to identify the *seismic risk,* or the probability that an earthquake higher than a given magnitude will occur during a certain time interval in the vicinity of the site. Design decisions depend upon the answers to a number of fundamental questions: What kind of structure is being built? How many years is it expected to remain in use? How serious would the consequences be if this structure did not withstand an earthquake of a given magnitude?

The answers to these questions determine the degree of safety that must be engineered into the design of the structure and thus dictate the method of geological and seismological investigation. For example, the consequence of a nuclear power plant failing—that is, the widespread release of deadly radiation into the environment—is unacceptable. As a result, federal regulations mandate that nuclear power plants be designed to remain functional during the ground motion generated by the *maximum credible earthquake (MCE).* The MCE is the largest earthquake that can be expected to occur within a 320-kilometer radius of the plant. The epicenter of the MCE is assumed to be located along the fault, or other known seismically active center, that lies closest to the site of the nuclear plant.

Designing a structure to withstand an MCE involves the *deterministic method* of seismic hazard analysis. Time is not a factor in the deterministic method; instead, the worst case scenario is considered regardless of when—if ever—the earthquake is expected to occur. As an example, suppose that a nuclear power plant is to be built at Site 1 shown in the figure. The seismologist might determine that the MCE is a magnitude 7 event occurring along Fault A, which is located 30 kilometers from the plant site. The seismic hazard to be expected at the site from this event is calculated to be the peak ground acceleration of 0.6 g (or 60 percent of gravitational acceleration). Thus, the engineer must design the structure to

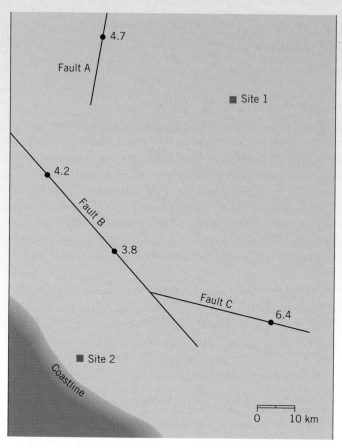

Hypothetical map of building sites, known faults, and magnitudes of recent earthquakes.

resist 0.6 g. More precisely, the structure is designed to resist the full spectrum of ground accelerations expected at the site.

The consequences of the collapse of nuclear power plants, hydroelectric dams, or hospitals are so horrendous that clearly every reasonable—or even unreasonable—precaution should be taken to ensure their safety. But do we need the same level of safety for a warehouse or parking garage? What about a supermarket or movie theater? If you think not, then you accept the proposition that for many structures the consequences of failure may not justify the added costs necessary to achieve the higher level of safety. In other words, every engineering decision includes the hidden question of *cost versus benefit.*

In designing for protection against earthquakes, construction costs tend to rise exponentially with each increase in expected magnitude. At the same time, the probabilities of high-magnitude earthquakes decline exponentially—the higher the magnitude of the earthquake, the rarer the event. The art of the engineer is to find the proper balance between safe design and cost. For this purpose, he or she takes into account the concept of *acceptable risk*.

Determining acceptable risk requires an answer to the question, What is the highest-magnitude earthquake that can reasonably be expected during the life span of the structure? The engineer then designs for the agreed-upon magnitude—or more precisely, for the intensity and type of ground acceleration that such an earthquake would set in motion at that particular site. This time-dependent approach is called the probabilistic method of seismic hazard analysis.

Let us consider the four steps in assessing risk using the probabilistic method, for a three-story office building to be built at Site 2 in the figure. The owner expects the building to last 30 years.

1. The seismologist begins by identifying the potential sources of earthquakes in the region surrounding the site and estimates the average number generated per year at each source. Extensive geological field work and inspection of historical earthquake records are usually required to establish a set of data that is as complete as possible.

2. Next, the seismologist calculates the strongest earthquake each source is likely to generate during the projected life span of the structure, based on records of yearly seismic activity at each source in the past. For example, he or she might conclude that there is a 50 percent chance that Fault B in the figure will generate an earthquake of magnitude 5 or higher during the next 30 years.

3. The seismic waves released during an earthquake lose energy, or attenuate, as they move away from the focus. Thus the next step taken by the seismologist is to estimate attenuation rates, or the decline of ground acceleration with distance, for each of the high-magnitude earthquakes calculated in step 2. One technique ties the attenuation rate to variations in the seismograms from scattered stations recording the same earthquake. Another approach uses historical accounts, such as modified Mercalli surveys or police and fire department reports, insurance records, newspaper accounts, and other documents. These methods are inherently uncertain, which makes step 3 the most inexact stage of the process.

4. The seismologist then integrates the data gathered in steps 1, 2, and 3 and determines the general probability that sometime in the next 30 years, this structure will be shaken by an earthquake whose magnitude exceeds the magnitude the building will be designed to withstand. If the probability is low—for example, 0.01 (one chance in a hundred)—the intended design is considered safe for that site.

Thinking About the Issues

This discussion of seismic hazards and risks raises some interesting issues concerning our reliance on experts. We—meaning the general public—hire seismologists, geologists, and engineers because "they know." But what is it that they know? Certainly not the timing of an earthquake, the location of its epicenter, or its strength. Nor do they know precisely the degree of ground shaking the earthquake will cause at the site, or precisely how the structure will respond to it. Despite their expertise, these things are unknowable given our present level of scientific understanding.

What these experts "know" is how to estimate risk and how to devise a rational plan for coping with the uncertainties of nature. Should the general public acknowledge this expertise to the point that we have no input into the decision-making process? What does cost-benefit analysis *really* mean? Who should decide that it is acceptable for a light manufacturing building or bowling alley to collapse in a certain magnitude earthquake—with the possibility that 50 or so people will die—because it is too expensive to design a safer structure? Who should decide that the location of a critical facility, such as a nuclear power plant or dam, is "safe" because the experts have determined that for any given year, the probability of a high-magnitude earthquake in the region is one in a thousand? Strip away the equations and the technical vocabulary, and you frequently find value judgments underneath.

Questions about the role of expert opinion do not apply exclusively to seismic hazards and risks, however. They also pertain to decisions concerning flood management or hurricane and tornado defense, and to such long-term problems as land conservation, toxic waste disposal, water pollution, acid rain, global warming, and other natural and human-induced environmental issues—many of which will be discussed later in this book.

THOUGHT QUESTIONS

1. If we cannot completely rely on experts, how can citizens make informed decisions?

2. How are important safety and planning decisions made in your community? How can the average citizen participate in the decision-making process?

3. How might the questions concerning our reliance on experts apply to personal issues, such as the relationship between a patient and a doctor?

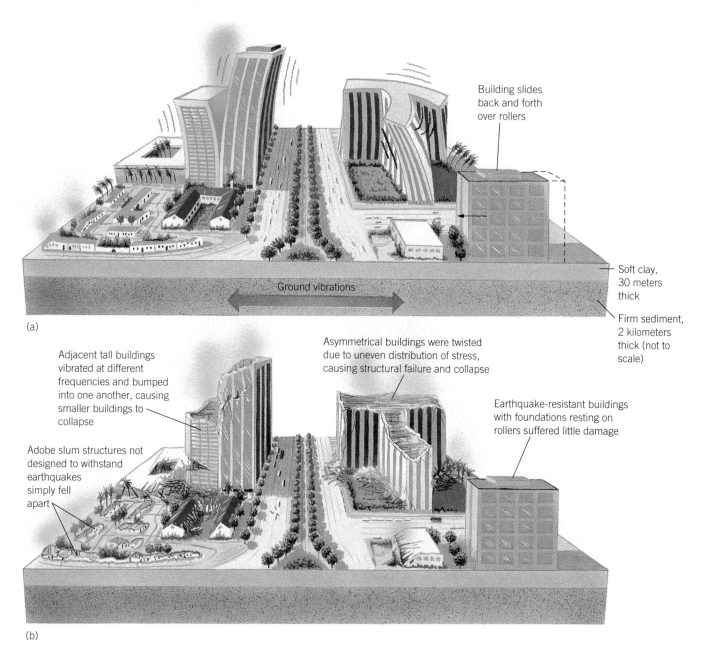

Building slides
back and forth
over rollers

Ground vibrations

Soft clay,
30 meters
thick

Firm sediment,
2 kilometers
thick (not to
scale)

(a)

Adjacent tall buildings
vibrated at different
frequencies and bumped
into one another, causing
smaller buildings to
collapse

Asymmetrical buildings were twisted
due to uneven distribution of stress,
causing structural failure and collapse

Earthquake-resistant buildings
with foundations resting on
rollers suffered little damage

Adobe slum structures not
designed to withstand
earthquakes
simply fell
apart

(b)

Figure 3.29 The effects of ground shaking on different types of structures during the magnitude 8.1 earthquake in Mexico City in 1985. (Adapted from William Stockton, "Lessons Emerge from Mexican Quake," *New York Times,* Nov. 5, 1985.)

large open spaces on the lower levels: corner buildings with high lobbies, hotels and office buildings with parking garages, and the like. Intense ground shaking caused the columns in these open areas to fail, because the upper floors of the buildings—packed with equipment, fixtures, and furniture—were less flexible and increased stress on the lower, open portions. Unfortunately, older buildings in California were not constructed with the benefit of this information. In the 1994 Northridge earthquake, the worst single disaster involved a three-story apartment complex; built over parking garages in the 1970s, it collapsed and killed 15 of its residents. Most of the freeway overpasses that collapsed were built according to pre-1971 standards and were scheduled for upgrading. Those that had been reinforced according to more recent standards did not collapse (Figure 3.30).

Buildings have different vibration periods: short buildings vibrate more rapidly than tall ones. In designing an earthquake-resistant building, engineers must minimize *resonance effects*—conditions that result when the building and ground vibrate at the same rate. Because of resonance, the building will sway with ever greater amplitude, a nightmarish effect. Without proper design precautions, a resonance condition can turn a low-magnitude earthquake into a major disaster.

Buildings between 5 and 20 stories are especially vulnerable. During the 1985 Mexico City earthquake, the 2-second periods of vibration of such buildings matched perfectly the natural vibration period of the lakebed sediment to which they were anchored. Consequently, they were subjected to ground accelerations equivalent to the acceleration of gravity (9.8 m/s^2)—a sideways thrust akin to being dropped from a vertical cliff. No ordinary structure can withstand this force; the buildings collapsed, and the floors pancaked one on top of the other.

No amount of planning can prevent catastrophe when a strong earthquake causes ground shaking beneath a densely populated city. Nevertheless, seismologists and engineers were unpleasantly surprised at the scope of the damage sustained during the magntitude 6.9 earthquake that struck Kobe, Japan, in January 1995, one day before the first anniversary of the Northridge disaster. The ground shaking, which lasted only 20 seconds, was especially intense because the shallow focal depth of the earthquake—only 10 kilometers—caused the energy released to be coupled more directly to surface structures such as buildings and elevated roadways. Although many were built according to the world's strictest design codes, over 50,000 structures collapsed, pitched over, or were damaged beyond repair. Over 5000 people were killed and 300,000 were left homeless. The total cost of the damage has been estimated at over $100 billion.

The lesson—and challenge—of Kobe is in how much we have yet to learn about earthquakes. Japanese scientists have been using signals from satellites and distant stars to measure minute shifts in the Earth's crust; they have drilled wells 2 kilometers deep and packed them with equipment to track movements at depth; and they routinely monitor oceanic faults in order to chart the grinding motion of the Philippine, Pacific, and Eurasian plates, the ultimate cause of the earthquakes that plague Japan. Yet the Kobe earthquake went unpredicted. Engineers also have a long way to go in planning for safety. Not only did many supposedly earthquake-resistant structures fail in Kobe, but in the months following the Northridge earthquake, it was discovered that the joints binding many earthquake-resistant buildings in Los Angeles may have sustained dangerous stress fractures.

Figure 3.30 Failure of freeway overpasses constructed prior to 1971 caused by the 1994 Northridge earthquake. Those overpasses built or retrofitted to more modern standards held up better.

STUDY OUTLINE

Earthquakes are vibrations triggered by the sudden slippage of rock along fault planes.

I. THE CAUSES OF EARTHQUAKES

A. The Reid-Gilbert **elastic-rebound theory** states that earthquakes result from the sudden release of strain energy that occurs when rocks rupture and lurch past one another. The motion is driven locally by **shear stress** generated, in turn, by regional shear, tensional, or compressional stresses.

B. Earthquake vibrations, or **seismic waves,** fan out in all directions from the earthquake **focus,** the point of initial rupture. The vertical projection of the focus onto the surface of the Earth is the earthquake **epicenter.**

II. SEISMIC WAVES

A. There are two basic types of seismic waves.

1. **Body waves** travel through the Earth's interior. The faster-traveling **primary (P) waves** alternately compress and expand the rocks they pass through. The slower **secondary (S) waves** cause the rock to vibrate at right angles to the wave path.
2. **Surface waves** travel mainly in the crust. They set the ground swaying horizontally while also causing it to rise and fall like ocean waves.

B. Seismic waves are detected by **seismographs.** A seismograph consists of a weight suspended by wires from a rigid framework embedded in bedrock. A sensor attached to the nearly stationary weight records vibrations of the framework on a rotating drum. The resulting wave pattern is a **seismogram.**
   1. Because seismic waves travel at different speeds, the time interval between the arrivals of P and S waves at a station is used to calculate the distance of that station from the epicenter.
   2. The compression and expansion pattern of the first P waves that arrive at different stations around an epicenter is used to determine the direction of fault movement.

C. The **Richter magnitude scale** uses wave amplitude as a measure of an earthquake's size. The scale is logarithmic; adjacent numbers differ by a factor of 10. The energy released by an earthquake increases 30 times for each **magnitude** number.

D. The modified **Mercalli intensity scale** measures the severity of an earthquake in terms of the level of destruction and panic it causes.

III. EARTHQUAKES AND PLATE TECTONICS

A. Plate boundaries are defined by seismic belts, the sites of 99 percent of all earthquakes.
   1. Shallow focal depths at the mid-ocean ridge and deep-focus earthquakes at the trenches are predicted by plate tectonics theory. First-motion studies confirm the fault movements predicted by the theory.
   2. As plates move by one another, the soft lower layers move smoothly, but the upper layers are subjected to the elastic-rebound effect. **Foreshocks** may occur as the rocks begin to crack. **Aftershocks** may follow as the rocks readjust.

B. The San Andreas fault is a transform boundary between the North American and Pacific plates. A bend in the San Andreas fault is transferring stress to the Los Angeles region. A previously unmapped thrust fault caused the 6.7 Northridge earthquake of 1994.

C. **Intraplate earthquakes** occur away from plate boundaries; nevertheless, they can be powerful and devastating to populated areas. Three theories relate their causes to plate tectonics: reactivation of ancient faults, stress transmitted from continental collisions, the doming of hot spots.

D. Other causes of earthquakes not related to plate tectonics include changes of groundwater pressure and glacial rebound.

IV. FORECASTING EARTHQUAKES.

Forecasting is used for long-range planning (such as building codes); prediction is used for immediate preparations (such as evacuation).

A. **Seismic gap** theory forecasts that great earthquakes will occur in the quiet segments of the seismic belts bordered by the epicenters of recent great earthquakes.

B. Paleoseismologists date past episodes of fault movement in order to determine the earthquake recurrence interval at a given locale along a fault.

C. Precursor studies focus on deviations from normal patterns in the vicinity of an active fault that warn of an imminent earthquake, such as changes in topography, groundwater levels, radon gas emissions, electrical conductivity, seismic wave velocities, and earthquake frequency.

D. **Tsunamis** are giant ocean waves generated by submarine earthquakes. More promising means of predicting them are based on their recently discovered relationship to certain long-period seismic waves.

V. **MINIMIZING EARTHQUAKE DAMAGE.** The extent of death and destruction in a region caused by an earthquake depends upon the concentration of the population and the location and types of structures built there.

A. Intensity surveys have shown the relationship between the type of subsurface and the levels of earthquake damage; for example, vibrations can cause **liquefaction:** the conversion of loose sediments from solid ground to mud.

B. Faulty design and defects in construction are the reasons most buildings collapse during an earthquake. Most dangerous are resonance effects—conditions that result when the building and ground vibrate at the same rate. If it is to withstand an earthquake, a building should be anchored to bedrock where possible and designed to flex and absorb shock.

STUDY TERMS

aftershock (p. 102)
body wave (p. 94)
earthquake (p. 89)
elastic-rebound theory (p. 91)
epicenter (p. 92)
focus (p. 92)
foreshock (p. 102)
intraplate earthquake (p. 105)
liquefaction (p. 112)
magnitude (p. 97)
Mercalli intensity scale (p. 100)

primary (P) wave (p. 94)
Richter magnitude scale (p. 97)
secondary (S) wave (p. 94)
seismic gap (p. 107)
seismic wave (p. 92)
seismogram (p. 95)
seismograph (p. 95)
shear stress (p. 91)
surface wave (p. 94)
tsunami (p. 110)

CRITICAL THINKING QUESTIONS

1. Using the elastic-rebound theory, describe what happens to the rocks at the focus just before, during, and just after an earthquake.
2. If you were a field geologist, what evidence would you search for to indicate recent fault movement?
3. What is the difference between earthquake magnitude and intensity?
4. Why do most earthquakes occur at plate boundaries?
5. Why are most earthquakes confined to the upper 20 or so kilometers of the crust?
6. How are the collision of India with Asia and earthquakes in China related?
7. What is the seismic risk of your present location? How safe or unsafe is the building in which you are currently living? Upon what factors would your answer depend?

CHAPTER 4

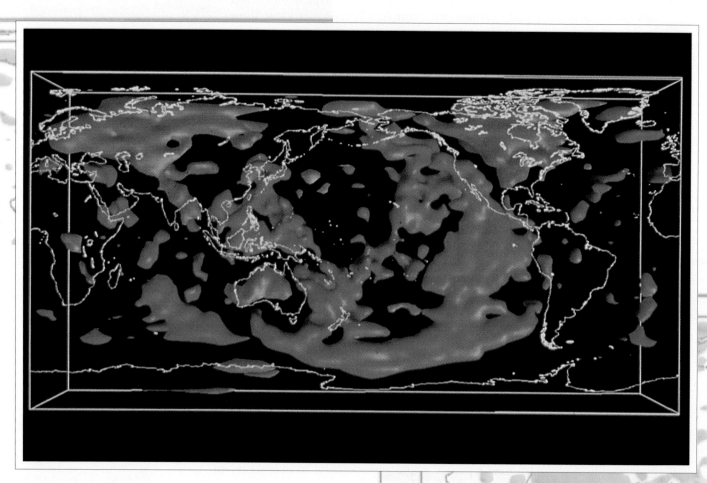

Interior of the Earth

Jules Verne notwithstanding, it is unlikely that anyone will ever journey to the center of the Earth. The deepest mine shafts are in South African diamond mines that penetrate about 3.6 kilometers below the surface. At this depth, temperatures are so high that elaborate cooling and ventilating systems must be employed to keep the miners alive. Direct sampling of the interior is also likely to remain limited. At present, modern techniques have enabled geologists to drill no deeper than about 15 kilometers. In comparison to the radius of the Earth, which is about 6370 kilometers, this amounts to less than a pinprick on the skin of an apple.

Despite these limitations, geologists have been able to gather a great deal of information about the interior from other sources. In this chapter, we will describe how geologists determine the composition, distribution, and condition of the materials that compose the Earth's mass and how they analyze the churning motions within the Earth's mantle that are responsible for the creation and destruction of the Earth's lithospheric plates. We will see that the Earth's internal dynamism can be traced to the origin of the Solar System and is a major reason that the Earth is a life-supporting planet.

Some Properties of Planet Earth

Our planet's density, moment of inertia, and magnetic field enable us to make important inferences regarding its interior without having to probe beneath the surface.

Consider first the Earth's average **density,** defined as the ratio of its mass to its volume. These two quantities were determined centuries ago: mass through experiments

density
The mass per unit volume of a substance.

◄ Seismic tomogram of the Earth's mantle from depths of 100 kilometers to 720 kilometers. This computer image, based on the analysis of seismic wave velocities, shows regions of rising hot rock (reddish color) and sinking cold rock (blue). The hotter regions of the mantle coincide with crustal regions of tectonic and volcanic activity (mid-ocean ridges and subduction zones), whereas the colder regions coincide with stable continental interiors far from plate boundaries.

involving Isaac Newton's law of universal gravitational attraction, and volume through knowledge of the Earth's dimensions and spherical shape. By the late eighteenth century, scientists knew that the Earth's density was 5.5 grams per cubic centimeter (or 5.5 times greater than the density of water). However, later measurements revealed that the granitic crust of the continents and the basaltic crust of the ocean basins have densities that are considerably lower than that of the Earth as a whole (2.7 and 3.0 grams per cubic centimeter, respectively). Apparently, then, the density of the Earth increases with depth, because somewhere beneath the surface there must exist matter denser than 5.5 grams per cubic centimeter to compensate for the light crustal rocks.

Another observation relevant here is that compression tends to pack most solids more tightly together, thereby increasing their densities. Could the rocks of the interior, therefore, be the same as the crust but denser because of the compression generated by the weight of overlying rocks? The answer is no, for compression alone would still leave the density of these rocks well short of the known density of the Earth. Consequently, we may infer that both the density and the composition of the Earth change with depth.

moment of inertia
A measure of distribution of mass within an object that determines the ease with which it rotates.

Next, consider the Earth's **moment of inertia** by imagining that you are a spinning ice-skater. Arms and legs extended, you spin slowly. But as you gather your arms and legs in close to your body, you spin faster and faster. Clearly, what changes is not your mass but the distribution of your mass with respect to your rotational axis—in other words, your moment of inertia. The rule is that the more concentrated the mass of a body is toward the center, the lower the body's moment of inertia is, and the easier it is to rotate.

Astronomers have determined the Earth's moment of inertia by observing the Earth's *precession rate*—that is, the rate at which its rotational axis traces a circular path when measured against the backdrop of the stars (Figure 4.1). Their observations indicate that the Earth's moment of inertia is 15 percent less than an identical sphere of uniform density. This result confirms the inferences made from the density of crustal rocks: that Earth is a planet whose mass is concentrated toward the center—that is, a planet whose density increases with depth.

magnetic field
The region surrounding a magnetized body that is subject to magnetic force.

Finally, consider what can be learned about the Earth's interior from its **magnetic field.** We can infer, for example, that the interior contains a great deal of iron, because iron is the only common material that is capable of generating a magnetic field as strong as the Earth's. The iron must be located in the center, because the magnetic field is symmetrical about the Earth; and the center of this symmetry—the place where the field is generated—is at the center of the Earth. This inference gives the added bonus of helping to account for the Earth's density distribution. The presence of the dense element iron in the center serves to balance the effect of lighter rocks near the surface.

Trying to examine the interior of the Earth using the approaches discussed so far is like trying to examine the organs of the human body without peering inside it—there is only so much we can learn by examining the outside. Just as we need an X ray or sonogram to see into a body, we need analogous means to probe the Earth's interior: seismic waves.

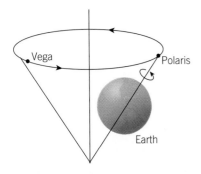

Figure 4.1 The Earth rotates about its axis while the axis traces a slow circular path with respect to the stars that serve as reference points. These motions prove that the Earth's moment of inertia is less than that of an equivalent sphere of uniform density. In other words, the Earth's density increases toward its center.

Seismic Waves in the Interior

We learned in Chapter 3 that an earthquake releases several types of seismic waves. Body (P and S) waves travel through the interior of the Earth, and surface waves travel near the circumference of the Earth, dying out with depth. We also learned that the velocities of these waves depend upon the elastic properties and densities of the materials through which they travel (Figure 4.2). Therefore, an abrupt change in the velocity of a body wave at a given depth signifies that the wave has intersected a boundary between rocks of different densities and/or elastic properties.

The velocities of *P* and *S* waves are dependent upon the properties of the materials through which they travel. They are calculated using the following formulas:

$$V_P = \sqrt{\frac{C + 4/3R}{\rho}}$$

$$V_S = \sqrt{\frac{R}{\rho}}$$

V_P = P wave velocity

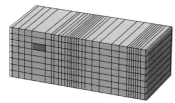

V_S = S wave velocity

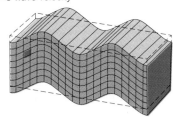

C is a measurement of the material's incompressibility, or resistance to volume reduction.

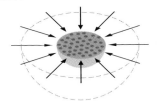

R is a measurement of the material's rigidity, or resistance to shear.

ρ is the density of the material.

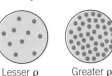

Lesser ρ Greater ρ

Figure 4.2 Important information about the properties of P and S waves. Notice that the velocity of an S wave depends upon the rigidity (*R*) of the substance through which it travels—that is, the resistance of the substance to shear. S waves can pass only through solids, because liquids offer no resistance to shear. The velocity of a P wave depends upon the incompressibility (*C*) of the substance as well as its rigidity. Thus a P wave is capable of passing through both liquids and solids.

Wave Refraction and Reflection

A phenomenon common to all waves—light, seismic, sound, and water—is that they will bend, or refract, upon entering a medium in which their velocities change (Figure 4.3). If a wave slows down upon entering the new medium, its path will bend away from the boundary surface; if it speeds up, its path will bend toward the surface (Figure 4.4). No **refraction** occurs when a wave enters a new medium at right angles to the boundary; although the velocity does change upon entering, there is no deviation in wave path.

A wave may also exhibit varying degrees of **reflection** upon striking a boundary. Figure 4.5 illustrates that reflection may be total or partial, depending upon the nature of the material at the boundary and the angle of the incoming wave. In partial reflection, a portion of the incoming wave energy is also refracted at the boundary, so that a single wave may give birth to two waves whose paths differ from the original.

refraction
The change of direction that occurs when a wave passes from one medium to another.

reflection
The return of a wave to its original medium upon striking the boundary of another medium.

Figure 4.3 The refraction of a light beam as it passes through a glass of water.

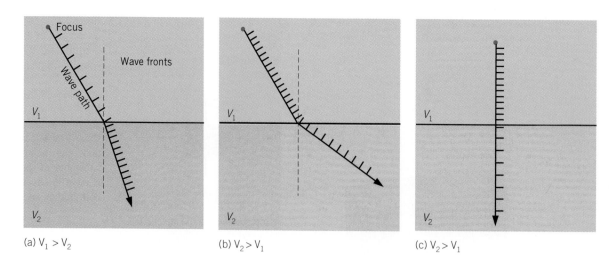

(a) V₁ > V₂   (b) V₂ > V₁   (c) V₂ > V₁

⌐⌐⌐⌐⌐ = Path traveled during equal time intervals (Higher
velocities indicated by more widely spaced ticks)

Figure 4.4 Refraction occurs when a wave changes velocity upon entering another medium. (a) A wave that slows down is bent away from the boundary surface. (b) A wave that speeds up is bent toward the surface. (c) No refraction occurs if the wave crosses into the new medium perpendicular to the boundary.

Of course, the Earth is not transparent—we cannot see the boundaries between layers, nor can we see the abrupt changes in wave velocity and path. But we can infer the existence of these features by examining seismograms from stations around the globe. The basic idea is that refraction and reflection alter the paths of the waves and their travel times. For this reason, the waves spreading from an earthquake focus may fail to arrive at certain stations and show up at others earlier or later than expected. By observing which stations were skipped, and by recording the precise arrival times of the waves at the other stations, researchers can determine the paths of waves, the depth of the boundaries, and the velocities of the waves as they travel through the Earth.

Detection of the Crust, Mantle, and Core

One way to follow the reasoning of seismologists in deciphering the Earth's internal structure and composition is to compare the actual patterns of seismic waves through Earth with a model of a hypothetical Earth whose composition is homogeneous (that is, uniform). Figure 4.6a shows that the waves spreading from an earthquake in a homogeneous Earth would travel in straight lines through the interior from focus to seismo-

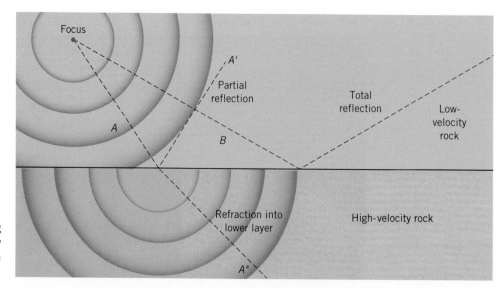

Figure 4.5 The energy of a wave striking the boundary of another medium may generate both reflection and refraction effects.

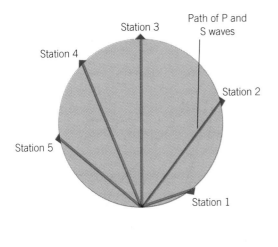

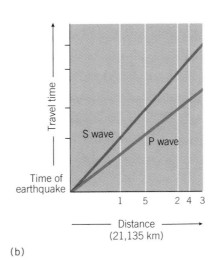

(a)   (b)

Figure 4.6 The homogeneous Earth model. (a) Seismic waves travel in straight paths from focus to recording station. (b) The greater the distance to the recording station, the deeper through the Earth the P and S waves must travel. The travel times to the various stations are plotted as straight lines on the time-distance graph.

graph stations; those stations farthest from the epicenter would record the deepest penetrating waves. Because the velocities of the waves would be constant, their travel times to the various stations would depend only on distance. Thus their velocities would plot as straight lines on a time-travel graph (Figure 4.6b).

Now look at the travel-time graph of seismic waves that travel through the *real* Earth (Figure 4.7a). The curving P- and S-wave lines indicate that these waves are arriving earlier than in our homogeneous Earth model; indeed, the farther the waves travel, the greater the gap is between them. Thus seismic waves in the real Earth speed up as they travel through the interior, indicating gradual changes in density and elastic properties with depth. The reconstruction of their paths, using data from many stations, is shown in Figure 4.7b. Notice the thick region between about 20 and 2900 kilometers in depth, where the waves curve smoothly and surface 105° from the epicenter. This enormous concentric layer of the Earth is called the **mantle.** It comprises about 80 percent of the Earth's volume and 67 percent of the Earth's mass. The mantle is bounded by the outer core below and the crust above.

Stations located between 105° and 142° of an earthquake epicenter receive no direct P and S waves. The blanked-out region is called the **shadow zone** (Figure 4.7b). Direct P waves are received beyond the shadow zone, but they are much delayed. Direct S waves, on the other hand, are simply not recorded beyond 105°. As the term *shadow zone* implies, these drastic changes are explainable only by the existence of a major obstacle at great depth that slows down and refracts P waves and completely blocks the S waves. This obstacle is the Earth's 2240-kilometer-thick **outer core.** The depth of the core is determined by two methods: (1) reconstructing the path of the deepest wave that travels undisturbed from focus to seismograph station (105°), or (2) recording reflections from the outer-core–mantle boundary.

The fact that P waves travel through the outer core whereas S waves do not leads geologists to believe that the outer core must be molten. Liquids lack the internal strength (or rigidity) to support the shearing motions by which S waves are propagated; that these waves are not detected beyond 105° of the epicenter means the interior is molten at outer-core depths. At the same time, a molten outer core explains why P-wave velocity decreases: as shown in the equations of Figure 4.2, one of the factors that adds to P-wave velocity—rigidity, or the resistance to shear—is lost upon entering a liquid. However, P waves that travel through the center of the Earth arrive at the opposite side earlier than would be expected if the core were completely molten. This observation, along with more complex reflection and refraction patterns, implies that the solid **inner core,** 2460 kilometers in diameter, rests like the hard pit in a soft peach at the very center of the Earth.

mantle
The layer of the Earth located between the base of the crust and outer core whose depth extends from about 20 to 2900 kilometers.

shadow zone
A region between 105° and 142° from the epicenter of an earthquake where no direct P and S waves reach the Earth's surface, owing to the properties of the outer core.

outer core
The upper layer of the Earth's core whose depth extends from about 2900 to 5140 kilometers. It is presumed to be molten.

inner core
The solid spherical center of the Earth about 2460 kilometers in diameter.

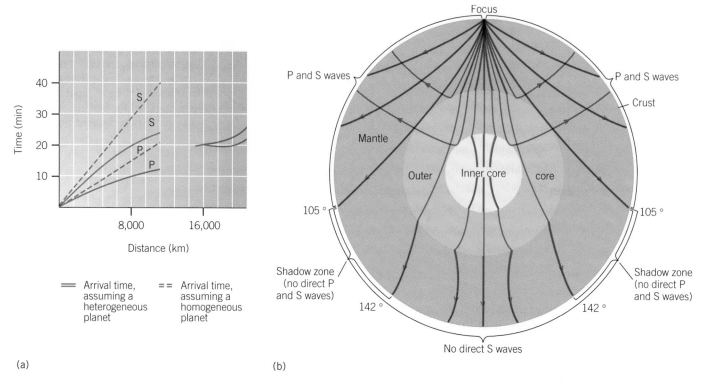

Figure 4.7 Seismic wave velocities in the real Earth differ from those expected in the homogeneous Earth model. (a) The arrival times of P and S waves for the real (heterogenous) Earth are curved, indicating that the waves speed up as they travel through the interior. The gap in the P-wave curve is caused by an obstruction, the Earth's outer core. (b) Each of the Earth's layers is characterized by distinctive seismic wave patterns. The region of smoothly curving ray paths (up to 105° from the focus) is the mantle. The presence of a shadow zone (from 105° to 142°) and P-wave reflections are evidence of a liquid outer core. P waves that arrive earlier than expected at stations located 180° from the focus are evidence of the existence of a solid inner core.

crust
The outermost layer of the Earth, about 2 to 50 kilometers thick, representing less than 0.1 percent of the Earth's total volume.

Smaller-scale travel-time curves give information that allows geologists to detect the boundary of the **crust** and mantle. In Figure 4.8(a), stations 1, 2, and 3 recorded P-wave arrivals at evenly spaced distances close to the epicenter of an earthquake. Because their arrival times plot as a straight line, we know that the waves reaching the three stations traveled with constant velocity. However, the first P waves reaching station 4 and 5 arrived *earlier* than expected, which is the reason for the abrupt change in the slope of the line in Figure 4.8a.

We can draw two conclusions from Figure 4.8(a). The first is that the waves arriving at stations 4 and 5 traveled a deeper route than those reaching stations 1, 2, and 3; the more distant the station, the deeper the P wave must penetrate the Earth in order to reach it. The second conclusion is that the P waves recorded at stations 4 and 5 traveled in higher-velocity rocks than those recorded at 1, 2, and 3; otherwise they would not have arrived earlier than expected.

Figure 4.8b is an interpretation of the arrival-time patterns in Figure 4.8a. Waves that arrive earliest at stations 1 to 3 have traveled the low-velocity (about 6 kilometers per second) route through the crust. Those arriving earliest at station 4 and beyond have been refracted into the **upper mantle,** where they gained speed (8 kilometers per second), and have outrun those taking the direct route through the crust. The boundary between the crust and the upper mantle is referred to as the **Mohorovičić discontinuity,** or **Moho,** in honor of its discoverer, the Croatian scientist Andrija Mohorovičić (1857–1936). The depth to the Moho is quite variable. It ranges from 2 to 10 kilometers beneath the ocean floor to an average depth of about 35 kilometers below the continents. It may reach depths of 70 kilometers under certain mountain ranges.

upper mantle
The outermost part of the mantle whose depth extends from about 20 to 670 kilometers.

Mohorovičić discontinuity (Moho)
The surface that defines the boundary between the Earth's crust and mantle.

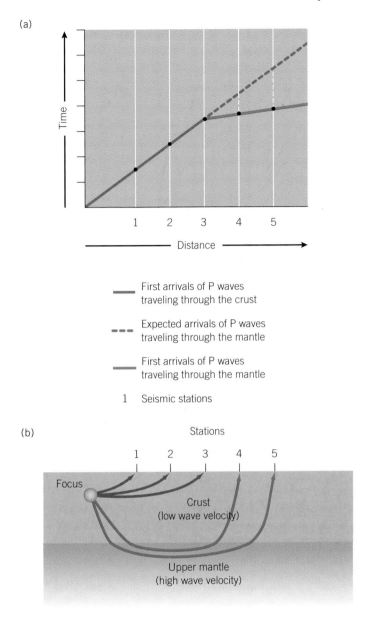

(a)

(b)

Figure 4.8 Detection of the crust-mantle boundary (Mohorovičić discontinuity). (a) The first P waves to arrive at stations 1 to 3 plot as a straight line, which indicates that they traveled at constant velocities through rocks of fairly uniform composition (the crust). The earlier arrival times at stations 4 and 5 indicate that those waves sped up along their paths in the mantle. (b) Reconstruction of the wave paths showing the crust-mantle boundary.

A Closer Look at the Earth's Structure and Composition

The depth-velocity curve of Figure 4.9 summarizes the Earth's structure and densities as we know them from the analysis of seismic waves.

The Lithosphere

Recall from Chapter 2 that the crust constitutes the upper part of a much thicker zone, the **lithosphere.** It is this zone that is divided into the rigid plates whose interactions are so closely connected with the occurrence of earthquakes, volcanoes, mountain building, and continental drift. These plates are able to slide across the globe because they rest on a relatively soft zone within the mantle—the asthenosphere. So it is fair to ask, What is the evidence of the existence of the lithosphere and the asthenosphere?

The main evidence rests on a general similarity of seismic wave velocities within each layer that contrasts markedly with the seismic wave velocities between them. We

lithosphere
A rigid zone of the Earth that includes the crust and a sliver of upper mantle and that rests directly on the asthenosphere.

Figure 4.9 The boundaries between zones are marked by abrupt changes in wave velocity and density. These changes reflect differences in the composition and/or physical properties of the rocks. Note that S waves do not travel in the outer core—strong evidence that the outer core is molten.

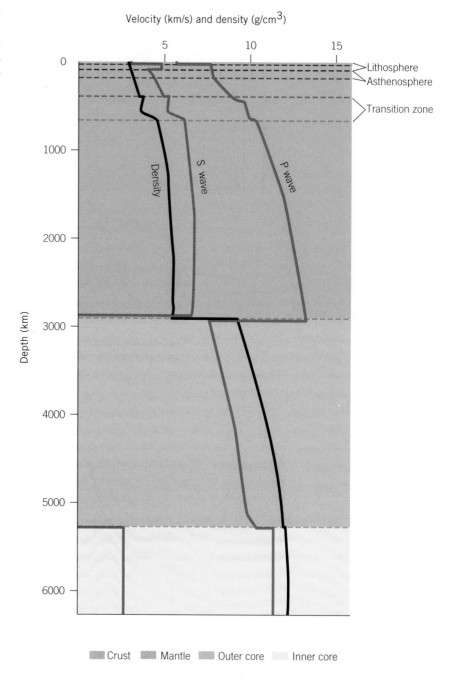

have just seen that there is a jump in velocities at the Moho, but the behavior of these waves nevertheless indicates that they pass through rigid rock from the Earth's surface to a depth of about 100 kilometers (Figure 4.10). We have, then, a relatively thin crust and a part of the upper mantle that display a more or less coherent set of physical properties—namely, rigidity and the ability to move together in large slablike units. Thus the lithosphere is largely defined by similarities in physical properties and contains subunits of different chemical properties: a granitic continental crust, a basaltic oceanic crust, and a thin slice of uppermost mantle that underlies both.

The Upper Mantle

We observed in Chapter 2 that the magma that erupts from beneath the central rift valleys of the mid-ocean ridges and builds the basaltic oceanic crust is derived from the upper mantle. This magma is *mafic* in composition—meaning that in addition to con-

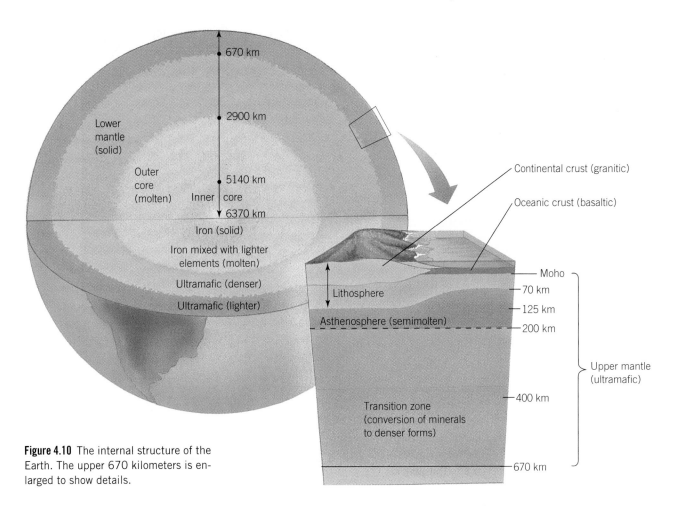

Figure 4.10 The internal structure of the Earth. The upper 670 kilometers is enlarged to show details.

taining lighter elements such as silicon, oxygen, and aluminum, it is fairly rich in the denser elements iron and magnesium. What might be the composition of the upper-mantle rock that gives rise to this magma?

One possibility is that the upper mantle is basalt that melts to yield mafic magma of the same composition. But when geologists experimentally subject basalt to upper-mantle pressures and pass seismic waves through it, they find that the wave speeds do not match the speeds observed in the upper mantle. However, if they conduct the same kinds of experiments on an *ultramafic* rock, one richer in iron and magnesium than basalt, the seismic waves in the sample match perfectly those observed in the upper mantle. Furthermore, geologists find that when they heat the ultramafic rock gradually, it melts in stages, yielding a mafic magma first before it converts completely to liquid. They therefore conclude that the upper mantle is composed of ultramafic rock that partially melts to form the mafic magma of the oceanic crust (see Figure 4.10).

This conclusion is further supported by study of the diamond-bearing rocks of South Africa. The diamonds are embedded within the ultramafic rock of the pipe-shaped intrusions called diatremes (Figure 4.11). The high pressures under which diamonds form suggest that the diatremes were brought up from great depths in the mantle—perhaps as deep as 240 kilometers.

The Asthenosphere

If the upper mantle is ultramafic, then what accounts for the decrease in seismic wave velocities between approximately 100 and 200 kilometers, the zone within the upper mantle known as the **asthenosphere** (see Figure 4.9)? Experimenters have shown that if they subject ultramafic rock to the pressures at that depth and raise its temperature to the point where a tiny fraction begins to melt, seismic wave velocities will match those

Note about mafic and ultramafic rocks: Mafic refers to rocks with a composition rich in iron and magnesium silicate. Ultramafic rock is so named because it is higher in iron and magnesium content than mafic rocks.

asthenosphere
A semimolten zone of the mantle just below the lithosphere whose depth extends from about 100 to 200 kilometers and is characterized by diminished seismic wave velocities and rocks that are close to melting.

Figure 4.11 The Big Hole diamond mine in Kimberly, South Africa, was formerly worked at a depth of 400 meters. The pit was excavated out of a pipe-shaped intrusion believed to have originated in the mantle. Analysis of the pressures and temperatures necessary to form diamonds suggests that they are formed about 240 kilometers beneath the surface.

present in the asthenosphere. The high temperatures lessen the rigidity of the rock, slowing the passage of P and S waves. The lithospheric plates rest directly on this relatively soft layer; without it, the plates would be welded to the lower mantle, and there would be no such thing as plate motion (see Figure 4.10). Located close to the surface beneath the mid-ocean ridges, the asthenosphere is also undoubtedly the source of the magma that builds the plates.

The Transition Zone and Lower Mantle

transition zone
The upper-mantle region, between 400 and 700 kilometers in depth, characterized by a series of steplike increases in seismic wave velocities and the conversion of minerals to denser forms.

Experiments show that the steplike velocity increases of the **transition zone** (between depths of 400 and 700 kilometers) are caused by the conversion of minerals in ultramafic rock into denser forms. In one important sequence, the internal structure of the common mineral olivine is transformed into a more tightly packed *perovskite* structure. In all, the various transformations increase the density of the mantle about 20 percent. The transformations also have the effect of increasing the rigidity and compressibility of the mantle, which accounts for the velocity changes. The gradual increase in wave velocities with depth beneath the transition zone suggests no sharp compositional changes in the **lower mantle** down to the outer-core boundary (see Figure 4.9).

lower mantle
The part of the mantle that extends from 670 to 2900 kilometers in depth.

The Outer and Inner Cores

P waves decline sharply and S waves die altogether upon encountering the molten outer core; within the solid inner core, the P waves speed up and the S waves begin again. The outer and inner cores are dense and therefore are probably composed mainly of dense metallic elements. The presence of the Earth's magnetic field is evidence that the metal is probably iron; however, shock wave experiments on possible core materials show that pure iron is a bit too dense for the outer core. A mixture of iron and a lighter element more closely fits the data. Some geologists believe this lighter element is sulfur, others opt for silicon, and still others choose carbon. The answer to the question is important. For example, carbon mixes with iron at a much higher temperature than sulfur does. If we can determine which element is present, we can estimate the temperature of the core far more precisely. This knowledge leads to a greater understanding of important processes such as mantle convection, which drives the lithospheric plates.

The Earth's Internal Heat

A major goal of geoscientists is to fully understand what drives the Earth's lithospheric plates with all of their related tectonic and volcanic activity. Another goal is to learn the secret of the Earth's magnetic field and the possible connections between the forces that generate the field and those driving the plates. To gain insight into these problems, researchers must be able to trace the flow of energy in the Earth—specifically, the transfer of heat from high- to low-temperature regions.

The Geothermal Gradient

The first step in tracing the flow of energy in the Earth is to determine how the temperature changes with depth—that is, the **geothermal gradient** (Figure 4.12).

Measurements taken in mine shafts and boreholes reveal that temperatures near the Earth's surface increase between 20 °C and 30 °C for every kilometer of depth. Geothermal gradients are generally higher in tectonically active regions, such as rift zones, than in currently inactive regions far from plate boundaries. In any event, these rates of temperature increase cannot be sustained for more than a few kilometers. If we take the average rate of increase to be 25 °C per kilometer, we would project a temperature of more than 150,000 °C at the Earth's center! An Earth that hot could not remain solid. Indeed, it would probably be a swirling mass of ionized gas. Thus we are sure that the geothermal gradient declines drastically in the upper mantle.

Even at lower rates of temperature increase, deep crustal and mantle rocks become blast-furnace hot within 10 to 20 kilometers deep; if subjected to the same temperatures at the surface, they would undoubtedly melt. What keeps them solid is the fact that they are under enormous pressure, and pressure raises the melting point of the rocks. Therefore, it is highly significant that the mantle begins to melt at depths between 100 and 200 kilometers. This zone is, of course, the asthenosphere, and experiments place its temperature at about 1150 °C.

The mineral transformations of the transition zone, between about 400 and 670 kilometers, mark the next great change in the mantle. Experiments show that for the reactions to occur at the depths indicated, temperatures must be in neighborhood of 1500 °C at the top of the transition zone and about 2000 °C at the bottom of the zone. Temperatures increase steadily toward the base of the mantle.

Estimates of the temperature at the outer-core–mantle boundary vary. Only in the past few years have techniques been developed to reliably duplicate pressure at this depth, and there is debate over the molten outer core's exact composition. We do know that temperatures must be low enough to keep the lower mantle solid and hot enough to keep the iron outer core molten. Recent experimental work places the temperature of the mantle–outer-core boundary at about 4500 °C and of the Earth's center at greater than 6600 °C. These temperatures are much higher than previously believed. If correct, the temperature at the center of the Earth is similar to the temperature at the surface of the sun. Whatever the generally agreed-upon estimate turns out to be, it is clear that the Earth's core generates more than enough heat to power the Earth's internal engine.

geothermal gradient
The rate of increase of the temperature in the Earth with depth. The gradient averages 25 °C per kilometer of depth in the crust.

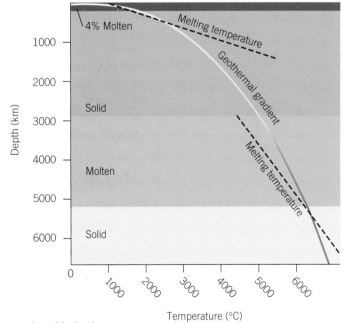

Figure 4.12 An estimate of how temperature varies with depth in the interior. Notice that the rate of increase is rapid in the crust, but declines in the upper mantle.

Asthenosphere Outer core
Mantle Inner core

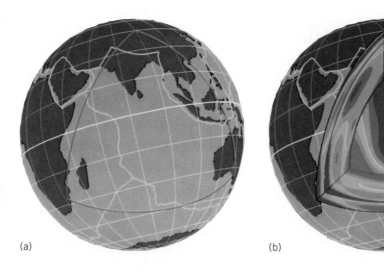

(a)   (b)

Figure 4.13 (a) The Earth. The yellow lines are plate boundaries in the region of the Indian Ocean. The surface location of the triangular cross section shown in (b) is marked by red lines. (b) A tomographic cross section of the Earth down to the outer core. The regions where seismic waves slow down, indicating high heat flow, are colored red; the cold regions, where seismic waves speed up, are blue. Note the correlation between hot and cold regions of the mantle and core. Also note that hot and cold regions of the mantle match well with tectonically active and inactive regions of the crust.

Seismic Tomography

Seismic tomography is to the study of the Earth's interior what a CAT scan is to the study of the interior of the human body. Both use a network of crisscrossing waves to construct a series of closely spaced cross sections or slices of the interior of the object. These slices are then stacked into a three-dimensional image. The major difference between them is that the CAT scan uses electromagnetic X rays, whereas seismic tomography uses seismic waves, which are similar in many respects to the ultrasonic waves used to image a fetus during a pregnancy.

seismic tomography
A computer-imaging technique based upon the analysis of the variations in velocities that occur along the path of seismic waves.

A travel-time curve such as the one we saw in Figure 4.7a merely records the *average* time that seismic waves take to reach a station located a given distance from the epicenter. However, a specific wave may arrive slightly early or late at that station, depending on the state of the materials encountered along its path. High-temperature materials, being soft, will slow the wave; low-temperature materials, being denser and more rigid, will speed the wave up. A single wave traveling through the mantle may not yield much useful information about velocity variations along its path. However, thousands of waves that originate from as many points form a dense crisscrossing network that outlines the relatively hotter and colder regions of the mantle.

Figure 4.13 shows a three-dimensional tomographic cross section of the Earth down to the outer core. First, notice that hot (red) and cold (blue) regions of the core match up with hot and cold regions of the lower mantle. Presumably, the lower-mantle rock, heated by the core, becomes less dense and slowly rises. Second, notice the high degree of correlation of crustal features with mantle heat-flow patterns. Subduction zones, rift zones, and other tectonically active regions are located over hotter mantle rocks, as we would expect. Tectonically quiet regions, such as continental and oceanic crust far from plate boundaries, sit over colder regions of the mantle.

Models of Mantle Convection

The pattern just noted strongly suggests that the mantle is heated from below by the core like a pot of thick tar sitting over a fire. So let us consider how heat circulates in such a pot of tar. The tar at the bottom of the pot absorbs heat, expands, becomes less dense than its surroundings, and rises buoyantly through the colder tar to the surface of the pot. There, it spreads, surrenders heat to the air, becomes dense, and sinks to the bottom of the pot, where it is reheated. This type of heat transfer, involving the movement of warmer materials to colder areas, is called **convection** (Figure 4.14).

convection
A circular heat transfer mechanism in which material heated from below rises because it is less dense than the cooler material above it. The warm material then surrenders heat, becomes denser, and sinks down to the bottom, where it is reheated.

The hot tar brought to the surface by convection will spread, cool rapidly, and harden into a thin slab. The slab is driven away from upwelling parts of the convection current by warmer, less viscous tar and toward the cooler downwelling parts of the current. If the slab is stiff and dense enough, it may sink beneath the surface, where it will be reheated. However, because convection is slow in the viscous tar pot, the surface of the tar will be covered almost entirely by slabs.

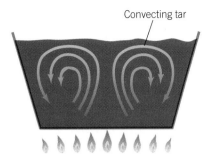

Heat is transferred from the massive churning pot of tar to the air through these thin slabs by means of conduction—that is, by atomic vibrations rather than flow, as in liquids. Conduction is efficient at first, while the slabs are thin, but grows less so as the slabs thicken. After a while, heat will be generated in the convecting tar faster than the slabs can conduct it to the surface. As the heat builds within the pot and temperatures rise, the tar becomes less viscous. Convection currents, now more active, break through the stiff slabs and release heat directly to the surface; they also bring fresh tar to the surface. The violence of the process recycles some of the older slabs; and after a time, a rough balance is reached between heat generation at the bottom of the pot and the heat released at the surface by conduction and convection. The balance will be maintained as long as there is no basic change in either the rate at which heat is supplied to the pot or the nature of the convecting-conducting tar.

One school of thought does indeed consider the mantle to act in a manner similar to the slowly churning tar pot described. In this view, which is called **whole-mantle convection,** hot and soft rocks at the base of the mantle—having received heat from the outer core—rise as **mantle plumes** and make their way to the surface, where they erupt to form hot spots and lava plateaus (Figure 4.15). Other plumes forge long cracks in the overlying crust (the mid-ocean ridges) and supply the magma that becomes the Earth's lithospheric plates. The plates cool as they spread and are subducted back into the mantle. Once in the mantle, being dense, they sink toward the core.

In support of whole-mantle convection, advocates point not only to the heat-flow patterns of the mantle such as we saw in Figure 4.13, but also to elaborate "mountains" and "valleys" at the outer-core–mantle boundary. They claim that seismic tomography shows that the mountains are located over hot regions of the outer core and argue that these thickened regions are absorbing heat from the core and will eventually rise.

Another group of geologists advocates a *two-layered model of mantle convection* (Figure 4.16). They point out that the Earth's internal structure is layered according to

Figure 4.14 Mantle convection has been compared to the motions that occur inside a pot of boiling tar. Heat supplied from below lowers the viscosity of the tar and causes it to rise slowly to the surface, where it cools and sinks to the bottom to be reheated. The "skin" that forms on the top is analogous to the Earth's lithosphere.

whole-mantle convection
The theory that the entire mantle circulates and mixes in the course of bringing heat from the outer core to the surface.

mantle plume
A pipe-shaped mass of heat-softened light rock that rises from the mantle toward the crust.

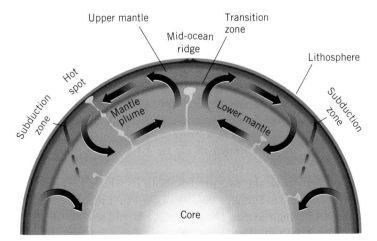

Figure 4.15 Mantle plumes play a key role in some models of whole-mantle convection. The plumes may rise to the surface as isolated hot spots or they may combine to initiate the crustal rifting that results in the formation of lithospheric plates.

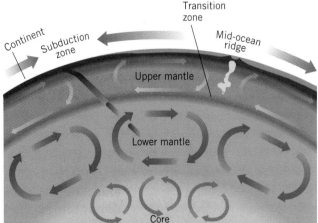

Figure 4.16 In a two-layered mantle convection model, heat is transferred from the core to the surface through a series of convection cells. However, no mixing occurs between the lower and upper mantle, as proposed in whole-mantle convection models. The outer core churns much more rapidly than the sluggish mantle.

ASIDE 4.1 *Mechanisms of Plate Motion*

While it is generally acknowledged that convection plays some sort of role in plate motion, we have yet to pin down the precise mechanism by which the plates move with respect to the soft, hot asthenosphere. The figure illustrates three proposed models of plate movement.

Ridge-push postulates that the plates are moved aside and toward the trenches by the injection of lava at the mid-ocean ridge. *Gravity sliding* emphasizes the difference in elevation between ridge crest and trench, and suggests that the plates simply slide downhill over a sloping asthenosphere, thus keeping a central rift valley open and allowing magma to rise. *Slab-pull* suggests that the hot plate is pulled from the ridge crest by the cold, dense part of the slab that descends beneath the trenches. It is rather like soaking the edge of a towel and letting it hang over the edge of a table; the towel will slide off the table because of the weight of the wet end.

Disguised in these models is the age-old chicken-and-egg problem. Does eruption of magma initiate plate motion, as in the ridge-push model, or does plate motion initiate eruption of magma, as in the other models? Each of these proposed mechanisms has strong points and drawbacks. Indeed, it may be that each contributes to plate motion.

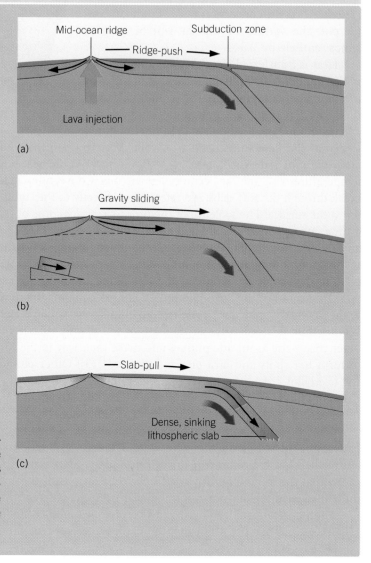

(a)

(b)

(c)

Hypothesized mechanisms of lithospheric plate motion. (a) Ridge-push: The wedge of hot rock beneath a mid-ocean ridge pushes the plates toward the subduction zone. (b) Gravity sliding: The plates slowly glide downhill under the influence of gravity, from the mid-ocean ridge toward the subduction zone. (c) Slab-pull: The dense portion of the plate that sinks into the subduction zone drags the rest of the plate with it.

density: light rock in the upper mantle, dense rock in the lower mantle. How, they ask, can rock rising from the lower mantle break through the rigid density barricade of the transition zone between 400 and 670 kilometers? They also point out that deep-focus earthquakes extend to the transition zone and no farther. Because these earthquakes originate in the lithospheric plates, they cite this as evidence that the plates can penetrate no deeper than the transition zone. Instead of sinking to the bottom, they probably smear out and blend with their surroundings like the chocolate in a huge marble cake. In the two-layered convection model, heat is transferred to the surface by means of the lower and upper convection currents shown in Figure 4.16, and the upper and lower mantle do not mix.

Attempts are being made to reconcile the two models, both of which are based on firm, if seemingly contradictory, evidence. Some geologists suggest that the difference in density between the upper and lower mantle may not prevent very cold and dense lithosphere plates from penetrating the boundary and sinking to the bottom. Recent advances in dynamic computer modeling of the interior suggest that descending plates do, indeed, stack up in the transition zone, but every few hundred million years or so, they break through the barrier to the outer-core boundary, where they are reheated and rise. However, computer models depend entirely on the quality of the information fed into them, and this information is constantly changing.

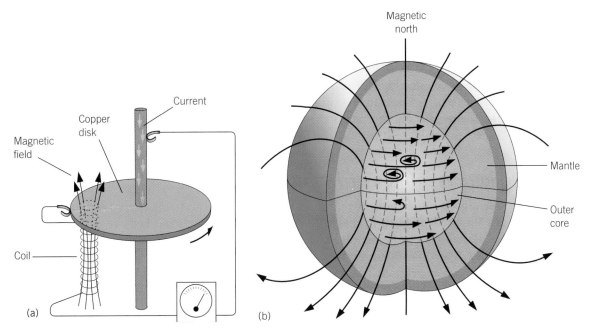

Figure 4.17 (a) The self-sustaining dynamo. A current in a coil produces a magnetic field; a rotating disk cuts the field and induces a current that is fed back into the coil, thus reinforcing the field. A constant input of energy is required to keep the disk rotating. (b) The Earth is probably a self-sustaining dynamo that works in a similar manner. The magnetic field is aligned parallel to the Earth's rotational axis and the dynamo is powered by the Earth's internal heat.

The Earth's Core and the Magnetic Field

Heat-driven outer-core motions probably play a key role in generating the Earth's magnetic field. Although the precise mechanism eludes them, most geophysicists believe that the outer core operates as a *self-sustaining dynamo* (Figure 4.17). Reasoning by analogy with motors and generators, geophysicists envision an electric current in the churning core that produces an intense magnetic field. This field, in turn, induces a current that feeds back into the core to reinforce the field. Meanwhile, the Earth's rotation keeps the magnetic field aligned in a roughly north-south direction. Each cycle of the "dynamo" dissipates a small amount of energy. If energy were not constantly supplied to churn the fluid core, the dynamo would eventually grind to a halt and the magnetic field would peter out. However, it is unlikely that the churning motions of the core are constant; they probably fluctuate in time and space because of friction or other effects. If they do, the polarity of the magnetic field will fluctuate and may even reverse.

A recent series of remarkable studies, based mainly on extremely detailed analysis of earthquake seismograms, suggests an outer core–inner core linkage in the generation of the magnetic field. One study found that seismic waves directed along the Earth's north-south axis travel 3 to 4 percent faster through the inner core than those that cross the inner core along the equatorial plane. This finding is intriguing because the inner core is composed of iron, and iron forms long, six-sided crystals that transmit waves most rapidly in the direction parallel to their long dimension. One possible conclusion is that the inner core consists of a single gigantic crystal of iron, the size of the Moon, whose long axis points north-south! Or perhaps the inner core is composed of billions of crystals all lined up in the same direction, similar to the grain in a block of wood. The crystal (or crystals) could have formed over billions of years by the slow solidification of the outer core, and the heat released in the process may serve as the source of energy for outer-core convection.

A second study focusing on the inner core found that its long axis is actually tilted about 10° to the Earth's rotational axis. This finding led to a third investigation, in which researchers used earthquake records going back many years to find that each day the axis of the inner core shifts ever so slightly eastward. The only logical explanation for this phenomenon is that the inner core spins slightly faster than the Earth as a whole.

The implications of these discoveries are enormous, as they provide our first real insight into the workings and structure of the inner core. Already, mathematical models have been developed that link the motion of the inner core to the generation of the magnetic field. These models propose that the inner core supplies the outer core with the energy necessary to power the self-sustaining dynamo, and that the rotation of the inner core stabilizes the magnetic field.

Sources of the Earth's Internal Heat

Identifying the sources of the internal energy that powers the Earth machine remains an unsolved problem in the geosciences. Until they *are* identified, we will never fully understand the way our planet—the most geologically active in the Solar System—works. Many possibilities have been suggested.

The heat released from the disintegration, or decay, of *radioactive isotopes* is an undoubted source of internal energy. These elements have been observed in the crust and in lavas that have erupted from the mantle. Upon extracting these elements and testing them, scientists found that they release heat as they decay to stable forms. Because their rates of decay are exceedingly slow and their concentration minor compared with that of other atoms present in rocks, their heat production is imperceptible over the course of a human life time. Rocks dissipate heat so sluggishly, though, that over millions of years, enough of it accumulates to fuel all the processes associated with plate tectonics.

However, a question remains about whether the radioisotopes are in the proper location to supply the heat for whole-mantle convection and the churning core motions that produce the magnetic field. Although they supply much internal heat, most radioisotopes combine readily with those elements concentrated in the granitic continental crust, and so they themselves are concentrated there. For this reason, the geothermal gradient is steep in the crust and levels off in the mantle. The concentration of radioisotopes apparently diminishes with depth, and they are scarce in the lower mantle. Because they do not combine with iron, they are probably not found in the core.

Other explanations of the Earth's internal heat are more conjectural. One suggests that the outer core is slowly solidifying—a process that releases heat. Another hypothesis relies on the recent discovery that the inner core rotates more rapidly than the rest of the Earth. In this model, the differential motion generates frictional effects that are dissipated as heat.

Internal Heat and the Origin of the Earth

An intriguing class of theories links the Earth's internal heat to the Earth's early history and suggests that the planet runs on leftover heat from that period. One well-established hypothesis is that the Earth formed by the accretion of particles or larger bodies from the same cloud of interstellar dust that produced the other bodies of the Solar System. Studies of meteorites indicate that some of these bodies were hunks of iron; others were rocks similar to the composition of the Earth's mantle. Presumably, these bodies accreted randomly to the primitive Earth so that it lacked a well-defined structure. There is also good reason to believe that the Earth formed as a relatively cold body, because its rocks contain compounds such as water, carbon dioxide, and ammonia that would have been driven off had the Earth formed at high temperatures.

However, if the cold accretion hypothesis is correct, then the Earth must have heated up later, because it is zoned according to density, having a heavy iron core, a less dense ultramafic mantle, and a light basaltic and granitic crust. This is precisely the structure we would get if the Earth had been molten and gradually cooled, and it is similar to the kind of separation that occurs in steel production. Thus there must have been heat sources that initiated melting of the original conglomeration of matter that made up

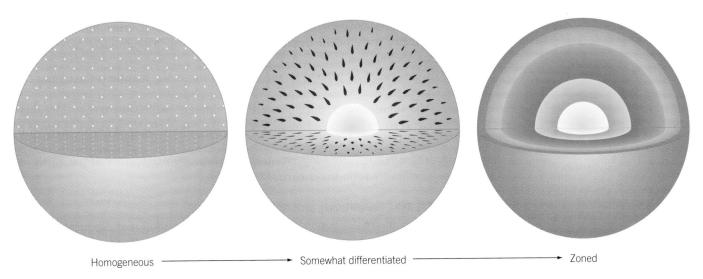

Homogeneous ⟶ Somewhat differentiated ⟶ Zoned

the primitive Earth. Possible sources include compression caused by the weight of over-lying materials, impact of meteors striking the Earth's surface, and, above all, the pres-ence of radioactive isotopes that decayed far more rapidly than those slowly decaying forms that remain today.

According to the cold accretion hypothesis, these heat sources increased the inter-nal temperature of the primitive Earth to the point where certain substances with low melting points—namely, iron mixed with sulfur—became molten and sank toward the center, where they formed the core (Figure 4.18). This sinking may have released suffi-cient heat to trigger melting of silicate compounds that later rose and wrapped around the core to form the mantle. Thus melting and zonation would have occurred billions of years ago, but enough heat remained to cause mantle convection and to drive the plates. Also expelled during this early period were the gases, water vapor, and carbon that later formed the Earth's atmosphere and oceans and eventually led to the appear-ance of living things.

Figure 4.18 The present-day zoned Earth probably evolved from an earlier, homogeneous state. Heat released from the decay of radioisotopes led to melting of iron sulfide trapped in the interior. The dense liquid sank to the center, forming the core and releasing huge quantities of heat. The silicate rocks surrounding the core melted and convected to form the mantle. The mantle is still convecting, although at a slower rate, and this convection drives the processes associated with volcan-ism and plate tectonics.

STUDY OUTLINE

The Earth's internal dynamism is responsible for the creation and destruction of the lithospheric plates. It is also a major reason that the Earth is a life-supporting planet.

I. SOME PROPERTIES OF PLANET EARTH The Earth's **density, moment of inertia,** and **mag-netic field** enable geophysicists to make important inferences regarding its interior.

A. The Earth's density increases and its composition changes with depth.

B. The Earth's center is composed mostly of iron, a dense mineral that is the only common material capable of generating the Earth's magnetic field.

II. SEISMIC WAVES IN THE INTERIOR An abrupt change in the velocity of a body wave at a given depth signifies that the wave has intersected a boundary between rocks of different densities and/or elastic properties.

A. As seismic waves intersect boundaries, **refraction** or **reflection** changes their arrival times at seismic stations. By comparing actual and expected arrival times, researchers can trace the paths of the waves.

B. If the Earth's interior were homogeneous, the paths of seismic waves would plot as straight lines. Their actual paths indicate, however, that the Earth is divided into structural and compositional zones. Changes in seismic wave velocities indi-cate that the Earth's interior is divided into the following boundaries and zones.

1. The boundary between the low-velocity rocks of the **crust** and the high-velocity rocks of the **upper mantle** is called the **Mohorovičić** discontinuity, or **Moho.**

2. Comprising 80 percent of the Earth's volume, the **mantle** is the largest layer.

3. The **shadow zone,** through which no direct P or S waves travel, is located between 105° and 142° of an earthquake epicenter. Its existence is proof of a molten **outer core.**

4. The early arrival of P waves that travel through the center of the Earth implies that the **inner core** is solid.

III. STRUCTURAL AND COMPOSITIONAL DIVISIONS OF THE INTERIOR

A. The **lithosphere** is a zone that contains subunits of different chemical properties—a granitic continental crust, a basaltic oceanic crust, and a thin slice of ultramafic upper mantle. The subunits of the lithosphere display a coherent set of physical properties—namely, rigidity and the ability to move together in large slablike units.

B. The rigid upper mantle is composed of ultramafic rock that partially melts to form the mafic magma of the oceanic crust. The **asthenosphere** is a zone of softer, warmer rock within the upper mantle over which the lithospheric plates slide.

C. The **transition zone,** an upper mantle region between 400 and 700 kilometers in depth, is characterized by steplike increases in seismic wave velocities and the conversion of minerals to the denser forms that compose the **lower mantle.**

D. The outer and inner cores are believed to be composed of iron plus some lighter elements yet to be determined.

IV. THE EARTH'S INTERNAL HEAT

A. The temperatures of the interior generally increase with depth, but the rate of increase, or **geothermal gradient,** differs in the various layers and zones. The gradient averages about 25 °C per kilometer in the crust and declines steeply in the mantle.

B. **Seismic tomography** is a research method using crisscrossing waves to determine variations of temperatures within the interior. There is tomographic evidence that tectonically active crustal regions are located over hotter mantle regions, and tectonically inactive crustal regions sit over colder mantle regions.

C. **Convection** is the means of heat transfer in materials that are heated from below; warmer materials rise, cool, sink, reheat, and rise again.
   1. In the **whole-mantle convection** model, the entire mantle convects by way of hot, rising mantle rocks, or **mantle plumes.**
   2. In the two-layered mantle convection model, the transition zone is a barrier between upper and lower mantle convection.

D. Three proposed mechanisms of plate motion are ridge-push, gravity sliding, and slab-pull.

E. Relatively hotter and colder regions of the lower mantle and outer core match up in seismic tomograms, suggesting that the churning, liquid outer core is heating the mantle. Heat-driven core motions may also generate the Earth's magnetic field. The intensity of mantle and core convection may control the Earth's magnetic-field reversals.

F. The source of the Earth's internal heat remains an unsolved problem in the geosciences. One proposed source is the decay of radioactive isotopes. Another hypothesis suggests the heat is left over from the formation of the Earth; another suggests that the Earth was cold when it formed but later heated up because of compression and radioactive decay. The energy was liberated by core formation added to the heat supply.

STUDY TERMS

asthenosphere (p. 129)
convection (p. 132)
crust (p. 126)
density (p. 121)
geothermal gradient (p. 131)
inner core (p. 125)
lithosphere (p. 127)
lower mantle (p. 130)
magnetic field (p. 122)
mantle (p. 125)
mantle plume (p. 133)

Mohorovičić discontinuity (Moho) (p. 126)
moment of inertia (p. 122)
outer core (p. 125)
reflection (p. 123)
refraction (p. 123)
seismic tomography (p. 132)
shadow zone (p. 125)
transition zone (p. 130)
upper mantle (p. 126)
whole-mantle convection (p. 133)

CRITICAL THINKING QUESTIONS

1. Spheres *A* and *B* have equal mass and volume. The mass of *A* is evenly distributed, whereas the mass of *B* increases toward the center. If both spheres are rolled down an inclined plank, which will reach the bottom first? Name the property responsible for the difference. How does this property add to our knowledge of the Earth's interior?
2. Sound cannot be transmitted by S waves for the same reason that S waves do not travel through the outer core. Explain.
3. How does the P- and S-wave pattern on the time-travel graph of Figure 4.7a prove that the Earth changes properties with depth?
4. Explain how the properties of the outer core cause the shadow zone.
5. Give two examples of abrupt changes of wave velocities in the mantle that do not necessarily indicate a compositional change. What accounts for the changes?
6. How may uncertainties regarding the exact composition of the outer core influence temperature estimates?
7. What evidence favors whole-mantle convection? Two-layered convection?
8. What accounts for the leveling off of the geothermal gradient beneath the crust?
9. Describe two possible sources of the Earth's internal heat.
10. Describe the possible relationship between the Earth's internal heat and the Earth's origin.

Minerals

The crust of the Earth is composed of igneous, metamorphic, and sedimentary rocks. Each rock is composed of smaller units distinguishable by their colors and shapes. Some are visible to the naked eye; others require a microscope. These constituents are called minerals. Thus a rock is simply an aggregate of one or more types of minerals. To determine the composition and structure of minerals is more difficult—requiring chemical analysis, X-ray, and other techniques—but it has been done (see Aside 5.1 on page 157). It turns out that minerals are composed of tiny particles called **atoms,** which are arranged in regular patterns. A **mineral** is defined as a naturally occurring, inorganic solid, having a definite chemical composition and orderly internal atomic arrangement.

All atoms are composed of the same basic subatomic building blocks: **protons, neutrons,** and **electrons.** The various elements that comprise minerals—such as carbon, oxygen, hydrogen, sodium, silicon, and uranium—are merely substances composed of atoms that differ from one another in the number and arrangement of these building blocks. We will see throughout this chapter that the differences between things often comes down to their organization.

Consider two well-known minerals, diamond and graphite (Figure 5.1). Clear, brilliant diamond, the hardest known mineral, can scratch virtually everything. Black, opaque graphite, one of the softest of minerals, can scratch virtually nothing; two of its uses are as a lubricant and as the "lead" in pencils. Minerals of more contrasting properties are difficult to find, and yet each is composed of the single element carbon. It is the arrangement of carbon atoms within diamond and graphite—their internal organization—that makes them different from one another.

atom
The smallest unit of an element that still retains the properties of the element; a dense, positively charged nucleus surrounded by negatively charged electrons.

mineral
A naturally occurring, inorganic solid with a definite chemical composition and orderly internal atomic arrangement.

proton
A positively charged particle in an atomic nucleus.

neutron
A particle in an atomic nucleus with a mass virtually equal to the proton's but with no electric charge.

electron
A tiny particle with a negative charge—equal to a proton's positive charge—that orbits the nucleus of the atom.

◀ This display illustrates the symmetry and color of some gem quality minerals. Included are tourmaline, aquamarine, morganite, heliodor, topaz, kunzite, spodumene, and citrine.

Figure 5.1 (a) Graphite and (b) diamond are composed of the same element, carbon; their contrasting properties are traceable to the differences in the arrangement of their carbon atoms.

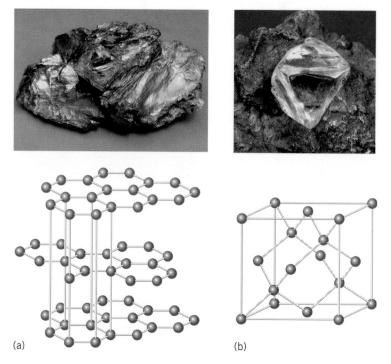

(a)   (b)

Organization, or how things fit together, is the central theme of this chapter. It is repeated, with variations, in our discussion of atoms and elements and of the composition, symmetry, internal structures, and various other properties of minerals.

One final point before turning to specific details: Albert Einstein was once asked if the laws of physics could ever perfectly describe reality. "Yes," he replied, "it is possible to completely describe Beethoven's Fifth Symphony in terms of equations for a series of air pressure pulses—but who would want to?" Similarly, X-ray analysis gives us an accurate depiction of the internal atomic structure of minerals; but when most geologists think of minerals like garnet, quartz, feldspar, or diamond, they do not think of X-ray photographs of symmetrically arranged dots. Reducing a mineral to its parts is not necessarily the same as understanding it or experiencing its beauty. A field trip with hammer, hand lens, and guidebook and a visit to the mineral collections of some of the great museums of the world are the best ways to gain such knowledge.

Atoms and Elements

element

A substance made up entirely of atoms of the same atomic number that cannot be decomposed into a simpler substance by ordinary chemical or physical means.

An **element** is a substance that cannot be further subdivided by ordinary chemical or physical means. There are 92 elements that occur naturally on the Earth, and 17 others have been manufactured in laboratories. Separately or in combination, elements form minerals, rocks, water, air, life—in short, the Earth as we know it. The discovery of the various elements and the identification of their physical and chemical properties are the result of centuries of labor by many scientists.

We know that elements are composed of atoms, but what holds atoms together? The answer lies in the concept of electric charge.

Electric Charge

Note: The electrical forces exerted by the charged rods are what is commonly called static electricity.

Vigorously rub a glass rod with silk (Figure 5.2). You will find that the glass acquires electrical properties; it is able to attract paper, for example. Now rub fur on a rubber rod. The rubber rod will also attract bits of paper and, in general, display the same electrical properties as the glass. But when the glass and rubber rods are brought close

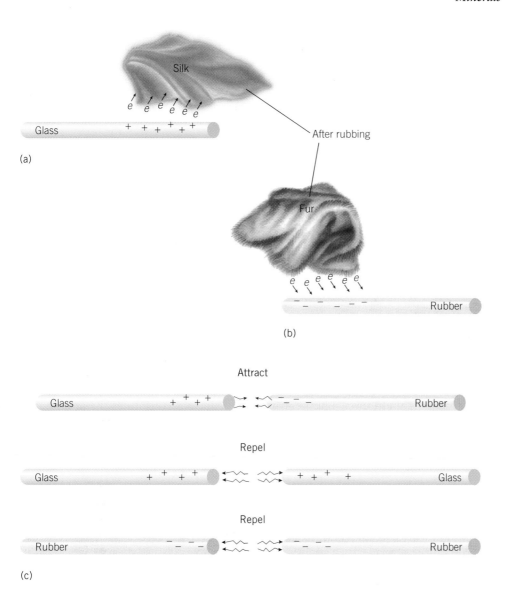

(a)

(b)

(c)

Figure 5.2 An electric charge demonstration. (a) When a glass rod is rubbed with silk, electrons are transferred to the silk, leaving the rod with a net positive charge. (b) When a rubber rod is rubbed with fur, electrons are transferred to the rod, giving it a net negative charge. (c) Opposite charges attract; like charges repel.

together, they will attract one another; whereas glass to glass, or rubber to rubber will repel. From these and similar experiments, we conclude that there are two kinds of electrical effects (or charges), and that like charges repel and unlike charges attract. About two centuries ago, Benjamin Franklin named the charge acquired by the glass *positive* and that by the rubber *negative*.

Today, we know that electric charge comes in equal units that are either positive or negative. Each proton carries a positive charge unit; each electron a negative charge unit. The equality between protons and electrons does not extend to their mass, however. Protons are 1800 times more massive than electrons.

Atomic Number and Mass

Most of the mass of an atom is confined to a dense knot of matter called the **nucleus.** Within the nucleus are the positively charged protons and neutrons, particles of nearly equal mass but of zero charge. Neutrons play the crucial role of overcoming the repulsive forces between protons, thus binding the nucleus into a tight structural unit. All atoms of a given element contain the same number of protons, called the **atomic number** of the element; in the most fundamental sense, this number serves to distinguish one element from another. For example, oxygen (atomic number 8) contains 8 protons; nitrogen (atomic number 7) and fluorine (atomic number 9) have 7 and 9 protons, respectively.

nucleus
The dense center of an atom composed of protons and neutrons. Nearly all of the mass of an atom is concentrated in the nucleus.

atomic number
The number of protons in the nucleus of an atom.

Figure 5.3 Three isotopes of carbon. Isotopes differ from one another in their atomic masses; each contains a different number of neutrons than other isotopes of the same element.

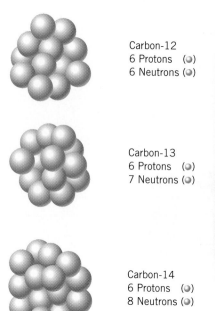

Carbon-12
6 Protons
6 Neutrons

Carbon-13
6 Protons
7 Neutrons

Carbon-14
6 Protons
8 Neutrons

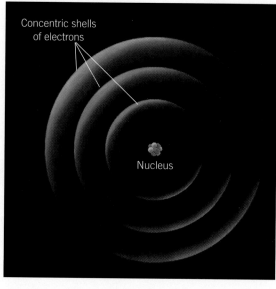

Concentric shells of electrons

Nucleus

Figure 5.4 In this simplified model of an atom, electrons are confined to concentric shells surrounding a dense nucleus of protons and neutrons.

atomic mass
The average mass of the atoms of an element; to a close approximation, the number of protons plus the number of neutrons in an atom of the element.

isotope
A variety of an element that differs from other varieties of the same element in the number of neutrons it contains and, thus, in its atomic mass.

radioactive decay
The spontaneous release of subatomic particles and energy from the nucleus of an atom.

The sum of the protons and neutrons in the nucleus constitutes the **atomic mass** of the atom. (Electron mass is so slight that it is generally ignored.) Most elements occur in a variety of atomic masses, for although the number of protons in the atoms of a given element is constant, the number of neutrons may vary. The varieties of a given element that differ in atomic mass are called the **isotopes** of that element. For example, the three isotopes of carbon (atomic number 6) have atomic masses of 12, 13, and 14 and are referred to as carbon-12, carbon-13, and carbon-14, respectively (Figure 5.3). (Appendix B displays the atomic masses of the elements in terms of the weighted contribution of each isotope's atomic mass.)

The isotopes of an element, although nearly identical in their ability to combine with other elements, may vary in geologically important respects. Certain isotopes are subject to **radioactive decay;** their nuclei are unstable, and they change into stable isotopes of other elements at fixed, measurable rates by emitting subatomic particles. For example, the *half-life* of uranium isotope U-238 is 4.5 billion years—that is, half of a quantity of uranium isotope U-238 decays to the lead isotope Pb-206 in 4.5 billion years. They are therefore extremely valuable "nuclear clocks" that allow us to date the age of rocks (which we will learn more about in Chapter 10, Geologic Time).

Atomic Structure

In the neutral state of the atom, the number of protons and electrons is equal; in other words, positive and negative charges are balanced. Electrons orbit at specific distances from the dense, positively charged nucleus; the farther away the orbits are from the nucleus, the greater the energy of the electrons within them (Figure 5.4). The paths of electrons are three-dimensional, so orbits are also called **electron shells.**

electron shell
A region surrounding the nucleus occupied by electrons having approximately the same energy.

Electrons fill the shells of elements of increasing atomic number in predictable sequence. In general, shells closest to the nucleus are filled first; then successively higher shells are filled (Figure 5.5). The closest shell holds a maximum of two electrons, and the outer shell can hold no more than eight. Helium (atomic number 2) has only two electrons, and they fill the innermost shell. The next largest element is lithium (atomic number 3), whose nucleus contains three protons. Three electrons are required to balance the charge; two occupy the first shell, and the third must occupy the next shell.

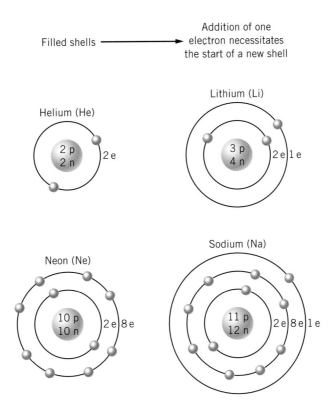

Figure 5.5 Electrons fill shells in an orderly sequence, with the innermost shell filling first. If additional electrons are needed to balance the charge after a shell is completed, these electrons occupy the next highest shell.

Because the outermost shell of any element can hold no more than 8 electrons, the electrons of neon (atomic number 10) must fill two shells. Neon, therefore, has a structure similar to helium in that its shells are completely filled. Sodium (atomic number 11), with 11 electrons, has two full shells plus a third shell where the eleventh electron orbits alone. In this respect, sodium is similar to lithium (atomic number 3), which also has a single electron in its outer shell. We will soon see that the number of electrons in an element's outer shell is a critical determinant of its chemical behavior.

The Periodic Table

The regularity of atomic structure serves as a basis for organizing the elements and displaying them on a chart called the **periodic table of the elements** (Appendix B). Figure 5.6 is a simplified version of the table, showing the structure of elements 1 through 20. The table is arranged so that all members of a vertical group (or family) contain the same number of electrons in their outer shell. For example, all the elements in Group I—lithium, sodium, potassium, and so on—have a single electron in their last orbit. The table is also arranged in horizontal rows, which specify the number of shells present. Each element in row 2, for example, has two electron shells surrounding its nucleus.

There is a striking correlation between the chemical and physical properties of the elements and their placement on the periodic table, as shown in Figure 5.6. Compare, for example, the chemical activity of neon (atomic number 10), fluorine (atomic number 9), and sodium (atomic number 11). "Nonactivity" is a good description of the activity of neon. It belongs to the family of *inert* elements, so called because they do not ordinarily participate in chemical reactions. Sodium and fluorine, on the other hand, are extremely active and participate in many reactions with other elements.

The major difference between the structure of neon and the structures of fluorine and sodium is that the outer shell of neon is filled. Fluorine is one electron short of being filled, and sodium is one electron in excess of it. Evidently, the chemically active elements are those with unfilled outer shells. When fluorine gains an electron or sodium loses one, the result in either case is an atom whose shells are filled. In fact, virtually

periodic table of the elements
A chart of the elements arranged in order of increasing atomic number and according to similarities of electron structure. As a consequence, the elements are divided into groups having similar chemical properties.

The reason for much of the complexity of the complete periodic table is that the electrons of heavier elements tend to occupy their outermost shell before the inner shells are completely filled. Thus more than one shell remains active. Consequently, heavy elements have complex chemistry. For example, iron will donate two electrons in some reactions and three in others.

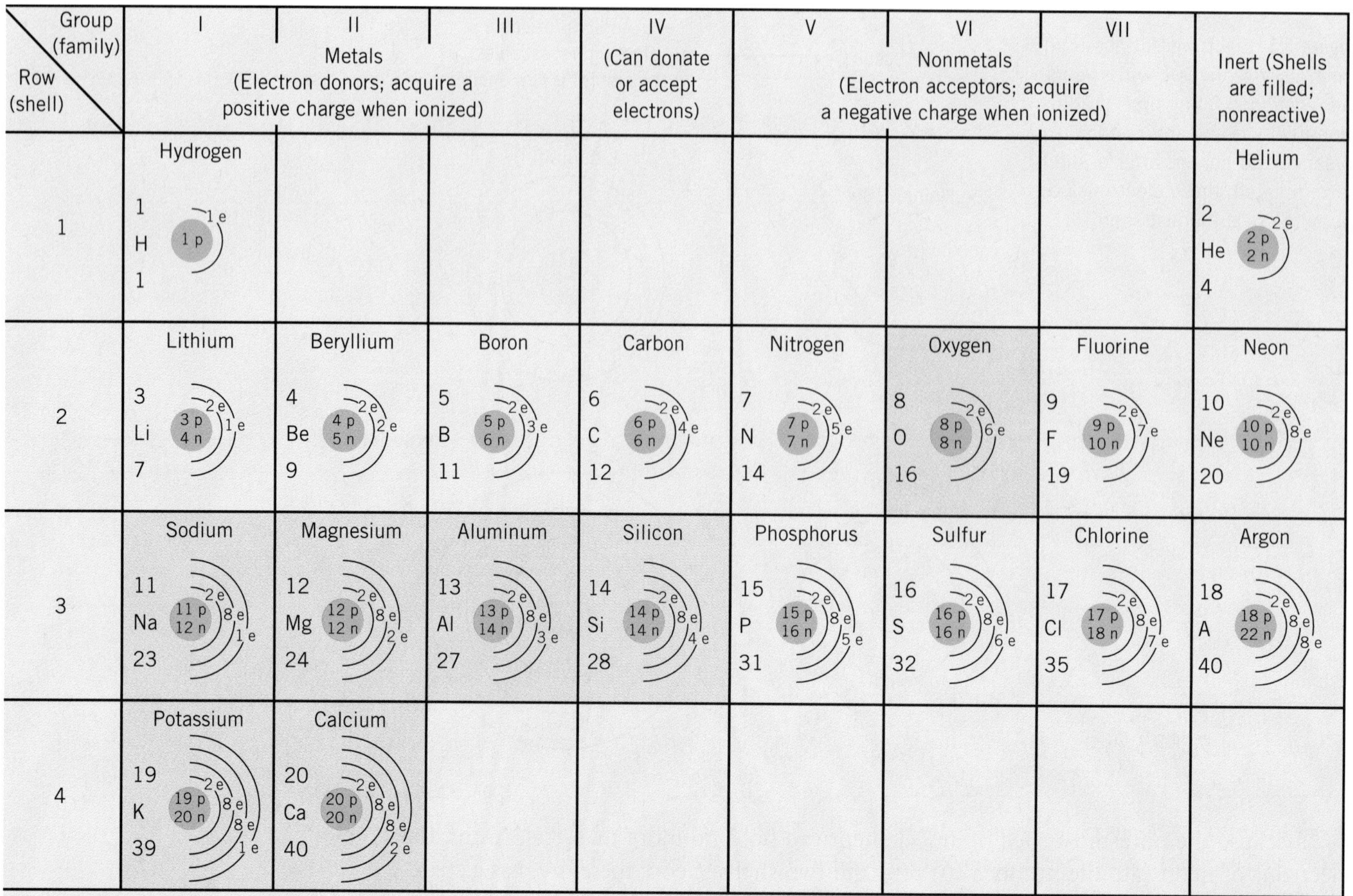

Group (family) / Row (shell)	I	II	III	IV (Can donate or accept electrons)	V	VI	VII	Inert (Shells are filled; nonreactive)
	Metals (Electron donors; acquire a positive charge when ionized)				Nonmetals (Electron acceptors; acquire a negative charge when ionized)			
1	Hydrogen 1 H 1 (1 p) 1 e							Helium 2 He 4 (2 p 2 n) 2 e
2	Lithium 3 Li 7 (3 p 4 n) 2 e 1 e	Beryllium 4 Be 9 (4 p 5 n) 2 e 2 e	Boron 5 B 11 (5 p 6 n) 2 e 3 e	Carbon 6 C 12 (6 p 6 n) 2 e 4 e	Nitrogen 7 N 14 (7 p 7 n) 2 e 5 e	Oxygen 8 O 16 (8 p 8 n) 2 e 6 e	Fluorine 9 F 19 (9 p 10 n) 2 e 7 e	Neon 10 Ne 20 (10 p 10 n) 2 e 8 e
3	Sodium 11 Na 23 (11 p 12 n) 2 e 8 e 1 e	Magnesium 12 Mg 24 (12 p 12 n) 2 e 8 e 2 e	Aluminum 13 Al 27 (13 p 14 n) 2 e 8 e 3 e	Silicon 14 Si 28 (14 p 14 n) 2 e 8 e 4 e	Phosphorus 15 P 31 (15 p 16 n) 2 e 8 e 5 e	Sulfur 16 S 32 (16 p 16 n) 2 e 8 e 6 e	Chlorine 17 Cl 35 (17 p 18 n) 2 e 8 e 7 e	Argon 18 A 40 (18 p 22 n) 2 e 8 e 8 e
4	Potassium 19 K 39 (19 p 20 n) 2 e 8 e 8 e 1 e	Calcium 20 Ca 40 (20 p 20 n) 2 e 8 e 8 e 2 e						

1. Members of groups (or families) contain the same number of electrons in last shell.
2. Each row contains the same number of shells.
3. The number above the element symbol is the element's atomic number; the number below is its atomic mass.
4. The eight most abundant elements of the crust (over 99% by weight) are oxygen, silicon, aluminum, iron, calcium, sodium, potassium, and magnesium. Iron is a metal and an electron donor. It is not on the simplified chart because it has a more complex structure. See complete table in Appendix B.
5. Electron donors (metals) are in Groups I–III; electron acceptors (nonmetals) are in Groups V–VII.
6. Carbon and silicon combine easily with metals and nonmetals; they can be thought of as positive in calculating formulas.

Figure 5.6 The arrangement of the first 20 elements in this chart illustrates the principles behind the complete periodic table of the elements in Appendix B.

every chemical reaction involves the loss, gain, or sharing of electrons in order to fill shells. Because all members of the same group on the periodic table have the same electron requirements for a filled outer shell, they will react similarly. For this reason, the chemical properties of members of the same group are closely related.

The tendency of atoms to fill their shells is not based on some mysterious property of atoms too difficult to understand. It is simply that the structures of atoms are more stable when the electron shells are filled; less total energy is required to hold electrons to the nucleus. As we will see, this tendency is the basis for the bonding of atoms.

Metals, Nonmetals, and the Periodic Table

The simplified periodic table (see Figure 5.6) helps us to understand reaction patterns. Elements in Groups I through III shed their extra electrons, since this is the most direct way for them to fill their outer shells. Elements having this ability are called *metals;* they donate electrons. Elements of Groups V through VII, especially those of low atomic number, fill their outer shells by accepting electrons; they are called *nonmetals.* Elements of Group IV occupy a midpoint between metallic and nonmetallic elements. Carbon

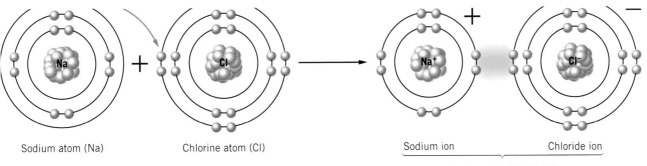

Sodium atom (Na)   Chlorine atom (Cl)   Sodium ion   Chloride ion

Sodium chloride (NaCl)

Figure 5.7 The ionic bond involves the transfer of electrons. The opposite charge of the resulting ions holds the ions together.

and silicon, for example, have four electrons to either donate or share in order to fill shells, which gives them great chemical flexibility. They combine in a variety of ways with a variety of elements. It is no coincidence that carbon is the key structural element of the atomic architecture of living things, and silicon, together with oxygen, is the key structural element of the Earth's crust. You may notice that helium, neon, and argon are classified as inert. The electron shells in this group are filled, making these elements inactive.

Bonding

When atoms combine, they form **chemical bonds.** The bonding may involve the same elements, for example, H_2, O_2, or N_2. Or it may involve several elements whose atoms combine to form **compounds.** For example, carbon dioxide, CO_2, is a compound consisting of a single carbon atom attached to two oxygen atoms. The bonding involves either transferring or sharing of electrons until all atoms have filled outer shells.

Ionic Bonds

Ionic bonds are accomplished through electron transfer. For example, in Figure 5.7, the single electron in the third shell of sodium has jumped over to the third shell of chlorine. This transfer leaves sodium with a net positive charge of 1 (written Na^+) and chlorine with a net charge of -1 (written Cl^-). The charged atoms are referred to as **ions.** Because opposite charges attract, sodium and chlorine ions bond, forming sodium chloride (NaCl), better known as table salt and less well known as the mineral halite.

In the process of ionization, atomic sizes change (Figure 5.8). Positive ions like sodium, having lost electrons, become smaller in radius than their neutral atoms. Negative ions like chloride, having gained electrons, generally increase in radius. The relative size of positive and negative ions influences the internal geometry of minerals, and this geometry in turn determines many mineral properties.

Covalent Bonds

Covalent bonds are formed by the equal sharing of electrons in the outer orbit of each element. In Figure 5.9, for example, the shells of four hydrogen atoms, each with one electron to share, combine with the outer shell of carbon, which has room for four electrons. Electrons orbit freely in all the outer shells. The result is the common gas methane (CH_4). Carbon dioxide (CO_2) and water (H_2O) are combinations of different elements held together by covalent bonds. Through covalent bonding, atoms of the same element can combine: oxygen (O_2), hydrogen (H_2), and nitrogen (N_2) are typical examples. These covalently bonded atoms tend to collect in the atmosphere as gases, but not all covalently bonded substances are gases. Diamond is an outstanding example of a covalently

chemical bond
The forces exerted between atoms that hold them together.

compound
A substance formed by the chemical combination of two or more elements in definite proportions and commonly having properties different from those of its constituent elements.

ionic bond
A chemical bond that holds two oppositely charged ions together through electron transfer.

ion
An atom with either a positive or a negative charge caused by the loss or gain of electrons.

covalent bond
A type of chemical bond in which electrons are shared by atoms.

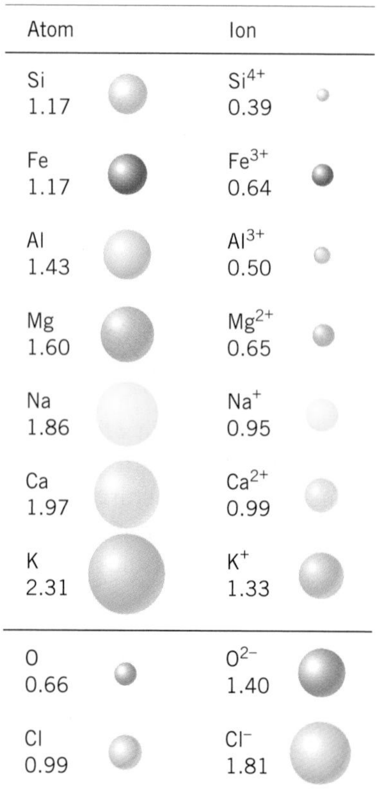

Figure 5.8 The radius of an atom changes when it is ionized. A positive ion has a smaller radius than the atom in a neutral state; a negative ion has a larger radius than the neutral atom.

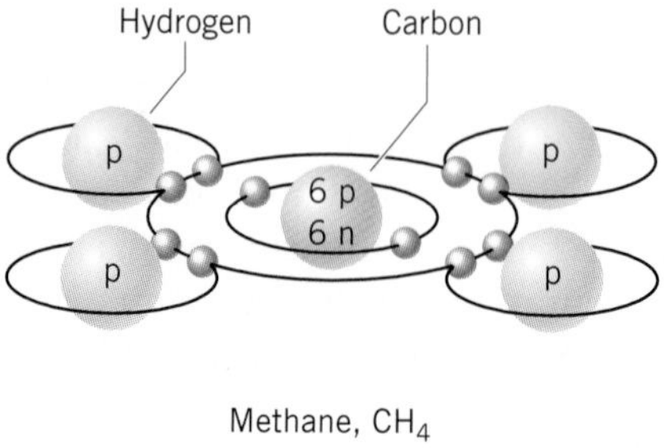

Atom	Ion
Si 1.17	Si^{4+} 0.39
Fe 1.17	Fe^{3+} 0.64
Al 1.43	Al^{3+} 0.50
Mg 1.60	Mg^{2+} 0.65
Na 1.86	Na^{+} 0.95
Ca 1.97	Ca^{2+} 0.99
K 2.31	K^{+} 1.33
O 0.66	O^{2-} 1.40
Cl 0.99	Cl^{-} 1.81

Figure 5.9 Methane is a typical covalently bonded molecule in which shells are filled by electron sharing rather than by electron transfer.

bonded mineral; its bonds are responsible for the perfectly symmetrical arrangement of its carbon atoms, and therefore, its great hardness and brilliance.

Metallic Bonds

metallic bond
A chemical bond created in electron-donating elements through the merging of electron shells.

Although metallic elements on the Earth's surface form ionic bonds with nonmetals, especially oxygen, they are also, on occasion, found in the pure or native state. In these cases, the metals form **metallic bonds.** In a manner analogous to the covalent bonding of nonmetals, the metallic elements combine by sharing electrons. However, metals have far more vacant spaces to fill in their shells than nonmetals have. With many electron vacancies to fill, and with each atom capable of sharing but a few electrons, many atoms are involved in the metallic bond. Atoms are thus packed so tightly together that their outer shells tend to merge. Outer electrons no longer belong to a particular shell but migrate freely from atom to atom within the mass. Positively charged nuclei of these metals exist in a "cloud" of loosely held electrons.

Metals in the native state have many familiar properties: they are good conductors of heat and electricity; they also tend to be dense, malleable, and opaque. All these properties stem from the metallic bond. The high electrical conductivity of metals, for instance, which is a manifestation of the flow of electrons, is the result of the greater freedom of electron movement in metallic bonds. The *malleability* of metals (their ability to be hammered into thin sheets) arises from the ability of closely packed atoms of the metallic bonds to roll over one another, like marbles. The metals' opaqueness stems from their tightly packed atoms, which concentrate many electrons in a small space; these electrons absorb incoming light energy.

Van der Waals Bonds

van der Waals bond
A weak bond between neighboring compounds or between atomic layers.

The atoms of a given compound may attract the atoms of a neighboring compound, forming **van der Waals bonds.** These bonds are not caused by electron transfer or sharing; rather, they are caused by electrons that cluster on the same side as they orbit the

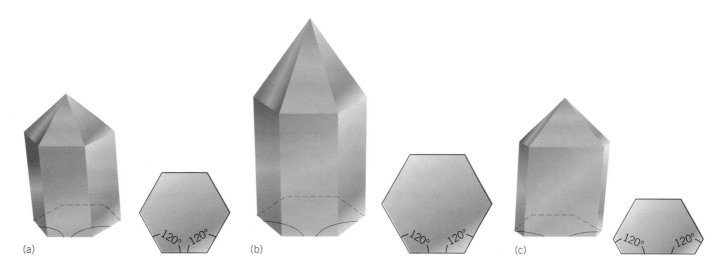

(a)   120° 120°   (b)   120° 120°   (c)   120° 120°

Figure 5.10 The angles between adjacent crystal faces of a given mineral (in this case, quartz) remain constant, despite superficial differences in the size and shape of individual specimens.

nucleus. One side of the atom is thus positively charged and the other side negatively charged for an instant. This polarized atom is able to attract neighboring atoms. But the electrons do not permanently cluster, and the billions of bonds throughout the substance randomly flick on and off. Therefore, the total energy of the bonds is weak. The weakness of van der Waals bonds is the reason many substances behave as they do— from the peeling of mica and the smudging of graphite to the melting of ice.

The Physical Properties of Minerals

Most minerals can be identified by inspection of their physical properties, which include crystal form, cleavage, fracture, hardness, color, streak, luster, and specific gravity.

Crystal Form

Long before the development of modern techniques for analyzing the atomic structure of minerals, scientists knew that if minerals were allowed to grow without interference (that is, in an open space), they would develop into **crystals,** objects whose regularity of form, or symmetry, is a manifestation of an orderly arrangement of atoms. The term *crystal* comes from the ancient Greek *kristolis,* meaning "clear ice." The Greeks were referring to transparent six-sided quartz crystals, which lined rock cavities and veins. The quartz reminded them of ice prisms hanging from branches, rock ledges, and cave entrances. Let us look at some of the properties of crystals.

Crystals are bounded by flat, planed surfaces called **crystal faces,** which are made of closely spaced atoms. If we take a number of quartz crystals of differing form—some short and stubby, others long and needlelike, still others displaying prominent pyramids—and measure the angle between neighboring faces in each crystal, we discover an interesting relationship: the angle between adjacent faces in all the specimens is the same, about 120 degrees (Figure 5.10). The *constancy of interfacial angles* is true for a given mineral regardless of the different appearance of particular specimens.

Most minerals crystallize when ions migrate through a fluid and bond with one another in orderly atomic arrangements. Under ideal conditions, a mineral will grow into one of the six perfect crystal systems described in Figure 5.11. However, the outward symmetry of crystals is usually destroyed by interaction with competing crystals during growth, by circulating fluids after formation, or by mechanical damage from tectonic forces. It is also possible for two or more minerals having identical compositions to belong to different crystal systems. Such minerals, called **polymorphs,** are discussed in Aside 5.2 on page 161. Even in these cases, though, the internal atomic order of the crystal remains intact.

crystal
A solid element or compound whose atoms display a definite, orderly atomic arrangement repeated throughout the solid.

crystal face
The smooth surface of an unbroken crystal.

polymorph
A mineral having different crystal structure from minerals of the same composition.

Figure 5.11 There are six crystal systems. Any given mineral displays a symmetry that belongs to one of the six systems.

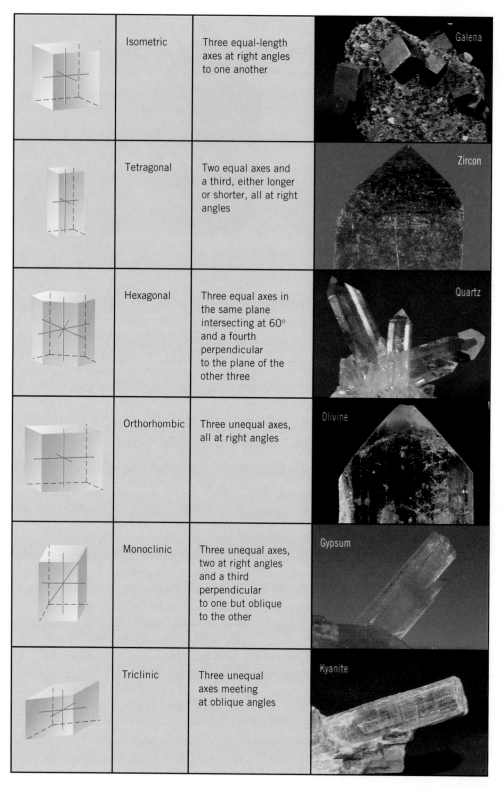

	Isometric	Three equal-length axes at right angles to one another	Galena
	Tetragonal	Two equal axes and a third, either longer or shorter, all at right angles	Zircon
	Hexagonal	Three equal axes in the same plane intersecting at 60° and a fourth perpendicular to the plane of the other three	Quartz
	Orthorhombic	Three unequal axes, all at right angles	Olivine
	Monoclinic	Three unequal axes, two at right angles and a third perpendicular to one but oblique to the other	Gypsum
	Triclinic	Three unequal axes meeting at oblique angles	Kyanite

Cleavage

cleavage

The tendency of a mineral to break along parallel planes of weak bonding.

Cleavage is the tendency of some minerals to break along definite parallel planes; it develops in the directions of the weakest bonds. Strong bonds occur in planes that are parallel to the crystal face, where atoms are closely spaced; the bonds that attach the planes together are weak. Thus minerals displaying cleavage break either in planes parallel to a crystal face or in planes that bear an angular relationship to it. Because cleav-

One plane	Basal (micas, chlorite, talc)
Two planes at right angles	Prismatic (feldspars)
Two planes at oblique angles	Prismatic (amphibole)
Three planes at oblique angles	Rhombohedral (calcite, siderite)
Three planes at right angles	Cubic (galena, halite, anhydrite)
Four planes	Octahedral (diamond, fluorite)

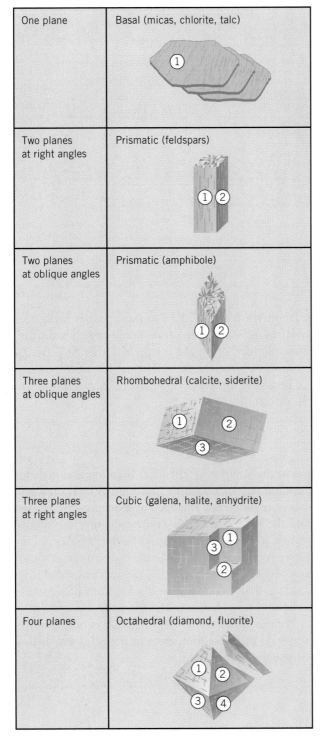

Figure 5.12 Common types of mineral cleavage. All cleavage planes either bear a constant angular relationship to a crystal face or are parallel to one.

age directions are controlled by the basic atomic arrangement and internal symmetry of the mineral, a given mineral will always display the same cleavage.

Figure 5.12 shows cleavage, which is often the key trait leading to the identification of some common minerals. Mica has perfect cleavage in one direction, which allows it to be peeled in thin onionlike layers. Feldspar cleaves in two planes that intersect at right angles. Galena, like halite (table salt), shatters into tiny cubes having three perpendicular cleavage planes. Rhombus-shaped calcite has three cleavage planes that do not meet at right angles.

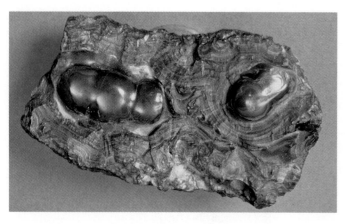

Figure 5.13 Typical conchoidal fracture in malachite.

Table 5.1　Mohs Hardness Scale

	1. Talc
	2. Gypsum
Fingernail ⟶	3. Calcite
Copper penny ⟶	4. Fluorite
Glass plate, knife blade ⟶	5. Apatite
	6. Feldspar
	7. Quartz
	8. Beryl (topaz)
	9. Corundum
	10. Diamond

Fracture

fracture
The manner in which minerals break other than along planes of cleavage.

As usually applied, the term **fracture** refers to the irregular broken surfaces of minerals that do not display cleavage when they break. Some minerals display smooth fractures; some exhibit rough, splintery, or jagged fractures. Other minerals, such as quartz and malachite, exhibit *conchoidal fracture,* the tendency to break along smoothly curved surfaces (Figure 5.13). Fractures usually develop where average bond strengths are approximately equal in all directions.

Hardness

hardness
The resistance of the surface of a mineral to scratching.

Hardness is related to the strength of chemical bonding and is defined as resistance to scratching. The common way of measuring hardness is by means of the Mohs hardness scale, which consists of ten minerals arranged in sequence from softest to hardest (Table 5.1). Any mineral on the scale will scratch a mineral with a lower number; minerals with the same hardness will scratch one another. Thus the hardness of any mineral may be determined by testing it against one of the minerals on the scale. A convenient reference is the hardness of a fingernail, about 2.5, which means that it will scratch gypsum and talc but not calcite. A copper penny, with a hardness of 3.5, will scratch calcite but not fluorite; a good steel knife blade has a hardness of 5.5 and will scratch apatite but not feldspar. Also, keep in mind that the Mohs scale is only relative, not quantitative. Diamond, for example, is probably thousands of times harder than, say, gypsum and is far harder than its neighbor, corundum.

Color

Although color is an obvious physical trait, it is not necessarily a reliable one. Often, many minerals display the same color; conversely the same mineral may occur in a wide variety of colors (Figure 5.14). For example, quartz (SiO_2) is usually clear, but you can also find varieties that are white (milky), black (smoky), green, pink, or purple (amethyst). However, calcite ($CaCO_3$) is also found in many of these colors. Color variety is usually caused by trace elements that amount to no more than a minute percentage of the atoms present, just as a drop of dye can color a pot of water. Because weathering almost invariably alters color, the true color of a specimen can best be observed on a freshly broken surface.

Some minerals almost invariably display the same color, determined by the presence of distinctive ions in their crystal structure. For example, malachite [$Cu_2CO_3(OH)_2$] is always rich green due to the presence of copper (see Figure 5.13). It

Figure 5.14 A collection of quartz specimens. Quartz, like many minerals, comes in a wide variety of colors.

Figure 5.15 Azurite is one of the few minerals that come in only one characteristic color.

is closely associated with azurite [$Cu_3(CO_3)_2(OH)_2$], which owes its distinctive blue color to the oxidation of copper ions (Figure 5.15). Still, color alone is not usually sufficient grounds for sure identification.

Streak

If you grind a mineral against a hard surface, it will powder; **streak** is the color of the powder. It is often a more consistent property than color, for powdering eliminates differences caused by the effects of minor impurities. The streak is usually observed by rubbing the mineral on a unglazed porcelain *streak plate*. A mineral harder than porcelain (between 6 and 7) will not streak. In some cases, a mineral's streak may not agree with its color in a hand specimen. For example, brass yellow pyrite (fool's gold) streaks black.

streak
The color of the powder of a mineral when scratched on a porcelain plate.

Luster

Luster is determined by the way the surface of a mineral reflects light. It is a subtle property dependent upon the ratio of reflected to absorbed light, the depth of light penetration, surface impurities and irregularities that scatter light, and other traits. Yet luster is often a more reliable identifying property of a mineral than color. Probably the most useful determination to make is whether the mineral has metallic or nonmetallic luster (Figure 5.16). Metallic luster is often displayed by metallic and metal sulfide minerals, such as pyrite (FeS_2) and galena (PbS), which gleam like stainless steel. Types of nonmetallic luster include pearly (some varieties of talc or opal), vitreous (glassy), resinous (like resin or pitch), silky, dull, or earthy (clay). Some minerals—for example, hematite (Fe_2O_3)—occur in both metallic and nonmetallic guises.

luster
The quality and intensity of light reflected from the surface of a mineral.

Specific Gravity

The **specific gravity** of a mineral is defined as the ratio of its weight to the weight of an equal volume of water. Because the density of water is one gram per cubic centimeter, specific gravity and density are numerically equivalent terms. For example, a mineral that weighs two-and-a-half times as much as an equal volume of water has a specific gravity (or density) of 2.5.

specific gravity
The density of a substance compared with the density of water.

(a)   (b)

Figure 5.16 Luster is the appearance of a mineral in reflected light. (a) Pyrite displays metallic luster. (b) Limonite displays a nonmetallic (earthy) luster.

There are devices for measuring specific gravity in the laboratory, but with a little practice, you can get a rough estimate of the specific gravity of a mineral by feeling its heft in the palm of your hand. Many minerals that appear superficially similar with respect to color, luster, and cleavage may be distinguished by this simple test.

As you might expect, the specific gravity of a mineral is controlled by the atomic mass of the elements that compose it and by the degree of atomic packing—that is, how closely spaced the atoms are. The specific gravity of halite, which is composed of light, widely spaced atoms, is 2.1. By contrast, galena, a densely packed lead sulfide, has a specific gravity of 7.6. In general, native metals and metal sulfides tend to have the highest specific gravities.

Other Properties

Minerals frequently exhibit a host of subtle properties that the geologist, like a good detective, learns to observe and interpret. The ability of a substance to transmit light is one of them. Minerals may be described as *transparent* if you can see through them; as *translucent* if they transmit light, but not well enough to permit seeing through them; and *opaque* if they do not transmit light at all. Clear calcite exhibits *double refraction*, which means that light, upon entering the crystal, divides and vibrates along separate planes; the result is two images (Figure 5.17). Try this experiment. Draw an *X* on a piece of paper. Place clear calcite over it. You will see two *X*'s. Rotate the calcite and one image will rotate with it. There are other physical properties of minerals as well. Halite, being salt, tastes salty. Magnetite, being magnetic, attracts iron. Talc has a smooth, soapy feel, as talcum powder does.

A mineral identification chart can be found in most geology laboratory manuals. Because no two minerals have exactly the same properties, you can use such a chart with a process of elimination to distinguish one mineral from another on the basis of luster, hardness, cleavage or fracture, streak, and so on. For example, a mineral having nonmetallic luster, a hardness greater than glass, no cleavage but conchoidal fracture, and colorless streak is probably quartz. Pyrite can generally be identified by its metallic luster, its brassy yellow color, cubic crystal form, and black streak. With practice, you should be able to identify good specimens of the common minerals, but keep in mind that such determinations may be ambiguous.

The microscope is as indispensable to mineralogy as it is to biology. High magnification shows features of minerals not otherwise discernible: minute cleavage planes, inclusions of

Figure 5.17 Polarization through double refraction in calcite. Light entering the crystal is bent (or refracted) and made to vibrate on separate planes; thus we receive two images of the crossing lines.

tiny crystals within larger minerals, and other clues to the origin and growth of the minerals, even of their chemical composition. The microscope thus leads to more precise identifications and, of course, wider knowledge of mineralogy in general. The optical mineralogist generally uses a polarizing microscope. Minerals or rock specimens are sliced to a standard thickness (usually 30 microns, or 0.03 millimeters) and glued to a glass slide; this slice is known as a *thin section.* When light pierces the thin section, it interacts with the minerals, producing a variety of telltale effects. For example, some minerals exhibit double refraction, as mentioned earlier. The refracting waves interfere with one another and form vivid colors and patterns unrelated to the color of the mineral in a hand specimen. The photomicrographs in Chapter 6 show what some minerals of igneous rocks look like under a polarizing microscope. A polarizing microscope is not useful for all minerals, however. Metallic ore minerals, such as galena, generally do not transmit light; distinguishing one from the other requires the use of a reflecting microscope.

Common Minerals of the Crust

In examining the abundance of elements in the crust, we are confronted with a striking fact: a mere 8 elements comprise 98.5 percent of the total mass of the crust. They are (in order of abundance by weight): oxygen, silicon, aluminum, iron, calcium, sodium, potassium, and magnesium. The other 84 or so naturally occurring elements amount to a mere 1.5 percent of the crust.

Table 5.2 illustrates that the abundance of oxygen in the crust is overwhelming. The lightest of the eight elements listed, it nevertheless accounts for nearly 46.6 percent of the weight of the crust. The largest of the common elements, it occupies 94 percent of the volume of the crust. Of all the atoms of the crust, 62.6 percent—nearly two-thirds—are oxygen. We often think of oxygen as a gas, as part of the air we breathe, or perhaps as a component of water; there is more oxygen by far in dense crustal rocks, however, than in the atmosphere and oceans combined. The oxygen ion, with a charge of -2, is the only negative ion among the eight elements listed in Table 5.2. A glance at the periodic table shows that the other seven elements occur as positive ions. Not surprisingly, then, the vast majority of minerals of the Earth's crust are compounds involving oxygen, either as oxides or as *radicals,* clusters of atoms that act as single ions.

Refer again to Table 5.2, and note that silicon is the second most abundant element of the crust and by far outstrips the remaining six. It accounts for 27.7 percent of the weight and 21.2 percent of all atoms of the crust. Therefore, it should also be no surprise that the majority of crustal minerals are **silicates,** compounds involving silicon and oxygen in the form of the silicate radical, $(SiO_4)^{-4}$.

silicate
A compound containing silicon and oxygen ions arranged as negatively charged ions.

Table 5.2 Abundances of the Common Elements

Element	Symbol	Percentage of Crust		
		By Weight	By Volume	By Atoms
Oxygen	O	46.6%	94.0%	62.6%
Silicon	Si	27.7	0.9	21.2
Aluminum	Al	8.1	0.5	6.5
Iron	Fe	5.0	0.5	1.9
Calcium	Ca	3.6	1.2	1.9
Sodium	Na	2.8	1.1	2.6
Potassium	K	2.6	1.4	1.4
Magnesium	Mg	2.1	0.3	1.8
All others		1.5	0.1	0.1

Table 5.3   Simplified Mineral Classification Chart

Chemical Group	Typical Minerals	Formula
Native elements		
Metals	Gold	Au
	Silver	Ag
	Copper	Cu
Nonmetals	Carbon (diamond, graphite)	C
	Sulfur	S
Oxides	Corundum (ruby, sapphire)	Al_2O_3
	Hematite	Fe_2O_3
	Uraninite	UO_2
	Ice	H_2O
Hydroxides	Limonite	$Fe_2O_3 \cdot nH_2O$
	Brucite	$Mg(OH)_2$
Sulfides	Pyrite	FeS_2
	Galena	PbS
	Chalcopyrite	$CuFeS_2$
Halides	Halite	NaCl
	Fluorite	CaF_2
Silicates	Feldspar	$Na,Ca,AlSi_3O_8$
	Olivine	$(Fe,Mg)_2SiO_4$
	Mica	$KAl_2(Si_3Al)O_{10}(OH_2)$
Carbonates	Calcite	$CaCO_3$
	Dolomite	$CaMg(CO_3)_2$
	Siderite	$FeCO_3$
Sulfates	Gypsum	$CaSO_4 \cdot 2H_2O$
	Barite	$BaSO_4$
Phosphates	Apatite	$Ca_5(PO_4)_3(OH,F,Cl)$

Aluminum is the third most abundant crustal element. As you might expect, the aluminum silicates constitute the most abundant crustal minerals. In fact, the vast bulk of the crust is composed of

1. the **feldspars:** aluminum silicates combined with varying proportions of sodium, calcium, and potassium ions;
2. the **ferromagnesian** minerals: silicate compounds of iron (Fe) and magnesium (Mg);
3. **quartz:** composed of silicon dioxide (SiO_2), which is technically classified as an oxide but obviously has much in common with the silicates.

Continental crust is overwhelmingly composed of feldspar and quartz, mostly included in the rock granite. Dense ferromagnesian minerals are relatively minor in the crust of the continents. Oceanic crust, in contrast, is composed overwhelmingly of feldspar and ferromagnesian minerals, mostly included in basalt and similar rocks. Little or no quartz is present. This fundamental difference in composition is the reason that continental crust is of lower density than oceanic crust (specific gravity of 2.7 versus 3.0).

Lighter continental crust tends to rise and denser oceanic crust to sink deeper into the mantle. Thus the major topographic feature of our planet—the difference in elevation between the continents and the ocean basins—is the direct result of the density dif-

feldspar
A common group of aluminum silicate rock-forming minerals that contain potassium, sodium, or calcium and display right-angle cleavage.

ferromagnesian
Category of silicate minerals rich in iron and magnesium.

quartz
A mineral composed exclusively of silicon dioxide, whose tetrahedra are linked in a framework structure.

ASIDE 5.1 *Peering Inside Minerals*

There are techniques that allow geologists to "see" inside minerals; the one most widely used is X-ray diffraction. In this technique, a beam of X rays of a single wavelength is passed through a narrow opening in a lead screen and directed at a sample. Figure 5A is a schematic diagram of the apparatus used to take an X-ray photograph of halite. The dots on the film show the arrangement of the atomic planes within the mineral. X-ray diffraction allows geologists to see that the internal architecture of minerals consists of atomic planes arranged in regular, geometric order.

A wealth of information lies hidden amid the patterns of dots representing the orderly spacing of atomic planes. For example, by measuring the distances between planes and the intensity of the deflections from the planes (that is, the brightness of the dots), crystallographers can estimate the size of the various atoms or ions, their electric charge, and their arrangement within the planes. This combination of traits makes the X-ray pattern of a mineral unique, a "fingerprint" that serves to identify it. Atoms within atomic planes can be located to an accuracy of a thousandth of an angstrom (10^{-13} meter)! Precision of this sort allows mineralogists to build a model of the crystal structure brick by atomic brick.

The electron microscope is an instrument used to determine the composition and crystal structure of minerals. The transmission electron microscope (TEM) determines the internal crystal structure by measuring the distance that electron beams travel through a specimen to a detector. The composition is determined by the nature of the radiation emitted by the atoms that the electrons strike along their paths. The scanning electron microscope (SEM) produces a three-dimensional surface image. The electron beam does not penetrate the mineral, but dislodges electrons from some point on the mineral surface. As the beam scans the surface, the distances that the electrons must travel to detectors are processed by computer into images that are displayed on a video screen (Figure 5B).

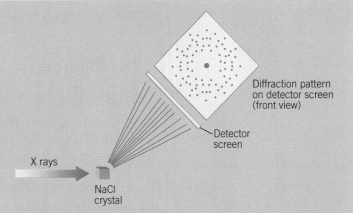

Figure 5A Simplified schematic of X-ray apparatus for determining mineral structure. A narrow beam of powerful X rays penetrates the mineral where it impinges on the internal atomic planes. Those planes, spaced a multiple of a wavelength apart and reinforcing the X rays, register as dots on the detector screen. The internal structure of the mineral can then be interpreted from the symmetry of the dot pattern. The mineral shown is halite.

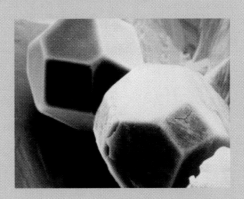

Figure 5B This scanning electron micrograph (SEM) captures gold crystals that are 32 microns (0.032 millimeters) in diameter. The minerals display complex isometric symmetry.

ference between continental and oceanic rocks. As we learned in Chapter 4, the ultramafic mantle is denser than both continental and oceanic crust. The term *ultramafic* refers to a group of rocks composed almost totally of the ferromagnesian minerals olivine and pyroxene. The most common ultramafic rock is peridotite.

The Classification of Common Minerals

There are other minerals in the crust besides the silicates, as shown in Table 5.3. The major mineral groups are typically named after the negative ion or radical present. The most common simple compounds are those in which metals combine directly with oxygen (the oxides and hydroxides); sulfur (the sulfides); and fluorine, chlorine, and iodine (the halides). The major radicals of the more complex minerals include the silicates, the carbonates, and the sulfates. Because the minerals within each group have similar chemical compositions, they often have similar physical properties and atomic structure. As a rule, they are identified through similar chemical tests.

The Structure of Silicate Minerals

There are two fundamental rules of mineral construction. The first is that ions must pack together so tightly that they touch. This packing arrangement is determined by the relative sizes of the positive and negative ions of the mineral. The second rule is that the mineral as a whole must be electrically neutral; the charges of the various ions must balance.

Let us first consider the packing arrangement of the silicates. X rays reveal that the positive silicon ion is quite tiny in comparison to the negative oxygen ion. In fact, only four oxygen ions can fit around the silicon ion. The structural arrangement that results can be regarded as a four-sided figure (a tetrahedron), with oxygen at the ends and silicon enclosed in the center (Figure 5.18). This simple unit, the **silicon-oxygen tetrahedron,** forms the skeleton of the silicates and of virtually the entire crust and upper mantle.

A single silicon-oxygen tetrahedron is incapable of satisfying the second rule of mineral construction—that the mineral as a whole must be electrically neutral. The silicon ion with a charge of +4 cannot balance the combined charge of the four oxygen ions, which is −8. To neutralize the excess negative charge, the tetrahedron forms linkages—either with other tetrahedra or with positively charged ions, such as metals. In this manner, the various silicate minerals are constructed.

silicon-oxygen tetrahedron
An arrangement in which four oxygen ions surround one silicon ion, forming a four-sided structure of negative charge.

olivine
A ferromagnesian silicate common in mafic and ultramafic rocks having an internal structure of isolated tetrahedra.

pyroxene
A ferromagnesian silicate common in mafic and ultramafic rocks having single-chain structure and right-angle cleavage.

amphibole
A group of double-chain ferromagnesian silicates whose cleavage planes intersect at approximately 120° and 60°.

(a) Tetrahedron

(b) Silicon-oxygen tetrahedron

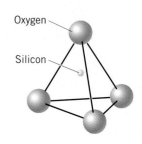

Oxygen

Silicon

(c) Silicon-oxygen tetrahedron expanded

Figure 5.18 Four oxygen ions surrounding a tiny silicon ion is the fundamental structural unit of silicate minerals. The shape of the unit suggests a pyramid, or tetrahedron. The net charge of the unit is minus four (−4).

Tetrahedral Linkages

The structural arrangements of the common silicate minerals are illustrated in Figure 5.19. In Figure 5.19(a), isolated tetrahedra are linked at their edges to positively charged metal ions. This is the structure of the **olivine** group, in which the metal ions are commonly iron (Fe^{2+}) and/or magnesium (Mg^{2+}).

Figure 5.19(b) shows two oxygen ions shared by adjacent silicon ions. As a result, tetrahedra link to form long single chains. The chains are attached by means of intervening metal ions, iron (Fe^{2+}), magnesium (Mg^{2+}), or calcium (Ca^{2+}). This is the structure of the **pyroxene** group.

In Figure 5.19(c), silicon ions that share three oxygen ions alternate with silicate ions that share two oxygen ions. The resulting structures are long double chains, the basic architecture of the **amphibole** group. Double chains are attached through bonds with metal ions.

In the *sheet silicates* (Figure 5.19d), tetrahedra are linked by means of three oxygen ions, and the linkages occur in the same plane. The fourth oxygen ion, pointing either up or down, attaches to metal ions such as aluminum, potassium, iron, magnesium, and calcium, and to hydroxide (OH^-). Sheets are then linked through van der Waals bonds.

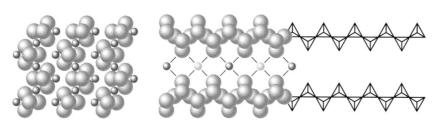

(a) Isolated tetrahedra   (b) Single chain

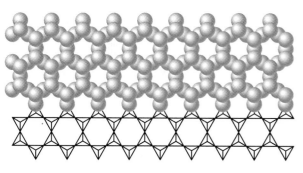

(c) Double chain

Figure 5.19 The structures of common rock-forming silicate minerals. (a) Isolated tetrahedra (olivine): the tetrahedra are separated in all directions by metal ions and their orientations are reversed in alternating rows. (b) Single chain (pyroxene): two oxygen ions of adjacent tetrahedra link to form a chain. (c) Double chains (amphibole): oxygen ions of adjacent tetrahedra link alternately in threes and twos to form a double chain. (d) Sheets (mica, clay, talc, and graphite): three oxygen ions align on a plane. The fourth points out of the plane. (e) Framework (quartz, feldspar): Tetrahedra are arranged in complex three-dimensional patterns.

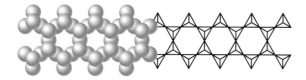

(d) Sheet

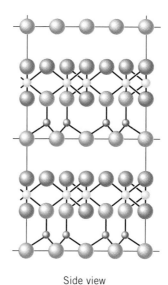

Side view

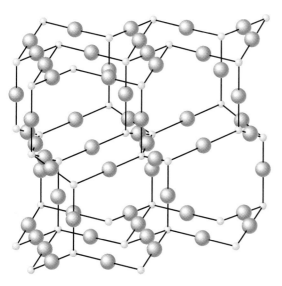

(e) Framework

OH   O   Al   Si   Fe   Mg

The result is a series of stacked planes of alternating tetrahedra and intervening ions—an atomic sandwich. The micas, clays, chlorite, and talc are some of the important sheet silicates.

Figure 5.19(e) shows that quartz (SiO_2) is composed solely of oxygen atoms linked to silicon atoms. Each oxygen is shared by adjacent tetrahedra. The resulting three-dimensional linkage is called a *framework structure*.

Properties Related to Internal Structure

We can readily trace many of the physical and chemical properties of the silicates directly to their atomic structures. For example, cleavage in the silicates is most pronounced in minerals whose internal structures display built-in planes of weakness. These planes are occupied by ions that link the major architectural elements of the mineral: the tetrahedral chains, sheets, and certain framework structures. For this reason, pyroxene (a single chain), amphibole (a double chain), and mica (a sheet) all display prominent cleavage. The cleavage planes of a mineral are parallel, because minerals are built of repetitions of the same pattern. Lack of cleavage is also traceable to the internal structure of minerals. Quartz and olivine fracture conchoidally, like thick glass, rather than along cleavage planes because neither mineral has a pronounced plane of weakness; the bonds are of approximately equal strength in all directions.

Scratching a mineral separates ions—that is, it breaks chemical bonds. But because silicate tetrahedra form especially strong bonds, most silicate minerals have hardness toward the upper end of Mohs hardness scale. However, not all silicates are hard; talc, kaolinite, chlorite, and mica can all be scratched with the fingernail. These minerals have sheet structure in common, and it is the weak van der Waals bonds linking the sheets that makes the minerals soft. The weak bond is also the reason that mica peels in thin sheets.

The chains and sheets of silicate minerals also allow storage of water in the form of hydroxide $(OH)^-$ radicals, which can fit between these structures. Thus some silicate minerals hold surprisingly large volumes of water; in fact, there may be more water stored in the silicate minerals of the crust and upper mantle than in the oceans. Clay minerals are especially efficient water sponges, which explains why they swell and become slippery when wet.

Variable Composition

Many silicate minerals display variable composition, a condition that can be traced to the internal atomic structure of the silicates. Recall that metal ions serve to hold individual tetrahedra together or to link large architectural units such as chains and sheets. The requirements for the job are proper charge, to balance the negative charge of the tetrahedra, and proper size, to fit between the chains, sheets, and tetrahedra without distorting crystal shape. Metals having these qualifications can freely substitute for one another. The result is a **solid solution** series, a group of chemically similar minerals in which certain ions substitute for one another in the chemical formula and, of course, within the structure of the mineral.

All the silicates shown in Figure 5.19, except quartz, form solid solution series to some degree. The olivine group is a good example. Magnesium (Mg^{2+}) and iron (Fe^{2+}) are of equal charge and are nearly the same size, and they easily substitute for one another in the structure of olivine. The composition of a given olivine mineral can therefore vary in any proportion between 100 percent fosterite (Mg_2SiO_4) and 100 percent fayalite (Fe_2SiO_4).

Iron, however, is heavier than magnesium and absorbs light differently, which makes the density and color of the minerals within the solid solution series change with composition. Magnesium-rich varieties are relatively light and are colored pale olive green.

solid solution
Variations in the composition of a mineral caused by the substitution of metal ions within the crystal structure.

ASIDE 5.2 *Polymorphs*

I n the introduction of this chapter, we contrasted the properties of diamond and graphite, two minerals of identical composition, the single element carbon. Diamond and graphite are classic examples of polymorphs, minerals of identical chemical composition but different internal crystal structure.

Whether carbon atoms are arranged as diamond or as graphite depends upon external conditions, notably pressure and temperature. Mineralogists have duplicated experimentally the ranges of pressure and temperature under which these minerals can form. Diamond is a high-pressure mineral, whereas graphite forms under less extreme conditions. In fact, diamond is found in close association with igneous rock, whose composition and other properties indicate that it was derived from the upper mantle. The experimental evidence bears this result out: diamond forms under pressures equivalent to about 240 kilometers of overlying rock. Pressures of this sort are responsible for the more compact mineral structure of diamond, whose density is 3.5. Graphite and other minerals formed nearer the surface are far less compact; the density of graphite is 2.2, for instance.

In Chapter 4, Interior of the Earth, we discussed the evidence for the existence of a transition zone between about 400 and 700 kilometers in the mantle. An abrupt jump in seismic wave velocities indicates a change in the elastic properties and densities at that depth. High-pressure experiments show that the chief mineral in the mantle, olivine, undergoes a series of polymorphic transformations at that depth (see the figure).

Polymorphs are useful indicators of the general environmental conditions existing at the time of their formation. As such, they aid in our understanding of important geological processes—for example, the pressure-temperature conditions existing in certain metamorphic environments, beneath the deep-ocean floor, or in a cooling magma.

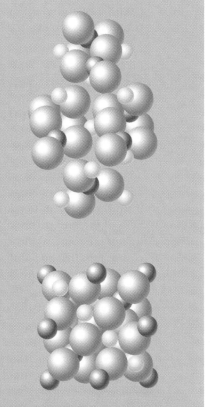

Compare the loosely packed structure of olivine (top) found in the crust and upper mantle to the densely packed structure of perovskite (bottom), a mineral of the same composition believed to be present in the lower mantle. The polymorphic transition occurs in the transition zone, at a depth of about 400 kilometers.

Iron-rich olivine is black and dense. Here, again, we see the relationship of internal structure to the physical properties of minerals: both color and density are determined by the ability of the structure to accept ions of a given charge and size.

Solid solution also accounts for the compositional variability of the crust's most abundant mineral groups, the feldspars. These aluminosilicates, the chief minerals of granite and basalt, compose more than half the volume of the crust. The feldspars are framework silicates, similar in structure to quartz but with the crucial difference that aluminum ions substitute for silicon ions in the center of some of the tetrahedra. Such substitution is possible because aluminum and silicon ions are about the same size and are neighbors in the periodic table of the elements. However, aluminum carries a charge of +3 and silicon carries +4; thus the introduction of aluminum requires the simultaneous introduction of positive metal ions to make up the charge deficiency. In this way, sodium, potassium, and calcium enter into the framework structure and attach to the tetrahedra. Of similar size, these metal ions comprise a solid solution series.

plagioclase
A group of feldspars in which calcium and sodium substitute for one another in solid solution.

Feldspars that are made up of solid solutions of calcium and sodium ions are called the **plagioclase** group. Composition within this group varies between 100 percent calcium feldspar and 100 percent sodium feldspar. Any given mineral in the series has a composition somewhere between these extremes.

Structure and Properties of Other Mineral Groups

We have concentrated on silicate minerals in this chapter because they compose the vast bulk of the crust and mantle; however, this emphasis should not imply that other groups are unimportant. The same principles that govern the atomic structure of the silicates apply equally to them. Charge must be balanced and distributed as evenly as possible throughout the mineral. Atomic packing arrangements must conform to the relative sizes of the ions involved. In all minerals, properties such as crystal symmetry, cleavage, fracture, hardness, and color depend upon internal atomic arrangement and chemical composition. Solid solution is found in many minerals and is not an exclusive property of the silicates. Figure 5.20 illustrates the structures of some other important groups: the carbonates, the sulfides, the oxides, and the halides.

Comparison of the silicates to the carbonates illustrates the importance of relative ionic size in determining the structure of minerals. A glance at the periodic table reveals that although carbon and silicon belong to the same group (IV) and have a charge of +4, carbon is a much smaller element. Because of this difference in size, only three oxygen atoms can fit around the carbon ion, whereas four can fit around the silicon ion. The silicon-oxygen tetrahedron, with its charge of −4, is able to form a large number of linkages, which accounts for the great variability of the silicate structures. The possibilities for the carbonates are more limited, however. Three oxygen atoms form a triangular unit with carbon at the center, and the overall charge of the unit is −2. This leads to the simple alternation of the negative carbonate unit with positive metal ions, such as calcium, magnesium, or iron, which results in a rhombic shape. Figure 5.20(a) illustrates the structure of dolomite, a carbonate mineral closely related to calcite. The formula for calcite is $CaCO_3$ and for dolomite is $CaMg(CO_3)_2$.

The Geologic Origin of Minerals

For all their myriad complexities and details, there are but four common ways that minerals form in the Earth.

1. Minerals crystallize from solutions of molten silicate rock (magma).
2. Preexisting minerals are altered by the atmosphere and surface water solutions.
3. Minerals precipitate from seawater, groundwater, or surface water solutions.
4. Preexisting minerals are transformed into new minerals while in the solid state.

The first of these origins leads to igneous rocks and includes the vast bulk of Earth materials—oceanic crust and much of the continents. The second involves weathering—for example, alteration by rainwater of feldspar to clay. The third also involves mineral formation by everyday processes that occur at or near the surface of Earth. Some examples are the precipitation of salt and gypsum resulting from evaporation of seawater and lakes and the extraction of calcium carbonate and silica by marine organisms. These minerals are considered sedimentary in origin. The fourth origin involves metamorphism, the application of intense heat and/or pressure to preexisting minerals; the new minerals form as an adjustment to these conditions.

The four origins appear to be distinct from one another; in the real world, however, minerals form as a result of processes that are often intimately related and difficult to

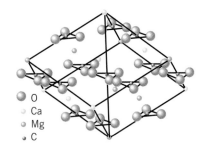

(a) Carbonates (dolomite)

O
Ca
Mg
C

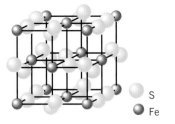

(b) Sulfides (pyrite)

S
Fe

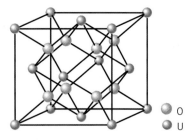

(c) Oxides (uraninite)

O
U

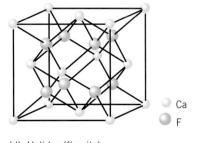

(d) Halides (fluorite)

Ca
F

Figure 5.20 Common structures of carbonates, sulfides, oxides, and halides.

define. Some minerals have multiple origins. To fully explore the origin of minerals is to examine igneous, sedimentary, and metamorphic processes. These important subjects are pursued in succeeding chapters.

STUDY OUTLINE

I. ATOMS AND ELEMENTS

    A. A **mineral** is a naturally occurring, inorganic solid, having a definite chemical composition and orderly internal atomic arrangement. A rock is an aggregate of minerals. The **elements** that comprise minerals are substances composed of **atoms** that differ from one another in the number and arrangement of their **protons, neutrons,** and **electrons.**

    B. An element cannot be broken down or subdivided by ordinary chemical or physical means. There are 92 naturally occurring and 17 manufactured elements.

    C. Atoms are held together by equal units of positive charge (carried by the protons) and negative charge (carried by the electrons).

    D. Most of the mass of an atom is in the **nucleus,** which contains the protons and neutrons—particles of nearly equal mass. Neutrons have zero charge and bind the protons tightly together.

        1. All atoms of a given element have the same number of protons, or **atomic number.** The number of protons and neutrons is the **atomic mass** of the atom.

        2. Most elements occur in a variety of atomic masses, called the **isotopes** of that element. Certain isotopes are subject to **radioactive decay.** Their nuclei are unstable, and they decay to stable isotopes of other elements.

    E. In the neutral state of the atom, the number of protons and electrons is equal. Regularity of atomic structure is the basis for the **periodic table of the elements.**

        1. Electrons orbit at specific distances from the nucleus. Their paths are three-dimensional, so orbits are also called **electron shells.**

        2. The first shell, which holds only two electrons and is filled first, is followed by successively higher shells, each of which may hold eight or more electrons; the outermost shell holds a maximum of eight. The number of electrons in an element's outer shell is a critical determinant of its chemical behavior.

II. BONDING. Elements whose outer shells are not filled are active and participate in reactions with other elements. **Chemical bonds** involve the loss, gain, or sharing of electrons in order to fill shells. Bonding of different elements results in **compounds.**

    A. Ionic bonds are formed by electron transfer. Charged atoms are called **ions.**

    B. Elements form **covalent bonds** by sharing electrons in their outer orbits.

    C. Metallic bonds are like covalent bonds but involve many atoms packed tightly together.

    D. Van der Waals bonds hold compounds weakly together.

III. THE PHYSICAL PROPERTIES OF MINERALS

    A. Unrestricted, minerals grow into symmetrical **crystals.** The planed **crystal faces** are made of closely spaced atoms.

    B. Some minerals display **cleavage,** the tendency to break along parallel planes.

C. Many minerals **fracture** to form smooth, jagged, or splintery surfaces.

D. **Hardness,** or resistance to scratching, is measured by the Mohs scale.

E. Some minerals can be identified by their distinctive color.

F. Some minerals can be identified by their **streak,** the color of the powder formed when they are rubbed against a porcelain plate.

G. **Luster** refers to the appearance of a mineral in reflected light.

H. **Specific gravity** is a mineral's weight relative to an equal volume of water.

I. Other mineral properties are transparency, opacity, and double refraction.

IV. COMMON MINERALS OF THE CRUST

A. Eight elements comprise 98.5 percent of the total mass of the crust: oxygen, silicon, aluminum, iron, calcium, sodium, potassium, and magnesium.

B. Most crustal minerals are **silicates,** compounds involving silicon and oxygen, the two most abundant elements of the crust.

C. Aluminum is the third most abundant element. The most abundant crustal minerals are the **feldspars, ferromagnesian** minerals, and **quartz.** Minerals of the ultramafic mantle are primarily ferromagnesian.

V. THE STRUCTURE OF SILICATE MINERALS.   The basic structure of silicate minerals is the tetrahedron: four oxygen ions around one silicon ion.

A. Tetrahedral linkages involve **silicon-oxygen tetrahedra** and intervening positive metal ions. Isolated tetrahedra are linked by positively charged metal ions as in the **olivine** group. Tetrahedra link to form single chains as in the **pyroxene** group. Tetrahedra of the **amphibole** group form double chains. Tetrahedra are linked to form sheet silicates, as in the micas. Tetrahedra that are linked to one another exclusively form framework structures, as in quartz.

B. Many chemical and physical properties of minerals are related to their internal structure. For example, the cleavage planes of micas and clays occur because of weak bonds between sheets, but quartz and olivine fracture conchoidally because both have bonds that are equal in all directions.

C. Many silicate minerals display variable composition. **Solid solution** series are groups of chemically similar minerals in which linking ions substitute for one another. Feldspars involving solid solution of calcium and sodium ions are called the **plagioclase** group.

D. **Polymorphs** are minerals of identical chemical composition but different internal structure. For example, diamond and graphite are two forms of carbon. Diamond forms under higher pressure than graphite.

VI. STRUCTURE AND PROPERTIES OF OTHER MINERAL GROUPS.   Other important mineral groups are the carbonates, the sulfides, and the oxides. The carbonates illustrate the importance of relative ionic size in determining the structure of minerals. Because only three oxygen units fit around a carbon unit, the structures of carbonate linkages are less varied than those of the silicates.

VII. THE GEOLOGIC ORIGIN OF MINERALS.   Minerals form in the Earth in four ways. Minerals crystallize from magma; minerals are altered by weathering; minerals are precipitated from seawater; or preexisting minerals are transformed into other minerals while in the solid state through metamorphism.

STUDY TERMS

amphibole (p. 158)
atom (p. 141)
atomic mass (p. 144)
atomic number (p. 143)
chemical bond (p. 147)
cleavage (p. 150)
compound (p. 147)
covalent bond (p. 147)
crystal (p. 149)
crystal face (p. 149)
electron (p. 141)
electron shell (p. 144)
element (p. 142)
feldspar (p. 156)
ferromagnesian (p. 156)
fracture (p. 152)
hardness (p. 152)
ion (p. 147)
ionic bond (p. 147)
isotope (p. 144)

luster (p. 153)
metallic bond (p. 148)
mineral (p. 141)
neutron (p. 141)
nucleus (p. 143)
olivine (p. 158)
periodic table of the elements
(p. 145)
plagioclase (p. 162)
polymorph (p. 149)
proton (p. 141)
pyroxene (p. 158)
quartz (p. 156)
radioactive decay (p. 144)
silicate (p. 155)
silicon-oxygen tetrahedron (p. 158)
solid solution (p. 160)
specific gravity (p. 153)
streak (p. 153)
van der Waals bond (p. 148)

CRITICAL THINKING QUESTIONS

1. Why are coal, glass, and manufactured diamonds not considered minerals?
2. Distinguish between the following:
   a. neutron—proton
   b. atomic number—atomic mass
   c. stable isotope—radioactive isotope
   d. electron shell—nucleus
   e. metal—nonmetal
   f. ionic bond—covalent bond
3. Explain the organization of the periodic table.
4. Why do comparatively few minerals in their natural settings display external symmetry? How would you prove they have internal symmetry?
5. Distinguish between the following:
   a. fracture—cleavage   b. color—streak   c. luster—translucency
6. Why are cleavage planes oriented parallel to crystal faces?
7. Why do some minerals display cleavage and others do not?
8. Why is graphite soft and diamond hard?
9. What is the relationship between the distribution of elements in the crust and the difference in elevation between the continents and ocean basins?
10. What are the two fundamental rules of mineral construction? How do they allow for the solid solution of the olivine and feldspar groups?
11. Why do the carbonates lack the structural complexity of the silicates?
12. What is a polymorph? How are polymorphs used as environmental indicators?

Igneous Activity

The Earth's crust is overwhelmingly igneous in origin, as a glance at a world map will confirm. There we see that most of the crust is oceanic, formed by the upwelling of magma from beneath the mid-ocean ridges. The oceanic crust is also very young. As we have seen in our discussion of sea-floor spreading, the entire oceanic crust has been recycled back into the mantle within less than 5 percent of geologic time. So from the dual perspectives of vigor and volume, it is fair to say that igneous activity is the Earth's most important geologic process.

The influence of igneous activity is not confined to the oceans. Too light to be subducted, the continental crust has grown through geologic time by the addition of igneous rocks. As the oceanic lithosphere descends beneath subduction zones, it generates fresh magma, which rises and, upon cooling, intrudes the continents and builds volcanic chains. Where erosion has stripped away the thin sedimentary veneer of the continents, the igneous rocks that formed beneath the surface are exposed. Some are billions of years old, and it becomes clear that deep-seated igneous rocks are the major architectural units of the continental crust.

Igneous Rock Bodies

From field studies and seismic evidence, as well as from the controlled melting of igneous rocks under laboratory conditions, geologists have learned that most **magmas** are generated through the partial melting of the upper mantle. The magmas then rise and spread from their source regions and, through a variety of means, intrude the crust (Figure 6.1). Most of them crystallize beneath the surface, forming intrusive rock

magma
Molten (hot-liquid) rock material, generated within the Earth, that forms igneous rocks when solidified.

◀ Photomicrograph of gabbro, a coarse-grained igneous rock, taken under polarized light. The long, rectangular crystals are plagioclase feldspar.

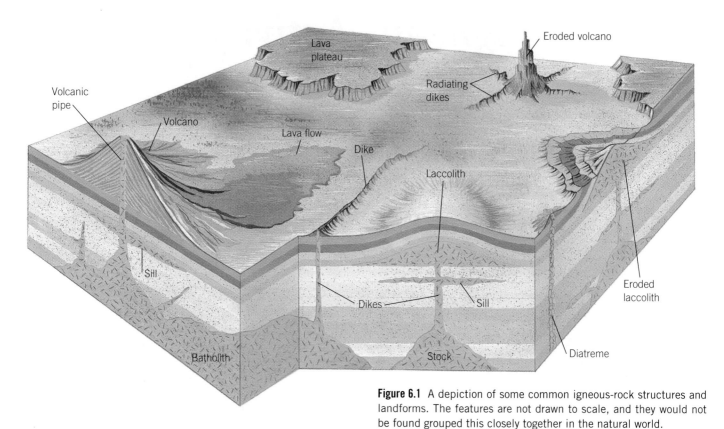

Figure 6.1 A depiction of some common igneous-rock structures and landforms. The features are not drawn to scale, and they would not be found grouped this closely together in the natural world.

pluton
An intrusive rock body.

concordant
Pertaining to igneous rock bodies that intruded and solidified parallel to the layers of country rock.

discordant
Pertaining to igneous rock bodies that cut across the layers of country rock.

sill
A tabular, concordant igneous intrusion.

laccolith
A mushroom-shaped, concordant igneous intrusion that has domed the overlying crustal rocks.

dike
A tabular, discordant igneous intrusion.

batholith
An igneous intrusion with a large mass, a surface area greater than 100 km², and no known floor.

bodies or **plutons** (after Pluto, the Greek god of the underworld). Processes connected to the formation of igneous rocks at depth are called plutonism.

Large volumes of magma may also rise through the crust and escape onto the surface—either relatively peacefully, as *lava,* or explosively, as fragments sent into the air. If the magma rises to the surface through a more or less localized vent, the lava and fragments will pile up around it. Repeated eruptions of this sort will build a large conical volcano. If, on the other hand, the lava pours onto the surface through a long fracture or fissure, it will flood the surrounding region. Many such episodes build a thick *lava plateau.* All these expressions of extrusive igneous activity go under the name of volcanism (after Vulcan, the Roman god of fire). Volcanism is the subject of Chapter 7.

Plutons are generally classified by their relation to the layering of the host or country rocks that they intrude. Plutons that are **concordant** form parallel to the layering of the host rock, whereas those that are **discordant** cut across the layering. Concordant plutons come in two main varieties: sills and laccoliths. **Sills** are flat, tabular bodies intruded parallel to the host rock layering (Figure 6.2). They can range from less than a meter to hundreds of meters thick; some extend for hundreds of kilometers, and others disappear in a few short steps. **Laccoliths** are mushroom-shaped bodies. Fed from a lower vent, magma rises and domes the overlying layers while it spreads laterally.

Discordant plutons cut across the layering of host rock; they include dikes, pipes, batholiths, and stocks. **Dikes** are tabular bodies (Figure 6.3). They may occur in a variety of patterns, some forming deep, irregularly shaped intrusions. *Pipes* are cylindrical bodies, intrusions that filled and solidified in the vents that once fed laccoliths, sills, or volcanoes high in the crust. *Radiating dikes* may spread from these pipes like spokes from a bicycle wheel, and still others may occur as concentric *ring dikes. Diatremes* are peculiar cylindrical, pipelike intrusions. Many of them contain diamonds—strong evidence that they are derived from the mantle, hundreds of kilometers below the crust.

Batholiths (*bathos,* meaning "bottom" in Greek; *lith,* meaning "rock") are enormous, complex rock bodies composed mostly of granite: they cover at least a hundred

Figure 6.2 A sill intruded parallel to the layering of the country rock.

Figure 6.3 A dike cutting across the layering of a shale formation in Grand Canyon National Park, Arizona.

square kilometers and have no visible floor. Many show signs that they were formed over millions of years by the repeated injections of relatively small magma bodies. Batholiths are major architectural elements of the continents. Some batholiths, such as the Sierra Nevada batholith in California, cover the greater part of entire states (Figure 6.4). Smaller bodies of similar origin are called *stocks*.

We will discuss the emplacement of igneous rock bodies in the course of this chapter.

The Formation of Rocks from Magma

A typical magma is a complex mixture consisting of molten silicate rock material, dissolved gases, solid minerals that have crystallized from the melt, and chunks of country rock incorporated within the liquid mass. The question is, How do solid rocks form from this mixture? To arrive at an answer, geologists begin with the fact that all rocks, including those that are igneous in origin, are essentially aggregates of minerals. By studying the texture of the aggregate—that is, the size, shape, and arrangement of the minerals—they can extract a great deal of information about the physical conditions that led to the formation of the rock.

Igneous Rock Textures

The texture of an igneous rock develops as minerals crystallize from a cooling magma. There are a few simple experiments, employing water and dissolved ions, that illustrate the origin of some important textural relationships—notably, how large and small crystals form. Take a beaker of hot water and dissolve blue powdery copper sulfate in it until the water cannot hold any more. Let the solution cool to room temperature. At this point, the solution is *supersaturated*, holding more copper sulfate than is normal for that temperature. Next, add a few tiny copper sulfate crystals, or seeds, that sink to the bottom. Let the solution stand for a few days and observe the results. You will find that the seed has grown into a large, symmetrical copper sulfate crystal and that the blue color of the water has lightened noticeably. Evidently, the copper and sulfate ions in solution were attracted to the seed and crystallized. We may conclude, therefore, that slow, undisturbed cooling of a solution that contains a few seeds tends to favor the precipitation of large, well-formed crystals.

Now, repeat the experiment, but this time add a number of seeds. You will find that a number of tiny crystals have precipitated. Next, try the experiment without adding

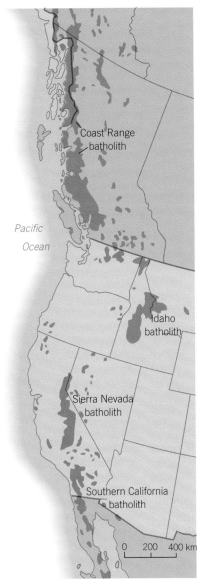

Figure 6.4 Batholiths cover wide areas of western North America.

Figure 6.5 Hand sample of a typical gabbro, showing phaneritic (coarse-grained) texture. The dark minerals are amphibole (hornblende), the lighter ones are plagioclase.

Figure 6.6 A typical basalt, displaying aphanitic texture and dark color. Compare with gabbro (see Figure 6.5), its coarse-grained equivalent.

phaneritic

An igneous-rock texture in which the mineral components are visible to the unaided eye.

aphanitic

An igneous-rock texture in which the mineral components are too small to be identified by the unaided eye.

glassy

The texture of an igneous rock having a high content of glass.

intrusive

An igneous rock derived from magma that solidified within the mantle or crust.

extrusive

Igneous rock formed from lava flows or pyroclastic materials that were spewed to the Earth's surface.

porphyritic

An aphanitic igneous-rock texture in which large crystals are embedded within an aphanitic or glassy matrix.

the seeds. Instead, cool the copper sulfate solution very rapidly and shake it at the same time. You will find again that tiny crystals have precipitated. Thus we may conclude that small crystals precipitate when there are many seeds present and that these seeds may form from a liquid that is agitated as it cools rapidly.

Although a magma is far more complicated than our simple copper sulfate solution, minerals will crystallize from it in an analogous manner. The major difference is that magma is a silicate solution capable of holding a far greater concentration of ions than water. Crystal growth within a magma begins with the precipitation of seeds. These seeds contain the basic structural plans of the minerals, and as ions migrate to them, they grow in size. Eventually, their surfaces interfere with those of other growing crystals. At this point, the crystals interlock and growth ceases. For this reason, igneous rocks typically display a mosaic texture of interlocking mineral grains.

In our experiments, we also saw that cooling rates control crystal growth. In a slowly cooling magma, relatively few seeds form; the minerals that grow around them are large. If the minerals grow large enough to be seen with the naked eye, we say the rock has a coarse or **phaneritic** texture (Figure 6.5). In a rapidly cooling magma, by contrast, numerous seeds compete for ions, and many tiny minerals crystallize; this gives the rock an **aphanitic** texture, meaning that the minerals are visible only under a microscope (Figure 6.6). Ultimately, if the magma cools too rapidly, ions are robbed of the energy to migrate to the seeds, and the magma solidifies without crystallizing. Glass, a smooth and hard substance, forms instead. It has the amorphous internal structure of a liquid, rather than the organized structure of minerals. Thus rapid quenching, or supercooling, produces a **glassy**-textured rock (Figure 6.7).

In the field, phaneritic textures are characteristic of **intrusive** rocks—those that cool beneath the surface. Large mineral crystals form because magmas are well insulated at depth and cool slowly. By contrast, aphanitic and glassy textures are characteristic of **extrusive** rocks—those that cool on the surface. Because the temperature contrast between air and lava is very high and heat loss is rapid, only small crystals form.

The link between cooling and depth of burial also provides an explanation for the origin of most **porphyritic** textures (Figure 6.8). Porphyritic rocks display a mixture of coarse-grained phaneritic and fine-grained aphanitic or glassy textures. In other words, rocks having this interesting texture contain minerals of two distinct sizes: large minerals embedded in a matrix of small minerals. As such, porphyritic rocks represent two generations of cooling. Picture a magma cooling slowly in a large chamber thousands of meters below a volcano. Large minerals characteristic of that depth crystallize in the

(a)

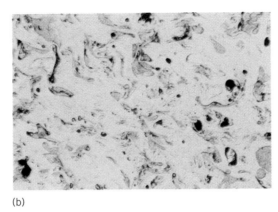

(b)

Figure 6.7 (a) A hand sample of the volcanic glass pumice. (b) This photomicrograph of pumice shows the vesicles and glass fibers that make it useful as an abrasive.

melt. Next, pent-up gases within the chamber burst through the plug that caps the volcanic vent. The magma then erupts onto the surface as lava, carrying the coarse crystals with it. As the lava cools rapidly, the aphanitic minerals that crystallize from it form the matrix of the porphyritic rock.

Up to this point, we have discussed the textures of rocks that cool either slowly or rapidly from a coherent body of magma. But that is not always the case. As we have mentioned, many volcanoes pass through an explosive phase, triggered by the escape of pent-up gases. The lava is literally torn to bits in the process, and clots ranging in size from minute droplets to small cars go flying out of the vent. On the way down, they cool; when the fragments land in a heap, they are cemented together. The rock that results has a **pyroclastic** texture (Figure 6.9). We will discuss pyroclastic rocks further in Chapter 7, Volcanism.

pyroclastic
Pertaining to rock material formed by an explosive ejection from a volcanic vent.

The Classification of Igneous Rocks

If the texture of an igneous rock points to the physical conditions at the time that the magma crystallized, then the mineral content of the rock suggests the origin and chemical evolution of the magma. Fortunately, of the more than 3000 minerals that exist, only 7 compose over 95 percent of all igneous rocks. They are the ferromagnesian minerals olivine, pyroxene, amphibole, and biotite; the aluminum silicate feldspars (both the plagioclase and potassium varieties); and quartz. To simplify matters still further, no one rock contains all these minerals; no more than 3 or 4 are commonly present, and they form characteristic groupings. This information forms the basis for classifying igneous rocks into *rock families,* which are named after their most common rock members.

Figure 6.8 A typical hand sample of rhyolite displaying porphyritic texture.

Figure 6.9 Pyroclastic texture is exhibited by this volcanic mudflow breccia, which consists of angular pyroclastic fragments of various sizes cemented in a volcanic mud matrix (from the 1963 eruption of Mount Agung, in Indonesia).

Figure 6.10 Peridotite, an ultramafic rock. This specimen consists almost entirely of the ferromagnesium mineral olivine.

Figure 6.11 Hand sample of serpentinite. Peridotite is often altered by hot-water solutions to serpentinite, a metamorphic rock.

The Peridotite Family

peridotite
A dark, coarse-grained, intrusive igneous rock composed mainly of olivine, small amounts of pyroxene, and little or no plagioclase.

Peridotite is a dark, coarse-grained intrusive rock composed mainly of olivine, with lesser amounts of pyroxene and containing little or no plagioclase (Figure 6.10). It is believed to form the bulk of the upper mantle. As such, it is either directly or indirectly the source rock of the magmas from which most other igneous rocks are derived. This important igneous rock is relatively rare in the crust, however. It sometimes occurs as inclusions in other igneous rocks of the oceanic crust or as layers in deep fracture zones of the oceanic crust. It is also preserved in slivers of oceanic crust that have been sheared off descending plates and thrown up onto the continents. Peridotite may often be altered to the soft, greenish rock *serpentinite* by hot-water solutions that penetrate the crust at mid-ocean ridges and other fracture zones (Figure 6.11). Because of its very high content of magnesium and iron, peridotite is often referred to as an *ultramafic* rock.

The Basalt-Gabbro Family

basalt
A dark, fine-grained, mafic, extrusive igneous rock composed largely of plagioclase feldspar, pyroxene, and olivine.

Basalt is a fine-textured, dark brown to black extrusive rock, composed primarily of calcium-rich plagioclase and pyroxene and, commonly, lesser amounts of olivine (see Figure 6.6). *Gabbro* is its coarse-textured, deep intrusive equivalent (see Figure 6.5). *Diabase,* whose texture lies between that of basalt and gabbro, often forms in intrusions that occur close to the surface. The ferromagnesian minerals remain important components of this family, though they are less plentiful than in the peridotite family. Therefore, basalt and gabbro are also known as *mafic* rocks.

Mafic rocks compose the entire oceanic crust—basalt forming the upper layers and gabbro forming the thicker internal zone upon which the basalt rests. The oceanic volcanoes of the Hawaiian Island–Emperor Seamount chain, the entire island of Iceland, and great oceanic plateaus of the sea floor are all composed of basalt. Vast outpourings of basaltic lava on the continents have built the Deccan Plateau of India, the Columbia Plateau of Washington, and similar features the world over.

The Andesite-Diorite Family

andesite
A dark, fine-grained, extrusive igneous rock composed of intermediate plagioclase and a ferromagnesian mineral.

Andesite is a gray, fine-grained volcanic rock consisting of plagioclase and a ferromagnesian mineral such as hornblende (amphibole) or biotite (mica). The plagioclase contains about equal amounts of calcium and sodium ions. *Diorite* is the coarse-grained equivalent (Figure 6.12).

The term *andesite* was coined in 1826 by pioneering field geologists for the rocks and lavas characteristic of the Andes Mountains of South America. About one hundred forty years later, geologists learned that the Andes range borders an active subduction

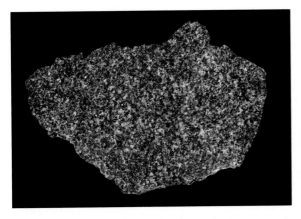

Figure 6.12 A hand sample of diorite, which is commonly lighter-colored than gabbro because it contains a higher proportion of sodium-rich plagioclase.

Figure 6.13 A typical granite displaying coarse-grained phaneritic texture. The rusty pink color is the consequence of an abundance of potassium feldspar. The colorless minerals are quartz.

zone, and indeed, rocks of the andesite family are typical of the volcanic island arcs and continental chains that border these zones.

The Granite-Rhyolite Family

Granite is a light-colored, coarse-grained intrusive rock consisting primarily of quartz, potassium feldspar, and/or sodium plagioclase (Figure 6.13). Ferromagnesian minerals such as hornblende or biotite are either absent or present only in minor quantities. However, this simple description does not do justice to the extremely varied textural and mineralogical complexity of granite. Members of the granite family run from the extremely coarse-grained *pegmatite* to the sugary-textured *aplite.*

Granite and its slightly more mafic variety *granodiorite* are the most common igneous rocks of the continental crust. Indeed, to a rough approximation, granitic rocks *are* the continental crust. It is also significant that granite is found only on the continents, whereas the oceanic crust consists entirely of basaltic rocks. Until recently, this division was never adequately explained. One of the triumphs of the theory of plate tectonics is that it provides this explanation (see Chapters 2 and 12).

Rhyolite is the extrusive equivalent of granite and is also generally confined to the continental crust (see Figure 6.8). Most of the violent volcanoes that border subduction zones involve either andesite, *dacite* (the extrusive equivalent of granodiorite), or rhyolite lava. Because of their extremely high silica content, members of the granite-rhyolite family are also referred to as *silicic* rocks.

granite
A coarse-grained, intrusive igneous rock containing quartz and feldspar (primarily potassium feldspar).

The Igneous Rock Classification Chart

Texture and mineral content form the basis for identifying and classifying igneous rocks, as shown in Figure 6.14. Using the chart, we see that granite is a coarse-grained (phaneritic) rock containing feldspar and abundant quartz—a mineral that in some rocks assumes the look of dull glass and in others looks like glass sparkling in the sun. Basalt, on the other hand, is a fine-grained (aphanitic) rock consisting of calcic plagioclase, pyroxene, and, commonly, olivine.

The classification chart shows that the igneous rocks vary systematically in mineral content, chemical composition, density, and temperature of crystallization. Note first the gradational nature of the mineral content. As we proceed from the peridotite end toward the granitic end of the chart (from right to left), olivine and pyroxene phase out and are replaced in sequence by hornblende and biotite. The ferromagnesians decline in abundance, and the percentages of feldspar and quartz increase. At the same time, the plagioclase feldspars change in composition from calcium-rich to sodium-rich varieties.

Figure 6.14 Mineral content and texture form the basis for classifying igneous rocks, as shown in this chart. The chart also illustrates systematic changes in the mineralogy, chemical composition, density, and crystallization temperature of the igneous rocks.

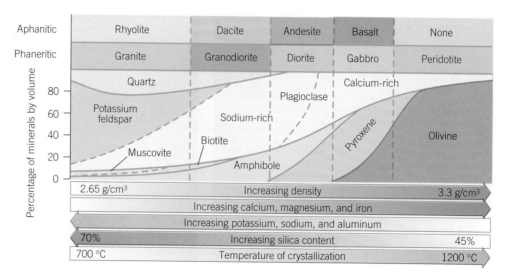

The changes in mineral content across the chart are broadly reflective of the chemical trends encountered as one rock type blends into another. Most significant is the increasing silica content, grading from about 45 percent at the peridotite end to up to 70 percent at the granitic end. Also notice that iron, magnesium, and calcium are important at the peridotite end, decline markedly, and are replaced by sodium and potassium toward the granitic end of the chart. These changes in mineralogy and chemistry are responsible for the marked decrease in density toward the granitic end of the chart; peridotite and basalt are heavy rocks, and granite and rhyolite are light rocks. Finally, note that the minerals at the peridotite end crystallize at about 1200 °C, and those of granitic rocks begin to crystallize at about 700 °C.

The Sequence of Crystallization of Igneous Rocks

When viewed under a microscope, igneous rock textures bear out the crystallization sequence suggested by the classification chart. The figures that follow show thin sections of some of the minerals of igneous rocks viewed under polarized light. They preserve the evidence of reactions, frozen in time, in which olivine converted to pyroxene, pyroxene to amphibole, and so on. The evidence is preserved because, in each case, the magma crystallized before the reaction proceeded to completion. Figure 6.15(a), for example,

Figure 6.15 Photomicrographs under polarized light of discontinuous reactions. (a) The brightly colored minerals are olivine. Each is encircled by a pyroxene reaction rim of a different color. (b) A pyroxene core is surrounded by an amphibole reaction rim. The grayish, rectangular minerals in both (a) and (b) are plagioclase.

(a)

(b)

Figure 6.16 Photomicrograph under polarized light of zoning in plagioclase feldspar. The layers or zones surrounding the calcium-rich core become progressively enriched in sodium and depleted in calcium toward the edge of the mineral. The reaction occurs continuously over the entire range of falling temperatures within the magma.

Figure 6.17 Photomicrograph under polarized light of granite with a complex mineral assemblage. The rectangular minerals with light and dark bands are plagioclase; the mottled gray ones are potassium feldspar; the bright-colored ones are biotite. The clear minerals showing shades of gray are quartz, the last to crystallize from the magma.

shows olivine cores encircled by pyroxene *reaction rims.* A feature of this type can occur only if the olivine first crystallized and reacted with the magma at a lower temperature, at which time its outer rim was converted to pyroxene. The same kind of reaction holds for the other ferromagnesian minerals of igneous rocks (Figure 6.15b). These reactions constitute a *discontinuous series;* they occur in steps, at specific temperatures, rather than continuously.

The plagioclase feldspars exist side by side with the ferromagnesian minerals in igneous rocks, and they also exhibit reactions frozen in time. Figure 6.16 shows a crystal of plagioclase displaying a series of rings or *zones* wrapped around a central, calcium-rich core. The succeeding zones mark a series of gradations in which the calcium content decreases and the sodium content increases. Clearly, the first plagioclase to crystallize was the calcium-rich core; as temperatures fell, it continued to react with the magma, exchanging calcium for sodium. Since there is no interruption of the reaction and there are no drastic conversions, we say that the plagioclase feldpars constitute a *continuous reaction series.*

Lastly, see the granite thin section of Figure 6.17. Quartz seems to exist as a formless mass between crystals of feldspar. This thin section shows that the last remaining liquid of the magma was pure silica, and it crystallized as quartz in the remaining spaces.

Figure 6.18 summarizes the sequence of mineral crystallizations expected from a typical basaltic magma. It is known as **Bowen's reaction series,** in honor of the Scottish-American scientist Norman Levi Bowen, whose *Evolution of the Igneous Rocks,* published in 1925, is a classic in the field of experimental geology. Note its similarity to the classification chart of Figure 6.14 and to the mineral sequences preserved in the frozen reactions we just described.

Bowen prepared glass samples whose compositions resembled simplified basalt magmas, and subjected them to high pressures and temperatures that simulated conditions of the shallow crust. Then he cooled his samples rapidly to freeze the reactions and see what minerals, if any, had crystallized from the artificial magma. He found that olivine crystallized at about 1200 °C; at about 1050 °C it became unstable, reacted with the magma, and converted to pyroxene. Indeed, by using glass magmas of the appropriate compositions and by adjusting the temperature-pressure relationships, he was able to reproduce all the relationships shown in Figure 6.18.

In Bowen's experiments, the minerals that crystallized maintained contact with the artificial magmas in a state of equilibrium. For this reason, the minerals and magma

Bowen's reaction series
A proposed sequence of mineral crystallization from basaltic magma, based on experimental evidence.

ASIDE 6.1   *The Palisades Sill: A Study of Layered Intrusives*

The Palisades sill of New Jersey is a classic illustration of Bowen's theory of fractionation. It is a sheer cliff of diabase—a coarse-grained equivalent of basalt—that stands like a fortress wall along the west bank of the Hudson River, directly across from New York City (Figure 6A). The sill, 300 meters thick, is exposed for approximately 80 kilometers along the river, and it serves as an excellent illustration of many of the processes we have described. Generations of New York and New Jersey college students have visited the sight, witnessing, 180 million years after the event, that a huge mass of molten magma had indeed sandwiched itself between nearly horizontal layers of country rock. Let us tag along on an imaginary field trip to this site. It will accomplish two objectives: to prove that the diabase was intruded as a sill and to examine the evidence that fractionation occurred.

Starting at the base of the cliff, we encounter soft, friable (crumbly) sedimentary rock: sandstones and shales (Figure 6B). The rocks are nearly horizontal, dipping gently to the west, and we are able to climb them easily, using the layers on the eroded cliff face like so many irregular steps. Within a few meters of the bottom of the sheer diabase cliff,

marked changes begin to show up in the sandstones and shales. Those rocks in direct contact with the diabase are fused to hard bricks, their colors glazed and indistinct. The intense heat of the magma baked the sediments as if they had been placed in a kiln. Directly on top of this *baked zone*, we encounter a *chilled zone* within the igneous rock itself. Within this zone, no more than a meter or so thick, the dark brown rock has the smooth texture of freshly broken chocolate. Evidently, hot magma came in contact with cold sedimentary rock, was quenched, and crystallized to a very fine-grained basalt. This conclusion is consistent with the notion that cooling rates are important factors in controlling the texture of igneous rocks. Xenoliths of shale and sandstone that broke off the host rock and floated into the magma are also encountered.

Toward the top of the diabase rock mass, west of the cliff face, we see that the diabase is covered by sedimentary layers identical to those that lie beneath it. Thus it is logical to ask: How do we know that the magma was intruded as a sill? In as much as the igneous rock is sealed between nearly flat sedimentary layers, how do we know that magma did not pour out onto the surface as a lava flow that was later

Figure 6A The sheer cliff of the Palisades sill along the New Jersey side of the Hudson River. The upper portion of the sill has been eroded and is not visible at this site; it is well preserved about one kilometer to the west.

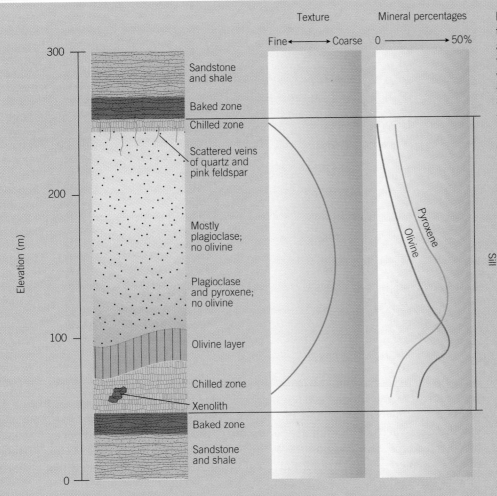

Figure 6B Textural and compositional changes in the Palisades sill and surrounding country rocks. Notice scattered quartz and feldspar veins toward the top of the sill and the concentration of dense, early-formed olivine toward the bottom. The fine-grained chilled zones at the upper and lower margins indicate that the magma cooled rapidly when it came in contact with the cold country rock. The coarsening of textures toward the center of the sill indicates a slower cooling rate there.

buried by younger sediment? The key piece of evidence is the existence of a second chilled and baked zone at the upper contact with the sedimentary rock. Had the magma poured onto the surface as lava, the upper sedimentary layers could not have been baked by the magma, for the simple reason that these layers did not then exist! Indeed, had the magma poured onto the surface, we would expect to find holes formed by escaping gas and perhaps a buried soil profile and erosion surface beneath the overlying sediment. But we find none of those things.

Having firmly established that the Palisades has been correctly named a sill, we return to the laboratory to analyze the rock samples we took when we climbed up the cliff. We find that the minerals change from bottom to top in a manner consistent with gravitational settling. The content of the rock fragments taken from the chilled zone consists of 1 percent olivine along with pyroxene and calcium-rich plagioclase. This finding indicates that olivine crystallized early in the cooling magma. About 4 meters above the chilled zone, we took samples from a soft, weathered olivine zone—a band of rock that is a meter or two thick consisting of 35–40 percent olivine. There is no olivine in the samples

we took above this zone in the great mass of the diabase. We can imagine that the early-formed olivine, being denser than the main body of the magma, sank toward to the bottom of the sill, until it was slowed by the cooling viscous magma above the chilled zone; it then accumulated to form the olivine zone.

Superficial examination of our rock samples does not reveal much else to prove that fractionation occurred in the sill, but detailed work reveals other features. The somewhat higher pyroxene content of rocks from the lower half of the sill suggests that these minerals also sank; but since many precipitated at lower temperatures than olivine and are not as dense, they could not concentrate as readily toward the bottom. The higher plagioclase content from the upper portion of the sill is consistent with the hypothesis that the late-crystallizing fraction of the magma consists of lighter elements. Although fractionation of the magma was not extensive enough to form rocks of intermediate to high silica content, we did find thin, scattered veins of quartz and sodium-rich feldspar toward the top of the sill. These veins probably represent the last remaining fluids of the molten rock.

Figure 6.18 Bowen's reaction series. This chart shows the experimentally determined sequence in which minerals crystallize from a cooling basaltic magma. The ferromagnesium silicates form a discontinuous reaction series and the plagioclase feldspars a continuous series. Potassium feldspar, muscovite (colorless mica), and quartz crystallize at the lowest temperatures.

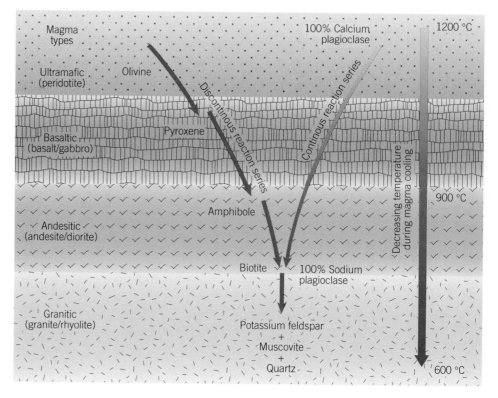

were able to freely react and exchange ions. He found that as the temperatures were lowered, more minerals formed and less liquid remained. The last drop of liquid had a composition very high in silica, and the minerals consisted of a mixture of pyroxene and plagioclase. When all the liquid was used up, the minerals that had formed, taken together, had the same composition as the original magma. The basaltic magma and the minerals that crystallized from it thus went through a series of interactions that resulted in a basaltic rock.

Bowen then considered what might happen if an early-formed mineral, such as olivine, is separated from the basaltic liquid. This mineral extracts a high percentage of iron and magnesium from the liquid, thereby shifting the liquid's composition toward the silica-rich end of the magmatic spectrum. A magma of this composition can remain liquid at lower temperatures than a basaltic magma, and more silica-rich minerals are able to crystallize from it. Therefore, the separation of early-formed ferromagnesian minerals and calcium-rich feldspar from a basaltic magma may prolong the life of the melt and produce a wide range of rocks. This process—in which early-formed crystals are prevented from reacting with the melt while changing the melt's composition through extraction—is called *fractionation*.

Early-Stage Mechanisms of Separation

Fractionation takes several forms in natural settings. *Gravitational settling* is effective in the fractionation of basaltic magma. It occurs as olivine and pyroxene, which are among the first minerals to crystallize, sink to the bottom of the magma chamber, out of contact with the liquid. These minerals collect to form peridotite. As in Bowen's experiments, their separation from the magma extracts a high proportion of iron and magnesium. Thus the remaining magma, enriched in lighter elements, will form basaltic and, in extreme cases, dioritic rocks. Aside 6.1 describes an example of fractionation through gravitational settling preserved in the 180-million-year-old Palisades of the Hudson River in New Jersey.

Though gravitational settling works well in extremely fluid magmas, geologists have come to question its effectiveness as a fractionation mechanism in most magmas. They doubt that the density differences between the early-formed minerals and the magma are great enough to overcome the magma viscosity, or resistance to flow. The situation is analogous to nuts mixed in a jar of honey; the nuts may be denser than the honey, but the honey will not let them sink. Geologists thus believe that other types of fractionation mechanisms may be more common. One such mechanism is *flow segregation*. As the magma rises in the crust, the early-formed minerals coat the walls of the enclosing country rock and are separated from the magma. Another mechanism, elegant in its simplicity, is the very act of crystallization itself. As we observed in our discussion of feldspar zoning and olivine reaction rims, only the surfaces of the minerals are subject to reaction with the magma and the early-formed cores of the minerals are effectively separated.

Late-Stage Mechanisms of Separation

Filter pressing is a separation mechanism that becomes increasingly effective as the magma evolves toward the solid state (Figure 6.19). In the late stages of cooling, the magma may crystallize into a mesh of hot crystals, trapping the residual liquid. As the mass continues to cool, the crystals settle and compact, squeezing out the remaining fluids like juice from orange pulp. Because these fluids remain late in the cooling of the magma, they are likely to crystallize as silica-rich minerals.

The late-stage silicate components of the magma are rich in water vapor and other volatiles. Superheated under high pressure, this hot mixture mobilizes and bursts through the soft, freshly formed igneous rock and out into the enclosing country rocks of the crust. There, it may blast out cavities and crystallize. Ions are able to migrate so rapidly in this environment that they form the huge crystals of coarse-textured pegmatite (Figure 6.20). These rocks are an exception to the general rule that coarse textures are associated with slow cooling. Finally, hot water is all that may remain of the original silicate magma. These

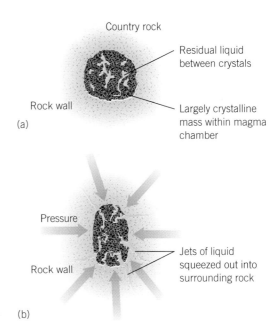

Figure 6.19 Filter pressing, during which residual liquids are squeezed from a nearly crystallized magma, is an effective late-stage fractionation mechanism.

Figure 6.20 A coarse-grained pegmatite dike cutting across country rock. Some pegmatites contain rare gems and elements that were retained in the late-stage fluids from which the igneous rock crystallized.

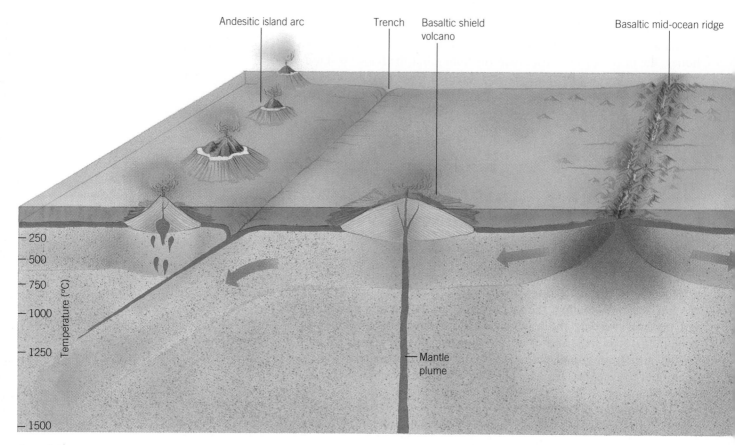

Figure 6.21 The structural settings of the various forms of igneous activity. Most of the activity occurs at divergent and convergent plate boundaries. Where they border island arcs, subduction zones generate andesitic magma; where they border continents, andesitic or granitic magma. Depending on the depth at which melting occurs, continental hot spots and rifts generate rhyolitic and granitic magmas (from crustal melting) or basaltic magma (from the mantle). Deep-seated mantle plumes feed oceanic hot spots.

hydrothermal solutions invade the cracks, the bedding planes, and the minute pores between the minerals of the surrounding country rock, all the while precipitating and dissolving elements along their paths as they cool.

The late-stage fluids of granitic magma are often enriched in rare elements that have remained in the liquid because their size or charge prevented them from fitting easily into the structures of most common minerals. Consequently, certain pegmatites and hydrothermal veins are the places to search for rare gems (such as ruby, emerald, sapphire, and tourmaline) and valuable elements (such as gold, silver, copper, and uranium).

The Geologic Settings of Igneous Activity

The global occurrence of igneous activity is confined primarily to plate boundaries or to intraplate regions above mantle hot spots (Figure 6.21). Each setting is characterized by magmas of distinctive composition and by distinctive styles of intrusive and extrusive activity. At divergent boundaries, basaltic magma is generated as the plates separate. The magma is derived from partial melting of the upper mantle, and it erupts from beneath the mid-ocean ridges to build basaltic crust. All this crust is recycled within about 200 million years as the oceanic lithosphere descends into the mantle at convergent boundaries.

Magmas display a more varied composition at convergent plate boundaries and are commonly richer in silica than the mafic magmas of divergent boundaries. Those plates that descend beneath the oceanic lithosphere generate magmas that are generally of andesitic composition. The magmas rise and build island arcs. Those that descend beneath continental crust generate both andesitic and granitic (silicic) magmas, which intrude the base of the continent. Most volcanoes that border subduction zones are

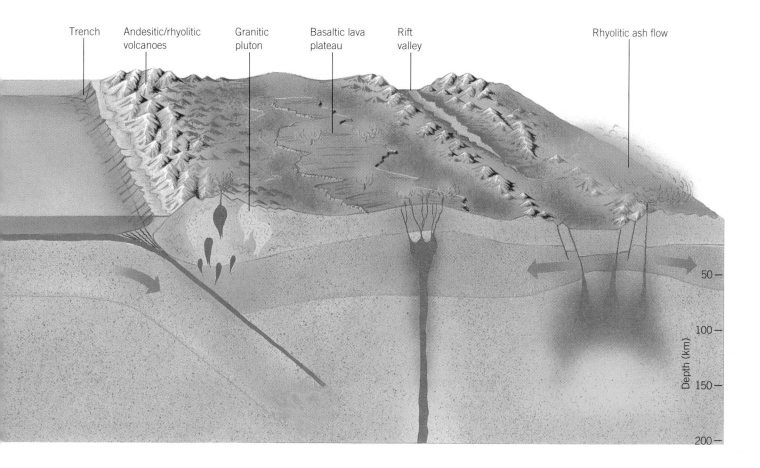

Trench   Andesitic/rhyolitic   Granitic   Basaltic lava   Rift   Rhyolitic ash flow
volcanoes   pluton   plateau   valley

Depth (km)

50
100
150
200

explosive, because their highly viscous silica magmas prevent gases from escaping. When pressures reach a critical point, the volcanoes erupt violently.

Mantle hot spots occur beneath both oceanic and continental crust. Those that occur beneath oceanic crust generate basaltic magmas and give rise to huge volcanoes. The Hawaiian Islands were formed over such a hot spot. Constructed layer by layer from the ocean floor by relatively peaceful liquid lava eruptions, the islands consist of some of the world's largest volcanoes, and they would be far larger if not for the fact that the Pacific plate upon which they rest is carrying them away from the hot spot. Those mantle hot spots that occur beneath continental crust have a more elaborate sequence of activity spanning millions of years. Early on, the hot spot melted the crust above it, which generated highly silicic and viscous magma that erupted explosively in huge ash flows. Later, the upper mantle was tapped directly, and basalt magma erupted.

Long fracture zones in the oceanic and continental crust constitute another major setting of igneous activity. These fractures extend to the mantle, where they tap voluminous sources of highly fluid basaltic magmas. Extensive lava plateaus are built of layer upon layer of basalt that erupts over millions of years. These fractures are associated with crustal rifting during the early stages of plate divergence.

Igneous Activity at Mid-Ocean Ridges

At the same time that Bruce Heezen and Marie Tharpe were mapping the global extent of the mid-ocean ridge system, other geologists were discovering that the oceanic crust is created by igneous processes within and beneath the ridge system. In the succeeding decades, geologists have studied these processes using seismic probes, deep-sea drilling, and submersible craft that descended to the rift valleys. In some locations, geologists have observed lava pouring out of cracks in the rift valley floor, and huge hydrothermal vents—called chimneys or black smokers—spewing mineral-rich hot water many

Figure 6.22 The layered structure of the oceanic crust and its relation to magmas beneath the rift valley of the mid-ocean ridge. The magma chamber is fed by partial melting of the asthenosphere. Olivine and pyroxene settle to the bottom of the chamber and form the peridotite of the upper mantle. Coarse-grained gabbro crystallizes from the main mass of the magma within the chamber. The basaltic dikes form as the crust separates and magma escapes the chamber through fractures. Basaltic magmas erupt through the dikes to the surface, cool rapidly when they come in contact with sea-water, and then congeal as pillow lavas.

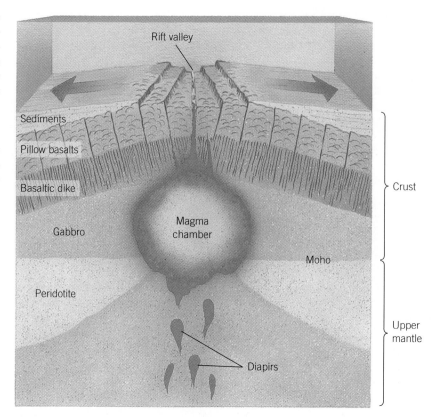

meters high. Magma lying at shallow depths beneath these locations provides the heat to drive these submarine hot springs. These on-site investigations are supplemented by inspection of slices of oceanic crust that have been uplifted and thrown onto the continents at subduction zone boundaries. The major result of these studies is the determination that the oceanic crust is basically a four-layered structure. Magma rises beneath the rift valley and crystallizes to form three of these layers; sediments form the fourth (top) layer (Figure 6.22).

Geologists have a reasonably good idea how this four-layer oceanic crust is constructed. First, basaltic magma, less dense than the surrounding mantle rock, rises beneath the ridge crest and accumulates in a space, or magma chamber, it has made for itself. There, it begins to crystallize in stages. Slow cooling forms the lower, massive gabbro layer. However, in the cooling stage, ultramafic olivine and pyroxene crystallize first and sink to the bottom of the magma chamber. These minerals form the peridotite of the uppermost mantle. While all this is happening, the crust continues to fracture and spread. The remaining magma forces its way through the vertical fractures in the gabbro and forms the sheeted basalt dikes. These dikes act as conduits for magma spilling onto the cold ocean floor, where it congeals rapidly to form the pillow basalts (Figure 6.23). As the solidified oceanic crust spreads away from the ridge, marine sediments gradually accumulate on the pillow basalts.

Magma production depends a great deal on spreading rates. In the Atlantic, where the plates spread slowly, magma chambers are few or nonexistent; consequently, oceanic crust production is low. Along the East Pacific Rise, on the other hand, spreading is so rapid that a full-fledged rift valley has not developed. For this reason, magma chambers are large and numerous, and produce a great volume of oceanic crust.

How is basaltic magma generated beneath the mid-ocean ridges? Recall that the asthenosphere is the zone in the mantle directly beneath the rigid lithosphere that is soft and close to melting. Seismic surveys reveal that it is normally located at depths of about 100 kilometers, but it rises to less than half that amount beneath the central rift valleys of the mid-ocean ridges. Indeed, the reduction in the speed of seismic waves at these

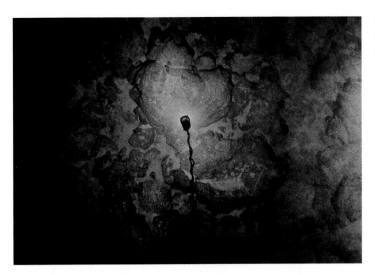

Figure 6.23 Recent lava flows on the crest of the East Pacific Rise, showing rounded pillow structures.

depths suggests that the asthenosphere may be 3 to 4 percent molten beneath the ridges. All that is needed is a slight change in the balance of conditions in the asthenosphere to generate basaltic magma.

Some years ago, geologists Alfred Ringwood and David Green of the Australian National Observatory conducted a series of experiments that suggested pressure reduction as a possible mechanism for the generation of basaltic magma beneath the rift valleys. They subjected samples of peridotite to the pressures and temperatures that exist 100 kilometers beneath the sea floor and they found that the samples remained solid. But when they slowly eased the pressure, the rocks yielded ever greater percentages of magma—up to 25 percent of their volume at the 1300 °C temperatures that prevailed. This phenomenon is an example of incomplete or partial melting (Figure 6.24).

Of equal significance, the composition of the magma generated in the Ringwood-Green experiment was similar to the composition of basalts observed to erupt at mid-ocean ridges. There is no surprise in this result; basalt has a much higher percentage of silica, aluminum, sodium, and potassium than peridotite. These are the low-melting point components of the rock, and we should expect them to liquefy earlier than the high-melting point components.

The attention of researchers has recently shifted to possible mechanisms of pressure reduction in the mantle. One hypothesis suggests that the oceanic crust is being pulled apart by the sinking action of the cold, dense parts of the plates at subduction zones. As the plates separate, fractures develop in the crust and upper mantle beneath the rift valley, which lowers the pressure on the asthenosphere. Another hypothesis suggests that rising convection currents lift plumes of hot, solid mantle rock close to the base of the crust. The mantle rocks, under reduced pressure at these shallow depths, then partially melt, releasing basaltic magma.

We should add that there are rivals to the pressure reduction hypotheses. For example, experiments on peridotite and other rocks show that wet samples melt at far lower temperatures than dry samples. Perhaps water seeping down from the oceanic crust beneath the mid-ocean ridges triggers partial melting. Actually, none of these explanations is mutually exclusive. All the suggested causes may play a role in partial melting—some being more important in a given location than others. Regardless of the exact cause, however, we know that under the high temperatures that prevail in the asthenosphere, the mantle generates basaltic magma.

Igneous Activity in Subduction Zones

Subduction zone magmas are even less accessible to observation than mid-ocean-ridge magmas, and geologists must rely more on experiments and conjecture to study them.

Figure 6.24 The geothermal gradient intersects the partial-melting curve at a depth of approximately 100 to 250 kilometers (the asthenosphere). Rocks are partially molten at these depths and serve as the source of the magma that forms the oceanic crust.

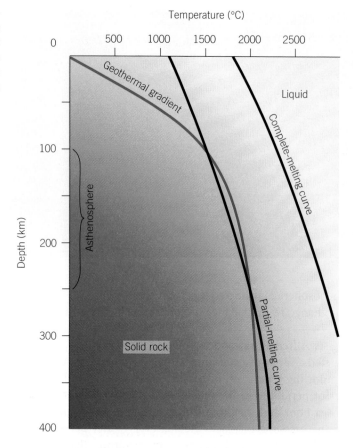

However, there is a central fact from which to start the investigations: subduction zone magmas are far richer in silica and other lighter elements than the mafic magmas of the central rift valley. The silica content of subduction zone magmas varies from intermediate (about 60 percent) to high (about 75 percent), and the rocks that crystallize from the magmas are reflective of these compositions.

Magmas of intermediate silica content give rise to diorite and its volcanic equivalent andesite, whereas highly silicic magmas produce granite and the extrusive rhyolite. Between these extremes are the most common of the high-silica rocks, granodiorite and dacite. Also significant is that the silica composition of subduction zone magmas generally increases from island arcs toward the continents. Intermediate magmas are common to island arcs and bordering continental ranges, but magmas of high silica content are almost exclusive to the continents.

Because subduction zone magmas form at lower temperatures than mafic magmas, it is tempting to imagine a simple system in which the lithospheric plates carry the basaltic rocks of the oceanic crust to the subduction zones, where they descend, partially melt at low temperature, and generate higher-silica magmas. This sequence may indeed happen to a small degree, but most geologists who have studied the process doubt that the cold subducting plates can absorb enough heat from their surroundings to melt appreciably; consequently, they seek other causes.

One explanation of the origin of subduction zone magmas involves water. As we have seen, it lowers the melting point of rocks at depth. The descending plates probably carry a great deal of water, some of it held by the sediments riding on the plates and some of it held by the minerals within the basalt. At depths beginning at about 75 kilometers, the pressures are so great that water is expelled from the plates to the hot, overlying mantle rocks. The melting points of the rocks are thus lowered enough to initiate partial melting, and given the lower temperatures that prevail in subduction zones, high-silica magmas are generated. The very-high-silica magmas that form the continental granites may themselves be generated by remelting of the granitic crust at the base of the continents.

The Ascent of Magma in Subduction Zones

How does magma rise in subduction zones and invade the crust? To answer this question, we must begin with the melting of mantle and lower crustal rock just above the subducting plate. The melting of source rock probably proceeds in stages. Early in the process, the rock is largely solid, with melting confined to the matrix between the grains.

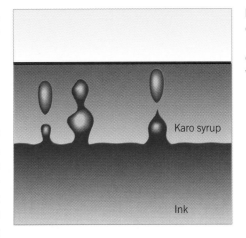

Figure 6.25 This experiment shows how density differences cause blobs of light liquid to rise through denser liquid. Geologists believe a similar process is at work in magma ascent.

Eventually, a point is reached at which the rock loses its integrity as a solid mass, even though much solid material is present in the melt. We may envision the magma at this point as having the consistency of slush (partially melted snow) within which the liquid and solid constituents move together as a unit.

For several reasons, this magma is less dense than an equal volume of surrounding rocks. Lighter, more silicic fractions of the source rock are liquefied first; if melting is total, then expansion occurs. Thus a given volume of the magma will weigh less than an equivalent volume of rock of the same composition, and it will have a lower density. The difference in density between the magma and the enclosing rock creates an unstable condition. In experiments simulating these circumstances, a layer of magma tends to bunch into a mushroomlike form, which buds off and rises buoyantly (Figure 6.25). The rising bud—which may reach tens of kilometers in diameter—is called a *diapir*. When it comes to rest in the crust, it crystallizes into a granitic or granodioritic batholith (Figure 6.26).

As they rise in the crust, diapirs encounter changing conditions that are reflected in the rock bodies they form and in their relationships to the rocks that enclose them. Those that come to rest deep in the crust (15–30 kilometers) have entered an environment in which temperatures and pressures are relatively high, and the enclosing rocks are relatively soft and yielding. At these depths, the metamorphic crustal rock and the

Figure 6.26 Half Dome and upper Yosemite Valley, California. Glaciers carved and smoothed a giant granitic batholith, leaving behind these spectacular features.

Apply and Decide 6.1

Radon, an Indoor Environmental Hazard

Each year, 136,000 people die of lung cancer in the United States. Some 25,000 of these deaths are caused by radon-222, a colorless, odorless, and tasteless, naturally occurring gas. Radon and smoking is an especially lethal combination—85 percent of the individuals who die from radon-related lung cancer are smokers.

Radon-222 is the product of the radioactive decay of radium, which in turn is a product of the decay of uranium, an element found in the crust primarily in granitic rocks.

Both radon and radium are links in the long decay chain that begins with the isotope uranium-238 and ends, finally, with the stable, nonradioactive lead isotope Pb-206. Because radon is an inert element like helium or argon, the atoms tend to collect in the air as a gas. Whereas uranium-238 has a half-life of 4.5 billion years, the half-life of radon is only 3.8 days. The element decays into polonium, which is itself radioactive and has a half-life of 138 days. When radon gas is taken into a person's lungs, it is in fact the polonium produced by the radon that injures lung tissue and causes lung cancer.

Radioactivity is commonly measured in picocuries (pCi). This unit of measure is named for the French physicist Marie Curie, who was a pioneer in the research on radioactive elements and their decay. One pCi is equal to the decay of about two radioactive atoms per minute. The Environmental Protection Agency (EPA) considers a radon level greater than 4 pCi per liter within a closed house to be unsafe. The agency estimates that several million homes in the United States exceed this level, and of these, several hundred thousand exceed 20 pCi/L. Inhaling this amount of radon gas is equivalent to smoking one pack of cigarettes per day.

The Geology of Radon

Uranium is the most massive naturally occurring element. When the Earth was in a molten state early in its history,

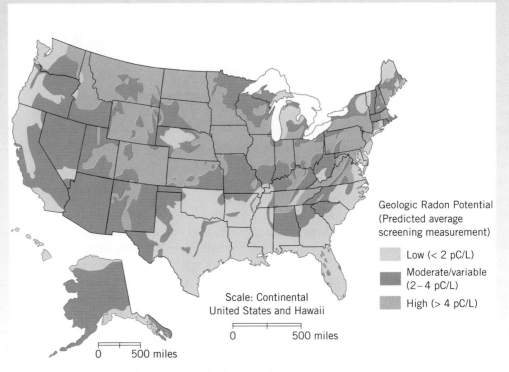

Figure 6C Distribution of radon potentials in the United States

most of the iron and other heavy metallic elements sank to the center of the planet and formed the core. Meanwhile, the lighter silicates in the melt rose to form the mantle and crust.

Because of its mass, much of the uranium sank along with the iron, indeed, the decay of this uranium may be the source of the heat presently escaping the core. However, uranium is in some respects chemically similar to the light elements potassium and phosphorus, in that it tends to form chemical bonds with aluminum silicate compounds in granitic magmas. Thus, some of the uranium was incorporated into upwardly migrating, high-silica magmas, and this is probably how it became concentrated in the granitic rocks of the crust.

Subsequent weathering, erosion, and metamorphism disbursed the uranium to other rocks of the crust. Today, all crustal rocks contain some uranium, although most contain just a small amount—between 1 and 3 parts per million (ppm).

Radon Formation and Movement

Just as uranium is present in all rocks and soil, so are its products radium and radon. When an atom of radium decays, it splits into an alpha particle—that is, a particle containing two neutrons and two protons—and a radon atom. As the alpha particle is ejected, the radon atom recoils in the

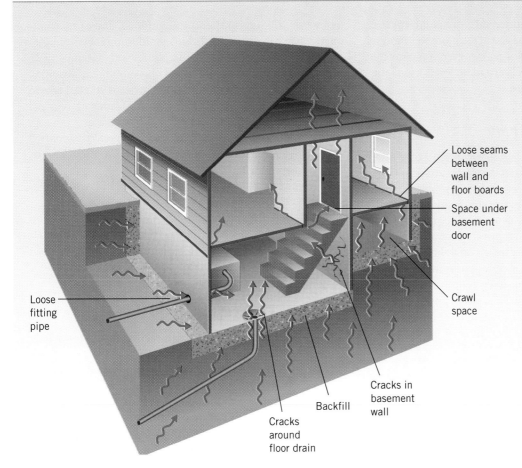

Figure 6D How radon penetrates a residential interior

because it is underlain by basalt lavas which are generally poor in uranium. Although it is not shown on this USGS map, regions of the Canadian provinces of Manitoba and Saskatchewan that lie just north of Minnesota and North Dakota have high radon potentials because they are situated on dry, highly-fractured clays. These fractures provide an efficient plumbing system for the radon to move through the ground.

Still, some houses in areas with a great deal of uranium in the soil have low levels of indoor radon, and other houses on uranium-poor soil have high levels of indoor radon. Clearly the amount of radon in a house is affected by factors in addition to the presence of uranium in soil and rock and the geology of the subsurface.

opposite direction, like a rifle recoiling when a bullet is fired. Some of the atoms are recoiled toward the grain's surface and out into the air that resides between grains or in a rock fracture.

Radon moves more rapidly through subsurface materials that are permeable, such as coarse sand and gravel, than through impermeable materials, such as clay. It also moves more quickly through air than through water; thus it travels longer distances in dry soil than wet soil before it decays. Consequently, homes in areas with drier, highly permeable soils and fractured or cavernous bedrock may have high levels of indoor radon.

Figure 6C shows the U.S. Geological Survey's analysis of the distribution of radon potentials in the United States. The radon potential varies with the uranium content of the bedrock and soil and with the bedrock structure of the region. For example, the Reading Prong of New Jersey and Pennsylvania, which is underlain by uranium-rich granites and metamorphic rocks and cut by numerous faults, has dangerously high radon potential. In contrast the Columbia Plateau of Washington and Oregon has moderate to low radon potential

Structural Factors in Buildings

Most of the radon that escapes the surface of the Earth enters the atmosphere, where it is such a small percentage of the total, it does no harm. However, if a building is present at an escape site, not only does the upwardly migrating soil air beneath the building move toward the foundation, but the surrounding soil air tends to flow toward the foundation as well (Figure 6D). There are three reasons for this lateral migration:

1. Increased permeability in the gap between the basement walls and the ground outside, which is "backfilled" with loose material from the construction site. Furthermore, having been shoveled, drilled, and sometimes even blasted with explosives, the backfill material itself generates radon.

2. Cracks, poorly sealed seams, utility entries, and uncovered soil in crawl spaces, all of which provide opportunities for the soil air to seep into the basement. From the basement, the air circulates into the interior of the house above.

3. The difference in air pressure between the soil and the house. Houses with poor ventilation, poorly

sealed foundations, and several entry points for soil air may draw as much as 20 percent of their indoor air from the soil. For this reason, even if the soil air has only moderate levels of radon, levels inside the house may be very high.

Radon in the Water Supply

Water in rivers and reservoirs usually contains very little radon, because the gas escapes from the water surface into the atmosphere. Radon in the water supply is a problem, however, for many small and medium size communities in North America where groundwater is the main source of tap water. Most small public water works and private domestic wells are closed systems in which the water travels only short distances to the end users. The radon has no opportunity to escape from the water, and with local delivery, transit times are too brief to allow the radon to decay. Thus radon travels into the home with the water supply and escapes into the indoor air as people take showers, wash dishes, etc. The areas most likely to have high levels of radon in groundwater are areas that have high levels of uranium in the underlying rocks.

Checking for Radon

You can measure the radon level in your house with a simple, inexpensive device that can be obtained from any hardware store. However, getting a reading is only the first step. You still need to know what the radon potential is—for the region, for the site on which your house has been built, and for the house itself. If the potential is high, then you should probably check your readings regularly. Check in the winter, when you put up the storm windows, and in the summer, when you turn on the air conditioning. Continue to check over the years as the house settles with age and cracks form in the foundation and spaces open around utility pipes.

You can obtain free information on steps you can take to minimize the radon danger in your home from your regional EPA office. Some recommended publications are:

A citizen's guide to radon. The guide to protecting yourself and your family from radon, 2nd ed. EPA 402-K92-001, 15 pages, 1992.

Home buyer's and seller's guide to radon. EPA 402-R-92-003, 32 pages, 1993.

Much of the information on radon in the discussion above was adapted from the U.S. Geological Survey paper entitled *The geology of radon,* which may be obtained from the Internet address: http://sedwww.cr.usgs.gov:8080/radon/georadon/3.html

THOUGHT QUESTIONS

1. To what extent is radon a natural hazard and to what extent is it a human-made hazard?
2. How might the users of water from private wells and local delivery systems process their tap water so that the radon in the water is diminished before it can escape into the indoor air of their homes?
3. Recall from Chapter 3, Earthquakes, that an increase in radon emissions from wells is a precursor of earthquakes. From your knowledge of how radon is generated and moves through the subsurface, explain why the emission of radon gas from a well might increase prior to an earthquake.

granite that crystallized from the diapirs form an intricate mixture called *migmatite,* which is present over a broad contact zone. Even within the main body of the batholith, the granite is intermingled with metamorphic rock. Exactly how emplacement occurs is not clear, but it involves much local melting and mixing. Some geologists suggest that this deep region of the crust serves as a source zone for silica-rich magma.

Diapirs emplaced higher in the crust (5–15 kilometers) encounter host rocks that are cooler and more rigid. At this level, it appears that they simply push aside the host rock. Therefore, contacts between pluton and host rocks are sharper than at lower depths, and the zone of mixing is not as extensive. Diapirs emplaced still higher in the crust (0–5 kilometers) encounter rocks at relatively lower pressures and temperatures. There, the cold, brittle crust tends to fracture. Consequently, the diapir rises by a combination of forceful injection, the shattering and cracking of host rock, and *stoping.* Stoping occurs when the roof of the enclosing rock breaks into huge chunks, which are assimilated as the magma eats its way higher into the crust. These fragments, called *xenoliths,* become stranded like islands within the igneous rock near the contact (Figure 6.27). In *caldron subsidence,* enormous blocks of overlying country rock, which may cover several square kilometers, partially sink into the viscous magma along concentric fractures. The magma intrudes into these fractures to form a series of ring dikes.

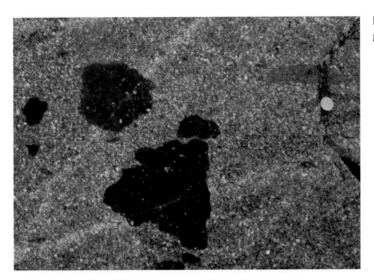

Figure 6.27 Basaltic xenoliths within granodiorite, Scotland.

Regardless of whether caldron subsidence and ring dikes develop, extensive fracturing of the shallow host rocks provides avenues for the magma to spread out into numerous channels and form dikes, sills, and laccoliths. At this level in the crust, the magma is under low pressure; and portions of it, fed from below, may be capable of bursting to the surface. The eruptions that result build volcanoes and other features (the subject of the next chapter). The high silica content of subduction zone magmas can make them extremely viscous, and their high viscosity prevents gradual escape of gases, so that pressures build until the magma erupts with great violence. As we will see in the next chapter, this condition makes subduction zone volcanoes extremely dangerous.

STUDY OUTLINE

The oceanic crust and most of the continental crust are igneous in origin.

I. IGNEOUS ROCK BODIES

  A. Igneous rocks begin as **magmas** generated from the melting of mantle and crustal rocks. Most magmas rise, intrude the crust, and crystallize to form rock bodies, or **plutons.** Some magmas escape to the surface as lava. Eruptions of lava can build conical volcanoes or thick lava plateaus.

  B. Plutons are classified on the basis of their relation to the layering of the host rocks that they intrude.
   1. **Concordant** plutons form parallel to the layering of the host rock either as flat, tabular **sills** or as mushroom-shaped **laccoliths.**
   2. **Discordant** plutons cut across the layering of the host rocks.
      a. **Dikes** are tabular bodies that cut jaggedly through layers of country rock.
      b. Pipes are cylindrical intrusions that filled and solidified in volcanic vents.
      c. Radiating dikes or concentric ring dikes may encircle these pipes.
      d. Diatremes are pipelike intrusions in which diamonds can be found.
      e. **Batholiths** are enormous granite plutons.

II. THE FORMATION OF ROCKS FROM MAGMA. A magma is a complex mixture of molten silicate rock material, dissolved gases, solid minerals, and chunks of country rock.

  A. One way to determine the conditions under which igneous rocks formed from magma is to study their texture. The large mineral crystals of **intrusive** rocks form when the magma cools slowly, undisturbed. Such conditions are found only at depth. The small crystals of **extrusive** rocks form under conditions of rapid, disturbed cooling at the surface.

B. There are five major types of igneous rock textures:
1. **phaneritic**—coarse-grained, large crystals
2. **aphanitic**—fine-grained, small crystals
3. **glassy**—smooth texture, no crystals
4. **porphyritic**—large crystals embedded in an aphanitic matrix
5. **pyroclastic**—cemented lava fragments

C. Igneous rocks can be grouped by mineral content into four families:
1. the **peridotite** family, consisting primarily of olivine and pyroxene and found in the upper mantle
2. the **basalt**-gabbro family, consisting primarily of pyroxene and calcium-rich plagioclase; the components of oceanic crust
3. the **andesite**-diorite family, consisting primarily of amphibole and sodium-rich plagioclase; associated with subduction zone volcanoes
4. the **granite**-rhyolite family, consisting of potassium feldspar, sodium-rich plagioclase, and abundant quartz; the main components of the continental crust

D. Igneous rocks arrayed on a classification chart display a number of gradational changes, from the peridotite toward the granitic end:
1. The percentages of the ferromagnesian minerals decrease, whereas the percentages of potassium feldspar, sodium plagioclase, and quartz increase.
2. The percentages of silica, sodium, and potassium increase, whereas the percentages of iron, magnesium, and calcium decrease.
3. The density of the rock decreases.
4. The temperature of magma crystallization decreases.

E. The microscopic analysis of mineral textures bears out the general sequence of crystallization suggested in the rock classification chart.
1. Reaction rims in the ferromagnesian minerals suggest a discontinuous reaction series. Zoning in the plagioclase feldspars suggests a continuous reaction. Quartz crystallizes last. Experimental work, summarized in **Bowen's reaction series,** lends support to the microscopic analysis.
2. Bowen's work also suggests that through fractionation, a single basaltic magma may produce many kinds of igneous rocks. The key to fractionation is that early-formed minerals are separated from the remaining liquid.
3. Fractionation mechanisms include gravitational settling, flow segregation, zoned crystals and reaction rims, and filter pressing.

F. The Palisades sill is an example of how fractionation can account for compositional differences in a single igneous rock body of basaltic composition.

III. **THE GEOLOGIC SETTINGS OF IGNEOUS ACTIVITY.**   Igneous activity occurs primarily at plate boundaries and above mantle hot spots. Each setting is characterized by magmas of distinctive composition and by distinctive styles of intrusive and extrusive activity.

A. At divergent boundaries, basaltic magma erupts and builds the layered igneous rock structure of the oceanic crust.

B. Where oceanic plates converge, andesitic magmas build island arcs. Where oceanic and continental crust converge, granitic magmas are generated. The high silica and gaseous content of subduction zone volcanoes causes them to erupt violently.

C. Hot spots beneath oceanic crust generate basaltic magmas that build huge, relatively peaceful volcanoes, such as those of Hawaii. Beneath continental crust, hot spots generate explosive ash flows and, later, basaltic lava flows.

D. Igneous activity in subduction zones produces the intermediate– to high–silica content magmas that form island arcs and continental mountain ranges. Water expelled from the descending plate may initiate partial melting of mantle rock, generating granitic magmas. They rise in the crust as diapirs to form batholiths.

STUDY TERMS

andesite (p. 172)
aphanitic (p. 170)
basalt (p. 172)
batholith (p. 168)
Bowen's reaction series (p. 175)
concordant (p. 168)
dike (p. 168)
discordant (p. 168)
extrusive (p. 170)
glassy (p. 170)

granite (p. 173)
intrusive (p. 170)
laccolith (p. 168)
magma (p. 167)
peridotite (p. 172)
phaneritic (p. 170)
pluton (p. 168)
porphyritic (p. 170)
pyroclastic (p. 171)
sill (p. 168)

CRITICAL THINKING QUESTIONS

1. Distinguish between concordant and discordant plutons.
2. What is the relationship between cooling rates and crystal size? How are these relationships illustrated in the textural changes from the bottom to the top of the Palisades sill?
3. By what mechanisms can fractionation produce rocks different in composition from the parent magma? By what mechanisms can partial melting produce a rock different in composition from the source magma?
4. Describe the evidence, viewed under the microscope, that supports Bowen's reaction series.
5. Describe the compositional changes that occur across the spectrum of igneous rocks from peridotite to gabbro, diorite, and granite.
6. What is the origin of magma beneath the mid-ocean ridges? Explain the process by which this magma cools to form the four-layered structure of the oceanic crust.
7. How do the magmas produced beneath subduction zones differ from magmas produced beneath the mid-ocean ridges? What processes may generate subduction zone magmas?
8. Explain the origins of batholiths and how they are emplaced in the crust.

CHAPTER 7

Volcanism

The Klamath Native Americans have lived in northern California and Oregon longer than the redwoods, and each generation has passed on the following story to the next.

> Long ago, the Chief of the Below World, who resided within a mountain called Lao Yaina, became enraged when a young maiden declined to marry him. He swore he would seek revenge by annihilating her people with the Curse of Fire, and he rose to the summit of Lao Yaina, flames streaming from his mouth. As the mountain trembled he hurled molten rocks through the air and unleashed a hot rain upon the forests that burned the trees to charcoal stumps. The people prayed for deliverance to the Chief of the Above World, who drove the Raging One back inside Lao Yaina. The top of the mountain sank behind him as he retreated into his subterranean lair. The next morning, the high peak of Lao Yaina had vanished.*

Geologists and anthropologists believe this legend recounts an enormous volcanic eruption that occurred in the Cascade Range of the Pacific Northwest. Lao Yaina is almost certainly Oregon's Mount Mazama, in whose collapsed center now sits the magnificent Crater Lake (Figure 7.1). The huge quantities of debris found along the flanks of the decapitated **volcanic cone** and in the surrounding region are records of the eruption. A charred moccasin was also found in the bottom layers of the debris. Radiometric dating of this artifact and of fragments of charred wood has established that the eruption occurred about 7000 years ago. Thus the legend constitutes an unbroken 7000-year-old verbal chain that links an ancient people with their modern descendants.

From projections across Crater Lake, geologists estimate that the summit of the volcano before the eruption stood 3700 meters above sea level and 1200 meters above the

volcanic cone
An accumulation of lava and/or pyroclastics around a volcanic vent.

---

*Adapted from Lawrence R. Kittleman's interesting article "Tephra," *Scientific American*, December 1979.

◀ A lava eruption on Mauna Loa. The Hawaiian Islands are constructed of innumerable flows of this sort.

Figure 7.1 Oregon's Crater Lake fills the caldera of the former Mount Mazama, whose summit was destroyed during an explosive eruption about 7000 years ago. In the center of the lake is Wizard Island, a small volcanic cone that formed within the crater around the tenth century A.D. Geologists have reconstructed the history of the Mount Mazama eruption by analyzing the layers of lava and debris ejected from the volcano.

present rim of the cliff encircling the lake. A thick deposit of thinly layered ash at the base of the volcano provides evidence that the first products of the eruption settled out of the air. Above these ash layers lies a crude jumble of pumice, ash, and shattered rock. From observations of modern volcanoes, we know that some of these sediments were carried by a ground-hugging fiery cloud that surged down the mountainside; other sediments crept slowly as viscous masses of mud. In all, 30 cubic kilometers of debris were ejected from the volcano, which is about a hundred times the volume ejected by the Mount Saint Helens, Washington, eruption of May 18, 1980.

Deprived of support from below, the summit of Mount Mazama collapsed in a heap along steeply dipping circular faults, leaving in its center a crater 9 kilometers in diameter. About the tenth century A.D., a fresh eruption occurred within the crater and built a small volcanic cone. Today, the cone is called Wizard Island of Crater Lake. The lake itself covers 50 square kilometers and is 600 meters deep in places. Sheer cliffs tower 150–600 meters above the lake.

We have presented two contrasting descriptions of the same event; one ancient and metaphorical, the other modern and scientific. The event is no less awesome in either version, but let us now pursue the scientific explanation of volcanoes and their important contributions to the geology of the Earth's crust.

The Anatomy of a Volcano

volcano
A vent in the surface of the Earth through which magma, gases, and rock fragments erupt; also the term for the landform that develops around the vent.

vent
A conduit through which magma rises to the surface.

The term **volcano** stems from *Vulcan*, the Roman god of fire. It refers to an outlet to the Earth's surface through which magma, gases, and fragments erupt and to the commonly conical landform constructed of these materials. Usually, the outlet or **vent** is cylindrical, having been shaped by swirling gases and rock fragments escaping the

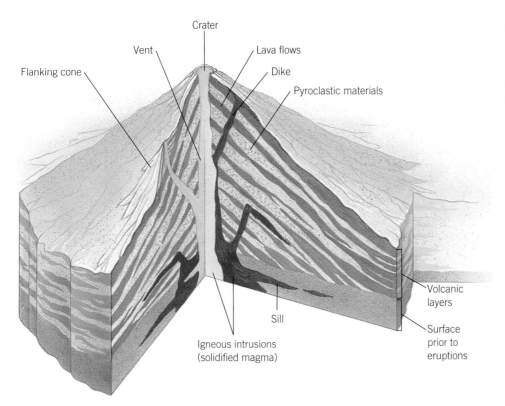

Crater

Vent

Flanking cone

Lava flows

Dike

Pyroclastic materials

Volcanic layers

Surface prior to eruptions

Sill

Igneous intrusions (solidified magma)

Figure 7.2 The internal skeleton of a volcano consists of a central vent and a network of dikes and sills filled with congealed lava. The cone itself is constructed of layers of lava and volcanic debris.

magma. During an eruption, most of the fragmental material accumulates close to the vent, as does the magma that pours out onto the surface as **lava.** Successive eruptions build the classic cone-shaped structure most people associate with the word *volcano* (Figure 7.2). The main pipe is generally located near the summit of the cone, where it opens to a bowl-shaped **crater,** or a much larger **caldera.** The crater can form from a violent eruption that blows the summit away or from the gentle piling of material on all sides of the vent. However, most craters are the result of collapse following an eruption, as in Mount Mazama. Large volcanoes are commonly fed through a network of vents that issue from their sides and create flanking cones. This elaborate plumbing system is filled between eruptions with congealed magma that provides the volcano with an internal skeleton. In active volcanoes, this skeleton serves as a pressure cap on the magma below.

Much remains to be discovered about the internal workings of volcanoes. No two volcanoes are precisely alike; and a single volcano, guided by rhythms that are as yet unknown, may not follow the same pattern each time it erupts. Nevertheless, there is little that is random about volcanoes. Every significant feature is linked to a causative chain: from tectonic setting, to magma composition, to eruptive style, to the physical state of the material ejected, to the shape and construction of the volcanic cone. In sum, volcanoes are excellent examples of the relationship between form and process in nature. Our purpose in this chapter is to explore this relationship.

lava
Magma that flows out onto the surface of the Earth; also refers to the rock body formed after the magma cools.

crater
A circular depression; a volcanic crater contains the vent or vents of the volcano.

caldera
A volcanic crater larger than 1 kilometer in diameter, usually formed by explosion or collapse.

Tectonic Settings of Volcanism

There are some 600 active volcanoes on the continents and islands of the world (Figure 7.3). That is, 600 have been observed by human beings to erupt during recorded history. Over two-thirds of these volcanoes constitute the *Ring of Fire* that borders the subduction zones of the Pacific. To the 600, add several thousand volcanoes that have not erupted in recorded history but whose lightly eroded cones indicate that they are of geologically recent origin and may spring to life again. Now add on the submarine volcanoes; 50,000 have been identified on the floor of the Pacific Ocean alone! How

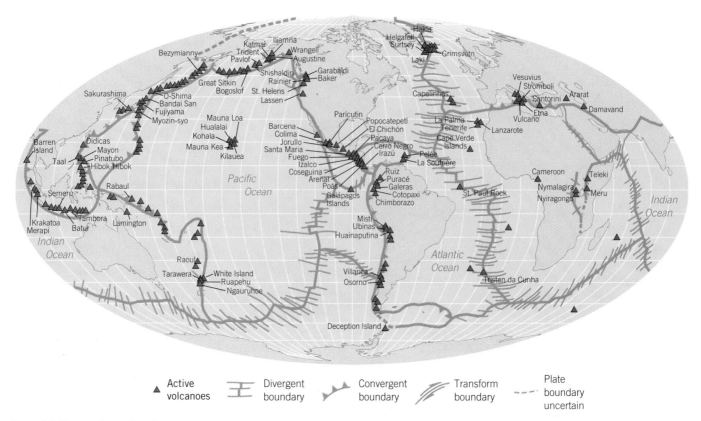

▲ Active volcanoes	⧧ Divergent boundary	⌁ Convergent boundary	⟋ Transform boundary	- - - Plate boundary uncertain

Figure 7.3 Most of the active above-sea-level volcanoes of the world are located at or near plate boundaries. They are so numerous along the margins of the Pacific Ocean that the region is referred to as the *Ring of Fire.*

many are active is difficult to say, but it is obvious that volcanic eruptions, frequently presented in the media as exotic events, are instead among the most common phenomena that shape the Earth's crust.

Most volcanic activity, or *volcanism,* occurs along diverging and converging plate boundaries—that is, along the rift valleys of the mid-ocean ridges and along the island arcs and continental margins that border subduction zones. However, the Earth's dynamism is not exclusively expressed by volcanism at plate boundaries. Hot mantle plumes rising from the deep interior can penetrate the crust far from plate boundaries. Intraplate hot spots result wherever they force or melt their way to the surface.

Plate interactions also control the composition of most of the world's volcanoes. The basaltic magmas that feed the mid-ocean ridge volcanism of Iceland are generated under conditions characteristic of divergent plate boundaries. The andesitic magmas that feed the volcanoes of the Aleutians and the Marianas are generated by the subducting plates that descend beneath these island arcs (ocean-ocean convergence). The subducting plates that pass beneath the continents (ocean-continent convergence) generate the andesitic and granitic magmas that feed coastal ranges such as the Cascades of the western United States and the Andes of western South America.

Subduction zone magmas contain a high percentage of silica and dissolved gases. Silica content controls the *viscosity* of the magma, that is, its stiffness or internal resistance to flow. Magma viscosity, which prevents the escape of pent-up gases, is in turn a key predictor of the explosiveness of a volcanic eruption. That is why subduction zone volcanoes are among the most violent and potentially dangerous.

The Mechanics of a Volcanic Eruption

The magma that fuels a volcanic eruption is generated by the melting of solid rock in the upper mantle or base of the crust. Because the molten liquid is hotter and less dense than the surrounding rock, it migrates upward through fractures. Prior to eruption, it

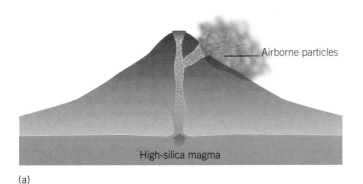

(a)

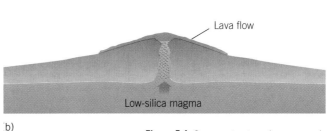

(b)

Figure 7.4 Gas content and magma viscosity affect a volcano's eruptive style. (a) Explosive eruption. Viscous magma retains gases under great pressure. When the magma plug rises higher in the vent, pressure is lowered and the gases are released suddenly and violently. (b) Peaceful eruption. As fluid magma flows from the vent, gases can escape gradually. Note also that how a volcano erupts affects the shape of its cone: the debris of explosive eruptions fall near the vent, forming a steep cone, but fluid lava flows away from the vent, forming a low, broad cone.

commonly collects in a large reservoir, or *magma chamber,* a few kilometers or so beneath the surface. The ascent of the magma from the chamber into the volcano proper is marked by earthquake tremors. They occur as the rising magma weakens fractures, melts the overlying rock, and moves through the newly created space.

Picture suddenly pulling the cap on a bottle of warm soda you have just shaken. Carbon dioxide bubbles surge rapidly to the surface and out the neck of the bottle, carrying the liquid along with them. Volcanologists believe a similar combination of pressure reduction and escaping gases triggers volcanic eruptions. The pressure is relieved by the fracturing and shattering of sealed vents at the summit as the magma migrates upward. The gases that are then released separate and initiate the actual eruption.

Magma viscosity plays a key role in the style of an eruption because it determines how the gases separate from the melt. A highly viscous silicic magma, for example, retards the expansion and passage of the gas through the liquid, so that pressure builds within the magma. Also, the gas escapes more slowly, so most of it is retained as the magma rises higher in the vent beneath the summit. Thus when separation does occur, it is likely to be rapid and the resulting eruption explosive (Figure 7.4a). In low-viscosity basaltic magmas, gases escape uniformly and gradually, so that eruptions are relatively peaceful (Figure 7.4b).

A magma's viscosity is determined by its silica content, as we have noted, and also by its temperature and the presence of dissolved gases. You may recall from Chapter 5 that the basic unit of silica (SiO_2) is the four-sided silicon-oxygen tetrahedron, a structure uniquely equipped to form strong linkages with neighboring tetrahedra. At the molecular level, these linkages stiffen the magma and retard its flow. Thus the higher the silica content, the more linkages that will form and the greater the viscosity of the magma will be. For this reason, a granitic magma (75 percent silica) is 10 orders of magnitude more viscous at the same temperature than a basaltic magma (50 percent silica)—a disproportionate difference for this change in silica content (Figure 7.5).

Raising the temperature lowers magma viscosity because it enables the silicon-oxygen tetrahedra to acquire the energy to break free of the forces exerted by their neighbors. On the other hand, cooling increases viscosity because it enables the tetrahedral linkages to grow longer and more complex. Dissolved gases play a dual role in magma viscosity. While dissolved in the magma, they break tetrahedral linkages, which enables the magma to flow more readily. But when the gases come out of solution and form bubbles, the magma stiffens rapidly.

Conditions at the time of eruption as well as the materials the magma encounters during ascent also influence the character of a volcanic eruption. Nowhere is this better illustrated than in typical sea-floor lava flows. Those that occur at great depths are

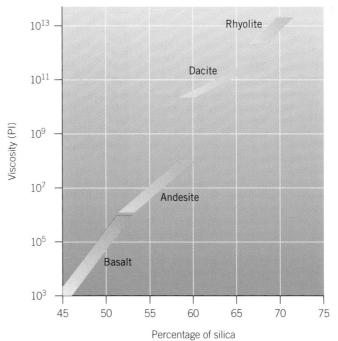

Percentage of silica

Figure 7.5 The viscosity of a magma varies exponentially with its silica content. At the same temperature, a rhyolitic magma (70–75 percent silica) is far more viscous than a basaltic magma (about 50 percent silica). (Adapted from Michael A. Summerfield, *Global Geomorphology,* New York: Longman Scientific & Technical, 1991, p. 111.)

Figure 7.6 This basaltic pillow lava, exposed on the face of a quarry wall, formed on the ocean floor. The rounded shapes were created when cold seawater came in contact with the molten lava.

Figure 7.7 A series of phreatic eruptions off the Icelandic coast in 1963 created the island of Surtsey. Steam and volcanic ash were propelled to great heights as cold seawater mixed with hot lava in the shallow volcanic vent.

pillow lava
A term applied to lavas of ovoid or pillow shape.

phreatic eruption
A volcanic eruption of steam, mud, or ash initiated by the contact of water with magma or hot rock.

seldom explosive, because the pressure of overlying water prevents gases from escaping the lava. Therefore, the lava flows smoothly and peacefully, and congeals to form ovoid **pillow lava** (Figure 7.6). On the other hand, **phreatic eruptions** involve the mixing of cold water and hot magma at shallow depths and are extremely explosive. One that occurred in 1963 created the new island of Surtsey off the coast of Iceland. The combination of hot magma and cold water that seeped into the vent generated steam of great propellant power (Figure 7.7). Two- and 3-ton lava bombs were shot hundreds of meters into the air for a period of weeks.

Magma reacting with country rock can also increase the violence of an eruption. The magmas beneath Mount Vesuvius of Italy gained carbon dioxide from the assimilation of limestone at shallow depths. The increased gas content probably contributed to the explosiveness of the eruption that destroyed Pompeii in A.D. 79. The El Chichon, Mexico, eruption of 1982 emitted a sulfur-rich plume that floated into the upper atmosphere and disturbed weather patterns in the Northern Hemisphere for more than a year. Much of the sulfur was probably derived from sedimentary deposits assimilated by the magma during its passage to the surface.

The Materials of Volcanic Eruptions

Volcanoes release three kinds of materials: lavas, gases, and fiery fragments. We will briefly describe these materials and the rocks that form from the lavas and fragments.

Lavas

In Chapter 6, we defined the two classes of igneous rocks: those that form from magmas that intrude country rock and cool within the Earth's crust, and those that form from lavas that erupt onto the Earth's surface. The two rock classes exhibit textural features reflective of their cooling histories. Intrusive rocks have coarse textures, because the magmas from which they were derived cooled slowly, allowing crystals to grow large. In contrast, extrusive rocks of the same composition exhibit fine-grained texture because

(a)

(b)

(c)

Figure 7.8 Common textures of volcanic glasses. (a) Obsidian exhibits smooth texture and conchoidal fracture. (b) Pumice is composed of innumerable glassy threads and tiny air holes. (c) Scoria superficially resembles a sponge, with large air holes.

the lavas cooled rapidly. The hot lava congealed when it met the cold air, and the flat shape of the lava flows exposed a greater surface area to cooling. Consequently, there was only time for microscopic crystals to form.

Mafic lavas form basalt, those of intermediate silica content form andesite, and silicic lavas (the equivalents of granodiorite and granite, respectively) form dacite and rhyolite. Though all lavas cool rapidly on the surface, differences in composition combined with widely fluctuating physical conditions create varied rock textures. If a lava loses heat too rapidly to form minerals, it may cool to a smooth glass. An example is **obsidian,** which lacks the organized internal atomic arrangement of a true mineral (Figure 7.8). Obsidian is usually a common feature of silicic lavas, whose high viscosities tend to prevent the migration of ions to crystal seeds within the melt. More commonly, volcanic glass is pockmarked with bubble cavities formed by escaping gases that failed to collapse in the viscous lava. Basaltic lavas exhibit large bubbles, which impart a spongy-looking texture to the cinderlike *scoria*. In silicic lavas, the bubbles are usually extremely numerous; and the light, frothy-looking, abrasive **pumice** is the result (see Figure 7.8). It is also common for the texture of a lava to consist of an intricate mixture of minerals and glass. Often, large crystals, or **phenocrysts,** are incorporated within an aphanitic or glassy matrix. Cooled lavas such as these are said to have *porphyritic* texture. As we described in Chapter 6, these minerals represent two generations of cooling. The phenocrysts precipitate at depth within the magma chamber, whereas the matrix forms rapidly on the Earth's surface.

Basaltic Lava Flows

Because of its relatively low silica content and high temperature, basaltic lava flows so easily from the vent that it is often referred to as runny lava. Individual flows have been traced for 200 kilometers in the Columbia Plateau of Washington and Oregon. When a thick basalt flow cools, it shrinks and develops **columnar joints.** The joints are a network of intersecting cracks that extend down the entire thickness of the flow and divide it into six-sided pillars (Figure 7.9).

Geologists distinguish between two kinds of basaltic lava flows, which bear Hawaiian names. **Pahoehoe lava** (pronounced pa-hoy-hoy) is billowy and undulating with a smooth continuous skin, but beneath this skin lies molten lava (Figure 7.10). As the lava pushes forward, it wrinkles the skin, causing it to take on a ropelike appearance. On occasion, the crust of the lava will thicken and grow rigid while the lava beneath continues to flow. When the supply of lava gives out, a hollow *lava tube,* sometimes over a kilometer long, is left behind.

Aa lava (pronounced ah′-ah) is the cooling viscous material that commonly forms at the front and sides of a river of pahoehoe lava (see Figure 7.10). It has a rough surface of blocks and fragments; partly congealed, red-hot lava lies beneath the rubble. If the main mass of the lava is still active, it may push the blocks forward or may churn

obsidian
A volcanic glass, either black or dark-colored, usually of rhyolitic composition, and characterized by conchoidal fracture.

pumice
A very light, cellular, glassy rock; often floats in water and is commonly used as an abrasive.

phenocryst
A relatively large crystal embedded in a finer-grained matrix.

columnar joints
Cracks that form as the result of contraction during the cooling of lava flows, dividing the lava into columns.

pahoehoe lava
A lava flow with a smooth, ropy surface.

aa lava
A lava flow with a rough, jagged surface.

Figure 7.9 Columnar jointing in the 60-million-year-old basaltic lavas of the Giant's Causeway, Northern Ireland. The parallel six-sided pillars were formed as the cooling lava contracted.

them over and plow them under in a tractor-tread motion. Pahoehoe and aa lava are identical in composition, but aa lava is the more viscous of the two, because gas has escaped from it.

Silicic Lava Flows

Lavas of very high silica content, being more viscous than basaltic lavas, generally do not flow far from the vent. Those that crystallize into aphanitic rock form rhyolite, the extrusive equivalent of granite. However, highly viscous flows cool to form obsidian. In the absence of explosive gases, the viscous lava may extrude from the vent—like toothpaste from a tube—forming a **lava dome,** or it may ooze a short distance down the flanks of the volcano (Figure 7.11).

lava dome
A convex structure of solidified lava extruded from the vent of a volcanic crater.

Figure 7.10 Dissolved water and gases can cause lavas of essentially the same composition to differ markedly in appearance and flow properties. Aa lava, deficient in dissolved gases, contains solid blocks of material and flows stiffly. Pahoehoe lava, richer in dissolved gases, flows smoothly and has a billowy or ropy surface.

Figure 7.11 Mount Saint Helens, Washington, in 1985. The large dome growing in the central crater is made of viscous silica that extruded from the volcano following the explosive 1980 eruption. Lava domes are a common late-stage feature of volcanoes having high silica content.

Pyroclastic Materials

The ancient Greek word for "fire" is *pyro,* and the word for "broken" or "fragment" is *klastos,* so the word *pyroclastic* accurately describes the materials that are expelled during the explosive phase of an eruption. Pyroclasts may be as small as dust particles or as large as trucks. They are classified according to size.

Ash, less than 2 millimeters in diameter, is not the residue of fire, as its name might imply. It is a tiny droplet of lava, composed of glass and mineral crystals, that cools in midair. **Cinders** and **lapilli** are pebble-sized fragments (2–64 millimeters) of the same composition as ash.

Blocks (defined as greater than 64 millimeters) are ejected as solid fragments. **Bombs** are the same size as blocks but are sent into the air in a semimolten state and are molded while in transit. Most bombs acquire the streamlined form that their name suggests. Upon landing, they may flatten like pizza dough. The enormity of some of the bombs that are propelled into the air during especially violent eruptions defies the imagination. Mount Cotopaxi of Ecuador, in full eruption, hurled a 200-ton bomb 14 kilometers from the vent!

Pyroclastic fragments cemented or welded together form **pyroclastic rocks.** They, too, are classified according to size. Cemented ash and lapilli-sized fragments form **tuff** (Figure 7.12a). If bombs, blocks, or other coarse fragments are incorporated within the tuff matrix, the rock is called **volcanic breccia** (Figure 7.12b). The cementing agents of pyroclastic rocks may be chemicals precipitated between grains, as with common sedimentary rocks, or the grains may simply weld together while semimolten.

Pyroclastic deposits ejected from a volcanic vent reach their final resting place by three means: air fall, surge, and mudflow. Air-fall deposits, or **tephra,** are fragments that settle out of the air; as a rule, they are well sorted. Large fragments land close to the vent, whereas finer materials are carried by winds and settle farther from the vent. The finest materials, volcanic dust, may actually circle the globe.

Surge deposits are ground-hugging pyroclastic materials propelled down the slopes of some volcanoes on a bed of hot steam. A **nuée ardente** (French for "glowing cloud") is a mixture of hot incandescent gases and fragments that can surge at speeds exceeding 400 kilometers per hour (Figure 7.13). Upon cooling, the gases escape and the fragments fuse together to form *ignimbrite,* or *welded tuff,* a rock as hard and massive as concrete.

Pyroclastic materials can also be transported as a muddy **lahar,** which is slower moving than a nuée ardente but can nevertheless outrun the inhabitants of a village downslope. The tops of large, steep volcanic cones are frequently covered with ice; others have well-forested slopes fed by abundant rainwater. Heat from the volcano combined with the shaking of the slopes during eruption mobilizes the waterlogged mixture of mud, soil, fragmented lava, and gases that pours rapidly into surrounding valleys.

ash
Pyroclastic material less than 2 millimeters in diameter.

cinder
Glassy, vesicular airborne fragments from 4 to 32 millimeters in diameter.

lapilli
Pyroclastic particles that range in size from 2 to 64 millimeters in diameter.

block
A pyroclastic fragment larger than 64 millimeters in diameter that is ejected in a solid state.

bomb
A clump of partially molten lava, ejected from a volcano, whose shape is streamlined in flight; larger than 64 millimeters in diameter.

pyroclastic rock
A rock of any size formed from the cementation or welding of volcanic fragments.

tuff
A rock of consolidated volcanic ash.

volcanic breccia
A pyroclastic rock composed of angular fragments that are larger than 64 millimeters in diameter.

tephra
A general term that refers to all airborne pyroclastic debris.

nuée ardente
A turbulent, ground-hugging, gaseous cloud erupted from a volcano.

lahar
A mudflow of volcanic material.

Figure 7.12 Two pyroclastic rocks. (a) Welded tuff (or ignimbrite), composed of cemented particles of volcanic ash. (b) Volcanic breccia, composed of coarse fragments in a tuff matrix.

(a)

(b)

Volcanic Gases

Volcanic gases dissolved in the magma under high pressure are released when the magma rises close to the surface. Carbon dioxide, water vapor, sulfur dioxide, hydrogen sulfide, hydrogen chloride, and nitrogen are the most common of these gases. Whereas lava accumulates relatively close to the vent, these gases, or *volatiles*, are dissipated into the atmosphere. Later in the chapter, we will discuss the possibility that carbon dioxide and fine sulfur dioxide droplets sent out during eruptions may markedly affect the Earth's climate.

Gases escape from lava in ways other than large eruptions. **Fumaroles** are vents that release hot gases (100 °C–1000 °C) either late in the history of a volcanic eruption or during the beginning phase. Some fumaroles emit gases that have directly escaped the magma; others—like those that once existed in the Valley of the Ten Thousand Smokes (Alaska)—release steam generated from water heated by buried yet still-hot ash deposits.

fumarole
A vent or ground opening that spews volcanic fumes or vapors.

Volcanic Structures and Eruptive Styles

As we have seen, there is a direct connection between magma composition and the eruptive style of a volcano. This connection also extends to the form and structure of the

Figure 7.13 A nuée ardente photographed during the 1980 Mount Saint Helens eruption. Because the hot steam reduces friction, these ground-hugging, fiery clouds can move down a mountainside at speeds of hundreds of kilometers per hour.

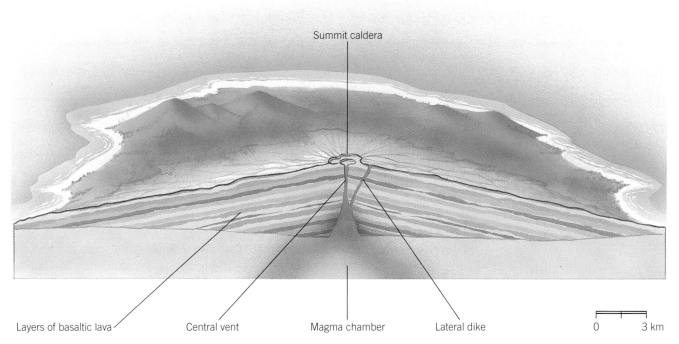

Summit caldera

Layers of basaltic lava   Central vent   Magma chamber   Lateral dike

0   3 km

Figure 7.14 A typical shield volcano is a broad, low cone built of innumerable lava flows.

volcanic cone. For example, high-silica magmas cause explosive eruptions that eject a high percentage of pyroclastic debris. These materials come to rest close to the vent at a high angle, thereby forming a steep cone. In contrast, the peaceful eruptions of fluid basaltic lavas spread far from the vent and form broad, gently sloping cones. We find, therefore, that magmas of similar composition produce similar types of eruptions and volcanic landforms.

Basaltic Volcanoes

Mid-ocean ridges. Oceanic hot spots. Certain continental fracture zones. These tectonic settings display broadly similar volcanic structures because they tap sources of extremely fluid basaltic magma derived from the mantle. The major difference among the structures is related to the shape of the vents through which the magma escapes to the surface. For example, volcanic cones are constructed around relatively localized and circular vents, whereas flat lava plateaus are built of lava that issues from a regional system of linear fissures in the crust.

Shield Volcanoes

Basaltic eruptions that occur on volcanic islands tend to be relatively peaceful. Mild explosiveness is confined to *lava fountains*—high arching sprays caused by escaping gas within the crater, or fissure, that can reach heights of 400 meters. During these displays, the eruptions thunder with a sound much like smashed crockery, as if a giant chef in a huge kitchen has gone mad. Following the display, a sheet of shimmering 1100 °C lava pours smoothly down broad mountain slopes at speeds of roughly 40 kilometers per hour, engulfing and smothering all in its path. An average flow forms a layer 10 kilometers long, 300 meters wide, and 12 meters thick and takes over three years to crystallize into a still formidably hot (750 °C) rock. Thousands of eruptions of this sort build **shield volcanoes.** Named for their broad, low profiles, shield volcanoes have slopes of less than 5 degrees (Figure 7.14). They are typical of mid-ocean ridge and oceanic hot-spot settings.

Thousands of eruptions of this sort have built the island of Hawaii. Located over the most active hot spot on earth, the island consists of five merging shield volcanoes

shield volcano
A broad, low-profile volcanic cone, commonly composed of basaltic flows.

Figure 7.15 The Hawaiian Islands are a chain of volcanoes that built up from the Pacific Ocean floor. As described in Chapter 2, all of the volcanoes were formed as they passed sequentially over a stationary hot spot that is now located beneath the big island of Hawaii. The volcanic islands were carried northwestward as the Pacific plate slid over the hot spot; thus the ages of the volcanoes increase steadily from Loihi to Niihau. (© 1986 Dynamic Graphics Inc.)

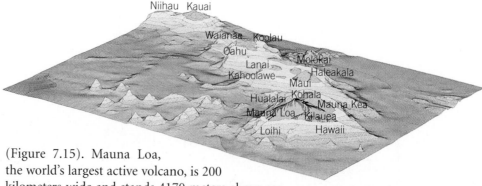

Figure 7.15 The Hawaiian Islands are a chain of volcanoes that built up from the Pacific Ocean floor. As described in Chapter 2, all of the volcanoes were formed as they passed sequentially over a stationary hot spot that is now located beneath the big island of Hawaii. The volcanic islands were carried northwestward as the Pacific plate slid over the hot spot; thus the ages of the volcanoes increase steadily from Loihi to Niihau. (© 1986 Dynamic Graphics Inc.)

(Figure 7.15). Mauna Loa, the world's largest active volcano, is 200 kilometers wide and stands 4170 meters above sea level. That does not tell the whole story, however, because its base rests on the Pacific Ocean floor, 5000 meters below sea level. So the full height of Mauna Loa is about 18.6 kilometers, taller than Mount Everest.

Kilauea is Hawaii's most active volcano. Earthquake analysis indicates that magma is generated by partial melting of mantle rock at a depth of approximately 50 kilometers. Prior to an eruption, magma rises to an accumulation chamber 1600–6400 meters beneath the summit. Lava escapes not only through the summit but also through fracture zones that radiate from the summit (Figure 7.16). Repeated forceful injections of magma into these fracture zones are splitting apart Kilauea's steep southern flank. When the cliffs become unstable, cracks develop and large blocks slump downward, shatter, and slide toward the sea. The impact of these landslides are often widely felt. Following an injection of magma in 1975, a slice of the cliff suddenly slipped 3.5 meters and set off a magnitude 7.2 earthquake. Submarine surveys have revealed that all the Hawaiian Islands are surrounded by sloping ramps of rubble, presumably built of innumerable landslides triggered in similar ways.

Cinder Cones

cinder cone

A steep-sided volcano formed by the accumulation of ash, cinders, and other debris close to the vent.

Cinder cones are steep-sided, beautifully symmetrical volcanoes built entirely of pyroclastic debris (Figure 7.17). Many cinder cones are the product of basaltic eruptions and can be found as satellite cones on the flanks of Hawaiian volcanoes, although they are

Figure 7.16 A fissure eruption along the eastern rift zone of Kilauea, Hawaii.

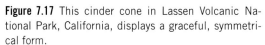

Figure 7.17 This cinder cone in Lassen Volcanic National Park, California, displays a graceful, symmetrical form.

Figure 7.18 Repeated basaltic lava flows have built the Columbia Plateau of Washington and Oregon.

also common on the continents. Cinder cones are exceptions to the general tendency for pyroclastic materials to be of high silica content. The reason is obscure but may be related to gas concentrations within the magma chamber. Seldom greater than 500 meters in height, they are minuscule in comparison to the great Hawaiian shield volcanoes or the subduction zone volcanoes of the Cascade or Andes ranges.

Fissure Eruptions and Lava Plateaus

If mafic magma issues from long fractures in the crust, rather than from a localized cylindrical vent, its fluidity will enable it to spread evenly as a **fissure eruption.** The lava behaves much like a stream overflowing its banks—hence its name **flood basalt.** Flow upon flow stacks up in thick accumulations called **lava plateaus.** The volume of the Columbia Plateau in Oregon and Washington, built out of hundreds of individual lava flows, is estimated at 200,000 cubic kilometers (Figure 7.18). Most of the lavas are about 15 million years old, but some of the eruptions have occurred as recently as 6 million years ago.

The Columbia Plateau is dwarfed by other continental lava plateaus, such as Russia's Siberian traps and India's Deccan traps. The term *trap* is derived from the Swedish *trappa,* for "stairs," and it refers to the pattern made by the eroded flat layers. In all, they constitute the greatest concentrations of volcanic material on the continents, and plateaus of similar size are found in the ocean basins. Most occur far from plate boundaries and appear as huge volumes of material in a localized area. This observation leads geologists to wonder whether there is a connection between flood basalts and the mantle plumes that create hot spots.

Figure 7.19 charts the position of flood basalts with respect to some known hot spots. You can see that the basalts may have been located at one time over the hot spots and that plate motions have since moved them away from these centers. Notice that the Columbia Plateau basalts are to the west of the Yellowstone (Park) hot spot. The two are lined up in the direction of plate movement; thus it is likely that the Columbia basalts were above the Yellowstone hot spot in the past.

fissure eruption
A volcanic eruption through a long fracture rather than a central vent.

flood basalt
Highly fluid basaltic lava produced during a fissure eruption.

lava plateau
An elevated, flat-topped region composed of a thick succession of horizontal lava flows.

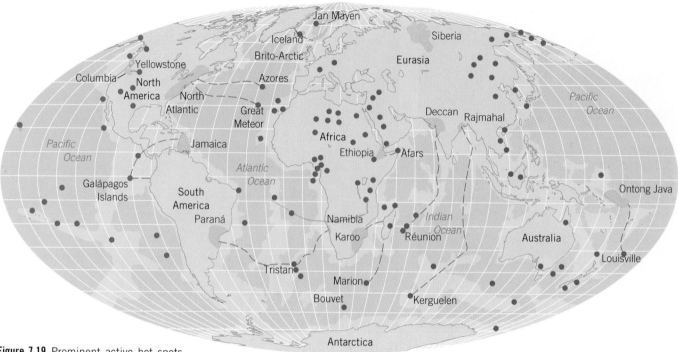

Figure 7.19 Prominent active hot spots and ancient lava plateaus of the world. Some geologists reason that the lava plateaus originated when lithospheric plates passed over the hot spots.

Andesitic and Silicic Volcanoes

The volcanic chains that border subduction zones, whether continental or island arcs, are fed by magmas of intermediate to high silica composition. The high viscosity and gas content of these magmas causes them to erupt explosively and, as a result, form volcanoes whose slopes and structures contrast markedly with those of shield volcanoes.

The high viscosity of subduction zone magmas traps gas bubbles and prevents them from expanding—a condition that intensifies as the magma continues to cool and rise close to the surface. Pressure within the magma may then build to such a level that the gases escape in an explosive eruption. Clots of molten magma and fragments of ancient volcanic rock are propelled into the air along with the gas. The coarsest debris falls closest to the vent, but the finest may spread over hundreds of kilometers. Following the escape of the pent-up gases, the eruption may enter a quiet phase, during which viscous magma oozes out the vent. The lava also congeals in the fissures and side vents that radiate from the central vent.

Many such cycles of violence and relative peace build a **composite cone** or **strato-volcano,** consisting of alternating layers of pyroclastic debris and viscous lava (Figure 7.20). Composite cones are much smaller in diameter than shield volcanoes, but they have steeper profiles and form some of the world's most majestic mountain peaks: Japan's Mount Fujiyama, Mount Mayon in the Philippines, Mount Cotopaxi of Ecuador, Italy's Mount Vesuvius, Indonesia's Krakatoa, and Mount Rainier of Washington, to name a few.

stratovolcano (composite cone)
A volcanic cone consisting of alternating layers of pyroclastic deposits and lava.

Figure 7.20 Mount Rainier, Washington, compared to Mauna Loa, Hawaii. Although the steep, composite cones of continental subduction zone volcanoes appear huge and rise to great heights, they are in fact dwarfed in comparison to oceanic shield volcanoes.

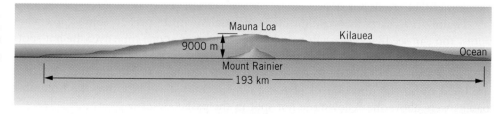

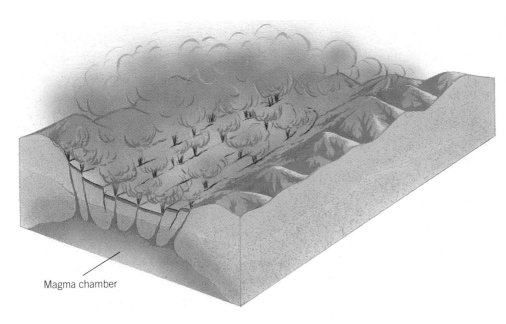

Magma chamber

Figure 7.21 Caldera eruptions such as those that occurred repeatedly in Yellowstone, Wyoming, and the Valles Caldera, New Mexico, are perhaps the most dangerous of all eruptions. Slowly rising plumes of viscous magma cause the crust to sink along circular fractures; at the same time, high-silica ash particles are released through the fractures. The ash forms a suffocating cloud that spreads over wide regions surrounding the caldera.

Subduction zone magmas of extremely high silica content intensify the cycle of eruption just described. Because these magmas are extremely viscous, they prevent the escape of gases until they rise within a few kilometers of the surface. When the gases are finally released, they are under extreme pressure. If their path to the top of the vent is blocked, they may instead blast through the sides of the volcano. A cataclysmic eruption then ensues.

Combine the explosiveness of a subduction zone volcano with the proximity of a population center and you have a formula that may result in huge losses of life. In 1902, Mount Pelée, on the island of Martinique in the West Indies, sent out threatening signals. For two weeks, minor flows and poisonous gas were observed escaping the vent. The emissions were strong enough to kill livestock and make people sick. The population of the nearby port city of Saint Pierre was terrified, but instead of evacuating the town, local politicians kept them there to ensure that they would vote in an upcoming election. Suddenly, the side of Mount Pelée blew open, triggering a nuée ardente. The surge, moving at 150 kilometers per hour, reached the nearby port city of Saint Pierre in a matter of minutes. Except for a badly burned shoemaker and a convict confined to an underground cell, the entire population of the city—approximately 30,000 people—perished in a matter of minutes.

Ash Flows

Some of the largest and most explosive eruptions have occurred in continental regions far from plate boundaries. These eruptions have not built conventional cones but, rather, immense sunken craters surrounded by ash fields (Figure 7.21). Yellowstone, Wyoming; the vast ash field of the San Juan Mountains in Colorado; the Valles Caldera of New Mexico; and the Long Valley–Mammoth Lakes region of California were formed in this manner.

Yellowstone National Park is situated over the Yellowstone hot spot, a rising mantle plume that initiated the slow melting of granitic continental crust. The park is named for the color of the rhyolite lavas that characterize the region. Rhyolite tuff deposits, composed of cemented needlelike shards of high-silica volcanic ash, are evidence of an extremely violent eruption (Figure 7.22). Although no volcanic cone is present, geologists have been able to trace the outlines of three sunken calderas in the park and surrounding regions. From the volumes of pyroclastics ejected, they have concluded that these eruptions were far greater than those of typical subduction zone volcanoes.

Mount Saint Helens: Case Study of a Volcanic Eruption

Mount Saint Helens is a moderate-sized volcano in the Cascade Range. Its highly silicic dacite magma is derived from the subduction of the Juan de Fuca plate beneath the North American plate (Figure 7.23). In May 1980, it became the most publicized and best studied volcanic eruption of its kind.

On Sunday morning, May 18, 1980, Mount Saint Helens appeared relatively calm. Earthquake activity had slackened over the past few days. With the exception of puffs of smoke rising from small vents in the crater—something that had occurred on and off since March 25—no other signs of trouble were visible. David A. Johnston, a 30-year-old U.S. Geological Survey volcanologist, had been working through the night 8 kilometers from the crater, laser-measuring the gas emissions and keeping watch on the general condition of the mountain. Since March 27, the north slope of the summit had been expanding steadily and ominously (Figure 7.24). "It's like standing next to a dynamite keg with the fuse lit. Only we don't know how long the fuse is," Johnston told a group of reporters who visited the site. As they took notes, the mountain trembled beneath their feet, and they could see chunks of ice breaking off the glaciers around the summit.

The state of relative calm continued up until 20 seconds after 8:32 A.M. that Sunday morning of May 18. Then Johnston saw the north face of the mountain blow open: "Vancouver! Vancouver! This is it!" Johnston cried to headquarters via his two-way radio. Those were his last known words, for he never got off the mountain. The full

Figure 7.22 A typically rhyolitic cliff in Yellowstone Park, Wyoming, composed of the ash deposits formed during caldera eruptions.

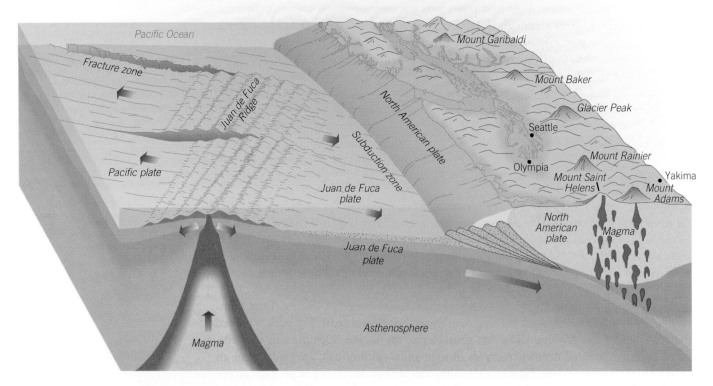

Figure 7.23 The regional tectonic setting of Mount Saint Helens in the Cascade Range. The Cascade volcanoes are fed by magmas that form as the Juan de Fuca plate descends beneath the North American plate off the west coast of the United States.

Figure 7.24 An ominous fractured bulge in the north slope of Mount Saint Helens grew steadily in the weeks before the eruption on May 18, 1980.

Figure 7.25 The May 18, 1980, eruption of Mount Saint Helens, seen here at 2:44 P.M., sent a mushroom cloud of dust and debris high into the air.

force of the eruption came roaring right at him in a lateral blast. "The trees were gone. David's trailer, the jeep, everything was blown away," said a geologist friend who visited the site of the Johnston work station by helicopter a few days later.

The very instant Johnston had relayed his message, two other geologists, Dorothy and Keith Stoffel, were flying 400 meters directly above the summit of Mount Saint Helens aboard a light reconnaissance plane. "The whole north side of the summit crater began to move instantaneously as one gigantic mass," Dorothy Stoffel recalled. "The entire mass began to ripple and churn without moving laterally. Then the whole north side of the summit started moving to the north along a deep-seated slide plane." The north slope, which had swelled over 24 meters in the months preceding the eruption, had just collapsed.

Within seconds following the landslide, an enormous gray cloud of ash and steam shot from the crater (Figure 7.25). The Stoffels had to escape by putting the plane into a steep dive to the south, just ahead of the cloud that threatened to engulf them. They saw lightning bolts flashing through it. Few people had ever come this close to an explosive volcano and lived to tell of it.

Ten seconds before the Stoffels and Johnston witnessed the collapsing summit of Mount Saint Helens, seismographs deployed around the volcano had registered a 5.1 magnitude earthquake centered 1.6 kilometers below the north flank of the cone. The tremor triggered the collapse of the oversteepened bulge of the volcano's north flank that the Stoffels observed. The collapse produced the largest landslide-debris avalanche recorded in historic time. The rocks and fragments were waterlogged with melted snow, groundwater, and hot steam ejected from the volcano. At speeds of 200 kilometers per hour, a 4-cubic-kilometer of mass of rocks and mud moved from the north slope 28 kilometers to the Toutle River.

Collapse of the north wall pulled the plug on the hot gases and steam contained in the magma reservoir. In the instant following the landslide, a lateral blast of ash and steam blew out the north slope of the mountain. Traveling at close to the speed of sound, it was strong enough to rip trees 2 meters in diameter up by their roots within a few kilometers of the crater. At a distance of up to 30 kilometers from the summit, prime Douglas firs were snapped like matchsticks. A truck was found overturned 26 kilometers away, the plastic parts melted. In all, the blast devastated 590 square kilometers of countryside.

The cloud of ash and steam that the Stoffels had dodged expanded into a mushroom cloud 25 kilometers high. Within hours, 10 centimeters of ash blanketed Yakima, Washington, 140 kilometers to the east. Spokane, Washington, 500 kilometers to the

Figure 7.26 The enormous crater left behind by the May, 1980, eruption, which blew out the top and north side of the mountain.

northeast, was so darkened that the automatic street lamps turned on and remained lit all day. Enough ash fell on Montana to spoil crops, and thin wisps of dust were traced across the country.

For five hours after the blast, a series of ground-hugging nuées ardente poured through the gaping hole in the north flank at speeds of 130 kilometers per hour, carrying ash and pumice downhill. At the same time, the glaciers of Mount Saint Helens melted and combined with centuries-old soil, rock fragments, and organic debris high up on the mountain. This dense fluid mass formed lahars that flowed down into the surrounding valleys.

The eruption took 400 meters off the top of the mountain and left a crater 750 meters deep (Figure 7.26). Less forceful eruptions have followed, but it seems that sufficient gas has been released to allow the magma to rise slowly rather than explosively. In the years since the eruption, the magma has been quietly building a 300-meter lava dome above the central vent (see Figure 7.11).

Forecasting Eruptions

Krakatoa, the great Indonesian island volcano that erupted August 29, 1883, unleashed huge tsunamis that killed over 36,000 people and set off a cannonlike roar heard as far away as central Australia. It is a sobering thought that this awesome volcano, whose fine dust turned sunsets red for months around the world and noticeably reduced global temperatures up to a year later, erupted unexpectedly after two centuries of quiescence. A century later, the science of volcanology had progressed to the level that the forecasts and warning bulletins of the Mount Saint Helens eruption saved many lives. The mountain and surrounding regions were cordoned off and emergency teams were alerted well in advance. Most of the 60 people who died in the eruption had received adequate warning to leave the area but chose to stay instead. As with earthquakes, however, there is a significant difference between forecasting and predicting the exact moment of an event.

Many steps are involved in forecasting volcanic eruptions. To begin, geologists draw upon their background knowledge of the broad geologic settings in which volcanoes erupt. Next, field studies, combined with radiometric dating in laboratories, help geologists identify which volcanoes have been most active in the past, their frequency of

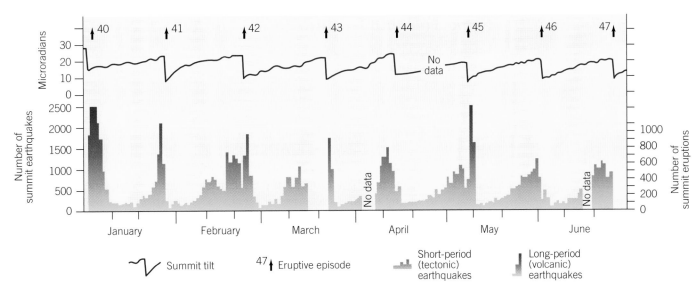

Figure 7.27 Determining the rhythm of the Kilauea eruption cycles. The first phase is characterized by summit inflation (measured in microradians) and short-period (tectonic) earthquakes, which indicate that the magma chamber is fracturing. The second phase is marked by long-period (volcanic) earthquakes, a signal that the magma is rising toward the surface. This phase culminates in eruption, accompanied by abrupt summit deflation.

eruption, and the type of activity they most likely can expect of them. For example, as we described at the beginning of the chapter, surveys of the volume and type of materials deposited on the flanks of a volcano help geologists to determine the power and explosiveness of its eruptions. Dating these materials establishes a volcano's rhythm—whether it erupts on the average of once each century or once every thousand years. The distribution of ejected material and its relation to local topography, such as valleys and ridges, suggests where material is likely to accumulate and, therefore, helps geologists to identify potential hazards. Work of this kind enabled geologists Dwight R. Crandell and Donald R. Mullineaux of the USGS to conclude, five years prior to the 1980 eruption, that Mount Saint Helens could erupt explosively within a few years. Studies such as theirs are not designed for short-range prediction but are useful in regional planning and in establishing research priorities.

Having decided that a volcano may erupt in the foreseeable future, geologists establish monitoring systems. Because Kilauea and Mauna Loa have stirred repeatedly for decades, the USGS maintains the Hawaiian Volcano Observatory (HVO) on the island. HVO geologists have placed seismographs strategically around the flanks of the volcanoes and toward their summits to track the location, severity, and frequency of earthquakes. They have also installed a network of *tiltmeters* to detect ground swelling or other distortions (Figure 7.27). Using tiltmeter and seismograph data, geologists have discovered that events preceding an eruption on Kilauea occur in a two-phase cycle of summit inflation and deflation.

In the first phase, the summit of Kilauea gradually inflates. Inflation is accompanied by an increase in the number of short-period earthquakes clustered around the summit. In this first phase, magma rises and accumulates in the reservoir beneath the summit, which causes the rocks to rupture as they are pushed aside. In the second phase, the summit deflates suddenly and sharply as the magma begins to leave the reservoir. These events change the earthquakes to a rumbling, long-period mode—which, HVO geologists have learned, is the preamble to the eruption. Close monitoring has given HVO the capability of forecasting a Kilauea eruption hours—even months—before the event.

Ground tilting and earthquakes are only two of the signs geologists monitor in the effort to forecast volcanoes. Other signs include changes in the composition and quantities of gases escaping the crater; changes in the heat escaping the crater; changes in groundwater composition and temperature; and changes in related igneous activity—for example, the appearance of geysers, hot springs, or steam vents (fumaroles) in the region. By integrating this information, geologists hope to improve their forecasting abilities.

Some Volcanic Hazards in the United States

Volcanic eruptions are not infrequent events that happen in exotic places. If we define an active volcano as one that has erupted at least once during historic times, it may surprise you to learn that the United States is the third most volcanically active nation. (Indonesia is first, with 140 potentially active volcanoes, and Japan is second, with over 70.) There are 53 active volcanoes in the United States, and the USGS has identified 35 that are likely to erupt in the foreseeable future. Most are situated in the Aleutian Islands, Alaska, and Hawaii, but there are hazards in the contiguous 48 states as well.

Besides Mount Saint Helens and Mount Mazama, there are 14 other volcanoes in the Cascade Range that have erupted in the relatively recent geologic past, and some of them are beginning to stir. Geologists believe that a few of them have reached a condition similar to what Mount Mazama was like just before the gigantic eruption that formed the Crater Lake caldera. The last major eruption of Mount Hood, for example, occurred 300 years ago, which may mean that it is long overdue for another.

The Mono-Inyo craters, a chain of small lava domes on the western front of the Sierra Nevada range, are the second-most active volcanoes in the 48 states. Eruptions occur every 200 to 300 years, and they are of the very dangerous silica–ash flow variety. The ash expelled during the last eruption incinerated entire forests. The Mono-Inyo craters are only a few miles from the sunken Long Valley, California, caldera, and these features are, in fact, related. About 700,000 years ago, Long Valley was created in a tremendous ash eruption.

Long Valley has proven that volcanology is still an uncertain science. On May 25, 1982, the USGS issued a "notice of potential volcanic hazard" for the Long Valley region. USGS scientists cited some compelling reasons for concern, such as a marked uplift of the caldera floor, formation of a new group of fumaroles in the area, and an increase in shallow earthquake activity. To date, however, no eruption has occurred. Nature's time scale is not always in harmony with our need for precision.

Volcanism and Climate

The atmosphere plays a dual role in regulating the Earth's temperature. On the one hand, minute dust particles and acid droplets in the upper atmosphere reflect incoming sunlight, which lowers the Earth's surface temperature. On the other hand, carbon dioxide and other gases in the lower atmosphere absorb the heat radiated from Earth's surface, which has a warming effect. Volcanoes release all these temperature-regulating substances. For this reason, many scientists suspect that the rate of volcanism may strongly influence the Earth's climate.

The June 16, 1991, eruption of Mount Pinatubo clearly illustrates the cooling effect initiated by the venting of huge quantities of dust and sulfur dioxide into the upper atmosphere. Sulfur dioxide readily combines with water vapor to produce sulfuric acid droplets, which are very effective in scattering sunlight. As these particles and droplets spread around the globe, average worldwide temperatures dropped by more than 1 °C and did not recover for the next two years (Figure 7.28). Similar temperature declines have been noted following most other major eruptions involving fine pyroclastic dust. One of the best known followed the Mount Tambora, Indonesia, eruption of 1815. The following year was so cold that it snowed in New England in July, and 1816 came to be known in North America and Europe as "the year without a summer."

Carbon dioxide's role in regulating Earth's temperature is well known, for it is a key component of the **greenhouse effect.** The hot sun sends out light waves that the relatively cold surface of the Earth absorbs and converts to infrared radiation. Carbon dioxide, along with water and methane, acts as a filter. It allows sunlight to pass through the atmosphere but absorbs some of the infrared energy that rises from the Earth's surface. The infrared energy eventually escapes to space, but in the delay, the Earth's lower

greenhouse effect
The warming of the Earth's atmosphere through the presence of atmospheric gases that absorb and re-radiate the heat rising from the surface.

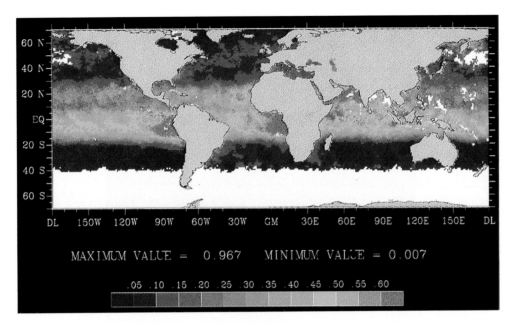

Figure 7.28 Measurement of the thickness of the dust blanket from the Mount Pinatubo eruption of 1991. The dust rose into the atmosphere and lowered average worldwide temperatures nearly two degrees for almost three years. (Courtesy of the National Oceanic and Atmospheric Administration.)

atmosphere and surface are warmed. The greater the concentration of carbon dioxide in the atmosphere, the more heat will be trapped and the higher Earth's temperature will be.

While it is difficult to trace the warming effect of the carbon dioxide released during a single eruption, geologists believe that volcanism cumulatively supplies the carbon dioxide that warms the Earth. From the presence of diamond in ultramafic rock derived from the mantle, we know that the mantle contains a great deal of carbon—and most magma is derived from the mantle. Volcanoes bring this carbon, combined with oxygen, to the atmosphere.

Some geologists see a connection between periods of intense volcanism and unusually warm periods in the Earth's past. For example, they note that the Mid-Cretaceous period, some 100 million years ago, was a time when the Earth's average temperature was perhaps 15 °C higher than at present. During that time, a number of mantle plumes also erupted and presumably brought carbon to the surface.

Notice that some researchers see volcanism as the cause of the Earth's cooling whereas others see it as the cause of Earth's warming. These views are not necessarily contradictory, since the cooling effects of a single eruption commonly last 2 to 5 years, whereas carbon dioxide lingers far longer in the atmosphere. Carbon dioxide release, therefore, should have the greater cumulative effect, and we should expect long-term warming. However, intense volcanism sustained over long time-periods might release enough acid droplets and dust to cause catastrophic declines in sunlight and temperatures. Some geologists have, in fact, suggested this mechanism as a cause of ice ages and of the sudden mass extinctions of plant and animal life that have occurred at irregular intervals throughout geologic time. The causal relationships between volcanism and climate are obviously complex; investigating them is a major goal of the geosciences.

Constructive Aspects of Volcanism

In closing, let us mention some of the constructive aspects of volcanoes.

- They are major landforms of the Earth's surface; without volcanoes, there would be no Hawaii or Iceland.
- A volcano may temporarily devastate the ecosystem that surrounds it, but it provides rich soils that strengthen the ecosystem that returns (Figure 7.29). Soils of volcanoes and lava plateaus are among the most fertile on the Earth.

Figure 7.29 The destructive effects of volcanism are short-lived, but the benefits are long-term. Here we see life reestablishing itself on the devastated flanks of Mount Saint Helens soon after the great eruption.

- The gases and particles that volcanoes emit are major regulators of climate. Most geologists agree that the oceans and atmosphere are the result of volcanism that occurred during the early stages of the Earth's formation.
- Valuable metals such as copper, gold, and mercury are brought close to the Earth's surface through volcanic vents and fractures.
- The near-surface magma bodies that supply volcanoes have enormous energy potential. The USGS estimates that geothermal heat of magma bodies within 10 kilometers of the Earth's surface in the continental United States could supply between 800 to 8000 times the energy the nation consumes each year! Iceland, New Zealand, and San Francisco have relied on geothermal power for years.
- Volcanoes are a priceless asset. Whether in repose or in eruption, they are beautiful.

STUDY OUTLINE

The eruption of Mount Mazama, which led to the formation of Crater Lake, Oregon, was witnessed by Native Americans 7000 years ago. Geologists have estimated the size of the eruption by analyzing the composition and distribution of the volcanic deposits. A **volcanic cone** in the center of Crater Lake is evidence of a later eruption.

I. **THE ANATOMY OF A VOLCANO.** A **volcano** is a **vent** through which magma escapes to the surface as **lava.** The flows of successive eruptions build the cone-shaped structure; collapse afterward leaves a **caldera** or **crater** at its center. Between eruptions, the magma hardens in the vents to form the volcano's skeleton.

II. **TECTONIC SETTINGS OF VOLCANISM**
    A. Most volcanism occurs along diverging and converging plate boundaries, but there are also intraplate hot spots, caused by mantle plumes.

    B. Plate interactions are the foundation of the following causative chain:
        1. The type of plate boundary controls the composition of the magma.
        2. Composition, temperature, and gas content control magma viscosity.
        3. The magma viscosity controls the degree of explosiveness of the eruption.
        4. The degree of explosiveness controls the shape and form of the volcano.

    C. This chain results in the following general relationships:
        1. Divergent plate boundaries yield low-viscosity basaltic magmas.
        2. Convergent boundaries yield high-viscosity andesitic, dacitic, and rhyolitic magmas.

III. **THE MECHANICS OF A VOLCANIC ERUPTION.**   Pressure reduction inside the vent and escaping gases trigger volcanic eruptions.
    A. The higher the silica content of the magma, the greater the viscosity. Viscous, high-silica magma retards gas expansion, which increases pressure, resulting in an explosive eruption. In low-silica, mafic magmas, gases escape easily, and eruptions are relatively peaceful.

    B. External conditions also affect the style of eruption. Peaceful sea-floor eruptions form **pillow lava.** The mixing of cold water and hot magma causes violent **phreatic eruptions.**

IV. **THE MATERIALS OF VOLCANIC ERUPTIONS.**   Lava, gases, and fiery fragments, when cooled, form extrusive igneous rocks with fine-grained, or aphanitic, textures.
    A. Basalt, andesite, dacite, and rhyolite, in order of increasing silica content, are the four major rock types formed from lava.
        1. If a lava cools rapidly, it forms a smooth glass, **obsidian. Pumice** is the end product of bubbly silicic lava. Large crystals, or **phenocrysts,** may be incorporated within an aphanitic or glassy matrix.
        2. Mafic, or basaltic, lavas flow easily away from the vent. A thick basaltic flow cools and develops parallel six-sided pillars, or **columnar joints. Pahoehoe lava** is billowy and ropelike. **Aa lava** is a rough surface of blocks and fragments.
        3. Silicic lava is viscous and remains close to the vent or extrudes from it as a **lava dome.**

    B. Pyroclastic materials come in different sizes. **Ash** is a minute droplet of lava; **cinders** and **lapilli** are pebble-sized fragments. **Blocks** are larger solid fragments; they are called **bombs** if they were rounded in flight while still molten. Fragments welded together form **pyroclastic rocks** such as **tuff, volcanic breccia,** and ignimbrite.

    C. Pyroclastic materials can be transported in three ways: as airborne pyroclastic debris (**tephra**), as a fiery cloud (**nuée ardente**), or as a mudflow (**lahar**).

    D. Volcanic gases, such as carbon dioxide, are expelled during a volcanic eruption. **Fumaroles** are vents through which gases or steam escape.

V. **VOLCANIC STRUCTURES AND ERUPTIVE STYLES**
    A. Low-viscosity, basaltic lava flows build broad, low **shield volcanoes,** such as those of the Hawaiian Islands, that erupt relatively peacefully. Exceptions are the pyroclastic eruptions of steep-sided **cinder cones.** Basaltic magma may also issue from crustal fractures in **fissure eruptions.** Cooled lavas form **flood basalts,** which build up over time into **lava plateaus.**

B. Viscous andesitic, dacitic, and rhyolitic lavas cause explosive eruptions and build steep **stratovolcanoes,** or **composite cones.** Explosive intraplate eruptions build broad, sunken calderas surrounded by ash fields.

C. The violent eruption in 1980 of Mount Saint Helens of the Cascade Range was typical for a moderate-sized, subduction zone volcano; the blast devastated 590 square kilometers of countryside and killed 60 people.

VI. FORECASTING ERUPTIONS

A. Dating volcanic deposits radiometrically helps geologists determine the eruption frequency of a volcano and the distribution of ejected material helps them to identify the hazardous zone.

B. Geologists monitor ground swelling and tilting, seismic patterns, and gas emissions to forecast eruptions.

C. There are 53 active volcanoes in the United States; the USGS has identified 35 that may erupt in the foreseeable future. Most are in the Aleutians, Alaska, and Hawaii, but some are located in the lower-48 states.

VII. VOLCANISM AND CLIMATE

A. In the **greenhouse effect,** carbon dioxide and other gases released by volcanoes absorb and reflect back the heat radiated from the Earth's surface, thus warming it.

B. However, the dust particles and acid droplets released by volcanoes reflect incoming sunlight, thus lowering the Earth's surface temperature. Although the cooling effects of a volcanic eruption are shorter-lived than its warming effects, some geologists have suggested that prolonged volcanism over long periods may have cooled the Earth enough to cause the ice ages and mass extinctions of the past.

VIII. CONSTRUCTIVE ASPECTS OF VOLCANOES.   Volcanoes are major landforms. They are the source of rich soils, mineral deposits, and atmospheric gases, and their geothermal heat is a potential source of energy.

STUDY TERMS

aa lava (p. 199)
ash (p. 201)
block (p. 201)
bomb (p. 201)
caldera (p. 195)
cinder (p. 201)
cinder cone (p. 204)
columnar joints (p. 199)
crater (p. 195)
fissure eruption (p. 205)
flood basalt (p. 205)
fumarole (p. 202)
greenhouse effect (p. 212)
lahar (p. 201)
lapilli (p. 201)
lava (p. 195)
lava dome (p. 200)

lava plateau (p. 205)
nuée ardente (p. 201)
obsidian (p. 199)
pahoehoe lava (p. 199)
phenocryst (p. 199)
phreatic eruption (p. 198)
pillow lava (p. 198)
pumice (p. 199)
pyroclastic rock (p. 201)
shield volcano (p. 203)
stratovolcano (composite cone) (p. 206)
tephra (p. 201)
tuff (p. 201)
vent (p. 194)
volcanic breccia (p. 201)
volcanic cone (p. 193)
volcano (p. 194)

CRITICAL THINKING QUESTIONS

1. How do geologists gauge the magnitude of an eruption that no one has witnessed?

2. Explain the relationship between a magma's silica content and the explosiveness of an eruption. What other factors may influence explosiveness? Which magmas tend to be most explosive? Which least explosive?

3. How does the theory of plate tectonics explain the Pacific's Ring of Fire?

4. Describe the mechanics of a violent volcanic eruption.

5. What kind of volcanic cone is associated with basaltic magma? How is its profile related to composition and eruptive style? What kind of volcanic cone is associated with a magma of high silica content? How is its profile related to composition and eruptive style?

6. Why are deep marine eruptions peaceful and shallow marine eruptions violent?

7. How is volcanism breaking the island of Hawaii apart?

8. Account for the differences in form of lava plateaus and shield volcanoes.

9. Describe the cycle leading to a Kilauea eruption. How is it monitored? In what ways were the warning signs of the Mount Saint Helens and Kilauea eruptions similar? How did they differ?

10. What kinds of geological evidence might you search for to support the hypothesis that volcanism has significantly affected the Earth's climate in the past?

Sedimentary Rocks

Some years ago, when archeologist Richard MacNiesh wanted to determine the origin of modern corn, he searched for an ancient garbage dump. After many years, he found the perfect site in Mexico's Tehuacan Valley, where Native Americans had discarded their refuse for thousands of years. An important attribute of the site was that its layers contained corncobs.

Working from the top down, MacNiesh found that the cobs were smaller in each succeeding layer. After four years of digging, he uncovered corncobs at the bottom of the site "no bigger than the filter tip of a cigarette—under a magnifying glass, one could see that they were indeed miniature ears of corn with sockets that had once contained kernels enclosed in pods." MacNiesh had discovered the ancestral wild grass from which all corn had descended and presumably the birth of Native American agriculture. Carbon-14 dating established this time at about 4000 B.C.

You can learn a lot about people from what they throw away. You can also learn a lot about the Earth from what *it* throws away. What the Earth throws away are **sediments,** fragmented particles weathered from preexisting rocks and transported and deposited by water, wind, and ice. Constituting only 7 percent of the crust by volume but covering 75 percent of its surface area, sediments and sedimentary rocks amount to little more than a thin coating on the hard igneous and metamorphic crust. Yet they are important for a variety of reasons.

Sediments and sedimentary rocks are the only Earth materials deposited at or near the surface under everyday conditions. As such, they contain the entire fossil record of life on the Earth. In them are recorded the composition, climate, and topography of former landmasses, as well as the physical, chemical, and biological conditions of oceans

sediment
Particles that have been mechanically transported by water, wind, or ice, or chemically precipitated from solution, or secreted by organisms, and deposited in loose layers on the Earth's surface.

◀ Layering is an important feature of sedimentary rocks, as illustrated by this sandstone exposure in the Paris Canyon Wilderness Area of Northern Arizona.

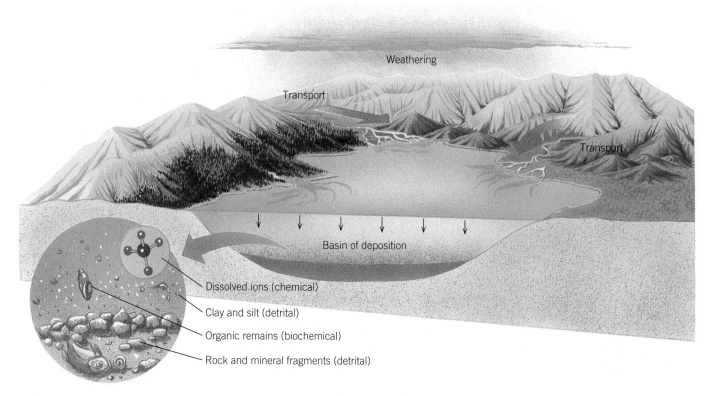

Weathering

Transport

Transport

Basin of deposition

Dissolved ions (chemical)

Clay and silt (detrital)

Organic remains (biochemical)

Rock and mineral fragments (detrital)

Figure 8.1 The components of a typical sediment are detrital, organic (biochemical), and chemical in origin. All result directly or indirectly from weathering, transport, and depositional processes.

that no longer exist. Without sediments and sedimentary rocks, we would have virtually no idea of the geography of the past, the nature of the Earth's past environment, or the life that inhabited that environment.

Not only do sediments retain the record of the past, but they also provide the means of arranging that record in chronological order. They are deposited in layers, and thus the oldest layer of a given sequence is on the bottom and is overlain in turn by successively younger layers. Because each layer was formed at the Earth's surface, we can chart the changes that have occurred at the surface over time by observing the changes in the sediments from bottom to top, using the fossils in the various layers as time markers. Recent events may also be evaluated in this manner. By carefully analyzing lake sediments, for example, we can date the accumulation of toxic chemicals and other pollutants in the environment and get a good idea of how the severity of the problem has changed in recent decades.

Sediments also have significant economic importance. Virtually every material useful to civilization is mined from them: groundwater, gold, copper, zinc, iron, lead, diamonds, limestone, sand, gravel, and clay. Oil, gas, and coal are exclusive to sedimentary rocks; in fact, coal *is* a sedimentary rock.

detrital sediment
Fragments derived from the weathering of rocks, transported by water, wind, or ice, and deposited in loose layers on the Earth's surface.

chemical sediment
A sediment composed of particles precipitated directly from water.

biochemical sediment
A sediment precipitated directly or indirectly by the activities of organisms.

Weathering, Transport, and Deposition of Sediments

Imagine a low-lying area of the Earth's surface that acts as a receptacle or basin for sediments, such as the ocean floor or a lake (Figure 8.1). What kinds of sediments can be found there? The most readily identifiable are the fragmental or **detrital sediments,** the land-derived gravel, sand, silt, and clay brought to the basin by streams or, perhaps, by winds or glacial ice. A typical sedimentary basin also contains precipitated particles formed from ions that were transported in solution. These sediments are **chemical** in origin if precipitated directly from the water, or are **biochemical** in origin if precipitated as a result of organic processes. Often, a single sedimentary deposit will contain fractions of all three sediment types: detrital, chemical, and biochemical.

The raw materials that compose both detrital and precipitated sediments are derived from three processes: **weathering,** the physical and chemical breakdown of exposed surface rocks; transport of the weathered materials to the basin; and **deposition** in the basin.

Weathering

Weathering causes both mechanical disintegration and chemical decomposition of bedrock. In mechanical disintegration, bedrock is broken into small fragments or into individual mineral grains, but the composition of the materials remains unchanged. It occurs when water that has seeped into rock fractures and soil pores freezes, expands, and wedges them apart; when temperature variations cause minerals to expand and contract unevenly; when release of pressure causes rock fractures to widen; and when roots get wedged between rocks.

In chemical weathering, water and dissolved ions react with solid rock and mineral fragments to produce materials of fundamentally different composition. Hydrolysis is one of the many intricate chemical changes that occurs. It involves the reaction of water—in particular, the hydrogen ions in water—with minerals. Although hydrolysis occurs slowly in pure water, in nature the reaction is greatly accelerated by the presence of dissolved carbon dioxide, which is derived from the respiration and decay of organisms in the soil. Water and carbon dioxide then combine to form carbonic acid, which adds chemical potency to the solution by increasing the number of hydrogen ions present.

In the hydrolysis of feldspar, the most abundant mineral of the crust, carbonic acid and water alter the mineral to insoluble clay residues and dissolved ions (potassium, sodium, calcium, and silica). The clay residues are transported as solid particles, and the ions are carried in solution. This pattern of hydrolysis holds for all silicate minerals, although the weathering products vary with the precise composition of the minerals. Feldspar, quartz, and a small percentage of mica or amphibole are the major constituents of granite—the igneous rock that forms the bulk of the continental crust. The weathering products of these minerals—quartz and feldspar fragments, clay, mica, magnetite, and dissolved ions—all end up, one way or another, in sedimentary rocks.

Transport and Deposition

Weathering prepares materials derived from bedrock for transport. As we saw in Chapter 1, it is the first step in the gravity-driven process of erosion that acts to level the continents. Erosion may act directly under the influence of gravity, as in landslides and rockfalls, or it may involve agents such as streams, glaciers, winds, and ocean waves and currents, all of which are capable of transporting sediments great distances from their sources. Streams carry the greatest sediment load by far. They deposit sediments in a variety of environments, where they are buried by later deposits and converted to sedimentary rocks. These environments include upland basins, continental margins, coastal plains, and the deep-ocean floor. Each of these broad environments is divided into myriad subenvironments. We will discuss them where appropriate throughout this chapter and the book.

Lithification

Lithification is the conversion of sediment to sedimentary rock, and **compaction** is the first step in the process. Successive layers of sediment exert pressure on the sediment

weathering
The physical and chemical alteration of rocks exposed to the atmospheric influences on the Earth's surface.

deposition
The gravitational settling of rock-forming materials by such natural agents as water, wind, or ice.

Note: As you will see in Chapter 13, Weathering and Soils, the production of raw materials for sedimentary rocks is only one aspect of weathering.

lithification
The conversion of sediment into rock through such processes as compaction, cementation, and recrystallization.

compaction
Reduction in volume of sediments resulting from the weight of newly deposited sediments above.

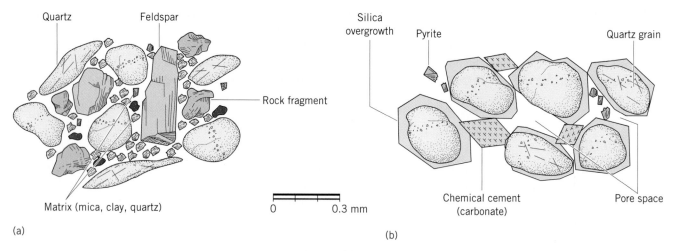

Figure 8.2 The grains of a typical detrital rock consist of a mixture of minerals and rock fragments. The grains may be bound together by (a) a clay and silt matrix or by (b) chemical cements.

matrix
The fine-grained material surrounding larger grains in a rock.

cementation
The process by which precipitates bind together the grains of a sediment, converting it into rock.

recrystallization
The formation of new crystalline mineral grains in a rock.

beneath, squeezing grains tightly together. In this manner, the underlying sediment may be reduced to a quarter of its original volume. Nevertheless, spaces remain between the coarse grains, and the spaces may be empty or filled with a **matrix.** The composition of the matrix determines the path that lithification follows. If the matrix is composed of clay and silt, the heating and drying induced by compaction will harden the matrix to a bricklike mass that embeds the larger fragments. However, if the matrix is filled with groundwater or saltwater saturated with silica, calcium carbonate, or iron oxides, these compounds will precipitate and bind the grains together in a process called **cementation** (Figure 8.2). During **recrystallization,** the mineral grains themselves act as cementing agents. Under pressure, the grains dissolve slightly at points of contact and then recrystallize in the voids between the original fragments. The matrix of a sediment need not fill all voids in order to bind the fragments. Many sedimentary rocks contain a large percentage of empty space.

Classification of Sedimentary Rocks

As with sediments, most sedimentary rocks may be divided into three broad categories: detrital, chemical, and biochemical (Table 8.1). The categories emphasize the origins of the rock components. The rocks are further classified within each of these categories on the basis of texture and mineral content. Texture, as you may recall from our discussion

Table 8.1 Classification of Common Sedimentary Rocks

Fragmental	Precipitated	
Detrital (classified by grain size)	Chemical	Biochemical
Breccia	Evaporites	Limestone
Conglomerate	Rock salt	Micrite
Sandstone	Gypsum	Chalk
Siltstone	Limestone	Coquina
Shale	Oolitic limestone	Reef rock
Mudstone		Chert (chalcedony)
		Coal

Table 8.2   The Wentworth Scale for Classifying Sediments by Size

Diameter (mm)	Particle	Rock
	Gravel	
Above 256	Boulder	Conglomerate and breccia
64–256	Cobble	
4–64	Pebble	
2–4	Granule	
	Sand	
$\frac{1}{2}$–2	Coarse sand	Sandstone
$\frac{1}{4}$–$\frac{1}{2}$	Medium sand	
$\frac{1}{8}$–$\frac{1}{4}$	Fine sand	
$\frac{1}{16}$–$\frac{1}{8}$	Very fine sand	
	Mud	
$\frac{1}{256}$–$\frac{1}{16}$	Silt	Siltstone, shale, and mudstone
$\frac{1}{2048}$–$\frac{1}{256}$	Clay	

of igneous rocks, refers to the size, shape, and arrangement of the components of the rock. This definition holds for sedimentary rocks also.

Detrital Rocks

As stated earlier, detrital rocks consist mainly of fragments weathered from preexisting rocks. The ancient Greek word for fragment is *klastos;* hence, detrital rocks are described as having **clastic texture.** However, clastic texture is so closely associated with detrital rocks that the term *clastic* is often used interchangeably with *detrital.* There are a number of characteristics a geologist looks for in analyzing the texture of a detrital rock. Among the most important are particle size, degree of sorting, and particle shape. We shall see that each has an interesting story to tell.

Perhaps the most obvious detrital textural characteristic is *particle size;* it is defined by the Wentworth scale (Table 8.2), which is the basis for classifying the common detrital rock types. Particle sizes range from boulders, cobbles, and pebbles down to sand, silt, and clay. Deposits consisting of sand or larger sizes are referred to as *coarse* sediments, whereas silt and clay are *fine* sediments. A rock formed largely of gravel is a **conglomerate** if the particles are rounded or a **breccia** if they are angular. A **sandstone** is formed of sand-sized particles and a *siltstone* of silt-sized particles. **Shale** and *mudstone* are composed of a mixture of the finest particles, clay and silt. The difference between them is that shale splits into thin layers, whereas mudstone is thickly layered and breaks into massive chunks.

Sorting refers to the range of particle sizes within the sediment and is another useful indicator of the differences between detrital rocks. Figure 8.3(a) shows a well-sorted sediment; its particles fall within a narrow range of sizes. It was taken from a typical beach, where it had been worked over by pummeling waves and currents. Contrast it with Figure 8.3(b), a poorly sorted sediment with a wide size range. It is *glacial till,* a sediment deposited directly by the ice of a melting glacier. These samples illustrate that sorting is related to the mode of sediment transport and deposition.

Particle shape tells a similar story, for whatever the shape of a fragment at its source, it will be modified during transport to the depositional site. Two measures of shape are the *sphericity* and *roundness* of the fragment. The first is a measure of the degree to which the particle approaches a spherical shape during transport, whereas roundness refers to the degree to which the sharp corners and edges of the fragment have been worn down or abraded.

clastic texture
Texture of a rock composed mainly of fragments of other rocks and minerals; most commonly used to describe detrital rocks.

conglomerate
A detrital rock consisting of rounded pebble-sized or larger fragments commonly set in a matrix of silt or sand.

breccia
A detrital rock consisting of angular pebble-sized or larger fragments commonly set in a matrix of silt or sand.

sandstone
A detrital rock consisting primarily of sand held together by a cementing agent.

shale
A fine-textured detrital rock with a layered structure resulting from compaction of clay, mud, or silt.

sorting
The process by which the agents of transportation (principally running water) separate sediments according to shape, size, and density.

(a)

(b)

Figure 8.4 Some common depositional environments of detrital rocks. Sorting, rounding, and sphericity generally improve with distance from the source. Turbidites are a major exception because fragment sizes are remixed in turbidity flows.

Figure 8.3 The degree of sorting is closely related to the transport history of a sediment. (a) Well-sorted beach sand transported by waves and currents. The finer particles have been sifted out. (b) Poorly sorted glacial till is deposited directly by melted ice; large and small particles had no opportunity to separate during transport.

Coarse Detrital Rocks

Detrital rocks are also classified by composition, and the composition is determined mainly by the degree of weathering at the source. Those minerals with the greatest resistance to weathering have the greatest chance of showing up in detrital sedimentary rocks. Therefore, quartz and clay are the chief constituents of detrital rocks. Appreciable amounts of feldspar, mica, and rock fragments are present in those rocks derived from a source region where weathering has been incomplete. Most detrital rocks also contain a small percentage of heavy minerals such as magnetite or other metals or metal oxides.

The mineralogy and texture of detrital rocks are the products of the composition of the source rocks and of their weathering, transport, and depositional history. The following simplified account describes the connections between these factors.

Consider the mountainous region shown in Figure 8.4. Erosion is rapid on the steep slopes, leaving scant time for mechanical and chemical weathering to work to completion. The rock fragments that weather directly from cliff faces accumulate as apronlike *talus slopes* at the base of cliffs, along with landslide and mudflow deposits. Under these circumstances, the weathering products have not traveled far, so there has been little opportunity for large and small particles to separate or for particles to become rounded. As a result, the sediments deposited this close to the source are likely to be poorly sorted and angular, with a high percentage of coarse fragments embedded in a matrix of clay and silt. The sediments that are buried and lithified here will form breccia (Figure 8.5).

The streams that emerge from narrow canyons of the mountainous region spread sediment as *alluvial fans* on the adjacent valley floors. Sorting is somewhat improved, and abrasion has rounded the fragments considerably. If fragments remain pebble-sized

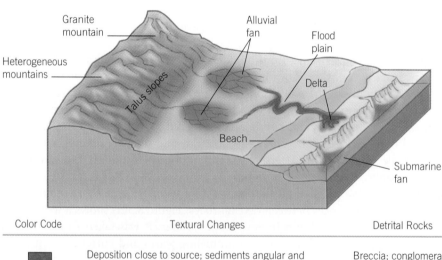

Color Code	Textural Changes	Detrital Rocks
	Deposition close to source; sediments angular and poorly sorted	Breccia; conglomerate; arkosic conglomerate
	Some sorting; some rounding of fragment edges; larger fragments mixed with clay and silt	Conglomerate; arkose; lithic sandstone
	Sorting, rounding, and sphericity improve; clay and silt start to be separated from coarse particles	Lithic sandstone
	Well-sorted, rounded, and spherical grains; clay and silt no longer present	Quartz sandstone
	Rock fragments of all sizes mixed with clay and silt in turbidity flows	Turbidite

Figure 8.5 Close-up of a typical breccia displaying angular fragments in a fine-grained matrix.

Figure 8.6 A quartz conglomerate. Compare the rounded pebbles in this sample with the breccia in Figure 8.5.

but are rounded, then the rock that results will be a conglomerate (Figure 8.6). Finer fragments are transported farther and more rapidly than coarse ones, so that sand-sized sediments are predominant at the outer fringes of the alluvial fans and in channels farther downstream. The type of sandstone that forms will depend on the composition of the source rock. Granitic mountains will yield granitic weathering products: sand-sized fragments of granite and grains of feldspar, quartz, and clay. A sandstone of this composition is called *arkose* (Figure 8.7). On the other hand, if the mountains consist of many rock types, the *lithic sandstone* derived from them will contain fragments reflective of the varied source region, in addition to quartz and clay. In both arkose and lithic sandstones, sorting is generally poor, with coarser fragments embedded in a clay and silt matrix.

Lithic sandstones are deposited in a broad range of environments: upland regions, downstream flood plains, transitional deltas where land and sea mix, and the continental shelf. However, their textures and compositions vary with distance from the source region. Roundness and sphericity improve as abrasion has more opportunity to shape the grains. Sorting improves as clay and silt are separated from the coarse particles, leaving behind a "cleaner" deposit—that is, one having less of a matrix. So the lithic sandstone deposited near sea level in deltas and flood plains or on the continental shelf is better sorted, better rounded, and cleaner than its upstream cousin.

Much of the sediment that streams carry down from the source region will sooner or later reach the beach. As waves wash the grains back and forth across the beach and nearby sea bottom countless times, the fine particles are sifted out and the coarser fraction remains in the beach zone. For this reason, beach sands are commonly well-sorted sediments.

The pounding waves also change the composition of the coarse sediments. Rock fragments composed of interlocking minerals are broken up. Minerals with structural weaknesses, such as feldspar, are relentlessly pulverized to powder. Soluble minerals, such as calcite, that escaped chemical weathering at the source are dissolved during transport. In fact, the only common source mineral that is chemically and physically tough enough to survive the relentless attack is quartz. When cemented, these well-sorted and well-rounded quartz fragments form *quartz sandstone* (Figure 8.8). Where the wave action is very intense, the relatively scarce gravels are left behind on the beach and the coarse sands are carried offshore. The well-rounded beach gravels then form a *quartz conglomerate.*

The sediments in the beach and near-shore environment are deposited in strips parallel to the coast: pebbles or sand-sized particles close to shore, finer sediments in the

Figure 8.7 Close-up of a typical arkose. The high percentage of pink feldspar, plus the angularity and coarse size of the fragments, suggest that the rock was deposited close to its granitic source region.

Figure 8.8 Close-up of a typical quartz sandstone. Note the similarity to the loose beach sand in Figure 8.3.

facies

The characteristics and appearance of a sedimentary rock unit whose conditions of origin differentiate it from neighboring units.

deeper waters farther out. These strips roughly coincide with the energy of the depositional environments, which tends to decrease from shallow water seaward. Each strip is composed of sediments whose mineralogy, texture, fossils, and the like distinguish it from neighboring units of the same age. Thus each is said to represent a different **facies** (Latin for "face" or "aspect"). The concept of facies is essential to the analysis of all sedimentary environments. In Chapter 10, Geologic Time, we will see how facies analysis can be used to trace ancient sea-level changes. Later in this chapter, we will discuss the facies of a coral reef.

The variations in rock type described thus far demonstrate that the composition and texture of most coarse detrital rocks are determined primarily by their weathering and transport history. A pure, well-rounded, and sorted quartz sandstone is the end-product of that history. Often, it is the result of several cycles of erosion. Geologists call it a *mature* rock (Figure 8.9). The other rocks we have described exhibit various stages of maturity. For example, breccia and arkose are *immature* rocks.

Turbidite Deposits   The sediments brought to the ocean by streams enter a new realm, where they are subject to submarine transport. Often, the sand and silt that would otherwise form lithic or quartz sandstone are swept by currents and waves to delta fronts and to the edge of the continental shelf, where they are deposited precariously on underwater slopes. Any disturbance—an earthquake, a storm, or simply too great a supply of

Figure 8.9 A detrital sediment attains maturity after being subjected to intensive abrasion, solution, and sorting during transport. The clay and silt matrix is sifted from the larger fragments, the fragments are rounded, and weak minerals are pulverized or dissolved. The product is commonly a well-sorted and well-rounded quartz sandstone.

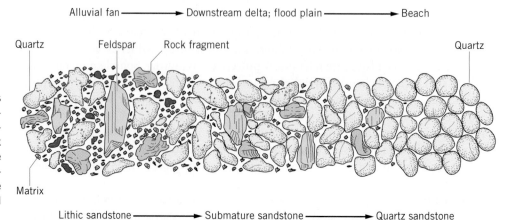

Figure 8.10 A turbidite bed, approximately 600 million years old, in North Wales, United Kingdom. Coarse fragments are embedded in a clay and silt matrix. The layers display graded bedding (see Figure 8.18).

sediment—may send dense flows of clay, mud, sand, and gravel down submarine canyons. The flows, which hug the sea bottom, emerge from the canyons, lose power, and spread their sediment as *submarine fans*. The sediments (and rocks) that result from these flows are called **turbidites** (Figure 8.10). They are poorly sorted, with coarse fragments embedded in a dense clay and silt matrix.

turbidite
The deposited sediment of a turbidity current; characterized by graded bedding and poor sorting.

Fine Detrital Rocks

Fine-textured sediment consists of minute clay minerals derived from the weathering of feldspar and tiny particles of quartz powder derived from the abrasion of larger quartz fragments. They tend to be poorly sorted because nature's sorting mechanisms are not sensitive to differences between extremely light, fine particles. The rounding process is also less effective, because the particles are so small that they are protected from abrasion and collision by films of water.

Because of their low mass, clay and silt particles do not sink easily. Slight turbulence can keep them in suspension nearly indefinitely, just as air turbulence can keep dust aloft. However, clay particles have surface charge and attract one another. When large clumps form, they sink.

Clays and silts are deposited in calm waters, which can occur in a great variety of settings: lakes, flood plains, deltas, marshes and estuaries, the continental margin, and the deep ocean. Where their deposition occurs along with coarse sediments, the clay and silt form the matrix of the coarse detrital rock. Where the coarse sediments are absent, the clay and silt particles form shale and mudstone. The difference between shale and mudstone is connected to the marine life on the sea floor. Where burrowing organisms stir the sediment, mudstone tends to form. In the absence of such organisms, thinly layered shale forms (Figure 8.11).

Precipitated Sedimentary Rocks

Precipitated sedimentary rocks can be divided into two categories: chemical rocks, which form through direct chemical or physical means, and biochemical rocks, which form through the work of organisms in the water. Chemical precipitation is initiated by the evaporation of water within which ions are dissolved, by the addition of excess ions, or by temperature and pressure changes that affect the capacity of the water to hold the ions in solution. Biochemical precipitation is initiated by organisms through metabolic activities. These activities alter the pressure, temperature, and concentration of ions in water. Most of the organisms use these ions to synthesize shell matter.

-Mudstone

-Sandstone

Shale

Figure 8.11 Alternating beds of shale, mudstone, and sandstone in the Moenkopi Formation, Arizona. Notice the differences in the layering and textures of these rocks.

The Limestones

limestone
The consolidated product of calcareous sand, limy mud, and/or crushed shells.

Limestone, composed of calcium carbonate, is the most common of the precipitated rocks. The majority of these rocks are biochemical in origin, having been derived from shell matter. Experiments prove that increasing water temperature and lowering pressure causes calcium carbonate to precipitate. Thus carbonate sediments are directly precipitated in the warm, shallow waters of tropical and subtropical seas but are rarely precipitated in great quantity at colder, higher latitudes or in very deep water.

As with detrital sediments, shells and mineral matter are subject to abrasion and mechanical transport. Limestones are deposited in a wide range of marine environments. Thus their texture and composition are reflective of their transport and depositional history.

Biochemical Limestones   Shells are the raw materials of biochemical rocks. For example, soft and fine-textured *chalk* consists of the shells of billions of microscopic organisms that settled in shallow water (Figure 8.12). So plentiful was it 135 to 65 million years ago that the entire geologic period was named for it—the Cretaceous (*creta* is Latin for "chalk"). England's famous White Cliffs of Dover are the classic exposure of chalk.

What happens to shells after the organism dies largely determines the properties of the resulting rocks. They may remain intact, or waves and currents may break them into large fragments, carbonate sand, and limy mud. Their degree of sorting depends largely on the intensity and duration of wave and current action. *Coquina* is composed of large, poorly cemented shell fragments (Figure 8.13). Limestones made of shells that have been subjected to intensive abrasion in the beach zone display properties similar to those of quartz sandstones formed in the same environment. The ground-up shells are sand-sized, are well sorted, and have a clear, chemically precipitated matrix.

Now consider the other environmental extreme, that of deposition in water devoid of current and wave action, such as a tidal flat or quiet lake. Under these circumstances, fine particles settle out of the water and are deposited as lime mud. Limestones having

Figure 8.12 England's famous White Cliffs of Dover, facing the English Channel, are an ancient chalk deposit composed of the carbonate shells of microscopic organisms.

this fine texture and composition are called *micrite*—a name that calls attention to the microscopic size of the particles. One famous micrite deposit, the Solenhofen limestone of Germany, was the site of a large lake 150 million years ago. The lake was so still, and the sediments so fine, that the remains of dead animals and plants that settled to the bottom are preserved to the most delicate detail. The Solenhofen contains one of the world's foremost "library" of flying reptiles and ancestral birds, including *Archaeopteryx*, which is considered by some paleontologists to be the first bird (Figure 8.23).

Most limestones are deposited in an environment between these energy extremes. They consist of a mixture of intact and fragmented shells embedded in a micrite matrix. In this respect, they resemble immature lithic sandstones, which consist of large rock fragments imbedded in a clay matrix.

Coral Reefs   Biochemical sediments predominate where there is a scarcity of land-derived sediments and where waters are warm and teeming with life. In these regions, shells of marine organisms deposited on the continental margins form *carbonate shelves*. Where warm ocean currents sweep abundant nutrients to shallow waters, conditions are ideal for growth of coral *reefs*. Anchored to the mainland or an island, the reefs are built of the myriad skeletal secretions of tiny colonial coral. The coral depends upon microscopic algae lodged in its gullet to help it digest food; because algae need sunlight for photosynthesis, the live coral must grow close to sea level, where light can penetrate.

Figure 8.13 Close-up of coquina, a limestone composed of large cemented shell fragments.

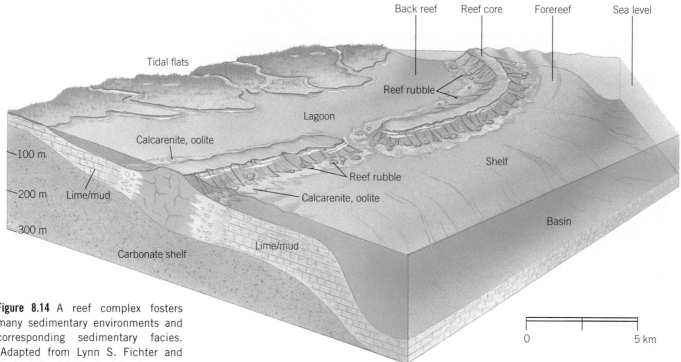

Figure 8.14 A reef complex fosters many sedimentary environments and corresponding sedimentary facies. (Adapted from Lynn S. Fichter and David J. Poché, *Ancient Environments and the Interpretation of Geologic History*, New York: Macmillan, 1979.)

Reefs are interesting not only for the variety of life that inhabits them but also for the variety of reef facies associated with them.

No two reefs are alike, but most share certain physical similarities (Figure 8.14). The *reef core*, the massive foundation, bears the brunt of breaking waves on the seaward side (the *forereef*), and it is often smothered in its own debris. Huge boulders of coral and other rubble plunge down the front of the reef core. Coarse carbonate sand and chemically precipitated oolites, the products of vigorous wave action and swift currents, are in turn deposited over this debris. The sands tail off to the fine lime muds and clays of the forereef and distant shelf.

Behind the reef core is the protected lagoon of the *back reef*. The sediments here are lime muds deposited in generally quiet waters, as well as oolite and limy sand that are swept in during storms. If the massive reef restricts circulation with the open ocean, and if the climate is dry, brines will concentrate in the back-reef lagoon, from which salts will precipitate. Because each part of the reef is distinct from the other parts that formed at the same time, each represents a different sedimentary facies.

Australia's Great Barrier Reef, 1500 kilometers long, is the largest living coral reef. A buried reef of similar dimensions, 150 million years old, runs several hundred kilometers offshore along the length of the eastern United States. It acted as a massive trap to sediments brought in from the continent and, in this way, was instrumental in building the continental shelf. Probably the greatest exposed fossil reef, the 220-million-year-old El Capitan of the Guadalupe Mountains, is currently stranded in arid West Texas (Figure 8.15). You can visit this spectacular structure and surrounding region and see all the features described above. Once you understand the origin of the rocks, you can almost hear the waves crashing in the desert.

Chemical Limestones   One interesting environment where limestone is directly precipitated is among the lagoons and shallow reefs of the Grand Bahama Banks. There we find deposits of tiny caviar-sized particles called *ooids*. Sectioning and observing an ooid under the microscope, we see that it consists of a minute sand-grain nucleus, around which are wrapped layers of calcium carbonate. Geologists reason that the calcite layers precipitated around the sand grain as it was tossed about by the waves in this high-

Figure 8.15 El Capitan, an ancient reef complex that is now part of the Guadalupe Mountains of west Texas and New Mexico. The steep cliffs are remnants of a massive reef core; the forereef and deep basin strata are in the foreground.

energy environment. *Oolitic limestones* are well sorted; the grains are held together by clear chemical cement (Figure 8.16).

Evaporites

Many **evaporites** are precipitated within the semienclosed basins of arid climates. In a typical setting, water within the basin is partially trapped by a shallow ledge or barrier, which closes off easy circulation with the open ocean. Heavy evaporation increases the salinity of the water in the basin to the point that it collects at the bottom as a dense brine from which the evaporites are deposited. The dense water creeps along the bottom and out into the open ocean, while the less saline water of the open ocean seeps into the basin to replace the salty water that escaped. It too becomes highly saline as it evaporates. As discussed in the Introduction to this book, some geologists believe that the thick layer of salt beneath the floor of the Mediterranean was deposited 6 million

evaporite
Deposit from the evaporation of aqueous solutions. The most common example is rock salt.

Figure 8.16 (a) Close-up of an oolitic limestone. (b) Photomicrograph of an ooid, displaying its delicate concentric structure.

(a)

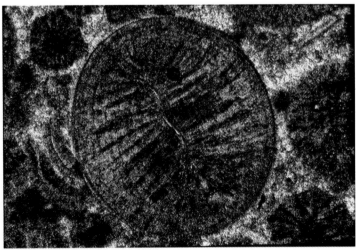

(b)

Figure 8.17 A bedded chert from the Franciscan formation, San Francisco, California. Deposits like these are believed to be derived from minute silica shells of microorganisms that were altered by heat and pressure after burial.

chert
A hard sedimentary rock formed from the lithification of biochemical silica.

coal
A combustible carbonaceous rock formed by the altered remains of plant matter.

beds (strata)
Visually distinguishable layers of sedimentary rock.

bedding plane
The plane that marks the boundary of each layer of stratified rock.

bedding
The stratification or layering of sedimentary rock.

graded bedding
Stratification in which particle size changes from coarse to fine from the bottom to the top of each bed.

years ago, when the sea was shallow and was cut off from the Atlantic Ocean.

If evaporation is extreme, salts precipitate in a bull's-eye pattern, with the least soluble on the fringes of the basin and the most soluble at the center. The sequence runs from carbonates (certain limestones and related rocks) to sulfates (gypsum) to halides (halite).

Thick evaporite deposits are spread over regions of New York, Pennsylvania, Ohio, Michigan, and Ontario. They are records of a shallow sea that covered the region some 400 million years ago. There are also evaporites of freshwater origin; the famous Bonneville Salt Flats of western Utah are the remains of an ancient lake that dried up; the Great Salt Lake is its descendent.

Chert

Some organisms use silica rather than calcium carbonate for their shells. Most biochemical silica is extracted from seawater by tiny single-celled *radiolaria* and by *diatoms,* an important single-celled alga. These shells rain on the ocean bottom and form layers of deep-sea ooze. Silica precipitates under the very same conditions that inhibit calcium carbonate precipitation, and oozes of this composition are found in polar seas and in regions where cold water rises to the surface from great depth. Upon lithification, the oozes form bedded **chert** (Figure 8.17). However, organisms with silica shells also exist in warm waters. If the shells are deposited along with abundant calcium carbonate, they will form impurities in limestone sediment and collect as nodules and geodes (structures we will soon discuss).

Coal

Coal is generally classified as a biochemical rock, although it is not precipitated by organisms but, rather, consists of the compressed, altered remains of vegetation. The typical environment where coal forms is a stagnant swamp. Dead plant matter is not decomposed by the oxygen-starved water but collects on the floor of the swamp. As it is buried deeper, compression and heat convert the plant matter first to *peat* and later to coal. Because coal is highly combustible, it is a valuable source of energy. As such, it is discussed more fully in Chapter 20, Mineral and Energy Resources.

Sedimentary Structures

No matter how they are transported, sediments ultimately settle to the sea bottom, lake bed, flood plain, and so on, under the influence of gravity. But sediments are never deposited continually in these environments. Streams overflow their banks onto the flood plain only seasonally, and shifting currents may prevent deposition on the sea bottom for a time. Temporary interruptions—which may vary from days to tens of thousands of years—allow earlier deposits to begin to compact before fresh sediments cover them. The compacted sediments form cohesive layers, known as **beds** or **strata.** The surface of the bed, called the **bedding plane,** lies parallel to the depositional surface and is the boundary between layers. **Bedding** is the most prominent structural characteristic of any sedimentary deposit and is an integral part of the depositional process.

Various types of bedding give clues to the environments in which they formed. In **graded bedding,** each layer displays a change in particle size—coarse to fine—from the bottom to the top of the bed (Figure 8.18). This grading indicates that the velocity of

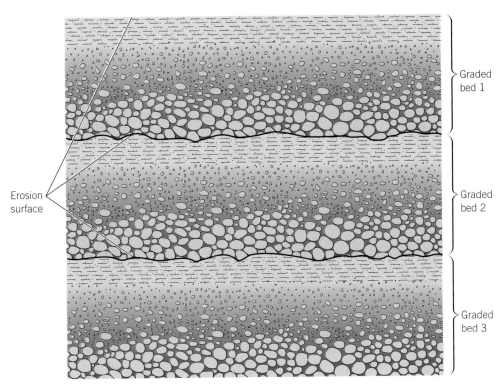

Erosion
surface

Graded
bed 1

Graded
bed 2

Graded
bed 3

Figure 8.18 Graded bedding in a typical turbidite deposit. The layers were deposited during separate flow events by slurries of gravel and silt that hugged the continental slope and spread over the deep-ocean floor. The coarse fragments settled to the bottom of the bed as the slurry lost energy.

the current carrying the sediment diminished. This condition occurs in submarine flows that deposit turbidites. Each turbidite bed represents a single flow event.

Sometimes, a rock outcrop has a thick lens-shaped layer of coarse sand and gravel surrounded by finer sediments, which is indicative of a buried stream channel (Figure 8.19). The channel was cut in the soft muds of a flood plain and later abandoned as the stream shifted course.

Cross-bedding refers to thin beds inclined at an angle to the main bed. Only the deposition of sediment carried by a moving stream of water or air can cause such a feature. Figure 8.20 shows a typical cross-bedded sandstone deposit, an ancient windblown sand dune. The cross-beds were formed as sand was blown up the gentle windward slope and rolled down the steeper lee slope. Shifts in the inclination of the cross-beds mark shifts in the direction of the wind that flowed over the sand dune; the average direction of cross-bedding reflects the average direction of the prevailing wind. The same principle applies to stream cross-beds and is a useful tool for interpreting the source region of the sediment. Cross-bedding can occur on a small or large scale—from fine layering

cross-bedding
Thin strata laid down by currents of wind or water at an oblique angle to the main bed.

Figure 8.19 An ancient stream channel, three meters thick, cut in even older sedimentary rocks, near Jesse Ewing Canyon, Utah. The channel is filled with conglomerate derived from coarse fragments deposited on the stream bed.

Figure 8.20 A spectacular 150-million-year-old example of a cross-bedded sandstone in what is now northern Arizona. These ancient sand dunes were probably formed in a coastal desert.

ripple mark
A corrugated form displayed in sedimentary rocks caused by currents of air or water that moved over the sediments prior to lithification.

mud crack
Polygonal crack formed by the drying and shrinking of mud, silt, or clay.

fossil
The remains, trace, or imprint of a plant or animal preserved in rocks.

within a single bed to the thick complex structures of huge deltas that form at the mouths of rivers entering calm portions of the ocean or lakes.

Ripple marks are formed by currents and waves that disturb the soft bottom where sediments have settled. The symmetrical ripple marks in Figure 8.21 indicate circular wave motion of the kind seen at beaches. Asymmetrical ripple marks result from currents flowing in one direction, such as stream and long-shore currents. Where sediment is deposited in water and then exposed to hot sun, **mud cracks** are caused by shrinkage when the sediment dries up (Figure 8.22). Tidal flats and seasonal lake beds are typical environments where mud cracks can be observed.

Fossils and Minor Structures

Many sediments contain **fossils,** the remains or traces of past life. Figure 8.23 illustrates the common modes of fossilization. Some are simple bedding-plane features, such as worm burrows, plant imprints, and animal tracks. Others are preserved bones, shells,

Figure 8.21 These ripple marks were formed by waves along a beach some 200 million years ago.

Figure 8.22 Mud cracks formed by the drying of sediment along the Colorado River.

Figure 8.23 Some modes of fossilization.

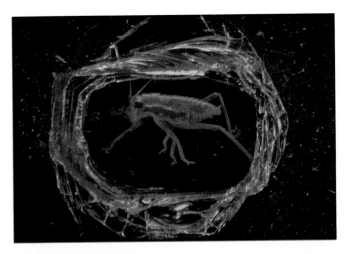

(a) *Direct preservation.* Forty million years ago, this grasshopper was trapped in a globule of tree sap that later hardened to amber.

(b) *Unaltered hard parts.* These rocks in Dinosaur National Monument, Utah, contain dinosaur bones.

(c) *Carbonization.* This 350-million-year-old seed fern has been altered to carbon, but so delicately that its structure is preserved.

(d) *Petrification.* The organic matter of this 200-million-year-old log has been replaced by silica, but the original tree-ring structure and bark texture have been preserved.

(e) *Molds and casts.* A mold is an impression of an organic structure, most commonly a shell. Sediment fills the mold, forming a cast in the shape of the organism. This near-perfect cast of *Archaeopteryx,* the oldest known bird fossil, was found in the 145-million-year-old Solenhofen limestone, Germany.

(f) *Tracks and burrows.* Organisms that live in or pass over sediments often leave behind fossil tracks or burrows. These vertical worm burrows, filled with mineral cements, were made 600 million years ago in muds that are now part of the sedimentary rocks of Scotland.

Apply and Decide 8.1

Figure 8A Mayan temples, Tikal II and Tikal I, rise above the rain forest of Guatemala Structures like these, and innumerable other artifacts, attest to a sophisticated and complex civilization that existed for over a thousand years.

Environmental Degradation and the Decline of Civilizations

Anthropologist T. Patrick Culbert recounts that when Spanish conquistador Hernando Cortés explored the lowlands of northern Guatemala (Central America) in 1525, he "encountered the same vast expanse of silent rain forest that covers the area today, and the few native settlements [he] passed were small and impoverished," as they remain today. It had not always been this way, however. The ruins of magnificent cities, with temples, pyramids, and courts fashioned of limestone blocks, are dispersed throughout the rain forest and surrounding regions like islands of white in a sea of green (Figure 8A).

Some of these cities supported populations of 30,000 to 80,000 people. To feed themselves as well as the city dwellers, Mayan farmers developed a number of innovative intensive agricultural systems, including irrigation, swamp drainage, and raised fields that were built above the low-lying, seasonally flooded lands bordering rivers.

Twentieth-century archeologists and anthropologists, applying modern radiometric dating techniques, determined that the Mayan buildings and other objects of art, science, and religion were created between 250 A.D. and 900 A.D. This classic period of flourishing Mayan culture was accompanied by massive population growth. Their numbers clearly began to rise dramatically around 3000 B.C. and may have exceeded 5 million people by 850 A.D. By 950 A.D., the popula-

tion had declined to the level encountered by Cortés six centuries later (Figure 8B, Graph 1). Archeological studies have confirmed that the decline was not a mere dispersal of Maya to surrounding regions. Instead, it was a true population collapse in which deaths far outweighed births.

What could have caused this complex society to implode so rapidly? Disease, food shortages, peasant revolts, warfare, and earthquakes have been suggested as sources of the decline. It is reasonable to assume that many interconnected factors were involved. In recent years, however, investigators have been paying close attention to *environmental disruption*—the fouling of one's own nest, so to speak—as the underlying cause. At this point, the Peten Lakes of northern Guatemala enter the picture.

The Peten Lakes are located in the heart of ancient Maya country. The layers of lake-bottom sediments that were washed in from the surrounding region span the rise and fall of the Mayan civilization. In the 1980s, geologists decided to investigate the environmental impact of the Mayan civilization by taking cores of lakebed sediment and dating the various layers using the radiometric carbon-14 method. Graphs 2 through 7 in Figure 8B will help us interpret the sedimentary record.

Note first the regional conversion from rain forest to low-brush *savanna* vegetation, as recorded by changes in the pollen, spore, seed, and wood content of the lakebed strata (Graph 2). The conversion occurred during the same period

Unfortunately for the Maya, the forest's root networks had served the useful purpose of soaking up rainwater and trapping loose sediment before it could be removed. Graphs 3, 4, and 5 illustrate the disastrous consequences of excessive land clearing: greater erosion of topsoil; higher lake sedimentation rates; and altered chemistry of the sediment—most significantly, a conversion from rich organic constituents to inorganic clays and silts. The conversion signals a decline in soil productivity, because it implies that many fields had been stripped of their organic, nutrient-rich topsoil.

Graph 6 shows a more direct human impact on the lake-bottom sediments—that is, a sharp increase in the amount of phosphorus derived from human feces. This poisonous influx, combined with increasing levels of turbidity, drastically reduced the biological productivity of the lakes, and consequently diminished the fish protein that the lakes supplied to the population (Graph 7).

In sum, the sediments of the Peten Lakes document an unmistakable history of environmental stress and degradation that closely correlates with the rise and fall of the Mayan civilization. Furthermore, recent paleoclimatic studies indicate that Mayan agriculture may also have been under stress from severe drought conditions that occurred during the same time interval.

We may envision the Mayan civilization of about 800 A.D. as a culture under stress from a burgeoning population, a deteriorating climate, and a rapidly degrading resource base. As Professor Culbert states, "the Maya collapse is an exemplary case of overshoot by a culture that had expanded too rapidly and had used its resources recklessly in an environment that demanded careful techniques of conservation."

Is there a lesson for us in the record of a sophisticated civilization that, having reached its peak, abruptly collapses? Could sudden climate change—perhaps human-induced global warming—be the straw that breaks the back of our modern world civilization, which is being stressed to the limit by overpopulation and environmental degradation? (For further discussion of this possibility, see Apply and Decide 17.1, Human-Induced Global Warming.)

THOUGHT QUESTIONS

1. Use your knowledge of the processes of weathering, transport, and deposition of sediments to explain why the layers of a lakebed may hold clues to the geologic, climatic, and biologic history of the surrounding region.

2. The upward swing of the Maya population growth curve (Graph 1) looks suspiciously like the growth curve of the entire human population over the past several centuries. Does it mean that civilization is headed for a cataclysmic decline? What other possibilities exist? What constructive steps may be taken to ensure a more positive destiny for the human species?

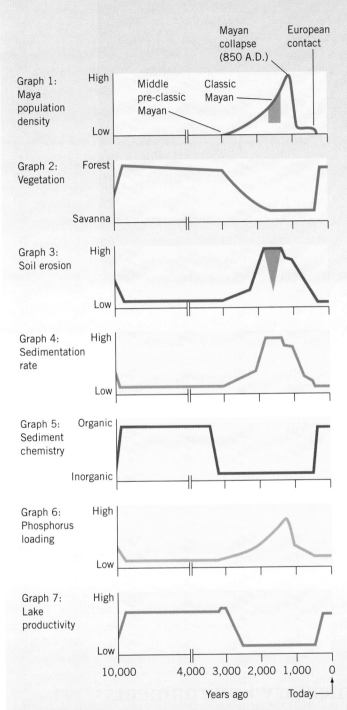

Figure 8B Data from the cores of Peten Lake Sediment

Source: M. W. Binford, et al., 1987, *Ecosystems, Paleoecology, and Human Disturbance in Subtropical and Tropical America,* Quaternay Science Review; 6, pp.115–128.

as the explosive Mayan population growth (see Graph 1) and was caused by slash-and-burn agriculture in which trees were cut down and burned and maize seeds were planted in the resulting ash. Savanna vegetation grew in the cleared spaces after the corn crops were harvested.

Figure 8.24 A geode is formed by deposits of silica that line the cavities in limestones and other rocks. This geode contains amethyst, a variety of quartz.

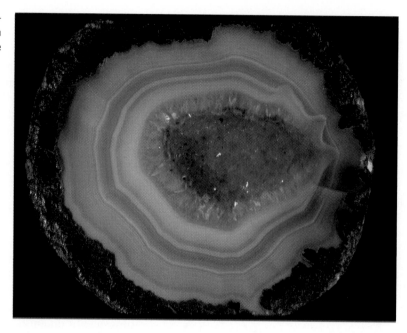

and other organic matter. Typically, they are either recrystallized shells or molds and casts of these shells. The orientation of the shells can be indicative of sea-bottom conditions when they settled. For example, shells arranged haphazardly indicate a calm sea bottom, for the slightest current or wave action will flip shells concave sides down.

Nodules and *geodes* commonly occur as masses of silica within limestone (Figure 8.24). The nodules lack an internal structure, but the geodes consist of cavities lined with silica, which serves as the base for beautiful crystals that grow inward toward the center. Geologists think that the silica was deposited throughout the limestone first, and circulating groundwater later concentrated it within cavities. There the silica dehydrated and formed gels that gradually hardened into a massive, waxy-looking rock called *chalcedony,* a variety of chert. The quartz crystals that form the chalcedony are so minute that special X-ray techniques are needed to detect them. Chalcedony forms the lining of the hollow geodes and the entire mass of the nodules. In some locations, such as the upper Mississippi valley, geodes and nodules weather out of the soft limestone bluffs that border streams and, like so many apples, lie scattered in the streambeds—to the delight of mineral collectors.

Concretions are rounded, solid bodies found within shale, sandstone, and limestone beds. Their composition is variable—silica, iron oxide, and carbonate material. The concretions originate when material precipitates around a rock fragment, shell particle, or other impurity and then hardens; finally, the entire conglomeration is cemented together.

Oceanic Sedimentary Environments

Although most sediment is deposited on the continental margins, appreciable quantities reach the ocean floor. As we discussed in earlier chapters, these sediments preserve the tectonic, environmental, and biological history of the deep-ocean basins. It is a complex subject, and we will touch upon only some of the highlights here. We will discuss three broad environments: the tectonically passive continental margin, the tectonically active margin, and the deep-ocean floor.

Passive Continental Margins

The east coast of North America is a good example of a passive margin, having been tectonically stable for several hundred million years. Detrital sediments predominate;

ASIDE 8.1 *Color as a Feature of Sedimentary Rocks*

Color is one of the first features one notices in a sedimentary rock (see the figure). Most colors are a function of the abundance of oxygen present in the depositional environment relative to the iron and organic matter present in the sediment. The coloring compounds are generally confined to the matrix and cementing agents between the larger particles and usually amount to a very small percentage of the rock (often less than 0.1 percent). Blackish colors in rocks are caused by organic compounds within the sediment that are normally broken down and converted to carbon dioxide when exposed to the oxygen. Black sediments accumulate in stagnant, oxygen-deficient environments such as swamps, certain lake bottoms, and semienclosed basins, such as the fjords of Norway, Alaska, and the Black Sea.

Bright red, orange, and yellow detrital sediments—often called redbeds—are formed above sea level under the oxygen-rich conditions typically found on flood plains, alluvial fans, and the exposed portions of deltas. The presence of hematite colors the sediment red, brown, and purple. Limonite is responsible for yellow and rust coloration.

Dull brown and gray, probably the most abundant colors of sedimentary rocks, are produced in a wide variety of environments—which, unfortunately, makes it difficult to interpret the colors. They may indicate mildly oxygen-rich or oxygen-poor environments or simply a sediment lacking iron or organic matter.

In analyzing color, it is important not to confuse the color of the weathered and oxidized surface of the rock with the true internal color of the sediment.

The striking colors of these sandstones in the Painted Desert, Arizona, are caused by oxidation of iron compounds that compose a small fraction of the rock.

indeed, in some places, the continental shelf seems to have been constructed from overlapping deltas that grew outward from the coast. As we go farther from shore, the sediments are finer and the continental slope is marked by silts and clays, in contrast to the coarse sands of the shelf. However, this relationship is complicated by many past sea-level changes. The most recent, which occurred during the last glacial advance that ended some fifteen thousand years ago, caused the sea level to fall about 90 meters and brought coarse sediment onto the continental slope. Since that time, sea level has been on the rise.

Moving farther out to sea, we observe large ripple marks and dune structures in sand and silt. Evidently, sediments are being moved along the bottom toward the shelf edge. There, strong turbidity currents that hug the ocean bottom carry sediments down the continental slope and deposit their load on the continental rise. The turbidite beds then mingle with, and gradually give way to, the fine clay and microscopic shells that settled out onto the deep-ocean floor. These deep-ocean sediments are thousands of meters thick close to the continental margins and form the nearly flat *abyssal plains* of the deep-ocean basin that have buried the basaltic oceanic crust.

Active Continental Margins

To understand depositional patterns on active margins, let us examine the west coast of South America. The Andes Mountain chain extends along the western edge of the continent and the deep Peru-Chile trench, located a few kilometers offshore, runs parallel to the coast. Oceanic crust is currently being subducted beneath that trench, and earthquake and volcanic activity is virtually unceasing. Under these circumstances, the continental shelf and slope are extremely narrow. Sediments are washed down into the trench, where they are trapped. Deep trenches rim much of the Pacific basin, in contrast to the passive Atlantic margin, which lacks bordering trenches. Thus turbidites in the active Pacific margin generally do not spread as far out to sea.

The Deep-Ocean Environment

Although deep-ocean sediments contain turbidites that have flowed down the continental slope and onto the abyssal plains, most sediments on the ocean bottom got there as the result of a steady drizzle of particles from the ocean surface. The particles are mainly of two kinds: *brown clay* and *deep-sea oozes*—the latter composed of the microscopic shells of minute organisms (Figure 8.25). Because the drizzle is exceedingly slow and fine, deep-ocean sediments accumulate at the rate of about 1 millimeter per thousand years. The distribution of brown clay and oozes over the vast expanse of ocean floor is largely controlled by variations in ocean depth, by deep-ocean circulation, by the kinds of organisms that thrive on the surface, and by movements of the lithospheric plates (Figure 8.26). The deep-sea oozes are classified on the basis of their composition. *Siliceous ooze* is derived mainly from organisms whose shells are composed of silica, and *calcareous ooze* mainly from the shells of organisms made of calcium carbonate.

Calcium carbonate oozes readily precipitate in warm carbonate-saturated waters in regions of low latitude (around the equator). But as they sink, they are dissolved by the high pressures and the cold bottom currents that originate off the Antarctic coast and creep along the ocean floor. The depth at which dissolving occurs, about 4300 meters, is called the *carbonate compensation depth,* or snowline, because the delicate white shells accumulate like snowflakes above it. Figure 8.26 shows that calcareous ooze predominates on the shallow mid-ocean ridges, which rise well above the snowline, but is generally absent on the deep abyssal plains.

Siliceous oozes, though present in all oceans, are generally less common than either calcareous ooze or brown clay. They are deposited where cold, nutrient-rich water predominates. The major accumulations occur off the Antarctic coast and in regions of

(a)

Figure 8.25 (a) Microscopic shells of foraminifera, magnified 50 times. Most calcareous oozes are composed of these single-celled amoebalike organisms. (b) Microscopic shells of diatoms, magnified 100 times. Most siliceous oozes are composed of these single-celled algae.

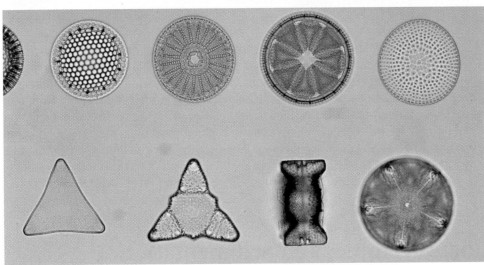

(b)

upwelling cold water—for example, off the western coasts of South America and Africa. Silica-producing organisms also thrive along continental shelves, but they are masked there by detrital or carbonate sediments. These siliceous sediments are the probable source of the silica of geodes, nodules, and concretions.

Brown clay, the most common deep-ocean sediment, comes from a variety of sources. Most of it is derived from the weathering of crustal rocks and is brought to the ocean by streams. Being very fine, the clay particles float to the open ocean far from shore before they settle. Windblown particles distinctive of the Sahara Desert have been traced far out into the Atlantic Ocean off the West African coast. Minute, but measurable, quantities of meteoric dust, formed when meteors burn through the Earth's atmosphere, have been found mixed with the brown clay of the North Pacific. In contrast to carbonate ooze, brown clay is found at all depths. It is the dominant sediment of the Pacific Ocean, most likely because the Pacific is the deepest ocean, and a higher percentage of its ocean floor lies beneath the snowline. Some geologists suggest that originally both carbonate oozes and brown clays accumulated, but the carbonate has since dissolved, leaving the brown clay residue.

Biochemical sediment
- Calcareous
- Siliceous
- Radiolarians
- Diatoms

Detrital sediment
- Deep-sea oozes
- Turbidites
- Glacial marine

Figure 8.26 The worldwide distribution of deep-sea sediments. Notice that calcareous sediments are associated with the mid-ocean ridge system, siliceous sediments with high latitudes and regions of upwelling ocean waters (the equatorial Pacific), turbidites with passive continental margins, and deep-sea oozes with abyssal plains far from the continents.

Manganese nodules are deep-ocean deposits directly precipitated from seawater (Figure 8.27). They consist of concentric layers rich in copper, cobalt, and nickel, as well as manganese wrapped around a hard rock nucleus. Geologists think that the nodules precipitated as a result of reactions between cold seawater and hot fluids rising from fractures along the mid-ocean ridges. The hot fluids—probably heated seawater that circulated through the fractures in a complex natural-plumbing system—leach elements from the ridge basalts and surrender them as they escape. The distribution of the nodules supports this idea. Though strewn throughout the ocean floor, they are concentrated mostly in the vicinity of mid-ocean ridges. Probably their wide distribution is at least partly explainable by the slow spreading of oceanic crust away from the ridges.

Plate Tectonics and Sedimentation

Plate tectonics plays a profound role in the distribution of past and present continental margin sediment. Active continental margins generally occur along plate boundaries, and passive margins occur far within the plates. As we have seen, these environments yield very different sediment patterns. Because plate tectonics is also a prime influence on the distribution of mountains and on continental topography in general, it exerts fundamental control on the courses that streams take to the sea. These courses, in turn, determine where the bulk of detrital sediment is deposited. Plate movements also control the rate of crustal uplift and, therefore, the rate of erosion and sedimentation.

The role played by plate tectonics in the distribution of deep-sea sediments is perhaps more subtle but equally profound. For example, layers of calcareous ooze are often found beneath brown clay and siliceous ooze on the deep-ocean floor below the snowline. The carbonate originally accumulated above the snowline on the flanks of the mid-ocean ridge. As the plates separated, and as the oceanic crust cooled and sank beneath the snowline, the calcareous oozes were covered and preserved by brown clay and siliceous ooze.

The migration of the oceanic crust and sediment away from the mid-ocean ridge crest brings us to an interesting point: since the oceanic crust is not permanent—being created at the ridges and destroyed at the trenches—the deep-ocean sediment deposited on it is not permanent either, at least not in its original condition. What happens to these former ocean sediments? Paradoxically, some are preserved on the continents. One way that deep-ocean sediment can be preserved is by being sheared off the oceanic crust as the crust descends into the mantle beneath deep-sea trenches. The sediment, together with slivers of oceanic crust, is highly deformed and slightly metamorphosed. This material, called a *mélange,* is often uplifted and thrust onto the landmasses bordering the trenches. The rocks of San Francisco are considered a classic example of a mélange. Other sediments become detached from oceanic crust during plate collisions and become attached to the continents as *accreted terranes.* Some of these terranes leave an excellent record of the ancient ocean floor. But, overall, the record of past deep-ocean sediments is confused and fragmented.

Figure 8.27 A manganese nodule, split open to display its internal structure.

The most complete record of oceans past comes from the deposits of shallow seas that invaded the continents repeatedly over geologic time. These deposits form the relatively thin veneer that covers the igneous and metamorphic rock of the ancient, gnarled continents.

STUDY OUTLINE

Sediments are fragmented or dissolved rock particles that have been transported by water or air currents and deposited in layers.

I. **WEATHERING, TRANSPORT, AND DEPOSITION OF SEDIMENTS**
   A. **Detrital sediments** are fragments weathered from rocks. **Chemical** and **biochemical sediments** are composed of ions that were weathered from rock, transported in solution, and precipitated either directly (chemical) or indirectly with the aid of organisms (biochemical).

   B. **Weathering** causes the mechanical disintegration and chemical decomposition of rock.
      1. In mechanical disintegration, bedrock is broken down into fragments by weathering agents such as rain, wind, freezing, and thawing.
      2. Chemical weathering alters the composition of rocks; for example, in hydrolysis, carbonic acid and water transform the mineral feldspar into insoluble clay residues and dissolved ions.

   C. Streams, the main agents of transport and **deposition,** carry sediments to lakes, flood plains, deltas, beaches, and the ocean floor.

II. **LITHIFICATION.**   There are three components to **lithification,** the process of converting sediments to sedimentary rocks.
   A. In **compaction,** layers are compressed by the weight of overlying layers.

   B. In **cementation,** mineral grains are held together by a **matrix** of precipitated ions.

   C. In **recrystallization,** dissolved edges bind neighboring mineral grains.

III. **CLASSIFICATION OF SEDIMENTARY ROCKS**
   A. Detrital, chemical, and biochemical sedimentary rocks are further classified by texture and mineral content.

   B. Detrital rocks display **clastic** (fragmented) **texture** and vary in the degree of **sorting** and rounding.

1. **Conglomerates** have large, rounded fragments; **breccias** have large, angular fragments.
2. **Sandstones** are formed of sand-sized fragments.
3. **Shale** is composed of the finest particles (clay and silt).

C. Coarse detrital rocks are the products of the composition of the source rocks and of their weathering, transport, and depositional history. As a rule, the farther from the source rock, the finer, better sorted, and better rounded the sediments.

D. **Turbidites** are detrital sediments that are subjected to further transport and deposition by undersea currents.

E. **Facies** changes allow geologists to analyze ancient sedimentary environments.

F. There are two categories of precipitated sedimentary rocks: those that are precipitated by chemical means and those that are precipitated biochemically through the aid of marine organisms.
1. The **limestones** are the most common of the precipitated rocks.
   a. Biochemical limestones are made of the shells of organisms. Coquina is formed from large shell fragments, chalk from minute fragments.
   b. Coral reefs are formed from the skeletal remains of marine organisms deposited on the continental margins, where ocean currents are warm.
   c. Oolitic limestones are precipitated in shallow waters.
2. **Evaporites** are formed from salts that settle out of the evaporating waters of semienclosed basins in arid climates. Some evaporites are chemical limestones, gypsum, and halite.
3. The biochemical silica shells of diatoms form bedded **chert**.
4. **Coal** consists of vegetation altered by burial, compression, and heat.

IV. SEDIMENTARY STRUCTURES

A. **Bedding** is the predominant structure of sedimentary rocks. The parallel layers are called **beds**, or **strata**. The surface of each bed is the **bedding plane.**
1. In **graded bedding,** each layer displays coarse to fine particles from the bottom to the top.
2. **Cross-bedding** is formed by moving wind or water currents and is laid down at an oblique angle to the main bed.
3. **Ripple marks** are formed when bedding is disturbed by currents or waves. **Mud cracks** form when the bed is exposed to air and dried.

B. Other sedimentary structures include **fossil** remains, masses of silica called nodules and geodes, and concretions, which are formed when chemicals precipitate around a rock fragment.

V. OCEAN SEDIMENTARY ENVIRONMENTS

A. On passive continental margins, detrital sediments are deposited such that the largest fragments are near the shore and the finest are far out on the continental slope.

B. Active continental margins are short and steep, and many sediments are subducted into the trenches.

C. In the deep-ocean environment, land-derived turbidites, siliceous and calcareous oozes, brown clay, and manganese nodules accumulate.

VI. PLATE TECTONICS AND SEDIMENTATION.
Plate tectonics controls sedimentation patterns. It also redistributes sediments by the movement of the oceanic crust. At the trenches, some sediments are subducted, some are thrust up onto the continents as mélanges, and some are scraped off onto the continents as accreted terranes.

STUDY TERMS

bedding (p. 232)
bedding plane (p. 232)
beds (p. 232)
biochemical sediment (p. 220)
breccia (p. 223)
cementation (p. 222)
chemical sediment (p. 220)
chert (p. 232)
clastic texture (p. 223)
coal (p. 232)
compaction (p. 221)
conglomerate (p. 223)
cross-bedding (p. 233)
deposition (p. 221)
detrital sediment (p. 220)
evaporite (p. 231)

facies (p. 226)
fossil (p. 234)
graded bedding (p. 232)
limestone (p. 228)
lithification (p. 221)
matrix (p. 222)
mud crack (p. 234)
recrystallization (p. 222)
ripple mark (p. 234)
sandstone (p. 223)
sediment (p. 219)
shale (p. 223)
sorting (p. 223)
strata (p. 232)
turbidite (p. 227)
weathering (p. 221)

CRITICAL THINKING QUESTIONS

1. Describe the evidence that would convince you that a given sequence of sedimentary rocks was deposited in each of the following environments:
   a. on land, close to a mountain range
   b. on an ocean beach
   c. on the ocean floor, far from shore
2. What rock types might you associate with the environments listed in question 1?
3. Describe the changes you would expect to see in a detrital sediment as it moves downstream from source to ocean.
4. How do the textures of quartz sandstone and glacial till differ? Describe the differences in terms of fragment size, degree of sorting, and particle shape. Explain the differences in terms of weathering, erosion, transport, and depositional history.
5. Imagine mountain ranges A and B, each composed of identical granite. Range A is located in an arid climate and range B in a humid climate. What differences would you expect in the weathering products yielded by each mountain range?
6. What features of a turbidite indicate its mode of deposition?
7. In what ways are coquina and chalk similar? In what ways are they different?
8. What conditions favor the formation of coal?
9. Why are carbonate oozes well preserved along the Mid-Atlantic Ridge in low latitudes but not in deeper waters of the same latitudes?
10. What conditions favor the growth of large coral reefs? What kinds of sediments are deposited in the forereef, core, and back reef?
11. How may the growth of a large reef change the conditions behind the reef so that evaporites are deposited?
12. Deep-ocean sediments are seldom found in an undisturbed state on the continents. Explain.

Metamorphic Rocks

The rock that became Michelangelo's magnificent sculpture *David* began life millions of years ago as limy mud on the bottom of a shallow sea that occupied a portion of what is now northern Italy. The mud was exceedingly fine, with few clay particles, quartz grains, or iron oxide compounds—what we call impurities. Upon burial, the mud was compacted and cemented into a dense, even-textured limestone. At that time, Italy was being squeezed against southern Europe to produce the Italian Alps, so the limestone deposit was subjected to much compression as well as heat from igneous intrusions in the region, and it recrystallized into a coarser-grained rock called marble. Prolonged erosion later exposed this marble, which was remarkably pure in color, texture, and composition and had few structural defects. For those reasons, architects and sculptors favored it for building ornaments and statues, and the marble was quarried for these purposes. Legend has it that Michelangelo took the particular block of marble that later became the *David* as a challenge. The block had been rejected by other sculptors because they considered it too thin and flawed for use.

This chapter describes metamorphism, the set of processes by which preexisting rocks are changed by the pressures, temperatures, and chemical conditions that prevail deep within the Earth. These changes produce rocks of markedly different mineralogy, texture, and structure—that is, metamorphic rocks.

Rocks remain in the solid state during metamorphism. What is mainly involved is change of form—a fact made clear by the Greek roots of the term: *meta*, "after" or "beyond," and *morphe*, "form." Metamorphism occurs at pressures and temperatures well above those prevailing at the Earth's surface but below those that cause melting. In most metamorphic transformations, the gross chemical composition of the rock changes

◀ A marble quarry in the Italian Alps (right). From a quarry similar to this one, Michaelangelo chose the block of metamorphic rock from which he sculpted his *David* (left).

Figure 9.1 The pressure-temperature range of metamorphism.

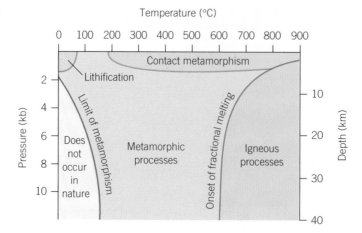

little, although reactions between minerals of the original rock may produce entirely new minerals.

The realm of metamorphism is inaccessible places: in the deep cores of mountain ranges, in the rocks surrounding igneous intrusions, in the fractured crust beneath mid-ocean ridges. In contrast to some igneous and sedimentary rocks, no metamorphic rock has ever been directly observed in the process of formation. It is the geologist's task to determine as closely as possible the nature of the preexisting parent rocks and the specific conditions responsible for their metamorphism. There are two components of this work:

1. Field studies of metamorphic terrains exposed through the erosion of overlying rock. These studies establish the mineralogical and structural features of the various metamorphic rocks and their variation in space and time.
2. Laboratory experiments on natural and synthetic rocks and minerals that attempt to simulate the conditions of metamorphism within the Earth.

Metamorphic rocks are found mainly in the core regions of the continents and mountain belts. Where exposed, they constitute 25 percent of the surface area of the continental crust. However, greater volumes of metamorphic rocks are blanketed by much younger sedimentary rock.

The Agents of Metamorphism

Heat, pressure, and chemically active fluids are the agents responsible for metamorphism. It will aid our understanding of the process to consider briefly the role played by each of them (Figure 9.1).

Heat

Heat is the energy source that powers metamorphic reactions. It raises the temperature of the rocks, which means that it causes the ions of the minerals to vibrate more rapidly. Chemical bonds are broken and the ions are forced to realign in combinations more in keeping with the high-energy environment. Thus new minerals are created as old ones are destroyed. In addition, heat mobilizes hot water and gases within the rock. The fluids act as a medium of exchange, dissolving and precipitating elements within the rock mass and, in general, accelerating the reaction rates. Heat also affects the mechanical properties of the rock by softening it and allowing it to deform plastically. Thus heat prepares the rock for the imprinting of characteristic metamorphic textures and structures.

Pressure

A body subject to pressure, in the form of a single force or a number of forces acting on it at the same time, is in a state of stress. If the forces are applied equally in all directions, then the body is said to be in a state of **lithostatic stress** (Figure 9.2a). In the Earth, this kind of stress is caused by the pressure of the overlying rocks.

Lithostatic stress squeezes rocks together and reduces their volume. Thus reactions in which the atoms of minerals recombine into denser forms are favored. We have mentioned two such reactions in previous chapters: the conversion of graphite to diamond in the upper mantle, and the conversion, through a series of steps, of olivine to perovskite in the transition zone of the mantle. High lithostatic stress, in conjunction with high temperature, also prohibits rocks from fracturing; instead, they deform plastically.

Although the degree of lithostatic stress, along with temperature, defines the general environment of metamorphism, nonuniform **compressive stress** is more directly responsible for many of the features associated with metamorphic rocks. For example, viselike pressure is exerted on rocks trapped between colliding lithospheric plates. Most rocks that are metamorphosed in this setting develop a pronounced layering, or **foliation,** perpendicular to the direction of maximum compression (Figure 9.2b).

lithostatic stress
The uniform stress in the Earth's crust, caused by the weight of the overlying rocks.

compressive stress
The stress generated by forces directed toward one another on opposite sides of a real or imaginary plane.

foliation
The arrangement of a rock in parallel planes or layers; in metamorphic rocks, caused by parallel alignment of the minerals.

Chemically Active Fluids

Fluids, especially water charged with dissolved gases, greatly accelerate chemical reactions because they have the ability to dissolve ions in regions of high pressure and temperature and to precipitate them in regions of lower pressure and temperature. Thus they are able to decompose old minerals and reconstitute new ones. These abilities increase enormously with depth, where the elevated pressures and temperatures convert the water to a superheated state, in which it is able to dissolve huge quantities of matter that are not ordinarily soluble at low temperatures and pressures. For example, quartz rarely dissolves on the Earth's surface, but it dissolves fairly easily in the presence of superheated hot water and dissolved gases. The superheated fluid also has great penetrative and propellant powers. Dissolved elements are commonly carried many kilometers from their source and disseminated throughout the rock mass.

The sources of the chemically active fluids that facilitate metamorphism are quite varied. The clays, the micas, and the amphiboles are among the many minerals that hold water chemically in the form of the hydroxide (OH^-) ion. These minerals release water if they are heated and subjected to high pressures. In addition, water is often held physically in the pore spaces and fractures of shallow rocks. The water in some sedimentary rocks may have originated as seawater trapped in the sediment at the time of deposition. In other instances, water is introduced from outside the rock body: from magma in the crust, from seawater that has penetrated fractured oceanic crust, or from groundwater that has filtered down from the Earth's surface. The dissolved gases and ions of metamorphic fluids are also derived from a

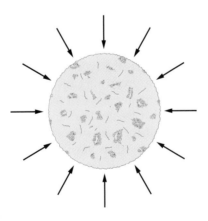

(a)

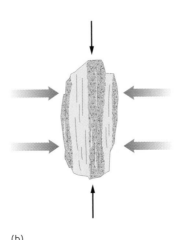

(b)

Figure 9.2 (a) Lithostatic stress results from pressure exerted equally in all directions. (b) Nonuniform compressive stress results in the parallel alignment of mineral grains, or foliation, a characteristic feature of many metamorphic rocks.

Figure 9.3 An outcrop of well-foliated and highly contorted metamorphic rock (gneiss), Painted Canyon, Mecca Hills, California. The pronounced layering reflects the segregation of minerals into light bands of quartz and feldspar and dark bands of biotite and hornblende. The structure, texture, and mineralogy of this rock are typical of regional metamorphism.

number of sources. Most are released from minerals subjected to high pressure and temperature. For example, carbon dioxide may come from the breakdown of calcite and shell matter.

Types of Metamorphism

Geologists classify the various types of metamorphism on the basis of the geologic processes responsible for them. **Regional metamorphism** is the most widely occurring type, and it is the type most commonly associated with the term *metamorphism*. It is induced in preexisting rocks during subduction and the plate collisions that produce fold-mountain ranges. In these mountain-building, or *orogenic,* environments, the rocks are invaded by the heat and fluids that escape from magmas rising from subducting plates, and they are subjected to enormous compressive stresses caused by the plates' squeezing together. Most regionally metamorphosed rocks exhibit a pronounced foliation developed in response to this stress (Figure 9.3). They occur as linear belts in the deeply eroded interiors of mountain ranges and are widespread in the core regions of the continents.

Contact metamorphism occurs where magma alters the cold country rock that surrounds an igneous intrusion or underlies a lava flow. The rock is baked, recrystallized, or otherwise changed through reactions driven by the infusion of heat and by the fluids mobilized within the magma and the rock itself.

Shear metamorphism typically occurs within active fault zones where one block of the crust moves with respect to another. High in the crust, where the lithostatic stress is low, rocks in the vicinity of the fault are subject to intensive grinding and are broken into rolled or rotated fragments, which are further pulverized to a fine powder. At deeper levels, the rocks are also subject to a great deal of plastic deformation and recrystallization. Faults and shear zones are generated along convergent, divergent, and transform boundaries. Therefore, shear metamorphism occurs in these settings.

Burial metamorphism is the change induced by the weight of overlying rocks. It occurs in the thick deposits of sedimentary basins at relatively low temperatures and is basically a continuation of the same processes that have converted the sediments into sedimentary rocks. These processes include the growth and hardening of the clay matrix, cementation, precipitation in pore spaces, and the dissolving and recrystallization of mineral grains at points of contact.

regional metamorphism
Metamorphism of an extensive area of the crust; generally associated with intensive compression and mountain building.

contact metamorphism
The transformation of rocks caused by heat escaping from an igneous intrusion or lava flow.

shear metamorphism
The transformation of rocks within the shear zone associated with active fault movement; mainly involves grinding and pulverizing high in the crust and recrystallization at deeper levels.

burial metamorphism
Metamorphism that results in response to the pressure exerted by the weight of the overlying rock.

Table 9.1   Mineral Content of Metamorphic Rocks

Category	Mineral
Minerals that crystallize from magma but also may form under metamorphic conditions	Quartz, feldspar, biotite, amphibole
Minerals most commonly, but not exclusively, formed under metamorphic conditions	Sillimanite, chlorite, garnet
Minerals that form only during metamorphism	Kyanite, andalusite, staurolite, talc

Shock metamorphism results from meteor impacts. Enormous pressures and temperatures are generated—and dissipated—in the short time interval during and immediately after the impact. Some rocks are fractured and pulverized; others are melted into droplets and fused into tiny glass beads; still others are recrystallized or converted to denser forms. Although meteorite impacts are difficult to find on the Earth's surface, those that are well preserved get much attention because of their suspected links to mass extinctions of the dinosaurs and other species. Evidence of shock metamorphism plays a prominent role in the identification of meteorite impact sites. Meteorite impacts and the resultant shock metamorphic features are well preserved on the Moon, Mercury, and Mars, which lack the Earth's efficient erosion and plate tectonic recycling systems.

The term *metamorphism,* strictly applied, refers to rocks that have not undergone a pronounced change of composition during transformation. However, many rocks do undergo significant change because of the invasion of fluids released by magmas or other sources deep in the crust. These fluids penetrate the rock, bathe the minerals, and then proceed to dissolve the original ions and replace them with others. The replacement process, which includes a change in composition, is called **metasomatism.** It is often so delicate that the original layering and other structures of the parent rock are preserved. **Hydrothermal metamorphism** is triggered by water that percolates through fractures above shallow magma chambers. The magma directly or indirectly heats the water, which, in turn, alters the host rock. Hydrothermal reactions of this type are extremely important in altering the highly fractured and porous basalt lavas of mid-ocean ridge rift valleys.

shock metamorphism
Changes in rock and minerals caused by shock waves from high-velocity impacts, mainly from meteorites.

metasomatism
A replacement process whereby the elements of a rock are exchanged with those of a magmatic fluid.

hydrothermal metamorphism
The transformation of rock through the action of high-temperature solutions.

Metamorphic Rocks

Metamorphic rocks contain information concerning conditions *within* the crust, just as sedimentary rocks contain information concerning the *surface* of the crust. Their textures and internal structures are a consequence of the deep crustal forces that have molded them. Their mineral content is a consequence of the pressures and temperatures that prevailed within the crust at the time of metamorphism (Table 9.1). In this sense, metamorphic minerals are sensitive pressure gauges and geological thermometers.

The Minerals of Metamorphic Rocks

Because rocks remain solid during metamorphism, nearly all reactions occur between minerals that have not melted. They are called *solid-state* reactions, in which two minerals normally stable at high temperatures react when they come in contact. In some solid-state reactions, ions diffuse across grain boundaries, recombine in new arrangements, and form a third mineral that grows at the expense of the original two. In other reactions, complex minerals may break down by solid diffusion into simpler forms.

Metamorphism often dries out rocks. Diffusion reactions commonly convert *hydrous* (water-rich) minerals to simpler *anhydrous* (water-free) ones. Table 9.2

Table 9.2 Typical Metamorphic Reactions

Dehydration

$$\begin{array}{ccc} \text{Muscovite} & \longrightarrow & \text{Potassium feldspar} + \text{Corundum} + \text{Water} \\ KAl_2AlSi_3O_{10}(OH)_2 & \longrightarrow & KAlSi_3O_8 \quad + \quad Al_2O_3 \quad + \quad H_2O \end{array}$$

Decarbonation

$$\begin{array}{ccc} \text{Calcite} + \text{Quartz} & \longrightarrow & \text{Wollastonite} + \text{Carbon dioxide} \\ CaCO_3 + SiO_2 & \longrightarrow & CaSiO_3 \quad + \quad CO_2 \end{array}$$

$$\begin{array}{ccc} \text{Dolomite} + \text{Quartz} & \longrightarrow & \text{Pyroxene} + \text{Carbon dioxide} \\ CaMg(CO_3)_2 + 2\,SiO_2 & \longrightarrow & (Ca,Mg)Si_2O_6 + \quad 2\,CO_2 \end{array}$$

Unmixing

$$\begin{array}{ccc} \text{Sodium feldspar} & \longrightarrow & \text{Jadeite} + \text{Quartz} \\ NaAlSi_3O_8 & \longrightarrow & NaAlSi_2O_6 + SiO_2 \end{array}$$

illustrates some typical *dehydration* reactions. The water driven off in these high-temperature reactions may recombine with other minerals at lower temperatures somewhere else in the crust. Table 9.2 also includes some reactions involving the carbonate minerals calcite and dolomite. Notice that these reactions are similar in form to dehydration, but carbon dioxide rather than water is driven off.

Diffusion reactions cause the growth of minerals through recrystallization, as in the conversion of limestone into marble. In this instance, the tiny calcite crystals of the limestone are transformed into the large, densely compacted crystals that characterize marble. Although technically the mineral composition of the rock remains unchanged, the original calcite structure of the limestone is obliterated, and the marble acquires a coarse texture. In the process, sedimentary features such as fossils are destroyed.

Polymorphic conversion is another means by which minerals are transformed during metamorphism. In a conversion of this type, the result is a mineral identical in composition to, but different in structure from, the original. Figure 9.4 illustrates the pressure-temperature ranges of three extremely important polymorphs common in metamorphic rocks of shaley origin: andalusite (dark, square to rectangular), kyanite (a blue, bladed mineral), and sillimanite (which usually occurs as needlelike brownish masses). Each has the identical composition of Al_2SiO_5.

The Textures of Metamorphic Rocks

The textures of metamorphic rocks—that is, the size, shape, and arrangement of their mineral grains—bear the stamp of the pressure-temperature environment in which recrystallization, mineral growth, and other reactions occur.

When rocks composed of granular minerals, such as quartz or calcite, are subjected to lithostatic stress, the minerals display no preferred direction of growth. As a consequence, the rocks develop a mosaic-tile pattern of interlocking equidimensional mineral grains—an *equigranular texture*. Rocks of this texture have a uniform, nonlayered appearance, as can be seen in many marbles and quartzites.

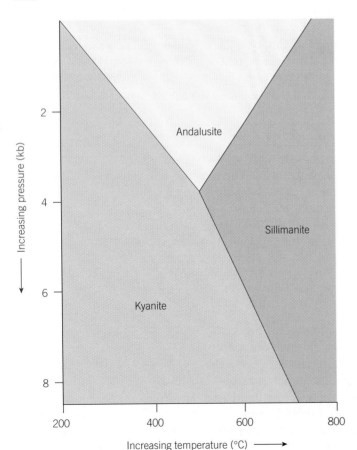

Figure 9.4 The pressure-temperature ranges of three polymorphs: kyanite, andalusite, and sillimanite. During metamorphism, a mineral such as kyanite may convert to sillimanite. This graph shows the range of temperatures over which this reaction (and similar ones) can occur. Note that there is only one point where the three minerals coexist.

Table 9.3 Common Metamorphic Rocks

Foliated Rocks		Nonfoliated Rocks	
Representative Parent Rock	Metamorphic Rock	Representative Parent Rock	Metamorphic Rock
Shale	Slate, phyllite, schist, gneiss, migmatite	Limestone	Marble
Basalt	Greenschist, blueschist, amphibolite (hornblende schist),* granulite,* eclogite*	Sandstone	Quartzite
Peridotite	Serpentinite	Basalt, gabbro	Amphibolite (hornblende schist),* granulite,* eclogite*
Rocks resulting from shear metamorphism	Fault breccia, mylonite	Country rock	Hornfels

*Can be either foliated or nonfoliated.

Most rocks subjected to the compressive stress of regional metamorphism display prominent parallelism of mineral grains. Minerals are sheared, rotated, and realigned at right angles to the direction of maximum compression. This action—plus growth of the platy minerals such as clay, mica, and chlorite—imparts a pronounced layering or foliation to the rock. Bladed minerals such as kyanite and sillimanite also line up with their long axes in the same direction, like swarms of parallel arrows. Marble and quartzite, composed of equidimensional calcite and quartz grains, respectively, tend to retain a superficially equidimensional texture even when subject to regional metamorphism. However, close inspection generally reveals subtle parallelism of structures even in these rocks.

The Classification of Metamorphic Rocks

Metamorphic rocks are classified on the basis of texture and mineral content, as are igneous and sedimentary rocks. As Table 9.3 shows, metamorphic rocks fall easily into foliated and nonfoliated categories.

Foliated Rocks

Increasing metamorphic intensity caused by elevated temperatures and pressures often leads to an increase in grain size and a general coarsening of the texture of foliated rocks. Various terms describe these changes. *Slaty cleavage* refers to the foliation of exceedingly fine-grained rocks whose platy minerals are of microscopic size. *Schistosity*, on the other hand, refers to the foliation of coarser-grained rocks with highly visible minerals. High metamorphic intensity may also produce banded *gneissic* rocks, which display pronounced mineral segregation, or banding. Layers of dark platy or elongated minerals like biotite, mica, and hornblende alternate with light layers of feldspar and quartz. **Slate, gneiss,** and **schist,** then, are among the most common foliated rocks (Figures 9.5, 9.6, and 9.7) At higher levels of metamorphic intensity, the melting points of some quartz and feldspar minerals are reached, and they form granitic fluids. The result upon cooling is **migmatite,** a complex mixture of banded gneiss and granite (see Figure 9.21 later in the chapter).

slate
A fine-grained metamorphic rock; mostly formed from the transformation of shale.

gneiss
A foliated metamorphic rock in which bands of granular minerals alternate with bands of flaky minerals.

schist
A foliated, coarse-textured metamorphic rock; most of the minerals display a pronounced parallelism.

migmatite
From the Greek migma, *meaning "mixture"; a complex mixture of metamorphic and igneous rock components (usually granite).*

Figure 9.5 A hand specimen of slate displaying closely spaced layering or slaty cleavage.

Figure 9.6 A hand specimen of biotite gneiss. Notice that the minerals are segregated into light and dark bands.

These rock names are generally made more specific by placing the names of key minerals in front of them: mica schist, garnet schist, sillimanite schist, hornblende gneiss, and so on. Some metamorphic rocks are referred to by color as a shorthand reference to their mineral content. *Greenschist,* generally the product of the metamorphism of basalt or gabbro, owes its color to the preponderance of chlorite and epidote (Figure 9.8). *Blueschist* derives its color from the metamorphic amphibole glaucophane. Hornblende schist is the higher-grade equivalent of greenschist.

Rocks subjected to shallow shear metamorphism are transformed to **fault breccia** (Figure 9.9). The rock consists of angular, fragmented mineral grains resulting from mechanical grinding and pulverization. Deeper in the fault zone, where temperatures are higher, rocks are subject to plastic deformation and are further ground down to a fine-grained, well-foliated *mylonite.* A compact rock with a streaky or banded structure, it is the product of extreme granulation and recrystallization (Figure 9.10).

fault breccia

A metamorphic rock consisting of angular fragments that are the result of the grinding and shattering action that occurs along active fault zones.

(a)

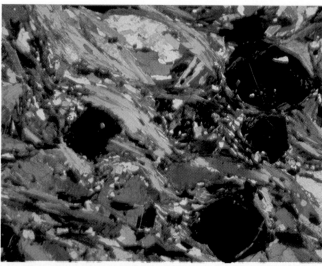

(b)

Figure 9.7 (a) A hand specimen of mica schist showing coarse texture and foliation. (b) The foliation is well displayed in this photomicrograph. The dark rounded minerals are garnet; the parallel bright-colored minerals, biotite; and the gray minerals, quartz.

Figure 9.8 Greenschist, which contains shiny green chlorite, green amphibole, quartz, and feldspar.

Nonfoliated Rocks

As described earlier, the nonfoliated rocks generally have an equigranular texture. They are distinguished mainly by their composition. **Quartzite** and **marble,** the most common types of nonfoliated rocks, are composed of interlocking quartz and calcite grains, respectively (Figures 9.11 and 9.12). They are the products of the metamorphism of sandstone and limestone. Greenstone, like the foliated amphibolite, is derived from the metamorphism of basalt or gabbro. It is composed of chlorite and epidote and is the low-grade, nonfoliated equivalent of greenschist.

Although equigranular textures generally imply mineral growth under lithostatic stress, heat is needed to trigger the reactions. Nearby magmas are the obvious heat source. Fieldwork bears out the fact that most equigranular textures are associated with rocks altered in contact zones that surround igneous intrusions. These rocks are given the special name **hornfels.**

quartzite
A metamorphic rock composed mainly of quartz, formed by recrystallized sandstone.

marble
A metamorphic rock consisting mainly of recrystallized calcite and/or dolomite.

hornfels
Fine-grained rock formed by contact metamorphism.

Figure 9.9 Fault breccia in the Grapevine Mountains, Death Valley National Monument, California. Notice the broken, sheared fragments—the products of mechanical grinding within the fine-grained matrix.

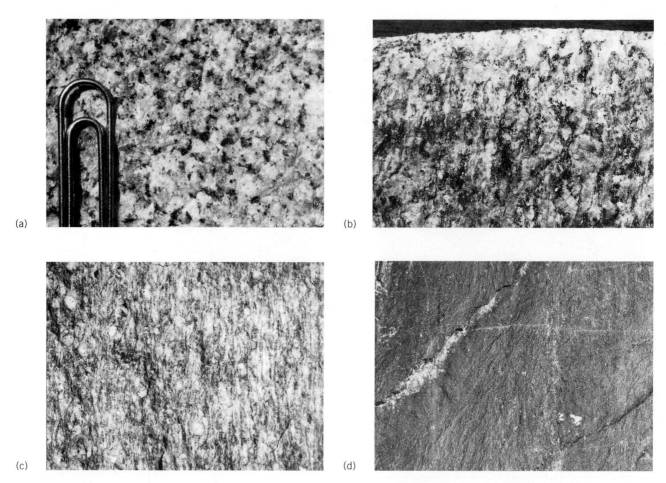

(a)

(b)

(c)

(d)

Figure 9.10 The transformation of granite into mylonite in the San Gabriel Mountains of California. Part (a) shows an undeformed granite at the periphery of a fault zone, while parts (b–d) show how the granite changes by degree into a fine-grained, well-foliated end-product.

aureole
An halolike region of contact metamorphism surrounding an igneous intrusion.

The Field Occurrence of Metamorphic Rocks

Like living things, metamorphic rocks are best appreciated in their natural habitat. In the following descriptions of common metamorphic rocks as they appear in the field, the textural, structural, and regional properties of the rocks will be emphasized more than their mineralogy.

Contact Metamorphic Rocks

Surrounding an igneous intrusion like a halo is a zone of transformed country rock called an **aureole** (Figure 9.13). In this display of contact metamorphism, the transfor-

Figure 9.11 Hand specimen of a typical quartzite.

(a)

(b)

Figure 9.12 (a) A coarse-grained pink marble from Tate, Georgia. (b) This photomicrograph (×40) of marble, taken under polarized light, displays a mosaic of interlocking mineral grains (calcite). This texture is typical of non-foliated rocks.

mation takes the form of textural and mineralogical changes that grow more pronounced toward the igneous rock, so there is little doubt that heat and, in some instances, fluids escaping the magma were the agents responsible for the changes.

The extent and intensity of contact metamorphism depend upon a number of factors: the quantity, temperature, and composition of the magma; and the composition, rigidity, and permeability of the country rock. In general, the effects are most conspicuous in shaley limestone. The effects are not as great in sandstones, which are composed of heat-resistant quartz. Igneous rocks that have crystallized at high temperatures are least affected.

The best place to view the effects of contact metamorphism is in the field. There we can walk out and see the changes in the country rock firsthand, and take samples back to the laboratory to examine under the microscope. Figure 9.13 is a typical contact aureole, within which we can trace the changes that occur from unaltered shale to a granitic intrusion.

The first signs of change are subtle, detected only through microscopic analysis. They involve recrystallization of clay to mica and an increase in the grain size of quartz (point X in Figure 9.13). Textures coarsen farther inward as we approach the granite, and for the first time, knots of new minerals are visible (point Y in Figure 9.13). The rock may thus acquire a spotted look. Some of the new minerals include the familiar quartz and biotite. Others are found only in metamorphic rocks—for example, the large, square, reddish brown andalusite and the greenish blue cordierite.

Close to the contact, recrystallization and mineral growth may have all but obliterated the bedding and other structures of the country rock (point Z in Figure 9.13). Through baking, the rock has acquired a hard, flinty look. Isolated larger minerals stand out amid the finer groundmass. Microscopic analysis reveals that the mineral grains have a mosaic pattern of interlocking boundaries with no discernible orientation. We have passed, by degrees, from unaltered shale into hornfels, a rock characteristic of contact metamorphism. There is little difference between the bulk composition of the hornfels and the shale. The major changes involve driving off water and carbon dioxide.

At this point, there are a number of possible endings to our hypothetical journey. In one version, contacts between the inner boundary of the aureole and the intrusion are razor-sharp. One can stand on the boundary between hornfels and the granite. In another version, the transition consists of an uneven zone where offshoots of igneous rocks—small dikes, sills, veins—cut into the aureole. Numerous xenoliths (broken pieces of country rock) are stranded like islands within the granite, and the country rock itself is contorted, as if compressed in response to the intense heat and force of the intrusion.

Figure 9.13 A granitic intrusion into shaley country rock forms a contact aureole of metamorphosed rock. Walking outward from point Z to point Y to point X, we can observe that the intensity of metamorphism, as manifested by textural and mineralogical changes in the country rock, decreases with distance from the granite. (Adapted from Myron G. Best, *Igneous and Metamorphic Petrology,* New York: W. H. Freeman and Company, 1982.)

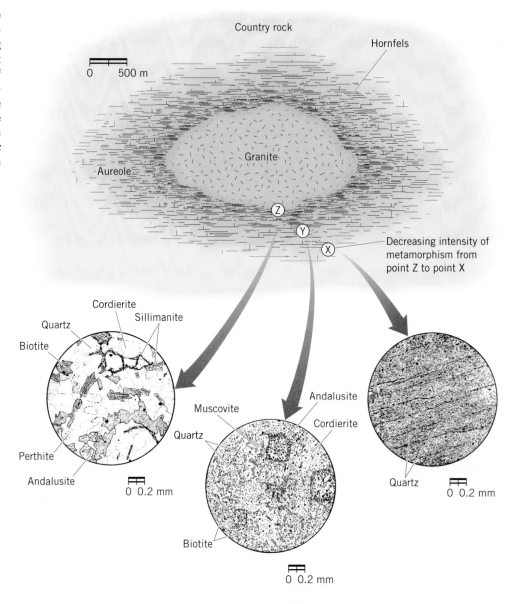

In still other zones, the transition of igneous rock proper into hornfels is so subtle that it is hard to locate the exact boundary between the two rock masses.

We find that contact metamorphic effects also vary with the composition of the host rock. For example, if pure quartz sandstone instead of shale is intruded by magma, it will be converted to quartzite. Few mineralogic changes occur, because quartz is stable at the high temperature of magma. However, the elevated temperatures recrystallize grains into an interlocking mosaic; the rock acquires a sugary texture. When struck with a hammer, it breaks sharply across the grains, rather than around them, as in sandstone (Figure 9.14).

However, if the host rock consists of impure limestone containing clay minerals and quartz, or if there has been interaction with high-silica magma, extensive mineralogic change occurs in the resulting marble. We saw one common reaction among the many that are possible in Table 9.2. Heat from the magma drives a reaction between quartz and calcite that forms *wollastonite,* a pearly white to gray calcium silicate mineral, similar to pyroxene. (Carbon dioxide is driven off in the process.) In fact, marble and limestone are so reactive with fluids derived from magma that they are often a mineral collector's paradise of rare specimens—rare because some of these minerals crystallize only during contact metamorphism of this type.

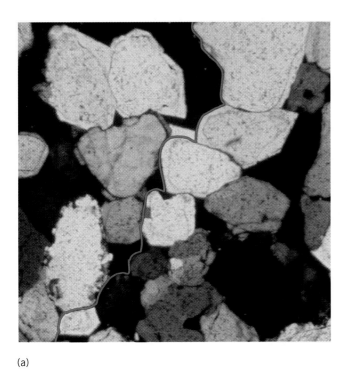

(a)

(b)

Figure 9.14 (a) Quartz sandstone. (b) Note the textural changes that occur when quartz sandstone is transformed to quartzite during contact metamorphism. The quartzite owes its mosaic texture to the fusion of recrystallized grains of quartz sand. The red lines show breakage patterns—between the grains in (a), and across the grains in (b).

Regional Metamorphic Rocks

Fold-mountain belts are commonly thousands of kilometers long and hundreds of kilometers wide. They are composed mainly of sedimentary rocks that were originally deposited as horizontal strata on continental margins, which in places have attained thicknesses greater than 10 kilometers. The mountains are formed where lithospheric plates collide. Strata caught between the plates are subjected to compressive stress and are deformed into a complex series of folds and faults. As streams strip the upper strata away, lessening the load on the mantle, the deep central cores of the mountains rise, and the streams cut into these as well. Eventually, the metamorphic core of the mountain belt is exposed (Figure 9.15).

Proceeding from the periphery of the mountain belt toward the core, we can often trace the transformation of sedimentary rocks into metamorphic rocks of increasing complexity. Here we are in the realm of the most widespread of metamorphic rocks, those formed as a consequence of deep burial, elevated temperature, and compressive stress. We will trace the changes in these regionally metamorphosed rocks, concentrating on the transformation of shale, the most abundant sedimentary rock exposed on the Earth's surface.

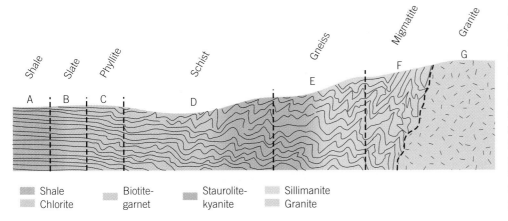

Shale | Slate | Phyllite | Schist | Gneiss | Migmatite | Granite

A   B   C   D   E   F   G

Shale
Chlorite

Biotite-garnet

Staurolite-kyanite

Sillimanite
Granite

Figure 9.15 A schematic representation of the textural and mineralogical changes that occur in shale when it is subjected to increasing regional metamorphism. These changes can be seen when we trace the rock toward the eroded core of a mountain belt. The shale is transformed by degrees into slate, phyllite, schist, and gneiss. With increasing heat and pressure, the gneiss partially melts, forming a mixture of granite and gneiss, or migmatite. Finally, the rock melts completely, forming granite.

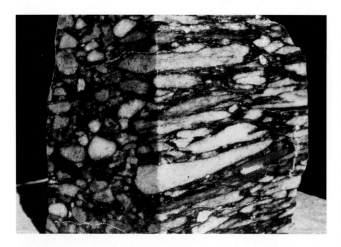

Figure 9.16 These elongated pebbles, found at the periphery of a fold-mountain belt, illustrate the role of compression in altering the original character of sedimentary rock.

Starting at the fringe of the mountain belt (point A in Figure 9.15), we note first that the folded rocks of the mountain belt have been squeezed in the direction of maximum compression and stretched in the direction of least compression. This process of squeezing and stretching is well demonstrated in conglomerates that contain elongated pebbles and in shales and limestones that contain distorted fossils (Figure 9.16).

We note, too, that distinct layering has developed at right angles to the direction of maximum compression. These layers, called slaty cleavage (Figure 9.17), are especially well displayed in the shales. Cleavage consists of a series of closely spaced parallel planes, along which microscopic grains of mica and chlorite (derived from clay) have grown in response to stress. The slaty cleavage grows more prominent in the shale as we move toward the core of the mountain belt. At the same time, bedding planes and other features of the shale grow faint, and it is the slaty cleavage that dominates the rock. We have now traced the original sedimentary shale into slate, a metamorphic rock (point B in Figure 9.15). The closely spaced cleavage planes, a type of foliation, impart a smooth parallelism to the slate. The rock is dense and "pings" when struck with a field hammer or tapped on a rock ledge.

Under the microscope, we see that slate is composed of minute mica flakes; chlorite, a green mineral similar to mica in its platy appearance and chemistry, except for its greater iron and magnesium content; and quartz. The chlorite and mica grow parallel to the cleavage planes. The basic pattern of change in regionally metamorphic rocks has now been established:

1. the reconstitution of older minerals into new minerals stable at higher temperature and pressure
2. the development of foliation in response to regional stress, and the growth of platy minerals, such as mica and chlorite, parallel to the foliation
3. the general coarsening of the rock textures.

Tracing the slate deeper into the core of the mountain range, we note that its slaty cleavage planes develop a silky sheen. The sheen signifies a coarsening of texture caused by the growth of mica and chlorite along the cleavage planes. By this time, the features of the original sedimentary rock are mostly obliterated. We have passed from slate to phyllite (Figure 9.18). Phyllite, in turn, grades into coarser-textured schist (point C to point D in Figure 9.15). The coarse texture is due to the growth and parallel alignment of constituent minerals, such as mica and chlorite. Quartz also is prominent. When struck with a hammer, this rock breaks readily in slabs along the foliation planes.

Figure 9.17 Slaty cleavage in a shale formation. Notice how the cleavage planes intersect the bedding planes, which have been folded by compression. Growth of minerals along the cleavage planes is responsible for much of the foliation displayed by regionally metamorphosed rocks.

(a)

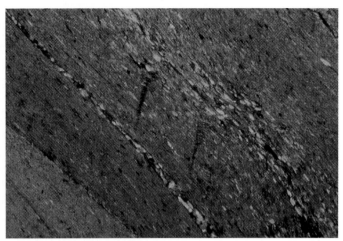

(b)

Figure 9.18 (a) A phyllite hand specimen showing pronounced foliation and silky sheen caused by the growth of mica. (b) This enlarged photomicrograph shows that the foliation of phyllite is caused by the alignment of minute mica and quartz grains.

The foliation of some schists may have a consistent orientation over a wide area. More commonly, however, schists are intricately folded (Figure 9.19). Large-scale folds in schists covering many kilometers are often reproduced in striking detail down to the microscopic level. The intricate fold pattern, transferred to virtually every segment of the rock, is clear evidence that in this region of great metamorphic intensity, the rocks have deformed plastically in response to compressive stress and elevated temperatures.

Farther into the mountain core, we find that the schist grades into gneiss, which is similar to granite in mineral content: mainly quartz, feldspar, and a dark mineral, either biotite, pyroxene, or hornblende (point E in Figure 9.15) Dark and light minerals are segregated into a new type of foliation having distinct alternating bands. Many of the bands in this gneiss are highly contorted in tight, intricate folds. When the rock is struck with a hammer, the bands do not split easily.

Here in the center of the vise, the core of the mountain belt, regional metamorphic effects blend with the effects of contact metasomatism, and metamorphic rocks blend with igneous rocks (point F in Figure 9.15). The gneiss gives way to a transitional zone of intricately mixed granitic and metamorphic rock, a migmatite (Figure 9.20). The migmatite, in turn, grades into a granite (point G in Figure 9.15). Evidently, temperatures

Figure 9.19 Intricate folds displayed in schist. These same folds are often reproduced on a regional scale.

Figure 9.20 A typical migmatite outcrop displaying a mixture of banded gneiss and granite.

inside the Earth reached the melting point of the metamorphic rock and transformed it to magma, which cooled to form granite. In other words, our journey from the fringes of the mountain belt, where we encountered sedimentary rock, to the core of the mountain belt, where we encountered the evidence that metamorphic rock melted to form granite, has been a journey around the rock cycle.

Metamorphic Zones and Facies

As we trace shale from the outer fringes of the metamorphic terrain toward the granite, we observe, in addition to textural and structural changes, systematic changes in the mineral content of the rocks. Certain minerals, such as chlorite, fade out and are superseded by others, such as biotite and garnet. These **index minerals** are formed by reactions with preexisting minerals at specific pressures and temperatures. They are therefore especially sensitive measures of *metamorphic grade,* the intensity of metamorphism.

If we map the first appearance of an index mineral in a region, we get an **isograd,** a line joining points of equal metamorphic grade. In the areas between successive isograds, the **metamorphic zones,** the mineral content of the rock remains constant. Proceeding from low-grade slate through the high-grade schist and gneiss toward the granite, we encounter the chlorite, biotite, garnet, kyanite, and sillimanite zones.

The concept central to the metamorphic zone is that the pressure-temperature conditions determine the mineral content of a rock of a given composition. Suppose, however, that we are dealing with basalt rather than shale, and we find that it has been metamorphosed to produce amphibole and epidote. How can we compare its degree of metamorphism with the metamorphic zones of the shale? Clearly, we need a measure of the degree of metamorphism of rocks having different compositions. It is here that the concept of **metamorphic facies** has proven useful.

Rocks belong to the same metamorphic facies if they have formed under the same set of pressure-temperature conditions regardless of their original compositions. Thus metamorphic rocks having very different mineral assemblages can belong to the same facies. For example, we find that a schist containing quartz and plagioclase feldspar belongs to the same facies as a gneiss containing garnet and sillimanite (Table 9.4).

The names of the facies reflect the minerals and common rocks that form at the experimentally determined pressures and temperatures shown in Figure 9.21. More important than memorizing these names is noting the various trends of metamorphism, as indicated by lines A, B, C, and D in Figure 9.21. Line A shows the path of contact

index mineral
A mineral that characterizes a given intensity of metamorphism, having developed under a specific range of temperature and pressure conditions.

isograd
A line on a map connecting points of equal metamorphic intensity; usually indicated by the first appearance of a given index mineral.

metamorphic zone
An area of equal metamorphic intensity between isograds in which the mineral content of rocks remains constant.

metamorphic facies
Mineral assemblages that have formed within a well-defined set of pressure-temperature conditions.

Table 9.4 Common Metamorphic Facies

Condition	Facies Names	Minerals		Common Environments
		Shaley Rocks	Basaltic Rocks	
Low temperature; low pressure	Zeolite	Clays, chlorite, quartz	Zeolites (hydrous silicates), calcite, chlorite	Low-grade regional metamorphism, hydrothermal metamorphism of sea floor, mild contact metamorphism
	Greenschist	Chlorite, muscovite, quartz	Epidote, green amphibole	Low-grade regional metamorphism, hydrothermal metamorphism of sea floor
	Amphibolite, muscovite	Quartz, amphibole, biotite, garnet, sillimanite	Quartz, plagioclase feldspar	Medium- to high-grade regional metamorphism (mountain belts at convergent boundaries)
Very high temperature; very high pressure	Granulite	Quartz, garnet, biotite, sillimanite, pyroxene	Quartz, pyroxene, plagioclase, garnet	High-grade regional metamorphism (mountain belts at convergent boundaries)
Low temperature; high pressure	Blueschist	Blue amphibole, quartz, chlorite	Blue amphibole (glaucophane), chlorite	Low-grade regional metamorphism and subduction zones
High temperature; low to moderate pressure	Hornfels	Andalusite, cordierite, mica, quartz	Pyroxene, quartz, plagioclase	High-temperature contact metamorphism adjacent to magmatic bodies
High temperature; high pressure	Eclogite	—	Garnet, pyroxene, kyanite	Deep metamorphism of igneous rocks, probably common in upper mantle

metamorphism. High temperature at relatively low pressures and depths produce various hornfels facies. Line B shows the path of metamorphism under high pressures and relatively low temperatures, as occurs in subduction zones. It produces the blueschist facies. Line C shows the path of metamorphism when both temperatures and pressures

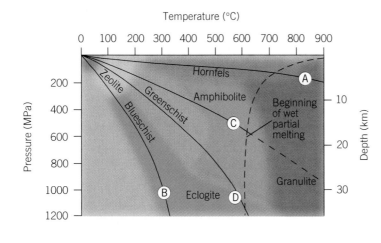

Figure 9.21 This graph illustrates the pressure-temperature relationships of the metamorphic facies described in Table 9.4. Lines A through D show common metamorphic pathways—that is, the facies changes to be expected from various combinations of increasing pressure and temperature.

Figure 9.22 The relationship to plate boundaries of the various types of metamorphism discussed in this chapter. Some types, such as shear metamorphism, can occur at convergent, divergent, or transform boundaries. Others, such as regional metamorphism, are confined to a single boundary type (convergent).

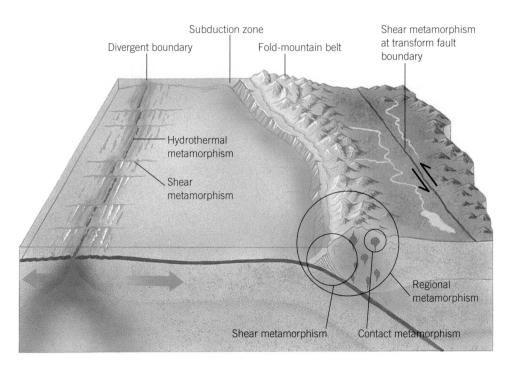

increase markedly, as in mountain belts formed by continental collision. It produces the classic regional metamorphism just discussed. Notice that line D, the geothermal gradient beneath the continents, follows a path similar to regional metamorphism—which suggests that the crust is basically composed of metamorphic rock.

Metamorphism and Plate Tectonics

Specific metamorphic rocks are closely associated with specific plate boundaries, in much the same way that specific kinds of igneous rocks and volcanoes are closely associated with these very same boundaries. Each type of plate boundary generates distinctive pressure-temperature-fluid conditions. Figure 9.22 is a plate tectonics model of the crust and upper mantle, with the principal sites of metamorphism labeled. Hydrothermal metamorphism occurs at divergent boundaries (mid-ocean ridges). Regional metamorphism occurs at subduction zones and in the mountain belts commonly associated with them. Within the subduction zone, rocks attached to the cold descending plate are subjected to relatively low-temperature but high-pressure metamorphism. In the bordering mountain ranges, the rocks of the continent or island arc are subjected to heating by rising magmas derived from the action of the subducting plate. Thus they are subjected to high-temperature but low-pressure metamorphism. Shear metamorphism occurs in the fault zones of convergent, transform, and divergent boundaries.

Metamorphism of the Sea Floor

Zones of intensive fractures develop as the oceanic crust slowly separates and spreads away from the mid-ocean ridge crests (Figure 9.23). The fractures penetrate deep into the crust and provide avenues for the intermittent rise of the magma lurking beneath the crests. The fractures are also an effective plumbing system. Cold seawater seeps deep into the crust along one set of fractures, where it is heated by the magma. It is then returned to the surface through another set of fractures that serve as hot-water vents. Submarine cameras have recorded spectacular plumes of hot water escaping from these chimneys, around which cluster complex communities of bottom life (Figure 9.24). Temperatures of over 300 °C have been recorded in the hot water.

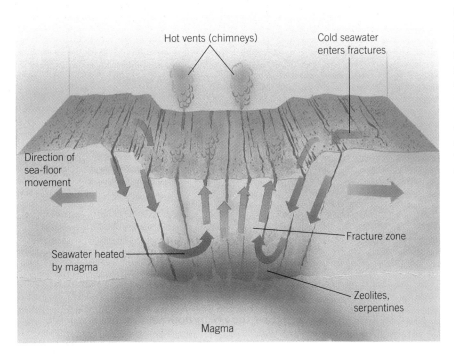

Hot vents (chimneys)

Cold seawater enters fractures

Direction of sea-floor movement

Fracture zone

Seawater heated by magma

Zeolites, serpentines

Magma

Figure 9.23 The environment of sub-sea-floor (hydrothermal) metamorphism. Fractures that develop within the central rift valley of the mid-ocean ridge act as passageways for seawater circulating within the crust. The cold seawater is heated by hot magma beneath the ridge crest. In the exchange between rock and water, the basaltic crust is converted to hydrous rocks, such as serpentinite, and metals extracted from the basalt by the water are concentrated high in the crust and near hot vents.

The combination of plentiful seawater from above and heat supplied from below acts as a potent agent for the hydrothermal metamorphism of oceanic crust, during which water-free minerals are transformed into hydrous, or water-bearing, forms. Dredge and core samples recovered from drilling reveal chlorite-rich greenstone, converted from basalt, and serpentinite, the soft, dark green rock composed of fibrous serpentine minerals derived from the conversion of peridotite. The alteration is most thorough along fracture linings where water penetrates. Microscopic analysis reveals that serpentine is derived from the alteration of olivine. Though sub-sea-floor metamorphism occurs mainly along the mid-ocean ridges, sea-floor spreading carries the rocks

Figure 9.24 A plume of hot water issuing from a submarine chimney.

Figure 9.25 A paired metamorphic terrain such as might be expected to form at a convergent boundary. Sediments and slices of oceanic crust trapped in the subduction zone are subjected to relatively high-pressure–low-temperature metamorphism (the greenschist and blueschist facies). Rocks of the bordering island arc or continent are affected by upwelling magma and are subjected to relatively low-pressure–high-temperature metamorphism (the amphibolite facies).

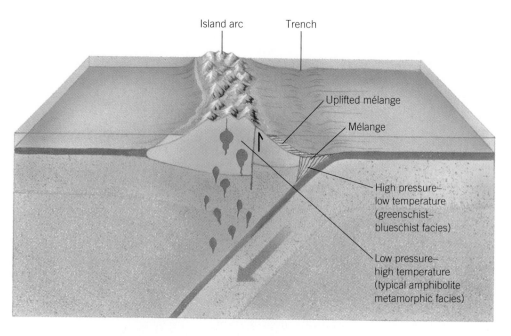

formed by this process far across the ocean floor and eventually down into the mantle at subduction zones.

The complex reactions between seawater and oceanic crust lead to the formation of many other minerals, including valuable ore-grade metal deposits and *zeolites*. The latter are aluminum silicates whose crystal structures contain large quantities of water easily expelled when heated. They have extensive industrial use as water softeners, gas adsorbents, and drying agents.

Subduction Zone Metamorphism

The subduction zones beneath the deep-sea trenches are regions of relatively high pressure but low temperature. Temperatures are depressed by the cold slabs of oceanic crust that descend into the mantle. However, intense pressure and shearing stress are generated as the subducting plate grinds against the bordering plate. Scraped off the descending plate as it passes beneath the trench are deep-ocean sediments, turbidite deposits, and slices of oceanic crust (called *ophiolite*). These materials are plastered onto the bordering plate as a chaotic, wedge-shaped mixture called a *mélange*. The rocks of the mélange are generally of the high-pressure–low-temperature greenschist and blueschist facies. But deep within the subduction zone beneath the trenches, the descending plate is subjected to the very-high-pressure conditions of the upper mantle. The basalt slab is converted to *eclogite,* a colorful rock composed of red garnet and green pyroxene but of the same chemical composition as basalt.

The descending plates also play a role in generating magmas that rise as diapirs and intrude the folded, mountainous sedimentary rocks of the continents and island arcs. The diapirs are, in effect, concentrated sources of heat delivered to the continents and arcs. This heat powers the high-temperature reactions of regional metamorphism that produce migmatites, sillimanite gneisses, and schists. These rocks pass into lower-grade phyllites and slates at the fringes of the mountain belt.

Because they are part of the same subduction process, the trends of the outer, low-grade rocks of the mélange and the high-grade rocks of the inner fold-mountain belt are parallel to one another. Together, they form a *paired metamorphic terrain:* low-grade rocks close to the deep-sea trench, high-grade rocks inland of the trench. Figure 9.25 shows a paired metamorphic terrain and associated volcanoes such as those that form the backbone of the Japan Islands. Interestingly, the rocks of the California Coast Range are largely low-grade mélange, whereas metamorphic rocks of the same age in the Sierra Nevada of eastern California are high-grade assemblages mixed with granite batholiths. Apparently,

California is similar to Japan in the sense that it consists of a paired metamorphic terrain, the result of subduction beneath an ancient trench that existed off the California coast.

STUDY OUTLINE

Metamorphism is the process by which the mineralogy, texture, and structure of solid rocks are changed by the pressures, temperatures, and chemical conditions that prevail deep within the Earth.

I. **THE AGENTS OF METAMORPHISM**

    A. Heat softens rocks and breaks ionic bonds.

    B. **Lithostatic stress** reduces the rocks' volume; nonuniform **compressive stress** causes parallel layering, or **foliation.**

    C. Chemically active fluids dissolve and disseminate elements throughout the rock mass.

II. **TYPES OF METAMORPHISM**

    A. **Regional metamorphism** is induced in preexisting rocks during subduction and the plate collisions that produce fold-mountain ranges.

    B. **Contact metamorphism** is the result of the baking and recrystallization of country rock by igneous intrusions.

    C. **Shear metamorphism** grinds and deforms rocks at transform boundaries and other faults.

    D. **Burial metamorphism** results from the pressure of overlying rock on buried sedimentary rock layers.

    E. **Shock metamorphism** results from meteor impact.

    F. **Metasomatism** is a companion process to metamorphism that does involve changes in composition. In it, ions are dissolved by fluids and replaced by other ions.

    G. **Hydrothermal metamorphism** alters rocks through the introduction of hot fluids.

III. **METAMORPHIC ROCKS**

    A. Because rocks remain solid during metamorphism, nearly all reactions are *solid-state*, in which ions recombine to form new minerals. Reactions often involve the loss of water and/or carbon dioxide. Where mineral compositions remain the same, their structures may change through recrystallization and polymorphic conversions.

    B. Rocks with platy or elongated minerals display foliation in response to nonuniform compressive stress. Rocks with equigranular minerals display non-foliated texture.

    C. Foliated rocks are classified on the basis of grain size and presence of banding. They range from the fine-grained **slate** to the coarser **schist** and the banded **gneiss. Migmatite** is a mixture of granite and gneiss derived from partial melting of the gneiss.

        1. The products of shear metamorphism are also commonly foliated. They include coarse, angular **fault breccia** and fine-grained mylonite.

        2. Nonfoliated rocks are classified by composition. **Quartzite** is the product of the metamorphism of sandstone; **marble,** of limestone.

        3. Rocks altered by igneous intrusions are called **hornfels.**

D. The field occurrence of metamorphic rocks follows a specific pattern.
1. In contact metamorphism, an **aureole** of altered country rock surrounds an igneous intrusion. Textural and mineralogical changes grow more pronounced closer to the center, illustrating that the heat and fluids of the magma caused the changes.
2. Regional metamorphic rocks are formed by deep burial, elevated temperature, and compressive stresses at convergent plate boundaries. The transformation can be traced from the periphery to the core of a fold-mountain belt; it involves a coarsening of texture, the growth of new minerals, and the development of foliation.

E. **Metamorphic zones** are areas where the mineral content of the rocks is constant. **Index minerals,** formed by reactions at specific pressures and temperatures, are measures of metamorphic intensity. Metamorphic zones lie between **isograds,** the lines that join points of equal metamorphic intensity.

F. Rocks characterized by mineral assemblages formed under the same metamorphic conditions belong to the same **metamorphic facies.** Depending on temperature-pressure conditions, a given parent rock can follow many paths of metamorphism.

IV. METAMORPHISM AND PLATE TECTONICS

A. Conditions at plate boundaries commonly control the kinds of metamorphism that occur in the crust.

B. Hydrothermal reactions occur at mid-ocean ridges, and they involve the alteration of oceanic crust to serpentinite through the action of seawater that penetrates the crust and is heated by subsurface magma. Sea-floor spreading widely distributes the rocks on the sea floor.

C. As a plate subducts, ocean sediment, turbidites, and slices of oceanic crust are transferred onto the bordering plate as a mélange. These areas are sites of low-temperature–high-pressure greenschist and blueschist facies.

D. Beneath the trenches, basalts of subducting plates are converted under high pressure to eclogite. Metamorphism in the island arcs or continents bordering subduction zones is influenced by the heat escaping rising granitic magmas. Rocks there are subjected to high-temperature–low-pressure metamorphism. Together, the low-grade rocks of the mélange and the high-grade rocks of the bordering mountain range constitute a paired metamorphic terrain.

STUDY TERMS

aureole (p. 256)	marble (p. 255)
burial metamorphism (p. 250)	metamorphic facies (p. 262)
compressive stress (p. 249)	metamorphic zone (p. 262)
contact metamorphism (p. 250)	metasomatism (p. 251)
fault breccia (p. 254)	migmatite (p. 253)
foliation (p. 249)	quartzite (p. 255)
gneiss (p. 253)	regional metamorphism (p. 250)
hornfels (p. 255)	schist (p. 253)
hydrothermal metamorphism (p. 251)	shear metamorphism (p. 250)
index mineral (p. 262)	shock metamorphism (p. 251)
isograd (p. 262)	slate (p. 253)
lithostatic stress (p. 249)	

CRITICAL THINKING QUESTIONS

1. If increasing pressure and temperature can transform a shale into a gneiss, do you think it is possible to convert a gneiss into a shale by decreasing temperature and pressure? Why or why not?

2. Suppose you were to trace shale deposits interbedded with basaltic lava into a region of steadily increasing metamorphic intensity. Furthermore, suppose you found that the shale had been transformed by regional metamorphism into sillimanite schist. Describe the mineral changes that occurred in the basalt.

3. Suppose you find the polymorphs kyanite, sillimanite, and andalusite in the same schist outcrop. What might their presence tell you about the pressure-temperature of the metamorphism?

4. Describe the conditions under which regional metamorphism, contact metamorphism, and metasomatism may merge.

5. Describe the plate boundary interactions that trigger sub-sea-floor, regional, and dynamic metamorphism.

C H A P T E R

10

Geologic Time

Geologists are historians of the Earth. They extract their information from rocks, in whose wide layers, like the pages of an enormous diary, are recorded the sequence of events that have affected the Earth's crust. Sedimentary rocks have time significance because they are deposited in layers, younger above older. Because these layers contain fossils that are distinctive of their age, rocks from different parts of the world can be compared and arranged in chronological order. From such arrangements, a scale of **relative time** has been constructed. Observing which layers have been disturbed by faulting, folding, metamorphism, igneous intrusion, or erosion also allows these events to be placed in sequence according to the relative time scale. Thus, Earth's history is pieced together layer by layer.

relative time
The chronological ordering of events without reference to their age in years.

The pioneering geologists of the eighteenth and nineteenth centuries noted evidence of mountains worn low, of coral reefs now stranded in deserts, and of complex and varied life-forms no longer living. They knew that the rocks held a long story within them, but they did not know precisely how long, for they had no way of determining **absolute time**—the ages of objects in years. Not until well into this century were radiometric techniques developed to date events beyond the recorded history of humankind. These techniques make use of the discovery that a given radioactive isotope decays into a specific stable isotope at a known rate, and that both of them can be extracted from rocks and measured. However, radiometric dating cannot be applied to all rocks and must be used in conjunction with methods of relative age determination.

absolute time
The ages of objects or events as measured in time units such as years; determined by radiometric-dating techniques.

Recognition of the immense span of geologic time—and, by implication, our minute share of it—is probably geology's key contribution to human knowledge. Its impact on our view of reality is comparable to the earlier realization that the Earth is

◀ A geologist examines an instant, frozen in time some two hundred million years ago when a dinosaur left tracks across a mudflat, now preserved as an ancient rock layer in the Painted Desert, Arizona. Behind the geologist are younger layers, worn back to expose the tracks; beneath the tracks are older layers. What can this sequence of strata reveal about the region's past?

more a speck of dust than the center of the universe. In this chapter, we describe how geologists tell time. First, we discuss how they place rocks, and the geologic events recorded in them, in chronological order (relative time). Then we explain how they use radiometric methods to determine the ages of rocks and geologic events (absolute time).

Relative Time

Relative dating concerns the chronological ordering of features and events without regard to their age in years. The general procedure is to work out the relative ages of the rocks in one locality or region and then to compare, or correlate, these rocks with those of other regions. The skills required depend ultimately on the few commonsense principles described in this chapter.

Deciphering Local Geologic History

How do geologists decipher the age relations and geologic history of a region? The first step is to identify the basic rock units, or **formations,** present. Formations are mappable bodies of rock that have distinctive mineralogical, structural, or textural characteristics that set them apart from the rock bodies with which they are in contact. Often, a formation consists of a single igneous, metamorphic, or sedimentary rock type, such as granite, slate, or sandstone. But it may also consist of a number of closely related rocks examined as a single unit—for example, a sequence of alternating shale and sandstone strata. The point is that rocks of a specific formation have a common origin.

The second step in deciphering the geologic history of a region is to place these formations in chronological order. Geologists begin by applying a few simple principles that relate the arrangement of rock bodies as observed in the field to their relative ages: original horizontality, superposition, and cross-cutting relationships.

Horizontality and Superposition

The character of a sedimentary rock depends upon its depositional environment. You may recall from Chapter 8 some of the many environments where sediments are deposited on today's Earth: alluvial fans, flood plains, deltas, the shallow continental shelves, and the deep-ocean floor. Though markedly different from one another in texture and mineralogy, these sediments all share an important trait: They are deposited in layers parallel to the Earth's surface—which, on our spherical planet, is taken to be a horizontal plane. This simple observation is the basis of the **principle of original horizontality,** which states that sediments are deposited in approximately horizontal layers; those found to deviate from this orientation have been disturbed at a later date.

Horizontality is a consequence of gravitation, in which all particles are attracted toward the Earth's center (the vertical direction) but strike the Earth's surface en route. Therefore, if a blizzard of particles settles out of water (or air), it will blanket the Earth's surface and form a horizontal layer. It is self-evident that the layer deposited next must rest upon the previously deposited layer. Thus, in a sequence of sedimentary rocks or lavas, each layer is younger than the layer beneath it and older than the one above it. This concept forms the **principle of superposition.** It holds for all layers deposited on the Earth's surface, no matter how thick the sequence, and it is the principle upon which nearly all relative-age relationships involving sedimentary rock are based.

Unconformities

The fact that superposition determines the order in which sedimentary rocks are deposited in a sequence does not imply that all layers have been deposited continuously

---

formation
A distinctive mappable rock unit.

principle of original horizontality
The principle that sediments are deposited in horizontal layers, parallel to the Earth's surface.

principle of superposition
The principle that in a sequence of sedimentary strata, the oldest layer is located on the bottom and is followed in turn by successively younger layers, up to the top of the sequence.

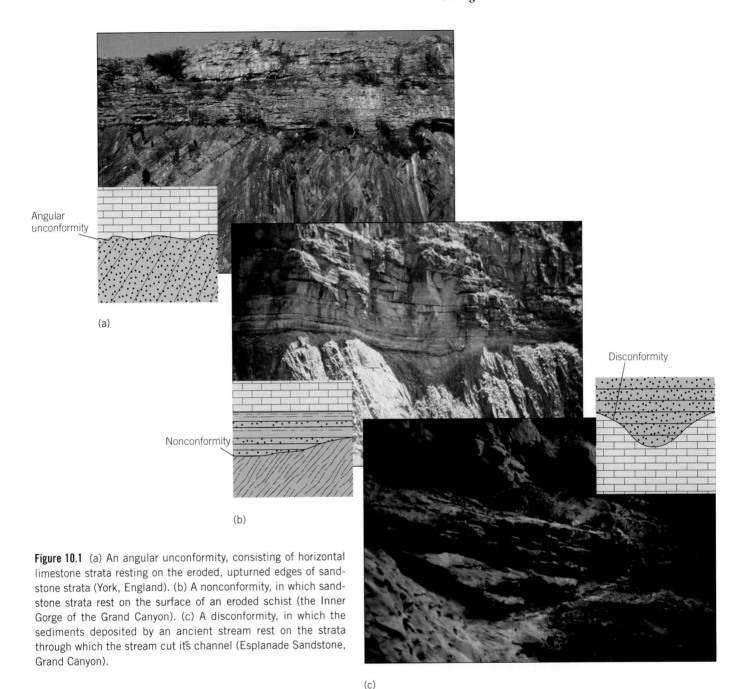

Angular
unconformity

(a)

Nonconformity

(b)

Disconformity

Figure 10.1 (a) An angular unconformity, consisting of horizontal limestone strata resting on the eroded, upturned edges of sandstone strata (York, England). (b) A nonconformity, in which sandstone strata rest on the surface of an eroded schist (the Inner Gorge of the Grand Canyon). (c) A disconformity, in which the sediments deposited by an ancient stream rest on the strata through which the stream cut its channel (Esplanade Sandstone, Grand Canyon).

(c)

over time. Indeed, the geologic record in any relatively large region is typically replete with gaps, as if pages were either torn from the book or were never included. A gap in the rock record is called an **unconformity;** it is identified by an erosional surface between rocks of markedly different ages, and it signifies a major depositional break between the rocks above and below that surface. Figure 10.1 shows three common unconformities, each of which represents a significant period of erosion.

The strata below the **angular unconformity** in Figure 10.1(a) had to have been folded, uplifted, and eroded prior to deposition of the strata above them. Missing from the record (the gap) is the time it took to accomplish these changes. The metamorphic rock below the **nonconformity** in Figure 10.1(b) formed deep within the crust, beneath several kilometers of rocks, all of which had to have been stripped away prior to deposition of the overlying strata. In this case, too, an immense time interval is missing from the record. The **disconformity** in Figure 10.1(c) shows that the underlying strata had to have been uplifted and eroded in order to produce the uneven surface upon which

unconformity
An erosion surface bounded by rocks of markedly different age and signifying a break in the geologic record.

angular unconformity
A surface formed by the deposition of sediments on the eroded, upturned edges of older, tilted strata.

nonconformity
An unconformity with stratified rocks above and igneous or metamorphic rocks below.

disconformity
An unconformity between parallel sedimentary layers.

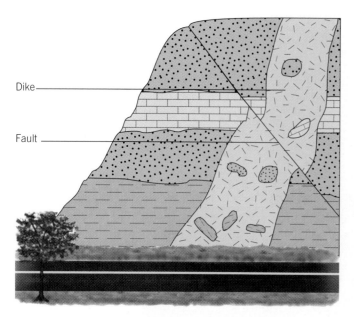

Dike —————

Fault —————

Figure 10.2 The principle of cross-cutting relationships applied to a diagram of a road cut. The dike, an igneous intrusion, cuts across— and is therefore younger than—the sedimentary strata. The fault displaces the intrusion and the strata and is younger than both of them.

the overlying strata were deposited. If they are of marine origin, then the underlying rocks must also have been depressed below sea level following the erosion period.

Although an unconformity is evidence of a major gap in the record, we have no way of knowing the magnitude of that gap—that is, how many years are missing—unless we can date the ages of the rocks above and below the unconformity. Determining numerical age requires radiometric-dating techniques, as we will see later in this chapter.

Cross-Cutting Relationships

Not all rocks are sedimentary, of course. Suppose, during fieldwork, we come across an outcrop like the one shown in Figure 10.2, in which an igneous dike cuts across a sequence of sedimentary rocks. Close inspection reveals that a halo of baked sedimentary rock surrounds the dike, and inclusions of

Figure 10.3 This panoramic view of the Grand Canyon, looking north, displays the rock formations from the Inner Gorge to the Canyon rim. Note the tilted sequence of strata to the lower right in the Inner Gorge (the Grand Canyon Supergroup) and the dark, upturned rock in the lowermost part of the Inner Gorge to the left (the Vishnu schist).

dislodged sedimentary rock (xenoliths) are found within the dike. This evidence tells us that the dike formed when magma intruded the formation and later cooled—in other words, it is younger than the strata. We notice also that both dike and strata have been displaced by a fault; so we conclude that the fault is younger than these features. In working out the sequence of events recorded at this outcrop, we have applied the **principle of cross-cutting relationships,** which states that a fault or intrusive body is younger than the rocks it cuts across or intrudes.

principle of cross-cutting relationships
The principle that an intrusion or fault is younger than the rock that it cuts.

Relative Ages of Grand Canyon Formations

The Grand Canyon's Inner Gorge is a classic example of how superposition, unconformities, and cross-cutting relationships may be used to unravel the geologic history of a region—in this case, the history contained in the ancient rocks that form the base of the Colorado Plateau (Figure 10.3). Figure 10.4 illustrates the stages in the history of the Inner Gorge.

Notice in Figure 10.4(a) that the dark rock formation composing the Lower Gorge is named the Vishnu schist. It is exposed where the Colorado River follows a zone of weakness created by a relatively recent fault. Schist, you may recall, is a metamorphic rock produced by the transformation of shale, a sedimentary rock composed of clay and silt. The schist is highly contorted, and metamorphic structures predominate; some of the original sedimentary character of the rock is preserved, however, and it is typical of what we find in today's oceans. Therefore, the first stage in the history we read from the rocks of the canyon walls was the deposition of clay and silt in an ancient sea (Figure 10.4b). The great thickness of the Vishnu implies that the ocean existed in this region for many millions of years.

In Chapter 9, we learned that schist forms at depths of several kilometers, under the high pressures and temperatures associated with the mountain building that occurs when lithospheric plates collide. So the next stage of history read from the canyon walls

Figure 10.4 The sequence of events inferred from the relationships displayed in the Inner Gorge of the Grand Canyon. (Adapted from *Geology Illustrated* by John S. Shelton, 1966.)

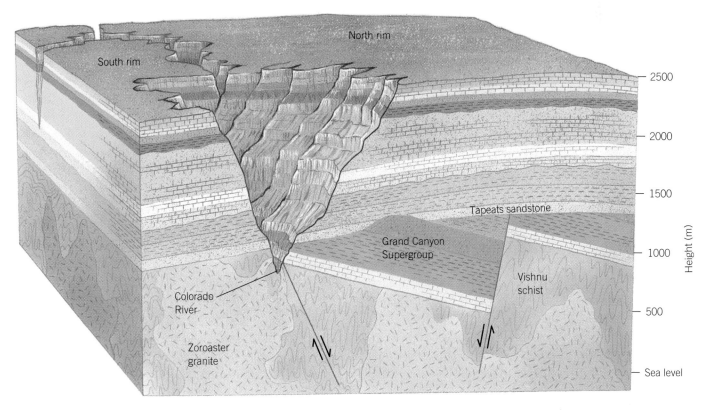

(a) Cross section of the rock formations of the Inner Gorge and the layers above it as they are today

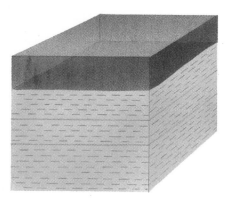

(b) Deposition of mud in an ancient sea; mud compacted to shale

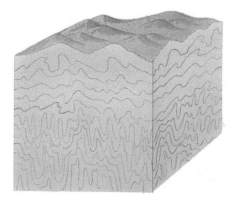

(c) Metamorphism of shale into schist in the core of a fold-mountain range

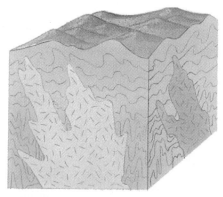

(d) Intrusion by Zoroaster granite into Vishnu schist

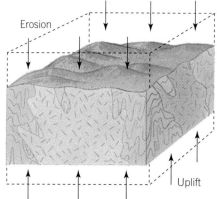

(e) Erosion of the overlying mountain range and vertical uplift of the Vishnu schist and Zoroaster granite to sea level

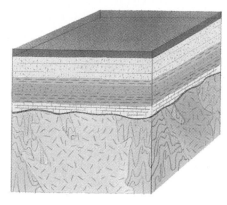

(f) Deposition of the Grand Canyon Supergroup on the Vishnu erosion surface

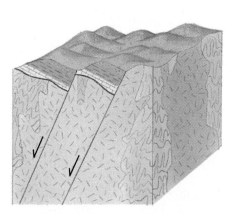

(g) Faulting uplifts Supergroup, forming a second mountain range

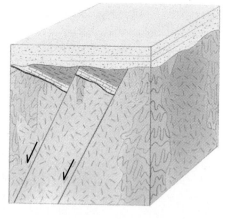

(h) Erosion of the fault block mountains to a flat surface, and the deposition of the Tapeats sandstone in a shallow encroaching sea was the metamorphism of the shale in the core of a fold-mountain range (Figure 10.4c). From the distribution of the Vishnu, which is found as far away as Las Vegas, it must have been an enormous mountain range that rivaled the Rockies in size.

Looking again at Figure 10.4(a), we see that prominent within the Vishnu are masses of pink Zoroaster granite that cut across the foliation planes of the schist and form dikes and irregular masses throughout the canyon. From cross-cutting relationships, we know that because the granite intrudes the Vishnu, it must be younger. So the third stage was the intrusion of the Vishnu by magma that later cooled to become the Zoroaster granite (Figure 10.4d).

Resting on top of these formations is a wedge-shaped mass of sandstone, shale, and limestone, well over a thousand meters thick, called the Grand Canyon Supergroup (see Figure 10.4a). The contact between these very different kinds of rocks is a nearly flat surface that cuts across the nearly vertical foliation of the schist. Ripple marks and mud cracks in the shale and sandstone of the Supergroup are evidence that the sediments were deposited in a shallow sea and intermittently exposed to air. But remember that these sediments were deposited on the Vishnu! How did the Vishnu get to the Earth's surface? Two things were required: erosion of the entire overlying mountain range down to the Vishnu, and vertical uplift of the Vishnu to sea level and above, thousands of meters from where it originated deep within the crust. These events constitute the fourth stage read from the Inner Gorge rocks (Figure 10.4e). The fifth stage was deposition of the Supergroup on the Vishnu erosion surface to produce a great nonconformity (Figure 10.4f).

Notice that the Supergroup slopes to the right where it abuts the Vishnu schist and is truncated by a huge fault that has displaced the strata by many thousands of meters (see Figure 10.4a). Since the fault is younger than all the rocks it displaced, a second period of mountain building marks the sixth significant stage in the history of the canyon (Figure 10.4g). This mountain building differed from the earlier event that created the Vishnu schist in that it involved normal faulting rather than folding. Mountains of this kind are tensional features associated with stretching and rifting of the crust. Present-day equivalents are found adjacent to the Colorado Plateau in the Basin and Range Province of Nevada.

We can also see that the tilted Supergroup is preserved only in the downthrown fault blocks and that the entire upper level of the Inner Gorge of the Grand Canyon is planed nearly flat (see Figure 10.4a). This ancient platform is overlain by a horizontal formation, the Tapeats sandstone, a typical shallow-water, near-shore deposit. Where it rests on the Vishnu schist, it forms a spectacular nonconformity. Where it rests on the Grand Canyon Supergroup, it forms an equally impressive angular unconformity. Thus we see that the final stage recorded in the rocks of the Inner Gorge was the erosion of the fault-block mountains to a flat surface and the deposition of the Tapeats sandstone in a shallow, encroaching sea (Figure 10.4h).

We have not yet discussed radiometric-dating techniques, but it is hard to resist skipping ahead to point out that the Vishnu schist is about 1.7 billion years old, and the overlying Tapeats sandstone is about 550 million years old. So about 1.2 billion years of the geologic record is missing at the nonconformity between these formations. That gap is four times greater than the time it took to deposit the entire sequence of Grand Canyon rocks above the unconformity—all the strata forming the cliffs and slopes to the canyon rim. The gap in the record is, in fact, twice as great as the entire time interval from the deposition of the Tapeats sandstone to the present day.

Correlation

Thus far we have seen that one of the first tasks for geologists studying a region is to identify the formations present and determine their chronological order. The next task is to correlate the formations with those of other regions. **Correlation** is the process of

correlation
The establishing of equivalence, either in age or rock type, of separated rock units.

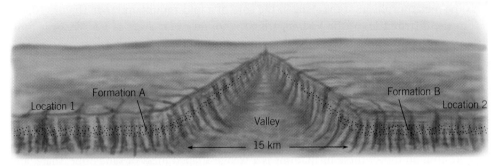

(a)

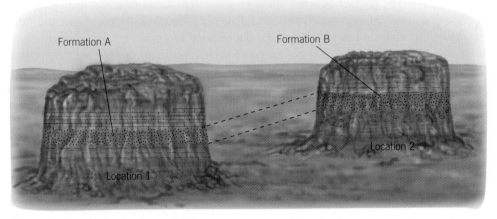

(b)

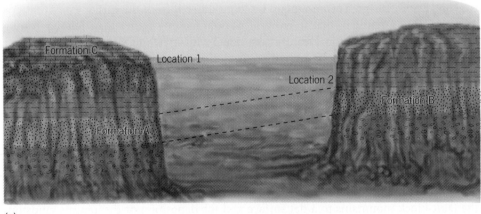

(c)

Figure 10.5 Three ways to correlate strata that are not distantly separated. (a) Lateral continuity. We can prove that formation A correlates with formation B by tracing A into B as we walk around the rim of the valley. (b) Physical similarity. These two isolated exposures are not widely separated, and their isolation was obviously caused by erosion. Thus the rocks of formations A and B, having the same distinctive characteristics, can be correlated by the similarity of their appearance. (c) Locations 1 and 2 are several kilometers apart. However, we can correlate the sandstone formations A and B on the basis of similarity of sequence—each is sandwiched between a conglomerate formation below and a shale formation above.

demonstrating equivalence, and the equivalence we are most concerned with here is age equivalence.

How could geologists prove that formation A of region 1 is equivalent to formation B of region 2 (Figure 10.5)? The simplest method is to physically trace the strata from A to B. Geologists call this method "walking it out," or more formally, proving *lateral continuity* (Figure 10.5a). This method is effective in regions where bedrock is contin-

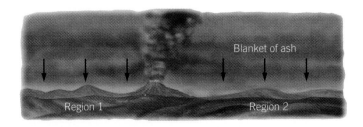

(a)

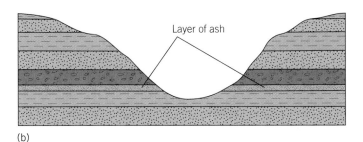

(b)

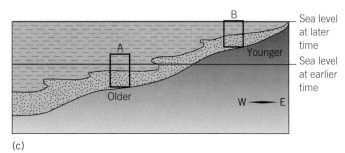

(c)

Figure 10.6 Correlating rocks on the basis of rock type alone does not necessarily prove that they are the same age. (a) A volcanic eruption emits ash that settles instantaneously on the surface as a continuous layer. (b) Therefore, we can assume this layer is the same age wherever we find it. (c) However, these shale and sandstone layers were deposited in a transgressing sea. Clearly the sequence at area A was deposited earlier than at area B; the rocks, though identical, are *not* the same age throughout their extent.

uously exposed—for example, in the bare cliffs of the Grand Canyon. However, most rocks are not continuously exposed, because they have been eroded or covered (Figure 10.5b). In those cases, geologists must rely on their *physical similarity,* provided that the outcrops are closely spaced, or their *similarity of sequence* (Figure 10.5c).

Correlation on the basis of physical characteristics has great value in oil, mineral, and groundwater exploration, as well as in civil engineering work—for example, in tunneling, securing foundations, or storing hazardous waste. It is also an essential step in organizing data to interpret geologic history. But it does not necessarily establish age equivalence between formations that are widely separated and lacking physical connection, because the same kinds of rocks may have formed repeatedly throughout geologic time. In these cases, physical similarity merely points to similarity in the processes that have formed these rocks.

In addition, formations correlated on the basis of physical similarity are not necessarily the same age throughout their extent because of shifting facies relationships (see Chapter 8). Consider an eruption that emits a thick layer of volcanic ash that blankets hundreds of square kilometers of countryside (Figure 10.6a). From the perspective of geologic time, the ash settles out of the air instantaneously; indeed, the resulting ash formation can be regarded as the same age throughout (Figure 10.6b). By contrast, the Tapeats sandstone and Bright Angel shale of the Grand Canyon in Figure 10.6(c) were deposited in a transgressing sea—one that advanced west to east over the land. Clearly, the sand-shale sequences at points A and B are physically equivalent but not age-equivalent. So the more separated two formations are, the less reliable are the correlation techniques we have discussed so far.

Stratigraphic Correlation and Faunal Succession

Toward the close of the eighteenth century, the surveyor and canal engineer William Smith made careful notes of the shells and impressions he collected at various levels of strata and over widely separated regions of the English countryside. On the basis of his observations, Smith proposed the **principle of faunal succession:** that each formation contains a unique assemblage of fossils and that the fossil assemblages succeed one another in an orderly and predictable fashion.

What had Smith correlated by comparing the fossils in strata of widely separated regions? Recall from superposition that younger strata overlie older strata in a sedimentary sequence. So the same must be true of the fossils preserved in the strata. Thus in matching fossil assemblages, Smith was correlating strata of the same relative age. He

principle of faunal succession
The principle that fossil organisms succeed one another in definite and recognizable order, so that rocks containing identical fossils are identical in age.

had discovered that fossils were a means of objectively dating the relative age of strata and, by extension, the events recorded in the strata.

Faunal succession constitutes material proof of organic evolution, for there is no other reason that the fossils in older strata should be different from, and yet related to, the fossils in younger strata and that fossils should succeed one another in an orderly fashion the world over. Furthermore, evolution is a one-way street—never have scientists observed mammals to devolve back to reptiles or multicellular forms to devolve back to single-celled organisms. Likewise, geologists have never found bones of ancestral horses, tigers, and primates beneath dinosaur bones in undisturbed strata. That would be akin to an archeologist finding pocket calculators and VCRs in King Tut's tomb. For these and other reasons, biologists and paleontologists consider evolution a fact of Earth history; only the specific whys and hows are debated today.

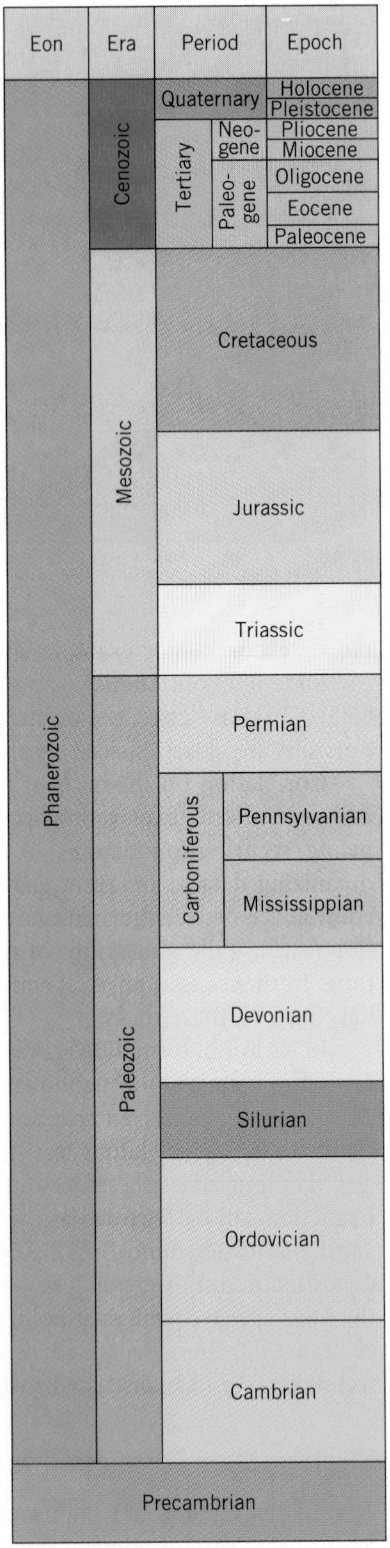

The Relative Geologic Time Scale

Superposition and the fossil record form the basis of the *relative geologic time scale,* a chart that lists the events of Earth history in chronological order, oldest on bottom, youngest on top (Figure 10.7). The construction of the time scale would have been simple if a region existed where a sequence of strata contained an uninterrupted record of the most ancient to the most recent fossil assemblages. These assemblages could then have served as a standard to date the fossils and events of other regions. Unfortunately, no such region exists, so geologists were forced to do the next best thing and piece together the assemblages of many different regions, mentally stacking them so that the time scale or column contains no time gaps (Figure 10.8). Strata from the various localities that contain these assemblages constitute *standard sections* for given intervals of the time scale. Like the reference volumes in a library, their fossil assemblages serve as a means to correlate rocks of other regions.

Correlation Tools: Index Fossils and Overlapping Ranges

If fossils succeed one another in recognizable order, each species must thus have existed for a certain time interval and then become extinct. The first and last appearance of each fossil therefore represents a fixed range on the geologic time scale. An **index fossil** is particularly useful in correlation

Eon	Era	Period	Epoch
Phanerozoic	Cenozoic	Quaternary	Holocene
			Pleistocene
		Tertiary — Neogene	Pliocene
			Miocene
		Tertiary — Paleogene	Oligocene
			Eocene
			Paleocene
	Mesozoic	Cretaceous	
		Jurassic	
		Triassic	
	Paleozoic	Permian	
		Carboniferous — Pennsylvanian	
		Carboniferous — Mississippian	
		Devonian	
		Silurian	
		Ordovician	
		Cambrian	
	Precambrian		

Figure 10.7 The relative geologic time scale. The scale is divided into eons, eras, periods, and epochs, in order of decreasing time magnitude.

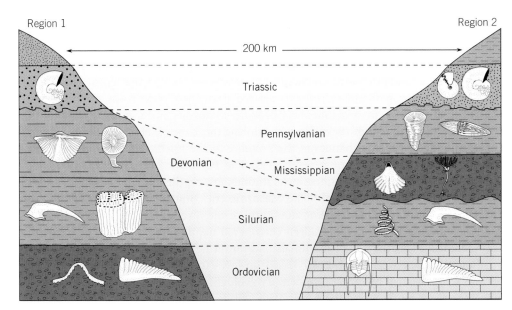

Figure 10.8 Similarity of fossil assemblages can be used to determine the relative ages of strata in widely separated regions. Between them, the two regions contain an unbroken record from the Ordovician through the Pennsylvanian periods.

because it has a narrow range but a wide geographical distribution. A narrow time range ensures precision in dating, and wide distribution makes such fossils useful far from the standard sections—as far away as separated continents. Fossil A in Figure 10.9 is an index fossil because it meets these specifications. Fossil B, however, has a broader time range and does not qualify as an index fossil.

Another way of correlating strata is to look for fossils whose ranges overlap. For example, there is a narrow time interval that fossils C and D in Figure 10.9 have in common. Therefore, the age of the stratum that contains both of them can be bracketed.

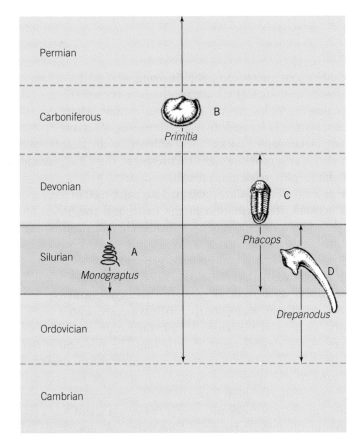

Figure 10.9 Correlation through the use of index fossils and overlapping ranges. *Monograptus* (fossil A) is an index fossil that inhabited a narrow interval of geologic time (the Silurian period). *Primitia* (fossil B), by contrast, has a wide range and is less useful for correlation purposes. *Phacops* and *Drepanodus* (fossils C and D, respectively) have different ranges, but because these ranges overlap, they can be used to bracket the age of rocks where both fossils are found.

The Terminology of the Time Scale

The relative time scale is divided into eons, eras, periods, and epochs, in order of decreasing magnitudes of time (see Figure 10.8). Sedimentary rocks of the **Phanerozoic eon** contain abundant, easily recognizable fossils—virtually the entire record of advanced life. (The name *phanerozoic* is derived from the Greek words meaning "visible life.") This eon is divided into the Paleozoic, Mesozoic, and Cenozoic eras, on the basis of the fossil content of rocks deposited during those times. The boundaries of these eras are marked by major changes in the presence of life-forms, including both extinctions and the development of new lineages. The **Paleozoic era** is the age of extinct shelled marine organisms, primitive fish, amphibians, reptiles, and plants. (*Paleozoic* is the Greek word for "ancient life.") The **Mesozoic era** ("middle life") is the age of reptiles (the best known being the dinosaurs) and of the invertebrate ammonite (the convoluted, ancient relative of the modern chambered nautilus and squid). The **Cenozoic era** ("recent life") is the age of mammals and flowering plants.

Boundaries between periods are based on less extreme change. The names of the periods were assigned by nineteenth-century geologists. Most were named for the places where the rocks representing these ages were first described or for distinctive rock types within these regions. For example, the Devonian period was named for rock strata exposed in Devonshire in southwestern England.

Periods are in turn divided into epochs. Although there are many epochs, only those of the Cenozoic era are listed in Figure 10.7. The Pleistocene epoch, for example, is the time of the ice ages—the most recent ice advance having retreated only 10,000 to 15,000 years ago. The lower boundary of the Pleistocene is marked by abrupt changes in microscopic marine fossils brought on by the onset of cold climates.

Geologists find it useful to distinguish between the abstract units of geologic time—eras, periods, and epochs—and the rocks deposited during those time units. For this reason, the rocks deposited during the Silurian period are referred to as belonging to the Silurian *system*. The rocks deposited during a lesser time unit, the epoch, are referred to as belonging to a *series*.

The Precambrian

The Cambrian period marks the onset of the Paleozoic era. Rocks of this era contain the first abundant fossils, of which trilobites are the best known (Figure 10.10). Cambria is the ancient name for Wales, the region in Great Britain where the standard section of the Cambrian system is located. In Wales, however, the Cambrian system rests on non-fossil-bearing metamorphic rocks. These older rocks are simply called **Precambrian** rocks, which is the origin of the term that refers to the time preceding the first appearance of abundant, easily recognized fossils.

Precambrian time is divided as follows: the Hadean eon (4600 to 3960 m.y.a.), which marks the interval between the formation of the Earth and the oldest known Earth rocks; the Archean eon (3960 to 2500 m.y.a.), which is characterized by primitive bacteria and algae; and the Proterozoic eon (2500 to 570 m.y.a.), which is characterized by the development of the soft-bodied, multicelled organisms.

Not until well into the twentieth century did radiometric studies reveal that the Precambrian comprises *about 85 percent of geologic time*. The immense history of life recorded in Paleozoic, Mesozoic, and Cenozoic rock is packed into a mere 15 percent of the Earth's history. Thus Precambrian rocks contain within them the bulk of Earth history and the story of the development of early life. The problem is that the fossil record is scant; much of it has been obliterated by metamorphism, mountain building, and erosion. Also, most of the organisms of that time had not yet evolved hard parts (shells, skeletons, plates, teeth), so there were few hard-part remains to be preserved as fossils. Geologists are currently at work unraveling the very ancient fossil record, which so far consists mainly of the minute remains of bacterial colonies, single-celled organ-

---

Phanerozoic eon
That part of geologic time represented by rocks in which the evidence of life is abundant.

Paleozoic era
The second of the geologic time eras, extending from the end of the Precambrian (about 570 million years ago) to the beginning of the Mesozoic (about 245 million years ago).

Mesozoic era
The era preceding the Cenozoic, extending from about 245 million years to 66 million years ago.

Cenozoic era
The most recent of the four eras of geologic time, beginning at the end of the Mesozoic era (66 million years ago) and extending to the present.

Precambrian
All geologic time prior to the beginning of the Paleozoic era (about 570 million years ago).

Figure 10.10 The trilobites *Modicia* (large) and *Ptychagnostus* (small) are index fossils of the Cambrian period. Rocks lying beneath Cambrian strata are of Precambrian age, an age from which fossils are generally scarce.

isms, and algal structures, stretching back nearly 4 billion years. For now, most of what we know about the ages of Precambrian rocks comes from radiometric methods that provide the basis for absolute-time determinations.

Absolute Time

Neither superposition, faunal succession, nor cross-cutting relationships enable us to date the age of rocks and geologic events in years. They merely enable us to arrange them in chronological order. To determine absolute age, we need geological clocks that record time at a steady pace over long periods without being affected by heat, pressure, chemical reactions, or other crustal processes. Only radioactive atoms—which are preserved in certain minerals and, in some cases, within organic remains, such as bone, wood, and shells—meet this requirement. Before we consider radiometric dating, then, let us briefly review atoms and radioactivity.

Modes of Decay

Recall from Chapter 5, Minerals, that atoms of the same element that differ in mass are referred to as *isotopes*. Some isotopes of certain elements are unstable, which means that they do not last forever. Their nuclei emit, or in some instances capture, subatomic particles, which changes their atomic number, their atomic mass, or both. (See Chapter 5 for review.) The process of emitting or capturing those subatomic particles is called **radioactive decay.** There are three principal modes of decay:

1. *Alpha decay,* in which two protons and two neutrons are emitted as an alpha particle (Figure 10.11a). The atomic mass of the isotope is thus decreased by 4, and the atomic number by 2. Because the atomic number is changed, the isotope is transformed into a different element.
2. *Beta decay,* in which a neutron decays within the nucleus to a proton and an electron: $n^0 \rightarrow p^+ + e^-$ (Figure 10.11b). However, the electron, or beta particle, escapes from the nucleus and is emitted. The atomic mass is not changed, but one proton is added to the nucleus; therefore, the element is transformed.

radioactive decay
The disintegration of certain isotopes by the emission of subatomic particles.

Figure 10.11 Three types of radioactive decay. In each type, an unstable parent isotope is converted to a stable daughter product.

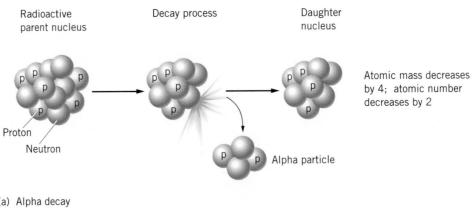

Radioactive parent nucleus   Decay process   Daughter nucleus

Proton
Neutron
Alpha particle

Atomic mass decreases by 4; atomic number decreases by 2

(a) Alpha decay

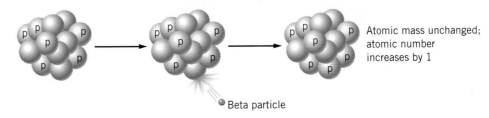

Beta particle

Atomic mass unchanged; atomic number increases by 1

(b) Beta decay

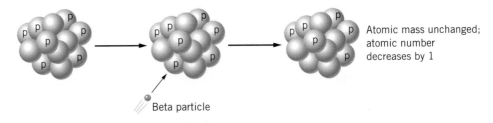

Beta particle

Atomic mass unchanged; atomic number decreases by 1

(c) Beta (electron) capture

3. *Beta (electron) capture,* which is the reverse of beta decay (Figure 10.11c). Here a proton captures an inner-shell electron to produce a neutron ($p^+ + e^- \rightarrow n^0$). The element is thus transformed because its atomic number has been decreased by 1. The atomic mass, however, is unchanged.

All radioactive decay, including electron capture, is accompanied by the release of energy in the form of heat. This heat is responsible for much of the Earth's high internal temperature.

As a result of emitting or capturing subatomic particles, a radioactive **parent isotope** is eventually transformed into a stable (nonradioactive) **daughter product.** The transformation may occur in a single step, or it may require a series of steps. In either case, parent and daughter form an identifiable pair within a sample. For example, the uranium isotope uranium-238 always decays to lead-206 after five alpha particle emissions, and rubidium-87 decays to strontium-87 after a single beta emission. For potassium-40, there are two daughter products: argon-40 and calcium-40; nevertheless, the percentage of each is well known and can be identified.

The Decay Principle

The fundamental principle of radiometric dating is that a given parent isotope will decay into a stable daughter product at a constant rate. The age of a sample, then, can be determined by knowing the decay rate of the parent and the amount of parent and

parent isotope
An isotope undergoing radioactive decay.

daughter product
An element formed from the decay of a parent isotope.

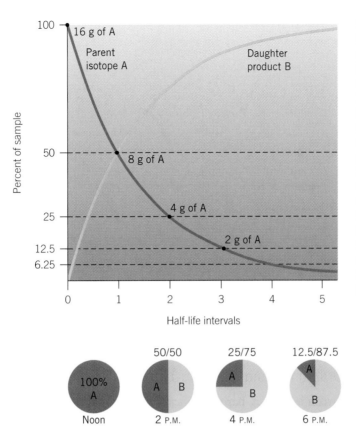

Figure 10.12 Hypothetical illustration of the half-life principle. A 16-gram sample of parent isotope A decays to daughter product B and has a half-life of 2 hours. If the sample is collected at 12 P.M., the end of the first half-life interval occurs at 2 P.M., and 50 percent of the parent remains; at the end of the second half-life interval (4 P.M.), 25 percent of the parent remains; and at end of the next half-life interval (6 P.M.), 12.5 percent remains. Daughter atom B accumulates at the same rate that the parent decays. Therefore, if we examine a sample containing 2 grams of parent isotope A and 14 grams of daughter product B, we will know that the sample is 3 half-lives, or 6 hours, old.

daughter in the sample. The number of atoms that decay is proportional to the number of atoms that remain. That is, if more atoms remain, more atoms decay; if fewer remain, fewer decay. The decay rate is expressed in terms of the **half-life** of the parent, which is the time it takes for half (or 50 percent) of the remaining parent to decay. Figure 10.12 illustrates how the half-life principle is used to determine the age of a sample.

half-life
The time required for half of a given parent isotope to decay to its daughter product.

Radioisotopes Useful in Dating

Table 10.1 lists the radioactive isotopes that are useful in dating ancient objects, together with their half-lives, daughter products, mineral occurrences, and effective time ranges. *Effective time range* is the interval over which a radioactive isotope yields useful dates. For example, the range of carbon-14, whose half-life is 5730 years, may stretch back 80,000 years; beyond that, the amount of carbon-14 remaining would be below the detection limits of our instruments. At the other extreme is uranium-238, which decays to lead-206 and has a half-life of 4.5 billion years. This isotope would not be useful in dating a 10,000-year-old lava flow because lead-206 would not have accumulated in measurable amounts. The effective range of uranium-238 begins with objects that are about 10 million years old and dips far back into the deep well of time. The effective time ranges of radioactive isotopes are not constant but expand with improved analytical techniques.

In general, the most reliable dates are derived from the minerals of igneous rocks, because radioactive isotopes are trapped within them at the time they crystallize from magma. Each crystallization is a clear-cut event. The parent-daughter ratio gives the time elapsed since the parent element began ticking away within the mineral. However, not every igneous mineral that contains radioactive isotopes is good for dating purposes; it must have remained impervious to weathering or physical damage throughout its history, so that no parent or daughter isotopes can have entered or escaped. Otherwise,

Table 10.1   Some Parent and Daughter Isotopes Used for Radiometric Dating

Parent Isotope	Daughter Isotope	Half-Life (years)	Occurrence in Minerals and Other Materials	Effective Range (years)
Rubidium-87	Strontium-87	47 billion	Muscovite, biotite, lepidolite, microcline, glauconite, whole metamorphic rock	>10 million
Uranium-238	Lead-206	4.5 billion	Zircon, uraninite, pitchblende	>10 million
Uranium-235	Lead-207	700 million	Zircon, uraninite, pitchblende	>10 million
Thorium-232	Lead-208	14 billion	Zircon, uraninite, pitchblende	>20 million
Potassium-40	Argon-40, calcium-40	1.3 billion	Muscovite, biotite, hornblende, glauconite, sanidine, whole volcanic rock	>100,000
Carbon-14	Nitrogen-14	5730	Organic materials, glacial ice, groundwater, ocean water	0–80,000

chemical analysis will yield a parent-daughter ratio that does not reflect the correct age of the mineral.

Uranium- (and Thorium-) to-Lead Dating

Zircon is a tough, hard, chemically resistant mineral that is a minor constituent of granite and other igneous rocks. Usually, it contains small quantities of uranium and thorium—both of which decay to lead—and it is widely used in radiometric dating. As in all uranium-bearing minerals, zircon contains closely related parent-daughter pairs: uranium-235 to lead-207 and uranium-238 to lead-206. Each pair has a different half-life, and each can be used to date the mineral.

Because uranium-235 decays more rapidly than uranium-238, lead-207 accumulates in the mineral more rapidly than lead-206. Thus the changing ratio between these daughter isotopes also reveals the mineral's age (Figure 10.13).

Figure 10.13 Because lead-207 and lead-206 are produced at different rates, the ratio of the two daughter products changes with time. Thus the ratio of the two lead isotopes extracted from a sample gives the sample's age.

Potassium-to-Argon Dating

Potassium is an abundant element, found in a wide variety of common minerals such as feldspar, mica, and amphibole and in the rarer sedimentary mineral glauconite. It consists of three isotopes: potassium-39, potassium-40, and potassium-41. Only potassium-40, which decays to argon-40, is radioactive. Because isotopes of the same element mix freely, potassium-40 is present in minerals containing potassium. Thus, the potassium-argon method has wide application in radiometric dating.

Rubidium-87–to–Strontium-87 Dating

Though relatively rare, rubidium and strontium are chemically similar to potassium and calcium, respectively. Both occur in a wide variety of common minerals, such as feldspar and mica. The 47-billion-year half-life of rubidium-87, which decays to strontium-87, makes it well suited for dating Precambrian rocks.

Carbon-14 Dating

The 5730-year half-life of carbon-14 makes this isotope extremely useful for measuring relatively recent events; in fact,

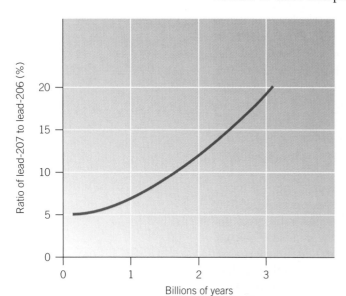

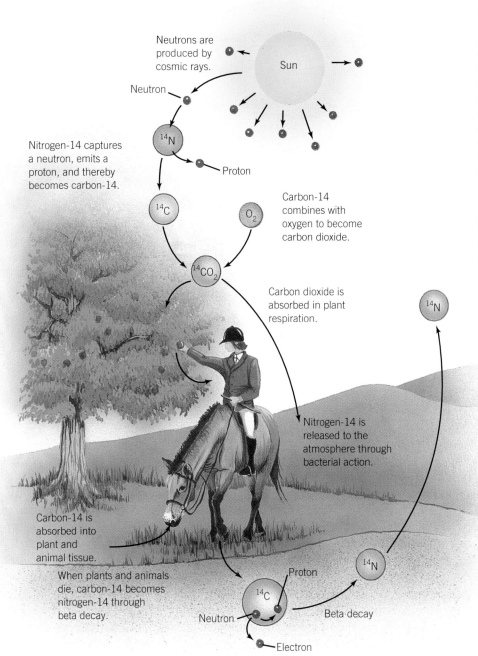

Neutrons are produced by cosmic rays.

Sun

Neutron

^{14}N

Nitrogen-14 captures a neutron, emits a proton, and thereby becomes carbon-14.

Proton

^{14}C

O_2

Carbon-14 combines with oxygen to become carbon dioxide.

$^{14}CO_2$

Carbon dioxide is absorbed in plant respiration.

^{14}N

Nitrogen-14 is released to the atmosphere through bacterial action.

Carbon-14 is absorbed into plant and animal tissue.

When plants and animals die, carbon-14 becomes nitrogen-14 through beta decay.

Neutron

Proton

^{14}C

^{14}N

Beta decay

Electron

Figure 10.14 Carbon-14 is produced from nitrogen-14 by cosmic ray bombardment. The carbon isotope then combines with oxygen to form carbon dioxide, which is incorporated into the tissues and shells of plants and animals. Through photosynthesis and respiration, there is constant interchange with the atmosphere, so the ratio of carbon-14 to other carbon isotopes remains constant while the organism is alive. However, upon death and burial, exchange with the atmosphere is cut off and the carbon-14 that decays back to nitrogen in the organism is not replaced. Therefore, the age of the organism can be determined from the half-life of carbon-14 and the amount of carbon-14 remaining.

we can go back approximately 75,000 years before it decays into quantities too small to measure. This time is sufficient for dating such things as

- the last glacial advance and retreat
- campsites of early Homo sapiens that allow us to trace migration routes
- historical documents made of skin (such as the Dead Sea Scroll parchments)
- climatic and sea-level changes dating back well into the last ice age
- plant and animal migrations
- volcanic eruptions
- deep-ocean circulation
- earthquake recurrence intervals

Carbon-14 is derived from nitrogen-14 through bombardment by cosmic rays in the upper atmosphere (Figure 10.14). But carbon-14 atoms immediately begin to decay back to nitrogen-14 atoms; as with all unstable isotopes, the number of atoms that decay

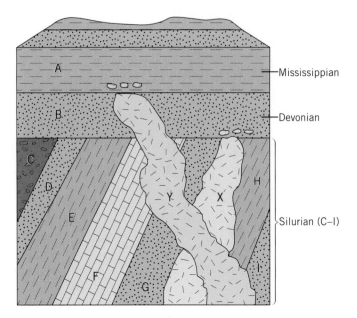

Figure 10.15 Cross-cutting relationships and radiometric dating are used to bracket strata whose relative ages were determined by fossil analysis. Through radiometric dating, we can determine that intrusion X is 400 million years old and intrusion Y is 350 million years old. Therefore, Layers C–I (of the Silurian period) are greater than 400 million years old. Layer B (of the Devonian period) is between 400 and 350 million years old, and Layer A (of the Mississippian period) is less than 350 million years old.

is proportional to the number of atoms present. Thus a balance is reached eventually between the carbon-14 that is manufactured and the carbon-14 that decays. At that point, carbon-14 in the atmosphere remains constant.

Carbon-14 is also in constant ratio to the dominant carbon isotope in the atmosphere, carbon-12. Because carbon-12 and carbon-14 are chemically identical, they combine with oxygen in the same ratio to form carbon dioxide (CO_2). Through photosynthesis and respiration, both are rapidly incorporated (or fixed) into the cells and tissues of plants and animals. And since interchange between organisms and atmosphere is open, the carbon-14/carbon-12 ratio within these living things equals that of the atmosphere.

Because photosynthesis and respiration end permanently with death, carbon-14 in a dead organism is cut off from the atmosphere. As it decays to nitrogen-14, the number of carbon-14 atoms in the organism declines. We can calculate the time since the organism died by comparing the percentage of carbon-14 relative to carbon-12 in the sample with the percentage of carbon-14 to carbon-12 in the atmosphere. For example, a drop of 50 percent means the sample is one half-life interval, or 5730 years, old.

Dating the Geologic Time Scale

Remember that the geologic time scale was worked out on the basis of superposition and fossil assemblages. How, then, do geologists date, in years, the boundaries between eras and periods of the scale?

One technique might be to analyze radioactive isotopes in the rock strata. Unfortunately, because extensive chemical weathering obscures their true age, sedimentary rocks do not lend themselves to accurate radiometric dating, so the next best thing is *bracketing*—combining the radiometric ages of igneous intrusions with the principles of cross-cutting relationships. Figure 10.15 shows that rock stratum B overlies intrusion X but is cut by intrusion Y. Clearly, B is younger than X and older than Y. If intrusion

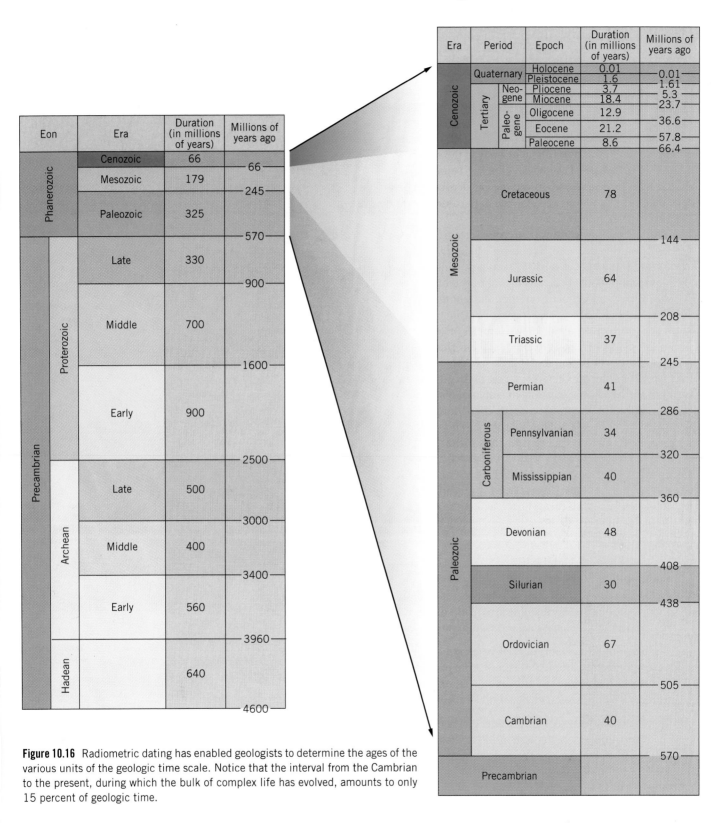

Figure 10.16 Radiometric dating has enabled geologists to determine the ages of the various units of the geologic time scale. Notice that the interval from the Cambrian to the present, during which the bulk of complex life has evolved, amounts to only 15 percent of geologic time.

X is dated radiometrically at 400 million years, and intrusion Y at 350 million years, then the age of B must fall somewhere between the two. If the fossils in B prove to be of the Devonian period, then we have an approximate date for that period. Obviously, the closer we can bracket the strata in future studies, the more closely we will be able to date the Devonian period. By applying these methods, geologists have dated the entire geologic time scale (Figure 10.16).

ASIDE 10.1 *Mystery at the K-T Boundary*

In the Apennines cliffs near the medieval town of Gubbio in northern Italy, you can climb to a non-descript exposure of limestone. The 6 meters of exposed strata would receive scant attention except for a particular 1-centimeter layer of clay that is sandwiched between two limestone layers about midway in the sequence. This innocent-looking layer of clay (the *K-T boundary*) marks the exact boundary between the Cretaceous (K) and Tertiary (T) periods—between the age of the dinosaurs and the age of the mammals.

Throughout the world you can trace certain fossil species upward in Cretaceous strata only to find that they are absent in strata above the K-T boundary. Dinosaurs are the most prominent of the extinct families, but of greater significance was the extinction of thousands of marine invertebrate species. In fact, late-Cretaceous extinctions eliminated about 25 percent of all known families of animals. Above the K-T boundary, very different life-forms are observed to have evolved and multiplied—notably the mammals, which during the Cretaceous were no larger than rats or squirrels, and the *angiosperms,* the modern plants, grasses, and trees. That there was mass extinction at the K-T boundary is a known fact. The *cause* of the extinctions, however, is the subject of a lively debate, which we will now describe.

In the late 1970s, a team of University of California scientists—Walter and Luis Alvarez, Frank Asaro, and Helen Michel—closely studied the ordinary-looking 1 centimeter of clay that marks the K-T boundary in Gubbio, Italy. Here is what they found:

1. The rare element iridium is several hundred times more abundant in the Gubbio clay than in normal clay or in the crust as a whole, but it closely matches the iridium content of some meteorites—rocky remains of extraterrestrial meteoroids that on occasion crash into the Earth.

2. The Gubbio clay contains distinctive glassy beads that were once molten droplets. We know from inspection of meteoroid craters on the Earth and the moon that such features are associated with meteoroid impact.

3. The Gubbio clay also contains quartz grains that display characteristic shock patterns identical to those found in meteoroid craters. These patterns have been duplicated in high-energy collision experiments that simulate meteoroid impacts.

Figure 10A The K-T boundary of Hojerup Church, Denmark. White Cretaceous chalk is overlain by gray, bedded, Tertiary limestone. The K-T boundary clay is found as a discontinuous layer in troughs between the two formations.

4. Features identical to those in the Gubbio clay were subsequently found in thin layers at the 95 other K-T boundaries located throughout the world (Figure 10A).

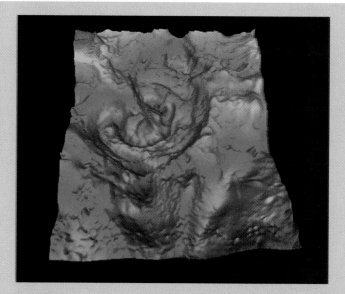

Figure 10B A computerized image of what is thought to be a large impact crater buried beneath the surface of the Yucatan Peninsula. The date of impact, the size of the crater, and the surrounding sedimentary deposits all suggest that it is the site of the collision of the extraterrestrial body that caused the catastrophic changes observed at the K-T boundary.

Upon examination of this evidence, Walter Alvarez and his colleagues proposed an explanation that would tie together the strange iridium content at K-T boundaries and the mass extinctions closing out the Cretaceous period: that a massive meteoroid struck the Earth 65 million years ago and raised an enormous fire cloud of ash and dust that rose to the upper atmosphere and circled the globe. The cloud blocked sunlight, caused acid rains, and catastrophically disrupted land ecosystems and ocean chemistry—to the point that a significant proportion of Cretaceous life died out instantaneously, or within a few years, at most. When the dust and ash settled, it formed a thin layer of sediment that blanketed the earth—the K-T boundary layer. Erosion, weathering, and burial have since destroyed the continuity of the layer, leaving only the scattered remnants we now observe.

It is relatively simple to estimate the diameters of the meteoroid necessary to form a worldwide layer approximately 1 centimeter thick and the impact crater it would produce. The numbers turn out to be 10 kilometers and 150 kilometers, respectively. Certainly, the credibility of the impact theory would be enhanced if a crater of the appropriate size and age were found. A leading candidate is the

Figure 10C A cloud of ejecta thrown up by the impact of fragment A of Shoemaker-Levy 9, the first in a series of impacts with Jupiter's atmosphere in July 1994. The ejecta cloud from this fragment alone is greater than the size of the Earth.

circular subsurface structure rimming the Yucatan coast of Mexico, shown in Figure 10B. Coarse marine gravel on nearby Cuba and as far away as the southern United States follows the arc of the crater, as if the impact caused a huge wave that tossed sediment on shore.

A number of geologists remain skeptical of the impact meteoroid-theory. Some challenge the evidence on technical grounds: Does the boundary layer show what the advocates of the meteor-impact theory claim it shows? Others offer alternative explanations: Could the K-T boundary layer have been deposited by the intense volcanism known to have occurred at the end of the Cretaceous?

The debate between the two camps is the latest flare-up in a long-running dispute centering on how to interpret the geologic record: Have geological changes been mainly gradual and of uniform intensity over the long term (the uniformitarian view) or have the most significant changes been sudden and violent, interspersed with long periods of incremental change (the catastrophist view)?

In 1994, we were able to observe firsthand the spectacular collision of the comet Shoemaker-Levy 9 with Jupiter. Figure 10C shows Earth-sized clouds sent up by fragments of the disintegrating comet as it penetrated the thick Jovian atmosphere. Had one of these fragments struck the Earth, it would have caused catastrophic extinctions. (For further discussion of meteoroid impacts, see Aside 11.1 The Geology of Meteoroid Impact Craters and Apply and Decide 21.1 Meteoroid Hazards.)

I n geology, we deal with radioactive isotopes that have half-lives of billions of years. Logically, we may ask: How can we be sure that their decay rates have remained constant through time? This question is certainly important, and it must be answered straightforwardly: the concept of constant decay through time is, indeed, an *assumption*. Although no one was present during the Precambrian to check out these "nuclear clocks," geologists nevertheless believe they have made a valid assumption.

First, there is experimental evidence that neither pressure, temperature, chemical reactions, nor any known crustal process affects nuclear decay rates in any measurable amount. Second, we can directly observe the decay of numerous short-lived isotopes, many of which have been created for medical use. All of these isotopes decay according to the half-life laws discussed in this chapter; there is no evidence that older, longer-lived elements play by different rules.

Third—and very telling—when a sample is dated by using two or more unrelated isotopes, all the dates are in very close agreement. If the assumption of constant decay rates through time were incorrect, there would be no reason—except extraordinary coincidence—for the agreement. The assumption of coincidence is difficult to sustain when thousands of samples have been subjected to numerous checks that are in agreement with one another.

Finally, older rocks, as determined by relative-dating methods, invariably yield higher daughter-parent ratios than younger ones. For example, the ratio is always higher in Precambrian than in Paleozoic rocks, in Paleozoic than in Mesozoic rocks, and so on.

The Age of the Earth

Radiometric dating has enabled us to determine the age of the oldest rocks that we can find. A few years ago, geologists thought that the 4-billion-year-old metamorphic and igneous rocks of Canada's Great Slave Lake region contained the Earth's oldest known objects. At present, that distinction belongs to Australia's Big Stubby deposit, which contains 4.3-billion-year-old zircons. Certainly, it is possible that we will find vestiges of older rocks, but geologists estimate the Earth's age as approximately 4.6 billion years. How do they justify this estimate if the Earth's oldest known minerals fall more than 300 million years short of this date?

The answer lies in the numerous meteorites they have dated. Originating in the asteroid belt beyond Mars, these meteorites are the rocky remains of meteoroids that strike the Earth's surface when their paths intersect the Earth's gravitational field. They are the oldest objects that have been physically dated, and the oldest of them invariably dates at 4.6 billion years. This result checks out using both uranium-lead and rubidium-strontium methods.

Lunar soils also yield dates of about 4.55 billion years; these are the oldest materials brought from the moon. So presumably, the largest planetary bodies of the Solar System all solidified at about that time.

This evidence does not prove the Earth's age, merely the age of the meteorites and moon. However, there are meteorites that contain no uranium but have a significant amount of lead-207 and lead-206 (the daughter products of uranium-235 and uranium-238). Since no uranium exists in these meteorites, this lead was probably incorporated in them at the time they formed, and their lead ratios have remained unchanged since then. The Earth's present lead-207/lead-206 ratio does not match these meteorites. Uranium has been decaying to lead-207 and lead-206 since the time that Earth formed, and the isotopes have accumulated at different rates owing to the differing half-lives of uranium-235 and uranium-238. Nevertheless, if we mentally reset the radiometric clock—that is, calculate how long it would take to change Earth's present lead ratio back to the ratio found in the meteorites—we come up with an interesting number: 4.6 billion years. In other words, it is probable that uranium became incorporated in the Earth about 4.6 billion years ago, no doubt along with the other elements.

Also present in the oldest meteorites, in far greater abundance than in the universe as a whole, are the decay products of short-lived isotopes. The original parent isotopes must thus have been derived from a local source, probably a newly exploded supernova that ejected matter into space. For these short-lived isotopes (with half-lives of approximately a few thousand years) to be incorporated into meteorites suggests that solid bodies formed less than a million years after the supernova explosion. In other words, the rocky materials of the planets—which are found to be about 4.6 billion years old—condensed rapidly from the dust cloud associated with the supernova explosion.

Some people argue on religious grounds that Earth is only 5000 to 10,000 years old, and geologists are often drawn into public debates as advocates of their scientific estimates. Sometimes, the position of geologists in this debate is misunderstood—for in the final analysis, geologists have no stake in how old the Earth is. They simply want to *know* how old it is! If the Earth is only 5000 years old, as some who take a literal interpretation of the Scriptures claim, so be it. However, the evidence points overwhelmingly to the contrary.

STUDY OUTLINE

Measurement of **relative time** is based on the fact that sedimentary rocks are deposited in layers, younger above older, and contain fossils that are of distinctive ages. Because a specific radioactive element decays into a specific stable element at a known rate, both can be used to measure **absolute time,** the age of objects in years.

I. RELATIVE TIME

  A. By determining the relative ages of rocks in one locality or region and then comparing or correlating them with those of other regions, geologists established the chronological ordering of geologic events in the Earth's history.

  B. Deciphering local geologic history
   1. **Formations** are bodies of rock with distinctive characteristics that set them apart from neighboring rock bodies.
   2. The **principle of original horizontality** states that sediments are deposited in layers parallel to the Earth's surface and, according to the **principle of superposition,** each layer is younger than the layer beneath it and older than the one above it.
   3. Geologists can learn a great deal about the history of rock formations from the way in which the rock layers make contact. An **unconformity** is an erosion surface between layers of markedly different ages.
     a. In an **angular unconformity,** strata rest upon the upturned edges of strata beneath.
     b. In a **nonconformity,** strata rest on eroded metamorphic or igneous rocks.
     c. A **disconformity** is an irregular contact between parallel strata.
   4. The **principle of cross-cutting relationships** states that a fault or intrusive body must be younger than the rocks it affects.

  C. Correlation is the process of demonstrating age equivalence. There are three methods of local correlation: lateral continuity, physical similarity, and similarity of sequence. Fossils are used to correlate widely separated rocks.

  D. The **principle of faunal succession** states that fossil assemblages succeed one another in an orderly fashion. Strata of the same relative age can be correlated by matching fossil assemblages.

1. Superposition and the fossil record form the basis of the relative geologic time scale, the events of the Earth's history listed in chronological order. Strata from various localities constitute standard sections for given intervals of the relative geologic time scale.

2. **Index fossils** are useful in correlation because they have narrow ranges but wide geographical distributions.

3. The relative time scale is divided into eons, eras, periods, and epochs, in order of decreasing magnitude. The **Phanerozoic eon** is divided into the **Paleozoic, Mesozoic,** and **Cenozoic eras.** The **Precambrian** comprises 85 percent of the geologic time scale.

II. ABSOLUTE TIME

A. To determine absolute age, geologists use radioactive atoms that decay at a steady pace over long periods.

B. The process by which unstable atoms emit or capture subatomic particles is called **radioactive decay.** There are three forms: alpha decay, beta decay, and beta (electron) capture.

1. As a result of radioactive decay, a **parent isotope** is eventually transformed into a stable (nonradioactive) **daughter product.** The age of a sample is determined by the ratio of parent to daughter and by knowledge of the decay rate.

2. The decay rate is expressed in terms of the **half-life** of the parent, which is the time it takes for half (50 percent) of the remaining parent to decay.

C. The effective time range is the interval over which a radioactive isotope yields useful dates.

1. Some radioactive isotopes used in dating are uranium-235 and uranium-238, which decay to lead-207 and lead-206, respectively; potassium-40, which decays to argon-40; and rubidium-87, which decays to strontium-87.

2. The 5730-year half-life of carbon-14 makes this isotope extremely useful for dating relatively young materials of organic origin.

D. Bracketing combines the radiometric ages of igneous intrusions with the principle of cross-cutting relationships to date the geologic time scale.

E. The 4.6-billion-year age of the Earth has been determined by dating meteorites from the Solar System and from lunar soil samples.

STUDY TERMS

absolute time (p. 271)
angular unconformity (p. 273)
Cenozoic era (p. 282)
correlation (p. 277)
daughter product (p. 284)
disconformity (p. 273)
formation (p. 272)
half-life (p. 285)
index fossil (p. 280)
Mesozoic era (p. 282)
nonconformity (p. 273)
Paleozoic era (p. 282)
parent isotope (p. 284)

Phanerozoic eon (p. 282)
Precambrian (p. 282)
principle of cross-cutting relationships (p. 275)
principle of faunal succession
  (p. 279)
principle of original horizontality
  (p. 272)
principle of superposition (p. 272)
radioactive decay (p. 283)
relative time (p. 271)
unconformity (p. 273)

CRITICAL THINKING QUESTIONS

1. In what sense does the principle of faunal succession derive from the principle of superposition?

2. In what sense does the principle of superposition derive from the principle of horizontality?

3. A zircon fragment is found within a granite pebble within a well-cemented conglomerate within a folded sequence of strata. The zircon is radiometrically dated at 1 billion years. What can you say about the age of the granite pebble? The relative age of the conglomerate? Of the folded strata? What sequence of events can you deduce from these dates?

4. Two rock sequences separated by thousands of kilometers contain no fossils in common. Does this discrepancy necessarily mean that their ages differ? Why or why not?

5. The Redwall limestone of the Grand Canyon is also found near Las Vegas. How do we know it is the same Redwall limestone?

6. What techniques discussed in previous chapters may allow the geologist to correlate strata not exposed to the surface?

7. Give three reasons that radiometric dates, when properly determined, are generally reliable.

8. There are geologists who disagree with both the impact and the volcanic eruption hypotheses of Cretaceous extinction. What might be their objections to these hypotheses?

9. Suppose that instead of Europeans, the Incas of South America had discovered the principles of relative dating and had applied them to the Earth. Suppose as well that they developed their own geologic time scale using their own names and standard sections. Would their scale differ from the present one? Explain.

10. Ash from an erupting volcano is deposited virtually simultaneously throughout its extent so that the age of the ash formation represents a distinctive brief interval of geologic time. What physical process involving sedimentary rocks would produce a rock formation whose age varies throughout its extent?

Rock Deformation

We live on the surface of a restless planet. Seismographs around the globe record several hundred thousand earthquakes a year, of which at least fifty thousand are strong enough to be felt. In rare instances, we can see fresh ground breaks following an earthquake. Eruptions of Kilauea on Hawaii are virtually continuous, and elsewhere large volcanoes spring to life at irregular intervals. Along the coast of southern California, beaches and wave-cut cliffs that are only a few thousand years old stand well above the present level of the sea. Similar signs of recent uplift or depression of the crust are found in other regions. Evidence like this proves not only that the Earth is currently an internally dynamic planet but also that the crust is warping and fracturing in response to this dynamism.

The evidence of Earth's past dynamism is preserved in the deformational features of rocks, in the contorted strata of mountain belts such as the Alps and Appalachians, in broad regional uplifts such as the Colorado and Tibetan plateaus, and in great faults such as the San Andreas of California. Most of these features are formed by the interaction of the Earth's lithospheric plates—namely, collision and subduction at convergent margins, rifting at divergent margins, and slide-by motions at transform margins.

This chapter describes the major deformational features of rocks: folds, faults, and joints, all of which are important to structural geologists because they are key elements in interpreting the architectural history of the crust. By determining the geometry of folds and faults in mountain belts, for instance, geologists are able to trace the direction of past plate collisions that have contributed to the growth of continents.

Folds, faults, and joints also exert major control over the topography and development of the Earth's landscape. Not only are they the major structural units of mountains, but they also profoundly influence erosion patterns. Streams seek out the zones

◄ Folded Paleozoic strata of the Northern Rockies near Borah Park, Idaho. Similar structures are typical of fold-mountain belts throughout the world.

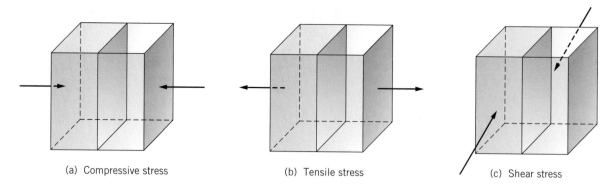

(a) Compressive stress   (b) Tensile stress   (c) Shear stress

Figure 11.1 The stress on an object is defined as the applied force divided by the area over which the force acts. The three basic types of stress are compressive, tensile, and shear.

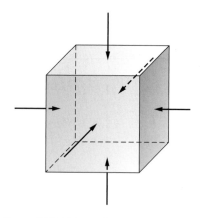

Figure 11.2 Lithostatic stress results when compressive stress is applied equally in all directions.

stress
The force applied to a plane divided by the area of the plane.

compressive stress
The stress generated by forces directed toward one another on opposite sides of a real or imaginary plane.

tensile stress
The stress generated by forces directed away from one another on opposite sides of a real or imaginary plane.

shear stress
Stress (force per unit area) that acts parallel to a (fault) plane and tends to cause the rocks on either side of the plane to slide by one another.

strain
The result of stress applied to a body, causing the deformation of its shape and/or a change of volume.

of weakness along faults, joints, and the belts of soft rock associated with folds. Differential erosion then leads to many of the ridge-and-valley patterns in the topography.

Underlying rock structure controls the occurrence of valuable mineral and petroleum deposits. Metal-rich seams, deposited by escaping magmatic fluids, line faults and joints. Oil and gas accumulate in certain fold structures. In many regions, rock structure controls the migration of groundwater in porous and permeable strata.

Civil engineers know that no dam, waste disposal site, tunnel, highway, or bridge should be constructed without detailed knowledge of the fold, fault, and joint patterns in the underlying rock structure. Lack of proper information may lead to design flaws that harbor the potential for disaster. Let us now take a close look at how rocks deform in response to stress and the various factors that govern their deformation.

Stress and Rock Deformation

Consider the three blocks of rock in Figure 11.1. To analyze the effect of the forces on these rocks, let us imagine an internal plane that divides the rocks as shown in the figure. Each force exerts **stress,** which is defined as the magnitude of the force applied divided by the area of the plane. The forces in Figure 11.1(a) exert **compressive stress,** because they tend to squeeze the rock. Typically, compression shortens, thickens, or buckles rocks. The forces in Figure 11.1(b) exert **tensile stress,** because they tend to pull the rock apart. Tension elongates and thins the rock. The forces in Figure 11.1(c) exert **shear stress,** because in acting parallel to the internal plane, they distort the shape of the rock. When subjected to shear stress, a cube is distorted to a rhomboid shape, and a sphere is made elliptical.

The blocks of rock in Figure 11.1 are subjected to unequal stress. The block in Figure 11.2, on the other hand, is subjected to equal stress in all directions. As mentioned in Chapter 9, such uniformly distributed lithostatic stress (or pressure) occurs in the Earth's mantle under many kilometers of rock.

The change in volume or shape of an object that results from stress is called **strain;** the particular strain caused by unevenly applied stress is expressed by a change in the object's shape. An increase in lithostatic stress, on the other hand, leads to volume reduction.

The Response of Rocks to Stress

Experiments have shown that the response of rocks subjected to stress takes three forms: elastic, ductile (or plastic), and brittle. The response in a given instance depends upon the type of stress applied, the temperature and pressure conditions, and the mechanical properties of the rock.

   If stress is light, a rock will deform in direct proportion to the stress; then when stress is removed, the rock will return to its original shape. This is an **elastic response.** However, above a certain level of stress, called the **elastic limit,** the rock may deform permanently in a **ductile response.** Strain is no longer in proportion to stress, and the rock will not return to its original shape once the stress is removed. On the other hand, stress continued beyond the elastic limit may cause a **brittle response,** in which minute cracks develop and the rock ultimately ruptures.

   Experiments in rock deformation aid our understanding of the behavior of rocks subject to stresses set in motion by geological processes. The propagation of seismic waves through rocks is an elastic response. The waves are vibrations—that is, small distortions or strains triggered by stresses well below the elastic limit of the rock. The rock deforms in response to the stress but returns to its original shape once the stress is removed. Slamming the palm of your hand on a table will trigger a similar response in the table.

   Ductile response is expressed by **folds,** which are permanent wavelike distortions in rocks. The stresses that created the folds may have ended millions or billions of years ago, yet the structures remain, frozen in time. In Chapter 9, we saw that in metamorphism, the shapes of pebbles, fossils, and other features are often distorted in folded rocks. Still further evidence of stress is the thickening and thinning of strata within the wavelike fold itself.

   The brittle response to stress is expressed by fractures. **Faults** are fractures in the crust along which the rocks on either side have been offset with respect to one another. Where such movement is absent, the fracture is called a **joint.**

elastic response
The deformation of a body in proportion to the applied stress and its recovery once the stress is removed.

elastic limit
The maximum amount of stress a material can withstand before it deforms permanently.

ductile response
The permanent deformation, without fracture, in the shape of a solid.

brittle response
The fracturing of a rock in response to stress with little or no permanent deformation prior to its rupture.

fold
Permanent wavelike deformation in layered rock or sediment.

fault
A fracture in bedrock along which rocks on one side have moved relative to the other side.

joint
A fracture in a rock, without noticeable movement along the plane of fracture.

Factors Influencing Rock Response

Two rocks may respond quite differently when subjected to similar stress. In Figure 11.3, for example, we see that a brittle rock (rock A) ruptures just beyond—or at—the elastic limit. Ductile rock B, on the other hand, exhibits permanent deformation beyond the elastic limit. Commonly, we can observe the two responses in the same rock exposure. Figure 11.4 shows an outcrop of folded sandstone and shale layers. The thickly bedded sandstone deformed in a ductile manner with little or no fracture, whereas the thinly bedded shale developed extensive fractures to the point that the original layering is obscured.

Figure 11.3 Two common responses to compressive stress. In a brittle response, rock A ruptures just beyond its elastic limit. In a ductile (or plastic) response, rock B is permanently deformed beyond the elastic limit. (Adapted from Marland P. Billings, *Structural Geology,* 2d ed., Englewood Cliffs, N.J.: Prentice-Hall, 1954.)

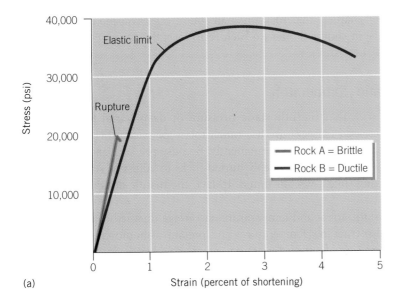

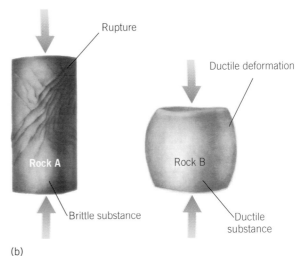

(a)

(b)

Whether a given rock responds as a brittle or ductile substance depends on a number of factors: the type of applied stress, the depth of the rock's burial, the temperature, the presence or absence of fluids, the homogeneity of the rock (that is, the degree of layering and impurities present), and time. Engineers make use of the fact that the response of a material to stress depends upon the nature of the stress applied. For example, concrete has great tolerance for compressive stress but ruptures easily when subjected to tension. Steel responds in the opposite manner; it is able to withstand enormous tension but buckles under moderate compression. So engineers reinforce the concrete in arches and other structures by inserting closely spaced steel rods in it. The concrete absorbs the compression, and the steel absorbs the tensional stress.

Most natural rocks behave much like that artificial rock, concrete: they have great resistance to compressive stress and display a wide range of plastic deformation. However, like concrete, they are weak and fail easily when subject to tensile stress. Figure 11.5 shows strata folded into an arch. The arrows indicate that the convex portion of the arch is under tension, and we can see fractures along the crest. The concave portion, however, is subject to compression; there are no fractures, but there is a thickening of the layers.

Recall from earlier chapters that rocks, like all solids, offer considerable resistance to shear stress. Liquids, in contrast, offer no such resistance; they change shape continuously in response to shear stress.

Depth of Burial and Temperature

Experiments have shown that the lithostatic pressure of deep burial fosters ductile behavior in rocks. Increased temperature has much the same effect; the elastic limit is lowered, and the rock tends to deform permanently. Because pressure and temperature increase with depth, we should expect deeply buried rocks to adjust to stress by ductile deformation, by "flowing" like a viscous liquid. In fact, the mantle does deform in this manner.

If increased lithostatic stress and temperature encourage ductile behavior, then the converse should be also true: low temperatures and pressures should encourage brittle behavior. With the exception of subduction zones, where cold brittle crust is forced into the hot upper mantle, failures by rupture—whether faults or joints—are common in the cool outer 10 to 15 kilometers of the crust.

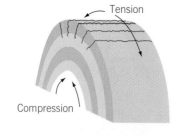

Tension

Compression

Figure 11.5 Tensile stress exerted on the convex side of an arch causes fracturing; compressive stress on the concave side leads to thickening of the strata.

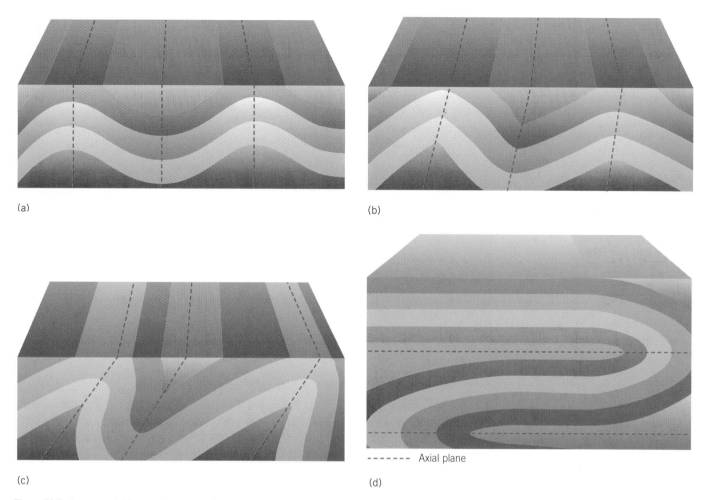

(a)

(b)

(c)

(d)

-------- Axial plane

Figure 11.8 Common fold types. The top of each block represents the surface pattern of the fold type after erosion has leveled its top layers. The front of each block shows the structure of the fold type in cross section. The dashed lines are the axial planes that divide the folds. (a) In symmetrical folds, the axial planes are vertical and the limbs of the folds dip in opposite directions at equal angles. (b) In asymmetrical folds, the axial planes are inclined and the limbs of the folds dip in opposite directions but at different angles. (c) In overturned folds, the limbs of each fold dip in the same direction. (d) In recumbent folds, the axial planes are horizontal.

Figure 11.9 Asymmetrical folds near Palmdale, California, in the vicinity of the San Andreas fault. The axial planes are inclined, dipping to the right. Notice also that the strata on the opposite sides of the axial planes dip at different angles.

Figure 11.10 A recumbent fold is an extreme type of overturned fold in which the axial plane is horizontal. Notice how the stratographic sequence is reversed in recumbent folds: Older strata overlie younger strata.

Plunging Folds

plunging fold
A fold whose axis is inclined rather than horizontal.

The folds we have discussed so far have axes that lie parallel to the Earth's surface—that is, horizontally. As a result, the outcrop pattern of eroded strata runs parallel to the strike of the fold axis. By contrast, a **plunging fold** is one whose axis is inclined to the Earth's surface. As a consequence, the outcrop pattern of the eroded strata converges toward what is called the "nose" of the fold (Figure 11.11). Because anticlines and synclines are folded in the opposite sense, the noses of the folds are oppositely directed. Anticlines close in the direction of plunge, and synclines close in the opposite direction. Because

(a)

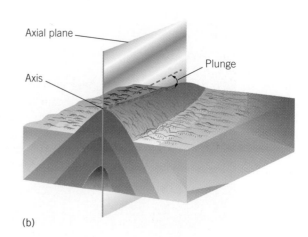

(b)

Figure 11.11 (a) A plunging anticline in the Zagros Mountains of Iran; the fold limbs converge in the direction of plunge. (b) The axis of the fold is inclined with respect to the Earth's surface.

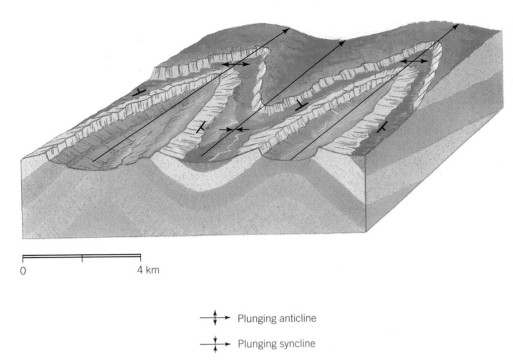

0   4 km

↕→ Plunging anticline

↕→ Plunging syncline

├ Strike and dip

Figure 11.12 The relationship between plunging anticlines and synclines. The "nose" of the anticline points in the direction of plunge, whereas the nose of the syncline points in the opposite direction. Also, it is common for deep valleys to erode in the cores of anticlines, leaving the synclines standing above them.

plunging anticlines and synclines interconnect in elaborate fold systems, the exposed strata follow a wavy pattern.

The plunging fold system, combined with the effects of erosion, dictates the topography of many mountain belts. Resistant strata stand out as ridges; valleys are carved out of weaker layers. Together, they form a zigzag ridge-and-valley topography that is characteristic of mountain belts. The Zagros Mountains of Iran (see Figure 11.11) are a young mountain belt. As might be expected, the highest elevations belong to the anticlines, which are convex-up folds. However, deep erosion may produce a topographic inversion, an interesting condition where the surface relief is out of phase with the underlying structure. Remember that anticlinal crests may be heavily fractured during folding, because the crests of convex-up structures are subject to tension. Thus they are easily eroded. Figure 11.12 shows a later stage of erosion in the Appalachians, a much older mountain belt than the Zagros. Anticlines are now the sites of deep valleys, while synclines form the high ground, standing above the anticlines as broad, flat-topped ridges.

The Appalachian Valley and Ridge Province is, in fact, a classic example of the control that rock structure exerts on erosion patterns and topography (Figure 11.13). Major streams, such as the Susquehanna River, cut through the hard-rock ridges perpendicular to the strike of the fold system. Tributaries work the belts of weak rock that parallel the resistant ridges, at right angles to the main stream. Indeed, this erosion is the reason that the ridges stand above the valleys. The long tributaries, in turn, develop short right-angle tributaries that flow straight down the hard-rock ridges parallel to the main stream. The entire right-angle drainage system resembles a vine on a trellis—hence the name, *trellis drainage pattern.*

Monoclines, Domes, and Basins

Not all folds form the repetitive pattern of anticlines and synclines that distinguish fold-mountain belts. Three important types that do not fit this pattern are monoclines, domes, and basins. A **monocline** is a single flexure, or bend, in otherwise horizontal rock strata. **Domes** and **basins** are rounded versions of anticlines and synclines.

monocline
A sudden steepening in an otherwise gently dipping strata.

dome
An anticlinal circular structure with rocks dipping gently away from the center.

basin
A synclinal circular structure with rocks dipping gently toward the center.

Figure 11.13 The Appalachians are an old, deeply eroded mountain belt. Judging from the thickness of the strata and the dimensions of the folds, the Appalachians probably stood higher in their youth than the Rockies or the Alps of today. Zig-zag ridges mark an elaborate system of plunging folds. Drainage is controlled by underlying structure and by differences in the resistance to erosion of the various rock formations. Major streams, such as the Susquehanna River, cut across the grain of the region, and their tributaries become established in the weak strata that trend parallel to the folds.

Monoclines are commonly formed by faults that cut brittle strata at depth but do not break through younger strata above (Figure 11.14). The older rocks are displaced along the fault plane, whereas the more ductile, younger rocks merely fold into the monocline. They are common structures of the Colorado Plateau, where strata deposited in shallow water have been uplifted thousands of meters without intense deformation. Monoclines are also formed by compression in the manner of anticlines.

Figure 11.15 is an aerial photograph of a typical dome, in which progressively younger strata dip away symmetrically from an older central core. Notice the topographic inversion—a valley occupies the center of the dome. As with the Appalachian anticlines, the apex of the fold has been subjected to deep erosion. This observation illustrates a general rule: One should not confuse underlying rock structure with topography—that is, the general configuration of the land surface. Although rock structure influences topography, the two are not the same. Topography is also a function of the resistance of rocks to erosion in a given climate, the age of the particular features, and many other factors.

The Black Hills of South Dakota are considered a classic dome. There, a Precambrian granite core is encircled by progressively younger strata that dip away from the

(a)

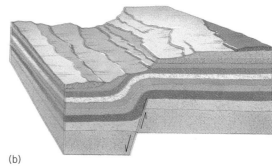

(b)

Figure 11.14 (a) A single flexure, or monocline, in the Cambrian sandstone of the Bighorn Mountains of Wyoming. (b) The flexure was formed by faulting in the rigid basement rocks beneath the sandstone strata.

core at relatively high angles. Whereas its topography is rugged, most domes and basins of the interior of the continent are far subtler structures, detectable only on a broad scale. For example, the rock strata that underlie the state of Michigan appear flat at any given locality. But when their outcrop patterns are studied on a regional scale, we see that Michigan is, in fact, a basin. Observe in Figure 11.16 the circular outcrop pattern of the formation, youngest toward the center.

The domes and basins of the midcontinent region are far from plate boundaries, and their origin is a matter of debate. Perhaps they are the distant ripples of plate collisions; perhaps they result from vertical faults unrelated to present plate collisions; or perhaps they are related to irregularities in the underlying basement complex.

(a)

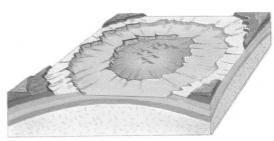

(b)

Figure 11.15 (a) Upheaval Dome, near the Utah Canyonlands. Notice how the strata dip symmetrically away from the up-arched center. (b) In a topographic inversion, a valley has been eroded in the core of the dome.

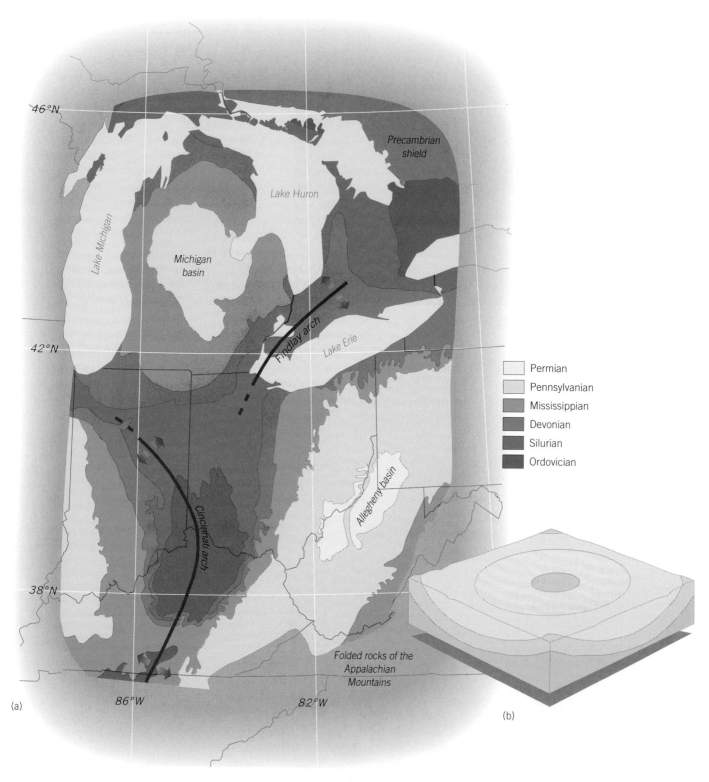

(a)

Permian
Pennsylvanian
Mississippian
Devonian
Silurian
Ordovician

(b)

Figure 11.16 (a) The mid-region of the North American continent consists of a series of broad, gentle domes and basins. (b) The strata of the Michigan basin dip so gently that the structure is discernible only through mapping on a regional scale. Notice the circular pattern of the rock formations, with the youngest formation at the center of the bull's-eye.

Fractures

Suppose, while standing, you were to hold a heavy suitcase in each hand. Plainly, your torso would be subjected to compression at the same time that your arms were subjected to tension. The parts of your body would respond according to their orientation with respect to the external forces applied. Rocks respond in similar fashion. Figure 11.17 shows the results of an experiment in which a rock of uniform composition has

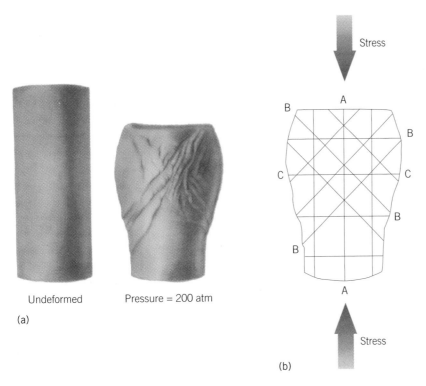

Figure 11.17 (a) This experiment was designed to show that an external force applied to a rock sample can cause a complex pattern of parallel fractures (or intersecting joint sets). (b) The vertical fractures (set A) are extension fractures. The diagonal fractures (set B) are the result of shear stress. The horizontal fractures (set C) are release fractures that developed after the force had been removed.

been subjected to external forces. The rock is subject to compressive stress in the direction of the external forces and extension at right angles to the forces. Planes midway between these extremes are subjected to shear. When its tolerance to these various stresses is exceeded, the rock will rupture. Extension fractures will form roughly parallel to plane A, and shear fractures roughly parallel to plane B. Upon easing of the forces on the rock, there will be a tendency for the compressed portion of the rock to expand, resulting in the release fractures parallel to plane C.

Joints

Natural fractures in rocks are called joints, and a group of parallel fractures is a **joint set.** As in the above experiment, joint sets bear a definite orientation to the applied force. Studies of joint sets help us understand regional topography as well as regional stress patterns. For example, we can see in Bryce Canyon (Figure 11.18) that joint sets intersecting at right angles divide the soft strata into rectangular blocks. This pattern is also common in areas of massive (nonbedded) igneous rocks of fairly uniform composition, such as parts of the Adirondack Mountains of New York. The joints and faults are zones of relative weakness. Streams may follow these joints and develop a pattern called *rectangular drainage.*

Not all joint sets are of tectonic origin. We saw in Chapter 7 that columnar joints develop as shrinkage cracks in cooling lava. In Chapter 13, we will see that granitic rocks peel in thin layers parallel to the Earth's surface when erosion removes the weight of the overlying rock.

joint set
A group of joints, generally parallel.

Faults

For most purposes, the classification of faults is based on the nature of the relative movement along the fault plane. Consider Figure 11.19(a), which is a block diagram showing a fault plane that dips at an angle of approximately 60°. Imagine you are a miner working in a shaft parallel to that plane. To use terms that were coined by Welsh coal

(a)

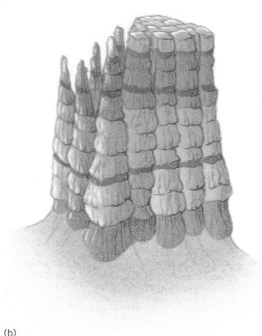

(b)

Figure 11.18 (a) These striking columns or "hoodoos" of Bryce Canyon, Utah, were formed from joint sets that intersected at right angles, dividing the soft limestone into rectangular blocks. (b) Weathering and erosion have widened the joints and rounded the columns.

miners in the nineteenth century, the block upon which you stand is the **footwall** (side A), and the block slanting diagonally overhead is the **hanging wall** (side B).

A **normal fault** is one in which the hanging wall has moved down with respect to the footwall (Figure 11.19b). Conversely, a **reverse fault** is one in which the hanging wall has moved up relative to the footwall (Figure 11.19c). In either case, displacement is along the dip of the slanting fault plane.

The normal and reverse faults in Figures 11.19(b–c) both dip at high angles (greater than 45°). However, there is an extremely important type of reverse fault in which the fault plane dips at less than 45° and most often at less than 15°. It is called a **thrust fault** because hanging-wall rocks appear to have been driven over the footwall rocks (Figure 11.19d).

Relative movement of the hanging wall in normal, reverse, and thrust faults is either up or down the fault plane. For this reason, all are referred to as **dip-slip faults.** In a **strike-slip fault,** however, displacement is horizontal—parallel to the strike of the fault plane (Figure 11.19e). The San Andreas is a right-lateral strike-slip fault, meaning that displacement is to your right as you face the fault from either side. For example, fences, roads, and similar features were offset in the right-lateral sense 3 to 4 meters during the 1906 San Francisco earthquake. Some formations on the West Coast have been offset hundreds of kilometers in the millions of years since the inception of the San Andreas fault. The Great Glen fault that divides Scotland is a strike-slip fault in which displacement is in the left-lateral sense; Loch Ness nestles in the long valley formed by the fault.

Figures 11.20 and 11.21 show examples of stratigraphic and structural displacements caused by faulting. Normal and reverse faults bring strata of different ages into direct contact (see Figure 11.20). This arrangement is contrary to the more common occurrence in which older strata pass beneath younger strata. Faults combined with ero-

footwall
The rock mass beneath an inclined vein or fault.

hanging wall
The rock mass overlying an inclined vein or fault.

normal fault
A fault in which the hanging wall has been moved downward relative to the footwall.

reverse fault
A fault in which the hanging wall has been raised relative to the footwall.

thrust fault
A low-angle reverse fault that generally dips at about 15° or less.

dip-slip fault
A fault in which the movement is parallel to the dip of the fault plane.

strike-slip fault
A fault in which the movement is parallel to the strike of the fault plane.

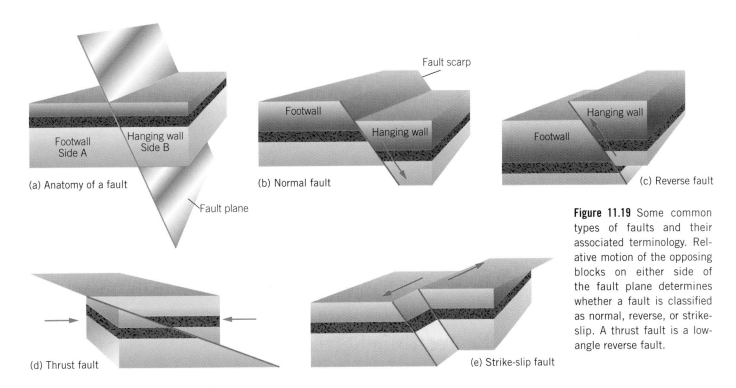

(a) Anatomy of a fault

Footwall Side A   Hanging wall Side B

Fault plane

(b) Normal fault

Fault scarp

Footwall   Hanging wall

(c) Reverse fault

Hanging wall

Footwall

(d) Thrust fault

(e) Strike-slip fault

Figure 11.19 Some common types of faults and their associated terminology. Relative motion of the opposing blocks on either side of the fault plane determines whether a fault is classified as normal, reverse, or strike-slip. A thrust fault is a low-angle reverse fault.

sion may also cause unexpected thickening, omission, and repetition of strata. In an extreme case, a fault may truncate the regional structure (see Figure 11.21)

Recognizing a fault and determining its relative movement are not always simple matters. Often, the topography in the vicinity of a fault can fool the geologist. The fault may be largely covered, or rock outcrops may be scarce, or the **fault scarp** on the upthrown block may have been eroded away. Indeed, the cliff may be located on the downthrown block rather than the upthrown block if the rocks on the downthrown side are resistant to erosion.

Close examination of the fault plane may also reveal the sense of fault movement. Often, the rock surface along the plane is polished and striated, so that the surface feels smooth in the direction of movement but rough in the opposite direction. The fault plane may also have been gouged and grooved during faulting, as if gone over by a gigantic rake. The grooves and gouges point in the direction of the fault movement. *Drag* in the strata on either side of a fault plane is another common feature that indicates the sense of fault movement (Figure 11.22). Because of friction, strata on the downside are deformed so that they dip away from the fault plane, whereas strata on the upthrown side dip toward it.

fault scarp
A cliff created by the movement along a fault.

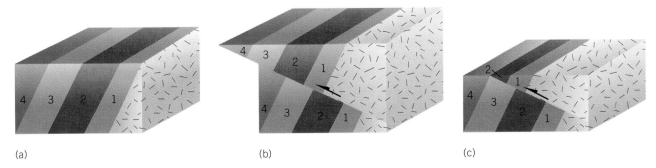

(a)   (b)   (c)

Figure 11.20 Faulting combined with erosion can disrupt the sequence of rock strata normally expected to be exposed on the surface. (a) Upturned layers 1, 2, 3, and 4 are visible at the surface. (b) A thrust fault slices and offsets the strata. (c) After erosion, layer 3 is no longer visible at the surface.

Figure 11.21 The fold pattern of this plunging anticline ends abruptly where it is truncated by a fault.

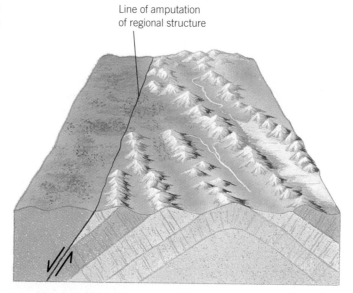

Line of amputation of regional structure

The Tectonic Setting and Topography of Faults

Each tectonic environment produces a distinctive pattern of deformation; these patterns, in turn, produce characteristic landforms. In the following sections, we will discuss the typical tectonic settings and topographic features of large-scale normal, thrust, and strike-slip faults.

Strike-Slip Faults

As we have mentioned, strike-slip faults result from shear stresses that displace rocks in the horizontal sense—that is, parallel to the Earth's surface and along the strike of the fault plane. Movements of this type typically produce zones of crushed rock more easily weathered and eroded than the rocks on either side of the fault. Strike-slip faults thus form long linear valleys that cut through the countryside. The valleys are commonly occupied by streams or by long narrow lakes and reservoirs. Streams crossing active faults are offset; even orchards, roads, and curbstones are displaced.

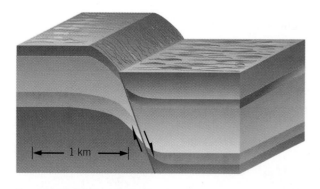

Figure 11.22 Friction caused by the motion of opposing blocks may cause drag in the strata close to a fault plane.

1 km

Long strike-slip faults can extend many hundreds of kilometers (Figure 11.23); among the longest is the Altyn Tagh fault of China. From the air or via satellite, it shows up as an enormous scar in the continental crust. The fault was formed as crust was displaced laterally in response to the collision of India and Asia. When traced, many modern strike-slips connect other large-scale structural features—for example, the offset segments of mid-ocean ridges. Indeed, such faults result from movement of the crust between these offset segments. In plate tectonics terminology, this type of strike-slip fault is referred to as a transform fault.

Normal Faults

Normal faults result from nonuniform tensile stresses that pull apart the crust, as occurs in the rift zones of the Earth's lithospheric plates. As the crust is stretched, some fault blocks drop relative to others along these planes to fill the excess space. Normal faults are the dominant structural units of Nevada, portions of its neighboring states, and northern Mexico (Figure 11.24). The region is referred to as the Basin and Range

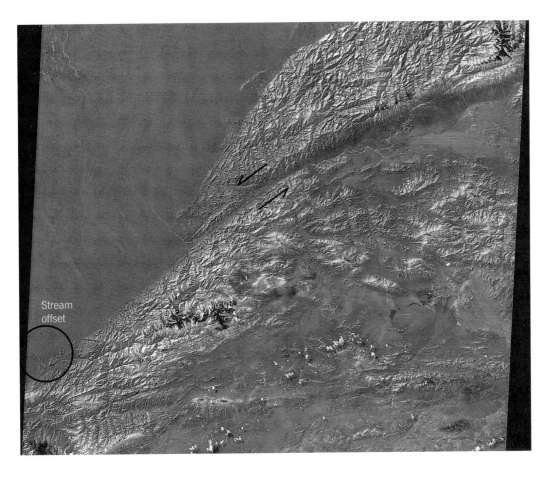

Stream
offset

Figure 11.23 A high-altitude, wide-aperture satellite photo of the spectacular Altyn Tagh fault, China. The displacement of rock strata and stream channels indicate that the Altyn Tagh is a left-lateral, strike-slip fault. It is one of many faults caused by the collision of India with Asia, which is forcing huge horizontal displacements in the Tibetan Plateau.

Province, a name that emphasizes the topography produced by these faults. The ranges are massive upthrown blocks called **horsts;** the basins are called **grabens.** The parallel structures extend north-south for hundreds of kilometers (Figure 11.25).

Several pieces of evidence point to a mantle heat source beneath the Basin and Range that is softening the upper mantle and deep crustal rock, and doming the overlying shallow crust. First, the age of various lavas and igneous intrusions associated with the faults prove that magma is generated beneath the Basin and Range. Second, field measurements

horst
An uplifted block bounded by normal faults on its long sides.

graben
An elongated, depressed block bounded by normal faults on its long sides.

Figure 11.24 The Hurricane fault is a normal fault that separates two major geologic provinces: the Colorado Plateau of Utah and Arizona and the Basin and Range of Nevada. The Plateau, with its eroded mesas and buttes carved out of flat-lying strata, is visible on the upthrown (or eastern) block. The Basin and Range, with its bold fault mountains and intervening valleys, is on the downthrown (or western) block. By measuring the distance between formerly connected strata, geologists have estimated that the vertical displacement along this segment of the fault was close to 2000 meters.

Figure 11.25 In the Basin and Range Province of the western United States, the crust is being stretched, causing some blocks (grabens) to drop relative to others (horsts) along normal fault planes.

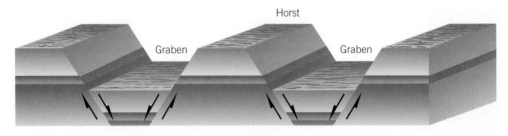

basin and range
A topography or landscape characterized by a series of tilted fault blocks forming longitudinal, asymmetric ridges or mountains with broad, intervening basins.

show that heat flow beneath the Basin and Range is extremely high. Third, studies of seismic waves passing beneath the Basin and Range reveal that the mountains have shallow roots. This observation runs contrary to the conventional theory of isostasy, which predicts that these mountains should have deep roots. It is also significant that the waves slow beneath the shallow roots, which is consistent with soft, hot rock high in the crust.

All this evidence presents a plausible explanation for the origin of **basin-and-range** features. As the softened upper-mantle and deep-crustal rocks expand and rise, they intermittently melt to produce magma. The rising of the soft rock also stretches the shallow crust, placing it under tension. Fractures develop and some blocks sink, but the expanding, hot mantle rock supports the weight of the fault-block mountains of the basin and range and prevents them from sinking and forming deep roots (Figure 11.26). Notice how normal faults, steep at the surface, can flatten with depth.

Thrust Faults

Thrust faults occur in regions subject to compressive stress, such as along colliding plate boundaries. Typically, such faults are an integral part of fold-mountain belts and, like them, are evidence of crustal shortening. Thrust faults tend to drive older rocks over younger ones (Figure 11.27). The cross section of the Canadian Rockies in Figure 11.28 illustrates that thrust faults often occur as thin sheets, in which fault planes overlap one another in series, like so many stacked shingles. Notice how the faults cause omission and out-of-sequence placement of the strata.

Figure 11.29 shows that thrust faults are capable of displacing strata great distances. Chief Mountain in Montana is an isolated remnant of a thrust fault separated by erosion from the main mass of the thrust sheet. The top of Chief Mountain consists of Precambrian strata about 1 billion years old. It rests on Cretaceous strata only 65 million years of age. It has been estimated that the Precambrian strata were driven a minimum of 25 kilometers. Some geologists believe the combination of folding and faulting is evidence that the crust has been shortened hundreds of kilometers in fold-mountain systems.

Thrust faults have long been of intense interest to geologists because of their importance to mountain belts. In recent years, however, they have been subject to even greater scrutiny. One reason is that geologists have recently been able to develop seismic wave

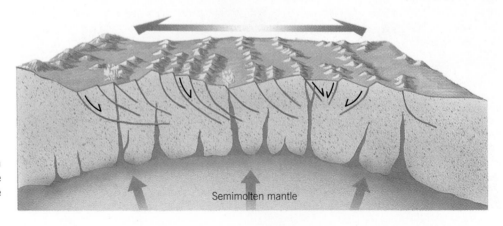

Figure 11.26 Magmas and semimolten rock rising from the mantle may be the cause of crustal stretching beneath the Basin and Range Province.

Figure 11.27 The classic Keystone thrust fault in the Spring Mountains of southern Nevada. Older Paleozoic rocks (the dark rocks above) have been driven from west to east over younger Mesozoic rocks (the light-colored rocks below).

techniques to probe deep crustal structures hidden beneath the surface. Especially important is what happens to large thrust sheets as we trace them in the direction of dip. Evidence is accumulating that they remain shallow in many mountain belts, such as the Appalachians and Brooks Range of Alaska. The shallow depth of the sheets suggests that slices of crust are pushed onto the continent during collisions with neighboring plates. We will have more to say about shallow thrust sheets in Chapter 12.

We should not leave the impression that faults are exclusively continental features; you may recall from Chapter 2 that faults are also common features of the ocean floor. Diverging plate boundaries in oceanic crust are marked by central rift valleys along the axis of mid-ocean ridges. The valleys are grabens, the product of normal faulting. Some plate boundaries in oceanic crust are sites of huge horizontal displacements—of strike-slip (transform) faults. Subduction of oceanic crust at convergent plate boundaries is a type of thrust faulting. In this case, the subducting plate, the one descending into the mantle, plays the role of footwall, moving down relative to the neighboring plate.

Figure 11.28 A cross section of the Canadian Rockies, which consist of a series of shallow, overlapping thrust sheets. For rocks to be driven over one another in this manner, considerable crustal shortening must have occurred. The westward dip of the thrust sheets is evidence that the compression drove the rocks from west to east.

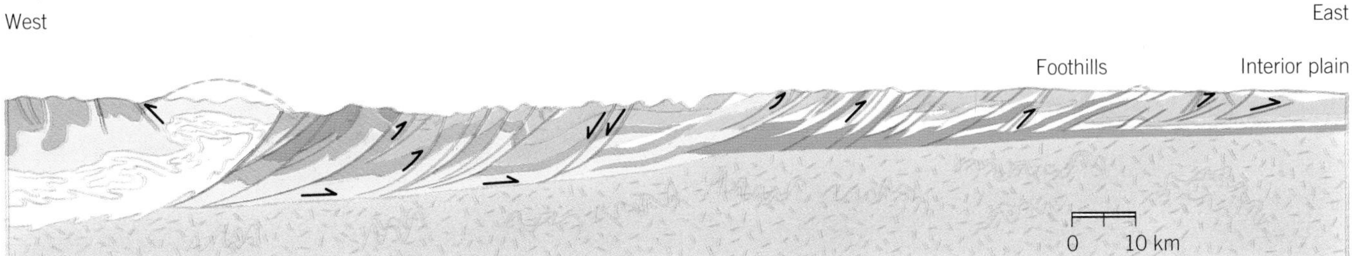

West

East

Foothills

Interior plain

0   10 km

ASIDE 11.1 *The Geology of Meteoroid Impact Craters*

Though Thomas Jefferson was open-minded about many subjects, the existence of meteoroids was not one of them. "Do we believe our common sense or the word of some Yankee professors?" was his response to a report that two Yale University researchers had retrieved rock fragments that had fallen from the sky over Connecticut.

The observations of the Yankee professors were, of course, correct. As discussed in Aside 10.1, many geology professors today are convinced that a large meteoroid impact 65 million years ago caused the mass extinctions that closed out the Cretaceous period. This theory, along with the realization that impact cratering is a major geological process affecting our neighboring planets in the Solar System, has motivated geologists to search for impact craters on Earth. Twenty years ago, these craters were thought to be rare, and their geologic effects rather trivial. Not so today, however.

Most meteoroids originate in the asteroid belt located between Mars and Jupiter. Orbiting the Sun within the belt are more than 1 million rock and iron fragments with diameters that range from a few millimeters to a few thousand kilometers. A *meteoroid* is formed when a collision sends a fragment of one of these bodies on an eccentric path. Sometimes, the path of a meteoriod intercepts the Earth's gravitational field. A *meteorite* is the rocky remains of a meteoroid that reached the Earth's surface.

Analysis of the frequency of impacts on our closest neighbor, the Moon, suggests that the Earth has suffered approximately 2400 large impacts over the past three billion years, with 720 of them occurring on the continents. Some 150 impact craters have been discovered on the continents thus far. It will be difficult to identify more than an additional few hundred, however, because many—perhaps most—have been destroyed by erosion and crustal recycling. Because of the prevailing dry climate, the Barringer (or Meteor) Crater of northern Arizona is probably the world's best preserved impact structure (Figure 11A).

It is now widely accepted that large impacts have produced profound effects. Here we will concentrate on the direct effects of meteoroid impacts—their landforms and associated features. In Chapter 21, Planetary Geology, we will discuss other issues, such as the origins of meteoroids, and the likelihood of large meteoroids striking the Earth in the foreseeable future.

The Energy of Impact

Probably millions of meteoroid particles shower the Earth every 24 hours. Most burn out in the atmosphere, and are observed as meteors or "shooting stars" in the night sky. Occasionally, boulder-sized meteoroids make it through the atmosphere; although they can flatten a car on impact, they have been slowed considerably by the time they reach the

Figure 11A Barringer (Meteor) Crater of northern Arizona.

Earth's surface. Meteoroids large enough to blast impact craters are a different story, because the atmosphere has little effect on them.

To survive the Earth's atmosphere and crash into the planet's surface with enough force to form a crater, a meteoroid must weigh at least 320 metric tons and travel at 15 to 30 kilometers per second—roughly five to ten times the speed of sound! The mass and velocity of the meteorite determine the *kinetic energy* available to do the work of excavating the impact crater; the greater their values, the greater the size and depth of the crater. The smallest meteoroid capable of blasting a crater would be about 50 meters in diameter.

The rule of thumb is that a meteoroid will blast a hole roughly ten times wider than its diameter, although actual craters may vary from this estimate by as much as a factor of ten. Because the ocean offers little resistance to a large meteoroid in a high-velocity impact, this ratio holds for both oceanic and continental crust.

The Structure of an Impact Crater

Though no one has ever witnessed an impact-cratering event firsthand, the landforms created by an impact are unmistakable (see Figure 11B). Most distinctive is the circular pit of the crater, which is bounded by an *outer ring* consisting of steep walls and an elevated *crater rim* that dips gently away from the center of the pit. The floor of the crater itself contains one or more circular fractures, or *inner rings*, that encircle the *core* of the crater. The core consists of a *central mound*, or peak, of elevated crater floor.

As shown in parts (a) and (b) of Figure 11B, the inner ring fractures are generated instantaneously by the enormous shock wave that spreads from the point of impact. In part (c), rebound from this wave causes the floor of the crater to rise, forming the central mound. A considerably longer interval of readjustment follows these rapid events. The crater walls stabilize by slumpage and other forms of mass wasting, which widens the crater and reduces its steepness. As indicated in part (d), the final circumference of the crater constitutes its outer ring.

A large-scale impact leaves behind other compelling evidence as well:

- *Ejecta,* consisting of particularized bedrock blasted out of the crater, are strewn about the surrounding region. If the impact occurs in a shallow sea or on a continental shelf, it may also trigger an enormous, circular wave pulse that tosses a ring of coarse sediment onto distant shores.

- *A zone of fractured and pulverized rock* may extend for hundreds of meters beneath the crater floor.

- *Meteorites* are identifiable as chunks of iron and nickel or rocks of characteristic mineral assemblages that are coated with a remelted and glazed crust.

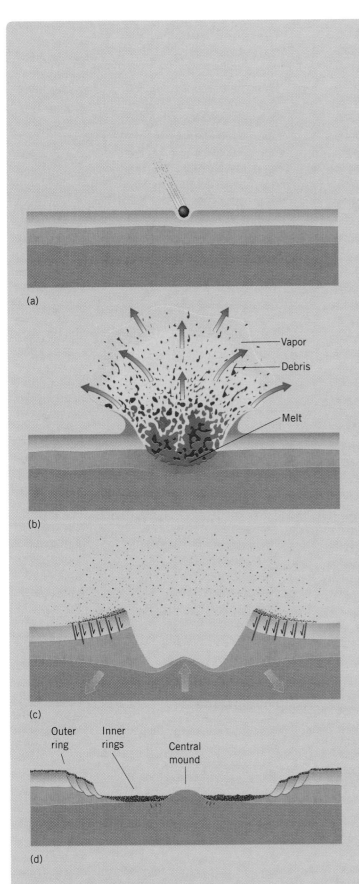

(a)

(b)

Vapor
Debris
Melt

(c)

(d)

Outer ring   Inner rings   Central mound

Figure 11B Impact of a meteoroid.

- *Distinctive minerals, melted rock, and shock structures* that form only during high-velocity impact (i.e., certain forms of quartz, glassy beads, and metal droplets) may be present.

- *Rare elements* may be found in the ejecta—for example, an iridium content that matches meteorites but no other materials formed on the surface of the Earth.

Unfortunately most craters, having formed millions of years ago, are deeply eroded. Only faint imprints of the rings, detectable by satellite imaging, remain. Other craters are not only eroded, but also buried, and thus are detectable only through geophysical means or drilling.

Effects of Impact Cratering

Many of the craters discovered so far have been implicated in the formation of a number of important geological features and historical events.

1. The inner and outer rings of a 85-kilometer-wide crater lie beneath Chesapeake Bay and the Delmarva Peninsula. At the center of the bull's eye lies a thick zone of fractured rock, churned-up sediments, and telltale shocked quartz grains. The 4-kilometer-wide meteoroid was roughly the size of the largest fragment of the Shoemaker-Levi 9 comet, which entered Jupiter's atmosphere in 1994. The meteoroid struck the Earth about 35 million years ago, and its impact probably created the depression in the continental shelf that later became Chesapeake Bay.

2. Geologists generally agree that Sudbury, Ontario, is the site of a two-billion-year-old impact crater, whose fractures tapped an ultramafic magma source beneath the crust. Sudbury is also the site of one of the world's largest and most valuable nickel deposits, which is found within the layers of the ultramafic igneous rock.

Sudbury raises an interesting question: To what extent are meteoroid impacts responsible for igneous activity? Large impacts on the Moon are known to have created the deep fractures in the crust through which basaltic lavas erupted, forming the smooth, flat regions of the Moon's surface called the lunar maria. Has the same process been at work on the Earth? Is it responsible for forming the huge lava beds, such as the Columbia and Snake River plateaus? Were some of the violent, ash-laden calder eruptions that have occurred throughout the Earth's history caused by meteoroid impact? It is a hypothesis some geoscientists are pursuing enthusiastically.

3. An important piece of evidence supporting the theory that a meteoroid impact led to the Cretaceous extinctions is a large, 65-million-year-old impact crater buried beneath the Yucatan Peninsula of Mexico. According to this theory, the extinctions were caused not by the force of the impact itself, but by secondary effects. First the atmosphere was heated to a boil by the fiery dust clouds and forest fires ignited by the impact. It was then cooled as the dust circled the Earth and blocked the sunlight. These effects, plus the torrential acid rains that followed, destroyed photosynthetic plants and marine algae, and generally disrupted the Earth's ecosystems to the point where the mass extinctions occurred in a matter of months.

Paleontologists who support the Cretaceous meteoroid theory are now asking the question, To what extent have other meteoroid impacts influenced the course of evolution through mass extinctions? This research question is worth pursuing, for if such impacts have caused extinctions in the past, Homo Sapiens may be on some future meteoroid's agenda. We will return to this subject in Chapter 21.

STUDY OUTLINE

The evidence of Earth's past dynamism is preserved in the deformational features of rocks: folds, faults and joints.

I. STRESS AND ROCK DEFORMATION

A. **Stress** is the magnitude of the force on a rock divided by the area of the plane. **Compressive stress** squeezes the rock from two opposite directions, **tensile stress** pulls it apart, and **shear stress** distorts its shape. Lithostatic stress applies equal forces in all directions.

B. The change in a rock's volume or shape that results from stress is called **strain**.

C. Rocks exhibit three modes of response to stress:
   1. A rock exhibits **elastic response** if it returns to its original shape once the stress has been removed.

(a)

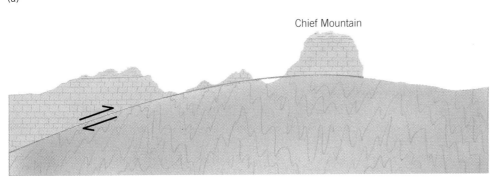

Chief Mountain

(b)

Figure 11.29 (a) Crustal shortening in the Northern Rockies as illustrated by Chief Mountain, of Glacier National Park, Montana. (b) Precambrian rocks of the mountain have been driven up to 25 kilometers to the east along the Lewis thrust fault, and have come to rest upon Cretaceous rocks. The mountain is called an outlier because erosion has separated it from the main mass of Precambrian rocks to the west.

2. Beyond the **elastic limit,** rocks can exhibit **ductile response,** in which the shape of the rock is permanently changed. The result is **folds,** which are wavelike distortions in the rock.

3. In **brittle response** to stress, rocks fracture. **Faults** are fractures in which the rocks on either side have been offset. If there has been no movement along the plane of fracture, the fracture is called a **joint.**

D. The high temperatures and pressures of deep burial encourage ductile response in rocks; low temperatures and pressures encourage brittle behavior. The presence of water in the cracks and pores of a rock expedites fracture. The composition of the rock and the amount of time over which stress is applied also influence rock response.

II. **THE MAP DEPICTION OF PLANAR FEATURES.**   **Strike** is the compass direction of a line along the intersection of an imaginary horizontal plane and a bedding plane. The **dip** is the acute angle between the bedding plane and the horizontal plane, measured in the vertical plane.

III. **FOLDS**   A fold may be divided into two parts, or limbs, by an imaginary surface called an **axial plane.** The line of intersection of the limbs and axial plane is the **fold axis.** An **anticline** is a fold with a convex-up orientation. A **syncline** is a fold with a concave-up orientation.

   B. Folds may be symmetrical or asymmetrical. In **overturned folds,** a limb is rotated so that older strata lie above younger strata, in a reversal of the normal sequence. In a **recumbent fold,** the axial plane is horizontal.

   C. The axis of a **plunging fold** is inclined; in a nonplunging fold, the axis is parallel to the Earth's surface.

   D. A **monocline** is a single flexure, or bend, in otherwise horizontal rock strata. A **dome** is a circular flexure that dips gently away from the center; a **basin** is a circular flexure that dips gently toward the center.

IV. **FRACTURES**
   A. A group of parallel fractures is a **joint set.** Joint sets that intersect at right angles divide rock into rectangular blocks.

   B. The classification of faults is based on the nature of the relative movement along the fault plane.
   1. When a rock is dissected by a fault that is diagonal to the Earth's surface, the underlying block is the **footwall** and the overlying block is the **hanging wall.**
   2. In **dip-slip faults,** movement is vertical along the dip of the fault plane. In a **normal fault,** the hanging wall has moved down with respect to the footwall. In a **reverse fault,** the hanging wall has moved up relative to the footwall. A **thrust fault** is a reverse fault in which the hanging wall overlaps the footwall.
   3. In a **strike-slip fault,** displacement is horizontal—parallel to the strike of the fault plane.
   4. If displacement is vertical, a cliff face, or **fault scarp,** is formed.

V. **THE TECTONIC SETTING AND TOPOGRAPHY OF FAULTS**
   A. Transform faults are strike-slip faults that result from movement of the crust between offset plate segments.

   B. Normal faults result from tensile stresses that pull the crust apart at divergent plate boundaries. **Basin-and-range** topography is produced by normal faulting. The ranges are massive upthrown blocks called **horsts,** and the basins are called **grabens.**

   C. Thrust faults are the result of compressive stress along convergent plate boundaries. They are an integral part of fold-mountain belts and are the result of crustal shortening.

VI. **THE GEOLOGY OF METEOROID IMPACT CRATERS.**   Impact craters are caused by the collision of meteoroids with the Earth's surface.
   A. The structure of an impact crater consists of an outer ring, crater rim, inner rings, core, and central mound.

   B. Impact cratering, once thought to be a minor occurrence on the Earth, is now known to be a major geological process that may have been responsible for a wide variety of events in Earth history. Such events include mass extinctions and the formation of lava plateaus, calderas, and deep basins, such as the one beneath Chesapeake Bay.

STUDY TERMS

anticline (p. 302)

axial plane (p. 302)

basin (p. 305)

basin and range (p. 314)

brittle response (p. 299)

compressive stress (p. 298)

dip (p. 301)

dip-slip fault (p. 310)

dome (p. 305)

ductile response (p. 299)

elastic limit (p. 299)

elastic response (p. 299)

fault (p. 299)

fault scarp (p. 311)

fold (p. 299)

fold axis (p. 302)

footwall (p. 310)

graben (p. 313)

hanging wall (p. 310)

horst (p. 313)

joint (p. 299)

joint set (p. 309)

monocline (p. 305)

normal fault (p. 310)

overturned fold (p. 302)

plunging fold (p. 304)

recumbent fold (p. 302)

reverse fault (p. 310)

shear stress (p. 298)

strain (p. 298)

stress (p. 298)

strike (p. 301)

strike-slip fault (p. 310)

syncline (p. 302)

tensile stress (p. 298)

thrust fault (p. 310)

CRITICAL THINKING QUESTIONS

1. List three examples of the Earth's current dynamism.
2. List three examples of the Earth's past dynamism.
3. Explain the difference between stress and strain.
4. A single, externally applied force may cause many kinds of stress in a rock body. Explain.
5. The nose of a plunging anticline tapers gradually, whereas the nose of a plunging syncline is blunt. Why?
6. How does the angle of dip influence the width of the outcrop pattern of rock strata?
7. Why do we call an anticline whose limbs dip in the same direction an overturned fold?
8. What kind of fault is associated with crustal shortening? With crustal extension? With crustal shear? Name three regions of the United States that display these features.
9. Give two examples of how faulting followed by erosion may influence the sequence of rock strata you encounter in the field.
10. Which feature of the oceanic crust resembles a graben? A strike-slip fault? A thrust fault?
11. Describe the distinguishing features of a meteorite impact crater. Why are impact craters easy to observe on the Moon but difficult to find on the Earth?

Plate Tectonics and Continental Crust

Astute observers have long noted the close connection between the Earth's highest and lowest places; between the ocean depths and the peaks of massive continental mountain belts. Leonardo da Vinci, for example, found that fossils in Alpine strata appeared similar to the living organisms of the Adriatic Sea, which led him to conclude that mountains were composed of rocks uplifted from the ocean floor. His discovery is applicable to all the world's mountain belts. You can find whale bones 15 million years old embedded in the mile-high submarine sediments of the Chilean Andes. When, in 1953, Sir Edmund Hillary became the first person to view the world from its highest peak, Mount Everest, his feet rested on a limestone reef—clearly the product of a tropical ocean of times past.

As we have seen, there is another interesting connection between the Earth's highest and lowest places—namely, that the present ocean basins are relatively ephemeral features that encompass only the most recent 5 percent of geologic time, whereas the continents contain traces of rocks 4 billion years old. The oceanic crust passes through relatively rapid cycles of birth, growth, and decline; but the continental crust endures. Too light to sink, it floats on the mantle surface and accumulates the records of "deep time."*

The dominant processes of continental evolution are accretion and synthesis, as juxtaposed to the divergence and spreading that characterize the evolution of the ocean basins. We will begin our exploration of the structure and evolution of the continental crust by examining the anatomy of its parts.

*As coined by John McPhee in his book *Basin and Range*.

◄ The world's highest peak, Mount Everest of the Himalayas, is composed of sedimentary rocks deposited in the long-vanished Tethys Ocean.

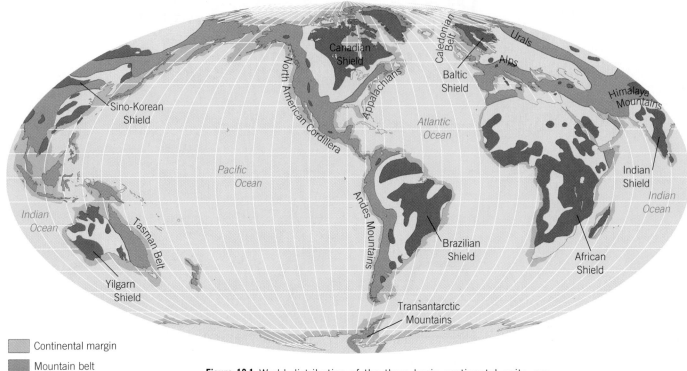

Continental margin
Mountain belt
Craton: Shield
Craton: Platform

Figure 12.1 World distribution of the three basic continental units: cratons, mountain belts, and shields. The theory of plate tectonics explains the origins of these features and their role in continental evolution.

Subdivisions of the Continental Crust

Though varied in detail, the continental crust consists of just three units, each with its own characteristic combination of age, rock types, and structure. The units are cratons, mountain belts, and continental margins (Figure 12.1).

Cratons

shield
A large region of exposed metamorphic and igneous basement rocks, generally having a gently convex surface and surrounded by a sediment-covered platform.

platform
That part of the craton consisting of essentially horizontal sedimentary strata overlying the older basement rocks.

craton
The stable core of the continental crust that includes basement rock, shield, and platform.

All the continents consist of a more or less central foundation of ancient *basement rock.* The exposed regions of basement rock, the **shields,** are so called because they are broad, low, and very gently arched (Figure 12.2). Their relatively subdued topography belies their structural complexity, however, for shields are composed of the oldest and most complex rocks on Earth. Basically, they are mixtures of granites, gneisses, schists, ancient sedimentary rocks, and interbedded lavas.

Vast regions of basement rock covered by flat-lying or gently tilting sedimentary strata constitute the interior **platform** of the continent (Figure 12.3). The younger platform strata cover the ancient igneous and metamorphic rocks like a thin blanket. For the most part, they were deposited in shallow seas that washed over parts of the Precambrian shields at intervals between 600 million years ago and the more recent past. These strata are not products of today's oceans, although it is clear that many were formed in a manner similar to present continental shelf deposits. Platform and shield together form the **craton,** the core of the continent that has remained tectonically stable and suffered little or no deformation for a geologically long time-period—perhaps hundreds of millions of years. Figure 12.4 is a generalized cross section of the North American craton, which includes the Canadian Shield. The Cambrian strata of the platform that rest upon the shield are undisturbed, which indicates that the craton has been stable for at least 600 million years.

Figure 12.2 A LandSat image of the Canadian Shield, in the vicinity of northeast Manitoba. Pink granitic rocks intrude darker-colored metamorphic rocks. Lakes, their basins scoured in bedrock by glaciers, appear dark and accentuate the regional structure.

Figure 12.3 An aerial view of typical farmland in the midwestern United States. The flat topography is underlain by sedimentary rocks of the interior platform.

Mountain Belts

Continents also contain *mountain belts,* most of which weave graceful arcs along the outer edges of the cratons. Most people think of mountain belts as simply terrain of higher elevation than the surrounding region. Geologists, however, think about mountains in terms of their structure—folds, faults, igneous intrusions, and metamorphism (Figure 12.5). Such features are often of greater significance than elevation, for they enable geologists to investigate the causes of mountain building. They also provide evidence of past mountain-building episodes, or **orogenies,** in regions that are no longer mountainous.

orogeny
The process of mountain building.

There are many kinds of mountains. Some are volcanic in origin, as we learned in Chapter 7. Others are primarily of tectonic origin and bear the distinctive marks of the kind of stress that created them. Fault-block mountains, for example, are the products of crustal extension and are associated with the rifting of divergent plate boundaries or

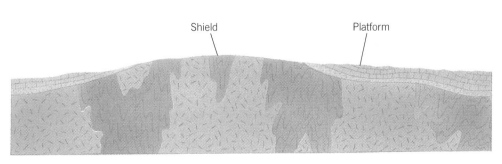

Shield   Platform

Figure 12.4 A cross section showing the relationship between complex shield and basement rocks and undeformed platform strata. (The vertical scale is greatly exaggerated.)

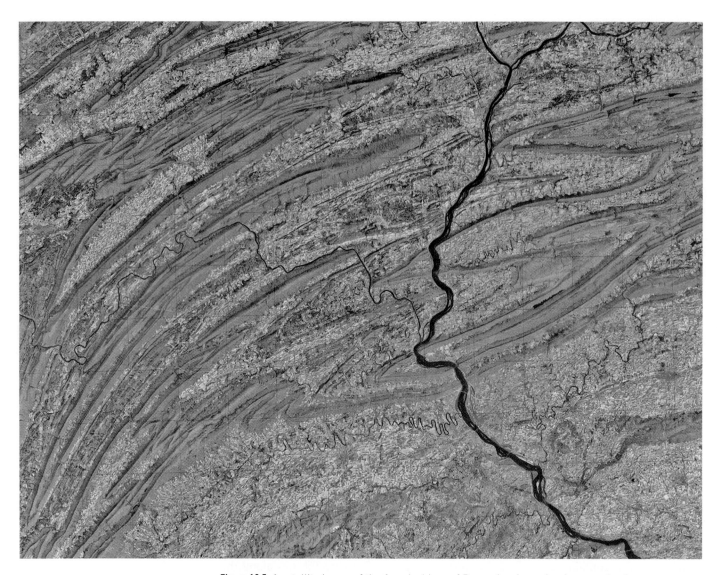

Figure 12.5 A satellite image of the Appalachians of Pennsylvania, a classic example of an extensively eroded fold-mountain belt. Resistant sandstone ridges delineate the zig-zag pattern of the plunging anticlines and synclines. The low-lying metamorphic rocks of the Appalachian-Piedmont are found to the southeast.

back-arc basins. In this category are the East African Rift Valley, the Mid-Atlantic Ridge, and the mountains of the Basin and Range of Nevada (discussed in Chapters 2 and 11).

Still other mountain belts are the products of compression generated at convergent plate boundaries; they are characterized by folds, thrust faults, granitic batholiths, and regional metamorphism. These are the mountains of the loftiest elevation and the most significant evidence of continent formation. When we examine these fold-mountain belts in detail, we find that they are commonly created over many millions of years and are the products of more than one episode of mountain building.

Continental Margins

Between the high-standing continental blocks and the deep-sea basins lie the continental margins, transition zones whose extent and shape are greatly influenced by their tectonic settings. As described in Chapter 2, active margins are commonly located close to plate boundaries, typically subduction zones, such as the one that runs directly offshore

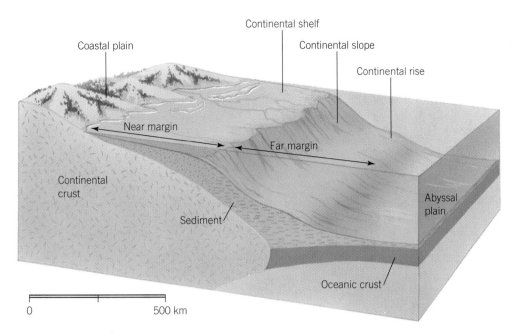

Figure 12.6 Anatomy of a passive margin. (The exaggeration of the vertical scale magnifies the ocean depth and makes the continental slope appear much steeper than it is in reality.)

of western South America. Some active margins, such as the Gulf of California, border rift zones. Passive margins are located far from plate boundaries in regions of little tectonic activity. The eastern margins of North and South America, for instance, are thousands of kilometers from the actively spreading Mid-Atlantic Ridge.

Passive Margins

A passive margin consists of a thick wedge of land-derived sediment that slopes gently seaward so that its tapered edge rests on oceanic crust. The wedge is often thousands of meters thick, yet examination of the sediment reveals that much of it was deposited at (or close to) sea level. Therefore, the crust must have subsided as the load accumulated. Passive margins exhibit broad, flat continental shelves and gently seaward-tilting continental slopes that trail down to great depths. Their lower boundaries are marked by a continental rise at the base of the slope, and the rise tapers into the abyssal plains of the ocean floor.

A passive continental margin is divided into two broad environment zones: the *near margin,* which consists of the shallow continental shelf and upper slope, and the *far margin,* which consists of the deeper slope and rise (Figure 12.6). Near-margin sediments, brought to the coast by rivers and glacial ice, are generally clean, well-sorted gravels, sands, and interbedded shales that reflect the work of breaking waves, currents, storms, and tides. Where detrital sediments are lacking, and where the water is warm, extensive coral reef and other carbonate sediments may cover the near shelf. Far-margin deposits are a mixture of fine clay, silt, and microscopic shells that floated far from shore; poorly sorted muds and sands deposited by swift undersea turbidity currents that flowed down the continental slope; and deep-ocean assemblages of brown clay, lava, volcanic ash, and biochemical oozes.

The Atlantic and Gulf Coast margin of North America is a classic example of a passive continental margin. Because the sea has retreated from its maximum level of the past, wide areas of the shelf are exposed today. This exposed region of the shelf is called the Atlantic and Gulf Coastal Plain. There the marine sediments, which were originally set down in predominantly shallow waters, cover ancient igneous and metamorphic basement rocks. The sediments tilt gently seaward, where they are overlain by younger strata. By taking drill cores and conducting seismic surveys, geologists have traced strata exposed on land far out onto the Atlantic shelf. Similar coastal plain features are found on the passive margins of other continents.

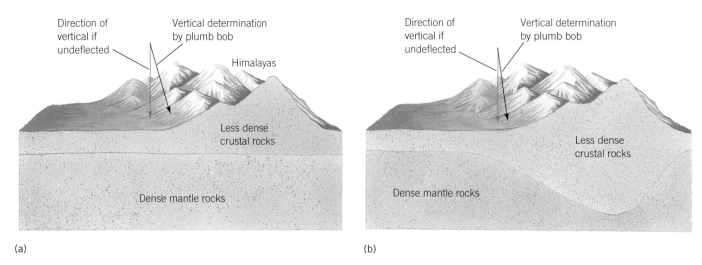

Figure 12.7 (a) If the Himalayas simply sat on top of the mantle, a plumb bob would be deflected by the gravitational attraction of the mountains and the denser mantle rocks beneath them. (b) The actual angle of deflection is found to be much smaller, which suggests that the lighter rocks of the mountains sink deep beneath the visible mountain peaks and displace denser mantle rocks. The deficiency of mass in the subsurface leads to weaker gravitational attraction.

Active Margins

An active continental margin has a steep gradient down to a deep trench and is located in a zone of tectonic instability. These factors combine to inhibit development of the elaborate features of passive margins. Instead, active margins are characterized by a narrow shelf and thick accumulations of sediment in the bordering trench. The sediments are brought to the trench by turbidity currents that are triggered by frequent earthquakes and rapid uplifts. It seems as though the vibrating crust were shaking the sediments down to the trench. Trenches that rim the Pacific Ocean act as traps that partially prevent the spread of sediment out onto the wider ocean floor.

There are two good reasons for studying the continental margins. First, these submerged regions of the continents comprise a significant percentage of the Earth's surface, so they are important in their own right. Second, the great fold-mountain belts of the world are built in large part of thick sedimentary wedges that resemble those found along today's margins.

Mountain Building and Plate Tectonics

In the course of surveying the subcontinent of India about one hundred fifty years ago, British engineer George Everest was surprised by the lateral pull of gravity on his instruments exerted by the nearby Himalaya Mountains. It was *less* than he and other scientists had expected, considering the great mass of matter that towered above the surrounding plain (Figure 12.7a). The cause of the discrepancy was the subject of spirited debate among geophysicists of the time. Gradually, a consensus emerged: the light crustal rock of the mountains extended down into the mantle, displacing the denser mantle rock and causing an overall deficiency of mass in the subsurface—hence, the weaker gravitational force (Figure 12.7b). Today, we have seismic evidence to support their conclusion that continental mountain belts like the Himalayas are composed of thickened crust. Viewed in cross section, their roots sink into the mantle upon which they float. Like any floating object, they rise where matter is eroded from them, just as the regions that receive the sediment are depressed in response. This state of balance due to buoyancy is called isostatic equilibrium, or *isostasy* (see Chapter 1).

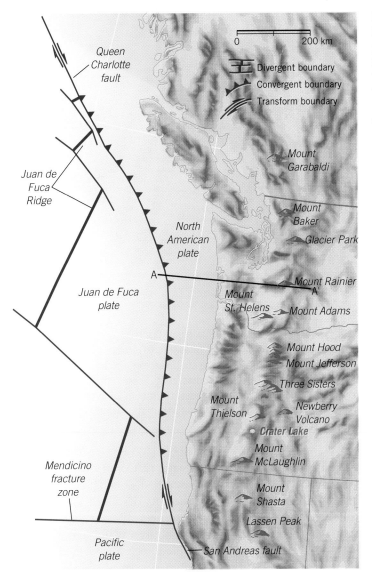

Figure 12.8 Subduction of the Juan de Fuca plate beneath the North American plate is responsible for three roughly parallel features: a sediment-filled trench, the coastal Olympic Range, and the Cascade Range farther inland. (See Figure 12.9 for an explanation of the cross section A to A'.)

In examining the Himalayas and similar mountains, we see that the reason for the crustal thickening is that the rock strata composing the mountains have been compressed perpendicular to the trend of the chain. This compression is evident in the structure of the chain, which is composed of folds and thrust faults. The thickened crust adjusts isostatically by sinking deeper while rising higher.

It took geologists over a century after George Everest's discovery to grasp the reason for the thickening and compression of the crust that creates mountain belts. When the theory of plate tectonics became accepted in the 1960s, it revolutionized the concept of mountain building, just as surely as it changed ideas about the origin of oceanic crust. In fact, oceanic crust and mountain building are united by plate tectonics.

Subduction and Mountain Building

Geologists learn about mountain belts by observing evidence of mountain-building processes at convergent boundaries and by reconstructing the events recorded in the deeply eroded rocks of ancient mountains. These studies reveal patterns consistent with the theory of plate tectonics.

The convergent boundary just off the coast of Oregon and Washington is a good place to study mountain-building processes (Figure 12.8). Here we see the spatial rela-

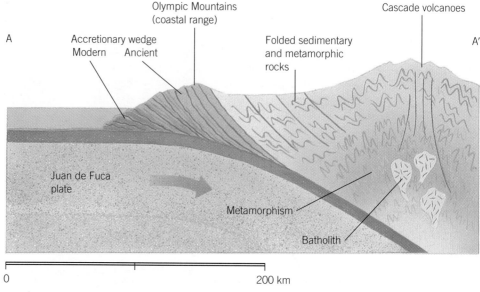

A

Olympic Mountains
(coastal range)

Accretionary wedge
Modern   Ancient

Folded sedimentary
and metamorphic
rocks

Cascade volcanoes

A'

Juan de Fuca
plate

Metamorphism

Batholith

0   200 km

Vertical scale is greatly exaggerated.

Figure 12.9 Cross section A to A' in Figure 12.8. Subduction of the Juan de Fuca plate has set a number of events in motion: (1) the plastering of an accretionary wedge onto the continent to form the Olympic Range; (2) the creation of magma bodies that rise in the crust, forming batholiths and the Cascade volcanoes; (3) the metamorphism of crustal rocks surrounding the batholiths; and (4) uplift, folding, and faulting of the overlying sedimentary rocks.

tionship among three parallel features: (1) an offshore subduction zone, (2) a coastal range made of sedimentary wedges, and (3) a higher inland mountain range with a core of granitic and metamorphic rock. As described in Chapter 2, this is a site of ocean-continent convergence. The dense Juan de Fuca plate is subducting beneath the light continental crust of the North American plate, setting a number of events in motion (Figure 12.9). The coastal Olympic Range was formed when the subducting plate plastered land-derived sediments, submarine lavas, and slivers of oceanic crust onto the edge of the neighboring North American plate as an **accretionary wedge** (also known as a **mélange** because of the chaotic structure of the rocks).

At depths of about 80 to 120 kilometers in the mantle, the descending Juan de Fuca plate has caused plumes of granitic magma to rise and intrude the continental crust above, forming batholiths or, where the magmas have reached the surface, the Cascade volcanoes. Deep in the continental crust, the heat released by the rising magmas has combined with the compression exerted by the converging plates to metamorphose the surrounding rocks. Shales and lavas have been converted to slates, schists, and gneisses. Higher in the crust, the same compression has thrown the rocks of the overriding continent into complex folds and thrust faults. The crustal thickening caused by the subduction, the emplacement of batholiths, and the compression has resulted in the rise of the complex mountain belt.

This generalized pattern of subduction, sedimentary accretion, and complex mountain building applies as well to the west coast of South America and to complex island arcs. The Japan Islands, for example, consist of several generations of volcanoes resting on platforms of granitic rock generated by the subduction of the Pacific plate. Sediments eroded from the islands and slivers of oceanic crust have been plastered back onto the islands, metamorphosed, uplifted, and eroded through several cycles. To geologists, Japan is a continent in miniature, a *microcontinent.*

accretionary wedge
A large mass of sediment and lava scraped off a descending plate that accumulates on the margin of a continent or island arc bordering a subduction zone.

mélange
A chaotic assemblage of accretionary wedge sediments and oceanic crust thrown up onto land at a subduction zone.

Mountain Building by Accretion

The Cascades, Andes, and island arc microcontinents illustrate the importance of subduction, magma generation, and compression in orogenesis. But these processes are not the entire cause of mountain building. In Chapter 2, we compared a lithospheric plate to a vast, slow-motion conveyor belt that moves from a mid-ocean ridge to a deep-sea trench. Transported to the trench by this belt is its cargo: continents and microcontinents, island arcs, extinct seamount chains, submarine lava plateaus, deep-ocean sediments, and continental margin sediments (Figure 12.10).

What happens to all this material when it is brought to a trench adjacent to a continent? For a fragment of continental crust embedded in the descending plate, the answer is straightforward. Because it is not dense enough to sink into the mantle, it must collide with the continent on the overriding plate. As for the denser materials, such as basaltic volcanoes and marine lava plateaus, it was first believed that they were ground up in the deep-sea trench and forced down into the mantle. However, recent studies have shown that although some of the material is forced into the trench, most of it is sheared off the descending plate and driven onto the continent on the overriding plate. In other words, the continent grows by the addition of incoming fragments of continental and oceanic crust, large and small. These fragments are called **accreted terranes.** Having been brought in by the oceanic conveyor belt, an accreted terrane is foreign to the rocks that surround it and is separated from them by faults. These faults are the "skids" along which the terrane was transported to its present position. Still, how are mountains created from accreted terranes?

First, recall that huge quantities of sediment—about 70 percent of the world's total—are deposited on continental margins. Much of this sediment accumulates on passive margins that later become active margins as the continents are carried toward convergent plate boundaries. It is not surprising, therefore, that most fold-mountain belts are composed mainly of margin sediments uplifted and folded by collision with accreted terranes.

Second, remember that all these events occur within the relatively narrow zone of plate convergence. Margin sediments and accreted terranes are likely to be distorted further by the relentless stress within that zone. A terrane may be shortened in the direction of compression but stretched, or elongated, at right angles to it. Pieces of the terrane may be sheared off, rotated, and moved great distances from where it originally landed. New sets of folds may be imposed on uplifted sediments and older terranes by later-arriving terranes. Subduction zone magmas rising from the base of the crust may intrude and metamorphose these assemblages. These adjustments pack terranes and sediments of the mountain belt tightly; they also bind accreted materials to the continent.

Accreted Terranes of Western North America

Accreted terranes extending from Alaska to Mexico compose approximately 70 percent of western North America (Figure 12.11). The truly intriguing challenge is trying to discover where they came from. For this task, we rely on essentially the same techniques developed by the pioneering geologists who argued the case for continental drift: analysis of the ancient magnetism, fossils, and structural or depositional features of the terranes. The task is complicated by the fact that most terranes have been distorted and

accreted terrane
A body of rock that is foreign to its surroundings and is bordered on all sides by faults.

Note: *The term* terrane *should not be confused with* terrain, *a general term for an area of land characterized by a particular feature, such as mountainous terrain or rocky terrain.*

Figure 12.10 (Pages 332–333) Continents and oceanic features in and around the Pacific. The map shows that the continents are a collection of terranes that were accreted before, during, and after the creation of Pangaea (see legend). Because of active subduction at the margins of the Pacific, the volcanic arcs, oceanic plateaus, and seamounts created after Pangaea have the potential of becoming accreted terranes. (Adapted from David G. Howell, "Terranes," *Scientific American,* November 1985.)

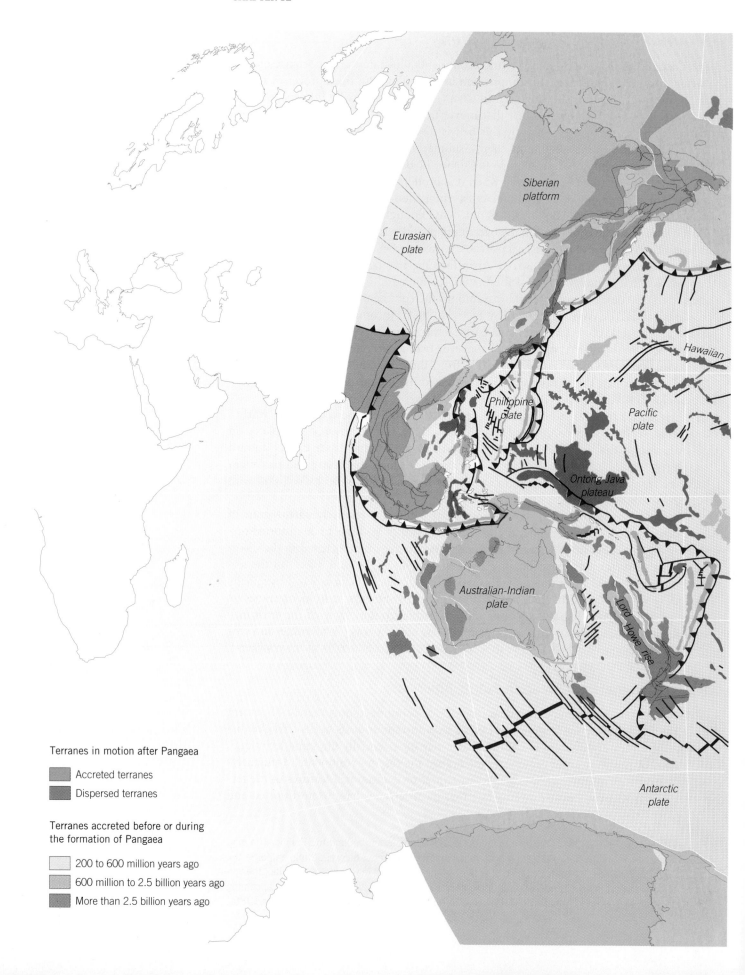

Terranes in motion after Pangaea

Accreted terranes

Dispersed terranes

Terranes accreted before or during
the formation of Pangaea

200 to 600 million years ago

600 million to 2.5 billion years ago

More than 2.5 billion years ago

Potential future terranes

Terranes now accreting or amalgamating

Volcanic island arcs

Oceanic plateaus

Basaltic seamounts and islands

Overlap sequences, igneous "stitching," and rifted continental margins covered by sediment

Divergent boundaries

Subduction zones

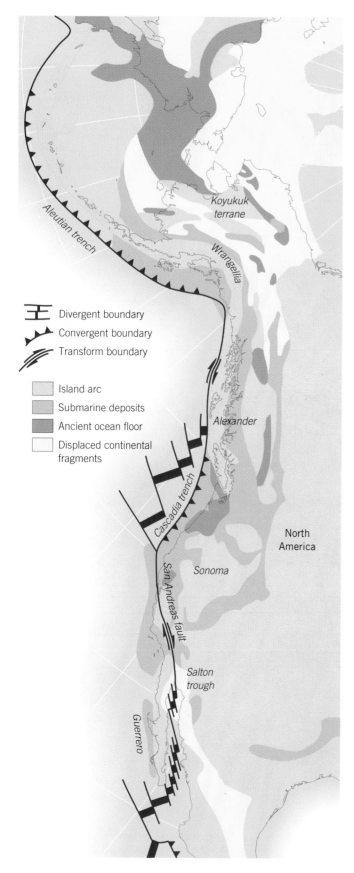

Figure 12.11 Most of western North America is an amalgam of accreted terranes.

altered after accreting to the continent, or docking. Consequently, researchers must design models that restore the terranes to their predocking condition. This job is similar to trying to read a torn and crumpled newspaper: first you have to straighten out the pages and paste them back together (Figure 12.12).

A typical terrane may consist of a sliver of basaltic oceanic crust—formerly a lava platform or seamount—buried beneath layers of oceanic sediment. Recall from Chapters 1 and 2 that as a lava cools, its magnetite crystals acquire the orientation of the Earth's magnetic field—an orientation that depends on the location of the lava. For example, magnetite crystals that form at the magnetic equator acquire a horizontal magnetization, while those that cool at the poles acquire vertical magnetization.

Thus the magnetic properties of the lava enable us to determine the approximate latitude at which the oceanic platform was formed. Furthermore, by dating the fossils in the sediments directly overlying the basalts, we can also determine the approximate age of the basalts. Therefore, by analyzing the terrane from top to bottom, we can trace its journey from point of origin to its present location.

One especially interesting terrane is Wrangellia, named for the Wrangell Mountains of Alaska. Geologists believe it originated near the equator some 200 million years ago as a large chunk of oceanic crust that traveled 9000 kilometers parallel to the west coasts of South and North America before docking in Alaska. Along the way, pieces of Wrangellia were sheared off, rotated, and embedded in other regions of North America; patches of it have been found in Oregon and British Columbia, among other places. This, by the way, is a conservative interpretation; some geologists, relying on similarity of fossils, believe Wrangellia originated in the South Pacific around the latitude of present-day Indonesia.

Alaska is a prime example of a continental region that is an assemblage of accreted terranes. Fragments were brought in and docked as the northern edge of the Pacific plate descended beneath Alaska and the Aleutian Islands. In all likelihood, northeast Alaska is the only bit of Alaskan crust that originated as part of North America, and even that is debatable. According to one USGS scientist, "You can stand in one spot and there will be three different terranes within two miles of where you are standing: continent, ocean deposit, and seamounts all smashed together." Another geologist describes Alaska as a "tectonic garbage heap."*

The Himalayan Orogeny

The term *accreted terrane* is generally applied to rock bodies that are small in relation to the host landmass. However, a subducting lithospheric plate may drag an entire continent

---

*Both quotations from Susan West, "Alaska, the Fragmented Frontier," *Science News* 119 (1981).

along with it. Large as it is, the continent is still less dense than the mantle and cannot sink. If the trench borders a continent on the overriding plate, the two continents have no choice but to collide. Obviously, the actual motion of convergence and collision is excruciatingly slow by human standards. As a result of the convergence, the sediments that were deposited on the intervening ocean basin and on the margins of the advancing continents are buckled and uplifted into huge fold-mountain chains.

Figure 12.13 shows the path of India across the ocean floor on its collision course with Asia that began during the Jurassic period some 150 million years ago. The journey was initiated in the wake of the fragmentation of Pangaea, which led to the modern configuration of the continents and ocean basins. A vast region of oceanic crust that was between Asia and India, having been subducted beneath Asia, no longer exists (Figure 12.14). Working backward from the fact of this enormous mountain range—uncrumpling the newspaper, so to speak—we note several pieces of evidence as we reconstruct the events that led to the Himalayan orogeny. The first is the presence of a belt of *ophiolites,* which are metamorphosed basalts, gabbros, and peridotites. Found in the Indus-Tsangpo valley just to the north of the Himalayas, they are slices of oceanic crust that were sheared off the descending Indian plate that subducted beneath Asia. Together with mélange sediments, they mark the boundary, or **suture zone,** between the colliding Indian and Eurasian landmasses. The granitic batholiths that intrude the Eurasian plate north of this zone were generated by the subduction of the Indian plate in advance of the collision (see Figure 12.14c).

The next piece of evidence is the tremendous thickness of the crust beneath the Himalayas, approximately 60 kilometers, almost double the average of the continents as a whole. The great thickness was caused by the underthrusting of the Indian plate beneath the Eurasian plate and by shortening due to compression. Too light to sink, the doubled-up mass of granitic crust has risen to great heights, in part because of isostatic adjustment and in part because the weight of the Himalayas is supported by the strength of the underthrust Indian plate. The thickened crust forms the platform that supports the Tibetan plateau north of the Himalayas. The plateau is itself higher than any mountain range of the United States. Finally, notice the large fold and thrust-faulted mountains south of the suture zone. The upper zone of the range is composed mainly of sediments that accumulated on the passive continental margin of India and on the deep-ocean floor during that continent's long journey. At greater depth are metamorphic gneisses and schists of the deep crust, also thrown into huge folds.

The Indian subcontinent continues to jam against Asia, squeezing, distorting, and faulting that continent over a wide region. Recall from Chapter 3 that the great earthquakes that plague China far to the northeast have been attributed to the stresses generated by this motion.

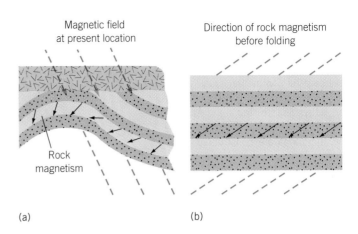

(a)   (b)

Figure 12.12 (a) The magnetization of most accreted terranes is distorted because of later deformation. (b) The geologist must correct for the deformation in order to reconstruct the original direction of magnetization.

suture zone
The narrow region that marks the juncture of two colliding blocks of continental crust.

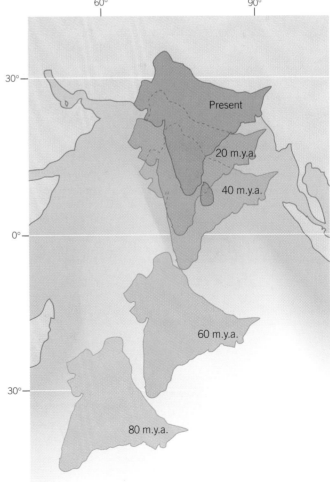

Figure 12.13 The path of India over an 80-million-year period, on its collision course with Asia.

Figure 12.14 The Himalayan orogeny. (a) India and Asia prior to collision. The lithosphere that subducted beneath Asia generated granitic batholiths that intruded the continental crust. At the same time, sediment was deposited in the Tethys Sea. (b) India and Asia collided, causing crustal uplift. (c) Today, the Himalayas are constructed of deformed Tethys sediment and granitic batholiths. The thickened continental crust is responsible for the great heights of the Himalayas and the Tibetan Plateau. Subduction of the Indian lithosphere has ceased.

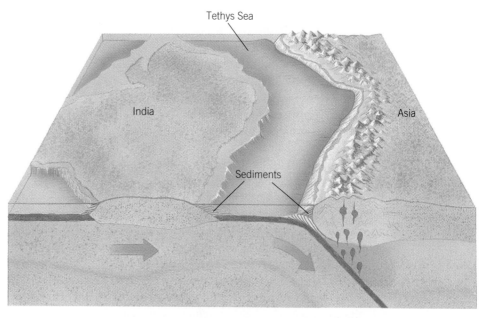

(a)

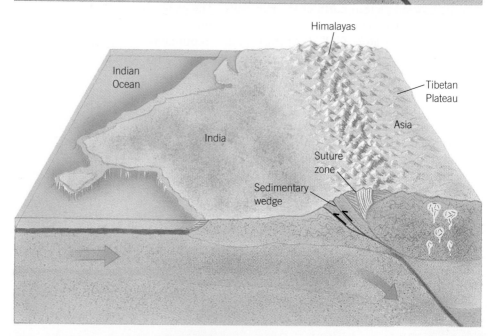

(b)

(c)

The Appalachian Orogeny

One wouldn't expect the Indian subcontinent and the rest of the Eurasian landmass to have a common history prior to the collision that welded them into a single unit. The same can be said of accreted terranes of any size and the continents in which they are embedded. Their ages, rocks, fossils, metamorphism, magnetism, and other features bear the stamp of the unique events that shaped them while they were separate. Geologists search for evidence of these differences in order to identify accreted terranes. For example, the fossils of the Carolina slate belt, a distinctive rock type, bear little resemblance to fossils of North America, but they match West African fossils of the same age.

The Appalachian mountain range extends from Newfoundland to Alabama and then continues a short distance below the Gulf Coastal Plain, where it is covered by younger sediments. Another mountain system, the Ouachitas, similar in age and structure to the Appalachians, extends across Arkansas, Oklahoma, and Texas down into northern Mexico (Figure 12.15). The two mountain systems are divided along their lengths and widths into a number of provinces that differ in rock type, structure, and age. Most geologists believe that these provinces were formed by a series of terranes that collided with the North American craton. The collisions, which occurred at irregular intervals, spanned the Paleozoic era between 570 and 250 million years ago. Each of the terranes that was added to the continent brought with it the rocks, structures, and fossils reflective of its history. The problem for the geologist is deciphering the complicated chain of events that created the mountains we see today.

Seismic reflection profiling provides strong evidence of the role played by accreted terranes in building the Appalachians. This technique employs large truck-mounted vibrators that send pulses through soil and rock strata deep into the crust. The vibrations reflect off various rock layers and structures and are picked up by arrays of receivers spread throughout the region. Huge amounts of data are collected and then processed by computer to form an image of the subsurface. Features identified from these images are interpreted by geologists to reveal faults, folds, and solidified igneous intrusions.

Especially relevant to this discussion is the seismic profile that runs from southeast to northwest along the southern Appalachians (Figure 12.16). The survey shows that the metamorphic Blue Ridge and Piedmont regions of the eastern Appalachians are each bounded by major thrust faults. While the existence of these faults was long known, their depth and amount of displacement had not been previously determined. Seismic profiling revealed the faults to be shallow, nearly horizontal structures that dip gently to the east and range from 5 to 15 kilometers in depth. In addition, the extent of displacement along the Brevard fault—which separates the metamorphic Blue Ridge Mountains from the sediments of the Valley and Ridge Province—is enormous, at least 250 kilometers. Moreover, seismic surveys reveal that sedimentary rocks lie beneath the shallow, nearly flat fault plane. These rocks are continuous with those exposed in the Valley and Ridge Province to the west, and the metamorphic and igneous rocks of the Blue Ridge

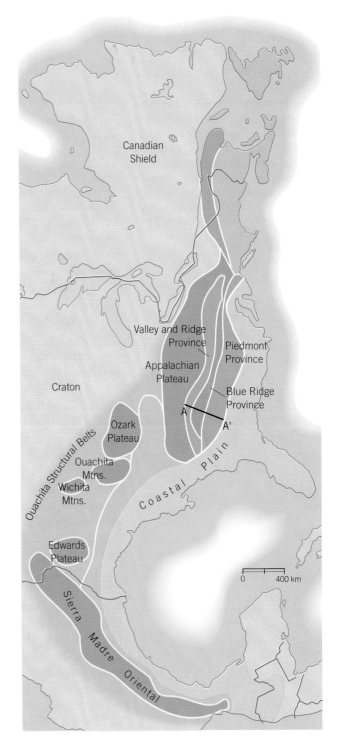

Figure 12.15 The Appalachian mountain belt (and associated structures) is a complex feature that trends from Labrador to Mexico. (See Figure 12.16 for an explanation of the cross section A to A'.)

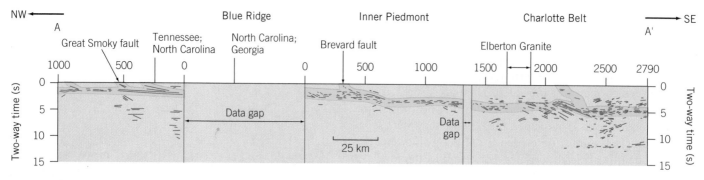

Figure 12.16 The structure of the crust beneath the southern Appalachians from A to A' in Figure 12.15, based on the analysis of artificially generated seismic waves. Notice the flat, relatively shallow (thrust) faults. They indicate that the rocks above them are decoupled from the deeper crust. (Adapted from Frederick Cook.)

and Piedmont rest on top of them. The only interpretation that makes sense of these data is that the older Blue Ridge and Piedmont rocks were driven westward over the younger sedimentary rocks.

Figure 12.17 employs the modern theory of plate tectonics to explain the origin of these huge faults. During the late Precambrian (800–700 million years ago), the ancestor of the Atlantic Ocean, the *proto-Atlantic*, opened and spread so that what are now the African and North American cratons were separated. However, several continental fragments and island arcs were left close to North America. The whole region probably resembled the configuration of islands, microcontinents, and true continents akin to those surrounding the Southeast Asian mainland today.

Throughout the Paleozoic era, the proto-Atlantic slowly closed, and the continental fragments collided with North America. What the collisions ultimately led to was the reassemblage of the continental crust into the landmass we know as Pangaea. The first great series of collisions, some 450 million years ago, involved relatively small microcontinents and island arcs. The second series involved large segments of what is now northern Europe. Finally, by the end of the Paleozoic era, the proto-Atlantic closed out, and Africa collided directly against North America, completing the enormous jigsaw puzzle. The total of these orogenies left behind an Appalachian mountain range that rivaled the Alpine-Himalayan belt of today.

Throughout the long history of mountain building, each newly arrived terrane bulldozed the one ahead of it farther inland. As a result, the ancient shelf sedimentary rocks of the continent were decoupled from the hard basement rock upon which they had been deposited and wrinkled into huge folds in the process. Sedimentary rocks to the west of the fold belt were gently uplifted with a slightly westward tilt; this thick wedge now constitutes the Appalachian Plateau (see Figure 12.15). Formerly a vast deltaic plain, it is constructed of the sediment eroded from the mountains to the east that were created by the collisions of the various incoming terranes.

Following the period of Appalachian mountain building, the supercontinent Pangaea fragmented; the modern Atlantic opened up; and North America, Europe, and Africa drifted toward their present positions. As described in Chapter 2, the breakup of the supercontinent was initiated by continental rifting. Sediment-filled grabens and lava flows caused by the rifting are preserved along the east coast, from Nova Scotia to North Carolina. New Jersey's Palisades Sill (described in Chapter 6) was formed of basaltic magma that erupted along an ancient rift valley. Indeed, each of the modern continents contains within it fragments that had previously belonged to other continents. Fragments of ancient Africa that were left behind in North America during the breakup include the Yucatan Peninsula, Florida, patches of the Appalachian-Piedmont, and parts of the southeastern New England states.

The Appalachians have been tectonically quiet for the past several hundred million years—which should be expected, because eastern North America has moved far from plate boundaries. As North America drifted westward from the Mid-Atlantic Ridge, new river systems were established, with their outlets in the present-day Atlantic Ocean. These rivers have stripped thousands of meters of sediment from the Appalachian

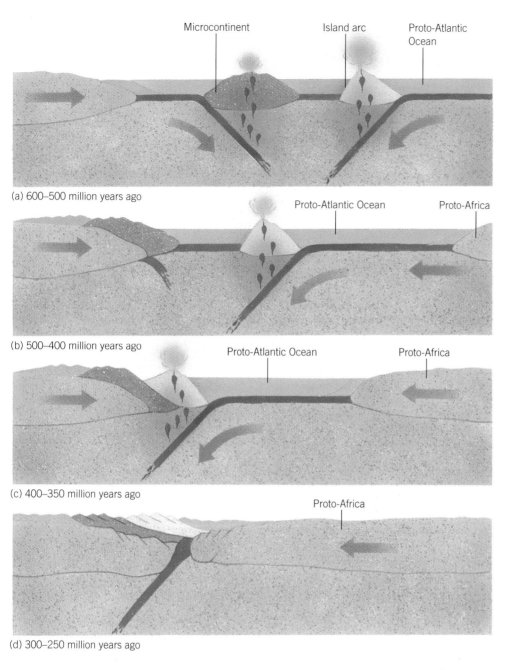

Microcontinent Island arc Proto-Atlantic Ocean

(a) 600–500 million years ago

Proto-Atlantic Ocean Proto-Africa

(b) 500–400 million years ago

Proto-Atlantic Ocean Proto-Africa

(c) 400–350 million years ago

Proto-Africa

(d) 300–250 million years ago

Valley and Ridge Province | Blue Ridge Mountains/ Piedmont | Carolina slate belt | Remnant of Africa | Atlantic Ocean

(e) Present

0   500 km

Figure 12.17 A schematic depiction of the origin of the southern Appalachians. The range was constructed over hundreds of millions of years during the Paleozoic era. A series of collisions involving microcontinents, and finally Africa, drove terranes that had arrived earlier westward along the thrust faults shown in Figure 12.16. Ancient shelf sediments, decoupled and deformed by the collisions, became the fold-and-thrust belt of the Valley and Ridge Province. Sediments eroded from the high mountains to the east created the deltaic rocks exposed in the Appalachian Plateau.

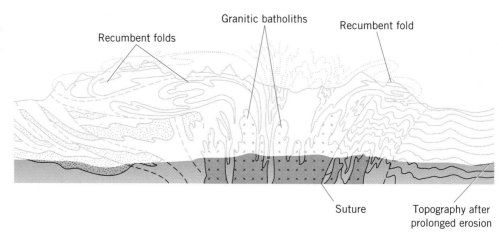

Figure 12.18 Shield-basement complexes are basically the remnants of former great mountain chains. Deep erosion has stripped most of the thick, elaborately folded sedimentary cover. Only the metamorphic rocks and batholiths of the deep crust remain. (Adapted from Robert H. Dott and Donald R. Prothero, *Evolution of the Earth,* 5th ed., New York: McGraw-Hill, 1994.)

Mountains. The present topography of the Appalachians reflects the effects of prolonged erosion. The Valley and Ridge Province east of the Appalachian Plateau delineates the anticlines and synclines of the fold system, but the ridges stand high merely because they are made of resistant sandstone, while weak limestones form the valleys. Erosion has dissected the great plateau itself so that its high remnants now stand as mountains themselves. To the east of the Valley and Ridge Province, the formerly rugged sections of the Appalachian-Piedmont are covered by the younger sediments of the Atlantic Coastal Plain. The seaward extension of this sediment composes the broad, passive Atlantic continental margin.

Growth and Evolution of the Continental Crust

The hard-rock cores of the continents, the cratons, are composed of deeply eroded microcontinents and larger crustal fragments. Preserved within them are the records of plate collisions and orogenies that occurred billions of years ago (Figure 12.18). Suture zones mark the ancient plate boundaries. Granitic batholiths and regionally metamorphosed sedimentary rocks and lavas mark ancient episodes of subduction. Rift zones filled with basaltic lava mark plate divergences. Each event left its imprint on those that occurred earlier. Thus the geologist seeking to interpret the history of a collision must first subtract from the rock record the impact of any later episodes of collision.

Figure 12.19 shows the age pattern of the ancient shields and mountain belts of North America. Proceeding from the margins inward, we encounter progressively older mountain belts and terranes: the Cenozoic Pacific coast ranges, the Mesozoic Cordillera, the Paleozoic Appalachians, the terranes of the Precambrian craton. We are led to the inescapable conclusion that our continent consists of a conglomeration of terranes, the younger ones welded to the older ones by subduction, compression, and later collisions. For this reason, geologist Paul Hoffman refers to our nation as "the united *plates* of America."*

Figure 12.19 The "united plates of America." The continent is mainly an amalgam of terranes welded together through subduction, compression, and later collisions. In general, the age of the terranes is younger toward the margins of the continent, but the pattern is disrupted by rifting and subsequent rearrangements.

Less than 900 m.y.a.
900 to 1200 m.y.a.
1200 to 2000 m.y.a.
2000 m.y.a. or greater
(Areas approximate)

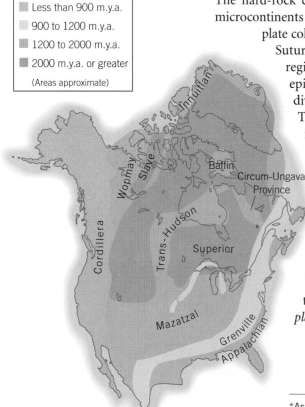

*As quoted in Robert H. Dott and Donald R. Prothero, *Evolution of the Earth,* 5th ed., New York: McGraw-Hill, 1994, p. 172.

So we see that plate tectonics elegantly explains the evolution of continents, just as it does the evolution of ocean basins. Consistent with the uniformitarian axiom that the present is the key to the past, the processes responsible for this evolution are readily observable on today's Earth. We can see how plate subduction produces simple island arcs (such as the Aleutians). We also see how these arcs grow into microcontinents (such as Japan), and then continents (such as North America), through complex volcanism, erosion, metamorphism, the addition of sediments, and the accretion of other terranes. Finally, we see how huge continental blocks riding on lithospheric plates may collide and form supercontinents (such as Asia/India).

Yet to proclaim that the present is the key to the past is not to say that the Earth was the same billions of years ago as it is today. One difference is that about 4 billion years ago, when the oldest known continental rocks were forming, the Earth's internal heat production was much higher than at present (Figure 12.20). At that time, the radioactive atoms responsible for the production of the heat were more abundant, for the simple reason that they had not yet decayed. Calculations based on the abundances of these atoms suggest that at that time the Earth generated four to six times the heat it currently produces.

Field evidence also supports the hypothesis of a warmer Earth billions of years ago. Early Precambrian rocks contain a unique assemblage of metamorphosed sediments and lavas preserved in *greenstone belts*. The lavas are of an ultramafic (peridotite) composition identical to the rocks of the upper mantle. No such lavas are produced today because their production would require total melting of the upper-mantle rocks from which lavas of this composition are derived. For this melting to occur, the upper mantle would have to be 300 °C hotter than it is at present. From the fact that the greenstone belts are very old—the youngest is 2.5 billion years of age—we may conclude that during the past 2.5 billion years, the Earth has ceased to generate the heat necessary to totally melt upper-mantle rocks.

If the Earth was much hotter billions of years ago, then it is reasonable to assume that the rates of volcanism, subduction, and plate production were much higher then. Accordingly, the continental crust must have formed at a much higher rate. Field surveys reveal that 70 percent of the continental crust was fully formed by the time the Earth was 2.6 billion years old and that only 30 percent has been formed in the 2 billion years since then. In other words, the rate of continent formation is slowing down, which implies that the Earth's internal processes, including plate tectonics, will someday grind to a halt. That event, however, may be several billion years away, because the Earth's production of internal heat is declining at an exceedingly slow rate.

In this chapter, we have envisioned the continents as collages of accreted terranes, swept together at irregular intervals by descending plates. Remember, however, that continental collages are not permanent works of art. The continents are continually being torn apart and reassembled in different forms. Crustal fragments that were once combined may later be separated by rifting or may be moved sideways over great distances by transform faulting; all are subject to the whims of the convection currents that drive the plates. The latest reshuffling of the crust, involving the assembly of the modern continents following the breakup of Pangaea, has yet to be completed. Crustal fragments are still fleeing toward Asia and North America around the rim of the Pacific and Indian oceans, as well as around the Mediterranean Sea.

The total volume of continental crust continues to grow. The subducting plates cause the production of huge volumes of magma that are added directly to the crust as granitic batholiths. Magma added to the oceanic crust through igneous activity at mid-ocean ridges and hot spots are accreted to the continents. Even sediments recently eroded from the continents will eventually find their way back. Through subduction, plate collisions, metamorphism, and igneous intrusion, these sediments will be pushed back and welded to the continents. As long as its supply of internal energy holds up, the Earth will go through many cycles of plate tectonics. The history of these cycles will be preserved in the hard-rock continental crust.

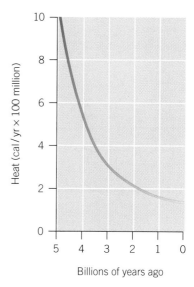

Figure 12.20 The decline of internal heat production over geologic time. The decline was steep during the first two billion years, but has long since leveled off.

STUDY OUTLINE

I. SUBDIVISIONS OF THE CONTINENTAL CRUST

A. The continental crust consists of three units: cratons, mountain belts, and continental margins. **Cratons** are the central, tectonically stable cores of the continents. They consist of foundations of ancient igneous and metamorphic basement rock, blanketed by **platforms** of sedimentary strata. The exposed regions of basement rock are called **shields.**

B. In mountain belts, geologists see folds, faults, igneous intrusions, and metamorphism as evidence of intense deformation that enables them to investigate the causes of past mountain-building episodes, or **orogenies.**
   1. Volcanoes are the products of the igneous activity associated with subduction zones.
   2. Fault-block mountains are the products of the crustal extension associated with rifting at divergent plate boundaries.
   3. Fold mountains are the products of compression generated at convergent plate boundaries.

C. The extent and shape of the continental margins are influenced by their tectonic settings.
   1. Passive margins, located far from plate boundaries, consist of thick, broad, sloping sediment wedges that are divided into continental shelf, slope, and rise.
   2. Active margins are characterized by narrow, steep continental shelves. Land-derived sediments are carried to the nearby trenches by turbidity currents.

II. MOUNTAIN BUILDING AND PLATE TECTONICS

A. Continental mountain belts are composed of thickened crust; their roots sink into the mantle upon which they float in a state of isostatic equilibrium. The crustal thickening is due to compression perpendicular to the trend of the chain.

B. Western North America, a typical site of ocean-continent convergence, is characterized by three parallel features: a deep-sea trench, an **accretionary wedge** forming a coastal range, and a higher inland fold-mountain range having a granitic/metamorphic core.
   1. Land-derived sediments, submarine lavas, and slivers of oceanic crust from the subducting plate are plastered onto the edge of the overriding plate as an accretionary wedge, or **mélange.**
   2. Plumes of granitic magma intrude the continental crust, forming batholiths or volcanoes. Thus heat and compression combine to metamorphose the country rock deep in the continental crust. Rocks closer to the surface are thrown into folds and thrust faults.

C. Continents grow by the addition of fragments sheared off the descending plate, or **accreted terranes.** Margin sediments and accreted terranes are distorted and uplifted by later-arriving terranes. Rising magmas may intrude and metamorphose these assemblages.

D. The Himalayan range resulted from the collision of continents. Slices of oceanic crust found in a valley north of the Himalayas and mélange sediments mark the **suture zone** between the colliding Indian and Eurasian landmasses.

E. Seismic reflection profiling provides strong evidence that the Appalachians consist of a series of terranes that collided with the North American craton between 570 and 250 million years ago.

1. Each newly arrived terrane bulldozed the one ahead of it farther inland. Ancient shelf sediments were wrinkled into huge folds and uplifted in the process.
2. The present topography of the Appalachians reflects the effects of prolonged erosion.

III. GROWTH AND EVOLUTION OF THE CONTINENTAL CRUST

A. Plate tectonics explains the evolution of continents, just as it does the evolution of ocean basins. Plate subduction produces simple island arcs that grow into microcontinents and then into continents through complex volcanism and metamorphism, the accretion of sediments, the arrival of other terranes, and continental collision.

B. Preserved within the continental cratons are the records of plate collisions that occurred billions of years ago; suture zones mark the ancient plate boundaries.

C. As long as the Earth has sufficient internal heat, the continents will continue to be torn apart and reassembled in different forms. Crustal fragments once combined may later be separated, and the total volume of continental crust will continue to grow.

STUDY TERMS

accreted terrane (p. 331)   orogeny (p. 325)
accretionary wedge (p. 330)   platform (p. 324)
craton (p. 324)   shield (p. 324)
mélange (p. 330)   suture zone (p. 335)

CRITICAL THINKING QUESTIONS

1. What evidence would prove that most of the sedimentary rocks of the Appalachian Valley and Ridge Province were deposited in a near-shelf environment?
2. What conditions would be necessary to change the passive Atlantic continental margin into an active one?
3. Looking at a world map, locate those regions other than the Himalayas that are the probable site of continent-continent convergences.
4. What sections of the North American continent show signs of becoming accreted terranes in the future?
5. How would you identify the line of convergence, or suture zone, between two continental blocks?
6. What kind of evidence might prove where an accreted terrane came from? How old it was? Its environment of formation?
7. Describe or sketch a sequence of events that may include ocean-continent convergence, accreted-terrane docking, and continent-continent convergence. What part of the world displays this sequence?
8. What is meant by the description of the continent as the "united plates of America"? What evidence supports the phrase?
9. What evidence suggests that the Earth's crust was much warmer 4 billion years ago than it is today?

Weathering and Soils

Forward-looking planetary scientists are now formulating plans to construct a permanent, inhabited station on the Moon. The self-sustaining station will function as a research center, serve as launching pad to Mars and other planets, and provide rest and recreation to weary astronauts and extraterrestrial real estate salespeople while their space vehicles are being serviced. All this is to take place indoors, in an oasis of greenery, water, and air.

But how do we get the air, the water, and the myriad other substances necessary to maintain an Earth-like colony on an airless, waterless Moon? Geochemists tell us that the answer lies in lunar rocks. They contain all the necessary elements—carbon, hydrogen, oxygen, nitrogen, sulfur, iron, phosphorus, calcium, magnesium, silicon, and so on. Unfortunately, the chemists have not yet developed practical techniques for extracting these elements, which are tightly bound in the crystal structures of lunar rock minerals.

In contrast to the Moon, the Earth possesses an atmosphere, water, and organisms. These agents act to physically disintegrate and chemically decompose the rocks and minerals of the land surface, thus freeing elements and rendering them more mobile. This process, called weathering, is so inconspicuous that its importance is often underestimated. Weathering creates the soil that makes plant life possible, and it releases to the ocean the nutrients necessary for the growth of marine algae. Algae and plants produce oxygen through photosynthesis; thus weathering is indirectly responsible for the air we breathe. Weathering also plays a key role in the carbon cycle—the circulation of carbon dioxide through the solid Earth, the atmosphere, the oceans, and living things (Figure 13.1). As we learned in Chapter 7, carbon dioxide regulates the Earth's temperature, for it is a key component of the greenhouse effect.

◀ Sugarloaf, the famous landmark of Rio de Janeiro, Brazil. The rounded form is the result of weathering, which causes thin concentric shells to peel away from the massive granitic rock.

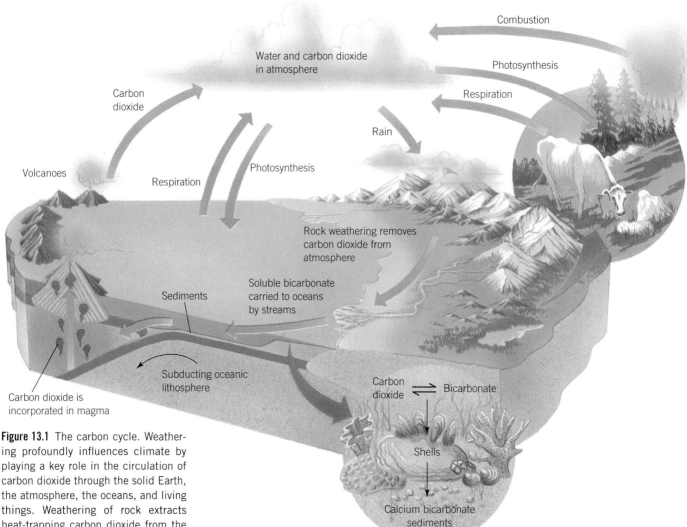

Figure 13.1 The carbon cycle. Weathering profoundly influences climate by playing a key role in the circulation of carbon dioxide through the solid Earth, the atmosphere, the oceans, and living things. Weathering of rock extracts heat-trapping carbon dioxide from the atmosphere and converts it to bicarbonate ions. Transported to the oceans by streams, the bicarbonate ions are temporarily stored in the calcium carbonate shells of corals, clams, microscopic plankton, and other marine organisms. Carbon dioxide is returned to the atmosphere via respiration, combustion, and subduction zone volcanism.

Weathering also provides many of the materials we use in our daily lives: aluminum ore; most iron and uranium ore; clay for pottery and bricks; and sand for porcelain ware and glass. We still coat our walls with the oxidized metal pigments derived from weathering, just as early humans used these pigments to decorate their cave walls—caves weathered from limestone formations. The list of things that weathering accomplishes for us is endless. The pages of this book are coated with a product of weathering—a clay extract, which acts as a preservative. Gold, discovered in the stream at Sutter's Mill in 1848, sparked the rush to California. Thousands of miners panned the rivers and creeks that flowed down from the Sierras to the east, and they combed the rugged hills for the elusive mother lode, the bedrock source from which the gold in the streams had weathered.

Weathering and Gradation

In time the Rockies may tumble,
Gibraltar may crumble,
They're only made of clay,
But our love is here to stay.

—George and Ira Gershwin

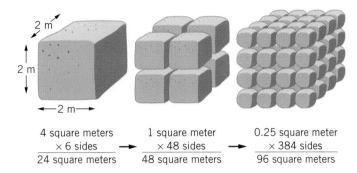

2 m

2 m

2 m

4 square meters
× 6 sides
24 square meters

1 square meter
× 48 sides
48 square meters

0.25 square meter
× 384 sides
96 square meters

Figure 13.2 By breaking a rock into smaller fragments, mechanical weathering increases the surface area that is subjected to chemical attack.

Allowing that George and Ira Gershwin were great songwriters, not geologists, they expressed the essence of an important fact: nothing exposed to the surface environment lasts forever. The Rock of Gibraltar does indeed crumble—although it is made of limestone rather than clay.

Weathering is the first stage in the vast redistribution process of **gradation,** in which erosion wears low the high portions of the land surface, and deposition builds up the low portions. In the course of gradation, the Earth's scenery is rearranged—indeed, to a large extent, created. Tributaries of the Mississippi River carve the valleys of the western slopes of the Appalachians and the eastern slopes of the Rockies. When it reaches the Gulf of Mexico, the Mississippi River constructs its delta out of the sediments excavated from these mountains.

Because most erosion occurs above sea level, and most deposition below, gradation works in the long term toward reducing the land surface of Earth to a featureless sea-level plain, with the planet seeming as monotonous as a billiard ball. Except in localized regions, however, this reduction has never happened, because the Earth is a dynamic planet. Tectonic, isostatic, and volcanic activity constantly uplift the surface and supply fresh material to the crust, so that the gradational agents of wind and water have never run out of work. We will see gradational processes at work in this chapter and in the six chapters that follow.

gradation
The balance between erosion and deposition that maintains a general slope of equilibrium, trending toward sea level.

Weathering Processes

Two major kinds of weathering take place on the Earth's surface. In **mechanical weathering** (disintegration), rock and mineral fragments are reduced to small sizes by purely physical means, without change in composition. In **chemical weathering** (decomposition), chemical reactions fundamentally alter the original composition of the rock.

There is often a reciprocal relationship between mechanical and chemical weathering. Mechanical disintegration accelerates the process of chemical weathering by exposing more surface area of the rock to chemical attack (Figure 13.2). In decomposing minerals, chemical weathering weakens the internal fabric of rock masses, leaving them susceptible to mechanical disintegration.

mechanical weathering
The combination of physical processes that disintegrates a rock without chemical change.

chemical weathering
The surface process that decomposes rocks and minerals through chemical reactions.

Mechanical Weathering

Mechanical weathering is accomplished through a variety of processes, from the freezing and thawing of water to the invasion of plant roots. All of them work to break rocks into smaller pieces.

Frost Wedging

Those of us who live in northern climates are well aware that pipes burst when water freezes in them over the winter. The reason they burst is that water expands about 9

Figure 13.3 These jagged-edged boulders in Finnish Lapland were formed through frost wedging, which shattered solid rock into smaller fragments.

percent upon freezing, which exerts enormous pressure against the pipe walls. Ice plays a similar role in rock weathering in any climate that allows ice to melt and freeze again. Water seeps into rock fractures, expands as it changes to ice, and dislodges rocks. The process is called **frost wedging.** At high elevations in mountainous regions, it can result in an enormous *boulder field* (also called *felsenmeer*, or "sea of rocks"), as shown in Figure 13.3. If ice wedging takes place along cliff faces, the dislodged blocks tumble down to the base of the cliff, where they accumulate. The newly exposed rock of the cliff face is, of course, subject to similar attack; and in this way, the mountain slowly crumbles.

frost wedging
A process by which water seeps into rock joints, freezes, and wedges the rock apart.

Salt Crystallization

Salt crystallization is effective in causing sandstone to disintegrate in arid climates. Fractures and pore spaces allow water to trickle through the rock. As the water evaporates, minute salt crystals are precipitated. As the crystals grow larger over the years, they pry apart sand grains and widen fractures. Eventually, larger fragments and blocks of rock are wedged loose from the cliff. Most of the action takes place at the base of the cliff, where salt tends to accumulate. In time, caves as large as amphitheaters will develop there. Caves formed in this manner provided shelter for the cliff-dwelling Indians of the American Southwest and for peoples of other cultures the world over (Figure 13.4).

Sheeting, Exfoliation, and Spheroidal Weathering

Miners and sandhogs who dig deep tunnels in hard rock do not have the world's safest jobs. Besides worrying about cave-ins, floods, blasting accidents, fires, and ventilation hazards, they must also worry about particularly nasty and unpredictable rock bursts. Rock bursts occur when freshly excavated rock experiences a sudden release of pressure. The rock expands, or "pops," often with a loud noise. Fragments that split along fractures go flying through the air.

In nature, rocks are exposed gradually by erosion and mass wasting; as the underlying rocks slowly expand, the release of pressure causes **sheeting,** or fracturing parallel to the rock surface. In **exfoliation,** a related though not identical process, concentric plates or scales are stripped from the exposed surfaces of large rock masses. The exact mechanisms are not well understood, but they involve physical and chemical processes

sheeting
A type of jointing parallel to the rock surface caused by pressure release. Similar to exfoliation.

exfoliation
The flaking or stripping of a rock body in concentric layers.

Figure 13.4 Ruins of the Anasazi cliff dwellings at Mesa Verde National Park, Colorado, where Native Americans constructed their homes in caves that were weathered from the base of thick sedimentary rock formations. Salt crystallization played a major role in the weathering process.

that widen fractures parallel to the exposed surface. Large exfoliated exposures of massive granite or similar rock form striking, rounded *exfoliation domes.* Stone Mountain, Georgia; Sugarloaf, Rio de Janeiro, Brazil; and the peaks of Yosemite, in the California Sierras, are classic examples of these features (see the photograph of Sugarloaf at the beginning of this chapter).

Spheroidal weathering is the peeling away of shells of decayed rock from isolated unaltered boulders and small rock outcrops. This type of weathering is caused by water that seeps into the fractures of jointed rocks and chemically weathers the minerals along the surfaces. Spheroidal weathering operates on a smaller scale than exfoliation but

spheroidal weathering
The peeling of small rock bodies into onionlike layers.

Figure 13.5 Spheroidal weathering of loose granite boulders in the Algerian Sahara desert.

produces similar rounded onionskin layers of decayed rock (Figure 13.5). Weathering molds rocks into spherical shapes even if they are not subjected to peeling action. The points of rectangular fragments are attacked from three directions at once, whereas the edges are attacked from two and the faces of the rock from one. Thus the points and edges weather more rapidly than the sides. Eventually, a spherical shape is attained.

Weathering Effects of Plants and Animals

Living organisms accelerate rock disintegration. Figure 13.6 shows the roots of a tree wedging between and enlarging cracks in the surrounding rocks and, with the aid of water and dissolved elements, reducing the rocks to fine particles. The root systems of plants and trees also extract nutrients from soil and act as a medium of chemical exchange, a topic we shall discuss later in this chapter. Burrowing organisms churn the soil, which exposes fresh rock and mineral fragments and allows air and water to penetrate to greater depth. Bacteria and fungi cycle elements through the soil and are responsible for the production of acids that are important to chemical weathering. Humans accelerate weathering by digging, blasting, excavating, and, in a variety of ways, uncovering rock and sediments that would otherwise be protected.

Chemical Weathering

The vast majority of chemical-weathering processes involve three major reactions: (1) direct solution, (2) hydrolysis, and (3) oxidation. These are descriptive terms that geologists find useful and are not necessarily the formal definitions used by chemists.

Figure 13.6 An eroded stream bank in the Lost Maples State Natural Area of central Texas exposes tree root systems in the process of disintegrating solid rock.

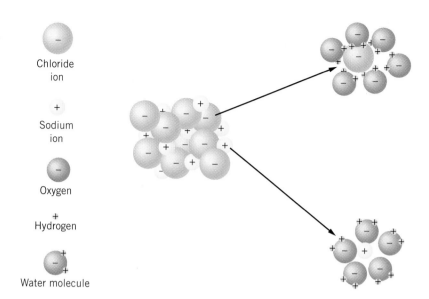

Figure 13.7 The dissolution of halite. The positive hydrogen ends of the water molecules attract the negative chlorine ions and the negative oxygen ends attract the positive sodium ions.

Chloride ion

Sodium ion

Oxygen

Hydrogen

Water molecule

Direct Solution

Direct solution is a reaction whereby mineral and rock materials are chemically attacked and dissolved. In pure water, reactions can be traced directly to the dipolar structure of the water molecule (see Figure 13.7). This structure allows the molecule to act as if it were a minute magnet whose north and south poles attract opposite charges and repel like charges. Figure 13.7 illustrates a well-known example: the dissolving of common salt (halite, NaCl). The positive hydrogen ends of water molecules dislodge the negative chlorine ions (Cl^-), and the negative oxygen ends dislodge the positive sodium ions (Na^+). Subjected to this attack, halite is thus dissolved. Direct solution works best on minerals in which the bonds between ions are relatively weaker than the attraction of these ions for the water dipole.

direct solution
The dissolving of rock or mineral materials.

Hydrolysis

Hydrolysis is an exchange reaction involving minerals and water. Even pure water contains a small but measurable amount of free hydrogen (H^+) and hydroxide (OH)$^-$ ions that are shaken loose through random collisions of water molecules. These ions are potent agents able to replace mineral ions and force them into solution. In the process, the mineral's atomic structure is converted to other forms.

Although hydrolysis works in pure water, the reaction rate is accelerated enormously if other sources of hydrogen ions are also present. Dissolved carbon dioxide (CO_2) supplies these ions. It is present in the atmosphere and soil, being a product of the respiration and decay of animals and plants. The carbon dioxide combines with rain and groundwater to form carbonic acid (H_2CO_3) that instantly separates into positive hydrogen ions and the negative bicarbonate ion (HCO_3^-).

hydrolysis
The reaction between minerals, especially silicates, and water.

$$H_2O + CO_2 \rightarrow H_2CO_3 \rightarrow H^+ + HCO_3^-$$

Hydrolysis involving dissolved carbon dioxide is the fundamental weathering reaction of the silicate minerals, which compose well over 90 percent of the Earth's crust. As an example, consider the hydrolysis of potassium feldspar, a representative of the most common mineral group of the crust.

$$2\ KAlSi_3O_8 + 2(H^+ + 2\ HCO_3^-) + H_2O \rightarrow Al_2Si_2O_5(OH)_4 + [2\ K^+ + 2\ HCO_3^- + 4\ SiO_2]$$

| Potassium feldspar | + | Carbonic acid ions | + | Water | Yields | Kaolinite (clay residue) | + | Potassium ion | + | Bicarbonate ion | + | Silica |

In solution

Table 13.1   The Products of Weathering Reactions Involving Hydrolysis of Silicate Minerals

Mineral	Generalized Composition	Insoluble Weathering Product	Dissolved Ions
Quartz	SiO_2	Quartz	Silica
Feldspars			
Plagioclase	(Na, Ca) Al silicate	Clay [$Al_2Si_2O_5(OH)_4$]	Sodium, calcium, silica, bicarbonate
Potassium	$KAlSi_3O_8$	Clay	Potassium, silica, bicarbonate
Amphibole, pyroxene, olivine	(Fe, Mg) silicate	Limonite ($Fe_2O_3 \cdot H_2O$)	Silica, magnesium, bicarbonate

First, note the overall reaction: potassium feldspar, in the presence of carbonic acid, is converted to an insoluble clay mineral. The reaction involves an exchange: the hydrogen ions (H^+) enter the clay, whereas the potassium ions (K^+) and silica (SiO_2) are released. The clay is incorporated in the soil and sedimentary rocks. The soluble potassium ions either react to form potassium bicarbonate ($KHCO_3$) or are taken up by plants, which use them as nutrients. The dissolved silica will either be carried away by groundwater or will precipitate as chert or flint.

This reaction follows the hydrolysis pattern of all silicate minerals: the mineral is converted to *insoluble residues* and *dissolved ions* (Table 13.1). The composition of these weathering products depends mainly on the composition of the mineral under attack. For example, feldspar contains abundant aluminum, which is not soluble; thus weathering produces clay mineral residue. Olivine, on the other hand, contains no aluminum, and its iron is insoluble under weathering conditions. Consequently, iron compounds, rather than clay, become the insoluble residue. Quartz is another example of special note, because it is chemically and physically resistant. Hydrolysis has little effect on it, except in extremely hot, humid climates and only after thousands of years of exposure.

Oxidation

oxidation
The process by which an element combines with oxygen.

Oxidation is a chemical reaction that results when elements combine with oxygen. Oxidation reactions are probably the form of chemical weathering most familiar to us. They turn wrought iron to orange rust and copper pennies green. Iron that remains in the soil or on exposed rock surfaces also rusts: it combines with oxygen to form the red mineral hematite or with water to form the yellow mineral limonite. Iron oxides and hydroxides such as hematite and limonite are responsible for the bright colors of many soils.

In certain oxygen-poor environments, such as stagnant swamps, iron combines with sulfur to form pyrite. If, at a later time, pyrite is exposed to oxygen, either directly through erosion or by circulating groundwater, the oxygen will replace the sulfur. The released sulfur may combine with hydrogen ions in the water to produce hydrogen sulfide (H_2S) or, if dissolved oxygen is abundant, sulfuric acid (H_2SO_4). Both are potent weathering agents.

Rates of Weathering

Rocks of different compositions and origins weather at different rates. Figure 13.8 lists the common igneous-rock–forming minerals in order of their increasing resistance to weathering. The arrangement closely resembles Bowen's reaction series, which gives the

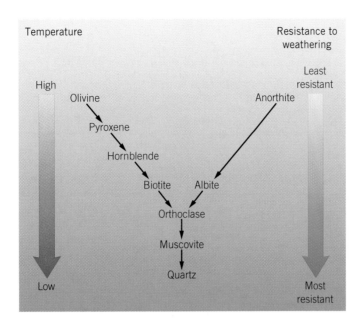

Figure 13.8 Minerals are weathered chemically in reverse order to the sequence of crystallization outlined in Bowen's reaction series (see Chapter 6). Minerals that crystallize at high temperatures are generally the most easily weathered; those that form at low temperatures are the most resistant.

general sequence of mineral crystallization during the cooling of typical basaltic magmas. We see that olivine, the mineral that precipitates from magma at the highest temperature, also weathers the most rapidly. Quartz, which precipitates at the lowest temperature, weathers at the slowest rate. In general, those minerals that form closest to surface conditions are the most resistant.

We can see now why quartz and clay minerals are the most common of the loose materials we usually observe on the beach, in streambeds, and similar places—they are the most stable minerals at surface conditions. Quartz is the chief mineral of sandstone and conglomerate. In combination with clay derived from feldspar, it forms the bulk of the thick sediment deposits found on the Earth's continental margins. Remember that granite is the most common rock of the continental crust, and that it is composed mainly of feldspar and quartz. Most of the carpet of loose materials and sedimentary rocks that cover the Earth's surface is derived directly or indirectly from the weathering of granitic rocks.

The Role of Climate in Weathering

Since this chapter is about weathering, it is appropriate to raise the question, What precisely is weather? It is simply the state of the atmosphere at any time or place. Many factors are involved: temperature, precipitation, humidity, cloud cover, suspended dust, and hours of daylight. These factors affect weathering because they control the delivery of solar energy and water that interact with Earth's surface materials.

Though a region's weather can prove quite variable on a daily basis, it displays a characteristic pattern over long time-periods. This long-term pattern is the region's *climate*. The weather pattern of a given region is long term because it is determined by a number of unchanging or very slowly changing parameters: the distribution of solar heat over Earth's surface, the planetary wind systems, the positions of the continents, the proximity of mountain ranges and ocean currents, altitude, and vegetation. Figure 13.9 illustrates the major climate zones.

There is an important relationship between climate and weathering. To establish it, we need a means of readily comparing the two. We may begin by noting that each of the world's climate regions displays a characteristic range of precipitation and temperature. The range of each climate can be shown graphically, as in Figure 13.10(a). A given climate's annual range of precipitation is plotted on the vertical axis and its temperature range on the horizontal axis. Our next step is to use the temperature-precipitation

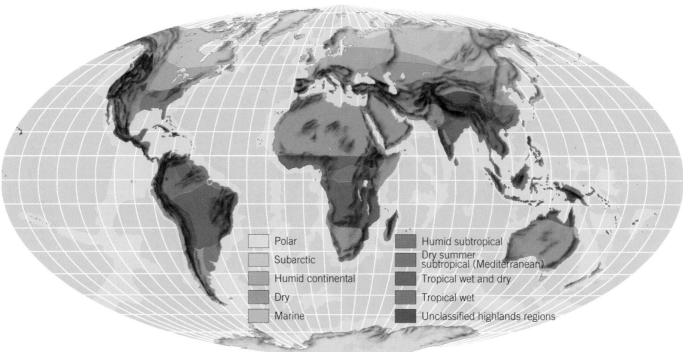

Figure 13.9 The major climatic zones of the world. (From Robert A. Muller and Theodore M. Oberlander, *Physical Geology Today*, 3d ed., New York: McGraw-Hill, 1984.)

graph of Figure 13.10(a) to depict the conditions under which the various kinds of mechanical and chemical weathering are most effective (Figure 13.10b).

The striking correlation between the two graphs dramatically illustrates the control that climate exerts on weathering. Chemical weathering is most intense in humid tropical environments, because reactions leading to mineral decomposition are rapid in the

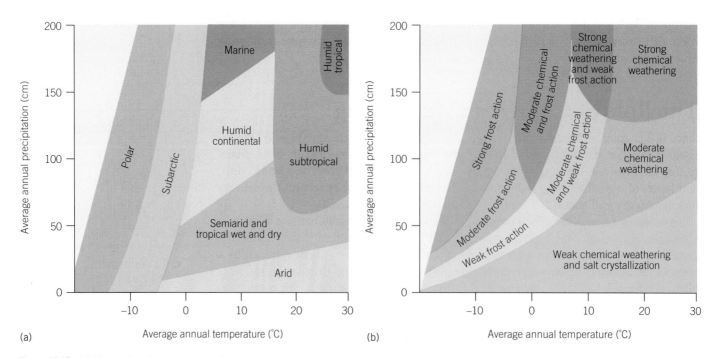

Figure 13.10 (a) The major climatic zones of the world exhibit characteristic ranges of temperatures and precipitation. (b) The temperature and precipitation ranges of the major weathering processes. The striking similarity of the two graphs illustrates the control that climate exerts over weathering. (From Robert A. Muller and Theodore M. Oberlander, *Physical Geology Today*, 3d ed., New York: McGraw-Hill, 1984.)

presence of abundant water and high temperatures. For this reason, soils are extremely thick in the humid tropics, where intense chemical weathering has proceeded without interruption for many thousands, perhaps millions, of years. In some regions, the soil thickness exceeds 100 meters.

If elevated temperatures and high rainfall accelerate chemical weathering in the tropics, then the depressed temperatures and low rainfall of polar and subarctic climates slow reaction rates enormously (Figure 13.11). Weathering rates are similarly slow in hot, arid climates, for there is simply not enough water available to generate intensive chemical weathering. Perhaps this has as much to do with the preservation of Egyptian mummies as the skills of ancient embalmers. However, it does not necessarily follow that because chemical weathering is weak in a particular climate, its effects are insignificant. Given sufficient time, it partially decomposes large quantities of feldspar and other silicate minerals and destroys rock fabric. Keep in mind that rainfall and temperature vary with altitude as well as with latitude. Snowfields and glaciers cap the peaks of the Andes Mountains along the equator. At these low latitudes, frost wedging provides most of the mechanically weathered debris carried by the Amazon to the Atlantic Ocean and is contributing heavily to the destruction of this South American mountain chain.

Differential Weathering and Climate

Differential weathering refers to rocks that weather at different rates in the same climate. Graveyards are excellent places to compare the weathering rates of various kinds of rocks. The tombstones are dated, and we need only observe the condition of rocks of the same age. A walk through a New England cemetery, or a cemetery in any other suitably humid climate, is especially revealing (Figure 13.12). Limestone and marble tombstones weather most rapidly because of their susceptibility to the dissolved acids in rainwater. The inscriptions of stones whose age is greater than about two hundred years are usually indecipherable. By contrast, slate tombstones remain clear and readable. Granite tombstones hold up longer than limestone, but they lose their gloss within a century. The feldspars appear dull and discolored, and the rock surfaces are somewhat pitted, perhaps due to the dissolution of more soluble dark minerals like biotite or hornblende. But the blocks are otherwise in good shape.

The long-term effects of differential weathering and erosion may be observed in Vermont. Here the major north-south valleys are underlain by limestone, marble, or other easily weathered rock. The rugged mountains that border these valleys are made of granite, gneiss, or similarly tough silicate igneous or metamorphic rock. The same pattern is maintained throughout the length of the Appalachian Mountain belt. For example, the Great Valley of the Appalachians, a striking feature that stretches from Alabama to Vermont, is underlain by soluble limestone approximately 500 million years old. Extending along the length of the valley on the eastern border are the resistant metamorphic and igneous rocks of the Blue Ridge and similar ranges. To the west, the Great Valley is hemmed in by hard, well-cemented sandstone ridges. Thus differential weathering provided a natural north-south transportation route. Some of the most important battles of the Civil War were fought for control of this valley.

Limestone and marble tombstones of the arid Southwest are in strikingly better condition than those of the same age in the rainy East. Again, climate is the key influence. The sparse moisture inhibits chemical weathering, allowing the mechanical properties

Figure 13.11 That chemical weathering rates slow significantly in a cold, dry climate is graphically illustrated by the well-preserved corpse of Chief Petty Officer John Torrington, buried on Beechy Island, Canada, far above the Arctic Circle. Torrington was 20 years old when he died, probably of pneumonia, in January 1846. His body was exhumed in 1984 by anthropologists in an effort to reconstruct the history of the polar expedition that cost him his life.

(a)

(b)

Figure 13.12 Differential weathering of gravestones, displayed in a Long Island, New York, cemetery. The lettering is more decipherable on (a) the slate stone, dated 1701, than on (b) the limestone marker, dated 1816.

of the rock to determine how rapidly it will disintegrate. Because limestone blocks are relatively compact and largely homogeneous, they offer greater resistance to temperature change, salt crystallization, and similar mechanical processes than many sandstones and shales. This resistance translates to sheer, massive limestone cliffs in arid regions, just about the opposite of the retiring aspect that limestone assumes in humid climates. Figure 13.13 shows the Redwall limestone of the Grand Canyon. The gentler slopes above and below the prominent Redwall are softer, more easily weathered shales. In fact, the entire canyon wall, with its differing slope angles closely keyed to various rock types, is a lesson in the effects of differential weathering in arid climates.

Soil Formation

The loose rock material that nearly everywhere forms the land surface and covers the bedrock of the crust is called **regolith,** from the Greek *regos* ("blanket") and *lithos* ("stone"). Except for minor amounts of volcanic ash and infinitesimally small quantities of meteoric dust, all regolith is derived from the weathering of bedrock. **Soil** is regolith that has incorporated organic matter and is capable of supporting vegetation.

The soil of any given locality is a dynamic body. Its condition results from the interplay of climate, rock material, groundwater, organic activity, topography, and time. It is no wonder, then, that soils are difficult to study and classify. However, the overall patterns of soil development are more easily understood.

The condition of a typical soil is determined by additions, subtractions, rearrangements, and recyclings of materials. Materials are added to the soil by weathering that reduces bedrock to clay, silt, and sand and that releases soluble elements for uptake by plant roots. Materials are subtracted from and rearranged within the soil body by rainwater, which percolates through the soil on its way to the water table. **Leaching** is the process by which soluble elements are transported downward in solution from the upper levels. Some are redeposited at depth along with clay and silt carried in suspension. Materials are also recycled and retained within the soil by complex plant, animal, and bacterial communities that live within and above the soil. The net result of their activities is called *nutrient cycling.*

regolith
All the fragmented and unconsolidated material overlying bedrock.

soil
Unconsolidated material capable of supporting vegetation.

leaching
The transfer in solution of organic matter and other elements from upper to lower soil levels.

The Soil Profile

The downward percolation of water through soil produces a characteristic layering. Each layer is called a **soil horizon,** and the totality of these horizons constitutes the **soil profile** (Figure 13.14).

Just below the vegetative mat of partially decomposed plant matter, called the **O horizon,** is the first soil layer, called the **A horizon.** This is the zone from which positive ions are leached and clays are removed. Because of the leaching, the layer tends to be porous with a sandy or silty texture. At the surface, its color is dark because it mixes with partially decomposed organic matter; lower down, its color generally lightens. The

soil horizon
A soil layer with physical and chemical properties that differ from those of adjacent soil layers.

soil profile
A vertical cross section of soil that displays all soil horizons.

O horizon
The top layer of soil consisting of decayed plant matter.

A horizon
The uppermost zone of the soil horizon.

Figure 13.14 A typical soil profile for temperate climates. The various horizons are the products of weathering, leaching, and redeposition by percolating groundwater.

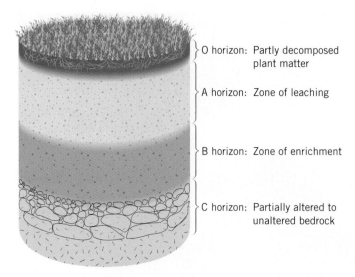

O horizon: Partly decomposed plant matter

A horizon: Zone of leaching

B horizon: Zone of enrichment

C horizon: Partially altered to unaltered bedrock

B horizon

The zone of the soil profile consisting of enriched clays and precipitates leached from the A horizon.

C horizon

The layer of weathered bedrock at the base of the soil horizon.

B horizon marks the enriched zone in which the positive ions that are leached from the A horizon are precipitated, and the clays carried in suspension are deposited. The accumulation of clay may produce a dense structure that is hard, physically tough, and difficult to plow. "Hardpan" is the common name given to this layer. The B horizon is characteristically richly colored. If calcium accumulates, as in desert climates, it is white. If iron accumulates, as in humid climates, it is dark red or orange.

A zone of partially weathered bedrock, the **C horizon,** lies beneath this enriched layer. It is formed by direct reaction between the downward-percolating groundwater and the rock. The C horizon grades into fresh bedrock below and closely resembles it, but appearances are deceiving: the feldspars have been weathered to clay, and the "rock" crumbles easily. Drillers aptly describe this zone as "rotten rock." As long as conditions are unchanged, the A, B, and C horizons will continue to thicken at the expense of the bedrock beneath.

Nutrient Cycling

Leaching impoverishes the A and O horizons. Thus, over the long term, it would inhibit plant growth were it not for a countervailing process that replenishes the soil: nutrient cycling. Nutrients extracted from the soil are returned to it when plants die or shed leaves. Leaves and other plant matter are broken down by bacteria, which produce humus, a black to dark brown organic substance. Humus mixes with clay to provide numerous bonding sites for nutrients such as potassium, sodium, magnesium, iron, and calcium ions. Because the nutrients are weakly held, plant roots can easily extract them. Thus soils with rich clay-humus complexes are generally highly productive, whereas those lacking sufficient humus tend to be infertile. There is a practical application here for gardeners who bag and discard fallen leaves and lawn cuttings. This practice interferes with the nutrient cycle by not allowing the leaves to naturally decompose to soil nutrients—driving the gardeners to spend money on fertilizers to replace them.

A given plant species may cycle some nutrients and exclude others. For this reason, agricultural systems—especially those that rely heavily on a single crop not native to the region's soil—often impoverish the soil for other species, including other crops, by altering the nutrient balance.

Climate and Soil Types

The two major soil-forming processes, leaching and nutrient cycling, depend upon the availability of water in the form of precipitation and energy in the form of solar heat.

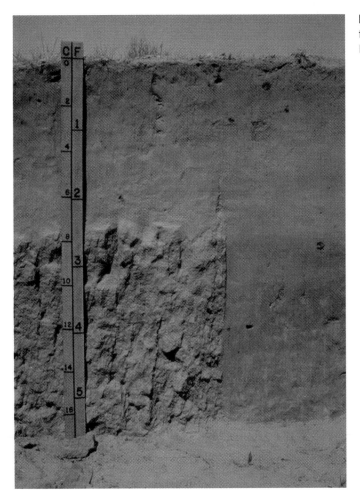

Figure 13.15 A typical laterite soil. Note the deep red color and thickness of the B horizon.

Since precipitation and solar heat are the key elements of climate, most soils, given sufficient time, reflect regional climate. As climate varies, so do soil types.

Soils of Tropical, Humid Climates

Brick red **laterite** (from the Latin *later,* meaning "brick" or "tile") soils develop in hot, humid climates that support rain forest or savanna vegetation (Figure 13.15). The drenching rains and high temperatures of these climates promote extensive soil leaching. Calcium, sodium, potassium, and magnesium ions are leached from the soil by the percolating rainwater and carried to the water table at greater depths. Large quantities of silica are also carried away. Much of the silica is derived from the decomposition of clay minerals, something that occurs only under extreme conditions of rainfall and heat. Left behind in the leached soil are oxides of iron and aluminum. The iron oxides produce the red color of laterite soils.

In some regions, these residual aluminum and iron oxides accumulate in commercially valuable concentrations. Bauxite, the chief aluminum ore, forms in this manner. Huge reserves of it are found in the deep laterite soils of Jamaica, among other places (Figure 13.16). The iron oxide ores of western Australia have formed in similar fashion. They constitute one of the world's richest iron deposits. Both ores are outstanding examples of how weathering makes things easy for humans by concentrating minerals through leaching and deposition. The metals have been freed from tightly bound silicate crystal structures, making them economical to extract.

Thick laterite soils support the magnificent rain forests of the Amazon, Central America, equatorial Africa, and Southeast Asia. Some of the world's poorest and most densely populated nations are also located in or near these rain forests. It is one of

laterite
A highly weathered red soil typical of tropical and semitropical climates, enriched in iron and aluminum oxides.

Figure 13.16 This bauxite, an aluminum ore, is the product of prolonged and intensive soil leaching in a humid tropical or subtropical climate.

calcification
The encrusting of a soil with alkaline compounds, particularly calcium oxides.

caliche
Desert sand or soil cemented with porous calcium carbonate.

nature's cruel ironies that these soils, capable of spawning luxuriant plant growth, will not generally support sustained agriculture or ranching; they are among the world's poorest soils for these purposes. High rainfall leaches the nutrients out of the A horizon. In addition, those nutrients that remain in the soil are snatched up rapidly by the fast-growing plants. We can readily predict the results of aggressive human intrusion into this environment. Clearing the land, planting crops, and allowing cattle to graze further impoverishes the soil, because it destroys the vegetative cover and short-circuits the recycling of nutrients in the soil by plant decay. It also accelerates leaching, erosion, and the development of hard laterite crust in the B horizon.

Soils of Arid and Semiarid Climates

Soils formed in arid and semiarid climates are as different from laterites as deserts are from tropical rain forests. Whereas laterites are residues, the products of extensive leaching, arid soils result from insufficient leaching, so that they retain highly soluble nutrients that would be carried away under more humid conditions.

Two factors work in arid climates to keep nutrients from escaping the soil: **calcification** and natural vegetation. Calcification is the process of enriching the soil in highly alkaline elements, especially calcium compounds. It works in the following manner: downward-percolating rainwater is evaporated before it can pass through the soil; indeed, it may actually be drawn back up toward the surface by capillary action. Left behind in the B horizon are very soluble ions that are normally carried away in rainier regions: sodium, potassium, magnesium, and calcium. These ions impart a light, chalky color and texture to the B horizon. Under extremely arid conditions, a hard calcium carbonate crust called **caliche** forms in the B horizon (Figure 13.17). Due to evaporation and capillary action, these highly soluble ions are plentiful in the A horizon as well.

The plants of semiarid climates, such as prairie grasses, must be tough to survive. They have dense root networks that enable them to withstand high winds and to soak up soil water before it trickles down or evaporates. The organic activity and decay associated with the roots produce clay-humus complexes, which retain nutrients close to the surface.

Arid conditions may therefore favor the development of potentially fertile soils, because some of them contain large stores of nutrients and humus close to the surface in the A horizon. All that is needed is sufficient rainfall to support vegetation and leach

Figure 13.17 A thick caliche cap on Mormon Mesa in southern Nevada.

away some of the alkaline elements. The proper blend of ingredients occurs in midcontinent regions of middle latitudes, where extremely arid conditions give way to a somewhat wetter climate capable of supporting grassland vegetation. These areas are the great grain-producing breadbasket regions of the world: the North American Great Plains, the Ukraine, Asia's Manchurian plains, and South America's pampas (Figure 13.18). Yet the balance between factors that make them productive is subtle. For example, fertile soils also exist in the dry fringes of these regions, and they are a great temptation to farm. The problem is that the rainfall trails off, making these dry areas prone to periodic long-term droughts. Overall, farming in the fringe areas has proven to be a losing proposition, both economically and in human terms.

Salinization is another problem in lands irrigated for agriculture in arid climates. The irrigation water soaks into the ground and raises the level of underground water until it seeps to the surface and evaporates, leaving behind a white, powdery salt residue (Figure 13.19). Little of value can grow in soil like this; the salty soil kills plants and soil bacteria alike. Agricultural scientists estimate that each year, salinization ruins hundreds of thousands of acres the world over (Figure 13.20). For example, salinization threatens major portions of California's San Joaquin Valley, which produces billions of dollars worth of crops annually. In the past, salinization has literally destroyed entire civilizations; 4000-year-old records from Mesopotamia (modern Iraq), for instance, describe abandoned farmland and mass migration as the land salted over.

Figure 13.18 A thick soil profile in highly productive farmland, North Dakota. The dark color indicates that the soil is rich in organic matter.

Soils of Cold to Temperate Climates

The soils of cold, subarctic climates support the great belt of pine forest that stretches across northern Canada and Europe. These pine forests have shaped the properties of the soils that support them by affecting the chemistry of the plentiful rainwater that filters through the fallen pine needles matting the forest floor. The water that filters through the acidic pine needles also becomes acidic. This water, in turn, leaches from the A horizon not only such soluble elements as sodium, calcium, potassium, and magnesium but also normally insoluble aluminum and iron oxides. The oxides are derived from the clay-humus complexes, which are decomposed by the acidic water. Left behind in the A horizon is a pale, ash-colored, silica-rich soil that is poor in nutrients. The pines have adapted by cycling few nutrients through the soil.

However, the clays, oxides, and organic matter leached from the A horizon are redeposited in the B horizon. They color that layer yellow, red, or black, which gives the entire profile a distinctive color banding. The soil-forming process we have described is called **podzolization.**

North of the pine forests, moisture and precipitation diminish and weathering is slow. Podzol soils grade into the poorly developed permafrost soils of the Arctic tundra. The ground surface is frozen much of the year, and nutrient cycles are greatly retarded. Only mosses, lichens, low shrubs, and other tenacious species can survive in this harsh environment.

To the south of the pine forest region, however, conditions are less restrictive. The vegetation changes to the mixed-forest type; trees and brush are more luxuriant; and roots sink deeper into the soil in search of greater quantities of nutrients. Nutrient cycling is more intense, and in general, the soils are richer than those farther north. Much of the mixed-forest soils of the mid-Atlantic states has been converted to highly productive farmland.

podzolization
The process of nutrient leaching of the A horizon that produces the ashy gray podzol soils of coniferous forests in cool, moist climates.

Apply and Decide 13.1

Acid Rain

An acid is a substance that releases hydrogen ions (H+) in solution. So the greater the concentration of hydrogen ions in solution, the greater the strength of the acid as measured on the logarithmic pH scale. The term acid rain is used to denote acid-polluted rain, but this term is a bit misleading, because natural rainwater is also somewhat acidic, having a pH of about 5.5, whereas the pH of pure, acid-free water is 7. The acidity of natural rainwater is derived from dissolved carbon dioxide, which combines with water to form carbonic acid. The rainwater that is destroying statuary and causing other environmental problems, however, has an acidity from 10 to 1000 times greater than this (Figure 13A). The high acid content of the polluted rainwater is derived from coal-burning electric utilities, other heavy industries, and automobiles. They all emit sulfur dioxide and nitrogen dioxide, which, when combined with water vapor in the atmosphere, form sulfuric and nitric acids. These acids are then precipitated downwind from their sources. Mountainous topography intensifies the problem. Winds that travel from the sources of the pollution are trapped by the mountains, concentrating the acid rain on windward slopes and valleys.

Highly acidic rain poses its greatest threats to lakes and forests. It damages foliage by direct contact and by leaching metal ions held in the clay-humus complexes of the soil and replacing them with hydrogen ions. The replacement starves the forest of important plant nutrients. To make matters worse, the acidic water kills the bacteria and fungi that recycle dead plant and organic matter back into nutrients. Consequently, the regenerative capacity of the soil may be severely damaged.

Through surface flow and groundwater seepage, acid rain eventually finds its way into lakes, lowering their pH. In addition, the water carries with it highly toxic aluminum ions leached from clay, where they are immobile under normal conditions. The combined acidic and toxic effects have a disastrous impact on insect, fish, and amphibian larvae and on the animals that feed on them. The aquatic life of entire lakes has been destroyed, leaving behind crystal-clear liquid deserts.

One way to attack the effects of acid rain is to neutralize the affected soil and water with buffers. Many of us employ the same tactic when we take an antacid to soak up hydrogen ions released by stomach acids or treat yellowed lawns with calcium carbonate (lime). So one solution is to add buffers to forests and lakes to restore their natural pH balance. Though this approach has been effective locally, it is obviously impractical, for we are dealing with a problem of continental proportions. Quebec's $24-billion forest industry is under threat, as are the Great Smoky Mountains and the

Figure 13A The damage of acid rain to statuary, and ultimately to our cultural heritage, is illustrated by the corrosion of this figure on the front of a church in Surrey, England.

Adirondack forest (Figure 13B). Over four thousand Swedish lakes are devoid of life, and almost half of the country's forests have been severely damaged. Yet these are but a few examples of the magnitude of the events set in motion by acid rain. As with most pollution problems, the only permanent solution lies at the source: steps *must* be taken to lower harmful emissions.

Figure 13B Acid fog, rising from the valleys to the west, has destroyed these trees high in the Great Smoky Mountains of Tennessee.

Finding a Solution

There are three possible approaches to lowering harmful sulfur emissions; each has drawbacks as well as benefits. One possible solution is to require power plants to install filtering and scrubbing devices in their smokestacks. However, these devices are expensive and their costs would be passed along as increases in local utility rates.

Another solution would be to require power plants to use low sulfur coal. In the United States, the source of this coal is the near-surface deposits of the Powder River basin of the Dakotas and Montana. However, low sulfur coal is extracted by strip mining, a technique that devastates extensive land areas and leads to ecological disruption and the threat of groundwater pollution.

A third possible solution is to replace fossil-fuel-burning plants with non-sulfur-emitting nuclear power plants. Nuclear power is inexpensive, clean, and plentiful. However its radioactive wastes are lethal and remain so for thousands of years. As we will see in Chapter 16, the questions of where and how to store nuclear wastes are major problems that have yet to be solved. (See Apply and Decide 16.1, Hazardous-Waste Disposal.)

THOUGHT QUESTIONS

1. How might you go about calculating the economic and environmental trade-offs involved in solving the acid rain problem? How might such costs compare to the long-term costs of leaving the problem unsolved?

2. Most acid rain is produced by point sources—that is, coal burning power plants. However, the pollutants that cause acid rain are spread by the winds to other states and even adjoining nations. What mechanisms might be available to the recipients of acid rain for reconciling their grievances?

Figure 13.19 Salt-ravaged agricultural land northwest of Victoria, Australia.

Other Soil-Forming Factors

Climate is the broad framework within which the soils of a region develop. In addition, there are local factors that influence soil formation, including the parent material of the soil, regional topography, organisms, and time.

Parent Material

The parent material from which the soil is derived determines the original composition of the soil. For example, limestone and basalt produce fertile soils in a humid climate because they contain large percentages of soluble ions that serve as nutrients for plants. Basalt and shaley limestone also weather to clay, which can aid the formation of clay-humus complexes. Quartz-rich sandstones, by contrast, contain a high percentage of materials that do not decompose easily and yield few soluble ions (for example, not much sodium or potassium). They produce acidic soils that can support only limited vegetation, such as pine forest.

Topography

A rule of thumb in agricultural nations is that the higher and steeper the slope upon which the farm is located, the poorer the farmer. Soils generally thin in the upslope direction, as do crop yields. The reason soils thin upslope is that water and weathered materials migrate downslope and accumulate at the base of the hill or in the valley. Thus soils are commonly thicker there. They are also more fertile, because the moister soils facilitate chemical weathering and nutrient cycling.

Another example of the importance of topography is shown in Figure 13.21. A strip of fertile land extends along the base of the San Andreas fault scarp. The relatively recent faulting disrupted the flow of underground water, which issues to the surface as springs. The steady flow reduces the severity of evaporation in this otherwise desert environment and modifies the soil.

San Joaquin Valley

Areas prone to salinization

Figure 13.20 The regions prone to salinization in the western United States include some of our most productive farmland.

Figure 13.21 This strip of vegetation follows the location of springs along a fault at the base of the San Andreas fault, Indio Hills, Imperial Valley, California.

Organisms

Burrowing animals, in addition to plants and bacteria, play an important role in weathering, soil formation, and nutrient cycling. By mixing, churning, and loosening soil, they allow for easy passage of soil water and nutrients to plant roots. Their churning motion also brings oxygen to the soil. Their respiration, digestion, and decay produces organic compounds, carbon dioxide, and other materials. These factors affect the composition of soil water, and, therefore, reactions involving rock and soil. Interestingly, the last book Charles Darwin wrote did not directly concern evolution but was devoted to earthworms! The creatures fascinated Darwin, and he concluded that their ceaseless activity was largely responsible for the rich soil of much of Great Britain.

Time

No bedrock can yield soil if there is not enough time for weathering, leaching, and nutrient cycling to occur. Simply contrast the deep, rich soil of ancient Hawaiian lavas with the thin or nonexistent soils of recent flows. On the other hand, some soils are productive merely because they have not had sufficient time to be leached. This is true of many river flood plains that support crops or other vegetation on soil not intrinsically rich in nutrients.

Just how rapidly does soil form? Again, this question is difficult to answer because it depends on climate, site, parent material, and organisms. Volcanic ash or sediment deposited in a humid, tropical climate is already partially decomposed, and thus much of its surface area is vulnerable to attack. It weathers rapidly, producing as much as 30 centimeters of soil per 50 to 100 years. In contrast, weathering of freshly exposed bedrock may take 10,000 years to produce the same amount of soil.

STUDY OUTLINE

I. **THE GEOLOGIC IMPORTANCE OF WEATHERING.**   Weathering is the first stage in the vast redistribution process of **gradation,** in which erosion wears low the high portions of the land surface and deposition builds up the low portions.

II. WEATHERING PROCESSES

A. Mechanical weathering reduces the size of rock and mineral fragments without changing their composition.

1. Water may seep into rock cracks and pores and then freeze. As the ice expands, it pries the rock apart in a process called **frost wedging.**

2. In arid climates, expanding salt crystals, precipitated from evaporating water in cracks and pores, pry the rocks apart.

3. In **sheeting,** caused by pressure release, and **exfoliation,** caused by physical and chemical processes, large rock exposures fracture and peel in layers. In **spheroidal weathering,** rocks and boulders peel in rounded, very thin layers.

4. Plant roots also penetrate rock cracks and pry them apart. Burrowing animals expose rock surfaces to the weather.

B. Chemical weathering increases the efficiency of mechanical weathering by weakening chemical bonds that hold rocks together.

1. In **direct solution,** the positive and negative poles of water molecules dislodge negative and positive ions in a mineral, thus dissolving it.

2. In **hydrolysis,** free hydrogen and hydroxide ions in water replace ions of silicate minerals and convert them to insoluble residues and dissolved ions.

3. **Oxidation** (commonly known as rusting) results when elements (such as iron) combine with oxygen.

III. THE ROLE OF CLIMATE IN WEATHERING

A. A region's climate is its long-term weather pattern. Climate is determined by the distribution of solar heat, winds, the positions of the continents, the proximity of mountain ranges and ocean currents, elevation, and vegetation.

B. Chemical weathering is most intense in regions that have abundant water and high temperatures. However, rocks of different composition weather at different rates under the same climatic conditions.

IV. SOIL FORMATION

A. Weathering disintegrates and decomposes rocks and minerals, thus rendering elements mobile. The loose rock material is called **regolith. Soil** is regolith that has incorporated organic matter and is capable of supporting vegetation. Thus weathering is responsible for the soil that makes plant life possible.

B. Soil formation involves two main steps.

1. Soluble elements needed by plants are released by the chemical weathering of the clays, silts, and sands that were mechanically weathered from bedrock.

2. Soil materials are rearranged by rainwater that percolates through the soil and leaches soluble elements down through the soil levels.

C. Leaching produces soil layering, or the **soil profile.** Each layer is called a **soil horizon.** The top mat of decomposed vegetation is the **O horizon.** The first soil layer from which nutrients are leached is the **A horizon.** The next layer, the **B horizon,** receives the nutrients. The **C horizon** is a layer of partially weathered bedrock beneath the B horizon.

D. Nutrient cycling through the soil is accomplished by complex interactions involving plants, animals, and bacterial communities in the soil.

E. Soil types vary with climate.

1. **Laterite** soils develop in tropical, humid climates. They support tropical rain forests in humid climates but are ill suited for agriculture.

2. Because of minimal leaching by rainwater, soils of arid and semiarid climates retain nutrients through the process of **calcification.** A hard calcium carbonate crust called **caliche** may form in the B horizon. Although arid-climate soils are fertile, salinization can ruin soils that are overirrigated.

3. Pine forests are well adapted to the nutrient-poor, acidic subarctic soils. Leaching causes colored banding of the soil profile in a process called **podzolization.**
4. Arctic soils are called permafrost because the ground is frozen nearly all year.
5. Soils of temperate-climate regions are fertile because of intense nutrient cycling.

F. Rain with a much higher than normal acidity, called acid rain, is destroying forests and lakes. The main producers of acid rain are electric utilities, heavy industry, and automobiles, whose sulfur dioxide and nitrogen dioxide emissions combine with water in the atmosphere to form sulfuric and nitric acids.

G. Other soil-forming factors
1. Soil composition depends on the composition of the parent materials from which the soil components were weathered. Topography also affects soils; in general, the higher the elevation, the thinner the soil.
2. Plants, animals, and bacteria are important in soil formation. The process takes time, possibly thousands of years to produce soil from bedrock.

STUDY TERMS

A horizon (p. 358)
B horizon (p. 358)
calcification (p. 360)
caliche (p. 360)
chemical weathering (p. 347)
C horizon (p. 358)
direct solution (p. 351)
exfoliation (p. 348)
frost wedging (p. 348)
gradation (p.347)
hydrolysis (p. 351)
laterite (p. 359)

leaching (p. 356)
mechanical weathering (p. 347)
O horizon (p. 357)
oxidation (p. 352)
podzolization (p. 361)
regolith (p. 356)
sheeting (p. 348)
soil (p. 356)
soil horizon (p. 357)
soil profile (p. 357)
spheroidal weathering (p. 349)

CRITICAL THINKING QUESTIONS

1. What is the role of gradation in the growth of continents, as described in Chapter 12?
2. How does mechanical weathering aid chemical weathering? How does chemical weathering aid mechanical weathering?
3. Why is it not surprising that the order of chemical weathering in silicate rocks is the reverse of Bowen's reaction series?
4. Discuss the reason for the correlation of climate with weathering patterns. Describe the kinds of weathering prevalent in the various climates.
5. Describe the development of a typical soil profile in a temperate climate.
6. Describe how leaching and nutrient cycling keep a soil in balance.
7. Why are tropical rain forest soils generally poor agricultural soils?
8. Contrast the soils of humid and tropical, temperate, and arid climates.
9. What factors other than climate affect soil?

C HAPTER

14

Mass Wasting

- An entire Canadian mountainside suddenly crashes down on a village nestled at its protective base. The event takes less than 3 minutes, and virtually no one gets out alive.

- A fragment works loose from the face of a sheer rock cliff in the Nevada desert. Ringing sharply, it skips downhill to the base of the cliff, where it joins several million similarly displaced relatives. Together, they form an apron 100 meters thick and 1 kilometer wide.

- The trees, the fences, and all the headstones in a New England graveyard tilt downslope. The soil carpet is clearly on the move, but at a stately rate of a millimeter per year.

This chapter is about how soil, boulders, and even large rock masses move from the top to the bottom of hills. It is about landslides, rockfalls, mud- and earthflows, slumps, soil creep, and similar occurrences. More formally, it is about **mass wasting,** the downhill movement of the Earth's materials under the direct influence of gravity.

Recall from Chapter 13 that weathering is the first step in the process of gradation in which erosion wears down the high portions of the land surface and deposition builds up the low portions. Mass wasting is also an integral part of the process of gradation, along with that most important gradational agent, streams (which we will come to in the next chapter). Weathering, mass wasting, and streams are as closely linked to one another as the men who quarry rock are to those who truck it away.

Figure 14.1 illustrates that weathering, mass wasting, and slope erosion supply most of the sediment that streams carry away. Conversely, by downcutting deeply into the land, streams create the sloping surfaces that make mass wasting possible. Other gradational agents also contribute to mass wasting by creating conditions of instability on slopes. Glaciers, groundwater, breaking waves, and even the wind all work to undercut and oversteepen slopes.

mass wasting
The downslope movement of rocks and soil caused by gravity alone, without the aid of a transport medium such as a stream, glacier, or lava flow.

◀ Unrelenting hillside creep is buckling this stone wall on a Berkeley, California, street. Although most mass wasting proceeds so slowly that it is scarcely noticeable in a short time frame, its cumulative effects are impressive.

Figure 14.1 The Yellowstone River occupies a small portion of the V-shaped valley it formed by downcutting into the land surface. The sloping valley walls are widening through mass wasting, which supplies most of the sediment that the stream transports.

Why do some materials cling to hillside slopes and others slide downhill? Why would a mountainside suddenly collapse? We begin this chapter by exploring some of the physical factors that affect materials perched on inclined surfaces in order to better understand the basic principles that underlie all forms of mass wasting. Next, we will describe the different forms that mass wasting may take, and finally, we will examine the role of human beings as both victims and initiators of mass wasting.

The Balance of Forces on a Slope

To understand why some rocks slide downhill and others remain stationary, place a palm-sized rock fragment on a sloping board. Tilt the board steeply so that the rock slides freely (Figure 14.2a). Now, let us analyze the various forces (pushes and pulls) on the rock. Foremost is the force of the Earth's gravitational attraction, known as the rock's weight. Its direction is always vertical (toward the center of the Earth), and that is how the rock would fall if not constrained by the board. Because of this constraint, the rock is instead pulled downward by the component of the weight that acts parallel to the slope. We will call this component the *driving force;* it is fundamental to all mass movement.

Next, put the rock back on the board and slowly lower the angle of slope until the rock is delicately balanced but remains stationary (Figure 14.2b). Since the effect of the driving force is to have the rock slide down the slope, there must be other forces acting on the rock in the opposite direction to restrain it. We will call these *resisting forces.* When the rock is delicately balanced, driving and resisting forces are equal; that is, the rock is in *equilibrium.* Any action that tips this equilibrium in favor of the driving force will cause the rock to move downslope.

The major resisting force is *friction,* the force generated when the rock rubs against other surfaces (in this case, the inclined board). The type of surface influences the amount of friction: the rock would slide down a smooth, oiled board at a lower angle than it does down a board with a roughened, sandpaper surface. Figure 14.2 also shows that tilting the board at a higher angle lessens the component of weight pressing the rock to the slope, thereby lowering friction; it also increases the driving force, which pulls the rock down the slope. Therefore, landslides are common on very steep slopes.

Shear Strength and Slope Stability

When oppositely directed forces act parallel to a plane running through an object, we say that the object is being subjected to *shear stress.* The shear stress exerted on a rock or any object perched on a slope tends to deform it in a plane parallel to the slope. We have employed the concept of shear stress when considering transform plate boundaries (Chapter 2), earthquakes (Chapter 3), and rock deformation (Chapter 11).

The resistance that a body offers to deforming forces is called its **shear strength.** The shear strength of a solid rock involves the forces between its atoms, whereas the strength of an aggregate of loose solids—the gravel, sand, clay, and rock fragments on

shear strength
The internal resistance of a body to shear stress, due to particle friction and cohesion, and moisture surface tension in pore spaces.

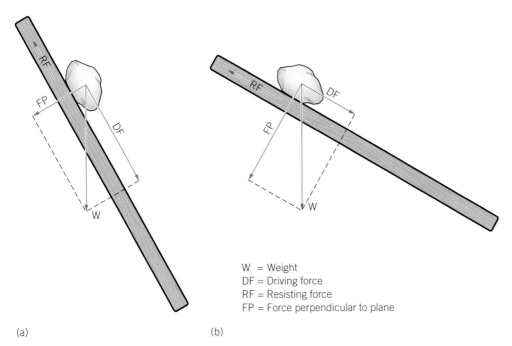

Figure 14.2 The forces acting on a rock placed on an inclined surface. The steeper the incline, the greater the component of the weight that acts parallel to the slope (the driving force). The boulder is restrained from sliding by resisting forces (that is, friction). (a) If the driving force is greater than the resisting force, the rock moves downslope. (b) When driving and resisting forces are equal, the rock is considered to be in a state of equilibrium and remains stationary.

W = Weight
DF = Driving force
RF = Resisting force
FP = Force perpendicular to plane

(a)   (b)

hillside slopes—is derived from the frictional forces between fragments and the cohesive forces between clay particles.

The origin of the cohesive forces between clay particles is electrostatic, similar to what causes clothes to cling after they come out of the dryer. The surfaces of the clothes are oppositely charged, which is what attracts them to one another. Because of their sheetlike crystal structure, clay minerals have relatively large surface areas upon which oppositely charged ions collect. The opposing charges attract other clay and silt particles, and large masses form, which resist deformation. Most movement is the result of *shear failure* along planes of weakness in sloping surfaces—that is, along fault planes, bedding planes, and joint sets.

Angle of Repose

So far, we have described the forces that act upon a single rock on a slope. When an entire hillside consists of unconsolidated materials in equilibrium, we say that the slope has assumed a natural **angle of repose.** The angle of repose reflects the natural balance between driving and resisting forces. That such balance exists is demonstrated in Figure 14.3. Notice that the pile of sand forms a cone with a maximum slope of about 30°. This is the angle of repose of the sand-sized fragments, and this slope angle is maintained tenaciously. Sand added to the top of the cone or removed from the base oversteepens the pile and causes the sand to slump or trickle downhill until the 30° angle is restored. If we try lowering the slope by subtracting sand from the top of the cone or adding it to the base, the sand that we then drop on the pile accumulates in such a manner that the slope angle of the pile is returned to 30°.

The angle of repose of natural slopes depends upon the size of the individual particles, their angularity, the degree of sorting, and whether they are wet or dry. Larger, more angular particles, such as gravel, tend to form slopes with steeper angles of repose than those of finer particles. Most natural slopes vary between 25° and 40° in steepness.

angle of repose
The maximum slope, or angle, at which a pile of loose material remains stable.

Slope Stability and the Role of Water

Small amounts of water can increase shear strength through surface tension. This effect occurs when water soaks into loose sediment and only partially fills the space between

Figure 14.3 A pile of loose material exhibits a characteristic angle of repose that depends on particle size, angularity, degree of sorting, cohesiveness, and other factors. A 30° angle of repose is typical of sand-sized particles.

the grains. The water molecules on the surfaces within the spaces form a membrane or web that clings to the grains and binds the loose sediment into a cohesive mass. Surface tension can create remarkably strong structures. Think of a sand castle on the beach. The damp sand forms almost vertical walls. When the water between sand grains evaporates, though, the dry walls crumble easily.

On the other hand, large amounts of water, from rain or a rising tide, may soak the sand and saturate the spaces between the grains. This destroys surface tension and transforms the sand-water mixture to mud. So it is on hillsides. Changes in surface tension both from overdrying and from oversoaking facilitate downslope movement.

Water affects slope stability in other ways. Rain or groundwater seeping into loose materials adds weight to a slope, which can itself trigger downhill movement. When water seeps between fracture and bedding planes, it exerts an outward pressure that lowers the net force acting perpendicular to the slope; this causes the overlying materials to slide along these planes. Most landslides are triggered by an abnormal increase in water pressure, by either natural or human-made means.

Water also affects the stability of hillside slopes by changing the properties of the materials that compose the slopes. We saw that a little added water makes sand more cohesive, but large quantities of water transform damp sand into mud. However, water's effect on clay is even more dramatic. Wet and dry clay have markedly different mechanical properties. Dry clay is hard and difficult to deform. By contrast, wet clay is soft, slick, and moldable, because the clay minerals absorb large quantities of water. Not only does this absorption greatly increase the weight of materials on a slope, but it also causes swelling that reduces the electrostatic attraction between clay minerals. Resisting forces are therefore lowered in wet clay, and easy slippage is the result. Try climbing a wet clay embankment!

Saturated wet clay and sand are vulnerable to liquefaction, a quicksand state in which particles lose cohesiveness and behave as a viscous mass, incapable of supporting loads. Heavy rains can trigger liquefaction. So, too, can earthquakes, which may send groundwater bubbling to the surface. As we mentioned in Chapter 3, this is a major cause of building collapse in low-lying, earthquake-prone regions reclaimed from bays, such as the Marina District of San Francisco.

Types of Mass Wasting

Mass movements of materials downhill take four general forms: fall, slide, flow, and creep. Because materials often exhibit more than one type of movement in their journey downslope, it is not always easy to classify a specific event.

Falls

fall

The free downward movement of detached rock fragments through the air from a cliff or steep slope.

Falls result when rock fragments are dislodged from a steep cliff face and fall unrestrained or when the face is undermined, causing it to collapse in a heap along vertical joints.

In Chapter 13, we discussed many of the mechanical- and chemical-weathering effects that can dislodge rock materials at high elevations: frost wedging, exfoliation, and salt crystal growth. Generally, these effects produce a steady, though scarcely noticeable, cascade of fragments. Occasionally, however, weathering produces sudden, dramatic rockfalls. Figure 14.4 shows the results of collapse along the Palisades sill of New Jersey, which runs along the west bank of the Hudson River. As you may recall from Chapter 6, the steep sill is composed of resistant diabase and is subdivided into massive columns by vertical joints. The joints are responsible for the steep, imposing appearance of the cliff. Because the thin zone of olivine-rich rock at the base of the sill weathers more easily than the rest of the sill, it leaves the huge vertical columns unsupported. As the joints

Figure 14.4 Rockfall boulders stacked at the base of the sheer cliff of the Palisades, New Jersey. The cliff is divided by vertical joints or fractures into massive columns. Not visible in the photo is an easily weathered layer at the base of the cliff (the olivine zone). Erosion of this layer deprives the columns of support, which leads to their collapse.

widen, friction is diminished, and the rock columns collapse under their own weight. In this manner, the Palisades are being worn steadily back from the Hudson.

The fragments dislodged in falls collect at the base of the cliff, forming wedge-shaped accumulations called **talus** slopes (Figure 14.5). The angular fragments of the talus slopes come to rest at steep angles of repose (about 35°), which are maintained as fresh talus is added to the slopes. Given sufficient time, the cliffs can choke in their own debris. More typically, however, fragments at the base of talus slopes are removed through stream erosion, which allows the slow downhill movement of fragments from the head to the base of the slopes. Thus, the form of talus slopes may remain constant, but over long time periods, the fragments that compose them change continuously.

talus
Coarse, angular rock fragments that collect at the base of the cliffs from which they were dislodged.

Figure 14.5 Talus slopes at the base of the Crowfoot Glacier in Canada. The slopes are composed of fragments weathered from the cliff face. The fragments were dislodged from loosely cemented sedimentary rocks, which accounts for their relatively small size. Contrast their size with the massive boulders at the base of the Palisades in Figure 14.4, which were derived from resistant igneous rock.

Figure 14.6 Two kinds of slides. (a) A rockslide involves the movement of large blocks of bedrock along a shear plane. (b) A landslide involves more or less unconsolidated rock and soil, although the term is often used in a more general sense to include many types of mass wasting.

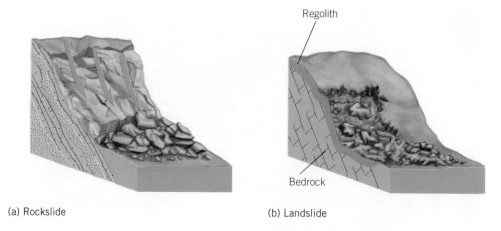

(a) Rockslide

(b) Landslide

Slides

slide
The rapid downslope movement along shear planes of more or less consolidated rock or fragmented materials.

rockslide
The rapid downslope movement along shear planes of a mass consisting mostly of large chunks of bedrock.

landslide
The rapid downslope movement along shear planes of a mass of rock fragments and soil.

The term **slide** generally refers to the mass movement of rock and soil downhill along a shear plane (Figure 14.6). More specifically, a **rockslide** is the movement of blocks of bedrock, and a **landslide** is the movement of a more or less consolidated mass of soil and rock fragments.

Rockslides and landslides may occur quite suddenly, although close inspection—usually after the fact—reveals plenty of warning signs. On October 1, 1963, a shepherd noticed that his flock of sheep refused to graze on the north slope of Mount Toc in the Italian Alps. Eight days later, 300 million cubic meters of rock and soil slid down the north slope at more than 90 kilometers per hour and plunged into the Vaiont Reservoir, which runs along the axis of a narrow valley (Figure 14.7). A chunk of mountainside nearly 600 meters high and 2.4 kilometers wide had collapsed and filled the reservoir with debris 150 meters above its former level. To cart the load away, a typical pickup truck would have had to make 500 million trips!

The landslide sent a huge blast of compressed air, water, and rock up the valley walls; the blast was strong enough to lift the roof off a house 250 meters above the reservoir. Like an enormous plunger, the slide also sent a wall of water 100 meters high over the top of the Vaiont Dam. There, it poured as a 65-meter wave to the Piave River 1.5 kilometers below the dam. The surging waters spread upstream and downstream along the Piave River for many kilometers and flooded the populated valley on either side. The lives of 2600 people were lost in an event that took about 7 minutes from start to finish.

The Vaiont Dam was newly constructed at the time and one of the world's highest, at slightly more than 265 meters. It sustained no significant damage, despite the tremendous stresses placed upon it—which is ironic, since it was the construction of the dam that set the disaster in motion. First, the dam had been built in an area of unfavorable geologic structure. The bedding planes of the limestone and shale formations dipped steeply toward the valley of the reservoir, as did prominent fault planes within these rocks. The limestone and shale strata, weakened by fracturing and weathering, were subject to shear failure.

Second, the groundwater level had been raised by impounding the water behind the dam. The water infiltrated the bedding planes, fractures, and pore spaces of the rock formations, further lowering the shear strength of the rock. Conditions had worsened in the weeks preceding the disaster. Drenching rains saturated the soil and seeped into the bedrock, thus adding weight to the slope. In the three years prior to the heavy rains, the rate of soil creep along the north slope of Mount Toc had been a few centimeters per week, but it jumped to 25 centimeters per day in late September. The day before the slide, the creep rate had reached 1 meter per day as the mountainside began to slide along the shear planes.

(a)

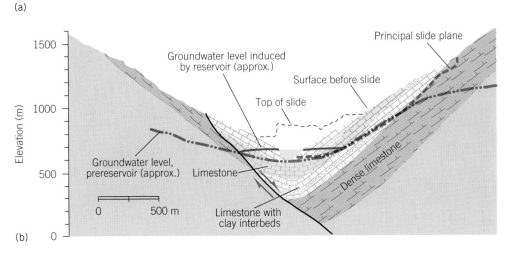

(b)

Figure 14.7 (a) This photo, taken shortly after the Vaiont dam landslide disaster, shows the enormous quantity of debris that filled the reservoir. The debris acted as a plunger, forcing a wall of water to spill over the dam and flood the communities downstream. The crest of the dam, visible in the photo, stands approximately 265 meters above the valley floor. (b) This geological cross section of the reservoir behind the Vaiont dam illustrates some of the conditions that led to the disaster: shear planes dipping steeply toward the valley; massive limestone formations interbedded with clay layers that served as weak support; and a rising water table that lowered friction along the slide surfaces.

These events were not lost on the engineers at the site. They suspected that disaster was imminent and sought to release water via the spillway of the dam. But the water released was more than offset by the rainwater washing down from the slopes. The Vaiont Reservoir and Dam had turned out to be a magnificently designed deathtrap.

Conditions Favoring Slides

The conditions that led to the Vaiont disaster are similar to those of most other slides. Analysis of the four well-studied slides illustrated in Figures 14.8 through 14.11 showed that each of them shared three characteristics of the Vaiont disaster.

Figure 14.8 (a) Turtle Mountain, site of the massive landslide that buried the town of Frank, Alberta, in 1903. The slide occurred along fractures that paralleled the steep cliff face on the right. (b) Regional cross section prior to the slide. Coal-mining operations in the valley undermined the support of the massive limestone formations of the mountain and widened the fractures that eventually led to the catastrophe.

(a)

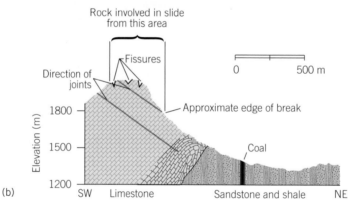

(b)

1. *Steep slopes and adverse geologic structures.* All the slides studied involved either bedding or fracture planes that dipped steeply downslope. This orientation provided shear planes for rocks to slide over. For example, observe in Figure 14.8 the steep dip of the fracture planes in the massive limestone cliff above the coal-mining village of Frank, Alberta—a catastrophe just waiting to happen.

2. *Weak layers that support heavy loads.* Shaley layers lying beneath thick strata or loose soil played a prominent role in the Gros Ventre (Figure 14.9), Palos Verdes (Figure 14.10), and Vaiont slides. Shale, especially when wet, lacks the shear strength of sandstone or granite. Like greased skids, shaley layers provide surfaces for massive rocks to slide over.

3. *A triggering mechanism.* Slides are initiated by natural processes and/or human activities: earthquakes, oversteepened or undercut slopes, waterlogged rock and soil, and altered groundwater levels.

Slides are triggered by natural factors and/or human activities. The 1903 Frank, Alberta, slide was set in motion by the mining of the steeply dipping coal seam in the valley. The excavations undermined support of the massive cliff, causing adjustments that ultimately set the slide in motion along fracture planes. The village and its inhabitants were buried in about 3 minutes. The Vaiont disaster also was caused by human

(a)

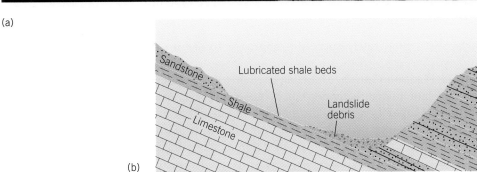

(b)

Figure 14.9 (a) Aftermath of the Gros Ventre slide. (b) The slide involved massive sandstone layers sliding over weak shale layers. The stage was set by the undercutting of the steep slope by the Gros Ventre River.

activity, construction of the dam, which was responsible for the raised groundwater levels that lowered the resistance of the mountainside strata to shear stress. Natural factors, in the form of drenching rains, provided the triggering mechanism.

Water pressure also played a key role in the Gros Ventre, Wyoming, slide of 1925. Water lubricated the underlying shales, allowing the massive sandstone layers above to slide downslope. However, the stage for the rockslide was set long before the event. As the Gros Ventre River cut through the sandstone layer at the base of the valley, it deprived the upslope sandstone of support. Only friction between shale and sandstone prevented movement. When friction was reduced by heavy rains, approximately 37 million cubic meters of mountainside collapsed, raced downslope at about 100 kilometers per hour, and smothered everything in its path.

Unlike the sudden catastrophes we have discussed so far, the Portuguese Bend slide of Palos Verdes, California, is a long-term, slowly moving event that was first noticed in the mid-1950s and continues to this day. Natural causes, aggravated by human activities, have set the slide in motion; over the years, rates of movement have ranged from about 0.5 to 2.5 centimeters per day.

Notice in Figure 14.10 that the rock formations of the Palos Verdes region dip gently toward the coast, where wave action has cut through the Portuguese formation, a cemented volcanic tuff. The formation rests on weak shales that provide a slide plane for the overlying rocks and other materials on the slope. Although the slide is accelerated by heavy rains, the increased load of housing tracts and septic tanks has added to slope instability. Pumping water out of the ground to prevent lubrication of the slide

(a)

Figure 14.10 (a) The Palos Verde region is an example of a complex, slow-moving slide acting over a large region. (b) Wave erosion is undercutting the Portuguese formation, which is sliding toward the sea along its contact with the underlying shale.

(b)

surface is a moderately successful approach to stabilizing the slope that has been employed recently.

The Madison Canyon, Montana, slide of 1959 (Figure 14.11) was triggered by neither rain nor human activities but by an earthquake. Prior to the earthquake, thick, steeply dipping dolomite strata supported the weight of even thicker, highly weathered metamorphic rocks and soils. In effect, the dolomite acted as a natural retaining wall. The earthquake shook loose blocks of dolomite along fractures high on the valley wall, causing the mountainside to collapse. The debris dammed the Madison River.

(a)

Figure 14.11 (a) Aftermath of the Madison Canyon slide of August 17, 1959. (b) Cross section showing the profile of the mountainside before and after the slide.

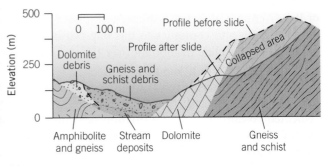

(b)

Slumps

Not all slides involve the downslope movement of bedrock along bedding planes or pre-existing fractures. A **slump** is a type of slide that occurs when intact blocks of rock or unconsolidated debris slide downward along concave planes. The blocks also tend to tilt backward during slump, so that rotation as well as downward motion takes place. Commonly, slumps occur on hills thickly blanketed with soil or on steep shoreline cliffs made of loose, poorly cemented materials.

The reason the glide planes of slumps are curved is not precisely known, but as with all slides, they are surfaces of shear failure. As in all mass movements, slumps are most common on slopes that are oversteepened, water-saturated, and undercut by streams, waves, or human activities. Earthquakes also set slumps in motion. Most of the damage that Anchorage, Alaska, sustained during the earthquake that nearly destroyed the city in 1964 has been attributed to slumps and other mass-wasting effects (Figure 14.12).

Precarious subsurface conditions exist beneath Anchorage. The city was built on an 18-meter layer of gravel supported by the thick Bootlegger Cove Clay. This water-soaked clay formation liquefied as the ground shook during the earthquake, a situation intensified by water pouring from broken water mains and pipes. The relatively stable gravel deposit and stiff upper portion of the clay slid over the liquefied clay beneath, toward Cook Inlet to the southwest. In the center of the city, the motion caused huge fault blocks to sink along fractures. At Turnagain Heights, the high bluff overlooking Cook Inlet, the ground split as the heights slid toward the inlet.

The cracked ground formed huge slump blocks that followed curved paths downslope (Figure 14.12a–c). Note the steep, curving *heel* of the slump and the *toe* of slope material that extends outward at the base. This arrangement reflects the balance of forces that brought the block to rest as it slid downward and rotated along the curved glide planes. Should the toe of the slump be removed—by natural erosion or by a poor road cut—the instability will trigger further slumpage.

Turnagain Heights was a fashionable suburb, but the slump blocks did not respect property; houses were severed and splintered. The panicked inhabitants were carried down in the dark on the slump blocks together with their shattered houses. It was sheer luck that few people were injured.

Flows

A **flow** is a water-soaked mass of loose rock, soil, and sediment that tends to behave as a viscous fluid; that is, it more or less sticks together as it moves downhill. However, the characteristics of a given flow depend on its water content, its particle size, the degree of particle sorting, the angle of the hillside slope, and the width of the flow channel. All these features make flows difficult to classify, but there are three main types: earthflows, mudflows, and debris flows.

Earthflows commonly consist of water-saturated clay, shaley fragments, soil, and similar materials that slide downhill as a viscous mass. During downslope movement, an earthflow may be confined to a channel of some sort. Once it reaches the valley floor, it spreads more like stiff molasses than muddy water. For this reason, earthflows end in distinctly curved boundaries or lobes.

Mudflows are less viscous and consist of finer-grained materials than earthflows. Containing up to 60 percent water by weight, these dense, well-lubricated masses can move very rapidly. As you may recall from Chapter 7, snowcapped volcanic peaks are particularly susceptible to a type of mudflow called a lahar. The peaks frequently contain thick deposits of pyroclastic debris set at a high angle of repose on steep slopes. During a volcanic eruption, the snow around the crater melts and begins to flow quickly downslope. At the same time, the shaking of the mountain loosens debris, which mixes with the flowing water. Huge volumes of mud may run down the flanks of the volcano at high speeds and smother all below it.

---

slump
Downslope movement of rock or loose debris along a concave plane.

flow
Downslope movement of loose rock and soil as a viscous fluid mass.

earthflow
Downslope movement of a loose mass consisting mainly of rock fragments and soil in a semifluid state.

mudflow
Downslope movement of a fluidized mass of clay and other fine-grained materials.

Figure 14.12 (a–c) Stages in the development of slump blocks in Turnagain Heights. The slumping was set in motion by the Anchorage, Alaska, earthquake of 1964. Notice the steep scarp, the curved slide plane, the heel and toe of the slide. All are characteristic of slump structures. (d) An aerial photograph of Turnagain Heights, showing the complex, curving fractures at the top of the scarp.

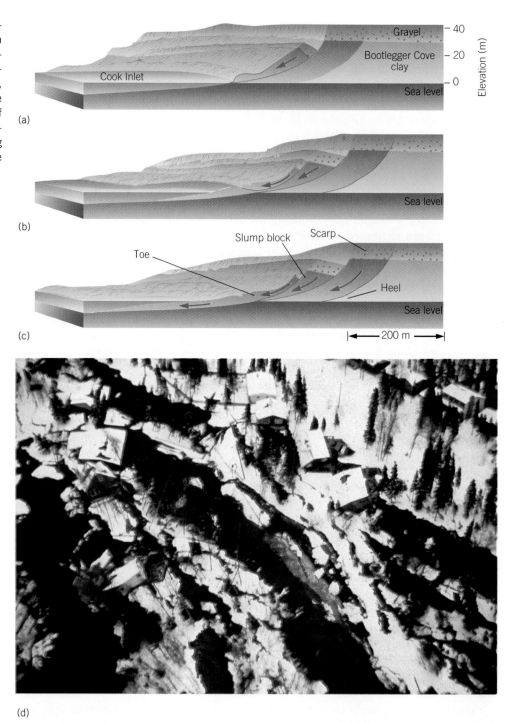

(a)

(b)

(c)

(d)

In fact, much of the damage and death in the Mount Saint Helens, Washington, eruption of 1980 was caused by lahars rather than by fresh lava flows. Also, the lahars triggered by the eruption of Mount Pinatubo in 1991 devastated huge portions of the island of Luzon in the Philippines. Green farmland and thriving villages were smothered in hot mud, which cooled to an ugly gray mass, as hard as cement. A far greater disaster, in human terms, occurred on November 13, 1985, when a lahar descended the 5400-meter slope of Nevado del Ruiz, Colombia (Figure 14.13). Channeled within the valley of the Lagunilla River, it was 15 meters high and traveling 70 kilometers per hour by the time it reached the town of Armero, 48 kilometers from the summit. The inhabitants did not stand a chance. Approximately 25,000 people were buried alive in the mud.

Figure 14.13 In 1985, this mudflow (or lahar) poured down the flanks of the Nevada del Ruiz in Colombia at 70 kilometers per hour. The town of Armero and its 25,000 inhabitants were buried in 15 meters of mud.

Debris flows are defined as moving masses of rock and soil in which more than half the materials are coarser than sand size. Depending upon the viscosity of the mass, debris flows can vary in speed between 1 meter per year to hundreds of kilometers per hour. In 1970, Yungay, Peru, bore the brunt of what is probably this century's most disastrous debris flow. Figure 14.14 illustrates the chain of events set in motion by a large

debris flow
Downslope movement of a viscous mass of rock and soil particles, more than half of which are larger than a sand grain.

Figure 14.14 The regional setting of the Yungay, Peru, debris flow of 1970. The scar in the face of Nevada de Huascarán at the right marks the site where a huge block of ice was dislodged from the mountain by a large earthquake. Upon impact, the snow and ice melted, thus setting in motion the debris flow that buried Yungay and neighboring towns along the Rio Santa at the base of the mountains.

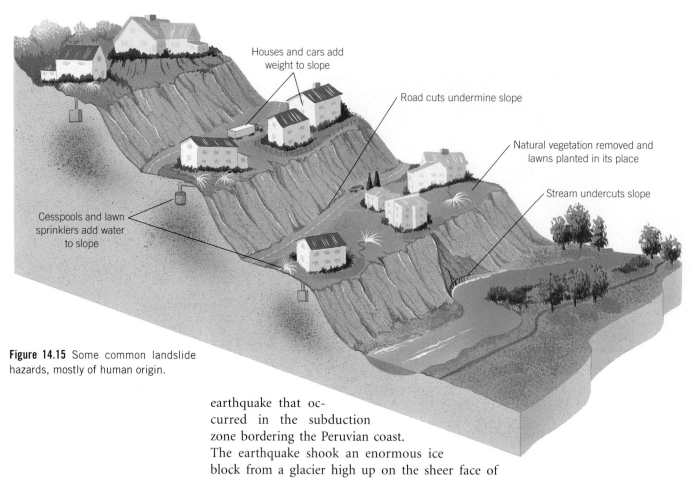

Figure 14.15 Some common landslide hazards, mostly of human origin.

Labels in figure:
- Houses and cars add weight to slope
- Road cuts undermine slope
- Natural vegetation removed and lawns planted in its place
- Stream undercuts slope
- Cesspools and lawn sprinklers add water to slope

earthquake that occurred in the subduction zone bordering the Peruvian coast. The earthquake shook an enormous ice block from a glacier high up on the sheer face of 6540-meter-high Nevado de Huascarán—a mountain that looms above the valley where the city is located. The ice plunged 1000 meters before striking the ground. The collision set in motion 80 million cubic meters of rock, melted snow, and mud that roared down the the valley of Rio Santa at 280 to 335 kilometers per hour. Well over 20,000 people—including the entire population of Yungay and the inhabitants of innumerable tiny villages—were killed.

Slumps, earthflows, mudflows, and their coarse-loaded cousins, debris flows, are common occurrences in arid climates. Lacking vegetation, the soil of hillside slopes in these climates is quickly saturated when beset by sudden, torrential rains. Of course, humans often unknowingly use their ingenuity to create these very same conditions in nonarid climates. People strip hillside slopes naked through deforestation, careless forest fires, or poor farming practices. They may then compound the problem by building houses and planting grass on the slopes, and watering the grass for hours on end with sprinklers. At some later date, they may express genuine surprise and consternation when the hillside runs away with their homes, their lawns, and all their possessions. Figure 14.15 shows some all-too-common ways that human activities can turn a stable slope into an unstable one.

Creep

creep

The imperceptibly slow downslope movement of soil and rock particles, mainly occurring in cold climates where water alternately freezes and thaws.

Creep is so slow that it can be measured in centimeters by the downward migration of fence posts, tombstones, and trees over decades or centuries (Figure 14.16). Yet on a worldwide basis, more material moves to the bottom of hills by creep than by any other form of mass wasting. The secret to understanding the enormous impact of creep is to grasp the cumulative impact of gravity acting over long time-periods. Whether it is the size of a sand grain or a boulder, a dislodged rock will roll downhill, not up. A soft,

Figure 14.16 Creep in soil and bedrock.

semifluid mass of soil will likewise sag downhill. Thus, nearly every event that loosens fragments or softens clay, silt, and sand contributes to downslope movement: the impact of grazing and burrowing animals, the splash of raindrops, the saturation effects of groundwater.

Creep processes are common at high latitudes and high elevations where water alternately freezes and thaws. As water seeps into the ground and freezes, its expansion lifts rock and soil at right angles to the sloping hillside surface. However, when the ice melts, materials are dropped vertically under direct influence of gravity. The result is a net downhill movement of the surface, scarcely noticeable but cumulatively important. The alternate wetting and drying of clay minerals, leading to expansion and contraction, produces similar downslope movements in all climates.

The freezing and thawing of water produces a number of striking landforms and other effects in cold climates. Most are associated with **permafrost,** a condition that develops when all the water within the pore spaces in regolith and rock remains frozen solid throughout the year. Permafrost covers 20 percent of the land area of the Earth, including 80 percent of Alaska. The frozen ground can reach depths greater than 600 meters in Alaska and Siberia. The thin soil blanket above the permafrost thaws each summer and becomes saturated, because the frozen ground beneath will not allow water to seep deeper into the regolith. This process is responsible for the frequently boglike, marshy conditions of arctic tundra landscape. Saturation is also responsible for **solifluction,** a form of creep in which the waterlogged soil slips slowly downhill and, like ripples in a rug, forms wavy *solifluction lobes* (Figure 14.17). Although the lobes give the appearance of great activity, solifluction is a slow process.

Permafrost presents difficult environmental and engineering problems that depend on soil type, regional drainage, local climate, and type of construction planned. At the root of most of these problems are the human-made disruptions in the natural regime that cause the subsurface ice to melt: construction that adds weight to the soil; the clearing of natural vegetation that alters the amount of solar heat absorbed by the soil; rumbling heavy machinery that compresses and disrupts the subsurface; and subsidiary roads that disturb the natural surface drainage. If a summer is unusually warm, the melting permafrost can cause extensive damage. The softened ground leads to land

permafrost
A condition of permanently frozen soil or subsoil occurring in cold climates.

solifluction
The slow downslope movement of waterlogged soil caused by repeated freezing and thawing in cold climates.

ASIDE 14.1 *Mass Wasting on the Ocean Floor*

Mass wasting does not stop at sea level. For example, rocky debris is spread over thousands of kilometers at the base of the Hawaiian Islands. Many geologists think that the debris is indicative of an ongoing process of island collapse. The relatively young volcanic islands are shaken by eruptions and subsurface adjustments. Fault and fracture planes abound, and waves undercut the island slopes. This is an ideal environment for landslides. As mentioned in Chapter 7, the southeastern coast of Hawaii is particularly vulnerable.

Marine geologists have detected many of the same features present on land at the base of the continental slopes: mudflows, debris flows, slumps, and slides. The most likely places to find them are similar to the types of locations where they are found on land: near oversteepened slopes, often in zones of active volcanism and tectonism where earthquakes set them in motion.

The greater our knowledge of submarine topogra-phy, the more we learn about submarine mass wasting. GLORIA (geologic long-range inclined asdic), a side-scanning sonar device, has proven to be a breakthrough in this regard. Figure 14A is a remarkable image that GLORIA made of the Bering Sea floor off the coast of Alaska. It is dominated by the 144-kilometer-long Zumchug Canyon. Geologists who have analyzed this region believe that the large block at the base of the canyon was dislodged from the cliff in the background, initiating the canyon erosion.

When GLORIA is towed from a ship, it emits a beam of sound waves that spreads over a 60-kilometer-wide track of sea floor (Figure 14B). Upon striking the sea floor, the sound waves are absorbed and reflected at varying intensities that depend upon the composition of the sea bottom. This information is processed by the computer to yield three-dimensional images. Hard rock surfaces that reflect a greater proportion of the wave energy show up as bright patterns, whereas the more absorbent soft sediments produce darker patterns.

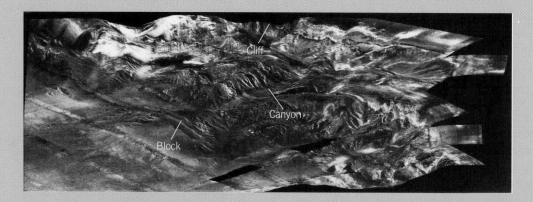

Figure 14A A mosaic of GLORIA images of the Bering Sea floor, showing rugged topography in the vicinity of the 145-kilometer Zumchug Canyon. The source of the huge block at the mouth of the canyon was probably the cliff in the background. The dislodgement and subsequent motion of the block may have led to erosion of the canyon.

subsidence, settling of structures, lateral slippage of load supports, landslides, and flows. When the ground refreezes in the winter, it triggers massive frost heaving, which, in turn, causes further stress on structures.

The ecology of arctic tundra has come under increasing stress as oil and other resources have been discovered near and within the Arctic Circle and as populations have spread north. Though the survival problems of tropical rain forests have received a great deal of attention, comparatively little has been given to the perils facing the starkly beautiful tundra landscape, underlain by permafrost.

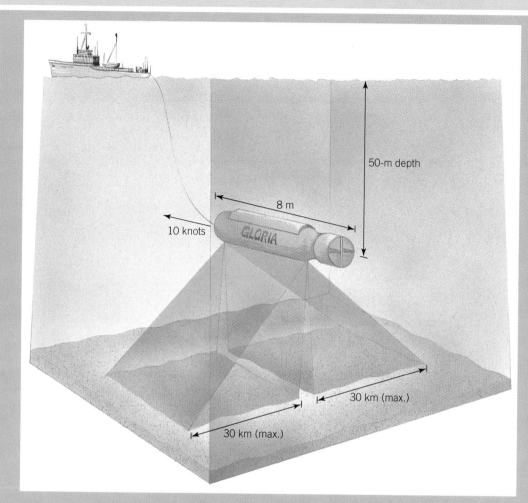

Figure 14B GLORIA, an acoustical scanning device, is used to construct three-dimensional images of the sea floor.

Predicting and Preventing Mass-Wasting Disasters

Most of this chapter has been devoted to describing how soil, rock fragments, and boulders move from the top to the bottom of hills. It is reasonable that we should now discuss ways to prevent them from burying us. The superficially simple solution—get out of the way—is not always possible, as the examples cited in this chapter and elsewhere illustrate. Assuming that it is possible to escape with one's skin intact, there is the further

Figure 14.17 Solifluction lobes of the Alaskan tundra.

problem of preventing or minimizing property loss to roads, utilities, buildings, and dams. These structures cannot get out of the way. We also have to contend with the slower, more subtle forms of mass wasting, such as soil creep. While not immediate threats, they often spell disaster on the installment plan—the ruination of farms and other valuable property.

Recognition of Hazards

Recognition of unfavorable sites begins with a detailed geological survey of the region. The objective is to prepare a landslide hazard map like the one shown in Figure 14.18. Frequently, maps of this kind serve as a basis for the enactment of zoning laws that halt construction in dangerous areas, institute stricter building codes, and regulate land use activities. At the least, the maps identify the risks of owning a home in landslide-prone regions.

In constructing a geological survey of the region, geologists search for all the conditions discussed in this chapter.

1. *The relationship of bedrock structure to topography.* Do the rock strata dip toward the sea, where wave erosion is active? Do they dip toward steep valleys, where streams undercut slopes?
2. *The condition of the bedrock.* Do massive rock and soil formations overlie strata of low shear strength? Is the rock highly fractured?
3. *The composition and steepness of hillside slopes.* Is the hill mostly bedrock? Or loose material? If the latter, what is the percentage of clay, sand, and boulders? Are these materials close to their angle of repose, or are the hillsides over-steepened? What is the shear strength of the materials on the slope? How will they be affected by heavy rains? Are the slopes vegetated or bare? Has there been evidence of recent slippage?
4. *The ground water conditions that underlie the slopes.* How do such structures as reservoirs or factories affect them?
5. *The tectonic history of the region.* What is the likelihood that earthquakes of a given magnitude will occur? How are the rocks and soil at specific locations likely to respond to severe ground shaking?

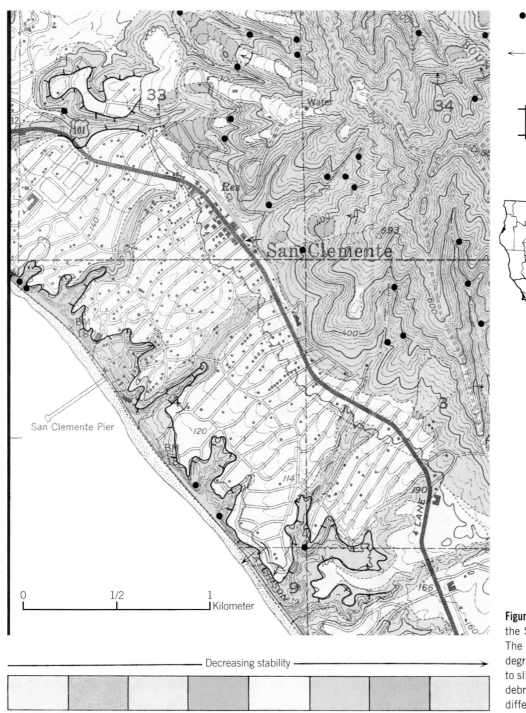

● Locations subject to sliding

← Relatively stable areas subject to overriding by landslide debris

Horizontal contact between rocks of different stability; tick marks point toward more easily eroded rocks.

← Decreasing stability →

Figure 14.18 A landslide hazard map of the San Clemente region of California. The slopes are color-coded to indicate degrees of stability, locations subject to sliding, areas vulnerable to landslide debris, and contacts between rocks of differing stabilities.

6. *The location of roads, dams, bridges, buildings*—indeed, whole towns—with respect to potential hazards.

Geologists are, of course, primarily concerned with being able to predict mass-wasting disasters. Telltale signs include nearly parallel, curving cracks high on steep slopes or recent excavations, fresh springs, and acceleration of the soil creep rate. A slope on the move is generally wavy and has few deep-rooted trees. The creep rate of centimeters per year often changes to a gallop as a full-fledged landslide develops. Thus, if cracks develop in your hillside home, you should have an engineering geologist check out the site, although there may be other explanations.

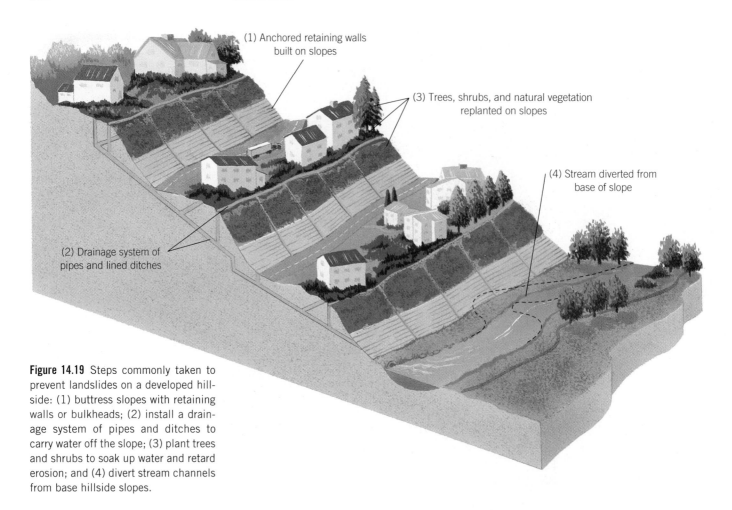

(1) Anchored retaining walls built on slopes

(3) Trees, shrubs, and natural vegetation replanted on slopes

(4) Stream diverted from base of slope

(2) Drainage system of pipes and lined ditches

Figure 14.19 Steps commonly taken to prevent landslides on a developed hillside: (1) buttress slopes with retaining walls or bulkheads; (2) install a drainage system of pipes and ditches to carry water off the slope; (3) plant trees and shrubs to soak up water and retard erosion; and (4) divert stream channels from base hillside slopes.

Prevention and Control of Mass Wasting

Once a potential landslide site has been spotted, how can the slope be modified so that it remains stable? Recall that on stable slopes, a balance exists between gravity and driving forces (which tend to pull material downslope) and frictional and cohesive resisting forces (which prevent movement). In theory, there are many ways to achieve stability (Figure 14.19). However, the engineering geologist must pick the most practical means available within the physical and financial constraints of the project, which is quite another matter. Also, some solutions are beyond the capacity of geologists and engineers. They require changes in zoning regulations and lifestyle habits. For example, if waterlogging is a problem, community waste may have to be treated in a separate facility rather than in septic tanks on the slope. Or perhaps lawn sprinkling will have to be rationed.

This discussion of mass-wasting control has focused on the hazards confronted in the United States and similar economically advanced nations. In parts of the world that are less developed economically, however, the prevention of slides and mudflows is an extremely serious matter affecting millions of people. Population experts forecast that by the year 2000, more than half of humankind will reside in huge metropolitan regions called "megacities." The most spectacular growth will occur in less-developed countries where recent migrants from the provinces frequently settle in the shanty towns that carpet the steep hillsides surrounding older central cities such as Lima, Peru; Rio de Janeiro, Brazil; or Tijuana, Mexico. These slums often lack sewers and plumbing and are bare of vegetation; a heavy rain can thus mobilize an entire hillside (Figure 14.20). As the mud and water pour downslope in torrents, they carry away virtually everything these impoverished people own. Preventing mudflows is a vital step toward improving their lives.

Figure 14.20 The aftermath of a mudslide in Puerto Rico. Slides such as these are tragedies wherever they occur, but their impact on impoverished communities is especially devastating.

STUDY OUTLINE

Mass wasting refers to the downhill movement of the Earth's materials under the direct influence of gravity.

I. **THE BALANCE OF FORCES ON A SLOPE.** The resisting force of friction acts to hold a rock on a slope stationary; in contrast, the driving force, the component of gravity that acts parallel to the slope, acts to pull the rock down the slope.

   A. The opposition of driving and resisting forces subjects an object on a slope to shear stress. Most mass movement is caused by shear failure.
      1. The **shear strength** of a solid fragment—its resistance to shear stress—involves the forces between its atoms. The shear strength of loose fragments is derived from friction between rock fragments and cohesion between clay particles. Most mass movements result when external forces exceed the shear strength of the material along fault planes, bedding planes, and joint sets.
      2. Small amounts of water between particles create surface tension, which also lends shear strength to the aggregate. But saturation makes clay and sand vulnerable to liquefaction, a quicksand state in which they behave as a viscous mass.

   B. When loose fragments form a pile, the steepness of its slope, or **angle of repose,** depends upon the size of the individual particles, their angularity, the degree of sorting, and whether the particles are wet or dry.

II. TYPES OF MASS WASTING

A. Falls involve the free fall of rock and soil fragments through the air, out of direct contact with the slope. They result from the undermining of sheer cliffs by weathering, erosion, wave action, glaciers, earthquakes, and human activity. The pile of dislodged fragments that collects at the base of a cliff is called **talus.**

B. Slides involve the downslope movement along shear planes of blocks of rock (**rockslide**) or rock fragments and soil (**landslide**). Three conditions favor slides.
   1. Steep slopes and adverse geological structures—in particular, the downhill orientation of bedding planes and rock fractures;
   2. Weak layers—for instance, shale or loose soil—that support heavy loads; and
   3. Triggering mechanisms, which can be natural (high rainfall or earthquake) or human-made (deforestation, building on steep slopes, dams that raise groundwater levels).

C. Slumps involve intact blocks of rock or unconsolidated debris that slide downward and tilt backward along concave planes. Slumps occur most frequently on hills with thick soil blankets and steep cliffs made of loose materials.

D. Flows involve the movement of rocks and soil as a fluid mixture. Flows behave differently depending on their water content, particle size, and sorting.
   1. Viscous **earthflows** commonly consist of water-saturated clay, shaley fragments, and soil.
   2. **Permafrost** is a condition of cold climates. The water in regolith and rocks remains frozen solid all year, but in summer, the soil layer thaws and becomes saturated. In **solifluction,** saturated soil slips slowly downhill, causing wavy solifluction lobes. Melting and refreezing cause *frost heaving*.
   3. **Mudflows** contain more water and are less viscous than earthflows.
   4. **Debris flows** are defined as masses of rock fragments, more than half of which are larger than sand grains.

E. Creep is a net downhill movement of rock and soil, scarcely noticeable but cumulatively important.

F. Mass wasting also occurs on the ocean floor. Slides, flows, and slumps have been detected at the base of the continental slopes. Sonar-scanning devices have greatly improved our knowledge of submarine topography and mass wasting.

III. PREDICTING AND PREVENTING MASS-WASTING DISASTERS.

Besides causing the loss of human lives, mass-wasting disasters cause extensive damage to roads, utilities, buildings, and dams. The slower forms of mass wasting—soil creep and slope erosion—can also cause the ruination of farms and other valuable property.

A. To recognize unfavorable sites and imminent hazards, geologists analyze the condition of the bedrock and its relationship to the topography, the steepness of hillside slopes, underlying groundwater conditions, the tectonic history of the region, and the location of human facilities with respect to potential hazards.

B. To control mass-wasting hazards and achieve slope stability, engineers can adjust the steepness of the slope, regulate or divert excess water, or artificially buttress the slope.

STUDY TERMS

angle of repose (p. 371)
creep (p. 382)
debris flow (p. 381)
earthflow (p. 379)
fall (p. 372)
flow (p. 379)
landslide (p. 374)
mass wasting (p. 369)

mudflow (p. 379)
permafrost (p. 383)
rockslide (p. 374)
shear strength (p. 370)
slide (p. 374)
slump (p. 379)
solifluction (p. 383)
talus (p. 373)

CRITICAL THINKING QUESTIONS

1. Discuss the physical factors that maintain slope equilibrium.
2. How can the presence of water improve slope stability? In what ways can water undermine slope stability?
3. Distinguish among falls, slides, slumps, and creep.
4. Discuss the factors common to all slides.
5. Discuss the mass-wasting problems of permafrost regions.
6. Describe five ways that human activities contribute to mass wasting.
7. Describe three types of physical evidence that a geologist looks for in identifying a landslide hazard.
8. Describe three methods of stabilizing a slope.

Streams

Each year, the Mississippi River transports about 300 million tons of North America to the Gulf of Mexico. This much silt, clay, gravel, and dissolved salts comprises enough sediment to fill a train nearly 6 million boxcars long—dramatic proof that "streams are the gutters down which flow the ruins of the continents."* The Mississippi spreads this sediment over its flood plain and delta and in this manner enlarges the Gulf states. Thus the Mississippi simultaneously adds to and subtracts from North America as it carries out the three major functions of all streams: erosion, transport, and deposition of sediment. In the process, new landforms are created, old ones are destroyed, and the configurations of continents are altered. **Streams**—to the geologist, this term means channeled flows of any size—are, in fact, the chief agents responsible for these gradational changes on the Earth's surface.

stream
Flowing water within a channel of any size.

As we saw in Chapter 1, streams are also key components of the hydrologic cycle, the interconnected set of processes that maintains the Earth's water balance (Figure 15.1). Each year, 496,000 cubic kilometers of water are evaporated from the Earth's surface, and the same amount is returned. However, the oceans surrender more water to the atmosphere than they receive from it, and more water falls on the continents than is evaporated. This trend creates a water deficit in the oceans and a surplus on land. Streams are the instruments by which the surplus water that falls on the land is returned to the oceans. Some of the water that falls on land infiltrates the surface to become part of the enormous bank of underground water that is so important to the life of this planet. Eventually, most of this underground water feeds into streams that return it to the ocean.

*This statement has been attributed to geologist Luna Leopold, a pioneer in the study of stream systems.

◀ This turbulent segment of the Big Thompson River in the Rocky Mountains is cutting a narrow gorge through resistant bedrock. Eventually the rapids will be eliminated and the stream will establish a smooth channel.

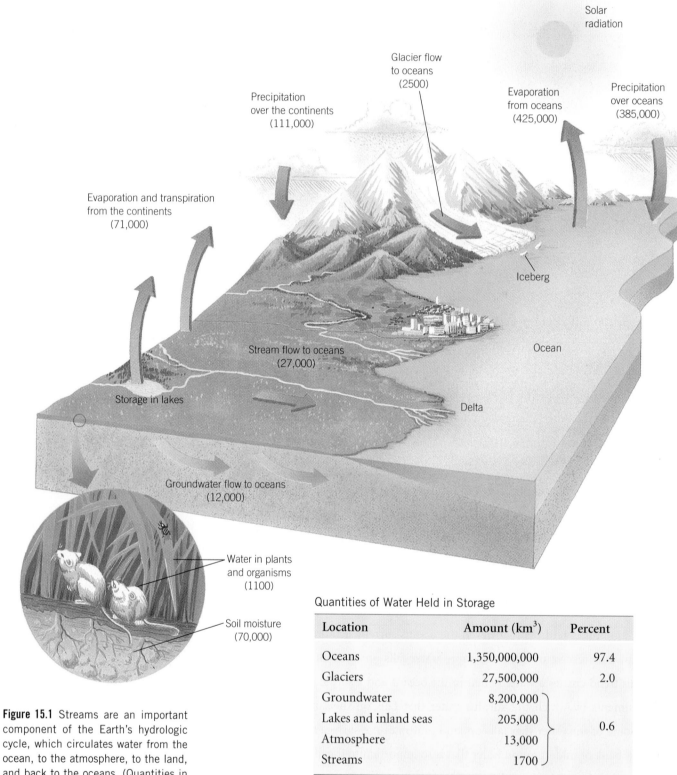

Figure 15.1 Streams are an important component of the Earth's hydrologic cycle, which circulates water from the ocean, to the atmosphere, to the land, and back to the oceans. (Quantities in parentheses are in cubic kilometers per year.

Quantities of Water Held in Storage

Location	Amount (km³)	Percent
Oceans	1,350,000,000	97.4
Glaciers	27,500,000	2.0
Groundwater	8,200,000 ⎫	
Lakes and inland seas	205,000 ⎬	0.6
Atmosphere	13,000	
Streams	1700 ⎭	

tributary stream
A stream that flows into a larger stream.

drainage basin
The total area that contributes water to a stream system.

A stream is best studied as a complex interdependent system—one that includes every branch, from the main trunk, to the **tributary streams** that feed it, to the most insignificant-looking hillside gully. The Mississippi system, for example, covers two-thirds of the United States; and every drop of water that falls within this area drains into the branches that feed into the main trunk. The total area drained by the Mississippi constitutes its **drainage basin** (Figure 15.2).

Figure 15.2 The major drainage outlets of the United States and Canada. A large part of the continent is drained by the Mississippi and its tributaries.

A stream system is, in fact, an "organism" remarkably sensitive to its environment. It adapts to environmental change by adjusting its course, gradient, channel, and velocity to the volume of water and sediment it conducts to the sea. The nature of these adaptations is the major theme of this chapter.

The Energy of Streams

Energy is the capacity to do work. Lift a rock above a given reference plane, such as a table, and you have given the rock a certain potential energy relative to the plane. Three factors determine the potential energy of the rock: its mass, elevation, and gravity. The same is true of all objects, including the water that constitutes streams. The total poten-

Figure 15.3 Some representative stream profiles drawn to the same scale. Although they vary a great deal in length and elevation at their sources, they all exhibit concave profiles. (Vertical scale is greatly exaggerated.)

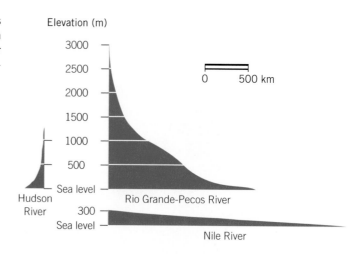

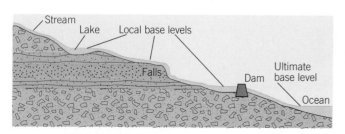

Figure 15.4 The base levels encountered by a stream, shown in profile. All base levels are local (and temporary) except sea level.

gradient
The steepness or slope over a specific length of a stream channel.

long profile
A cross section of a stream channel showing the gradient from the source to the mouth of the stream.

base level
The limiting level below which a stream cannot erode its bed.

ultimate base level
The lowest possible level to which a stream can erode its channel; with rare exceptions, sea level.

local base level
A transitory, local feature, such as a lake or resistant rock layer, below which an inflowing stream cannot erode its channel.

discharge
The volume of water moving past a particular point in a stream over a given time interval; the product of the cross-sectional area of the channel and the stream velocity.

tial energy per year of the Earth's streams—that is, the power they have available to do work on the land—has been calculated by geologist Arthur Bloom from knowledge of the average elevation of the continents and the mass of water that streams bring to the ocean each year. It amounts to about 12 billion horsepower, or the equivalent of supplying each square kilometer of the Earth's surface with a one-horsepower machine that scrapes away at rock and soil 24 hours a day, every day of the year. Over millions of years, the cumulative effects of this ceaseless work would be to reduce the land surface of the continents to featureless, sea-level plains, were it not for the tectonic and volcanic activity that constantly uplifts the surface and supplies fresh material to the crust.

As they flow downhill, streams convert their potential energy into motion. The velocity of the stream is governed by the steepness of its slope, or its **gradient**—that is, by the change in the altitude of the channel per horizontal length of the stream course. All other factors being equal, the steeper the gradient, the more rapidly the stream flows. It is significant that the gradients of all streams are steep at the source, or head, and taper to gentle slopes before they enter the sea. Thus when viewed in cross section throughout their entire length, they exhibit a concave, longitudinal profile, or **long profile.** Although their slopes are concave, stream profiles nevertheless vary from one another in form. Compare, for example, the steep gradient of the Hudson River with the nearly flat gradient of the Nile River shown in Figure 15.3.

A stream's profile is controlled by the lowest level to which it can erode its channel—its **base level** (Figure 15.4). For a stream that empties into a below–sea-level inland basin, such as the Jordanian Dead Sea, the basin acts as the base level. However, the ultimate destination of the overwhelming majority of streams is the sea. For this reason, sea level is considered their **ultimate base level;** lakes, dams, flat valleys, or stream junctions serve as **local base levels.** We will see that local base levels often exert greater control over streams than the more distant sea level.

Stream Discharge

The more water that flows through the channel in a given period, the greater the energy available to the stream. Therefore, an extremely important measure of a stream is its **discharge,** or the volume of water that flows past a monitoring station in a given time interval (usually 1 second). Discharge, measured in cubic meters per second, is the product of the cross-sectional area of the channel and the stream velocity (see Figure 15.6).

At a given locality, stream velocity, channel width, and channel depth vary along with discharge, which is commonly controlled by seasonal rainfall patterns and snow melts. Because the main trunk of a stream system must transmit all the water funneled

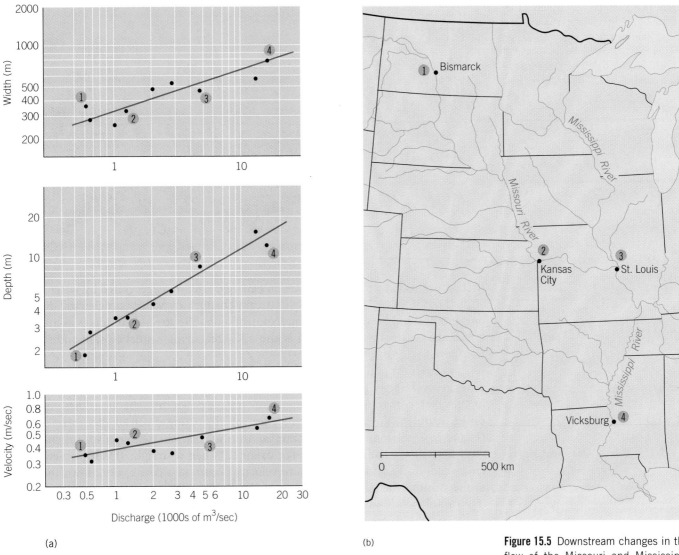

Figure 15.5 Downstream changes in the flow of the Missouri and Mississippi rivers. The graphs show that channel dimensions (that is, width and depth) and velocity increase from head to mouth. These changes accommodate the increased discharge. (*Source:* After L. B. Leopold and T. Maddock, Jr., USGS Professional Paper 252, 1953.)

into it by its tributaries, discharge tends to increase in the downstream direction. To accommodate the added discharge, the stream channel changes downstream in much the same way that it changes at a specific location during the rainy season: stream velocity increases along with channel width and depth (Figure 15.5). These latter changes lower friction and facilitate flow, because a lesser proportion of the water is in contact with the channel surface (Figure 15.6).

Stream Flow and Channel Erosion

If all the potential energy of a stream were converted to motion, most streams would rush to the ocean at the speed of a waterfall. This does not happen, however, because

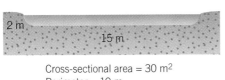

Cross-sectional area = 30 m²
Perimeter = 19 m

(a) Wide, shallow channel

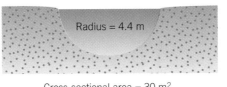

Cross-sectional area = 30 m²
Perimeter = 13.7 m

(b) Semicircular channel

Figure 15.6 (a) The shallow channel and (b) the deep channel have approximately the same cross-sectional area (30 m²). However, the deep channel conducts water more efficiently than the shallow channel does because less water comes in contact with the channel walls.

Figure 15.7 Laminar flow occurs at low velocity in a frictionless channel. Because natural stream channels are not frictionless, stream flow is nearly always turbulent, displaying random variations on the overall direction of flow.

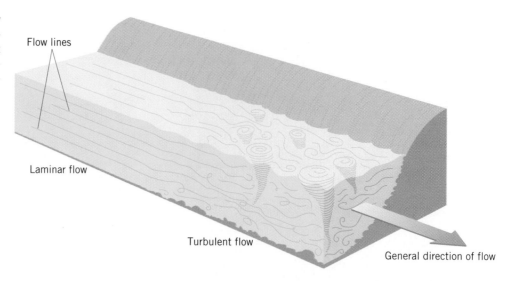

Figure 15.7 Laminar flow occurs at low velocity in a frictionless channel. Because natural stream channels are not frictionless, stream flow is nearly always turbulent, displaying random variations on the overall direction of flow.

laminar flow
A type of flow in which water moves in straight paths parallel to the channel.

turbulent flow
A type of flow in which irregular, chaotic motions are superimposed on the general current direction.

pothole
A smooth, bowl-shaped depression in the bed of a stream caused by abrasion when turbulent currents circulate stones or coarse sediment.

bed load
Large or dense particles that are transported by streams on or immediately above the streambed.

friction drains away most of the stream's energy. Friction is generated by water rubbing against the sides and bottoms of the stream channel, by water particles colliding with one another, and by water dragging sediment along the streambed. It acts as an important brake on stream velocity and is also a key factor in channel erosion and sediment transport.

In experiments using colored dyes and smooth glass tubes that minimize friction, water at low velocity moves in parallel paths as **laminar flow.** But if velocity is increased or the channel is roughened, laminar flow breaks down into chaotic eddies and swirls called **turbulent flow,** which has the effect of dissipating much of the stream's energy (Figure 15.7). Stream velocity, friction, and turbulence vary within a stream channel in ways that affect channel erosion. For example, when a straight stream is viewed in a cross section, the zone of maximum velocity occurs in a core region down the center of the stream, where friction is the least (Figure 15.8).

Turbulence, however, is greatest where the velocity *changes* most abruptly, which is just above the channel bottom; this condition ensures its maximum erosion. First, the swifter-moving water above the streambed creates a pressure difference that lifts particles off the bottom through hydraulic action; second, the turbulence keeps the particles suspended. The moving sand and gravel particles also contribute to the erosion of the stream channel. Their work is most evident where the particles, spinning rapidly in whirlpools, have scoured cylindrical **potholes** in a bedrock channel (Figure 15.9).

Transport and Deposition

Geologists refer to all of the materials transported by a stream as its load. Larger, heavier particles are dragged and skipped along the bottom as **bed load.** Fine particles car-

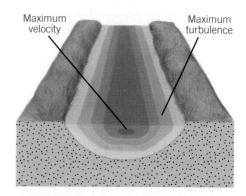

Figure 15.8 The regions of maximum turbulence and maximum velocity in a straight stream channel. Maximum turbulence occurs where the velocity change is most rapid.

Figure 15.9 These potholes in the stream channel were abraded by gravel and sand that were trapped in swirling eddies and whirlpools.

ried within the mass of the stream make up the **suspended load,** and ions carried in solution are the **dissolved load.**

Most dissolved load is brought to streams by groundwater seepage into the channel. The ions in solution are derived from chemical weathering of rock and soil. Thus the dissolved load of a particular stream is directly related to the degree of chemical weathering in the drainage basin. Each year, streams in the 48 contiguous United States carry a dissolved load that amounts to an average of 35 percent of their total load.

Most of the solid load of a stream is probably transported in suspension. This load frequently colors the stream or makes it opaque. The Colorado River was named by Spanish explorers for the brilliant red silts it carried in suspension before its flow was regulated by twentieth-century dams. Table 15.1 shows the amounts of suspended load

suspended load
The fine particles carried in the mass of a stream and kept aloft by turbulence.

dissolved load
The portion of a stream load carried in solution.

Table 15.1 Suspended Loads Transported by the Earth's Major Stream Systems

River	Country	Average Discharge at Mouth (cubic meters per second)	Length, Head to Mouth (kilometers)	Area of Drainage Basin (thousands of square kilometers)	Average Annual Suspended Load (millions of metric tons)	Average Annual Suspended Load (metric tons per square kilometer of basin
Amazon	Brazil	180,000	6300	5800	360	63
Yangtze	China	22,000	5800	1900	500	260
Mississippi	United States	18,000	6200	3200	310	97
Irrawaddy	Burma	14,000	2300	430	300	700
Bramaputra	Bangladesh	12,000	2900	670	730	1100
Ganges	India	12,000	2500	960	1450	1520
Mekong	Thailand	11,000	4200	800	170	210
Nile	Egypt	2,800	6700	3000	110	37
Missouri	United States	2,000	2300	640	140	210
Colorado	United States	200	2300	640	140	210

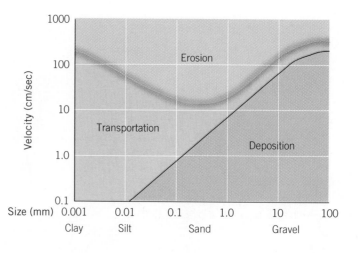

Figure 15.10 The velocities necessary to erode, transport, and deposit sediment vary with particle size. The greater the particle size, the greater the velocity necessary to transport the particle in suspension. Because fine particles are cohesive and stick to the stream bed, it takes nearly as much stream velocity to lift clay and silt as it takes to lift coarse gravel. (*Source:* After F. Hjulstrom, 1935.)

brought to the ocean each year by the Earth's major stream systems. It is interesting to note that the Mississippi's suspended load, though great, is not exceptional relative to the loads of other great rivers when measured in tons of suspended load per square kilometer of drainage basin. Nor are the suspended loads carried by the Amazon or Mekong rivers especially high. These streams drain humid regions containing vegetation that inhibits rapid runoff and erosion. The humid climates facilitate intense chemical weathering, which produces the high dissolved load of these rivers. In contrast, streams such as the Colorado and Ganges occupy dry drainage basins. The sparse vegetation allows for the erosion of huge quantities of sediment, which increases the suspended load.

Two useful measures of a stream's ability to transport solid load are its capacity and competence. **Capacity** refers to the *quantity* of sediment the stream can carry past an observation point in a given time span. Clearly, it is proportional to discharge—the water volume flowing through the channel. **Competence** refers to the maximum sediment *size* the stream can carry. The velocity of flow determines the competence of a stream: the higher the velocity, the greater the stream competence.

The experimental work summarized in Figure 15.10 shows that the stream velocity necessary to transport a particle in suspension may differ markedly from the one required to lift it from the streambed. Transport depends upon the **settling velocity** of the fragment, the rate at which it sinks through the water. If the stream's velocity is greater than this sinking rate, the fragment will stay in suspension; if less, it will sink.

Fine silt and clay are transported at such low settling velocities that, theoretically, they may remain in suspension indefinitely; for this reason, they are the major constituents of suspended load. Coarse fragments, in contrast, have such high settling velocities that stream turbulence cannot keep them in suspension for long and they are therefore transported as bed load. Heavier particles roll and slide along the bottom, but lighter bed load particles may travel by **saltation,** in which grains bounce and skip along the bottom. These modes of transport result in a natural sorting of sediments during their journey downstream. Fine suspended load moves with the velocity of the water, whereas the coarser bed load that is dragged along the bottom lags behind. In this manner, the percentage of suspended load increases downstream.

The Graded Stream

As we have seen so far, the function of a stream is to transport all the water and sediment funneled through its channel. If given enough time, the slope and channel of the stream adjust so that transport is accomplished with maximum efficiency. A stream that has reached this state is said to be in a graded condition. In a **graded stream,** slope and channel are so adjusted that the stream has just enough energy to transport its load; there is no excess energy present to erode the channel, nor is there a deficiency of energy

capacity
The transporting ability of a stream as measured by the quantity of sediment that the stream carries in a given time interval.

competence
The transporting ability of a stream, as measured by the largest-sized particles that it can carry.

settling velocity
The velocity required for a suspended particle of a given size to sink to the bottom of the stream channel. If the stream velocity is greater than this value, the particle will remain in suspension; if less, it will sink.

saltation
Transport by streams in which bed load particles are moved in short skips and bounces.

graded stream
A stream whose slope and channel are so adjusted that it has just enough energy to transport its load; no excess energy is present to erode the channel, nor is there a deficiency of energy that leads to deposition of sediment in the channel.

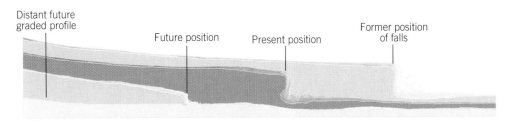

Distant future
graded profile

Future position

Present position

Former position
of falls

Figure 15.11 A waterfall is an irregularity of the stream profile and as such will eventually be destroyed. Energy is concentrated at the head, or knickpoint, of the falls. As the falls are worn back, their height is reduced until the irregularity vanishes and the smooth stream profile is restored.

leading to deposition of sediment in the channel. This ideal state of equilibrium is rarely achieved and even more rarely maintained. Because the discharge and the velocity of streams are variable—especially over short time periods—streams are constantly either eroding their channels or depositing sediment. Over a sufficiently long time span, however, a stream can come close to a graded condition by striking a rough balance between erosion and deposition. In this respect, a graded stream is analogous to a region's climate: the weather may vary from day to day, but the overall pattern remains constant from year to year.

If a stream's equilibrium is disturbed, specific actions are triggered to restore it to the graded state. For example, a sudden increase in discharge (caused, perhaps, by melting snow and spring rains) may temporarily throw a stream out of grade, creating greater turbulence and scouring of the stream bottom. However, this erosion reduces the stream gradient and deepens and widens the channel to accommodate the increased discharge. These changes lower the stream's velocity until there is energy just sufficient to transport the load, thus restoring the graded condition.

Other factors also affect the equilibrium of streams. Often, they must cut through rocks of varying resistance to erosion or through land surfaces disrupted by glacial action, tectonic forces, or human meddling. All these factors cause irregularities in the long profile by creating such features as waterfalls, rapids, dams, and lakes. Eventually, however, these features will be destroyed, because they interfere with the efficient transfer of energy throughout the stream system.

Waterfalls and Lakes

Waterfalls have many causes, but no matter what their origins are, they share a common property: all are located at a *knickpoint,* a place where the stream profile steepens abruptly, causing a vertical drop. Because energy is concentrated there in the form of high water-velocity, abrasion, and turbulence, a waterfall is a temporary feature in the course of a stream. The falls wear back and reduce the knickpoint, causing it to retreat upstream until it merges with the smooth, concave profile of the stream, at which time the fall is destroyed (Figure 15.11).

Figure 15.12 shows two waterfalls, New York's Niagara Falls and California's Yosemite Falls. Nearly a kilometer wide and 50 meters high, Niagara Falls is capped by a thick, resistant layer of rock, the Lockport dolomite, beneath which are soft shales. At the base of the falls, turbulence is eroding the soft shales that support the dolomite, and it has worn the falls back 1 kilometer since its origin some 8000 to 10,000 years ago. In the near geological future—perhaps as little as 50,000 years from now—the falls will probably be worn back and reduced until it becomes just another smooth stretch of the Niagara River.

A narrow sliver compared with Niagara Falls, the 700-meter-high Yosemite Falls obviously has a long way to go before it is reduced to a quiet stretch of stream. This waterfall is cut in resistant granites and gneisses and empties into a deep valley oversteepened and carved by a large glacier that has since melted. In spite of these obstacles, the tendency still exists for Yosemite Falls to wear back the formidable cliffs and produce a smooth, concave profile. The process will simply require a long period of time, perhaps several million years.

(a)

(b)

Figure 15.12 (a) Niagara Falls. (b) Yose-mite Falls.

Lakes are a departure of another kind from the ideal graded condition. They are places where a stretch of the stream profile flattens before steepening once again, which is the reverse of what occurs in a waterfall (Figure 15.13). In contrast to waterfalls, lakes are places where the potential energy of the stream is stored rather than used, because a portion of the stream's water is ponded at higher elevation than found downstream. Engineers build hydroelectric dams with artificial lakes behind them to tap this energy, and they construct controlled waterfalls at the outlet of the dams to convert the energy to electric power.

Two Contrasting Stream Types

Graded streams are alike in the sense that, over a long time span, they transport their loads with no net channel erosion or deposition. However, they may differ from one another in channel characteristics and other features, because they attain the graded condition in response to differing requirements of sediment load and discharge. To illustrate this point, we will describe the channel, gradient, and landforms produced by two contrasting stream types: the braided stream and the meandering stream.

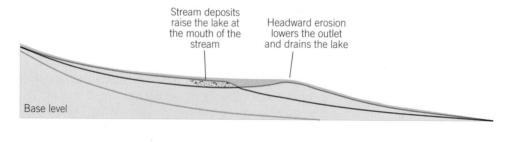

Stream deposits raise the lake at the mouth of the stream

Headward erosion lowers the outlet and drains the lake

Base level

—— Initial profile
—— Profile after elimination of lake
—— Graded profile

Figure 15.13 Stages in the destruction of a lake and reestablishment of the graded stream profile.

Figure 15.14 A braided stream draining an alluvial fan in the Jago River, Alaska. Notice the sand bars among the twisting, branching stream channels.

The Braided Stream

Braided streams are typically associated with the **alluvial fans** that accumulate at and spread from the base of dry mountain ranges. A fan develops where a mountain stream, no longer confined to its narrow channel, deposits its coarse sediment load in a large, sloping, apronlike structure at the point of emergence onto the plain. Braided streams twist their way through the fan and carry sediment from the fan's apex to its fringes, thus enlarging it. In addition, mud- and debris flows contribute to the fan's growth by spreading sediment down its sloping surface. Braided streams also develop in front of melting glaciers, where water flowing out from the ice carries a tremendous quantity of coarse debris with it (see Chapter 17).

A braided stream derives its name from the way its main channel divides into a complex series of interwoven branching and merging subchannels (Figure 15.14). Tear-shaped islands as well as bars of sand and coarse sediment lie between the subchannels, giving the entire stream a braided appearance when viewed from above. Seemingly smothered by its own debris, the braided stream is, in fact, a highly competent earth-moving machine. The braided form is an efficient response to the large cargo of coarse bed load sediment that must be moved by traction and saltation. The numerous shallow subchannels provide maximum contact between the coarse sediment and the meager amount of water available to transport it.

braided stream
A stream that divides into branching and intertwining subchannels separated by islands or sandbars.

alluvial fan
A gently sloping, fan-shaped mass of sediment deposited by a stream where it issues from a narrow canyon onto a plain or valley floor.

The Meandering Stream

As they approach base level and their gradients lower, most streams begin to **meander**—that is, their channels form wide loops in their course downstream. Like the undulations of a giant snake, the loops slowly migrate downstream as they wind back and forth across the nearly flat valley floor. Meandering streams may be as large as the Mississippi or as small as the Animas River, shown in Figure 15.15.

Viewed from the air, winding streams may leave the impression that meanders are random, haphazard features. However, in the late 1950s, geologist Luna Leopold and his research team at the U.S. Geological Survey conducted field studies that provided convincing evidence to the contrary. They determined that, although the typical meandering stream has a large suspended load, it also has more than adequate discharge to carry

meander
As a verb: to flow in a winding, sinuous course. As a noun: a loop created as a stream winds back and forth across a valley.

Figure 15.15 The Animas River of Colorado. In this view we see many of the features of the meandering stream: broad flood plain, oxbow lakes, meander scars, and point bar deposits. Notice that the point bar gravels are slightly off-center from the meanders in the downstream direction.

cutbank
A steep slope caused by erosion of the outer bank of a meander.

point bar
A low, crescent-shaped deposit of sand and gravel developed on the inner bank of a meander where the water velocity is low.

the load. Lacking sufficient coarse fragments to abrade and downcut the streambed and having, in most instances, a channel floor already close to base level, the stream expends its excess energy laterally as it winds back and forth across its valley.

Leopold and other geologists have studied meander formation in stream models and natural systems. The process begins with a straight channel having an initially flat bed of sediment, which is soon transformed into alternating *riffles* and *pools*, suggesting a wave pattern. Riffles are mounds, and the water becomes shallow and flows over them swiftly. Pools are pockets of deeper, quieter water. Exactly why these features develop in a streambed is not known. However, we do know that whenever a fluid flows over a solid, an aggregate of solids, or a denser fluid, frictional drag tends to cause wave-like features along the surface of contact. (Think of wind-driven ripples in a pond or sand dunes.) In any case, riffle and pool spacing tends to be uniform, about five times the width of the stream channel.

A meander is initiated where the water flowing swiftly over a riffle deflects toward a stream bank upon entering a pool, thus eroding a small bend in the channel (Figure 15.16). The water is then deflected to the opposite bank, where it strikes the channel wall, creating a similar bend. Continued deflection shifts the maximum water depth and velocity toward the outer channel walls and increases turbulence there as well. The outer channel wall is then eroded through impact, slumping, and undercutting, thereby forming a steep **cutbank.** At the same time, the complementary process of deposition occurs opposite the cutbank. Sediment is funneled to the low-energy inner bank downstream, where it forms a **point bar.**

Because the meandering stream has a gentle gradient, maximum cutbank erosion occurs at the most downslope portion of the bend, slightly below the midline of the

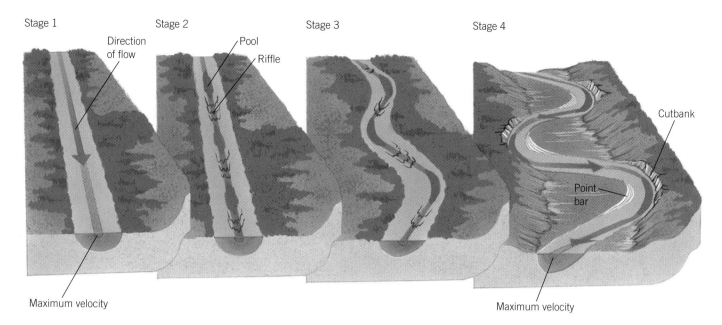

Figure 15.16 This sequential diagram shows how a straight channel evolves into meandering one. A key step involves the development of riffles and pools.

meander. This seemingly trivial effect has great consequence, because it forces the meanders to migrate downstream (Figure 15.17). The meanders do not move with equal speed, however. Some are delayed because of cutting through materials that are difficult to erode, such as compacted clays or sediments tightly bound by vegetation. The delay allows the upstream meander to catch up and break through the short segment of land between the loops. For a while, the channel will divide, one portion of the stream taking the long route, the other taking the **meander cutoff.** Eventually, the stream will be entirely diverted through the cutoff, which possesses the steeper gradient. The stranded portion of the meander will be isolated, forming an **oxbow lake.** In time, the lake will silt up, leaving only a *meander scar* (see Figure 15.17).

meander cutoff
A new channel created when a stream takes the shorter route across the neck of a meander.

oxbow lake
A crescent-shaped, abandoned meander channel isolated from the stream channel by a meander cutoff and sedimentation.

The Flood Plain

As the meandering stream winds back and forth across its valley, it simultaneously widens the valley through erosion and builds up the valley floor through deposition. Erosion occurs where the stream undercuts the valley walls, attacking one side and then the other as it pursues a winding course. Eventually, this lateral erosion wears back the hillside slopes and widens the valley floor. At the same time, the stream spreads channel and point-bar deposits across the valley floor. Periodically, when discharge becomes too great for the channel to handle, the stream overflows its banks. Layers of silt are deposited with each successive overflow, supplementing the channel deposits. This combined depositional process is called **aggradation.** Over thousands of years, aggradation builds a thick sediment blanket, the **flood plain,** upon which the stream flows.

Although aggradation is the process by which all flood plains are built, not all flood plains are deposited on level surfaces formed by lateral stream erosion. The flood plains of great meandering streams are often spread over lowlands that were created by large-scale tectonic movements of the crust. For example, recall from Chapter 2 that the lower Mississippi River follows an ancient continental rift that has since been reactivated and is now an active fault zone.

During floods, streams deposit the greatest portion of their loads close to their banks. Over the years, thick wedges of sediment accumulate there (see Figure 15.17). These wedges are called **natural levees,** and they serve as partial barriers against future floods. The flood plain tapers gently from the natural levees toward the main valley walls, forming subtle depressions along the base of the valley walls. Drainage is often poor in these regions and vegetation is thick; swamps are common. A meandering

aggradation
The process of building up a surface by deposition.

flood plain
Layers of sediment beds deposited across a valley as a stream periodically overflows its banks and meanders laterally.

natural levee
A ridge of sand and silt that parallels the stream channel and that is deposited over time, when the stream overflows its banks during floods.

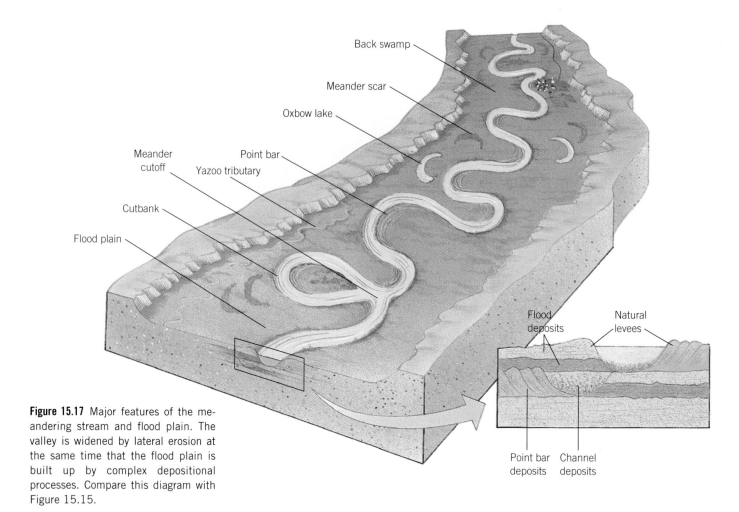

Figure 15.17 Major features of the meandering stream and flood plain. The valley is widened by lateral erosion at the same time that the flood plain is built up by complex depositional processes. Compare this diagram with Figure 15.15.

tributary stream may wander for quite a distance through these depressions, blocked from joining the main stream by the levees. Such a stream is called a *yazoo* tributary, after the small river that parallels the Mississippi River and eventually joins it at Vicksburg.

In terms of the graded-stream concept, the flood plain is an integral part of the stream system. During floods, water and sediment are released from the stream channel to the flood plain to relieve drainage system overload. The storage is temporary, because flood waters seep back to the channel through the flood-plain sediment. Sediment is returned to the channel when the meandering stream erodes cutbanks in the flood plain. The bulk of the sediment remains in long-term storage, however, awaiting a basic change in the energy of the stream system.

 The Debate Over Flood Control

Throughout human history, people have inhabited flood plains, which constitute some of the world's richest farmland. It is ironic that the same natural instrument that makes flood plains valuable by enriching them with nutrients also destroys farms, homes, and human lives. That instrument is, of course, floods. In the past, the way to prevent floods appeared simple enough: construct artificial levees on top of the natural levees to contain the rising water within the stream channel.

The traditional approach to flood control worked reasonably well for normal flooding, and the response to the occasional major floods that breached the levees was to build higher levees. However, in recent years, geologists and engineers have come to the conclusion that traditional flood control measures may have backfired and intensified the severity of flooding.

Figure 15.18 The Mississippi River flooded this Illinois town in July 1993.

The Upper Mississippi Valley Floods of 1993

Throughout the late spring and summer of 1993, the jet stream—the high-altitude air current that steers weather systems—looped farther to the south than is normal for that time of year. Why it did so, and why it remained locked in that pattern for so long, no one knows. Whatever the reason, it brought cold, dry air down from Canada over the drainage basins of the upper Mississippi and Missouri rivers. It also brought warm, humid air up from the Gulf of Mexico. Along the front, where the air masses clashed, there was rain. The unrelenting rains—actually, a series of storm systems that drenched the same midwestern states repeatedly—soon saturated the subsurface, and the runoff swelled the streams.

A flood occurs when the discharge exceeds the capacity of the stream channel to contain the water flowing in it. Engineers can measure the magnitude of the resulting flood by surveying the percentage of the flood plain covered when the stream overflows its banks. How often we can expect floods of similar magnitude to occur determines the recurrence interval of the flood—whether it is a ten-year flood, a fifty-year flood, or so on. By this criterion, the recurrence interval of the 1993 Mississippi floods was on the order of centuries; the floods spread over portions of the upper Mississippi Valley that had not been covered since the time when recordkeeping had begun. However, the high discharge of the upper Mississippi River—11 times normal and 4 times the flow of Niagara Falls—was easily accommodated south of Saint Louis. There, the channel of the Mississippi River broadens to receive the discharge of the entire drainage basin, covering 31 states, and the contribution of the upper Mississippi had little impact.

Although the 1993 flooding of the Mississippi River wreaked havoc on flood-plain communities, it was a natural event, similar to others that occurred throughout the long history of the river. After all, without floods, flood plains would not exist. The economic loss, misery, and deaths from floods are of human origin—resulting from the population density and the concentration of human activities on the flood plain. In these terms, the floods of 1993 were of catastrophic proportions—48 lives and $12 billion lost (Figure 15.18). The catastrophe has renewed an old debate among scientists and the public at large over the effect that human interference has had on the natural flow of the Mississippi River system—and, by extension, other river systems. The main source of conflict involves the impact that the construction of 5800 kilometers of dikes and levees has had on the river.

Figure 15.19 How a levee fails. Powerful flow tears away at the levee walls on the river side. Water under high pressure may seep beneath the levee embankment and boil up on the flood plain side, undermining the levee and causing it to collapse. Finally, the levee may collapse of its own weight as water saturates the structure and turns it to mud.

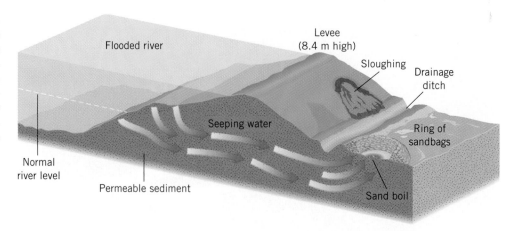

Those who support the construction and maintenance of levees point to the development of vast tracts of land for agricultural, commercial, and residential purposes, all made possible by the levees. Those who disagree accuse their opponents of fostering a false sense of security among flood-plain residents and, more important, of intensifying flooding by constructing the levees.

How may a levee failure intensify flooding? When a stream's flow is constricted, the sediment it would normally spread across the flood plain is deposited instead in the channel. This deposition raises the level of the channel above the flood plain and thus requires engineers to add height to the levee. Indeed, heights of 12 meters or more have become common along the upper Mississippi. The river at or near flood stage roars downstream within its confined channel at high speed, and it works to destroy the levees by a variety of means (Figure 15.19). The powerful flow tears away at the levee walls on the river side. Water under high pressure seeps beneath the levee embankment and boils up on the other flood-plain side. This water undermines the levee, which can collapse at various points along its length. Saturation of the levee is another threat, for it causes slumping of the levee on the flood-plain side. When the levee is breached, it is like a dam bursting; a high-pressure wave of water spreads out across the countryside at great speed. The erosive power of the floodwaters is therefore enhanced. In contrast, a stream that is allowed to flood naturally does so from the lower height of the stream bank, and its ability to erode is reduced. Furthermore, the levee and other structures retard the drainage of floodwaters back to the river channel. This leaves the land flooded longer than in a natural flood.

It is an underappreciated fact that flood plains are instruments of flood control, which they accomplish by storing and slowing floodwaters and by reducing floodwater heights. For this reason, the states hardest hit by the 1993 flood were the ones that had converted their wetland flood plains to intensive agriculture at the same time that they leveed and channeled the streams. Over the years, Missouri has lost 87 percent of its wetlands, Iowa 89 percent, and Illinois 85 percent. But the altered flood plains and stream channels are merely components of the much larger problem of the transformation of the entire drainage basin of the Mississippi. What was once water-absorbent forest and prairie now has been converted to runoff-enhancing farms and paved-over suburbs and cities. More rainwater goes directly to the stream channels and it gets there faster, thereby intensifying flood conditions.

The undesirable effects of confining a river system within levees and straightened channels were demonstrated again during the catastrophic floods in Washington State, Oregon, and northern California that were caused by record rainfalls early in 1997. The discharge of the Sacramento River—next to the Mississippi, the nation's most leveed stream—could not be contained. The river's flood plain was inundated, decimating some of California's most fertile farmland.

Figure 15.20 The straightened and restricted channel of the Kissimmee River of Florida. An attempt is underway to restore the river to its former meandering course.

Most experts agree that the present system of flood control leaves much to be desired. But they also agree that there is no going back to a state of pristine nature. In 1991, a federal task force reported that, nationally, 17,000 communities occupied our nation's flood plains, which amounted to 10 million households and $390 billion in property. Having learned some painful lessons from major floods, states and communities across the nation are beginning to pursue "soft" alternatives to the "hard engineering" solutions of dikes and levees. They are preserving the few stretches of flood plains left within cities and suburbs, buying back additional wetlands from private owners, and passing zoning regulations that discourage development of flood-prone areas. Some of these open spaces serve as parks and farmland between floods. Some states are planning for unleveed stretches that allow the river elbowroom to flood naturally. This approach will probably save lives and money in the long run. The National Research Council has recommended that we immediately initiate an effort to restore the ecology and normal flow patterns of 640,000 kilometers of the nation's 5.2 million kilometers of streams. Restoration of Florida's Kissimmee River has top priority (Figure 15.20). As a flood control and land development measure, its original meandering channel has been straightened into an 83-kilometer canal, at a cost of $32 million. Unfortunately, the project destroyed 16,000 square kilometers of wetlands. An estimated $370 million will be needed to put the bends back in the river and revive the ecosystem.

Deltas

When a stream enters an ocean or lake, its velocity is checked and much of its sediment load is deposited near the mouth of the stream. If the rate at which the sediment is

Figure 15.21 A satellite image that reveals the classic triangular shape of the Nile Delta. The city of Cairo is at the apex of the delta. The red color reveals the extent of agriculture in the fertile delta region.

delta
A triangle-shaped landform built up over time where a stream enters a calmer body of water and deposits its sediment load.

distributary
One of a system of small channels that carry water and sediment from the mainstream channel and spread them over the delta surface.

supplied exceeds the capacity of the waves and currents to carry the load away, sediment will accumulate and form a more or less triangular structure, with the river mouth at the apex. The name of the structure, **delta,** is from the Greek letter delta, Δ. It was first used by Herodotus in the fifth century B.C. to describe the Nile River deposits in the Mediterranean Sea north of Cairo (Figure 15.21). Nevertheless, variations in local currents, wave action, and shoreline configurations prohibit many deltas from achieving the idealized triangular shape.

A delta derives its form from the branching **distributary** channels, which flow from and divide the main trunk of the stream, bringing sediment to the delta front. The mouth of a stream entering a quiet lake is a good place to observe the formation of a delta (Figure 15.22). As the stream enters the lake and deposits sediment, a broad-layered platform that slopes into deep water is gradually constructed. These layers are the *foreset beds* of the delta. This platform becomes a base upon which a complex mixture of river and shallow-bay sediments are deposited. They form the *topset beds* of the delta. At the same time, fine clays and silt are carried to deeper, calmer portions of the lake and settle to the bottom. These *bottomset beds* grade into the foreset deposits in the direction of the delta and merge imperceptibly with the nondeltaic sediments of the middle of the lake. The delta continues to grow as topset-foreset-bottomset deposits reach ever farther into the lake. These three structural units form distinctive *deltaic cross-beds.*

A delta can be thought of as a battleground between two opposing forces: the stream supplying sediment, and the water body that redistributes the sediment through wave, current, and tidal action. The simple delta just described was deposited in still water, and obviously the sediment-carrying stream was completely dominant. In the ocean, waves and currents may not even allow the delta to form at all or may greatly reduce it. For example, Atlantic currents sweep half the Amazon River sediment load 1600 kilometers to the northwest and deposit it near the Orinoco River Delta in the Caribbean Sea. For this reason, the Amazon's delta is small relative to the great river's size.

Many deltas have built-in cycles of growth and retreat. During the growth phase, the delta pushes seaward for a time by depositing a thick wedge of sediment on the continental shelf. Coarse beach sands are thrown up by wave action on the delta front, while flood-plain and marsh sediments are deposited behind it. As the delta grows, the sedi-

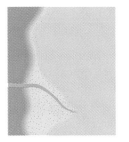

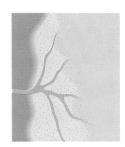

(a)

Figure 15.22 Stages in the development of a simple lake delta. In the ocean, strong waves and currents add complexity to this basic structure.

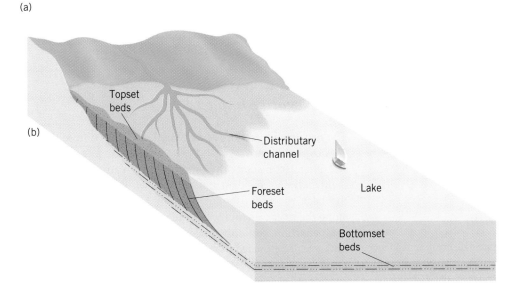

(b)

ment compacts of its own weight, and the crust beneath the shelf may buckle slightly in response to the sediment load. At the same time, extension of the delta lessens the stream gradient, which leads to a decrease in sediment supply. These events trigger a gradual flooding of the delta surface and the onset of the retreat phase. As the sea washes in, the beaches are beaten backward, and the encroaching ocean spreads shallow marine muds over the flood plain and marshes. But the growth cycle of the delta is eventually renewed, because the marine invasion temporarily increases the stream gradient and therefore the capacity of the stream to deliver sediment.

The great delta of the Mississippi River contains within its layers and on its surface an amazing record of advances, retreats, and shifting distributary channels (Figure 15.23). It is where most of the annual 300 million metric tons of sediment that the river carries away from the interior of North America ends up. Through analysis of core samples and other means, geologists have determined that the river has been depositing sediment at its present outlet for about 500 years. Although the delta is growing in front, little sediment is supplied to the rest of the vast deltaic plain. Its flanks are submerged and under attack. The bird's-foot shape of the modern delta is caused by the shallow marine invasions of the low marshland between the relatively few and widely spaced distributaries.

Figure 15.23 shows that the Mississippi River has shifted outlets many times over the past 7000 years, first to one side of the broad deltaic plain, then to the other. Progressive outbuilding has overextended the modern delta to the point where the Atchafalaya River, which branches off the Mississippi above Baton Rouge, offers a much shorter route to the sea. Indeed, the flood history of the past few decades has proven that if the Mississippi had its way, it would shift channels and flow through the Atchafalaya. New Orleans would be bypassed, a new delta would be established at the mouth of the Atchafalaya River, and the present bird's foot would gradually be worn back. In fact, the present course of the Mississippi is maintained artificially through the

The Environmental Impact of Large Dams

Since 1950, more than 36,000 dams of at least 15 meters in height have been constructed around the world. At 221 meters in height, the Hoover (or Boulder) Dam of Nevada, which spans a deep, narrow gorge of the Colorado River, was by far the tallest dam in the world at the time of its completion in 1936; it impounded the world's largest reservoir, Lake Mead, which holds 37 billion cubic meters of water. Today, however, the Hoover Dam is exceeded in height by at least 20 other dams, some of whose reservoirs hold four times the volume of Lake Mead.

A large dam exerts a major impact on the geology and ecology of the region that depends on it, although this impact varies with the regional environmental circumstances. The Aswan High Dam of Egypt, constructed in 1964, provides an ideal case study, giving us thirty years of observations with which to assess its impact.

The Impact of the Aswan Dam

The striking satellite image of Figure 15.21, page 410, demonstrates the dependence of Egypt on the Nile River. The country's entire population of more than 60 million people, as well as its arable land, are confined to the Nile Delta and to narrow strips on either side of the river; the rest of the country consists of empty desert. For this reason, the dam has a major impact, not only on the river and the surrounding environment, but also on the life of every Egyptian citizen—for better or for worse.

Benefits of the Dam

The purposes of a large dam are to prevent floods, generate power, and provide a steady, reliable source of water. When measured against those goals, the Aswan High Dam has been a successful project. Because excess discharge is stored behind the dam in Lake Nasser Reservoir, the delta's agriculture is no longer at the mercy of the yearly floods that would occur when the Nile drainage basin farther south was deluged by monsoon rains. Instead, the delta is irrigated year-round, and crops are harvested three times each year; in contrast, crops were harvested only once each year prior to the dam's construction. Intensified irrigation and a controlled supply of water have also allowed the cultivation of one million additional acres reclaimed from the desert. The waters of Lake Nasser have nursed the nation through potentially catastrophic periods of drought, when the headwaters of the Nile did not deliver the water volume upon which Egypt depends.

The waters impounded behind the dam store a formidable amount of gravitational potential energy. When converted to electricity, this energy source supplies about one-third of Egypt's power needs—power necessary to expand the nation's industrial capacity. In addition, Lake Nasser, which is roughly the size of Rhode Island, has enabled some of Egypt's population to relocate from the overcrowded delta to the sparsely settled south. Increased agricultural production, flood control, water regulation, electrical power generation, and resettlement potential—these advances are major gains for a desperately poor nation whose population has doubled since the dam was first conceived in the 1950s.

The Negative Impact

The negative impact of the dam has been more diffuse than the positive impact.

First, consider the disruptions caused by the construction of the dam and Lake Nasser. To make room for the lake, the entire populations of ancient communities along the river were uprooted and scattered, with little or no compensation. Not only modern-day culture was destroyed, however. The ancient Egyptians had built many of their tombs and shrines on the banks of the Nile, which they considered sacred. With financial help from a few foreign governments and many international organizations, a few structures were hauled inland and saved; nevertheless, much of Egypt's rich heritage was drowned in the waters of the new lake.

Second, consider the numerous environmental problems caused by the dam's interference with the normal stream flow. The impoundment of water in Lake Nasser has caused the groundwater table to rise upstream of the dam. Downstream, the water table has fallen so precipitously that many wells have run dry and others close to the shore of the Mediterranean Sea are regularly plagued by saltwater contamination.

The dam's impoundment of water has also drastically altered the Nile's gradient. With Lake Nasser now serving as a regional base level, the velocity of water flowing into it has been checked. As a consequence, the Nile deposits behind the dam virtually all the sediment load it used to carry to the delta and the Mediterranean Sea—which limits the effective life of the dam as the reservoir fills with sediment. Each year the capacity of the lake to hold water shrinks a little more. Just as surely as the river built the delta, it will eventually fill in Lake Nasser.

At the same time, the Nile Delta, which was built from the sediment delivered year-round by the Nile for millions of years, is now being starved. With no new sediments laid down over the old, the delta is steadily subsiding and the Mediterranean Sea is encroaching upon valuable crop land. In some places the sea is reclaiming the delta at the

Figure 15A Satellite image of Lake Nasser (the broad dark region on the left), the Nile River (flowing toward the right), and the Aswan Dam at their juncture. The narrow strip of arable, populated land along the river stands out in sharp relief against the uninhabitable desert on either side.

astounding rate of 100 meters per year. Furthermore, the delta's formerly rich soils are no longer being replenished by the yearly layer of nutrient-rich silt furnished by the Nile floods.

Delta farmers must now purchase commercial fertilizer, which is manufactured in new fertilizer plants—enterprises that, ironically, consume much of the electric power produced by the dam. The nutrients that the Nile brought to the Mediterranean Sea supported a thriving marine ecosystem that included sardine, mackerel, shrimp, and lobster and enabled the existence of a thriving fishing industry. That industry has since dried up at the cost of thousands of jobs and the loss of a significant source of protein for the Egyptian people.

The floods that washed over the Nile Delta every year prior to construction of the dam provided another invaluable—and free—service. They flushed away the salts that were precipitated from the delta's water-logged, irrigated soils. Without the cleansing floods, the salts now accumulate in the topsoil in a ruinous process called salin-

ization. Salinization has cost the Egyptian people three-quarters of the productivity they gained by reclaiming one million additional acres from the desert.

Also flushed out of the delta soils during the yearly floods were the snails that transmit shistosomiasis, a disease deadly to humans. Since year-round irrigation has replaced the natural flooding process, the snail population has increased exponentially, as has the spread of snail fever through the Egyptian population.

The Costs of Large Dams Versus the Benefits

As with all large dams, the key question regarding the Aswan Dam is, Are the costs worth the benefits? As with most dams, the answer varies according to whom you ask, because those who reap the benefits are not necessarily those who pay the costs. In the case of Egypt, the costs have generally been borne by the people displaced by Lake Nasser, by those trapped on the deteriorating regions of the delta, and by those who depend on the Mediterranean fishing industry for their living. Individuals who own or work the newly productive strips of land have clearly benefited, as have those who rely on the dam's power for their businesses.

Some of the effects associated with the Aswan Dam reflect the particular circumstances of Egypt—few national populations are as dependent as the Egyptians are upon a single river and its delta. Other effects, especially some of the environmental changes, are common to large dams everywhere. All large dams trap sediment in their reservoirs, for example. The capacity depletion rate is about 2 percent per year in the tropics, so the effective life span of such dams—most of which cost half a billion or more dollars to build—is on the order of 30 to 50 years. All dams starve the downstream flood plain and delta of sediment and nutrients, disrupt water tables, and cause disturbing changes in the regional ecology. Nearly all flood ecologically and culturally valuable stretches of river. In addition, dam construction invariably produces unanticipated or underestimated side effects peculiar to the local geology, climate, ecology, and patterns of human settlement (for example, see the discussion of the Vaiont Dam disaster, page 375).

Thought Questions

1. How would the impact of the Aswan Dam differ if Egypt's population were half its present size and twice as wealthy?

2. In what ways are the effects of dams and levees on streams similar and different?

investment of millions of dollars, intensive labor, and sophisticated engineering. The Atchafalaya River and surrounding lakes are used mainly as controlled safety valves to handle the excess discharge of the Mississippi River at high-flood stage and to limit flooding of New Orleans.

Stream Rejuvenation

stream terrace
An elevated, shelflike surface upon which the stream formerly flowed; typically, ancient flood-plain remnants abandoned as the stream cut downward and established a course at a lower level.

incised meander
A deep, narrow, winding stream valley caused by the combination of regional uplift and stream downcutting.

Streams that receive an infusion of energy from uplift, sea-level retreat or increased discharge due to long-term climatic changes are said to be rejuvenated. They put their fresh energy supply to work lowering the stream gradient. For a meandering stream that flows on a flood plain, the method of accomplishing this task is to cut deeply into the thick flood-plain sediment. Left behind on either side of the newly incised channel are **stream terraces,** the flat-topped remnants of the former flood plain (Figure 15.24). Complex terrace patterns commonly consist of many levels, like staircases. Rising above the main level of the flood plain, they are often well drained, have fertile soil, and usually escape flooding. Active tectonism may produce spectacular examples of rejuvenation. Figure 15.25 shows **incised meanders** along the course of the San Juan River of the Colorado plateau. The river, cutting rapidly downward, is keeping pace with the rapid uplift of this plateau.

Drainage Patterns

slope wash
Unchanneled water that moves downhill; also, the rock and soil that are transported by the water.

stream divide
The boundary that separates adjacent stream valleys.

Viewed from the air, streams and their tributaries etch characteristic drainage patterns into the landscapes that they drain. A large pile of earth left uncovered around a construction site is an especially good place to observe the development of a drainage pattern in miniature. Watch carefully the way water runs downslope during a heavy rain. If rain comes down hard and fast enough, it soon exceeds the capacity of soil and sand to absorb moisture, and the excess water runs off the pile in thin sheets called **slope wash.** Downslope, the sheets break into turbulent ribbons of water called *rills,* caused by local slope irregularities and slight differences in composition of the materials. The rills scour sediment and erode channels in the earth pile, creating small stream valleys. The hill crests separating neighboring valleys are the **stream divides,** and the areas encompassed within the divides are the drainage basins. Rainwater striking the pile within a given drainage basin descends downslope to the stream channel.

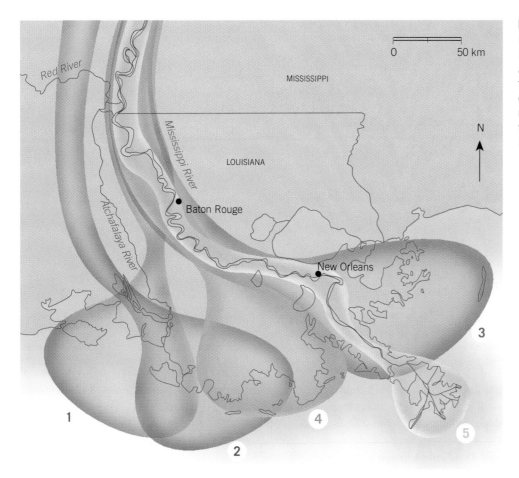

Figure 15.23 Some of the shifting outlets of the Mississippi River over the last few thousand years. Notice that the Atchafalaya River offers a shorter route to the Gulf of Mexico than the present course of the Mississippi does. Elaborate measures are employed to prevent the Mississippi from bypassing New Orleans, one of the nation's leading ports.

With time, an intricate *stream system* will evolve that consists of a main trunk and branching tributaries. Its drainage basin will then include all the drainage basins of streams that belong to the system. Thus all the water and sediment of the tributaries will be transported off the pile through the main stream. The stream system is in a constant state of adjustment. With each rain, the main stream cuts deeper into the pile, which forces the tributaries joining the main stream to follow suit. As the streams cut deeper into their channels, the heads of the tributaries gain the energy to eat farther back into the earth pile through **headward erosion** and to develop many branches of their own. Eventually, this system and competing systems drain the entire surface of the pile. In this manner, the pile is eroded and—given enough rainfall—is eventually flattened.

Drainage patterns to a large extent reflect underlying rock structures and differential erosion rates. In our example, the pile is essentially homogeneous and exhibits tree-like **dendritic drainage.** It is characteristic of regions underlain by materials of uniform composition or by horizontal layers whose properties do not change laterally (Figure 15.26a). Dendritic drainage is a pattern of random development based on rainfall and regional slope. However, much of the Earth's surface is underlain by rocks of variable composition and structure. Streams carve channels in belts of easily eroded rock and in zones of faulting and fracture, in which rocks have been mechanically crushed and weathered. Thus geologists are often able to identify underlying rock structures and compositions from inspection of drainage patterns shown in aerial photographs and maps.

headward erosion
The lengthening of a stream or gully by erosion at the head of the stream valley.

dendritic drainage
A stream drainage pattern resembling a branching tree that is common to regions underlain by materials of uniform composition or horizontal layers.

Figure 15.24 Stream terraces incised in the flood plain of the Strathmore River, Scotland. Point bars and cutbanks are well displayed along the course of the meandering stream.

trellis drainage
A drainage pattern in which the main stream cuts across the regional structure (typically folded or tilted strata), and tributaries follow parallel belts of weak strata that lie perpendicular to the main stream.

radial drainage
A stream pattern that radiates like the spokes of a wheel from the cone summit of a volcano.

Trellis drainage is characteristic of regions underlain by parallel belts of tilted strata, having differing resistances to erosion (Figure 15.26b). Main streams follow the regional slope and cut across the strata. Tributary streams then erode valleys in the weaker strata. The entire system resembles the pattern that vines make on a trellis. Trellis drainage is well displayed in the gently inclined strata of the Atlantic Coastal Plain and in the deeply eroded folds of the Appalachian Mountains of Pennsylvania. *Annular drainage* is similar to the trellis drainage of fold-mountain belts but is primarily associated with domes and basins (Figure 15.26c). The annular ring pattern develops because erosion of these structures produce circular rather then parallel rock exposures. Another pattern, organized like the spokes of a wheel, is the **radial drainage** common to the flanks of volcanoes (Figure 15.26d). It is formed by streams that flow outward in all directions from

Figure 15.25 The incised meanders of the San Juan River, Utah, are the result of stream rejuvenation. Rapid uplift has energized the stream, which has responded by downcutting its winding channel.

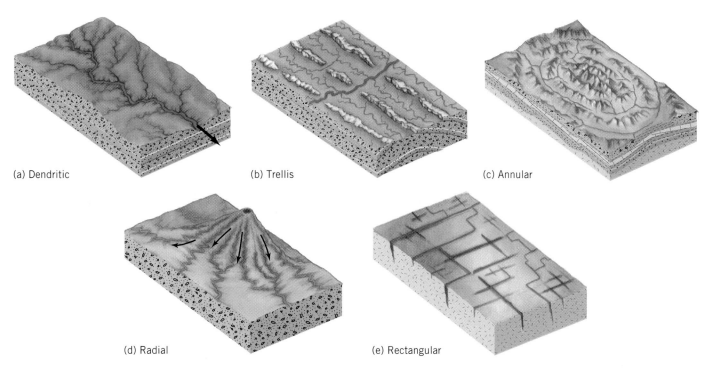

(a) Dendritic

(b) Trellis

(c) Annular

(d) Radial

(e) Rectangular

Figure 15.26 The relation of five common drainage patterns to underlying rock structure.

the cone summit. The stream valleys originate at the volcano base, and work upslope by means of headward erosion.

Often, fractures and faults form the major structural elements of massive rocks of homogeneous composition—commonly granite. These features may intersect at right angles and provide zones of weaknesses for streams to exploit. A pronounced **rectangular drainage** pattern develops where the stream system incises channels along the joint and fault planes (Figure 15.26e).

rectangular drainage
A drainage pattern commonly found in homogeneous igneous and metamorphic rocks in which tributary streams display right-angled bends that follow joints and faults.

Reorganization of Drainage (Stream Piracy)

As they work backward into their upland regions through headward erosion, the drainage divides between the tributaries of the same and neighboring stream systems are narrowed. The stream having the steeper gradient possesses a distinct advantage over its neighbor in offering a shorter, more rapid route to base level. Eventually, the steeper stream may undercut the divide between the two and divert through its own channel the headwaters of the neighboring stream. Appropriately enough, the process is called **stream piracy** (Figure 15.27). Often, an abandoned valley or *wind gap* exists where the diverted headwater of the pirated stream used to flow.

Stream piracy is the way drainage systems adjust in order to achieve maximum energy efficiency in the course of transporting water and sediment to the sea. As tectonic and climatic changes occur, the stream adjusts accordingly, and stream piracy is one type of response.

stream piracy
The diversion of a stream into another stream that has a steeper gradient.

Streams and Plate Tectonics

The reversed course of the Earth's largest stream, the Amazon River, provides a spectacular example of the influence of tectonic forces on drainage patterns. Today, the Amazon

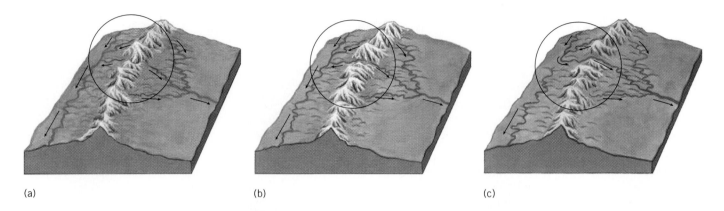

(a)   (b)   (c)

Figure 15.27 Drainage reorganization through stream piracy. (a) The streams flowing down the east side of the mountain range have steeper gradients than streams on the west side. (b) Streams on both sides of the mountain range cut into the mountains in headward erosion. (c) When their channels meet, the higher-gradient stream diverts or captures the headwaters of the stream with the gentler gradient.

system drains eastward into the Atlantic Ocean (Figure 15.28). Its headwaters flow down the eastern slopes of the Andes Mountains, which run like a spine along the western edge of South America. But 100 million years ago, the Andes Mountains did not exist, nor did the Atlantic Ocean. The super-continent Gondwanaland had begun to fragment, but South America and Africa were still joined. Flood-plain and deltaic sediments along the western slopes of the Andes, in what is now Peru and Colombia, prove that the Amazon drained westward into the Pacific Ocean, and this evidence is supported by the many organisms in the present Amazon whose origins are traceable to the Pacific, not the Atlantic.

As the fragmentation of Gondwanaland proceeded, the South American plate drifted westward and collided with the eastward-spreading Nazca plate, which initiated the uplift of the Andes Mountains. About 15 million years ago, the rising Andes blocked drainage to the west, and an inland lake developed over the vast lowland of what is today the Amazon basin. But as we have mentioned, lakes are temporary. Several million years ago, one of the streams that flowed to the Atlantic lengthened its course through headward erosion and provided the lake with an outlet to the east. The stream captured not only the waters of the lake but also the drainage systems of all the streams that had flowed into the lake. This reorganized and redirected drainage constitutes the present Amazon River system.

Figure 15.28 The reversal of the Amazon River is an example of how plate movements and mountain building may alter drainage patterns.

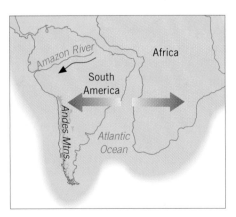

(a) 100 million years ago

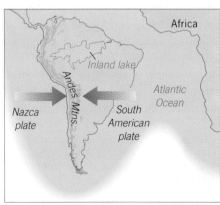

(b) 15 million years ago

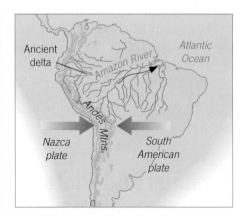

(c) Present

> ## STUDY OUTLINE

Streams are defined as channeled flows of any size; their three major functions are erosion, transport, and deposition of sediment. In the hydrologic cycle, water circulates from ocean to atmosphere to land and is returned to the ocean by streams. A stream system consists of the main trunk, **tributary streams,** and **drainage basin.**

I. THE ENERGY OF STREAMS. As streams flow downhill, they convert their potential energy into motion. The steeper the stream **gradient,** or slope, the more rapid the flow. All streams have concave **long profiles.** A stream cannot erode its channel below **base level. Local base levels,** such as lakes, are eventually destroyed by the stream. Sea level is the **ultimate base level.**

    A. As stream **discharge** increases, so do stream velocity, channel width, and channel depth.

    B. A stream flowing at low velocity through a glass tube will exhibit **laminar flow.** However, in nature, friction drains away most of a stream's energy. As friction is increased, streams move in **turbulent flow.** Maximum turbulence occurs just above the channel bottom, ensuring maximum erosion. Particles spinning in whirlpools may scour **potholes** in the channel.

    C. Materials are transported by streams in three ways: dissolved ions in **dissolved load,** fine particles in **suspended load,** and larger particles in **bed load.**

        1. In humid regions where chemical weathering is high, streams have greater dissolved loads. In arid regions, streams have low dissolved loads and high suspended loads.

        2. **Capacity** is the amount of sediment that a stream can carry, and **competence** is the size of particles that it can carry.

        3. If a particle's **settling velocity** is low relative to stream turbulence, it is transported in suspension. Heavier particles are rolled along the bottom; lighter ones are skipped along in **saltation.**

II. THE GRADED STREAM. A **graded stream** is one that has just enough energy to transport its load without eroding its channel or depositing sediment.

    A. Variable conditions are constantly throwing a stream out of grade; increased erosion and deposition act to return the stream to grade, as do changes in channel dimensions.

    B. Irregular features in the stream profile, such as waterfalls and lakes, are destroyed over time by channel erosion at high points and deposition at low points.

III. TWO CONTRASTING STREAM TYPES

    A. Where a stream must transport excessive bed load under low discharge, it develops a series of interwoven subchannels that give it a braided appearance. **Braided streams** commonly form in **alluvial fans** at the foot of desert mountains.

    B. As they approach base level, most streams form **meanders,** or loops. A meandering stream has a large suspended load but adequate discharge to carry it.

        1. A meander is initiated where the water is deflected as it flows over a series of riffles and pools. **Cutbanks** and **point bars** form on the outer and inner channel walls.

        2. Meanders tend to migrate downstream. The segment of land between the loops may be breached, forming a **meander cutoff.** The stranded portion of the meander forms an **oxbow lake.**

IV. THE FLOOD PLAIN

A. A meandering stream flows over a self-constructed **flood plain.** The flood-plain depositional process is called **aggradation.**

B. Deposits on the sides of the stream channel, called **natural levees,** serve as partial barriers against future floods.

C. The debate over flood control continues.
  1. Flooding of the Mississippi River in 1993 was a natural event, but the devastating results may have been intensified by human interventions designed to protect against floods.
  2. Supporters of the levee system of flood control point to the economic benefits of flood protection; opponents claim that confining the river forces it to grow higher and flow faster, which increases flood severity when the levees are breached. Softer approaches to flood control are being investigated.

V. DELTAS

A. When a stream enters an ocean or lake, its sediment load is deposited near the mouth and accumulates in a triangular structure called a **delta.** The sediment, which is brought to the front by branching **distributary** channels, is deposited as foreset, topset, and bottomset beds.

B. Many deltas have built-in cycles of growth and retreat. The Mississippi River, for example, has shifted outlets many times in the past.

VI. STREAM REJUVENATION.   A stream that is revived through tectonic uplift or increased discharge will cut into its flood plain and leave behind **stream terraces** that mark the former level of the plain. Tectonic uplift and stream downcutting also form **incised meanders.**

VII. DRAINAGE PATTERNS

A. Water from heavy rainfall runs downslope in thin sheets called **slope wash.** The water collects in turbulent ribbons called rills, which erode channels, creating small stream valleys. The crests between channels form **stream divides.** In time, channels join together to form tributaries and a main trunk. The tributaries develop branches of their own through **headward erosion.**

B. Stream systems take the following patterns.
  1. **dendritic drainage:** a treelike pattern characteristic of horizontal strata of homogeneous rock
  2. **trellis drainage:** a vinelike pattern found in fold-mountain belts
  3. **radial drainage:** a pattern that radiates from a volcanic crater
  4. **rectangular drainage:** a pattern that forms along joint and fault planes

C. **Stream piracy** occurs when a steeper stream undercuts a divide and diverts the waters of a neighboring stream through its own channel.

VIII. STREAMS AND PLATE TECTONICS.   When landmasses separate or collide, drainage patterns are altered and may be reversed, like the Amazon River, whose ancestor flowed west to the Pacific.

STUDY TERMS

aggradation (p. 405)
alluvial fan (p. 403)
base level (p. 396)
bed load (p. 398)
braided stream (p. 403)
capacity (p. 400)
competence (p. 400)
cutbank (p. 404)
delta (p. 410)
dendritic drainage (p. 415)
discharge (p. 396)
dissolved load (p. 399)
distributary (p. 410)
drainage basin (p. 394)
flood plain (p. 405)
graded stream (p. 400)
gradient (p. 396)
headward erosion (p. 415)
incised meander (p. 414)
laminar flow (p. 398)
local base level (p. 396)

long profile (p. 396)
meander (p. 403)
meander cutoff (p. 405)
natural levee (p. 405)
oxbow lake (p. 405)
point bar (p. 404)
pothole (p. 398)
radial drainage (p. 416)
rectangular drainage (p. 417)
saltation (p. 400)
settling velocity (p. 400)
slope wash (p. 414)
stream (p. 393)
stream divide (p. 414)
stream piracy (p. 417)
stream terrace (p. 414)
suspended load (p. 399)
trellis drainage (p. 416)
tributary stream (p. 394)
turbulent flow (p. 398)
ultimate base level (p. 396)

CRITICAL THINKING QUESTIONS

1. Describe the hydrologic cycle. Emphasize the role of streams in your discussion.
2. Use the terms *energy* and *base level* to explain why streams cannot erode their channels below sea level.
3. Explain how particles are lifted from the stream bed. Use the terms *turbulence* and *hydraulic action.*
4. How does climate affect the amount of dissolved and suspended load in a stream?
5. Waterfalls and lakes are temporary features in the life of a stream. Explain.
6. Describe the downstream changes in channel dimensions, discharge, and sediment load in a typical stream. How might they explain how a stream changes from a braided to a meandering pattern?
7. Explain what happens to a stream when it reaches a lake or ocean. You may want to include the following terms in your discussion: delta, channel, sediment, distributary, outlet, deltaic bed.
8. Describe how a straight stream becomes a meandering stream.
9. Describe the development of a stream system. You may want to include the following terms in your discussion: sheet, rill, channel, valley, divide, drainage basin, tributary, main stream.
10. Describe the extent of human interference with the Mississippi River system. In your opinion, has the interference been generally beneficial or harmful? What are some possible alternatives to present practices?

CHAPTER 16

Groundwater

Underground water makes up about 95 percent of the world's supply of freshwater, exclusive of glacial ice. To gauge the enormity of this hidden resource, consider that the volume of water trapped within a half mile of the continental surface is 40 times greater than the water residing in all rivers and lakes. However, these reserves are not inexhaustible, nor are they evenly distributed. Often, the demand is greatest where supply is least, such as in arid regions brought under agricultural cultivation.

Underground water—extracted at a rate of 340 billion liters per day—provides one-fifth of the water used in the United States. It is a fragile resource subject to contamination from agricultural, industrial, and municipal sources. Approximately 25 percent of the supply of underground water considered usable for drinking in the United States is contaminated. This percentage is likely to increase, because contamination has not been brought under control despite tougher cleanup laws.

Underground water is a major component of the hydrologic cycle. It feeds streams and lakes and contributes heavily to the surface runoff that returns water to the sea. It also plays a central role in rock weathering and erosion. Soluble ions that are dissolved by underground water from rocks are transported directly (or via streams) to the ocean. Thick deposits of limestone are particularly susceptible to dissolution, which forms most of the world's spectacular caves. The other side of the chemical coin is that underground water precipitates elements, leading to the formation of economically valuable mineral deposits and to such fascinating features as Arizona's Petrified Forest.

Like the hot water in a home heating system, underground water also serves as the working substance in heat exchange between magmas at depth and the Earth's surface.

◀ Groundwater springs issuing from cliffs in Grand Canyon National Park, Arizona.

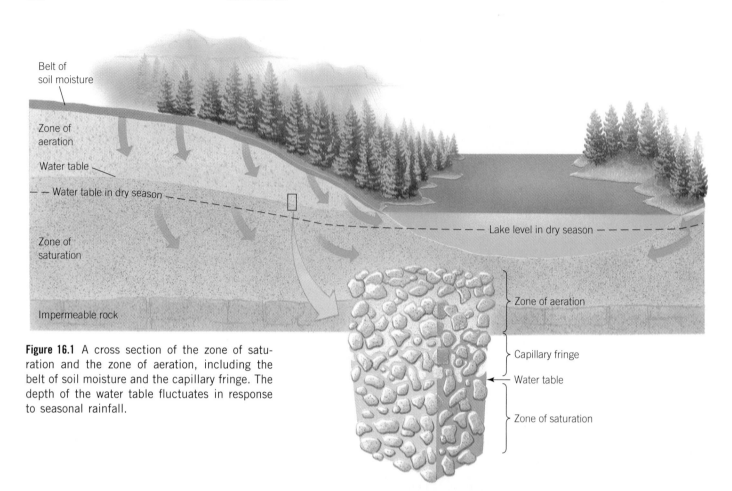

Belt of
soil moisture

Zone of
aeration

Water table

— — Water table in dry season — — — —

Zone of
saturation

Impermeable rock

Zone of aeration

Capillary fringe

Water table

Zone of saturation

— — — Lake level in dry season — — — — — — —

Figure 16.1 A cross section of the zone of saturation and the zone of aeration, including the belt of soil moisture and the capillary fringe. The depth of the water table fluctuates in response to seasonal rainfall.

This action is responsible for geysers and hot springs. Increasingly, we are learning to harness heated underground water as a largely untapped supply of geothermal energy.

This chapter describes how water accumulates underground and migrates through the subsurface; how it is extracted through wells; and how it does the geologically important work of forming caves, shaping the landscape, preserving fossils, and creating hot springs and geysers. The role of humans in polluting this valuable natural resource will also be examined.

Accumulation of Groundwater

Water that strikes the land surface as rain either evaporates, runs off directly as overland flow, or seeps into the ground. The amount of seepage in a given location depends upon a number of factors: the degree of slope; the frequency, intensity, and duration of rainfall; the absorbing and transmitting capacity of the subsurface; and the vegetation present. The rainwater that does seep into the ground percolates beneath the surface through the interconnected spaces between soil, sediment, and rock particles of the **zone of aeration.** Eventually, the water reaches a depth where movement is retarded. There, it accumulates and completely fills the voids between solid materials, forming a **zone of saturation** (Figure 16.1). Many hydrogeologists distinguish between the general term *underground water*, which encompasses all water beneath the surface, and the more specific term **groundwater,** which refers to water in the zone of saturation.

The upper limit of the zone of saturation, called the **water table,** fluctuates with the seasons, rising during periods of high rainfall and sinking when rainfall is scarce. Thus the zone of saturation, a natural reservoir, plays a key role in regulating the regional water economy. By slowly releasing water to streams during dry spells and storing it during heavy rains, it lessens the severity of droughts and floods.

zone of aeration
The subsurface zone between the water table and the surface; it retains little moisture except within the belt of soil moisture and the capillary fringe.

zone of saturation
A subsurface zone where groundwater accumulates and completely fills the voids in sediments and rocks.

groundwater
Underground water within the zone of saturation, below the water table.

water table
The surface marking the upper limit of the zone of saturation.

Though largely dry, the zone of aeration does contain two significant accumulations of water: the **belt of soil moisture** directly beneath the surface and the capillary fringe just above the water table. The belt of soil moisture is formed when a small portion of the water that seeps beneath the surface is trapped in the upper few meters of soil by sand, silt, clay, and organic matter. Plants draw on the soil moisture in this thin belt; the presence of this water is the reason many species survive dry spells. But the soil is extremely loose and pockmarked with voids left by decayed roots and burrowing animals, which allows most rainwater to percolate down to the zone of saturation.

The **capillary fringe** is the wet halo that extends upward about a meter or so from the water table. It results from the adhesive forces between mineral grains and water molecules that work against gravity to draw the water up from the saturated zone into small openings between grains. You can observe the same kind of thing by dipping the edge of a napkin into a glass of water. The boundary between the wet and dry portions of the napkin will rise above the water surface.

belt of soil moisture
The thin layer of moisture just beneath the land surface; the uppermost subdivision of the zone of aeration.

capillary fringe
The thin belt just above the water table within which moisture is drawn up by surface tension and partly fills the voids in sediments and rocks.

Porosity and Permeability

If you fill a container with common sand and then pour water into the sand all the way to the top, you may be surprised to find out just how much water the sand absorbs—up to 25 percent of the volume of the container. What you have measured is the sand's **porosity**—that is, the ratio of voids to the total volume of the sample.

porosity
The ratio of open space to total volume of a subsurface material.

The porosity of loose material depends mainly on three factors. The first, *sorting,* is the degree of uniformity of particle size. A poorly sorted gravel and silt mixture has less open space than pure gravel (Figure 16.2a). The second factor, *packing,* is a measure of how tightly the particles are arranged; the tighter the arrangement, the less the porosity (Figure 16.2b). The third factor is the *shape* of the particles. For example, spherical particles leave more open space than nonspherical ones (Figure 16.2c). Note that particle size is not on this list; it does not directly affect porosity. However, natural sedimentation processes tend to sort smaller sediments more effectively than larger ones; so, for example, clay is typically twice as porous as gravel, even though the voids in the gravel are far more apparent.

Typical soil is extremely porous because of its tendency to form clumps and because of its root systems and animal burrows. A soil's degree of porosity is an important component of its fertility, which is one reason that farmers plow the soil before they plant. The porosity of consolidated sedimentary rocks is affected by the three factors listed above and by the material that cements the particles. Where particles are partially cemented, many open spaces may remain; however, if cement fills the voids between particles, then obviously porosity is much reduced. If the cementing agent is soluble, then groundwater will dissolve it. Hard, massive rocks—like granite, basalt, and certain kinds of limestone—would appear at first glance to lack any space for water. However, because these rocks are frequently fractured, they may hold great

Low porosity ⟵⟶ High porosity

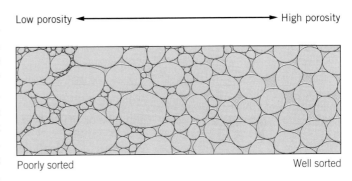

Poorly sorted   Well sorted

(a) Sorting

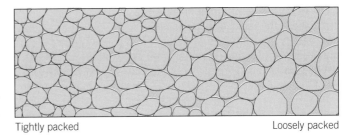

Tightly packed   Loosely packed

(b) Packing

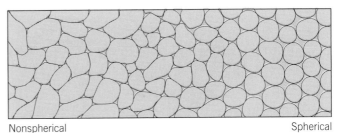

Nonspherical   Spherical

(c) Particle shape

Figure 16.2 In general, the particles of highly porous (coarse) sediments are (a) well sorted, (b) loosely packed, and (c) spherical. These characteristics maximize the open space within the sediment.

Table 16.1   Porosity and Permeability of Various Materials

Material	Porosity	Specific Yield	Specific Retention	Permeability
Soil	55%	40%	15%	Good
Clay	50	2	48	Poor
Sand	25	22	3	Excellent
Gravel	20	19	1	Good
Limestone	20	18	2	Poor to excellent
Sandstone	11	6	5	Poor to very good
Granite	0.1	0.09	0.01	Very poor

Source: Adapted from USGS Water-Supply Paper 2220, p. 9.

volumes of water. Basalt and other lavas are often quite porous owing to the holes left behind by escaping gas, lava tubes, and other features.

Although porosity is an important measure of the space available for groundwater, it does not specify how much water we can tap from a given host material, for thin films of water adhere to particle and rock surfaces. Because this water migrates slowly, if at all, it is not easily tapped by wells and therefore does not constitute a ready supply of water. Extraordinary measures such as blasting are often necessary to extract this water. *Specific yield* refers to the percentage of water that will drain under the influence of gravity; *specific retention* refers to the percentage of water that adheres to the rock surfaces.

Knowledge of the **permeability** of sediments and rock samples—that is, the ease with which they transmit water—is also important to our understanding of groundwater accumulation and movement. Permeability is measured by the volume of water that will move through a given cross-sectional area of material in a given period of time, subject to a given pressure difference. Table 16.1 compares the porosity, retention, yield, and permeability of various sediments and rocks. Again, note that clay, with a porosity of 50 percent, usually yields only 2 percent of its water. The balance is held within the small openings between grains and within the crystal structure of clay itself. By contrast, gravel yields most of the water that it holds.

The contrasting retentive properties of gravel and clay illustrate that while a permeable substance must also be porous, porous substances are not necessarily permeable. For example, the volcanic rock pumice contains so many open spaces that it floats in water; the spaces are not connected, however, and so water cannot penetrate it. Thus permeability depends upon the size, arrangement, and interconnectedness of openings,

permeability
The capacity of a material to conduct water; measured by the volume of water that will move through a cross-sectional area in a given period of time, subject to a given pressure difference.

Figure 16.3 A lens of impermeable strata above the regional water table may trap groundwater as it percolates downward, thus creating a perched water table.

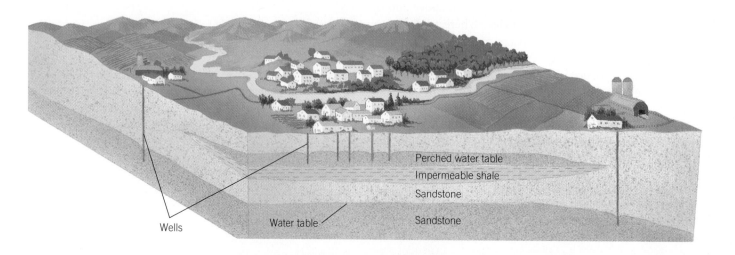

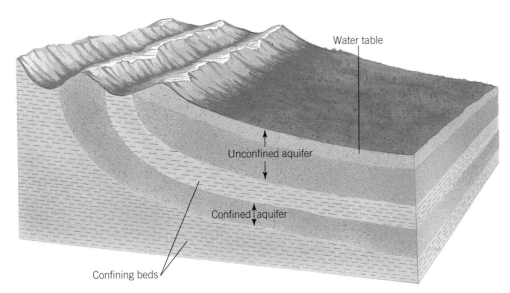

as well as on porosity. Unfractured granite, shale, and clay are impermeable; most soils, cavernous limestone, gravel, and sandstones are relatively permeable.

The porosity and permeability of rock and loose material frequently vary horizontally and vertically, and these changes affect groundwater accumulation and movement. Figure 16.3 shows what happens when downward-percolating groundwater encounters an impermeable barrier, or **confining bed**—typically, a lens of clay within a sand or gravel deposit, or a shale layer within sandstone or conglomerate—before reaching the zone of saturation. Groundwater accumulates directly over the lens and forms a **perched water table** above the main zone of saturation. Perched water tables are often extensive, and wells sunk in them can extract large volumes of water.

confining bed
An impermeable layer adjacent to an aquifer that prevents escape of groundwater from the aquifer.

perched water table
The upper limit of a local groundwater body stranded above the regional water table by an impermeable layer of rock or sediment.

Aquifers

Groundwater is contained within porous and permeable layers called **aquifers** (Figure 16.4). Where rainwater is able to percolate directly from the surface to the zone of saturation, it is contained within an *unconfined aquifer,* so named because the water table can rise or fall freely. However, aquifers are often sealed between confining beds, and this arrangement prevents local rainwater from percolating down to it. The groundwater in these *confined aquifers* originates in another region where the aquifer is exposed at higher elevation and receives precipitation directly. The groundwater in a confined aquifer is generally under considerable pressure and often will rise to the surface when wells are sunk. Wells like these, in which groundwater rises above the level of the aquifer under its own pressure, are called **artesian wells.** We will return to the subject of wells later in the chapter.

aquifer
A permeable body of subsurface sediment or fractured rock that conducts water.

artesian well
A well under sufficient hydrostatic pressure to force the water to rise above the top of the aquifer.

Movement of Groundwater

The term *water table,* taken literally, can be misleading, for it implies that the top of the zone of saturation forms a level surface beneath the countryside. Instead, the water table follows the outline of the topography, although in a more subdued and less detailed fashion. It is generally higher beneath hills than it is beneath valleys.

In lowland areas, the water table may intersect the land surface, forming swamps and marshes. More typically, it intersects stream channels. Streams that gain groundwater in this manner are called **effluent** (or gaining) **streams** (Figure 16.5). In arid regions, the water table may lie beneath the stream channel, in which case the

effluent stream
A stream that intersects the water table and gains water from the zone of saturation.

Figure 16.5 An effluent stream is fed by groundwater. Note the curved path of the groundwater migration from high to low elevations.

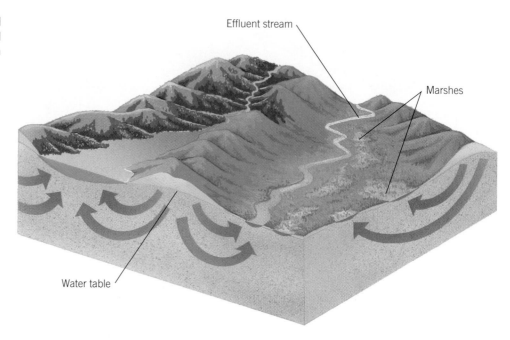

influent stream
A stream that lies above the water table and loses water to the zone of saturation.

spring
A place where the land surface intersects the water table and groundwater seeps or flows naturally out of the ground.

hydraulic gradient
The slope of the water table, measured by the difference in elevation of any two points along the water table divided by the horizontal distance between them.

influent (or losing) **stream** feeds the water table (Figure 16.6). At higher elevations, the water table may not mirror the topography because of abrupt changes in rock permeability, faulting, or other factors. In these instances, the water table may intersect the land surface where groundwater issues directly onto the surface as a **spring** (Figure 16.7).

Like all continuous bodies of water, groundwater in the zone of saturation flows from those regions where the water stands high to those where it is low. The velocity of flow is determined by the permeability of the material through which it flows and by the **hydraulic gradient.** The hydraulic gradient, or the slope of the water table, is equal to the difference in elevation of any two points on the water table divided by the horizontal distance between them.

The relationships that govern groundwater flow were explored in 1856 by the French engineer Henry Darcy, who derived the fundamental equation of groundwater movement (Figure 16.8). Known today as Darcy's law, it states that the velocity of ground-

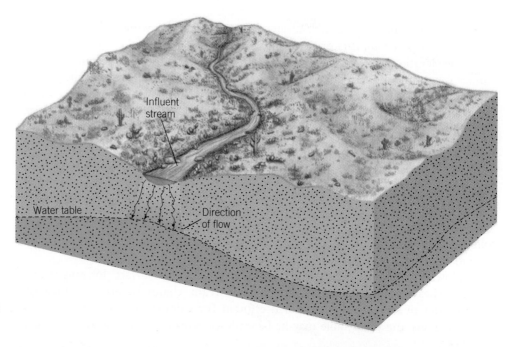

Figure 16.6 An influent stream feeds the water table.

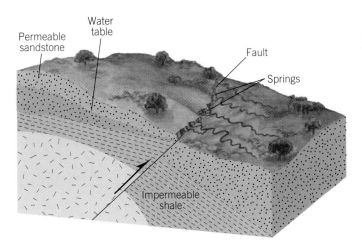

Figure 16.7 One of the numerous subsurface conditions that may lead to the occurrence of springs: faulted strata cause the water table to intersect the land surface.

water flow is equal to the hydraulic gradient of the water multiplied by a constant that represents the permeability of the sediment or rock through which the water flows. Groundwater moves in arcs through the subsurface; the deeper it penetrates before surfacing, the more slowly it flows. Sometimes, deep groundwater can take thousands of years to surface.

The normal flow of groundwater is from **recharge areas** (the broad uplands between stream valleys that receive rainfall and soak up groundwater), through the zone of saturation, to **discharge areas** (commonly, streams, lakes, and marshes). In time, the configuration of the water table reflects the balance between the rate of recharge, the rate of discharge, and the porosity and permeability of the subsurface. If these factors remain fairly constant, so will the configuration of the water table.

recharge area
The area—mainly the broad upland between stream valleys—that receives precipitation and adds water to the zone of saturation.

discharge area
The area—primarily stream channels—that receives groundwater from the zone of saturation and conducts it away.

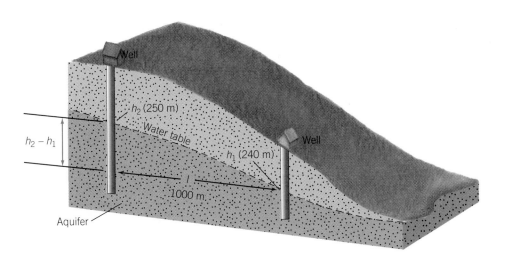

To calculate the hydraulic gradient *(H)*:

$H = \dfrac{h_2 - h_1}{l} = \dfrac{(250 - 240)\ m}{1000\ m} = 10$ meters per kilometer

To calculate the velocity of flow using Darcy's law:

$V = K \times \dfrac{h_2 - h_1}{l}$

where *K* is a constant that expresses the hydraulic conductivity (or permeability) of a given host material.

If an aquifer has a hydraulic conductivity of 1000 meters per day (mpd):

$V = 1000\ mpd \times \dfrac{(250 - 240)\ m}{1000\ m}$

$= 10\ mpd$

Figure 16.8 Determining the hydraulic gradient and the velocity of groundwater flow using Darcy's law.

Figure 16.9 Wells and the water table. (a) Pumping leads to a general lowering of the water table, which forms cones of depression surrounding the wells. (b) The graph shows that the water table falls during a pumping period and rises when pumping ceases; however, unless water is put back into the ground, the level of the water table gradually declines.

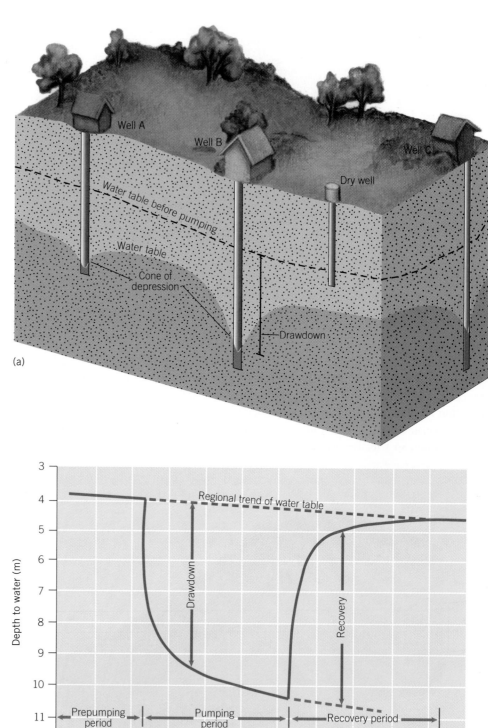

(a)

(b)

Wells

Wells that extract water from an unconfined aquifer are known as *water-table wells*. As points of artificially created discharge, they depress the water table in the vicinity of the well. Thus a **cone of depression** is produced in the aquifer's water table as water is drawn into the well from all directions (Figure 16.9).

cone of depression
A concentric indentation in the water table that develops around a well.

As Figure 16.9 illustrates, *drawdown* is the vertical difference between the water table before pumping and the level in the well during pumping. Note that the initial draw-down rate is extremely rapid as the cone develops, but it levels off when a stable condition is reached. Also note that even a stable cone of depression will slowly widen with time, because as the water table is lowered, the groundwater that is extracted must come from greater distances. The widening cone means that the hydraulic gradient between the top and the bottom of the cone is being reduced. This is bad news, because energy must be supplied by the pump to compensate for energy loss from the declining gradient. Water then becomes increasingly expensive to pump from the ground.

Long-term pumping at high rates from a large well can produce a cone of depression with a radius of 3 kilometers or more. Frequently, the cone may extend beneath neighboring properties, causing wells there to run dry. Lawsuits have been fought over this issue—water is the one thing that crops, livestock, factories, and people cannot do without. Where demand is high, many closely spaced wells are sunk. Cones of depression intersect and lead to a general lowering of the water table, to everyone's detriment.

Artesian Wells

Early in this century, railroad planners faced a serious problem: to find a reliable supply of water that could feed the huge steam-driven locomotives that moved goods and people across the arid South Dakota plains. Planners turned to the United States Geological Survey for help, and it is interesting to see how the USGS geologists solved the problem. They began by investigating the geology of the Black Hills dome of South Dakota. As we learned in Chapter 11, the Black Hills consist of a core of ancient igneous and metamorphic rocks encircled by a thick sequence of sedimentary strata. At one time, the strata covered the entire core region, but they have long since been eroded. As with all domal features, strata dip away from the core and are found at great depth far from the uplift.

Prominent in the sequence of strata is the Dakota sandstone, a porous, permeable formation that was deposited when a broad, shallow sea covered the western United States approximately 100 million years ago. In the vicinity of the Black Hills, the Dakota sandstone forms a thick ridge that is sandwiched between impermeable shales. On the basis of these relationships, survey geologists predicted that a 1000-meter-deep well drilled several hundred kilometers to the east—a location convenient to the railroad— would tap the Dakota sandstone. The prediction proved correct when water gushed from the new well.

The Dakota sandstone is the key element of a classic artesian well system—one in which well water rises above the level of the confined aquifer without the need for pumping (Figure 16.10). The Dakota sandstone is a porous and permeable aquifer with impermeable confining beds above and below the aquifer that prevent groundwater escape. The Black Hills recharge area receives water at a greater elevation than the out-let or discharge well. Therefore, water at the discharge well is under enough hydraulic pressure to rise toward the surface.

If it were not for friction, water spouting from wells drilled in the aquifer would rise to the same elevation as the aquifer at recharge. Because of friction, however, the level to which artesian well water is able to rise decreases with the distance of the well from the recharge area. The sloping line $(X - X')$ in Figure 16.10 indicates how this level changes with distance. If the land surface at a particular location lies beneath this line, water will reach the surface under its own power. Should the land surface lie above the line, water will rise only partially to the surface, and pumping is required to bring it up the rest of the way.

The Dakota sandstone and similar confined aquifers are the major water sources for numerous cities, towns, and farms of the Plains states. First tapped in the 1870s, they probably had more influence on the settlement of the American West than the covered

Figure 16.10 The conditions necessary for an artesian well in a confined aquifer. Well B is far below the height of the recharge area, so water gushes to the surface under hydrostatic pressure. Were it not for friction, the water at this location would rise to the level of the recharge area, but in actuality, it reaches level *X* – *X'*.

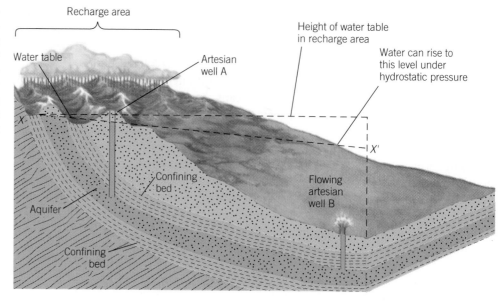

wagon. Artesian systems also are important water sources in the upper midwestern states and in the humid Atlantic and Gulf states.

Well Extraction Problems

As we mentioned at the beginning of this chapter, groundwater reserves are not inexhaustible. Excessive extraction can lead to land subsidence, saltwater encroachment, and loss of hydraulic pressure.

Land Subsidence

Discharge from an artesian well triggers a pressure drop surrounding the well similar in form to the cone of depression in a water-table well. However, the drop in an artesian well is commonly more rapid and far deeper and extends over a wider area, because water in a confined aquifer plays an important structural role. The weight of overlying strata is supported by the framework of the particles that compose the aquifer and by the fluid pressure of the water in the pore spaces between particles. As the well draws water from the confined aquifer, the particles distort and pack together more tightly, squeezing water from the pore spaces as from a sponge. The framework can no longer support the load and begins to collapse, causing the land surface above to subside, or sink. Thus excessive extraction of water from a confined aquifer may lead to **subsidence** of the land. Subsidence is accelerated if there is additional surface load in the form of buildings, roads, and other structures.

Subsidence may cause many problems. Subsurface utilities, such as water and drainage pipes, are commonly ruptured, and electrical installations are damaged. Foundations may shift, causing shear fractures in the supports, warped doors and windows, and other things that make buildings uninhabitable. Cracks can develop suddenly in roads, and roadbeds can sag. If subsidence occurs near coastal regions, the land surface may actually drop below sea level, which accelerates coastal erosion, shifts stream courses, and causes flooding.

Land subsidence can be seen in the San Joaquin Valley of California, where groundwater extraction to feed this rich agricultural region has led to a lowering of the land surface by 9 meters in 50 years (Figure 16.11). Mexico City bears the dubious distinction of displaying some of the most disastrous effects of land subsidence, because it rests on soft lakebed sands, silts, and clays. Excessive pumping of water from these aquifers has caused homes to sink as much as 9 meters, turning the intervening streets into ele-

subsidence
The sinking of an area of the Earth's surface upon compaction or dissolution of subsurface materials, often due to groundwater extraction.

Figure 16.11 The result of groundwater pumping and subsurface settlement in the San Joaquin Valley, California. In 1955 the land surface stood where indicated on the pole.

vated highways. In fact, the entrance to the magnificent Teatro de los Artistas is now on what was designed to be the second floor; the building has sunk to the point where the former entrance hall is now below street level. One of Mexico's most sacred religious shrines, the Shrine of Our Lady of Guadalupe, has tilted so dangerously over the years that it can no longer be used and a new cathedral has been built.

Saltwater Encroachment

In coastal regions, groundwater in the zone of saturation floats on saltwater that has seeped in from the ocean and penetrated the strata. The density relationship between the two water bodies is such that for every 1 meter that the water table rises above sea level, the freshwater sinks 40 meters below sea level (Figure 16.12). Therefore, a 1-meter well drawdown of the water table results in a 40-meter rise of the saltwater beneath the well. Unrestrained pumping thus brings saltwater up into the well very quickly.

Artesian wells are particularly vulnerable to **saltwater encroachment,** because of the rapid, deep, and widespread pressure drops in confined aquifers. As saltwater works its way up in the aquifer, it contaminates wells ever farther inland; it is thus a major problem along some reaches of the Atlantic Coastal Plain. Needless to say, saltwater contamination makes water useless for drinking and industrial purposes; it also ruins machinery and the wells themselves. It is not necessary for seawater to actually invade

saltwater encroachment
The displacement in the zone of saturation of freshwater by saltwater.

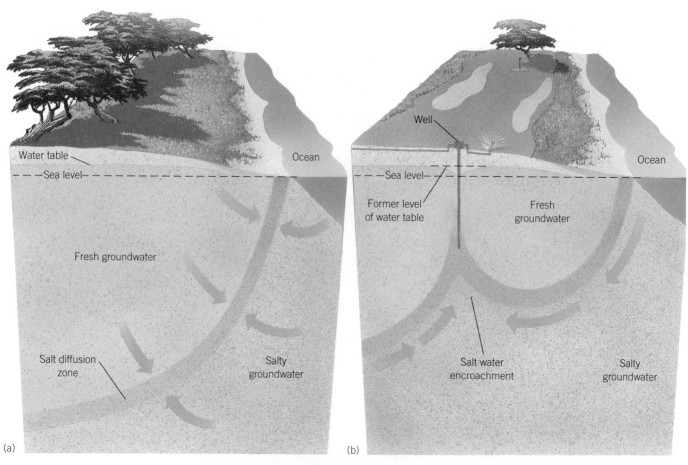

Figure 16.12 Saltwater encroachment and well contamination in a coastal region. (a) Because saltwater is denser than freshwater, it passes beneath and lifts normal groundwater. (b) A slight drawdown of freshwater due to pumping causes a great rise in the saltwater level beneath the well.

wells for contamination to occur; just the upward diffusion of salt is sometimes sufficient to contaminate freshwater.

Loss of Hydraulic Pressure

Groundwater extraction from unconfined aquifers in most cases does little long-term harm as long as the rate of extraction does not exceed the rate of recharge. If it does, though, the water table will be lowered, with undesirable results. Either deeper wells will need to be drilled, or existing wells must be pumped at higher rates. Both solutions adversely affect the cost of extraction and are long-term drains on the economy of the region.

Serious pressure problems develop where confined aquifers are overtapped. A well-known case is the Ogallala aquifer, a 65-meter-thick, porous deposit that spreads beneath the arid Texas Panhandle, western Kansas, eastern Colorado, and much of Nebraska (Figure 16.13). The entire region underlain by the Ogallala aquifer is responsible for feeding 40 percent of the nation's cattle and generating $30 billion a year in agricultural products. The aquifer is vital to at least 5 million cultivated acres of the Texas Panhandle and New Mexico. Although the Ogallala is recharged to some extent by present-day rainfall, most of the water that it contains was added during the wetter climate of the previous ice age. The aquifer is like a bank account from which money is withdrawn but not deposited. Over 150,000 wells are currently tapping the Ogallala aquifer at 10 times the rate of recharge, and the zone of saturation is shrinking drastically. Should the practice continue, this essentially nonrenewable aquifer will be depleted in less than 30 years. Because no other source of water is available to replace it, the loss of the aquifer would have a disastrous effect on the economy and ecology of the region.

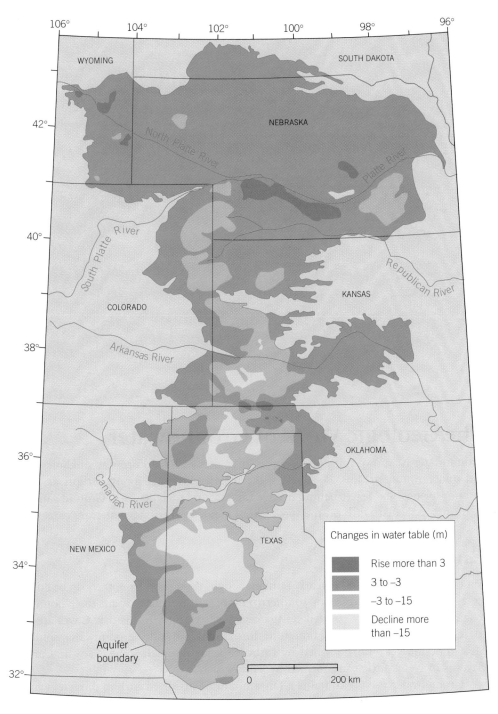

The best way to attack the problems of excessive extraction is for the users to treat the water and put it back into the ground, through either recharge basins or recharge wells. Although these practices are expensive, they maintain the level of the water table and slow land subsidence and saltwater encroachment.

While it may appear that the key to solving groundwater problems is technology, there are instances where the unrestrained and uncoordinated use of technology actually multiplies problems. Technological solutions are especially likely to backfire when modern devices are introduced to less-developed nations without regard to prevailing ecosystems or local customs. For example, the well-intentioned introduction of small pumps in the arid African Sahel region temporarily made life easier for the semi-nomadic tribes of the region. It also led to a lowering of the water table, which, combined with overgrazing by cattle, multiplied the original drought problems.

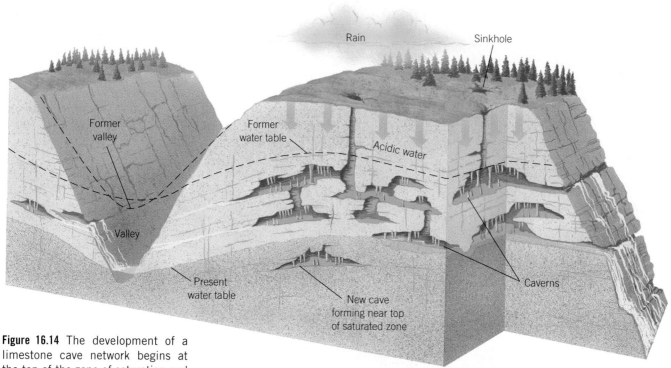

Figure 16.14 The development of a limestone cave network begins at the top of the zone of saturation and follows the lowering of the regional water table.

cave

A natural cavity beneath the surface of the Earth, usually formed by the dissolution of limestone by groundwater; large caves are also called caverns.

The Geologic Work of Groundwater

Groundwater is a powerful geologic agent. It creates such landforms as sinkholes, caves, and underground streams. It preserves the record of extinct forests, erupts as geysers and hot springs, and leaves behind diverse mineral and rock deposits. Groundwater does its most spectacular work on the thick limestone deposits of ancient seas.

Caves

Ten thousand years ago, when early humans sought protection from the grip of Europe's ice age, they made the caves of southern France and northern Spain their home. The magnificent murals they painted on the cave walls endure to this day. **Caves** are open spaces dissolved out of thick limestone formations by acidic groundwater solutions. Most limestone caves are formed by carbonic acid, although some, like the Carlsbad Caverns of New Mexico, are formed by sulfuric acid.

Groundwater usually contains a fair amount of dissolved carbon dioxide, which it derives from the atmosphere and decayed organic matter in the soil. Together, water and carbon dioxide form a weak carbonic acid solution that decomposes minerals. Limestone, composed mostly of calcium carbonate, is particularly vulnerable to acid attack. The hydrogen ions of the carbonic acid displace the calcium ions of the limestone, producing bicarbonate ions and free calcium ions. Both attach themselves to water molecules, and in this way, the limestone is dissolved.

$$H_2CO_3 \quad + \quad CaCO_3 \quad \rightleftharpoons \quad Ca^{2+} \quad + \quad 2\,HCO_3^-$$

| Carbonic acid | Calcium carbonate (calcite in limestone) | Calcium ion in solution | Bicarbonate ions in solution |

There is no clear agreement among geologists as to precisely how limestone caves are formed by carbonic acid. The long-standing theory is that their formation is a two-stage process that originates just below the water table and is orchestrated by changes

in regional base level. In the first stage, groundwater rich in carbonic acid trickles down to the water table and occupies the fractures and bedding planes of the limestone (Figure 16.14). As the limestone is dissolved, the planes and fractures enlarge in all directions, forming water-filled cavities that follow the slope of the regional water table.

The second stage of cave formation is initiated as stream valleys are lowered by erosion. This also lowers the water table, because groundwater flow (like surface runoff) is controlled by the level of the stream that it enters. The drop in the water table exposes the solution cavities to the air and extends the solution process to greater depth. Over time, the falling regional base level leads to an intricate, many-storied network, complete with weirdly shaped passages and vast, unexpected rooms that delight explorers.

The lowered water table is also responsible for two of nature's delightful creations: **stalactites,** which hang from the roof of a cave, and **stalagmites,** which grow up from the floor (Figure 16.15). Their development begins with a vertical crack in the limestone cave roof. Water supersaturated with carbon dioxide seeps down through the crack. However, the carbon dioxide escapes when it reaches the open-air cave, and calcite is precipitated as the water drips through the crack. Because the opening is maintained, the center of the stalactite is a hollow tube, like a soda straw (Figure 16.16). At the same time, thin concentric layers accrete to the stalactite's surface by precipitation from groundwater that runs down over it. By these means, the stalactite grows longer and

stalactite
A calcite deposit that projects from a cave roof, precipitated from groundwater supersaturated with calcium carbonate.

stalagmite
A cone-shaped calcite deposit, growing up from a cave floor, precipitated from groundwater supersaturated with calcium carbonate.

Note: The terms stalactite *and* stalagmite *are easily confused; try associating the* c *in* stalactite *with* ceiling, *the* g *in* stalagmite *with* ground.

Figure 16.16 Cave straws in a cave in Puerto Rico.

sinkhole
A circular surface depression that occurs when the roof of an underground cave collapses or dissolves.

wider. Meanwhile, the dripping groundwater falls to the floor of the cave. As the water evaporates, a thick mound accumulates and forms the stubby, wider stalagmite beneath the stalactite. The porous, banded limestone of stalactite and stalagmite columns is called *travertine;* the rock also precipitates from films of water that coat cave walls and from the carbonate-rich water of hot springs (Figure 16.17). Because of its exceptional beauty, travertine is in great demand as a finishing stone for the interiors of buildings.

Underground caves that are stranded in the zone of aeration by water-table lowering are subject to weathering and erosion. Their roofs are unsteady and frequently collapse for lack of support. If collapse undermines the land surface, the land will subside, forming a large, circular depression, or **sinkhole.** Frequently, sinkholes fill with groundwater to form deep lakes. Numerous sinkholes give the countryside of Florida (as well as other parts of the southeastern United States) a pockmarked appearance. While they provide scenic and recreational benefits, as well as habitats for wildlife, they also cause major hazards for homeowners and engineers. It is difficult to tell when or where a sinkhole will suddenly open (Figure 16.18). Urban sprawl has increased groundwater pumping and a corresponding lowering of the water table, which has, in turn, accelerated the formation of sinkholes.

Karst Topography

The Karst region of Slovenia along the eastern shore of the Adriatic Sea is characterized by many of the features discussed so far: thick limestone deposits with numerous sinkholes and intricate cave networks extending thousands of meters below the surface. Their depth is evidence that in the past, sea level was much lower than it is at present.

Figure 16.17 A travertine deposit in the Minerva Terraces, Yellowstone National Park.

The entire subsurface has become so pockmarked with caves, sinkholes, and interconnected channels that streams no longer run on the surface but drain underground through the water table. The porous topography takes on an arid aspect, even though rainfall may be plentiful.

Regions of the world that have features similar to those just described are said to display **karst topography.** The Yucatan Peninsula and portions of Florida, Indiana, and Kentucky all exhibit relatively early stages of karst development, in which most of the land surface has not yet been eroded, despite the proliferation of sinkholes and caves. The Kwangsi Province of China is a classic example of late-stage karst topography (Figure 16.19). The once-thick limestone strata have been reduced to an eerie landscape of stacks and towers. Full-fledged karst topography takes millions of years to evolve.

karst topography
Topography formed by the intensive dissolution of underlying limestone bedrock, featuring sinkholes, intricate cave networks, and diversion of surface drainage underground.

Figure 16.18 A sinkhole, formed from the sudden collapse of limestone bed-rock, near Orlando, Florida.

Figure 16.19 Late-stage karst topography, south of Auilin, China. The towers are the remnants of thick limestone formations dissolved away by groundwater.

Far from being a geological oddity, karst topography covers about 15 percent of the Earth's land surface. Because it is characterized by underground drainage, the potential for the spread of pollution is enormous. For instance, about 20 percent of the United States' freshwater drains through karst topography. Unfortunately, uncounted tons of domestic, agricultural, and industrial wastes have been pumped down into cave networks, converting them to sewers and polluting their waters, which were once symbols of purity. Bowling Green, Kentucky, for example, is situated on karst topography. *Time* magazine reported that on several occasions during the 1980s, benzene and other chemical fumes rose up from the caves, endangering homes and schools. Fortunately, cave specialists from Western Kentucky University were able to use their knowledge of groundwater flow to identify the sources of the pollution.

Preservation by Groundwater

In east central Arizona, near the southern edge of the Colorado Plateau, you can visit the ruins of an ancient forest. Logs of 40 different species of conifer trees, some 60 meters long and 6 meters wide, are scattered across the landscape, along with their roots, leaves, and seeds. You can see the scars of forest fires, insect holes, and rot in the logs. Most of the logs are not found where the trees grew; they were washed down hillside slopes and quickly buried in muddy backwaters that protected them from decomposing. The truly amazing thing about these logs is that they are made of silica (Figure 16.20). They are part of the famous Petrified Forest of Arizona, some 220 million years old, and the logs are preserved in the Triassic Chinle formation of the plateau.

Figure 16.20 A petrified tree trunk, 200 million years old, in the Chinle formation, east central Arizona. The log was buried and converted to silica by circulating groundwater.

The logs were converted to silica by groundwater that circulated through the Chinle formation after leaching silica from beds of volcanic ash directly above the logs. Close inspection reveals that two types of conversion were involved. In the simplest case, *replacement,* the internal structure of the log has been destroyed and the wood replaced by silica precipitated in the form of chalcedony, jasper, or agate. Only the external form of the log is preserved.

The more intricate form of conversion involves *permineralization,* in which the silica is deposited in wood cells with such delicacy that details of the cells and other internal structures are preserved. If you were to soak a chunk of the petrified log in hydrofluoric acid, the silica would dissolve and leave behind a residue of soft, 220-million-year-old wood. No one knows exactly the conditions that favor one process over the other. Probably, they are related to the rate at which groundwater circulated through the logs; in fact, you can see both kinds of conversion in the same log. The Petrified Forest is a spectacular example of fossilization by groundwater. Most fossils the world over are preserved through replacement or permineralization.

Geysers, Hot Springs, and Fumaroles

Chapter 9 described what happens when underground water comes in contact with hot rock or magma beneath the sea floor. When cold seawater trickles down through fractures along oceanic rift zones, it absorbs heat from the magma and rises back up to the ocean floor as hot-water chimneys. **Geysers** and related features are land analogies to these submarine chimneys (Figure 16.21). Most geysers erupt at intervals, although not necessarily with the clocklike efficiency of Old Faithful in Yellowstone National Park, Wyoming.

The eruptive cycle of a geyser begins as water filters down through fractures and passageways to a heat source at depth (Figure 16.22). The complex fracture network in effect forms a vessel or container. Water at the bottom that is in contact with the hot rock is under great pressure, which allows the water temperature to rise well above the normal 100 °C surface boiling point. Over time, the temperature reaches a critical point where the water at depth vaporizes to steam despite the pressure. The expansion causes water at the surface to spill over in a sort of preliminary event, reducing pressure on the entire water column, and allowing the superheated water in the column to vaporize

geyser

A periodic eruption of hot water and steam, caused by the heating of pressurized groundwater in a network of underground channels in contact with hot rock or magma.

Figure 16.21 A geyser in full eruption, Iceland.

hot spring
A spring whose waters have been heated above body temperature (36.7 °C) by hot rock or magma.

fumarole
A vent from which volcanic gases and water vapor escape.

to steam and erupt violently. After the pressure is released and the water column cleared, the water of the geyser trickles back down through the cracks to repeat the cycle. Left behind on the surface near the geyser vent is thick, encrusted silica called *geyserite.* Similar in composition to quartz, but lacking an organized internal structure, geyserite precipitates from the hot water.

Although geysers are the most spectacular manifestation of the interaction between groundwater and subsurface magma, there are others. Where spring waters have come in contact with magma or hot rock before issuing from the ground, they form **hot springs**—some warm enough to bathe in, others hot enough to boil eggs in. If magma-heated underground water intersects volcanic gases, the mixture may erupt through a vent in the surface as a steaming **fumarole.** Geysers, hot springs, and fumaroles are common in such volcanically active regions as New Zealand, Iceland, Alaska, northern California, Wyoming, and Montana. In many parts of the world, these features have been tapped as a source of geothermal energy. In fact, fumaroles in northern California supply much of the energy needed to run the utilities of San Francisco. Iceland and New Zealand also have been tapping subterranean heat for years; and recently, Hawaiian utility companies have begun to harness some of the enormous energy released by the island's active volcanoes.

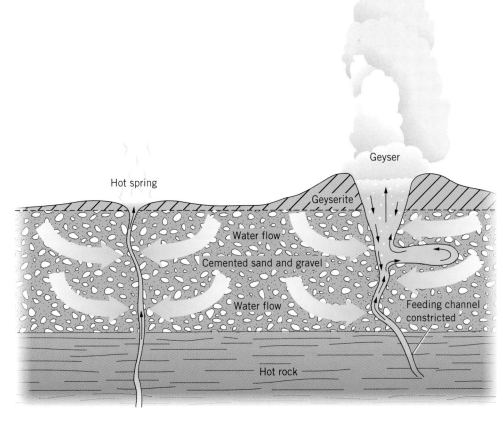

Figure 16.22 Two ways in which groundwater interacts with magma or hot rock: hot springs and geysers.

The Quality of Groundwater

Totally pure water is found only in laboratories; underground water contains many naturally dissolved substances. More often than not, groundwater quality depends on the composition, the concentration, and (in some places) the interaction of these dissolved substances. The common characteristic of *hardness* is caused by relatively high concentrations of calcium, magnesium, and iron salts that are dissolved in water; they can impart an unpleasant taste to the water and prevent soap from lathering. Hard water also gunks up machinery; it forms scales in boilers and hot-water pipes.

The calcium and magnesium ions in hard water are derived mainly from the solution of limestone, dolomite, and gypsum and, in part, from the solution of metamorphic and igneous minerals like olivine and pyroxene. Regions with underlying carbonate and sulfate sediments tend to have hard water. Often, chemical treatments that extract or filter these ions from the water are required.

Groundwater Pollution

Groundwater pollution is broadly defined as any deterioration in water quality. Salt-water contamination of freshwater aquifers that results from excess well pumping fits this definition. However, the main cause of groundwater pollution stems from waste disposal on or beneath the land surface. Industrial chemicals, radioactive wastes, heavy metals, pesticides, and other substances seep—or are purposely injected—into the ground. The main culprits are factories, nuclear power plants and weapons facilities,

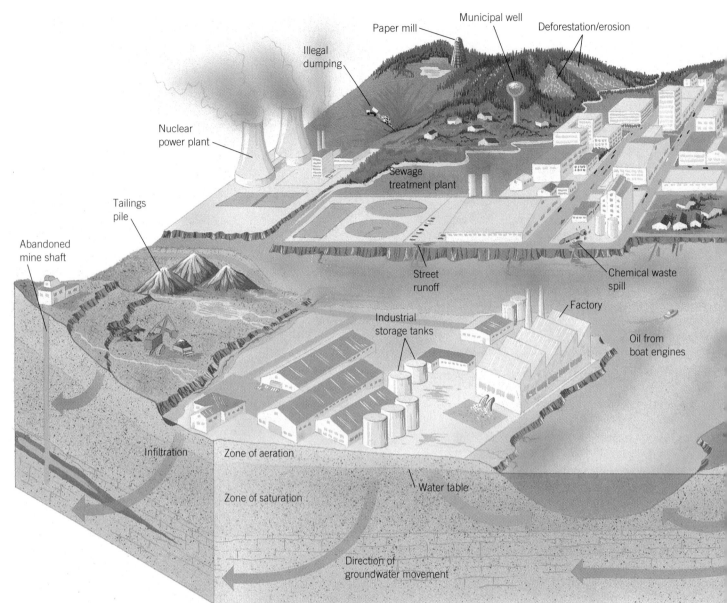

Municipal well

Paper mill

Deforestation/erosion

Illegal
dumping

Nuclear
power plant

Sewage
treatment plant

Tailings
pile

Abandoned
mine shaft

Street
runoff

Chemical waste
spill

Factory

Industrial
storage tanks

Oil from
boat engines

Infiltration

Zone of aeration

Water table

Zone of saturation

Direction of
groundwater movement

Figure 16.23 Some of the myriad sources of groundwater contamination. Sources such as crop dusting with pesticides spread contaminants over a wide surface area. Others, such as leaky industrial storage tanks, are localized point sources. Upon percolating through the subsurface, the pollutants migrate according to the regional slope of the water table. Streams, lakes, and wells are at risk.

farms, landfills, buried gasoline tanks, and septic systems (Figure 16.23). As populations grow and as industrial and agricultural activities multiply, the extent and intensity of waste disposal also increase. Approximately 25 percent of our nation's aquifers are unusable, the percentage expanding to nearly 100 percent in some states.

Once polluted, a groundwater system may take centuries and incredible sums of money to purify. On the national scale, billions of dollars are already being spent on purification. A more serious problem is that while some contaminants are evident because they are obnoxious in taste, odor, color, or the like, others are silent killers. If these poisons go undetected, over the long run, they can cause disease in humans, animals, and plants. As with so many environmental hazards, it is often difficult to trace the exact cause of illness (or death) to its source.

Figure 16.23 illustrates that pollutants are released to the subsurface from point sources—septic tanks, leaking gasoline storage drums, and industrial injection wells—and over broad areas, such as cropland that is sprayed with pesticides. Point-source and broad-area release occur in both urban and rural settings and have two kinds of effects on the environment: (1) local and usually short-term and (2) pervasive and usually long-term.

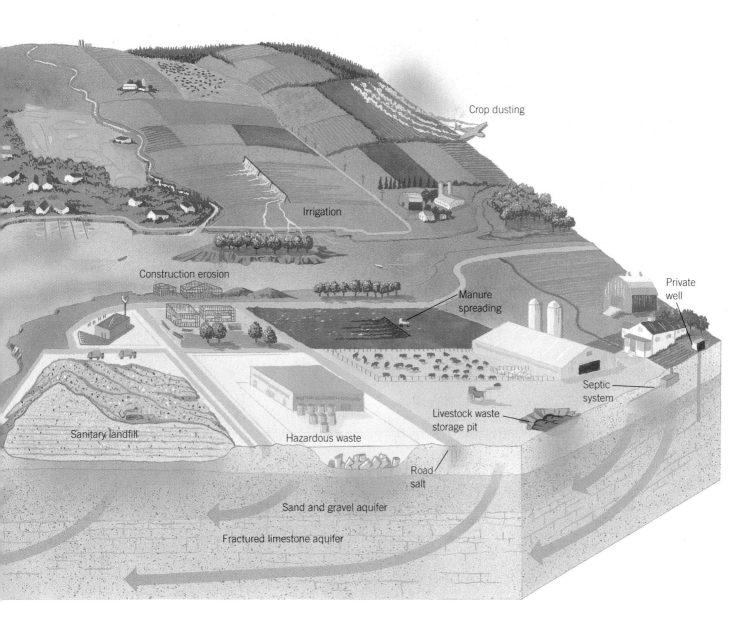

Take, for example, a typical point-source polluter: a septic tank located a short distance uphill from a well. Sinking a well in this setting is asking for trouble, since the pollutants will migrate toward the well. Nevertheless, the source of pollution can be pinpointed quickly, and the tank or the well can be moved. Some point-source problems are not as easily corrected because the pollutants migrate surprising distances. In addition, their slow rate of creep through the subsurface makes them difficult to locate in areas of intense land use.

Long-term pollution problems involve a slow decline in water quality and are often reflected by changes in community health. Unfortunately, water lives up to its reputation as the universal solvent where pollutants are concerned—virtually all are soluble in water, at least to some degree. Table 16.2 shows the maximum concentration allowed for a selection of inorganic contaminants in drinking water by the U.S. Environmental Protection Agency. Note that it does not take a high concentration for some elements to be considered harmful.

Because pollutants migrate slowly, some problems are the harvest of mistakes planted decades earlier. Thus groundwater quality continues to deteriorate despite current enlightened pollution control and treatment policies. The migration of pollutants

Apply and Decide 16.1

Hazardous-Waste Disposal

A hazardous waste is any solid, liquid, or gas that is harmful to living things and can escape into groundwater, air, or soil; that is toxic (DDT, dioxin, lead compounds, PCBs); that is corrosive (strong acids or bases); that can catch fire (waste oil, organic solvents, PCBs); or that can explode or release poisonous fumes (cyanide solvents). Radioactive wastes from nuclear power plants and weapons facilities are included because they are known to cause cancer. There are three approaches to preventing groundwater pollution from hazardous wastes. Solid wastes are sealed off from the surrounding environment in *secure landfills*. Liquid wastes are treated through *surface evaporation ponds*, if they are lightly contaminated, or through *deep-well injection*, if they are concentrated. In this nuclear age, we have yet to devise a generally accepted approach to the disposal of radioactive wastes. One method currently being considered by the federal government is deep storage.

Figure 16A illustrates the secure-landfill method of solid-waste disposal. An excavated pit is lined with thick, impermeable clay and plastic sheets that serve a similar function. Drainage in the vicinity of the site is adjusted to prevent erosion or leaching out of the structure. These approaches were used at the toxic-waste dump site in Emille, Alabama, which is among the largest of the nation's landfills. It opened in 1977 and was designed to store dangerous chemicals in 55-gallon drums for thousands of years. Unfortunately, within seven years, dangerous chemicals were already leaking from the site.

Lightly contaminated liquid wastes are treated through surface evaporation ponds. The untreated water is conducted through pipes into the pond, where it is allowed to evaporate. The solid residue is then carted off to a secure landfill. The chief advantage of this method is that it is inexpensive. But the potential for leakage is high.

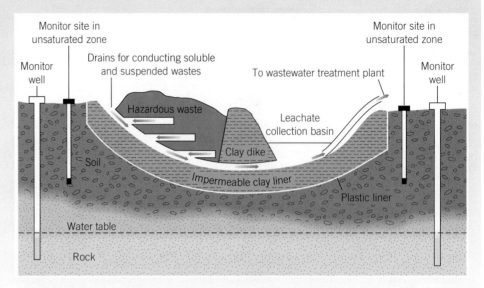

Figure 16A Some of the features installed to prevent groundwater contamination in a secure landfill site used for solid waste disposal.

Deep-well injection is the current method for disposing of highly contaminated liquid hazardous wastes, such as mercury, arsenic, and various nonbiodegradable synthetic compounds. They are pumped through impermeable strata into highly fractured igneous and metamorphic rocks or porous sedimentary formations, well below the zone of saturation. The assumption is that the impermeable strata will act as a seal that prevents the liquid pollutants from rising and polluting the groundwater. Some of these wastes can remain hazardous for thousands of years, so injection sites must be carefully chosen. For example, the region should have low earthquake probability to keep the subsurface seal intact. However, critics remind us that deep-well injection itself triggered a series of sharp earthquakes (magnitude 3 to 5) near Denver in 1964 (see Chapter 3). In addition, there are the many dangers of equipment and human failure involved in the complex process of storing, transporting, and pumping the wastes thousands of meters below the surface.

Over 100,000 tons of radioactive waste are stored on the sites of the world's nuclear power plants, and this number does not include radioactive wastes stored in military installations. There is no agreement among experts on how to dispose of these wastes over the ten thousand years required for

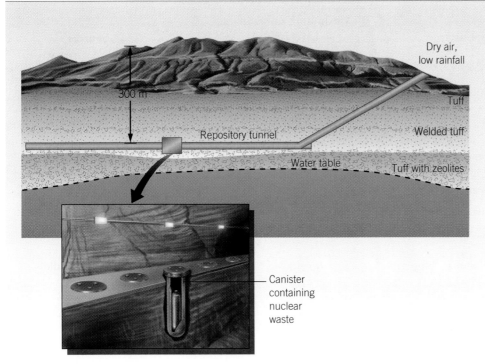

Dry air, low rainfall

300 m

Tuff

Repository tunnel

Welded tuff

Water table

Tuff with zeolites

Canister containing nuclear waste

Figure 16B Measures taken to assure safety from groundwater contamination at the proposed nuclear waste disposal site at Yucca Mountain, Nevada. This location has been selected because of its low level of rainfall. The waste will be converted to solid form (berylium-silicate) and stored in secure steel canisters buried beneath a mountain of impermeable and absorbent welded tuff, but well above the water table. (Adapted from DOE's *Yucca Mountain Studies,* U.S. Department of Energy, 1992.)

local solutions to this very long-term, international problem.

No Easy Solutions

Yucca Mountain and similar remote storage sites have been chosen in part as a reaction to the public's "NIMBY" syndrome: not in my backyard. Who can blame people for not wanting a hazardous waste material site in their community? There are one hundred temporary nuclear waste storage facilities located around the United States. The residents of those communities are fighting to get the wastes sent on their way to their final storage sites. But sending them off leads to another problem—transportation; the wastes would have to be trucked across country. Not surprisingly, communities en route are declaring, "Not through my backyard!"

As the battle rages on, those temporary storage sites are becoming more and more dangerous. Each was built to last only until arrangements could be made to move the waste from a local nuclear power plant or military facility to permanent storage. But most of the wastes still reside in those "temporary" facilities—some over half a century old.

them to decay to safe levels. All agree, however, that should these lethal products of nuclear fission contaminate groundwater reserves, the results would be catastrophic. U.S. government plans call for a national deep-storage site beneath Yucca Mountain, near Las Vegas, Nevada (Figure 16B). Advocates of this site argue that the rock types—volcanic tuffs—are good absorbers of radioactive-decay particles and that the dry regional climate, combined with the location of the water table well below the storage site, should ensure safety. The opposition continues to raise many objections; evaluations of groundwater flow patterns, rock fracture, and seismic (earthquake) risk, for instance, have been questioned. Recently, government officials announced that the site would not be ready to open for at least another twenty years or perhaps longer. In the meantime, the federal government has turned to state governments for short-term,

Thought Questions

1. What relationships exist between the problem of nuclear waste storage and transportation and the problem of pollution, resource depletion, and land degradation caused by fossil fuel burning power plants (see also Apply and Decide 13.1, Acid Rain)?

2. In a democratic society, conflicts are commonly resolved through compromise. Discuss ways in which communities might compromise and choose the best of imperfect solutions to the nuclear waste problem. How might such a process of resolution apply to other environmental problems?

Table 16.2 Drinking Water Standards of the U.S. Environmental Protection Agency

Contaminant	Health Effects	Maximum Contaminant Level (mg/L)	Sources
Arsenic	Dermal and nervous system toxicity effects	0.05	Geological, pesticide residues, industrial waste, and smelter operations
Cadmium	Kidney effects	0.01	Geological, mining, and smelting
Lead	Nervous system damage; kidney effects; highly toxic to infants and pregnant women	0.05	Leaches from lead pipes and lead-based solder pipe joints
Mercury	Central nervous system disorders; kidney effects	0.002	Used in manufacture of paint, paper, vinyl chloride; geological
Nitrate	Methemoglobinemia ("blue-baby syndrome")	10	Fertilizer, sewage, feedlots, geological
Selenium	Gastrointestinal effects	0.01	Geological, mining

Source: U.S. Environmental Protection Agency, 1977.

is governed by their intrinsic properties and the properties of the subsurface. One extremely important property of a pollutant is its density compared with the density of groundwater. Consider the leaking gasoline storage tank and a salt pile used to melt ice at the filling station in Figure 16.24. Because gasoline is less dense than water, it seeps down and collects at the top of the water table, where it spreads as a thin film. Salt, on the other hand, dissolves in rainwater and forms a concentrated brine that is denser than freshwater. This brine sinks beneath the water table until it is blocked by impermeable beds at the base of the zone of saturation. Eventually, the salt spreads into the groundwater or is drawn up into wells.

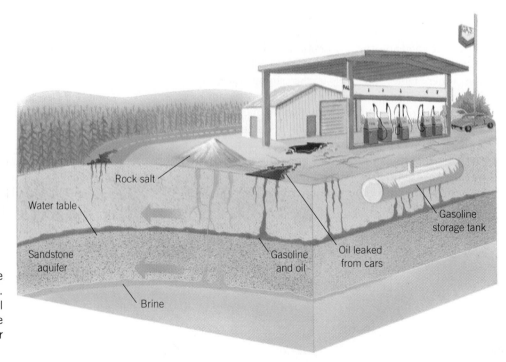

Figure 16.24 Gasoline, oil, and brine contamination in a filling station. The relatively light gasoline and oil form a film on the water table. The denser brine displaces fresh water in the aquifer.

Pollutants travel with groundwater, so their dispersion rates depend on the porosity and permeability of the subsurface. In sand or gravel, point pollutants spread in a widening cone, their concentration varying with time. Minute amounts arrive first, and amounts increase rapidly as the mass of groundwater arrives. Concentrations diminish slowly as the source is exhausted, but dispersion is intensified if leakage is continuous.

If given sufficient time, some natural processes will diminish groundwater pollution. Biodegradable compounds eventually break down, for instance, and radioactive elements with a short half-life eventually decay. Most important, the subsurface has a remarkable capacity to scrub groundwater clean. Sand and gravel are natural filters that remove harmful particles, and the oxygen-rich zone of aeration acts to decompose many potentially dangerous compounds. Clays are also very effective natural absorbers of metal ions and organic compounds that can be dissolved in groundwater. However, they are not permanent repositories of pollutants and can release them back into the groundwater. Also, the subsurface through which groundwater moves is of finite size, so its purifying capacity can be overloaded. Finally, there are some highly soluble industrial compounds, such as methyl mercury, for which there is no natural defense.

STUDY OUTLINE

Underground water supplies most of the world's usable freshwater.

I. **ACCUMULATION OF GROUNDWATER.** When rainwater seeps into the ground, it passes first through the **zone of aeration,** which contains little moisture except in the **belt of soil moisture,** directly beneath the surface. The water fills the voids in sediments and rocks, forming a **zone of saturation.**

A. The term **groundwater** refers specifically to water in the zone of saturation, and the upper limit of that zone is the **water table.** The **capillary fringe,** where water rises up through capillary action, is located just above the water table.

B. The **porosity** of loose materials depends on three factors: sorting, packing, and particle shape. The porosity of consolidated rock is also affected by the material that cements the particles and by the degree to which the rock is fractured.
1. **Permeability** is a measure of the ease with which materials transmit water. It depends upon the material's porosity and upon the size, arrangement, and interconnectedness of openings.
2. When groundwater cannot reach the zone of saturation because of an impermeable barrier, it forms a **perched water table.**

C. **Aquifers** are porous and permeable strata that hold and release groundwater. In an unconfined aquifer, the water table can rise or fall freely. A confined aquifer, sealed between impermeable **confining beds,** receives water from a distant, unconfined source.

II. **MOVEMENT OF GROUNDWATER**

A. The water table tends to mirror the topography. Where it intersects lowland surfaces, lakes or marshes are formed. Streams that gain water from the water table are called **effluent streams.** In arid regions, **influent streams** lose water to the water table. Due to faulting or other irregularities, the water table may be exposed and flow out as a **spring.**

B. Darcy's law states that the velocity of groundwater flow is equal to the **hydraulic gradient** of the water multiplied by a permeability constant. The flow of groundwater is from **recharge areas,** where precipitation replenishes the groundwater supply, to **discharge areas** (streams, lakes, and marshes).

III. WELLS

A. Wells sunk in unconfined aquifers produce a **cone of depression** in the water table. Drawdown is the difference between the levels of the water table before and after pumping.

B. An **artesian well** will flow without the aid of pumping if the recharge area lies above the elevation of the well. Because of friction, however, this effect decreases with the well's distance from the recharge area.

C. Well extraction problems
1. Extraction from a confined aquifer may lower its water pressure and cause its framework of particles to compact. The result may be land sinking, or **subsidence.**
2. Excessive pumping of coastal plain aquifers may cause **saltwater encroachment.** A slight freshwater drawdown leads to a marked rise of saltwater into the well.
3. Overuse of a confined aquifer increases the expense of extraction because of falling hydraulic pressure. Excessive extraction can be offset somewhat by treating the water and returning it to the ground through recharge basins or wells.

IV. THE GEOLOGIC WORK OF GROUNDWATER

A. Caves are open spaces dissolved out of limestone or other soluble rock by acid solutions. They form at or beneath the water table and are enlarged as the regional water table is lowered.
1. Groundwater and carbon dioxide combine to form carbonic acid, which decomposes limestone. In the process, the water table is lowered, leading over time to exposed cave networks. Water supersaturated with carbon dioxide that drips from cave ceilings precipitates calcite to form **stalactites** and **stalagmites.**
2. Some caves, such as Carlsbad Caverns, are formed by the action of sulfuric acid.
3. When an underground cave collapses, it may form a circular depression, or **sinkhole,** on the surface.

B. Karst topography is formed by the intensive solution of underlying limestone bedrock. It features sinkholes, intricate cave networks, and diversion of surface drainage underground.

C. Buried logs in the Petrified Forest were converted to stone when they were bathed in groundwater that had leached silica from volcanic ash.

D. Geysers and **hot springs** occur when water filters down through fractures and contacts hot rock or magma. **Fumaroles** are vents through which volcanic gases and vapor escape. In some regions of the world, they are tapped as a source of geothermal energy.

V. THE QUALITY OF GROUNDWATER. Groundwater quality depends upon the type and quantity of naturally dissolved substances. Groundwater pollution—broadly defined as any deterioration in water quality—stems mainly from waste disposal. Once polluted, a groundwater system may take centuries to purify.

A. Point-source pollution is local and short-term; broad-area release is pervasive and long-term. A pollutant's migration is governed by its density and the porosity and permeability of the materials through which it moves.

B. There are three major ways to store hazardous wastes, none of which is problem-free.
1. Solid wastes are sealed off in secure landfills.
2. Liquid wastes are treated in surface evaporation ponds.
3. Liquid wastes may also be pumped into subsurface rocks by deep-well injection.

C. A debated proposal for disposal of radioactive wastes is deep storage.

STUDY TERMS

aquifer (p. 427)
artesian well (p. 427)
belt of soil moisture (p. 425)
capillary fringe (p. 425)
cave (p. 436)
cone of depression (p. 430)
confining bed (p. 427)
discharge area (p. 429)
effluent stream (p. 427)
fumarole (p. 442)
geyser (p. 441)
groundwater (p. 424)
hot spring (p. 442)
hydraulic gradient (p. 428)
influent stream (p. 428)

karst topography (p. 439)
perched water table (p. 427)
permeability (p. 426)
porosity (p. 425)
recharge area (p. 429)
saltwater encroachment (p. 433)
sinkhole (p. 439)
spring (p. 428)
stalactite (p. 437)
stalagmite (p. 437)
subsidence (p. 432)
water table (p. 424)
zone of aeration (p. 424)
zone of saturation (p. 424)

CRITICAL THINKING QUESTIONS

1. Explain how the zone of saturation helps regulate the regional water economy during a one-year period.
2. A porous substance is not necessarily permeable. Explain.
3. Diagram and label the conditions necessary for an artesian well.
4. Describe three major well extraction problems.
5. Why does lowering the hydraulic gradient of an aquifer make pumping water more expensive?
6. Explain the origin of a limestone cave.
7. Describe four features associated with karst topography.
8. Explain the similarity between sea-floor chimneys and geysers.
9. Distinguish between point-source and broad-area pollutants. Make a list of each type.
10. What are the three major hazardous waste disposal methods? What are the advantages and disadvantages of each?

CHAPTER 17

Glaciers and Climate

Most of the world's ice is confined to two locations: Greenland, in the Northern Hemisphere, and Antarctica, in the Southern Hemisphere. The Greenland ice sheet is nearly a kilometer and a half thick, and it covers an area roughly the size of California, Arizona, New Mexico, Nevada, Utah, and Colorado combined. The Antarctic ice sheet is even more impressive (Figure 17.1). Concealing a continent the size of the United States and Mexico, it is 7.5 times as extensive as the Greenland ice sheet. The crushing weight of the ice has depressed the bedrock surface of Antarctica hundreds of meters. In places, the ice is nearly 5 kilometers thick. Together, the Antarctic and Greenland ice sheets keep hidden 10 percent of the land surface of the Earth.

Those giant deep-freeze lockers, the Antarctic and Greenland ice sheets, exert major control over the Earth's climate by extracting heat from the overlying air and surrounding ocean. For this reason, the ice sheets are closely monitored for climatic trends. Also studied closely for what they reveal of climatic trends are core samples extracted from the ice. They contain bubbles of *fossil air*—samples of past atmospheres trapped within the ice layers. The cores are, in fact, a reference library of the Earth's ancient atmosphere; their layers record hundreds of thousands of years of climatic history.

Is the Earth getting warmer? If so, a 5 °C increase would melt the glaciers and raise sea level at least 65 meters—enough to drown the land occupied by two-thirds of the world's population. Figure 17.2 shows what North America would look like under these conditions. Is the Earth getting colder? If so, a 5 °C drop would return the world to where it was 18,000 years ago, when glaciers covered 30 percent of the Earth's land surface and drastically altered its geography and environment (Figure 17.3). Although this

Note: If the Antarctic ice sheet were melted at a steady rate, it could feed all the rivers of the world for 750 years.

◄ Glaciers of the Fairweather Range, Glacier Bay, Alaska.

Figure 17.1 The Antarctic ice sheet. To gauge the scale of this continental glacier, the Ross Ice Shelf (brown-tinted area below center) is roughly the size of Texas.

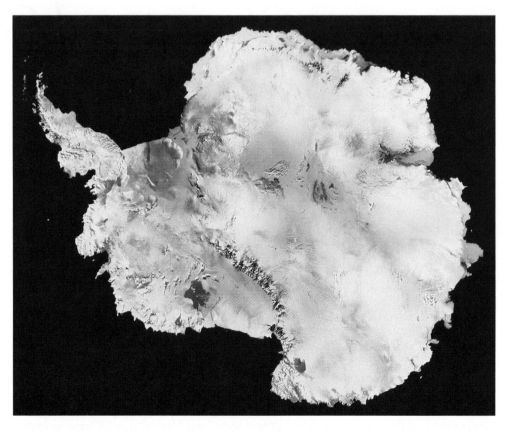

vast ice sheet was indeed impressive, there is clear evidence that its appearance was not a unique event. Glacial advances and retreats have been the pattern over at least the past 3 million years, and probably longer.

Evidence in support of the theory that ice sheets periodically covered large portions of the Earth's land surface comes, first, from observing today's glaciers in action and the geological features they produce and second, from observing that vast regions, now ice-free, contain deposits and landforms derived from glacial action. Glaciers have sculpted the mountain peaks of the Canadian Rockies, the Cascades, the Andes, the Alps, and the

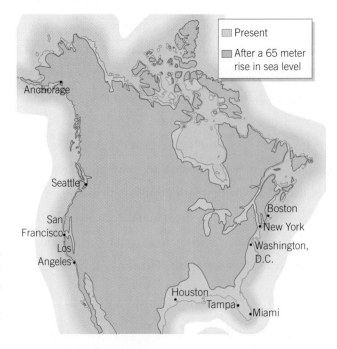

Figure 17.2 The effect of a 65 meter rise in sea level on North America if the Earth's present ice sheets were to melt. The major population centers of the East and Gulf Coast states would be submerged.

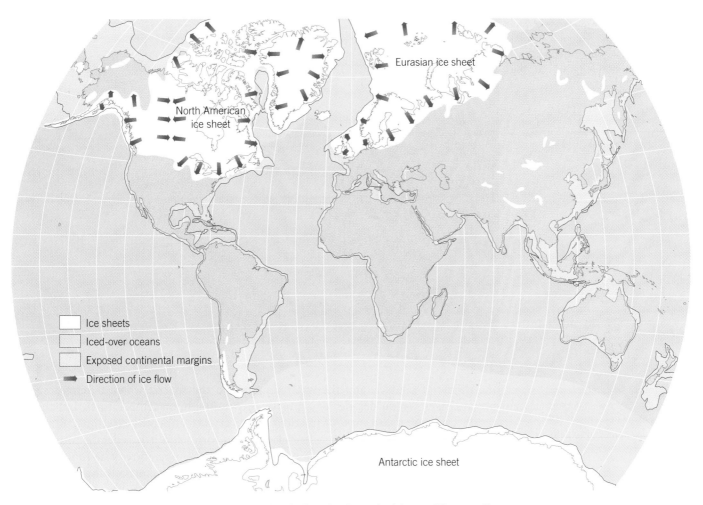

Figure 17.3 Global geography during the last glacial age. The growth of vast ice sheets lowered the worldwide sea level and exposed the continental shelves. Australia and Indonesia were one; Japan, Malaysia, and many Southeast Asian islands were joined to the mainland; and the Red, Black, and Caspian seas were dry. Humans migrated across the exposed Bering Strait from Asia to settle in North America.

Himalayas. The Vikings sailed out of the ice-carved fjords of Norway. Glaciers scoured and scraped clean the bedrock of eastern Canada and, upon melting, bequeathed a wilderness landscape of lakes, marshes, and rearranged stream courses (Figure 17.4). Also left behind by the melting ice were the stony soils of New England, Canada, and Northern Europe. Meltwater carrying fine silts and sands from the glaciers built the heavily farmed plains of Iowa, Central Europe, and China.

The Formation of Glaciers

A **glacier** is a large, slowly moving mass of ice. To understand how one forms, we must consider a key feature called the **snowline**—the altitude above which there is permanent snow. Above the snowline, temperatures are so low that snow forms into large snowfields; within these fields, the snow is converted to glacial ice. The conversion begins when the freshly fallen snow of these fields is changed through compaction into hard little ice pellets, or **firn.** The firn is buried under later snowfalls, and the added weight of the overlying snow subjects it to great pressure. Pressure is greatest where the grains make contact and least in the open spaces between the grains. The ice then melts along

glacier
A large ice mass formed on land by the compaction and recrystallization of snow that survives from year to year and shows evidence of present or past flow.

snowline
Altitude above which the snow is permanent.

firn
The pelletlike form assumed by snow in its transition to glacial ice.

Figure 17.4 Ice-sculpted topography of the Canadian Shield. Glacial scouring and lake-filled depressions impart a distinctive grain to the region.

ice sheet or **continental glacier**
A glacier more than 50,000 square kilometers in area that spreads in all directions, unconfined by underlying topography.

ice cap
A glacier less than 50,000 square kilometers in area; either dome-shaped, blanketing the summit of a mountain mass, or plate-shaped, covering a flat landmass, such as an Arctic island.

alpine glacier or **valley glacier**
A glacier in mountainous terrain that flows downslope in a valley previously eroded by a stream.

piedmont glacier
A glacier formed by the joining of two or more mountain glaciers as they flow out onto the lowlands at the base of a mountain range.

the points of contact, and the water migrates to the spaces, where it refreezes. The process takes several decades or centuries, depending upon temperatures, but the net result is a dense, fused mass of interlocking ice crystals at depths exceeding 50 meters.

The snowline defines just where glaciers can form, and it is generally found at or near sea level toward the polar regions. There, sea-level temperatures year-round are low enough to permit the growth of ice masses of continental proportions that spread outward in all directions regardless of the topography. Ice masses of this magnitude, such as those presently covering Greenland and Antarctica, are called **continental glaciers,** or simply **ice sheets.** Glaciers similar to ice sheets but having areas less than 50,000 square kilometers are called **ice caps.** Ice caps may cover mountain peaks or low-lying Arctic islands.

The snowline rises toward the equator, although its exact elevation is influenced by such factors as sunlight, temperature, snowfall, and wind conditions. Away from the polar regions, glaciers are confined to high altitude, the mountains. Mountain glaciers, which are more formally called **alpine** or **valley glaciers,** occupy former mountain stream valleys (Figure 17.5). Alpine glaciers occur at all latitudes, including the equator, as in the case of Mount Kilimanjaro. **Piedmont glaciers** are created when several valley glaciers flow out onto the broad lowlands at the front of mountain ranges and merge.

The Movement of Glacial Ice

As the ice in a valley glacier thickens by the addition of new snow, it begins to deform under its own weight. Ever so slowly, it flows, slips, and slides downhill. Geologists measure this motion by hammering a straight line of pegs into the ice across the valley and noting how the line changes in subsequent years (Figure 17.6). Over time, the line gradually deforms into a curve that bends downhill, which proves that ice velocity is greatest at the valley center and least along the valley walls. Clearly, friction with the valley walls and bottom retards ice flow, just as it retards stream flow.

zone of fracture
The rigid outer layer of a glacier where the ice fractures as a result of stress caused by glacial movement.

Measurements like these show that the mechanics of glacial motion depend upon the response of the ice to stresses induced by gravitational and frictional forces. The upper 35 meters or so constitute the **zone of fracture,** in which the ice responds rigidly, storing stress until it cracks (see Figure 17.6). The stress results from the downhill pull

Figure 17.5 Valley glaciers of the Wrangell Mountains in Alaska occupy former stream valleys. Note the rounded cirques at the heads of the glaciers and the knifelike arêtes that divide the glaciers. The numerous crevasses are caused by stress generated in the glaciers as they slide downslope.

of the ice at depth and from friction with the valley walls. It causes a series of fractures, or **crevasses,** to develop that extend in arcs across the width of the glacier. The crevasses open and close as the glacier creeps down over ledges in the valley floor.

Crevasses do not extend below about 35 meters, because at that depth, the ice deforms under its own weight in the **zone of plastic flow.** That is, rather than fracturing, the ice changes shape permanently and continuously in response to pressure generated by the weight of the overlying ice. In this zone, the ice flows—but slowly (see Figure 17.6). The squeezing is greatest at the head of the glacier where the ice is thickest. Constrained by the valley walls and floor, the ice is forced to flow downhill toward the **glacial terminus** (or front), like toothpaste squeezed out of an open tube.

Motion in the zone of plastic flow can be measured by drilling holes greater than 35 meters deep into the ice and observing how the holes deform and bend with time (see Figure 17.6). Again, the overall similarity between ice motion and stream flow is striking. For example, just as water velocity increases with height above the streambed, so ice velocity increases with height above the valley floor. In each case, the flow velocity increases in a rising curve, reaching its maximum about halfway up.

Laboratory experiments on the response of ice to pressure and microscopic analysis of ice samples taken at various points within the glacier shed light on the complex movement in the zone of plastic flow. Ice crystals high in the glacier snowfield tend to be small and relatively spherical, with their internal planes pointing every which way. But downhill, the ice crystals merge and align in planes that parallel the valley floor, enabling the ice to glide downhill, layer upon layer, in a shearing motion. Near the glacial front, where the valley slope lessens and the velocity of the ice is retarded, the stacked layers curve upward.

crevasse
A fracture in the rigid outer layer of a glacier caused by glacier movement.

zone of plastic flow
The inner mass of a glacier where the ice moves without fracturing.

glacial terminus
The extremity or outer margin of a glacier; the glacial front.

Figure 17.6 (page 458) The major structural units of a typical valley glacier. Note the various ways the glacier moves downslope: fractures that open and close near the surface, plastic flow at depth, and basal slip along the bottom. The rate of movement at the surface is greatest toward the center, as shown by the progression downslope of a line of stakes. Curving of drill holes indicates motion in the zone of plastic flow. The glacier slides downhill on flattened glide planes that stack up at the glacial terminus. A terminal moraine is deposited as the glacier melts, and the sediments deposited by meltwaters escaping the glacier form an outwash plain.

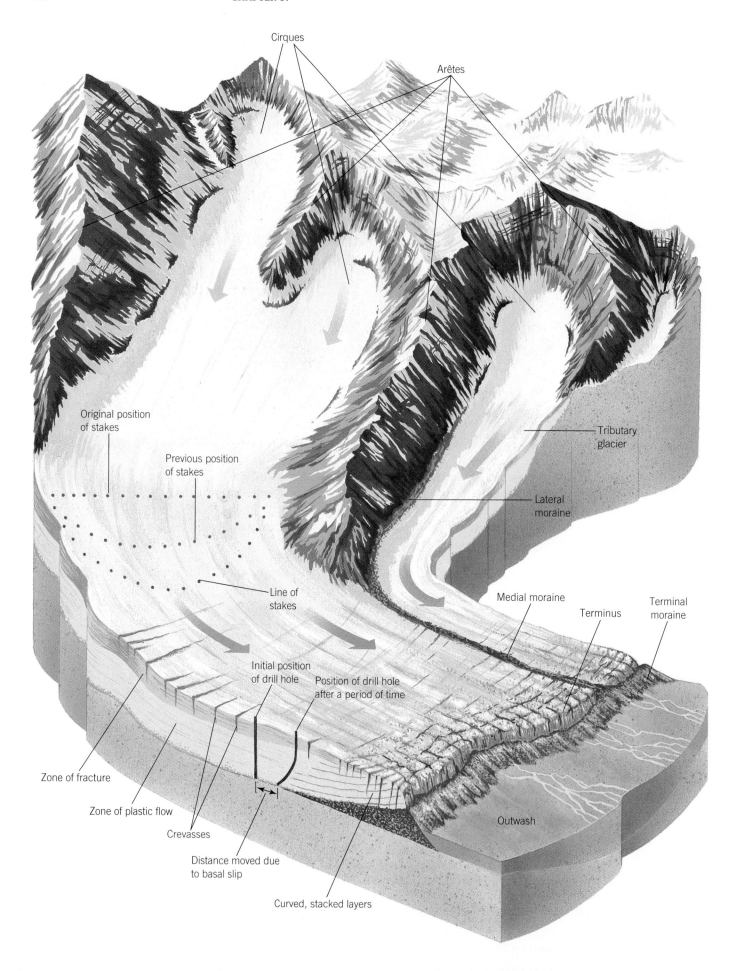

Cirques

Arêtes

Original position
of stakes

Previous position
of stakes

Tributary
glacier

Lateral
moraine

Line of
stakes

Medial moraine

Terminus

Terminal
moraine

Initial position
of drill hole

Position of drill hole
after a period of time

Zone of fracture

Zone of plastic flow

Crevasses

Distance moved due
to basal slip

Curved, stacked layers

Outwash

Glacial motion seems to involve both sliding layers and melting. We all know that ice melts under pressure and refreezes as soon as the pressure is released—that is the principle behind making a snowball. So as the ice moves under pressure, it alternately melts and refreezes in the downhill direction. Should the ice meet an obstacle in its path—say, a protruding rock or a bend in the valley—it may melt along the uphill portion of the path where the ice is squeezed against the obstacle and refreeze downhill from the obstacle.

The melting and refreezing of the ice speed up movement in the zone of plastic flow. The process also supplies the water that collects on the valley floor and mixes with sand and gravel that has attached to the base of the glacier, creating a slippery slush. Like a greased skid, this slush allows the glacier to decouple from the valley floor and slide en masse downhill. This action is called **basal slip** (see Figure 17.6). On occasion, the decoupling is very sudden and dramatic; glaciers have been recorded moving downhill at up to 20 meters per day. Movement of this kind is called a **surge.**

The great ice sheets, which flow from thick interiors to thin coastal margins, are not as well understood as valley glaciers. Teams of geologists are currently measuring them and combing satellite images to advance our understanding of these complex masses. It appears that the Greenland and East Antarctic ice sheets, for the most part, move exceedingly slowly—on the order of centimeters per year in their interiors. Contributing to this slow movement are low regional slopes and low year-round temperatures, which makes the ice more rigid and binds it to bedrock. On the other hand, portions of the West Antarctic ice sheet have been observed to surge on the order of 3 meters per day.

The Glacial Budget

Glaciers can be said to operate like a business. They have income and expenditures and, over a period of time, show profits and losses. Income for a glacier means, of course, snow **accumulation** converted to firn and then to glacial ice. Expenditures mean ice loss through melting, evaporation of water that did not refreeze, and *sublimation*—that is, the passing of ice directly into the vapor state (which is the reason ice cubes tend to shrink in the freezer). Such expenditures go under the name of **ablation,** or wastage (Figure 17.7).

For a typical valley glacier, accumulation exceeds ablation above the snowline, that is, where the glacier forms. However, ablation exceeds accumulation below the snowline, because temperatures are higher at the lower elevations and melting is greater. If, over a given time period, net accumulation above the snowline exceeds net ablation beneath it, ice is produced faster than it is wasted. This is a profit or growth situation, and the glacial terminus will advance from its previous position. If the situation is reversed—that is, if ablation exceeds accumulation—the glacial front will retreat up the valley, although ice will continue to move from the head to the terminus.

The balance between glacial advance and retreat can be delicate indeed. For example, in 1991, about 1 centimeter of windblown African dust settled on the ice margin of the Similuan Glacier in the Italian Alps. Some of the sunlight normally reflected by the ice was instead absorbed by the dust, heating the glacier and causing the ice to retreat several meters in a matter of months. What is unique about this instance of glacial retreat is that released from the ice was the preserved corpse of a man who had met his accidental death at that very spot 53 centuries ago (Figure 17.8)!

Ice Calving

The Antarctic ice sheet continues to grow even though very little snow falls on it (just 5 centimeters per year); because it is so cold, there is almost no melting. However, this growth is offset by ice that is lost to the sea along the glacier margin. The glacial ice flows from the interior to the edges of the continent, where it extends far out to sea as broad, flat **ice shelves.** Attached to the land on one end, the shelves terminate in steep-

basal slip
Movement in which a glacier decouples from the valley floor and slides downslope.

surge
A brief period of rapid glacial flow.

accumulation
The mass of ice gained by a glacier because of snowfall.

ablation
The loss, or wastage, of ice from a glacier through melting, evaporation, and calving.

ice shelf
A sheet of thick ice terminating in steep cliffs. One end remains attached to the land; the other end extends out into the sea, where it floats.

Figure 17.7 The glacial budget. On a yearly basis, accumulation exceeds ablation above the snowline; the reverse is true below the snowline. The glacier will advance if net accumulation exceeds net ablation and will retreat if the opposite is the case.

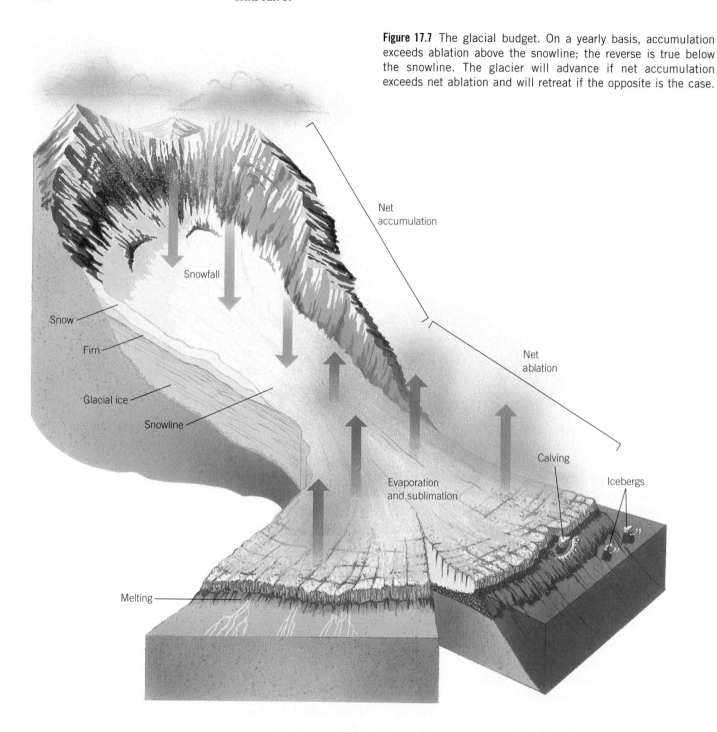

iceberg
A block of glacial ice floating on a body of water, with 80 percent or more of its volume below the surface.

calving
The process through which blocks of ice break off from ice shelves and float out to sea as icebergs.

walled ice cliffs—starkly beautiful sights. Ice shelves in the ocean behave like an ice cube in a glass of water, which floats with most of its volume submerged. The ice shelves grow until they lack a firm base of support on the seaward end. When they reach depths that allow 80 to 90 percent of their volume to sink underwater, the edges of the ice shelves calve—that is, they break into huge blocks and float out to sea as **icebergs** (Figure 17.9). Because melting is minimal on the vast Antarctic ice sheet, **calving** is the chief means by which the ice sheet is ablated.

Calving also occurs on the edge of the Greenland ice sheet, the ice caps of the Canadian Arctic Islands, and the numerous mountain glaciers that empty into the sea, but it is most impressive in the ice shelves of Antarctica. The Ross Ice Shelf, for example, is the size of Texas; icebergs, some twice the size of Rhode Island, calve from the shelf edge and float out into the Ross Sea.

Figure 17.8 This well-preserved corpse of a stone-age traveler was entombed in the Similuan Glacier of the Austrian Alps 53 centuries ago. It was uncovered in 1991.

Erosional Features of Glaciers

The geologic work of glaciers has a definite pattern, whether confined to mountain valleys or spread over continents. The ice erodes outward from its center of accumulation and transports the eroded material to the terminus, where it is deposited. As the melting glacier retreats toward its source, it leaves behind a countryside characterized by an intricate mixture of erosional and depositional features.

Two relatively simple processes are responsible for carving the wide variety of erosional features associated with glaciers. The first, **plucking,** is related to the fact that ice generally melts under pressure when squeezed against solid-rock surfaces. The meltwater trickles into fractures in the rock and refreezes. The water expands upon freezing, which exerts pressure and breaks off the rock fragments—some the size of a van, others the size of a sand grain. The rock fragments are dislodged (or plucked) and incorporated in the glacier as it slowly moves toward the ice front. The glacier then uses the

plucking
An erosion process in which meltwater trickles into rock fractures, refreezes, and expands, breaking off rock fragments.

Figure 17.9 An iceberg calving from the face of the Margerie glacier in Glacier Bay National Park, Alaska.

Figure 17.10 Glacial striations in a rock outcrop in the Van Horn Range in Alaska. Mapping striations on a regional scale enables geologists to trace the direction of ice movement.

abrasion
The mechanical wearing or grinding of rock surfaces by friction and impact of rock particles transported by wind, ice, waves, running water, or gravity.

cirque
A deep, curving, steep-walled depression scooped out of the bedrock at the head of a valley glacier.

glacial striations
Parallel grooves and scratches that were gouged in the bedrock by rock fragments attached to the bottom of the ice as the glacier slid over it.

roche moutonnée
A glacially sculpted knob of rock, gentle and smooth on its upstream side; steep and rough on its downstream side.

fragments to grind away the surfaces it moves against. The process is called **abrasion,** and it is the second way that the ice erodes its surroundings.

Plucking is largely responsible for the characteristic signature carved at the head of valley glaciers, the **cirque** (see Figures 17.5 and 17.6). A cirque is a natural amphitheater—a deep, curving, steep-walled depression scooped out of the bedrock. To investigate firsthand how a cirque forms, we must lower ourselves by ropes down the narrow crevasse, or *bergschrund*, that separates the steep wall of the cirque from the snow and ice at the head of the glacier. We observe all sizes of blocks of rock that are in various stages of dislodgment from the ice-coated cirque wall. Turning to the ice cliff formed by the head of the glacier, we see rocks that have already been engulfed in the ice and snow. Evidently, snow and ice are ripped loose from the cirque wall, along with their cargo of rock, and the whole collection is plastered onto the glacier as it slowly creeps downhill.

The plucked fragments serve as abrasives on the bottom and sides of the glacier, removing loose materials and gouging, polishing, and crushing bedrock. Figure 17.10 shows typical features of a rock outcrop subjected to glacial scouring. Note, first, that the outcrop has been scraped clean of soil. Parallel grooves and fine scratches, called **glacial striations,** were gouged in the outcrop by rock fragments attached to the bottom of the ice as the glacier slid over it. Finer silt and sand particles in the ice, mixed with slush, acted as sandpaper to give the rock its finer polish. The polish is smooth to the touch in the direction of ice movement but rough in the opposite direction, like planed wood. And like the fine sawdust created when wood is sanded, rock dust, or *rock flour,* was generated by the grinding and polishing of the rock surfaces and was washed away by streams of meltwater.

Next, look at the streamlined shape of the outcrop in Figure 17.11. The up-glacier side slopes gently, a result of abrasion and polishing; whereas the down-glacier side is steep and rough, caused by glacial plucking. This glacially sculpted knob of rock is called a **roche moutonnée,** from the French, meaning "sheep rock."

Ice-Sculpted Mountain Topography

Glacial plucking and scouring produce spectacular alpine scenery. Unlike a stream, which physically occupies only a minute percentage of the valley it has carved, a glacier may fill the entire valley, so that erosion extends far up the valley walls (Figure 17.12).

Figure 17.11 A roche moutonnée in the island of Skye, Scotland. The streamlined shape of the outcrop indicates that the ice moved from right to left.

Just as tributary streams feed water into the main stream, tributary glaciers feed ice into the main valley glacier. The ice originates at the headwaters of former streams, where, as we have seen, glacial plucking produces cirques. Expansion of the cirques narrows the divides between glaciers to knife-edged **arêtes.** In places, the glaciers expand their cirques headward to produce isolated peaks of rock, or **horns,** which stand like massive towers above the surrounding valleys.

As the ice flows and slides relentlessly downhill, the V-shaped stream valleys are converted into characteristic U-shaped **glacial troughs**—steep-sided, broad, and deep. In the process, the main glacier may amputate the former tributary stream valleys; it may also excavate a far deeper trough than the tributary glaciers dug. Thus when the glaciers melt, the tributary valleys are left stranded at a higher elevation than the main valley, forming spectacular **hanging valleys** that run along the upland margins of the main U-shaped glacial trough.

The rainwater that falls on this glaciated region will eventually erode these features and reestablish the dominance of stream erosion in shaping the landscape. Until then, running water accommodates itself to the reality of what the ice has accomplished. Cirques become cold, crystal-clear lakes, or **tarns.** Unevenly scoured bedrock forms the basins of **paternoster lakes** that are sometimes connected by a series of cascading waterfalls and rapids.

Mountain glaciers extending to a coast often cut their U-shaped valleys below sea level. When the ice in the valley finally melts, the sea invades these steep-sided troughs, forming **fjords.** These deep, narrow arms of the sea are prominent features of high-latitude seacoasts bordered by glaciated mountain ranges: the coasts of Norway, Greenland, Alaska, and British Columbia in the Northern Hemisphere, and Antarctica, Chile, and New Zealand in the Southern Hemisphere. Aside from being objects of scenic splendor, many fjords make great natural harbors.

Depositional Features of Glaciers

Glacial deposits are called **drift**—a term held over from the time, not so many years ago, that they were believed to have been laid down during Noah's Flood. The two vehicles for depositing drift are meltwater and ice. Meltwater allows sediment to separate

arête
A divide between adjacent valleys, narrowed to a knife-edged ridge by the backward erosion of the valley walls.

horn
A steeply carved rock peak, formed by the walls of three or more cirques.

glacial trough
A U-shaped valley carved by a mountain glacier in what was previously a V-shaped stream valley.

hanging valley
A glacial valley, formed in what was previously a tributary stream valley that was cut off and left stranded at a higher elevation from the main glacial valley.

tarn
A lake in a cirque.

paternoster lake
A small lake in a depression carved in a glacial valley floor.

fjord
A deep, narrow arm of the sea, formed when a glacial trough that extended to the sea became submerged after melting of the ice.

drift
Glacial deposit of rock fragments.

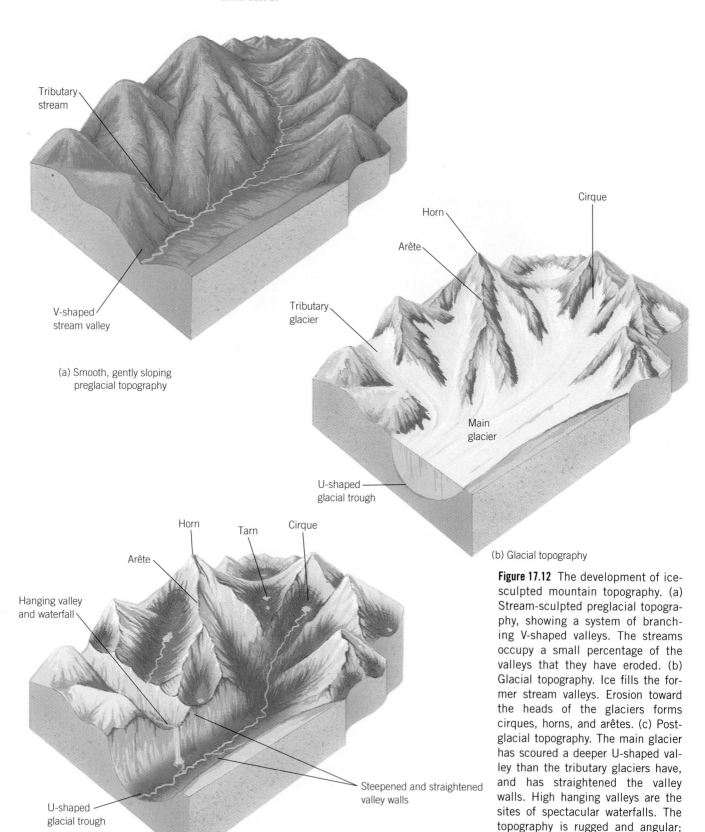

(a) Smooth, gently sloping preglacial topography

Tributary stream

V-shaped stream valley

(b) Glacial topography

Horn

Arête

Cirque

Tributary glacier

Main glacier

U-shaped glacial trough

(c) Steep, angular postglacial topography

Horn

Tarn

Cirque

Arête

Hanging valley and waterfall

Steepened and straightened valley walls

U-shaped glacial trough

Figure 17.12 The development of ice-sculpted mountain topography. (a) Stream-sculpted preglacial topography, showing a system of branching V-shaped valleys. The streams occupy a small percentage of the valleys that they have eroded. (b) Glacial topography. Ice fills the former stream valleys. Erosion toward the heads of the glaciers forms cirques, horns, and arêtes. (c) Postglacial topography. The main glacier has scoured a deeper U-shaped valley than the tributary glaciers have, and has straightened the valley walls. High hanging valleys are the sites of spectacular waterfalls. The topography is rugged and angular; postglacial streams are dwarfed by their glaciated valleys.

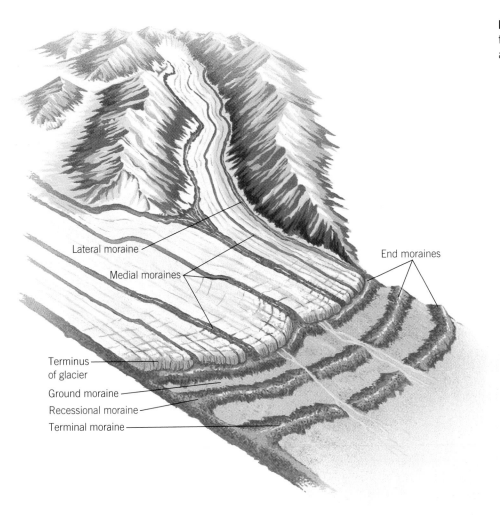

Figure 17.13 Glacial moraines reflect the various erosional, depositional, and flow patterns of the glacier.

Lateral moraine

Medial moraines

End moraines

Terminus of glacier

Ground moraine

Recessional moraine

Terminal moraine

according to size. Hence, the resulting deposits are sorted and layered, forming **stratified drift.** Sediments deposited directly by ice, called **till,** are neither sorted nor stratified, because all fragments, from the fine clay to boulders the size of trucks, are locked in place within the moving ice and dumped wherever the ice happens to melt.

Moraines

Moraines are landforms composed of till (Figure 17.13). Like other sedimentary landforms—sand dunes, beaches, deltas—a moraine reflects the manner in which the sediment was brought to the depositional site and the environment in which the sediment was deposited.

   Let us consider the situation at the front of a glacier whose budget is in balance—which is where ice accumulation equals wastage. Because ice is being produced at the same rate as it is being destroyed, the glacier can neither advance nor retreat. But as we have seen, the ice *within* the glacier continues to move to the front, where it melts. Deposited at the front of the glacier is the debris, or till, that the ice brought with it. Should the ice remain stationary for a number of centuries, a thick pile of till called an *end moraine* will accumulate. If the stationary front is, in fact, at the line of maximum glacial advance, then its end moraine is called a **terminal moraine.**

   Now let us assume that the budget of this glacier becomes unbalanced for several centuries by a warming climate, during which time melting exceeds ice accumulation. In this case, the ice margin steadily recedes toward the head of the glacier, leaving a rough blanket of till called a **ground moraine,** the most common and widespread type

stratified drift
Deposits left by streams or pools of glacial meltwater that are sorted and layered according to size.

till
Unsorted, unlayered drift deposited directly by glacial ice.

moraine
A landform composed of till left behind by a retreating glacier.

terminal moraine
A thick pile of till deposited at the line of maximum glacial advance.

ground moraine
A rough blanket of till that accumulated under the glacier, exposed as the terminus recedes toward the head of the glacier.

Figure 17.14 A glacial erratic in Tripod Rock Reserve, New Jersey.

of glacial deposit. If the glacial budget temporarily comes into balance during the long period of retreat, it may produce another stationary ice front, along which a **recessional moraine** will accumulate. Depending on the number of times the ice stalls, a series of recessional-moraine ridges may form parallel to the terminal moraine of the glacier. End and ground moraines are features of both continental and mountain glaciers.

A valley glacier, however, abrades and plucks the valley walls as well as the valley floor. This oversteepens the walls, causing landslides, slumps, and rockfalls. As this debris collects at the sides of the glacier, it forms a **lateral moraine.** Should lateral moraines merge, as when two tributary glaciers enter the main valley, they form a **medial moraine.** These moraines ordinarily are not well preserved because they are destroyed by meltwater during glacial retreat.

When a glacier melts, it leaves boulders strewn across the glaciated landscape. We know that they are ice deposits because they are too large to have been transported by running water, even a huge flood. Called **glacial erratics** (Figure 17.14), they commonly do not match the underlying bedrock, which makes them useful for what they reveal about the direction and mechanics of ice movement. The way to use them is to note carefully their distinctive features—mineral content, fossils, texture, color—and trace them boulder to boulder back to their bedrock source. In many instances, that source is local, perhaps a kilometer or two away. In other instances, ice carried these boulders hundreds of kilometers. For example, boulders of granite from Canada, weighing several tons, today rest peacefully on the limestone plains of Iowa.

In places, the ground moraine forms smoothly rounded, elongated hills called **drumlins** (Figure 17.15), whose long axes parallel the direction of ice flow. Their origin is uncertain. They may have been molded by the base of the glacier moving over previously deposited moraines. Averaging about 30 meters high and 1500 meters in length, most drumlins appear in groups or swarms. From the air, a group resembles a school of whales, their broad backs breaking the water surface. Some ten thousand drumlins are found in western New York State alone. They are also locally prominent landscape features of Wisconsin, Minnesota, Ontario, and New England. While not one

Figure 17.15 A drumlin field along Copper River, Alaska.

of the Earth's major landforms, the modest drumlin has had its rendezvous with history: Bunker Hill, where in 1775 Boston citizens fired on British soldiers in a battle that sparked the American Revolution, is a drumlin.

Landforms of Stratified Drift

The landforms of stratified drift are composed of glacial meltwater deposits. These waters issue from continental and valley glaciers, so it is no surprise that similar landforms are common to both—though they may differ in size and frequency of occurrence (Figure 17.16).

The meltwater issuing from a glacial terminus carries a choking load of coarse bed load sediment. Unable to move this load all at once, the stream deposits most of it within the channel and assumes a braided pattern (see Figure 17.16). As discussed in Chapter 15, this is the most efficient way to move coarse bed load in a low-volume stream. Over time, the stream's constantly shifting channels will deposit sediment in front of the retreating glacier, forming a broad, flat **outwash plain** (see Figure 17.16). As a valley glacier retreats, it leaves behind a long, narrow body of outwash called a *valley train*. Like all stream deposits, outwash plain and valley train sediments are sorted and layered. Coarse gravels are found closest to the glacier front, sands and finer materials farther away. Because most of the load is deposited closest to the glacier, outwash plain deposits thin as they slope away from the terminal moraine. As may be expected, the outwash plains of ice sheets are far more extensive than those of valley glaciers.

If we trace outwash plains back toward the source of the glacier, we find that they often connect with long, snakelike winding ridges called **eskers** (Figure 17.17). The eskers of continental glaciers may reach impressive dimensions: over 500 kilometers long, including gaps, with heights of 300 meters. Typically, however, they are far more modest, having lengths of a few kilometers and heights of perhaps 10 meters. That eskers are composed of poorly stratified gravel and sand and that they connect to outwash plains suggest that they were formed by streams escaping through tunnels at the base of the melting glacier.

There are many other opportunities for meltwater to deposit sediment within and along the margins of a melting ice sheet or valley glacier. As the ice recedes, its sediment forms mounds, knobs, or short ridges above the surrounding countryside. These

outwash plain
A broad, gently sloping sheet of stratified sand and gravel sediment deposited by meltwater streaming out of the front of a glacier.

esker
A snakelike ridge, up to 500 kilometers or longer, of roughly stratified gravel and sand, left behind when glacial ice melted.

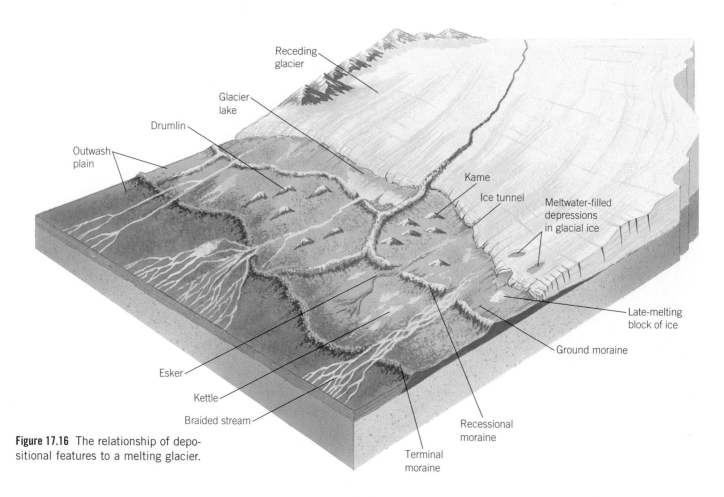

Figure 17.16 The relationship of depositional features to a melting glacier.

kame

A mound or short ridge of stratified sand and gravel deposited by water streaming under or trapped within glacial ice; from the Scottish word meaning "steep-sided monument."

kettle

A depression in a moraine or outwash plain formed when a large block of ice was isolated from the retreating glacier and buried in the drift, melting afterward.

elevated features are collectively called **kames,** from the Scottish word meaning "steep-sided monument." *Kame terraces* are long, narrow ridges of sediment that were deposited by streams that flowed between the ice and the walls of valley glaciers.

Kettles are depressions in moraines and outwash plains formed when blocks of late-melting ice were isolated from the retreating glacier. They often become rounded and

Figure 17.17 This winding ridge of gravel and sand is an esker in Campbellsport, Wisconsin. It was deposited in an ice tunnel by water escaping a melting glacier.

Figure 17.18 Kame and kettle topography south of Big Delta, Alaska. The mounds are kames; the lakes are kettles. This topography typically forms over a broad zone at the terminus of a retreating ice sheet.

secluded lakes. Henry David Thoreau's Walden Pond in Massachusetts is such a feature. Pitted *kame-and-kettle topography* is especially common along the ice margin, with its array of mounds, ridges, depressions, and small lakes (Figure 17.18).

Glacial Lakes

At the outset of this chapter, we mentioned that 18,000 years ago, North America was a very different place than it is today. Figure 17.3 shows an ice sheet that covered nearly all of Canada and large portions of the United States, all of which are ice-free today. This slowly advancing ice sheet scoured bedrock; upon melting, the retreating ice deposited tons of glacial drift over the exposed earth. Left behind was a new landscape, consisting of disrupted drainage patterns and numerous lakes.

Much of the drift that blankets Ohio, Indiana, Illinois, Michigan, Wisconsin, and Minnesota was derived from soft bedrock excavated by the last North American ice sheet from the deep basins that contain the Great Lakes. We can verify this observation by analyzing rock and soil fragments in the drift and comparing them with the bedrock farther north. Like a huge bulldozer, the ice scoured deep depressions and dumped the debris around the periphery.

A resource of inestimable value, the Great Lakes contain over 30 percent of all the fresh liquid water on the Earth's surface. It is perhaps difficult to imagine a feature of such magnitude being a mere ten thousand or so years old. However, the Great Lakes basins were filled by the North American ice sheet. As the ice retreated, the streams draining the Great Lakes flowed first to the Gulf via the Mississippi, then to the Atlantic via the Hudson, before finally flowing out to the Atlantic via the Saint Lawrence River.

The Great Lakes, Finger Lakes, and others, though formed by glaciers, survive because they are regional low points fed by present-day streams (Figure 17.19). But many glacial lakes have not survived because the ice sheet formed their margins, either in whole or in part; when the ice melted, the lakes drained. Left behind were lakebed sediments, as well as ancient beaches that defined their shorelines. Glacial Lake Agassiz, which spread over enormous areas of central Canada, Minnesota, and North Dakota, was the most extensive of the glacial lakes, larger than any of the Great Lakes. Its legacy is some of the richest farmland on the continent. The name Lake Agassiz is well chosen,

Figure 17.19 Satellite image of Finger Lakes, New York. The lakes occupy former stream valleys that were oversteepened by the ice sheet that covered the region about 15,000 years ago.

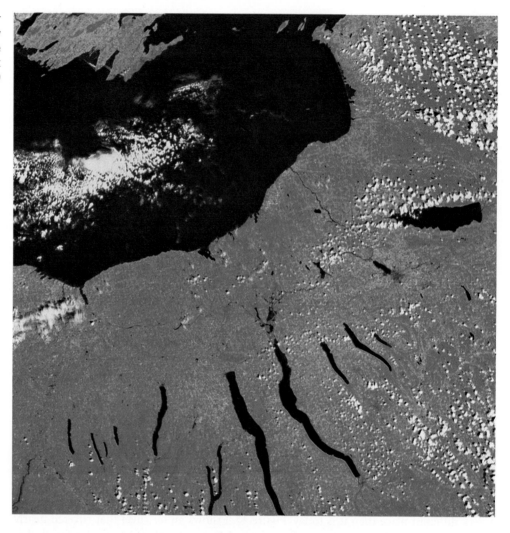

being named for Swiss-born geologist Louis Agassiz (1807–1873), who was the champion of the theory that enormous ice sheets had in the past periodically advanced and retreated over large areas of Europe and North America. Like many revolutionary scientific ideas, his theory was regarded with great skepticism when he proposed it in 1837. Many scientists of his time lacked the imagination to visualize an Earth vastly different from the one before their eyes. Their attitude is reminiscent of the outraged and mocking response Alfred Wegener received in the early twentieth century to his theory of continental drift.

The climate that produced the ice sheets was also responsible for the formation of lakes farther south. Dry regions, such as Utah, Nevada, and Southern California, were in fact heavily forested or had lush grassland supporting a fauna whose abundance rivaled the plains of Africa. The melting of valley glaciers—or simply abundant rainfall—fed water into the lowlands forming large **pluvial lakes.** Lake Bonneville, Utah, of which Great Salt Lake is a puny descendant, was one of them. Its shorelines are visible in the mountains high above the Bonneville Salt Flat, where the lake evaporated as the climate turned arid.

pluvial lake
A body of water in a nonglaciated region formed by abundant rainfall due to the cooler climate caused by the growth of ice sheets.

Glacial Lake Deposits

During the waning stage of a glacier, ice often impounds meltwater and prevents its escape. Lakes then develop at the edge of the ice margin. In the past, some of these lakes were enormous, the size of whole states. When the ice retreated, the water drained,

Figure 17.20 Varves deposited in a glacial lake. A light and dark layer together represents approximately one year. Geologists are able to date ice retreats by counting the varves in lake deposits left behind by the ice.

leaving behind various stratified lake deposits: gravelly and sandy kame deltas, beach sediment, and stranded shoreline gravel that record former water levels.

Toward the center of the lakes, coarse sediment grades into fine-textured, delicately layered **varves** (Figure 17.20). Each varve consists of a layer of light, silt-sized rock flour and a layer of dark clay. The rock flour is a summer deposit, brought in by milky streams when melting was active. The clay settled slowly to the bottom during more peaceful winter conditions when the lake had frozen over and wind turbulence was not a factor. Therefore, a single varve records approximately one year of time. Varve counts indicate that it took about twelve thousand years for a vast ice sheet to shrink from its maximum point of advance in what is now the United States back to Northern Canada.

Glacial and Interglacial Ages

Northern Hemisphere glaciation began at least 3 million years ago during the Pliocene epoch and has continued into the period most associated with glaciation, the **Pleistocene epoch.** Each ice advance, or **glacial age,** has been followed by a period of ice retreat, or **interglacial age.** Right now, we are living in the Holocene interglacial age. Geologists have arbitrarily set its beginning at the onset of the rapid sea-level rise that resulted from melting of the last ice sheet about 11,000 years ago.

varve
Two layers of sediment deposited in one year's time in a glacial lake: one layer of light sand and silt deposited in summer and one layer of dark clay deposited in winter.

Pleistocene epoch
An interval characterized by widespread glacial advances and retreats that began 2 million years ago and ended about 11,000 years ago.

glacial age
An interval of geologic time in which glacial ice sheets advance.

interglacial age
An interval of geologic time in which glacial ice sheets retreat.

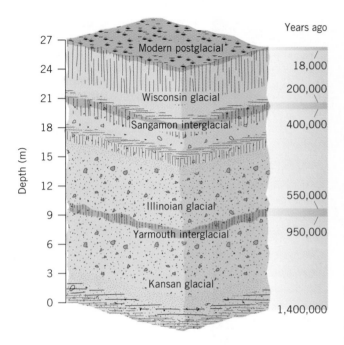

Figure 17.21 Geologists use the principle of superposition to establish the age relations of past glacial advances and retreats. A ground moraine deposit signifies that an ice sheet formerly covered the region. A buried soil profile above the moraine—often containing logs or pollen—indicates a warming period. Radiometric dating establishes precise dates.

How have geologists been able to piece together these ages? The traditional evidence of ice advances and retreats comes from land-based field surveys of glacial deposits and other features of climatic change. Figure 17.21 shows the kinds of evidence useful in glacial studies. The most important clues come from glacial till. As the figure shows, till deposits are stacked one on top of another. The highest—and youngest—till deposit contains the least weathered rock fragments. Tills at progressively lower levels show progressively higher degrees of weathering. Each till deposit represents an ice advance—most ancient on the bottom, most recent on top. Between the till deposits are ancient soil profiles, each marking an interglacial age. The profiles show evidence of pollen grains, seeds, root systems, and buried logs, making it clear that the soil nourished extensive forests. The profile may in turn be covered by glacial outwash and by nonglacial stream, lake, or bog deposits—all of which are evidence of glacial melting and warmer climates.

Although land-based fieldwork has continued to yield important information, emphasis has shifted to the analysis of minute fossils from marine sediments that are extremely sensitive to the changes in climate that accompanied ice advances and retreats. Certain species of amoebalike marine organisms called *foraminifera* are very useful for these purposes. Some species are confined mainly to cold surface waters; others are found primarily in warmer waters. Each species has a distinctive, identifiable shell; when these organisms die, their shells rain in a steady drizzle to the ocean bottom, where they are buried in clay and silt.

By taking a core of marine sediment and counting the abundance of shells of the temperature-sensitive species in various levels of the core, we can determine the fluctuations of water temperature over time. Intervals in the core where the cold-water species predominate mark low seawater temperatures; intervals where warm-water species are abundant mark high temperatures (Figure 17.22). Presumably, the onset of low seawater temperatures signifies a glacial age; the warmer temperatures, an interglacial age.

Marine cores containing foraminifera shells can be used in another way to track glacial advances and retreats, because the varying oxygen isotope content of the shells reflects cooling and warming climates. When seawater is evaporated, the lighter ^{16}O isotopes escape into the atmosphere at a slightly higher rate than the heavier ^{18}O isotopes. The excess is returned to the ocean through rainfall and the hydrologic cycle. During a glacial age, when snow falls, these excess ^{16}O isotopes become locked up in glacial ice, which leaves the oceans—and, therefore, the shells—correspondingly richer in ^{18}O and poorer in ^{16}O. During an interglacial age, the ice melts, and the excess ^{16}O is returned to the ocean. Therefore, the shells deposited during this warmer period contain greater amounts of ^{16}O than those deposited during the glacial age. Thus the changing percentages of ^{16}O relative to ^{18}O in the shells at various levels of marine cores are the mirror image of ice advances and retreats.

There is generally close agreement among the dates of ice advances and retreats determined from land-based field studies, from the variations of cold- and warm-water species, and from the oxygen isotope ratios found in marine cores. These marine studies, conducted in the 1960s and 1970s, place glacial ages at approximately 100,000-year intervals. We will see that there is a close match between these intervals and earlier theories of the causes of the ice ages.

Causes of the Ice Ages

Ice ages are an expression of climate, and much of what we know about them comes from the study of past and present climates. The climatic conditions that have led to

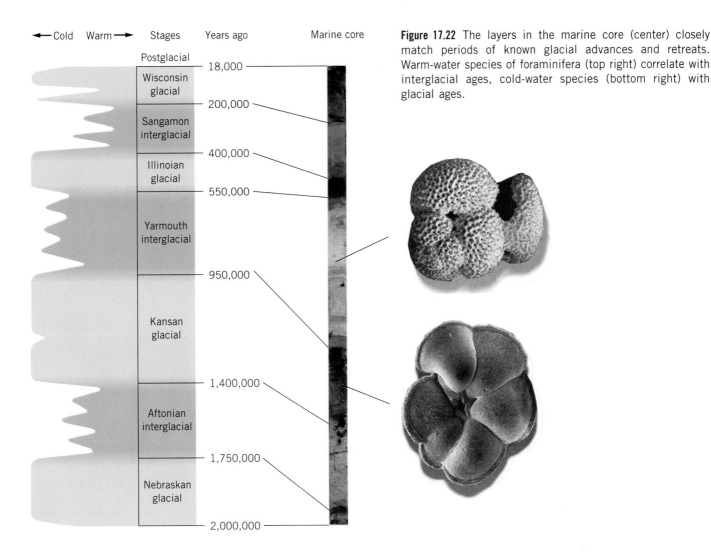

Figure 17.22 The layers in the marine core (center) closely match periods of known glacial advances and retreats. Warm-water species of foraminifera (top right) correlate with interglacial ages, cold-water species (bottom right) with glacial ages.

glacial and interglacial ages are the result of a multitude of factors. The three major factors we discuss are:

1. the movement of the Earth's tectonic plates and associated volcanic and crustal uplift
2. long-term cyclical variations in solar energy received by the Earth
3. the condition of the Earth's atmosphere—most importantly, its carbon dioxide and dust content

Plate Tectonic Effects

The rocks and sediments of the mid-Cretaceous period, 120 million years ago, preserve evidence of a world warmer by 15 °C than today's—a world in which dinosaurs spread throughout the globe feeding on lush tropical vegetation or on animals who fed on it. The mid-Cretaceous warming is an interesting subject; increased volcanism has been suggested as the cause. Our point here is that the Earth has been gradually cooling since the Cretaceous to temperatures that today support the Greenland and Antarctic ice sheets and that a short time ago supported glaciers three times as extensive (Figure 17.23).

Two effects of the movement of the Earth's lithospheric plates are responsible for the pronounced decline in average world temperature since the Cretaceous period, a decline that made the advent of the ice ages possible. The first is the slow drift of the continents toward high latitudes. The second is the general elevation of the continents due to volcanism and mountain building.

Figure 17.23 This graph shows that worldwide temperatures have cooled considerably since the Cretaceous period, some 80 million years ago.

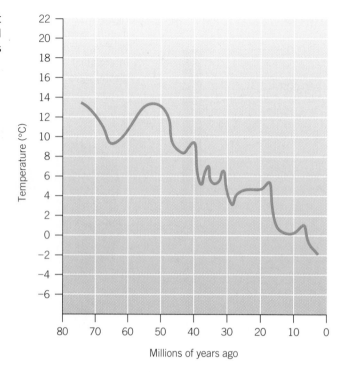

Drift of the Continents Present ice sheets are concentrated in, and have spread from, high latitudes. High latitudes are favored because temperatures generally decline toward the poles. However, continents must also be present in high latitudes to support the ice; apparently, only rarely in geologic history have the continents been in the proper position to support glaciers.

With this idea in mind, refer to Figure 17.24. It shows that some 275 million years ago, an ice sheet spread over large portions of the great Southern Hemisphere landmass of Gondwanaland that was centered around the South Pole. Since then, Gondwanaland has fragmented into Africa, South America, Australia, New Zealand, and India, all of which carry with them the evidence of the ice sheet.

Crustal Uplift During the Cretaceous period, not only were the continents in lower latitudes, but their average elevation was lower as well. We know this because marine deposits of that age cover vast areas of the now emergent landmasses. The Cenozoic era (our era), which followed the Cretaceous, has been a time of active tectonism. The enormous uplifts of the Himalayas, the Tibetan Plateau, and portions of the Rockies and Andes mountains have occurred during this time frame. Yet even slight crustal uplift can raise elevations enough to lower temperatures by a few degrees in surrounding regions. Uplifts far from ice sheets may affect climate in subtler ways. For example, they may alter global wind patterns, forcing frigid arctic air farther south. The cold air may lessen melting at the margins of continental glaciers, thus causing their advance. In addition, recent work has suggested that massive uplift affects weathering rates, which, in turn, affect the release of carbon dioxide to the atmosphere. We will soon see that variations of atmospheric carbon dioxide correlate closely with ice advances and retreats.

Clearly the drift of the continents to high latitudes, combined with crustal uplift and mountain building—both of which are controlled by the movement of the Earth's lithospheric plates—have contributed heavily to a lowering of worldwide temperatures during the past 120 million years. However, the plates move too slowly to cause periodic glacial and interglacial ages. Neither is there any evidence that they oscillated back and forth to cause the alternate warming and cooling necessary to trigger advances and retreats of the ice. What mechanism could be responsible for these rhythmical changes?

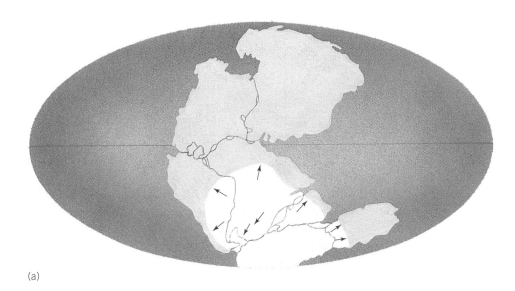

(a)

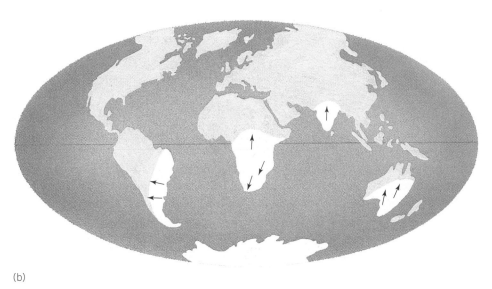

(b)

Figure 17.24 The concentration of the continents at high latitudes favors the growth of ice sheets. (a) Ice sheets covered much of the Southern Hemisphere 275 million years ago. (b) The drift of North America and Eurasia to polar regions led to the growth of ice sheets in the Northern Hemisphere during the latter part of the Cenozoic era. Evidence of the Southern Hemisphere glaciation remains in the scattered continents.

Variations in Earth's Orbital Geometry: The Milankovitch Theory

One mechanism that might trigger glacial advances and retreats is cyclical variations of the Earth's orbit and rotational axis that control the amount of solar energy reaching Earth's high latitudes. The ice would advance when incoming solar heat is reduced and retreat when these latitudes are warmed again. Most glacial geologists agree that a sustained drop of about 3 °C in average annual temperatures is sufficient to trigger an ice advance. This is the approximate difference in mean temperature between Boston and Washington, D.C. The entire difference in average temperatures between peak glacial and interglacial ages is probably on the order of 5 °C.

We are already familiar with one form of cyclical variation of solar heat over the Earth's surface that is controlled by orbital geometry—the seasons. Summer occurs in the Northern Hemisphere when, because of the tilt of the Earth's axis with respect to its orbit about the Sun, the Sun's rays strike most directly above the equator. Winter occurs when the rays strike most directly south of the equator. (Of course, the timing of the seasons is reversed in the Southern Hemisphere.) In the 1920s and 1930s, Serbian astronomer and meteorologist Milutin Milankovitch (1879–1958) proposed a concept similar to the seasons to explain ice ages. He suggested that other, more subtle orbital characteristics were responsible for slight, longer-term variations in the

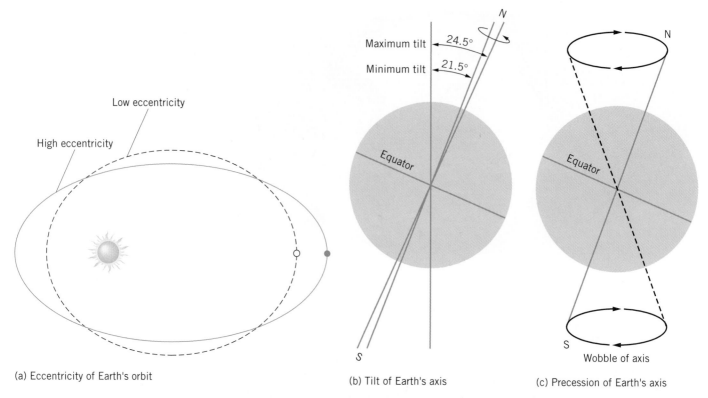

Figure 17.25 Three periodic variations in the Earth's orbital and spin characteristics contribute to the Milankovitch cycle. (a) Orbital eccentricity. (b) Axial tilt. (c) Axial precession. An ice age may be initiated when these motions coincide periodically, thus reducing the solar energy that reaches the Northern Hemisphere during summers.

(a) Eccentricity of Earth's orbit

(b) Tilt of Earth's axis

(c) Precession of Earth's axis

distribution of solar heat over the Earth's surface—variations sufficient to control ice advances and retreats. These variations are caused by the gravitational tug on the Earth by the other planets and the Moon as the Earth spins like a top and revolves around the Sun. Three gravitational effects are evident.

1. *Eccentricity of the Earth's orbit.* Figure 17.25(a) shows that the Earth periodically departs from a nearly circular orbit to a more elliptical one. Maximum departure occurs about once every 100,000 years. This effect periodically brings the Earth closer to and farther away from the Sun than at present and causes a slight ebb and flow of the heat received by the Earth over time.

2. *Tilt of the Earth's axis.* The Earth's rotational axis also tilts with respect to the plane of the Earth's orbit about the Sun. At present, the angle of tilt is 23.5°, but it varies (nods) between 21.5° and 24.5° every 40,000 years or so. The greater the angle of tilt, the more sunlight that reaches the polar regions in summer and the less in winter (Figure 17.25b).

3. *Precession of the Earth's axis.* In addition, the Earth's axis precesses, or wobbles, slightly, repeating each wobble about every 26,000 years. This precession changes the date when the Earth is closest to the Sun and, therefore, varies slightly the intensity of the sunlight striking polar regions in summer and winter (Figure 17.25c).

Milankovitch proposed that these cyclical motions periodically reinforce one another. When the combined effects warm the Northern Hemisphere slightly, they trigger an ice retreat. When they combine to reduce temperatures in the Northern Hemisphere, they trigger an ice age. Apparently, an ice advance correlates with a slight reduction in the heat reaching northern latitudes during the summer months, which leads to less snow melting and an overall ice surplus.

Unfortunately, when Milankovitch first proposed his hypothesis, it suffered a fate similar to that of many advanced ideas; seemingly impossible to prove or disprove, it was largely ignored. However, attitudes changed decades later when researchers discovered

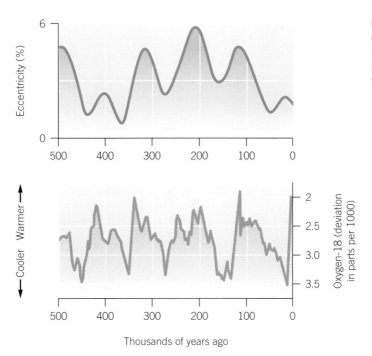

Figure 17.26 Warm interglacial periods and cold glacial periods are measured by the oxygen isotope ratio in marine foraminifera (bottom). They closely correlate with the Milankovitch cycle (top).

that the oxygen isotope content of foraminifera in deep-sea cores accurately reflects the timing of glacial advances and retreats predicted by Milankovitch (Figure 17.26).

That glacial advances and retreats correlate with variations in solar heat as predicted by the Milankovitch cycle is indeed an important contribution to our understanding of the origin of glaciers. But it is far from a complete explanation, for the cyclical heat variations are too weak to induce the temperature extremes necessary to cause glacial and interglacial ages. Other factors must serve to amplify the effect. There is also evidence that glacial advances and retreats lag behind the Milankovitch cycle, which suggests that the Earth's climate systems react sluggishly to slight ebbs and flows in solar heat. We need to know more about the interactions among these various components of climate, and their linkages to incoming solar heat before we can fully understand glaciers.

Variations in Atmospheric Carbon Dioxide and Dust

Earth's temperature is very sensitive to the percentage of atmospheric carbon dioxide present, because the carbon dioxide traps the heat rising from the Earth's surface. In this respect, the carbon dioxide acts like a blanket that warms you by slowing the dissipation of the heat your body radiates. This well-known role of carbon dioxide and other gases in regulating the Earth's temperature is called the greenhouse effect. Sunlight is able to penetrate the atmosphere to warm the Earth's surface, but carbon dioxide delays the escape of infrared radiation (that is, heat) rising from the surface. The greater the concentration of carbon dioxide, the more heat that is trapped and the higher the Earth's temperature will be; the lesser the carbon dioxide content, the lower the temperature.

Because the greenhouse effect is known to regulate present-day climates, the carbon dioxide content of ice cores drilled in the Greenland and Antarctic glaciers constitutes highly significant evidence of past climatic changes. The deepest cores, extracted from Antarctica, date back more than 160,000 years. Each layer of the core contains trapped bubbles of "fossilized air"—tiny samples of the atmosphere that existed when the layer formed. By measuring how the carbon dioxide varies layer by layer in a core, scientists are able to trace the fluctuations of atmospheric carbon dioxide over the time span covered by the core. By analyzing the oxygen isotopes in these layers, they are able to trace atmospheric temperature variations over that same time span. (^{16}O increases by a certain amount for each 1 °C drop in temperature.)

Figure 17.27 Variations in atmospheric carbon dioxide and temperature over the past 160,000 years are closely matched. This correlation suggests that carbon dioxide is a key regulator of the Earth's temperature.

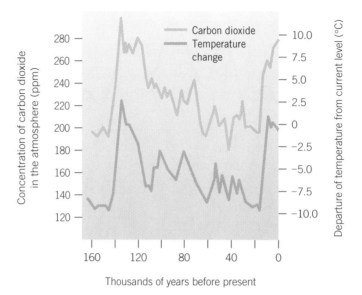

Data from the cores suggest that a strong correlation exists between variations in atmospheric carbon dioxide content, glacial and interglacial ages, and the Milankovitch cycle. The low temperatures of the ice ages correlate with low carbon dioxide—over 30 percent less than the present amount. The high temperatures of interglacial ages show correspondingly high carbon dioxide content. In addition, the data show that over the past 150,000 or more years, carbon dioxide buildups have preceded the onset of global warming, whereas declines have signaled cooling periods (Figure 17.27).

Having confirmed the relationship between carbon dioxide and the Earth's temperature, the next question must be, What controls the carbon dioxide content of the atmosphere? The answer is complex, as you might imagine, but there is no doubt where to search for it—the ocean. This enormous water body contains 50 times more carbon dioxide than the atmosphere and exchanges the gas with it, through the sea surface. These exchanges are controlled by the temperature, chemistry, biology, and circulation of the ocean, as well as by atmospheric conditions. Any slight change in these factors can add or subtract carbon dioxide to or from the atmosphere and thus affect climate in a big way.

Atmospheric Dust   Greenland and Antarctic ice cores contain other important information. Trapped in their layers are dust particles that had settled out of the atmosphere at the time of deposition. Again, the correlation with the Milankovitch cycle is striking. Low dust volumes match warming periods; high dust volumes match cooling periods. Dust particles in the upper atmosphere—most importantly, sulfate droplets—both reflect and absorb sunlight and therefore cause worldwide temperature reductions. As discussed in Chapter 7, volcanic eruptions are the chief source of these particles.

 # Future Climate Change

Is the Earth warming or cooling? This is an interesting question aside from practical considerations, because the answer depends upon the natural contributions to climate we have been discussing so far and the "unnatural" contributions made by industrial civilization. The latter effect is widely known today as *human-induced global warming.*

Taking the long view, there is little doubt that the Earth will grow increasingly colder and drier. We are currently in a warm, wet interglacial age that began approximately 6000 years ago. However, judging from the length of previous interglacial ages and the Earth's present position in the Milankovitch cycle, temperatures will probably decline

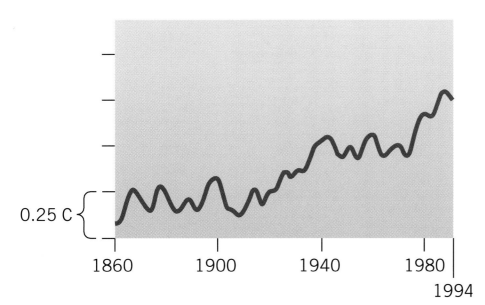

Figure 17.28 An estimate, based on meteorological records, of global temperature changes from 1860 to 1994. Although temperatures fluctuated from year to year, the overall increase was only about 0.7 °C.

within the next few thousand years; in approximately 20,000 years, they will reach full-scale glacial conditions.

Over the short run (that is, the past 150 years), worldwide temperatures have been on the rise (Figure 17.28). Geological evidence suggests that this trend is affecting Earth systems.

1. The world's alpine glaciers are in general retreat, a trend that has accelerated during the past 15 years. So is the Arctic Sea ice, which has shrunk 2 percent in the past 30 years. Large portions of the immense Ross and Larsen ice shelves that fringe Antarctica are disintegrating, and temperatures on the Antarctic Peninsula have risen nearly 3 °C over the past half-century.
2. Coral reefs of the Pacific and Caribbean have bleached white from the death of the algae upon which they depend. Many marine biologists believe that the deaths are attributable to rising water temperatures.
3. Over the past five years, sea level has risen an average of 2 centimeters per year worldwide—more than double the rate observed for the twentieth century. This increase correlates with the melting of the glaciers, which returns water to the sea, and with the warming of the sea surface, which causes water to expand slightly.

Although these changes fall well within the range of natural variation, geoscientists generally agree that the short-term temperature spurt is human-induced. They trace global warming to two types of human activities.

The first category includes activities that accelerate emissions of greenhouse gases: *carbon dioxide* from cars, factories, electrical power plants, and burning of forests; *methane* from natural gas wells, cattle, and rice cultivation; *nitrous oxide* from agriculture; and *chlorofluorocarbons* (CFCs) from refrigerants and other sources. As these gases are added to the atmosphere, they trap a higher percentage of the heat radiated from the Earth's surface.

The second type of activity contributing to global warming is the ongoing destruction of the world's great forests and other ecosystems, which decreases the amount of carbon dioxide that is extracted from the air via photosynthesis. Most geoscientists reason that the average global temperature is rising because heat-trapping gases are being added to the atmosphere more rapidly than they are being extracted. Recent predictions place this temperature increase within the range of 2 to 5 °C by the end of the next century. This range incorporates varying estimates of the amounts of greenhouse gases that human beings will add to the atmosphere—the lower, more optimistic number of 2 °C derives from predictions that emission rates will not increase, while the higher number of 5 °C assumes substantial increases.

Figure 17.29 This display, which reflects recent GCM simulations, shows a global pattern of climatic change that resembles the real-world pattern observed over the past 50 years.

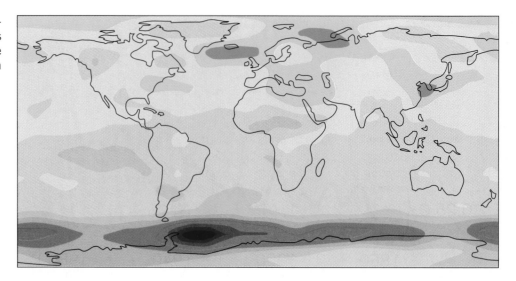

Geoscientists are gaining confidence in their predictions through the employment of increasingly sophisticated general circulation models (GCMs). Such computer programs are built of interconnected, physics-based equations that simulate—in simplified form—the complex relationships among the myriad factors responsible for climate. They are designed to process enormous quantities of past and present climatic data from the Earth's surface and atmosphere and then predict changes in regional and global climates over time. Although a detailed discussion of these climate models lies beyond the scope of this book, a simple example may suggest the complexities involved in developing a GCM.

Consider the influence of clouds on climate. Rising temperatures are known to increase the amount of moisture evaporated from the surface and therefore increase the cloud cover. In contrast, decreasing temperatures tend to decrease the cloud cover. However, clouds have both a warming and a cooling effect on the Earth. On the one hand, they warm the Earth by absorbing infrared waves rising from the planet's surface and reradiating the heat downward. On the other hand, clouds cool the Earth by reflecting sunlight that would otherwise be absorbed at the surface. Thus an accurate GCM must be able to calculate the *net effect* of an increase or decrease in the cloud cover on global temperatures over the long term.

A useful GCM predicts not only the average temperature increase for the Earth as a whole, but also—and more importantly—how temperature and rainfall will vary over the Earth's surface, and during the course of a year. First, the model is fed only "natural data" exclusive of human-induced greenhouse gases and other emissions. This step establishes a baseline from which to trace the effects of human-induced global warming. Second, the human contribution is added to the model. Finally, the model is tested by seeing how well it can replicate the known record of climate change over the time interval covered by the data.

Figure 17.29 shows the results of recent computer model simulations of the effects on the Earth's surface temperatures of human-induced carbon dioxide and sulfate aerosols. The carbon dioxide emissions, of course, *raise* worldwide surface temperatures. However, you may recall from our discussion of volcanoes that sulfate aerosols scatter sunlight and *lower* surface temperatures. Thus the aerosols act to partially offset the carbon dioxide increase. The significance of the simulations is that they match closely the pattern of climate change actually observed over the Earth's surface during the past 50 years. Such models give geoscientists confidence that their predictions of human-induced global warming are essentially correct.

Apply and Decide 17.1

Human-Induced Global Warming

While most scientists are becoming alarmed at the prospect of global warming, a few continue to wonder what the fuss is about. They remind us that natural climatic fluctuations lasted for centuries in the recent past; yet the human race survived prolonged periods of both relative warmth and relative cold. For example, we know that about 900 years ago, the climate in northern Europe during the Medieval Warm Period (1100–1375) was milder than it is today. Writers of the time described a Britain so mild that wine grapes flourished and a Greenland so warm that the Vikings farmed the now-frigid coast. A century or two later, during the Little Ice Age (1550–1850), Europe became so cold that the British held their winter carnivals on the frozen Thames River and the Greenland ice sheet extended back toward the coast.

Although an hysterical response to the evidence of global warming may not be called for, neither is complacency. During the Medieval Warm Period the world's population probably numbered fewer than 600 million people. Today's population is closer to 6 billion, and is welded together by a technological civilization infinitely more complex than it was centuries ago. The following are some of the changes we can reasonably expect as a result of global warming:

• We can expect a higher degree of weather variability than at present. As more water vapor and energy are transferred to the atmosphere, hurricanes and other storms are likely to grow more violent. The seasonal contrasts between land and sea temperatures will increase. Midcontinental droughts and heat waves will intensify in summer; in winter, a reverse cooling effect will bring more frigid, longer lasting cold spells.

• The highest latitudes are expected to receive the sharpest temperature increases. Although small regions around the north Atlantic and northern Eurasia may grow colder, virtually all GCM's predict that temperatures in the higher latitudes will increase—from 6 to 14 °C according to some models. Increasing temperatures at high latitudes may accelerate the melting of the ice caps. The degree of melting remains a point of disagreement among scientists. Indeed, there are those who think that the glaciers will actually grow for a time because of increased precipitation. However, if sea level should rise at close to current rates, as the Intergovernmental Panel on Climate Change (IPCC) has estimated, the oceans will rise between 0.5 and 1.5 meters over the next century. Some extreme estimates place sea-level rise at 6 meters.

• Temperature increases may also cause the North American and Eurasian rain belts to shift significantly northward. The reason for the shift is that most of the rain is generated when warm moist air from the subtropics clashes with dry, cold polar air along constantly fluctuat-ing boundaries or fronts. In North America the current "line of battle" extends across the northern Plains States, but with global warming, the warm and cold air masses are likely to clash at higher latitudes, well into Canada.

If the rain belts migrate northward, we can expect the Plains States to become hotter and drier. Such a drop in the average yearly rainfall is very likely to change this region back into a desert. Yes, back. If you dig below the soil cover of these states, you will find extensive windblown sands and other evidence that 10,000 to 15,000 years ago, deserts stretched across the now fertile region. The Sand Hills region of western Nebraska was in fact a dune field larger than any in the present Sahara Desert.

As with most rapid change, there will be winners as well as losers to the latitudinal creep of the warm climatic belts. The humid tropics and subtropics are expected to broaden, bringing rain and renewed fertility to the north African deserts and the surrounding semiarid zones, which today are periodically famine-stricken (see Chapter 18, Deserts and Winds).

• We can also expect a weakening of the global wind systems as the temperature contrast between the equator and the poles diminishes. Weakening of the wind systems will in turn lessen the efficiency of wind-driven ocean currents that transport heat poleward from the equator. For example, the Gulf Stream carries water warmed in the Caribbean up along the eastern coast of North America to the North Atlantic. In some GCMs, the Gulf Stream falters, turning Great Britain and coastal Scandinavia as cold as Greenland.

• We can expect major ecological changes. Global warming will endanger those species unable to evolve swiftly in the face of rapid environmental change. Animals that have long life spans, relatively few offspring, and specialized food and habitat requirements—for example, most large land mammals—will simply not be able to adapt to such rapid change and may die out in numbers reminiscent of the great Cretaceous extinctions. On the other hand, rodents and insects, being omnivorous and highly adaptable, will probably prosper.

The Impact of Global Warming on Human Society

The physical and biological effects of global warming will have major economic and social implications for that most prolific and adaptable of large mammals, *Homo Sapiens*. The changes will affect three important domains: agricultural production, coastal habitation, and disease control.

Agricultural Production.   The agricultural potential of a region depends upon its climate, soil type, and soil condition.

Figure 17A During spring melt, a tabular iceberg is dislodged from an Antarctic ice shelf. The extent to which global warming will melt the Antarctic and Greenland ice sheets, thereby raising sea level, remains an open question.

Many regions of the continents are unsuitable for anything beyond small-scale farming, while other regions have become worldwide suppliers of their agricultural produce. The United States, for example, is the world's largest grain exporter, and most of the grain is grown in the midwestern Plains states. However, as we discussed in Chapter 16, Groundwater, in many regions of the Plains states agriculture is already under environmental stress because of inconsistent rainfall and rapidly depleting aquifers. Further drift toward drought conditions will have devastating economic consequences for both farmers (a small percentage of the U.S. population) and grain consumers (an enormous percentage of the world population).

The rains that now water the grain fields of the Plains states will shift northward, so the logical question is: Won't Canada, the recipient of the shifting rains, eventually be able to take up the agricultural slack? This answer is: Probably not. Much of the soil of northern Canada is underlain by stony glacial deposits and is thin and poor. It is doubtful that the region could ever match the productivity of the present corn and wheat belts, where glacial outwash, nourished by thousands of years of rainy climate, has become the thick, rich soil blanket of the midwestern U.S.

Coastal Habitation.   The prediction of a one meter sea-level rise doesn't seem impressive until you consider that up to 70 percent of the world's population lives along coasts. A sea-level rise of a meter to a meter and a half would displace tens of millions of people, contaminate the freshwater wells and storage facilities of coastal communities, and intensify inland flooding during storms. The United States alone would lose some 66,000 square kilometers of coast. The most extreme estimates place sea-level rise at a catastrophic 6 meters. In that case, nearly all the world's coastal cities, farming belts, deltas, and flood plains would drown. The sea would inundate such places as Bangladesh, Venice, the Bahamas, Samoa, the Argentine Pampas, and most of Florida and the Gulf states.

Disease Control.   Global warming will facilitate the spread of a host of nasty tropical diseases to higher latitudes and altitudes. For example, the mosquito that carries malaria will be able to migrate to regions currently too cold for it to survive. IPCC scientists estimate that an average global temperature increase of approximately 3 °C will add 50 to 80 million malaria cases per year to the 300 million per year already occurring.

Global Warming as a Political Issue

Any time there are conflicting economic interests, there tend to be conflicting points of view. If you doubt the veracity of this statement, remember how long, in spite of a mountain of evidence, the tobacco companies continued to deny that smoking causes lung cancer. The global warming debate also has its political aspects, as anyone who has listened to the noted climatologist, Rush Limbaugh, knows.

Compare, for example, the stance that the coal, oil, and natural gas industries might take versus the stance that the insurance industry might take toward the evidence of global warming. (Hint: on the one hand, remember that the burning of fossil fuels is a major cause of global warming. On the other hand, if you haven't already guessed the relationship between the insurance industry and global warming, remember that Hurricane Andrew, which struck Florida and Louisiana in 1992, generated 15.5 billion dollars in insurance claims.)

THOUGHT QUESTIONS

1. Atmospheric CO_2 emissions have increased 13.6 percent since 1986. What steps might the United States take to reduce emissions? Would reducing emissions help, hurt, or have no effect on the economy?

2. Consider the economic and environmental impact of global warming on your current hometown. In what ways do you think the weather and climate will change? How will those changes impact on the local flora and fauna? On the local economy? On your livelihood? On the adult lives of your children?

Glacial ice covers 10 percent of the Earth's land surface, with most of it concentrated in the Greenland and Antarctic ice sheets. Glacial advances and retreats, accompanied by profound climatic fluctuations, have been the pattern for at least 3 million years. Glacial advances cause the sea level to fall, which exposes the continental shelves. Glacial retreats cause the sea level to rise, which floods the coastal regions of the continents. Glaciers are important agents of erosion, transport, and deposition.

I. THE FORMATION OF GLACIERS

A. A **glacier** is an ice mass that survives from year to year. Its formation begins with the compaction of snow into ice pellets called **firn.** The **snowline** is the elevation above which there is permanent snow. Glaciers of continental proportions are called **continental glaciers, or ice sheets. Ice caps** are somewhat smaller. Away from the polar regions, in areas of high elevation, **alpine** or **valley glaciers** occupy former mountain stream valleys. **Piedmont glaciers** form on lowlands where several valley glaciers merge.

B. As an alpine glacier flows downhill, ice velocity is greatest at the valley center and least along the valley walls.

  1. The top rigid layer, or **zone of fracture,** cracks under the stress of movement, forming **crevasses.** Below, in the **zone of plastic flow,** the ice flows under the pressure of its own weight toward the front, or **glacial terminus.** Melting along the valley floor allows the entire glacier to slide downhill in **basal slip** or in more rapid **surges.**

  2. Because ice sheets form at high latitudes, they are colder and more rigid than alpine glaciers. Ice sheets flow slowly outward in all directions from the centers of accumulation.

C. Like a business that has income and expenditures, the glacial budget is a balance between **accumulation** of snow and **ablation**—water loss through melting, evaporation, and **calving.**

  1. If accumulation exceeds ablation, the glacier grows and advances. If ablation exceeds accumulation, it shrinks and retreats.

  2. When a glacier reaches the sea, it extends outward as an **ice shelf,** where chunks break off, or calve, forming **icebergs.**

II. EROSIONAL FEATURES OF GLACIERS.

Glaciers erode rock by **plucking,** in which ice melts, seeps into rock fractures, refreezes, and breaks off rock fragments. Through **abrasion,** the ice polishes and grinds bedrock.

A. A mountain glacial system fills the former stream valley and all its tributaries, eroding V-shaped stream valleys into U-shaped **glacial troughs.** The main glacier may cut off the tributary glacier valleys, forming **hanging valleys. Tarns** and **paternoster lakes** fill the cirques and depressions left by a glacier in the valley floor. **Fjords** are seawater-filled glacial troughs that extend to the coast.

B. Bowl-shaped **cirques** are formed by the **plucking** of a valley glacier headwall. A bergschrund is a crevasse between the glacier and the cirque headwall. Between cirques are knife-edged **arêtes** and rock peaks called **horns.**

C. Parallel grooves in rock, caused by abrasion, are called **glacial striations.** The fine debris of abrasion is called rock flour. A **roche moutonnée** is a knob of rock smoothed by abrasion on one side and steepened by plucking on the other.

III. DEPOSITIONAL FEATURES OF GLACIERS.   There are two types of glacial deposits, or **drift: stratified drift** and **till.**

    A. Till consists of unsorted, unstratified rock fragments (including boulder-sized **glacial erratics**) that are transported by glacial ice and dropped at the terminus in mounds called **moraines. Terminal** and **recessional moraines** form at the foot of the glacier; **lateral** and **medial moraines** are formed along the sides. A **ground moraine** is a blanket of till left behind by a receding glacier. In places, the ground moraines of continental glaciers feature rounded, elongated hills called **drumlins.**

    B. Meltwater flowing in braided streams from the glacial terminus deposits sorted and layered **stratified drift,** leaving landforms such as broad, flat **outwash plains** and snakelike ridges called **eskers.** Other stratified-drift landforms are mounds called **kames** and depressions called **kettles.**

IV. GLACIAL LAKES

    A. The last North American ice sheet excavated the Great Lakes and other new lakes.

    B. The climate that caused the ice sheets was also responsible for large **pluvial lakes,** formed in regions that today are arid.

    C. Alternating layers of rock flour and clay sediments are deposited in glacial lakes that are formed when a receding ice sheet temporarily impounds meltwater. Counts of these layers, or **varves,** are used to date ice sheet advances and retreats.

V. GLACIAL AND INTERGLACIAL AGES

    A. Great ice sheets have occurred ten or more times during the past 2 to 3 million years. The most recent interval of glaciation began during the **Pleistocene epoch.** Each ice advance, or **glacial age,** has been followed by a period of ice retreat, or **interglacial age.**

    B. Timing of advances and retreats is estimated by carbon-14 dating and other radiometric methods and by analysis of the oxygen isotope content in climate-sensitive fossils of marine organisms.

    C. The following factors have led to glacial and interglacial climate conditions:

        1. movement of the Earth's tectonic plates from low to high latitudes and the associated elevation of the continents due to volcanism and mountain building;

        2. the Milankovitch cycle: long-term variations in solar energy received by the Earth; and

        3. the condition of the Earth's atmosphere—most importantly, its carbon dioxide and volcanic dust content.

VI. FUTURE CLIMATE CHANGE.   Future climate trends will depend on natural and human-induced effects.

    A. Temperatures in the long-range future will probably decline; however, over the short run, worldwide temperatures have been on the rise.

    B. Many experts now agree that dangerous global warming is being caused by two areas of human activity: increased emissions of greenhouse gases, such as carbon dioxide, and deforestation. Recently developed climate models bear out their concerns.

STUDY TERMS

ablation (p. 459)
abrasion (p. 462)
accumulation (p. 459)
alpine glacier (p. 456)
arête (p. 463)
basal slip (p. 459)
calving (p. 460)
cirque (p. 462)
continental glacier (p. 456)
crevasse (p. 457)
drift (p. 463)
drumlin (p. 466)
esker (p. 467)
firn (p. 455)
fjord (p. 463)
glacial age (p. 471)
glacial erratic (p. 466)
glacial striations (p. 462)
glacial terminus (p. 457)
glacial trough (p. 463)
glacier (p. 455)
ground moraine (p. 465)
hanging valley (p. 463)
horn (p. 463)
iceberg (p. 460)
ice cap (p. 456)

ice sheet (p. 456)
ice shelf (p. 459)
interglacial age (p. 471)
kame (p. 468)
kettle (p. 468)
lateral moraine (p. 466)
medial moraine (p. 466)
moraine (p. 465)
outwash plain (p. 467)
paternoster lake (p. 463)
piedmont glacier (p. 456)
Pleistocene epoch (p. 471)
plucking (p. 461)
pluvial lake (p. 470)
recessional moraine (p. 466)
roche moutonnée (p. 462)
snowline (p. 455)
stratified drift (p. 465)
surge (p. 459)
tarn (p. 463)
terminal moraine (p. 465)
till (p. 465)
valley glacier (p. 456)
varve (p. 471)
zone of fracture (p. 456)
zone of plastic flow (p. 457)

CRITICAL THINKING QUESTIONS

1. Explain how it is possible for ice to advance from the head to the terminus of the glacier even as the glacier retreats.
2. What factors affect the budget of a glacier? Under what circumstances will a glacier advance or retreat?
3. Assume that sometime in the future, all the world's glaciers will have melted. If you were living at that time, what evidence preserved on land would suggest that ice sheets had once existed?
4. What physical evidence preserved on land and in marine cores suggests that the great North American ice sheet periodically advanced and retreated?
5. Explain how cores of glacial ice have been used to chart variations in the Earth's climate over the past 160,000 years.
6. Name and explain at least three factors that may be involved in climate regulation.
7. Why do many scientists suspect that humans are contributing to global warming?

C HAPTER 18

Deserts and Winds

S and and deserts go together in the popular imagination, a marriage arranged by impressionable travelers to antique lands and by artistic film directors. David Lean's film *Lawrence of Arabia* (1963) shows one exotic scene in which magnificent sand dunes stretch to infinity like enormous waves in a dry ocean. In the foreground—in stark contrast to the white sands—a caravan of nomadic warriors on camels inches across the screen.

Most desert regions, however, are not covered by windblown sand. Running water—and not wind—is the chief gradational agent of deserts, and great sand dunes, though certainly impressive, occur only under specialized conditions. Perhaps the most impressive fact about deserts is that they occupy 19 percent of the land surface of the Earth. Semidesert steppe regions account for another 15 percent. Far from being exotic, either arid or semiarid conditions are common to one-third of all land (Figure 18.1).

Another popular myth to dispel is that all deserts are hot. Deserts are defined by lack of precipitation, not by temperature. Temperatures in the Sahara Desert, which can rise to well over 40 °C during the day, can easily drop below 4 °C at night. Midlatitude deserts and steppes can get brutally cold in winter—ask a Wyoming rancher or read the accounts of soldiers who fought on the Russian front during World War II. Because deserts are defined on the basis of annual precipitation, portions of Antarctica, Greenland, Siberia, northern Canada, and Alaska must be included. We do not usually think of these regions as deserts, though, because of their low temperatures and the effects of the ice sheets that either still cover them or have recently receded. Having discussed these regions in Chapter 17, we devote this chapter to the deserts and semiarid regions that fit the more conventional meaning of the term. We will look at the origins of deserts, the topographic features of deserts, the work of desert winds, desert water

◄ A striking desert scene at the foot of the Black Mountains, Death Valley, California. The blue-green colors are caused by the oxidation of mineral-laden groundwater seeping to the surface.

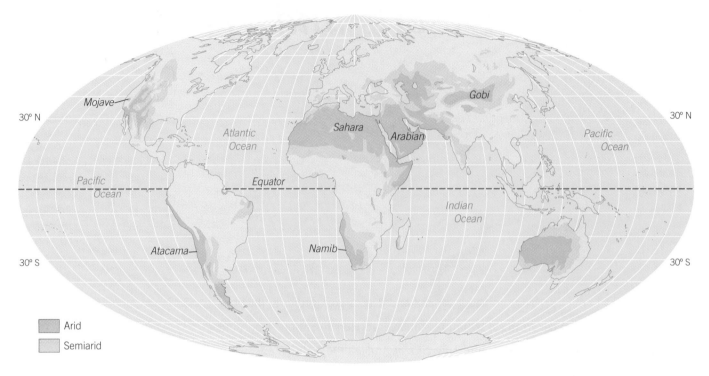

Arid

Semiarid

Figure 18.1 The worldwide distribution of arid and semiarid lands. Notice the concentration of deserts within belts approximately 30° north and 30° south of the equator.

desert

A region with an average precipitation of less than 25 centimeters, too dry to support more than sparse vegetation. (Climatologists define deserts more complexly on the basis of temperature-precipitation-evaporation ratios.)

resources, and the human contributions to desertlike conditions—a serious problem, as we shall see.

The Formation of Deserts

Lack of abundant, consistent rainfall (less than 25 centimeters per year) is the irreducible feature of all **deserts** and is responsible for all that is unique to them—including the fact that the land surface is too dry to support more than a sparse plant cover. Three factors, acting separately or in combination, create the arid conditions. The first concerns the placement of the continents with respect to the worldwide wind system; this leads to subtropical deserts. The second concerns the influence of topography on rainfall patterns, resulting in rain shadow and midlatitude deserts. The third concerns the proximity of the land to certain cold ocean currents, producing narrow coastal deserts.

Winds

The origin of deserts is intimately connected to the planetary wind system, which controls the delivery of moisture to the continents. So we will pause briefly to describe the factors that cause winds before examining how winds interact with the land to form deserts.

Winds are air movements and are controlled, either directly or indirectly, by the uneven distribution of solar heat over the Earth's surface. The uneven distribution results from Earth's spherical shape, which causes the Sun's rays to strike the surface most directly at the equator and least directly at the poles. Thus the equator receives more heat than the poles, and temperatures generally decline poleward—although seasonal variations, the location of landmasses, and ocean circulation complicate matters.

Though our intuition may tell us otherwise, the atmosphere does not draw its heat directly from the Sun. Rather, the Sun's rays penetrate to the Earth's surface, where they are absorbed; the warmed Earth *then* heats the atmosphere from below. That is why air temperatures decline with altitude. The air in contact with the surface of the Earth at the equator is warmed more than air over other regions; as it is warmed, it expands,

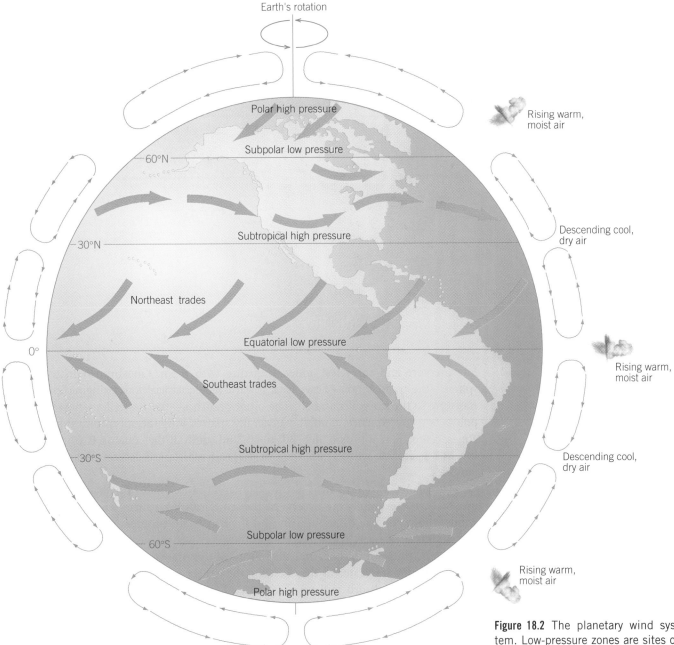

Earth's rotation

Polar high pressure

Subpolar low pressure

60°N

Subtropical high pressure

30°N

Northeast trades

0°

Equatorial low pressure

Southeast trades

Subtropical high pressure

30°S

Subpolar low pressure

60°S

Polar high pressure

Rising warm, moist air

Descending cool, dry air

Rising warm, moist air

Descending cool, dry air

Rising warm, moist air

Figure 18.2 The planetary wind system. Low-pressure zones are sites of converging winds and high rainfall. High-pressure zones are sites of diverging winds and low rainfall. The Earth's rotation causes the winds to deflect to the right in the Northern Hemisphere and to the left in the Southern Hemisphere.

becomes less dense, and rises. The rising air creates a surface zone of low pressure, known appropriately as the *equatorial low* (Figure 18.2). This warm air holds huge quantities of moisture, because a large portion of the solar heat that strikes the equator is used to evaporate water from the ocean. As the rising air cools, however, it releases the moisture as rain. Thus the vast rain belts of the world are located at the equator, as are the tropical rain forests.

As air at the equator continues to rise, it eventually hits a "lid" of even lighter air, known as the stratosphere. Unable to rise farther, it spreads—one loop heading north, the other south. At about 30° latitude north and south of the equator, the air has cooled sufficiently to sink and pile up in huge mounds, forming the *subtropical high-pressure zones*. These sinking air masses have two important effects. First, as they pile up, they compress and therefore warm the air at the surface, which increases the air's capacity to hold moisture; so water is evaporated from the surface. Second, since these air masses are mounds of high pressure, they generate diverging winds that flow toward regions of

Figure 18.3 Rain shadow deserts are formed on the leeward side of mountain ranges. Rising winds release their moisture on the windward slopes, causing these slopes to be heavily forested. Descending winds on the lee slopes of the range are dry.

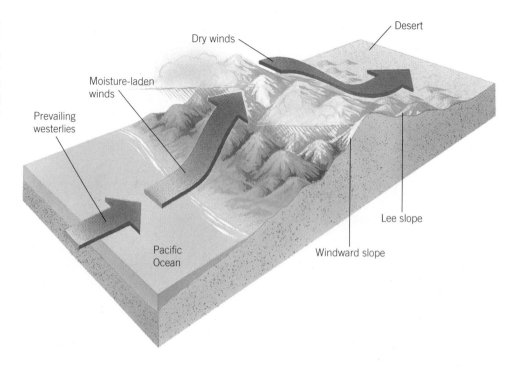

rain shadow
A dry region on the lee side of a mountain range where rainfall is significantly less than on the windward side; the region is said to be "in the rain shadow" of the mountain range.

low pressure and carry the added moisture with them. One branch in each hemisphere—the trade winds—flows back toward the equatorial low, bringing its supply of moisture to the tropical rain belt. The other branches—the westerlies—flow toward the poles. As a consequence, rainfall is generally sparse along the subtropical high-pressure zones, centered at approximately 30° north and south of the equator. If you look at Figure 18.1 again, you will see that the world's great subtropical deserts are located at these latitudes.

The planetary wind belt is also instrumental in forming **rain shadow** deserts (Figure 18.3). For example, the deserts of the American West form where moisture-laden winds coming off the Pacific Ocean find their eastward progress blocked by a series of high mountain ranges. The winds cool as they rise over the peaks; their capacity to hold moisture is thus diminished, and they drop rain on the western windward slopes. The now-dry winds are heated as they descend the eastern lee slopes of the mountains. Thus their capacity to absorb moisture is increased. The combination of having little moisture to give and an increased capacity to hold it ensures that deserts will develop east of the mountains. One need only compare dry Nevada with wet northern California to see the influence of the Sierra Nevada Mountains on the creation of deserts. Farther north, the Olympic Peninsula and the west coast of Washington constitute the rainiest portion of the United States, whereas central Washington is desert. The latter is in the rain shadow of the Olympic-Cascade mountain ranges.

The creation of deserts does not always involve low latitudes or mountain ranges adjacent to the ocean. For example, the great Gobi Desert of China is located at mid-latitude in a deep continental basin within the vast interior of Asia. Winds there are simply exhausted of moisture before reaching the basin, having passed over the Himalayas, the mountains of Afghanistan, and the Tibetan Plateau.

An interesting question in connection with North American rain shadow deserts is, Why do the winds blow from west to east across California? The answer is that the paths of the winds curve, as viewed from the perspective of the Earth's rotating surface. This apparent deflection is called the **Coriolis effect.** Viewed from above the North Pole, Earth resembles a carousel that spins counterclockwise at the rate of one rotation every 24 hours. Objects at the equator have higher velocities than those closer to the poles, because in the same 24 hours, they must complete a larger circle than those closer to the poles. So an object that starts out traveling north from the equator will have a greater

Coriolis effect
The apparent deflection of moving objects, such as winds, viewed from a rotating Earth; deflection is to the right of the initial path in the Northern Hemisphere, to the left in the Southern Hemisphere.

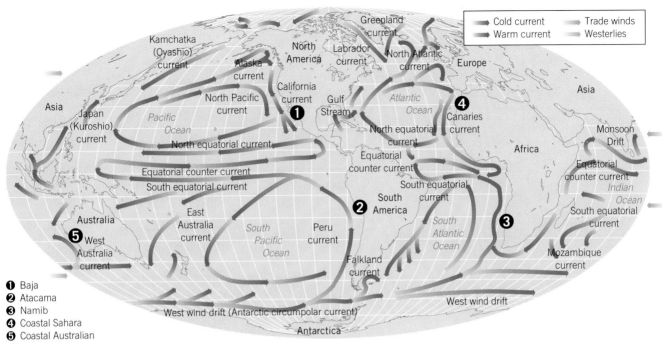

❶ Baja
❷ Atacama
❸ Namib
❹ Coastal Sahara
❺ Coastal Australian

Figure 18.4 Circulation of the wind-driven surface currents. The world's coastal deserts are generally located on the western margins of the continents, adjacent to the cold water currents.

eastward velocity than the land it moves over, and it will therefore curve to the east. Conversely, objects traveling south from the equator will curve in the opposite direction (to the west).

In sum, the Coriolis effect in the Northern Hemisphere causes the winds to curve to the right of their initial path (in the Southern Hemisphere the deflection is to the left), so Northern Hemisphere winds that start out blowing north from the subtropical high develop an eastward deflection. Those blowing eastward from the Pacific encounter the mountains of the American West and produce the rain shadow deserts just described.

The planetary wind system also plays a major role in forcing the ocean currents to flow in enormous circulation loops (Figure 18.4). These currents deliver a huge supply of heat from the equator to high latitudes. Most of the heat transfer occurs along the western boundaries of the oceans, and by the time the currents reach the eastern boundaries, they are cold. Those that run parallel to the western coasts of South America and Africa are responsible for the coastal deserts there. The cold Benguela current runs directly offshore of the Namib Desert, for example (Figure 18.5). The cold water cools the air above it, causing fog and lending the illusion of plentiful rainfall to follow. The cool air, however, contains very little moisture. The cold water causes the air mass to sink and spread, and as it sinks, it warms up. The warm air, with its increased capacity to hold moisture, dissipates the fog, leaving the land dry.

Plate Tectonics and Deserts

In discussing the origins of deserts, we must not overlook one extremely important long-term factor that does not show up on weather maps: plate tectonics. The motions of the great lithospheric plates that compose the Earth's crust have brought the continents to latitudes where diverging winds make rainfall scarce. The placement of the continents also contributes to the division of the ocean currents into circulation loops; for example, it forces currents to flow parallel to the west coasts of Africa and South America—a condition favorable to the formation of coastal deserts. Faulting, regional compression, and volcanism—all plate tectonic effects—have created the mountain barriers to the wind that lead to rain shadow deserts.

Figure 18.5 The Namib coastal desert of western Africa; the cold Benguela current is just offshore.

Prior to the uplift of the Sierras about 15 million years ago, for example, Nevada's climate was much like present-day central Florida and Louisiana—humid and swampy. Sediments deposited in Nevada at that time reflect this environment, and younger sediments deposited during the uplift of the mountains record the change to arid conditions. The Earth's geological record is replete with the history of lush tropics gone dry and vice versa.

Weathering and Erosion in Arid Climates

Sparse rainfall has an impact on every aspect of the erosion process in desert regions. First, the rate of chemical weathering is greatly retarded, and many rocks and minerals are not decomposed to the extent common in temperate and humid climates. Instead, mechanical weathering and incomplete chemical weathering produce a blanket of loose, porous material that covers the bedrock. Water trickles through these partially decomposed rocks and minerals during infrequent rains and dissolves soluble elements. However, water also evaporates rapidly in the desert, and these elements are precipitated near the surface rather than carried away, as in more humid climates. The result is highly alkaline soil, often colored pink, white, and buff. In the extreme, hard white masses of caliche, a calcium carbonate precipitate, accumulate at or near the surface. With the aid of bacterial action, loose fragments and bedrock surfaces may also acquire a strikingly dark-colored coating of **desert varnish,** an encrustation of clay, manganese, and iron oxide (Figure 18.6). The alkaline, granular nature of the soil, combined with the lack of water, accounts for the sparse, widely spaced vegetation of deserts. Sparse vegetation in turn greatly accelerates erosion and is thus responsible for many features of desert landscapes.

desert varnish
A dark, shiny, enamel-like film of iron oxide and manganese that coats the surface of rocks in desert regions.

Figure 18.6 Dark coatings of desert varnish on gravels.

Another factor that affects erosion in arid regions is the nature of the water flow. Although some of the great rivers of the world cross deserts, most of them receive the greater part of their water from regions of high rainfall. The headwaters of the Nile, for example, originate near the humid equator. The Colorado River is nourished from melting snows north of the Colorado Plateau. However, streams that originate within deserts leak water to the water table rather than being fed by it. Furthermore, there is little vegetation in the drainage basins to absorb water and release it slowly to streams, so desert streams must rely almost solely on the rainfall that occurs at widely spaced and irregular intervals. In addition, the high evaporation rate returns a great deal of water to the atmosphere soon after it falls. Under these conditions, the streams are said to be intermittent; their channels conduct water during periods of high rainfall and are left dry during intervening periods. Losing water to both atmosphere and subsurface, many **intermittent streams** simply disappear along their course (Figure 18.7). Their dry beds—known as *arroyos*—are themselves striking features, with nearly vertical channel walls that can rise 30 meters above flat channel floors. With walls that shield against wind, sun, and rain, an arroyo is often seen by novice campers as the ideal spot to pitch their tents. But experienced desert campers avoid arroyos in the rainy season, when a sudden storm can send a flash flood roaring down the dry stream channel.

intermittent stream

A stream whose channel conducts water during periods of high rainfall and is dry during intervening periods.

Figure 18.7 An arroyo cut by an intermittent stream. Notice the steep canyon wall.

The combination of sparse vegetation and sporadic water flow allows slopes in arid regions to be eroded steeply and rapidly. Particles are easily dislodged from the slope because little or no vegetation binds them there. Erosion is accomplished mainly by slope wash, extensive gullying, and mudflows and earthflows during floods. Even mild downpours cause extensive erosion, and the brief but torrential rains, common to many deserts, trigger flash floods that transport huge quantities of sediment. Floods are especially effective erosional agents in deserts having mountainous peaks upon which snow accumulates in winter and melts in spring.

In deserts, extensive and rapid erosion strips hillsides naked and exposes bare bedrock cliffs. This erosion accounts for the steep, angular, and generally rugged-looking topography of deserts. The lack of covering vegetation on the cliffs allows us to see the different erosion rates of the rock strata that form the cliffs. Resistant strata, such as well-cemented quartzite, basaltic lava flows, and massive limestone, form steep cliffs. Easily eroded shales and weak sandstones form gentler slopes. Cliffs with alternating strata of varying composition display varying slope angles.

Eroded fragments that collect at the base of desert cliffs are not colonized by vegetation, so they too are easily removed by the intermittent slope wash and do not accumulate to the same extent as they would in humid climates. Thus the hillside slopes of arid regions retreat at relatively constant angles. In contrast, the slopes in humid regions become gentler as they are eroded. Vegetation lessens the effect of slope wash by soaking up water, and fragments remain close to the base of the hill, which reduces the hill's angle of slope. We can think of slopes in arid regions as being worn back and slopes in humid climates as being worn down (Figure 18.8).

Structural Influences on Desert Topography

Deserts exhibit a wide variety of erosional and depositional features that are dependent upon the bedrock structure of the region and upon the tectonic activity currently at

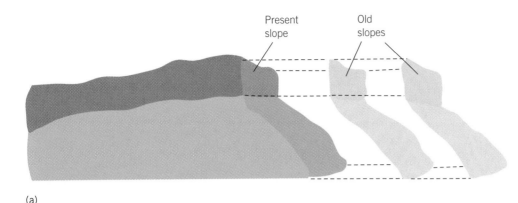

(a)

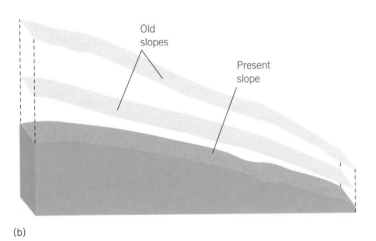

(b)

Figure 18.8 Slope erosion in arid and humid climates. (a) In an arid climate, slopes tend to be worn back at a more or less constant angle. (b) In a humid climate, slopes tend to be worn down, causing a reduction in the slope angle.

work. As illustrations, we will discuss two adjacent rain shadow deserts of the American West: the Colorado Plateau and the Basin and Range Province of Nevada (Figure 18.9). Each is the product of complex interactions involving the North American and Pacific plates.

The Colorado Plateau

The Colorado Plateau is an elevated region of relatively flat land, with a steep drop to lower ground on its western and southern borders. It is composed of sedimentary rock thousands of meters thick, and it is the nearly horizontal layering of this strata that is responsible for the flatness of the plateau. The entire region reminds one of a gigantic layer cake, sliced with steep canyons (Zion, Bryce, the Grand Canyon), from which the layers have been peeled back by mass wasting and stream erosion, exposing the different levels of the plateau as a series of steep cliffs. These cliffs consist of hard sandstone or limestone strata capping weaker shales. The pace of cliff retreat is not uniform, however, because erosion is more active in zones of weakness caused by fractures and slight differences in rock composition.

Left behind by the retreating cliffs are isolated erosional remnants. Resistant rocks that cap these features protect the softer strata beneath them from erosion and, at the same time, are undermined by the erosion of the soft strata. Depending on their size, the remnants are called *mesas, buttes,* or *chimneys* (Figure 18.10). Mesas (the Spanish word for "tables") encompass many square kilometers; as they are eroded, they evolve into the smaller buttes. Buttes, in turn, are reduced to smokestack-shaped chimneys. The

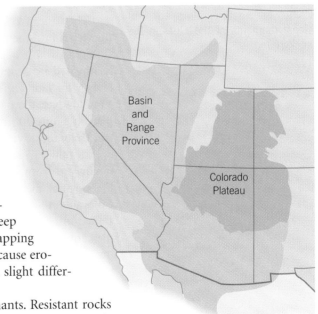

Figure 18.9 A map of the western United States showing the Colorado Plateau and the Basin and Range Province.

Figure 18.10 Erosion of the flat-lying sediments of the Colorado Plateau. The mesas, buttes, and chimneys visible in the foreground are erosional remnants of the cliffs in the background.

empty spaces between these remnants are testament to the extensive erosion of the plateau.

The Colorado River and its tributaries—including every branch, gully, and rivulet that feeds into the main stream—are responsible for most of the plateau's erosion. The Colorado, whose headwaters are nourished by the snowy peaks of the Rocky Mountains, has incised its Grand Canyon by exploiting structural weaknesses in the plateau—several major fault zones that lie along the axis of a broad regional arch.

However, there is much more to the story of the Colorado River and the Colorado plateau. Some 30 million years ago, the plateau was at a lower altitude, and the ancestral Colorado River flowed sluggishly, east-west, across to destinations unknown. However, 5 million years ago, the Gulf of California opened to the south. Streams working northward from the gulf diverted the Colorado's flow to the south, and this is where the story of the Grand Canyon began. Having a shorter, steeper route to the sea, the Colorado cut into the plateau with renewed vigor. At the same time, the plateau continued to rise, so that the river sawed down through the layers of the plateau to create the canyon in about 4 million years. This is an extremely rapid rate, geologically speaking, if you compare it to the age of the 2-billion-year-old rocks at the bottom of the gorge.

The arid climate of the plateau is conducive to the formation of some of nature's most striking features—*natural arches* and *natural bridges.* The arches formed from weathering and wind erosion of the lower portion of an exposed cliff face, leaving behind the upper portion (Figure 18.11). The natural bridges were created where the entrenched meander loops of a desert stream intersected by eroding through the lower layers and not the upper strata. Because chemical weathering is less intense, the internal fabric of the rock has not crumbled to the extent prevalent in humid climates.

Figure 18.11 Natural arches, dubbed "North and South Windows," in Arches National Park, Utah.

Having greater internal cohesiveness, the rocks in arid climates are better able to maintain arch and bridge structures.

Basin-and-Range Topography

Sheltered on the west by the high Sierra Nevada, the Basin and Range Province of the southwestern United States and northern Mexico is a classic rain shadow desert. As we learned in Chapter 11, the province is named for its rugged, bold mountains (horsts) and wide, intervening valleys (grabens), which extend north-northwest in hundreds of roughly parallel belts. The trend of the mountains and valleys follows major faults, formed as the crust has been stretched in an east-west direction. Unlike the crust beneath fold-mountain belts, the continental crust beneath the Basin and Range Province is thin, and it continues to stretch even thinner. The tectonic activity, which began about 30 million years ago, involves large-scale movements of the Pacific and North American plates and continues on an enormous scale. Geologists calculate that the mountains of the province are rising at the rate of 30 centimeters every 500 years.

The combination of tectonism and desert conditions has produced the distinctive structural, erosional, and depositional features of basin-and-range topography (Figure 18.12). Because the various regions within the province differ in age, we are able to trace the evolution of this topography through time. The basic pattern is one in which the basins and ranges begin as isolated erosional and depositional units and are gradually

Figure 18.12 Racetrack Playa in the Cottonwood Mountains, west of Death Valley. Block faulting has created the prominent basin-and-range topography. A playa occupies the sediment-filled basin in the foreground.

bajada
Alluvial fans that coalesce to form a continuous apron along the base of mountains.

playa
A flat area at the bottom of a desert basin that becomes an intermittent (playa) lake in the wet season.

linked to a regional stream system that connects to the ocean. As the mountains are uplifted, they are also eroded, and the basins are filled in.

In the early stages of basin-and-range topography, mountains uplifted by steep faults are separated by broad valleys. Occasional thunderstorms cause torrential rains that pour down the mountain slopes and gash narrow canyons in the bare cliffs. Water is funneled into these dry canyons during subsequent cloudbursts, and the high-volume water pulses scour the canyons, carrying a heavy load of coarse sediment to the flat valley floor. These short-lived streams—some more properly called mudflows—lose carrying power at the point where they emerge from the narrow canyon, and they spread, depositing their coarse sediment as gently sloping alluvial fans.

The fans coalesce to form a **bajada,** a continuous apron that runs along the base of the mountains. The bajadas grow outward and slope away from the mountains, leaving a shallow basin between them. After it rains, runoff from the intermittent streams accumulates in the basin, producing a temporary body of water called a playa lake. The lake dries up during the long interval between rains, leaving behind a dry lakebed, or **playa.** Thus unlike the Colorado Plateau, large areas of the Basin and Range Province exhibit internal drainage, because the intermittent streams have no outlet to the sea.

There are two major consequences of the erosion of the block mountains and the deposition of sediment in the adjacent basins. The first is that the basins fill with sediment. How much sediment? Death Valley, California, the lowest point in the United States at 85 meters below sea level, may provide a fair example. The present depth of the original valley floor is 2104 meters below the present floor. The difference between the two levels, 2019 meters, is the thickness of the sediment washed into the Death Valley basin, which clearly is not yet filled.

The second consequence of the erosion of the block mountains is that the formerly wide mountains are worn back. In time, all that are left of them are narrow divides

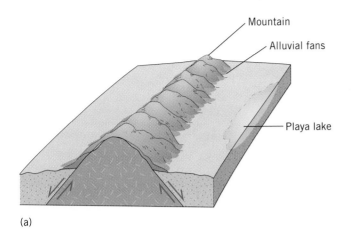

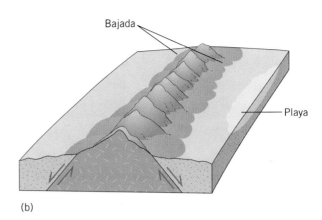

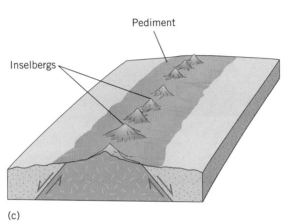

Figure 18.13 Stages in the erosion of a fault-block mountain in an arid climate. (a) Recent uplift produces bold mountains and alluvial fans at the mouths of steep canyons. (b) As the mountain is worn back, the fans coalesce to form a sediment ramp, or bajada. (c) Continued wearing back of the mountain creates an erosional ramp or rock pediment covered by a thin sediment veneer that is a continuation of the bajada. The divides of the mountains are breached, leaving isolated inselbergs.

capped by isolated remnants, **inselbergs** (Figure 18.13). The narrow divides form the crests of broad, gently sloping erosion surfaces called **pediments.** They extend from the former cliff lines and are created as the cliff retreats. In many places, a thin coating of sediment laid down on the pediments merges with the alluvial fans.

Some regions of the province are at a later stage in the evolution of basin-and-range topography in that erosion and deposition are no longer confined to local basins. This is how external drainage is established: as a higher basin fills with sediment, intermittent streams from the adjacent lower basin breach the mountains between them by headward erosion. Drainage from the higher basin is now funneled to the lower one. The process continues until the drainage systems of a series of basins are joined. The system is then captured by a main stream from outside the region, which has cut back through the low divides between valleys. Because the main stream empties into the sea, it is capable of eroding the topography of the basins and ranges to much lower levels. Not only are the hard-rock mountains eroded, but the alluvial fans and the playas of the basins are eroded as well. The layered playa sediments of soft clays, silts, and salts are dissected into a maze of gullies, channels, and shallow canyons—a typical badlands topography. Except for southwestern Arizona, exterior drainage has yet to be established in other parts of the Basin and Range Province.

inselberg
An isolated knob or hill, remnant of a heavily eroded mountain in the desert regions; from the German meaning "island mountain."

pediment
A broad erosional surface formed by running water that slopes gently away from a receding mountain front; the bedrock may be exposed or thinly covered with debris.

The Geological Work of Wind in Arid Climates

Winds are moving fluids. They erode, transport, and deposit sediment in a manner similar to that other moving fluid in contact with the Earth's surface, running water. Like streams, winds transport large, heavy particles as *bed load* close to the ground; small,

Note: You may find it useful to read (or reread) the discussion of load transport on pages 398–400 of Chapter 15, Streams.

Figure 18.14 A dust storm in Saskatchewan, Canada.

light particles are carried as *suspended load* within the main body of the fluid. Bed load particles are rolled, dragged, or pushed along the ground surface in *traction* or are skipped along in short leaps and bounces in *saltation.*

The size of the particles carried either as bed load or suspended load varies with the velocity of the moving fluid, so that winds, like streams, are efficient agents for sorting sediment. As with streams, the volume and size of sediment that winds are able to carry increase markedly with increased velocity. However, air lacks water's density and, as a result, lacks the buoyancy to transport particles larger than coarse sand at the wind velocities present on the Earth's surface. Rarely are these sand-sized particles lifted more than a meter or so above the ground in saltation, whereas fine dust particles may be carried in suspension great distances before they settle out. Particles of African clay fall on Miami, and Gobi Desert dust has been detected in the soil of Hawaii.

Features of Wind Erosion

Winds are most effective in the erosion of loose sediment. That is why wind erosion is a major factor in desert locations, which lack the vegetation and moisture to bind sediments to the ground. The removal of loose materials by wind is called **deflation.** Often left behind by the fleeing winds are broad fields of **desert pavement**—coarse fragments too heavy for wind transport, which protects the material below from further deflation. In regions of thick sand accumulations and strong winds, deflation produces *blowouts,* deep saucer-shaped depressions.

It is easy to underestimate the erosive work of wind if readily identifiable landforms like blowouts are missing. To identify former land surfaces, we must search for remnants, such as exposed tree roots that stand above the land surface, or mounds of soil protected by barriers to wind erosion, such as large boulders. Perhaps the most vivid means of appreciating the erosive capability of winds is to view a dust storm in action (Figure 18.14). A sizable one can transport millions of tons of sediment.

Although winds play a minor role in the direct erosion of solid rock, their effects are nevertheless locally important in deserts. Torrents of saltating sand are capable of cutting down telephone poles and fence posts at their base. This natural sandblasting

deflation
The removal of loose materials by wind, which often results in the lowering of the land surface.

desert pavement
The remaining pebble- to cobble-sized rock fragments covering a desert surface after removal of lighter fragments by wind.

Figure 18.15 A large wind-sculpted ventifact. Notice the faceted planes.

action abrades and pits rock outcrops close to the ground surface and converts rock fragments and boulders into angular, faceted **ventifacts** (Figure 18.15).

ventifact
A rock that has been polished and faceted by the action of windblown sand.

Features of Wind Deposition

Wind deposits are often so diffused that they are mixed with other sediments and soils and are thus not identifiable as massive features. Two exceptions are loess deposits and sand dunes.

Loess Deposits

Thick deposits of fine windblown dust are known as **loess.** Loess is a collection of angular, silt-sized rock and mineral fragments bound loosely by clay. When eroded, the loess forms vertical cliffs (Figure 18.16). The cohesiveness of loess also enables people to easily dig caves out of the cliffs. The insulated caves make comfortable habitats and were used for that purpose in central China and other regions. Loess, rich in nutrients, makes excellent agricultural soil. It underlies the midwestern U.S. corn belt, the wheat fields of eastern Washington State, and the vast farmlands of China.

loess
A thick deposit of fine windblown dust consisting of unstratified silt-sized calcium carbonate and bits of clay, which has the ability to maintain nearly vertical walls despite its weak cohesion.

The distribution and age (about 15,000 years) of loess deposits suggest that in many places, they are derived from glacial deposits. Typically, they are found in areas that border regions covered by ice during the last glacial advance, such as the Mississippi Valley and central Europe. They thin downwind from glaciated regions, which is strong evidence for their glacial origin. It is doubtful, however, that all loess is derived from glacial deposits. Some accumulates from windblown dust from deserts. The extensive, thick deposits of northern China, for example, have blown in from the Gobi Desert to the west. The great Yellow (Hwang-Ho) River of China erodes arid regions blanketed by loess. The color of the silt and clay loess particles carried away gives the river its name.

Sand Dunes

Dunes are sandpiles formed where winds that carry abundant sand lose velocity because of an obstacle, depression, or friction with the land surface (Figure 18.17). In time, the sand itself acts as a barrier to the wind, trapping newly arrived sand. Although they come

dune
A mound or hill of windblown sand.

Figure 18.16 Loess deposits, Shaanxi, China. The cohesiveness of these fine sediments allows them to be excavated for habitation.

slip face
The steeper downwind (or leeward) side of a dune.

in a variety of sizes and shapes, most dunes are variations of a single plan, having a gentle windward slope and a steep downwind or leeward slope, also called the **slip face.**

It is not difficult to see how this basic architecture develops: wind gradually drives sand up the windward slope, but the slip face is protected, in a wind shadow. Sand simply cascades down the slip face under the influence of gravity and comes to rest at the natural angle of repose for sand, about 30°–35°. Additional sand supplied to the slip face triggers slumps and small slides, which serve to maintain that angle. If not pinned down by vegetation or moisture, the dune will slowly migrate downwind, with sand eroded from the windward face being tossed over to the leeward slip face. Because the winds are constantly shifting the position of the sand within the dune, sand dunes display a complex cross-bedding in which steeply dipping layers formed on the slip face alternate with gently dipping layers of the windward face.

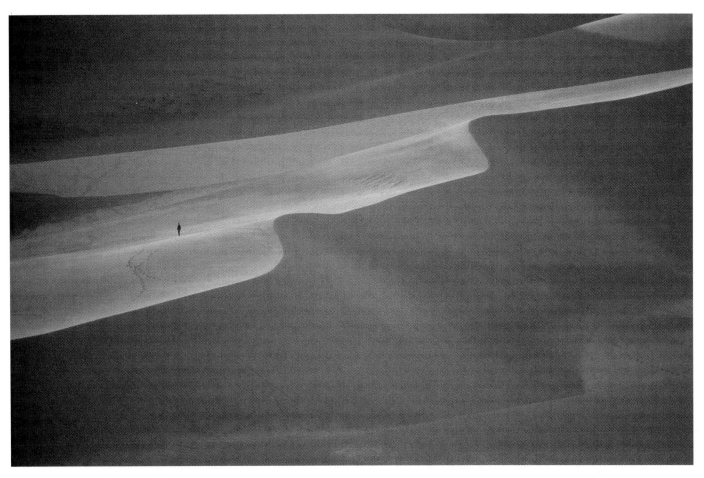

So distinctive are dune cross-beds that their occurrence in ancient sediments is indicative of dry climates at the time and place that the sediment was deposited. The bold cliff of the Navajo sandstone that overlooks Zion National Park, Utah, is an extensive rock formation of Jurassic age (about 150 million years old). Because the Navajo formation exhibits cross-bedding nearly identical to present-day windblown sands, we can assume it had a similar origin (see Chapter 8).

Sand dunes come in a variety of forms, depending upon the direction, strength, and constancy of the winds, sand supply, and vegetation. Wind direction is most important in determining the basic orientation of the dune and serves as a means of broadly classifying dunes. The long axis of a **transverse dune** is at right angles to the prevailing winds (Figure 18.18a). However, the long axis of a **longitudinal dune** lies parallel to the prevailing winds (Figure 18.18b).

Transverse dunes usually occur as a series of thick, nondescript ridges in regions of abundant sand. Some take the form of two of nature's more pleasing creations: the horned, crescent-shaped *barchan* and *parabolic dunes* (Figures 18.18c–d). Both develop in the face of steady winds that blow in an unchanging direction, but each tends to form in a different environment. Barchans generally develop on rocky or otherwise barren desert floors where sand supply is limited (Figure 18.19). Parabolic dunes tend to form in wetter climates than barchans, such as along beaches (see Figure 18.18d). The horns of the barchan crescent point downwind because the wind pushes their thin ends farther than their thicker centers. However, the ends of the parabolic dune tend to be anchored by surrounding vegetation, so their horns point upwind.

Longitudinal dunes are long and narrow (see Figure 18.18b). They are particularly well developed in the deserts of North Africa and the Near East, where they are called *seifs* (pronounced "safes"). *Seif* is Arabic for "sword," which the dunes resemble. Strong crosswinds, superimposed on the prevailing winds, give them a crested, tapering shape.

transverse dune
A long ridge of sand standing at right angles to the prevailing wind.

longitudinal dune
A long ridge of sand standing parallel to the prevailing wind.

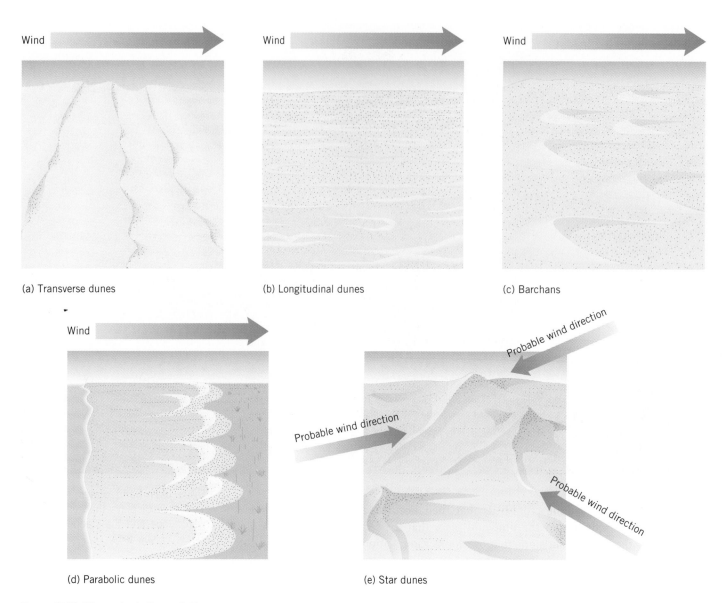

(a) Transverse dunes

(b) Longitudinal dunes

(c) Barchans

(d) Parabolic dunes

(e) Star dunes

Figure 18.18 The orientation of the major dune types relative to the prevailing wind direction.

These gigantic dunes can rise 200 meters high and stretch 300 kilometers in length, although most are about 100 kilometers long. Star dunes are neither longitudinal nor transverse (Figure 18.18e). They are probably formed by winds that shift regularly with the seasons, but their exact origin remains obscure.

Vast areas of the world's deserts are covered by fields of enormous sand dunes, which raises an interesting question: How did such huge quantities of sand come into being? Satellite imaging coordinated with fieldwork on the ground has led to the discovery of buried stream channels beneath the dune fields of the Middle East, which is clear evidence of a wetter climate in the past (Figure 18.20). We may hypothesize that the wetter climate accelerated chemical weathering, which over time converted large quantities of rock to fine-grained sediments. When the climate grew arid and the vegetation vanished, the wind carried away the finer particles and only the sand remained.

Water Resources in Arid Climates

If deserts are so dry, where do the people who live in them find water to survive? Our answer begins with a brief review of the geologic occurrence of water in desert and semidesert environments.

Figure 18.19 Barchan dunes in the Namib Desert, western Africa. The gentler windward slope forms along the convex side.

Most obvious are the streams that flow through deserts. You may recall from Chapter 15 that the population of Egypt is mainly confined to the narrow fertile strip on either side of the Nile River and to the great delta of the Nile north of Cairo (see Figure 15.21). Water is obtained directly from the river itself and from high-water-table wells along the banks. Intricate canals and irrigation ditches divert water during floods, especially on the delta, so that the maximum amount of land is brought under cultivation.

In many arid regions, water comes from confined aquifers (discussed in Chapter 16). Like the surface streams that flow through deserts, the recharge areas of these aquifers are in distant regions with higher altitudes and rainier climates. The water migrates slowly through the porous and permeable strata, which are sealed between impermeable layers, into the arid regions, where it is tapped by artesian wells. Because of the difference in elevation of the recharge and discharge areas, the water rises under hydraulic pressure. However, as we saw in our discussion of the Ogallala sandstone, these

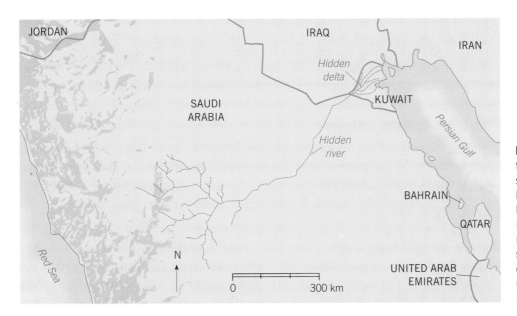

Figure 18.20 Radar used to penetrate the sand dunes located this hidden stream channel and delta in the Arabian desert. The main channel may have been nearly 5 kilometers wide; its delta covered much of what is now Kuwait. These features are strong evidence that a more humid climate prevailed in the past. (Adapted from the *New York Times*, March 30, 1993.)

aquifers are not an unlimited source of water, and overuse has led to serious depletion problems.

Other sources of water are the springs and shallow wells commonly found along the fringes of alluvial fans where sediments are capable of holding considerable quantities of water, like a huge sponge. Active faults may sever the water table of the fans and bring the upthrown side of the table to the surface. In these cases, springs may issue along the traces of the fault scarps. This type of water resource influenced the location of Salt Lake City, Utah, and Las Vegas, Nevada. Today, these growing cities face shortage problems as they continue to extract water at rates faster than it is replaced by rainfall. If wind deflation extends to the depth of the water table, it can also expose sources of groundwater. The water table then intersects the land surface and forms a spring or seep, producing a desert oasis.

Desalinization is the purification of seawater through distillation and other means to produce freshwater. It has become an increasingly popular supplement to water resources in wealthy, water-short countries of the Middle East. It is three to four times more expensive than conventional sources, but costs may be lowered by improved technology; for example, the waste heat from electrical power plants can be used to distill the water in a desalinization plant. However, desalinization will continue to be expensive because of the huge quantities of water involved. Furthermore, there are prohibitive energy costs of pumping the water uphill and distributing it over wide regions.

 Large-Scale Water Transfer

What happens when the population of an arid region exerts demands for water that exceed the local supply? Since the days of the construction of the Roman aqueducts, the only feasible solution has been to import water from far away. But importation creates further problems of a technological, ecological, and political nature.

Consider, for example, what it takes to slake the seemingly insatiable thirst of southern California. Enormous quantities of water are diverted 1000 kilometers from rainy northern California and Owens Valley in the foothills of the Sierra Nevada. From the overused Colorado River, water is diverted 200 kilometers west to the Mojave Desert. These water transfers have transformed a virtual desert into one of the nation's most populated regions and most powerful economies. But this monumental accomplishment has come at a high price. The water transfers are bitterly opposed by many residents of the source regions and by people throughout the state who are sensitive to the complex ecological and economic issues raised by the water transfer. For example, many Californians are outraged that shipping the water south from northern California may deprive the rich farms and wetlands of the Sacramento Delta of sediment and water—leading to the drying and salting up of the state's most fertile region. In 1995, a compromise limiting the amount of water transferred to the south was reached. This agreement is expected to ensure the survival of the wetlands and delta.

The Disappearing Aral Sea

Pumping water from wet to dry regions may not be an ideal policy, but transporting water from one dry region to another can lead to calamitous results. The second-largest inland body of water in Asia, the Aral Sea, is drying up—in 30 years, it will disappear—and not from natural causes. The lake and the 30 million people who depend on it are being sacrificed to cotton.

Figure 18.21 shows that the shoreline of the Aral Sea has receded since 1960 as the flows of the Amu Darya and Syr Darya rivers that feed the lake have been diverted to irrigate 7 million hectares of cotton, rice, and melon fields. The drying of the lake has markedly altered the region's climate. Formerly, the huge lake exerted a moderating influence on the seasonal temperature swings that are common to the interior plains of

It is not city dwellers who use most of southern California's water. Ninety percent of the water transfer is used to irrigate some of the world's richest cash-crop farmland, which supplies 25 percent of the United States' fruits and vegetables, as well as cotton, rice, soybeans, and sugar beets for export.

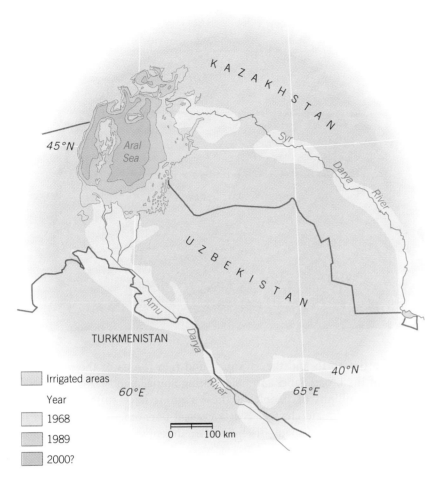

Figure 18.21 Water diverted for irrigation from the Amu Darya and Syr Darya rivers has led to the shrinkage of the Aral Sea.

central Asia. But as the lake has dried up, winters have become much colder, summers have become warmer, and dust storms have become more frequent.

The Aral has no outlet. With its supply of freshwater diverted, it simply evaporates in the dry, windy climate (Figure 18.22). The lake has lost 40 percent of its surface area since 1960. Winds whip across vast stretches of former lakebeds, now exposed salt flats. The salts are carried as far away as the Arctic. Close to the lake, the airborne salts have spawned a dramatic increase in throat cancer in children. Respiratory diseases are on the rise in adults. The quality of the water pumped from wells has dropped along with lake levels, leading to the spread of intestinal diseases and the decline of livestock.

The lake's economy is in ruins. The once-thriving fishing industry is gone because the water has grown too salty to support native species. Fish from the Arctic are frozen and shipped across the continent to keep the canneries of the Aral Sea working. Although the water diversion has been carried out to increase agricultural production, the agricultural program has been only partially successful. Intensive irrigation has led to salinization that is threatening the region's valuable cotton crop—the reason for diverting the water in the first place. Russian environmentalists, freed from the restrictive policies of the past, have taken up the cause of the Aral Sea and brought it to the attention of the world. There are plans to reverse the abuses, but it remains to be seen whether they will be put into action or achieve their desired effect.

On the Fringe of the Desert: The Sahel

Deserts grow and shrink over millennia, centuries, decades, or even years. Particularly sensitive to change are the broad semiarid regions bordering arid deserts. A striking example is the Sahel (Arabic for "fringe") of Africa, the transition zone between the

Figure 18.22 A ship stranded in the desert that was once the Aral Sea.

Sahara Desert to the north and the savannah and rain forests to the south. For many years, portions of the broad Sahel and neighboring Somalia and Ethiopia have suffered drought, and the millions of people of these regions have suffered famine. Is this drought a long-term climate change? Is it a cyclical phenomenon? Or is it a result of human activity? Scientists of many disciplines are urgently investigating the problem.

The Contribution of Climate

Figure 18.23 graphs the fluctuating periods of rainfall and drought in the Sahel since 1950. The sparse rainfall of the Sahel is derived from the clash between the moist winds that blow over the land from the tropical Atlantic Ocean and the trade winds pushing southward (Figure 18.24). In the summer, the trade winds are warmed and rise off the land while relatively cooler, moist air creeps in from the Atlantic to take its place. The air masses interact over a broad, slowly shifting front, causing rain. The rainfall occurs almost exclusively from June through September, peaking in August. The dry season of

Figure 18.23 Fluctuating rainfall in the Sahel. Twenty years of higher-than-normal rainfall (1950 to 1970) were followed by 20 years of drought (1970 to 1990).

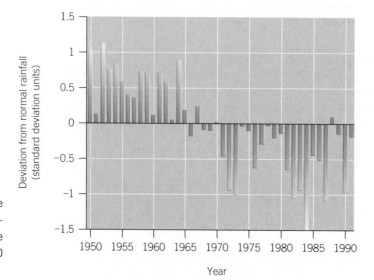

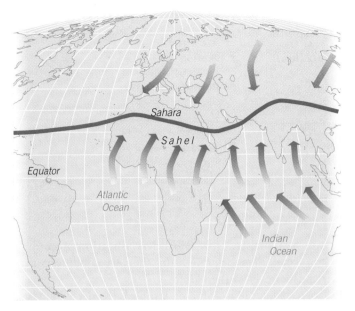

(a) Summer

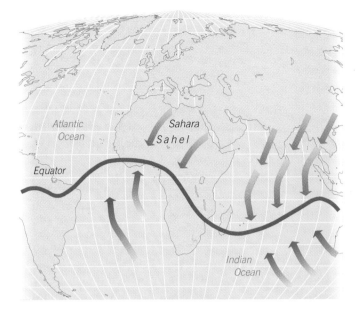

(b) Winter

Figure 18.24 Seasonal wind patterns that control rainfall in the Sahel. (a) In the summer, moist winds from the Atlantic blow onto the land, bringing on the rainy season. (b) In the winter, these winds are blocked, initiating the dry season.

the Sahel is initiated later in the year when the Sun passes well below the equator. This shifts the trade winds southward, and they block the moisture-laden Atlantic winds before they reach Africa. This seasonally reversing wind pattern is called the *monsoon effect.* There are indications that Sahel droughts occur when the trade winds blow farther south over the continent than is usual in summer, thus eliminating the Sahel rainy season.

More than high rainfall is needed for crops to grow and famine to be avoided. The rain must also fall at the right time and be of a duration and constancy to allow seeds to germinate and crops to mature. Early and ample rainfall is particularly important; lack of it proves a good predictor of famine. It is no wonder that human existence in the Sahel is fragile, resting on the vagaries of fluctuating water and air currents over the Earth's surface.

The Human Contribution

Scientists agree that the severity and duration of Sahel droughts have been worsened through the destruction of the vegetative cover by overgrazing cattle, chopping down trees and brush for firewood, and overplanting in an effort to sustain failing crops. These practices have reduced the capacity of the soil to retain the moisture that is derived from the monsoon rains. As vegetation disappears, the soil becomes rock-hard; when the rains come, the water simply runs off the sunbaked surface.

Dry soil inhibits rainfall in a variety of ways. It obviously supplies little moisture to the atmosphere through evaporation. Less obviously, it raises the air temperatures above the ground, because the sunlight normally employed to evaporate soil moisture is used instead to heat the atmosphere. The heated air is then capable of holding greater quantities of moisture; being able to hold more moisture, it yields less rainfall.

Perhaps more subtle is the inhibiting role played by the destruction of natural vegetation in the actual production of raindrops. The drops begin as ice crystals, which form around a minute nucleus, or "seed," of lipoprotein particles shed by bacteria onto leaves during the decay of plant matter. The particles are blown by stiff winds up to the clouds. Therefore, destruction of plant cover eliminates the organic seed particles needed for rain.

Until recently, there was consensus among scientists that human activities in the Sahel create desert conditions. This destructive process, in which human activities

Table 18.1   Migration of the Boundary Between the Sahara (North) and the Sahel (South)

Migration Period	Distance (km)	Direction
From 1980 to 1984	240	Southward
From 1984 to 1985	110	Northward
From 1985 to 1986	30	Northward
From 1986 to 1987	55	Southward
From 1987 to 1988	100	Northward
From 1989 to 1990	77	Southward

desertification

Declining productivity of arable land due to mismanagement, leading to a state resembling a desert.

accelerate the growth of deserts in marginal lands, is called **desertification.** According to this view, desertification of the Sahel, once begun, creates the very conditions that sustain it, and the Sahara Desert pushes ever farther south.

Is the desertification model correct? Like all hypotheses, it must be subjected to tests. From 1980 to 1990, scientists used satellite imagery to trace the southern boundary of the Sahara and see whether it was expanding or shrinking. They discovered that the Sahara had both expanded and contracted during this time and had exhibited no clear trend (Table 18.1). These findings do not support desertification. However, in human terms, the fact that permanent deserts have not been created in the Sahel by human activities is small comfort, because the cycle of land degradation has created what amounts to deserts—soils that are so dry and poor that crops cannot grow in them. No one can control the whims of the monsoon winds, but the human cycle of poverty must be broken if the cycle of land degradation is to be broken. Overgrazing, destruction of plant cover, and general misuse of the land are characteristic of poverty-stricken people trying to survive in a physical environment that cannot sustain their numbers (Figure 18.25).

In summary, drought is not the cause of the mass starvation in the Sahel. Volumes have been written on the misdirected government policies, warfare, and brutalities that have stripped many of the people of the Sahel of the capacity to feed themselves. Recent mass starvation in Somalia illustrates this point.

Figure 18.25 The human effects of desertification in the Sahel.

The human toll of land degradation is certainly not unknown in the United States. Dust bowl conditions milder than those of the Sahel, but not different in kind, affected the western American Plains states during the Depression years of the 1930s. Prolonged drought initiated the destruction of farmland, the uprooting of entire populations, and the mass migration of rural peoples from their homelands. The general origin of the dust bowl was much the same as in the Sahel of today—a combination of natural forces and human practices that destroyed vegetative cover and prevented rain. Consult John Steinbeck's *Grapes of Wrath* and the songs of Woody Guthrie, which describe the human impact of the dust bowl better than any scientific report.

STUDY OUTLINE

One-third of Earth's land surface is arid or semiarid. Running water, not wind, is the chief gradational agent of deserts. Not all deserts are hot, and only a small percentage of Earth's total desert area is covered by sand.

I. **THE FORMATION OF DESERTS**
   A. A **desert** is a region with less than 25 centimeters of rainfall per year and whose land surface is too dry to support more than sparse vegetation. The formation of deserts involves
      1. placement of the continents with respect to the worldwide wind system (subtropical deserts);
      2. the influences of topography on rainfall patterns (rain shadow and mid-latitude deserts); and
      3. the proximity of land to certain cold ocean currents (coastal deserts).

   B. The origin of deserts is connected to the planetary wind system.
      1. As air at the equator converges, it is warmed, rises, and releases its moisture, creating the tropical rain belts. The dry air cools, sinks, and diverges just north and south of the equator, causing subtropical deserts.
      2. As warm winds blowing from the ocean rise over a mountain range, they cool and release their moisture on the windward side, creating a **rain shadow** desert on the lee side.
      3. The deflection of winds as they blow over a rotating Earth is called the **Coriolis effect.**
      4. Cold ocean currents stabilize air masses and cause coastal deserts.

   C. Plate tectonics contributed to the formation of deserts by moving the continents to latitudes where rainfall is scarce, and by creating the mountain barriers to the wind that led to rain shadow deserts.

II. **WEATHERING AND EROSION IN ARID CLIMATES**
   A. In arid climates, mechanical and incomplete chemical weathering produce a blanket of loose material over the bedrock. Rapid water evaporation causes a highly alkaline soil. Mineral encrustations, called **desert varnish,** collect on rock surfaces.

   B. **Intermittent** streams are dry between periods of rainfall. Most erosion is accomplished during floods. Different erosion rates of resistant and weak rock strata, combined with lack of vegetation, account for the rugged topography of deserts. Slopes in deserts retreat at constant angles, whereas slopes in humid regions become gentler.

   C. The Colorado Plateau and the Basin and Range Province illustrate structural influences on desert topography.
      1. The Colorado Plateau is an elevated region of relatively flat land composed of nearly horizontal layers of sedimentary rock formed into cliffs by mass

wasting and stream erosion. Isolated mesas, buttes, and chimneys form picturesque erosional remnants. The Grand Canyon was created by the simultaneous rising of the plateau and increased downcutting of the river.

2. The Basin and Range Province is a rain shadow desert. Its topography is caused by a combination of crustal uplift, erosion, and deposition.

   a. Alluvial fans coalesce to form **bajadas. Playas** are dry lakebeds. The Basin and Range exhibits internal drainage, because its intermittent streams have no outlet to the sea.

   b. Deposition causes the basins to fill with sediment, and erosion causes the mountains to be worn back, leaving narrow divides with sloping erosion surfaces called **pediments** and remnants called **inselbergs.**

III. THE GEOLOGICAL WORK OF WIND IN ARID CLIMATES

A. Like streams, winds transport heavier particles as bed load and lighter particles as suspended load. The amount of sediment and the size of the particles carried in suspension vary with wind velocity.

B. Removal of loose materials by wind is called **deflation. Desert pavement** is the remaining coarse fragments too heavy for wind transport. In regions of thick sand, deflation produces deep, saucer-shaped blowouts. Wind abrasion converts rocks into faceted **ventifacts.**

C. Features of wind deposition

1. Thick deposits of windblown dust are called **loess.** When eroded, they form vertical cliffs.

2. **Dunes** are formed where winds carrying sand lose velocity. Dunes, which migrate slowly downwind, have a gentle windward slope and a steep leeward slope, or **slip face.**

   a. Shifting winds cause cross-bedding, in which steeply dipping layers formed on the slip face alternate with gently dipping layers of the windward face.

   b. **Transverse dunes** lie at right angles to the wind. Some related types are crescent-shaped: barchan dunes point downwind, and parabolic dunes point upwind.

   c. **Longitudinal dunes** lie parallel to the wind.

   d. Star dunes are probably formed by regularly shifting winds.

3. The sands of vast dune fields were probably created in the same regions during earlier periods of high rainfall and weathering.

IV. WATER RESOURCES IN ARID CLIMATES

A. Water is obtained in desert and semidesert environments from streams, confined aquifers, springs in alluvial fans, oases, and techniques such as desalinization—the purification of seawater.

B. When demand for water in an arid region exceeds supply, water may be imported from other regions. For example, water has been diverted from rainy northern California to dry southern California. The Aral Sea, in an otherwise semiarid inland region of Asia, is drying up and will soon disappear because of water diversion.

V. ON THE FRINGE OF THE DESERT: THE SAHEL.   The Sahel is a broad semiarid region bordering the Sahara Desert where drought and famine have been a severe problem for many years.

A. Drought probably occurs in the Sahel when the trade winds move farther south than usual, blocking the monsoon winds during the summer.

B. The severity of Sahel droughts has been worsened through the destruction of the vegetative cover, which reduces the capacity of the soil to retain moisture.
    1. Research findings do not support the theory of **desertification**—that destructive human activities accelerate the growth of deserts in marginal lands. However, the cycle of land degradation has created what amounts to deserts: soils that are so dry and poor that crops cannot grow in them.
    2. The same combination of natural forces and human practices caused the dust bowl conditions that affected the western American Plains states during the Depression years of the 1930s.

STUDY TERMS

bajada (p. 498)	intermittent stream (p. 493)
Coriolis effect (p. 490)	loess (p. 501)
deflation (p. 500)	longitudinal dune (p. 503)
desert (p. 488)	pediment (p. 499)
desertification (p. 510)	playa (p. 498)
desert pavement (p. 500)	rain shadow (p. 490)
desert varnish (p. 492)	slip face (p. 502)
dune (p. 501)	transverse dune (p. 503)
inselberg (p. 499)	ventifact (p. 501)

CRITICAL THINKING QUESTIONS

1. Explain why most of the world's deserts are found in the subtropics at approximately 30° north and south latitudes.
2. Explain the formation of a rain shadow desert.
3. Running water and not wind is the chief gradational agent of deserts. Explain.
4. Describe the process by which the internal drainage of the Basin and Range Province will evolve into an external drainage system.
5. Contrast the wind conditions that form transverse and longitudinal dunes; barchan and parabolic dunes.
6. The hills in humid climates wear down, whereas those of arid climates wear back. Explain.
7. Discuss the physical and social advantages and disadvantages of large-scale water transfer.
8. Describe the interaction of natural forces and human intervention that has led to starvation in the Sahel region of Africa.

Coasts and Shoreline Processes

Consider the following facts.

- Many of the world's great cities are seaports with natural harbors. These cities are the destination of ships carrying the food, raw materials, and manufactured goods of world trade. (Unfortunately, the coastal ocean and near-shore ocean bottom have become the sewers and dumping ground for these cities and their factories. Just about everything nobody wants ends up there.)
- The life cycles of countless marine and land organisms, an immense food supply, are tied to tidal marshes, estuaries, and beaches. (Yet these are the very areas where human intervention is most prevalent.)
- An increasing number of Americans prefer to live near the shore, either permanently or when on vacation. (Yet many shorefront homeowners are losing their life savings to the sea, as the beach or cliffs erode out from under them.)

In this chapter, we describe the physical processes that affect shorelines and coasts: the work of waves, currents, tides, sea-level changes, and tectonic forces. We also consider the effects of human intervention, since in many parts of the world, the shore is shaped as much by the clash of human interests as by the crash of waves. We begin with a discussion of wind-driven waves because they are the principal agents that affect the shore on a daily basis.

Wind-Driven Waves

Drop a stone in a still pond and watch the ripples spread to the shore. Their motion is deceptive, for although the ripples expand in ever-widening arcs across the water surface, if you closely observe a leaf floating on the water, you will see that the water itself rises and falls in a nearly stationary orbit (Figure 19.1). The high point of the orbit is the *crest* of a ripple; the bottom is the *trough*. What you are really seeing when you watch the ripples spread is the transfer of energy via vibrations from the stone to the

◄ Beach homes under attack during an Atlantic storm.

Figure 19.1 The components of a wind-driven wave.

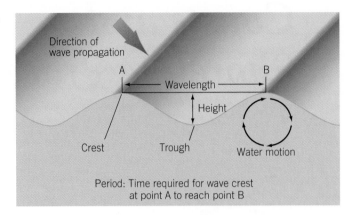

Direction of wave propagation

A ← Wavelength → B

Height

Crest　　Trough　　Water motion

Period: Time required for wave crest at point A to reach point B

water wave
A repetitive circular or elliptical motion in an ocean or lake, causing the water surface to rise and fall.

shore. These vibrations are called *waves*. Energy is transferred from the deep sea to the ocean shore through **water waves** as well, except that the energy is imparted to the water by winds that act on a vastly greater scale than the stone dropped in the pond.

Wind-driven waves and associated currents are the chief means by which the sea erodes the land, builds beaches, and transports sediment along the coast. Most of this chapter is devoted to explaining this work. Figure 19.1 shows the anatomy of a wind-driven wave. The distance between wave crests is the *wavelength,* and the time it takes for successive crests to pass a given point (that is, the time for one complete vibration) is the *wave period.* The vertical distance between the crest and the trough of the wave is the *wave height.*

Most wind-driven waves are generated during storms by friction between the moving air and the sea surface. Their height is dictated by the velocity and duration of the wind as well as the distance over which the wind is in contact with the sea surface. In the storm center, the sea surface is a chaotic mixture of wavelengths, periods, and heights. But if we observe carefully the pattern generated by the storm, we can see that the confused waves are transformed into a regular series of ripples as they move toward shore. We are witnessing the effects of wave interference close to the center of the disturbance and the effects of wave dispersion at greater distances.

Close to the storm center, wave patterns are dominated by *wave interference*—that is, the addition or subtraction effects of many waves to produce composite waves (Figure 19.2). The chaotic wave surface tosses ships about like toys in a bathtub. It is not uncommon for waves in a stormy sea to reach heights of 6 meters; severe storms have generated waves over 35 meters high. Far from the storm center, the pattern changes to *wave dispersion,* in which the long waves outrun the short. As with runners in a marathon, the farther a wave train travels from the source, the more the waves separate.

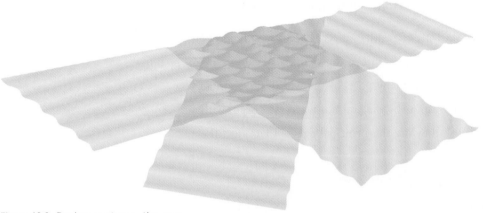

Figure 19.2 During a storm, the contributions of many waves that differ in length and period produce a chaotic sea surface.

Waves in Shallow Water

The greater the height of a wave in the open sea, the greater its energy. However, because only the sea surface makes contact with the wind, wave energy—and therefore wave motion—decreases with depth (Figure 19.3).

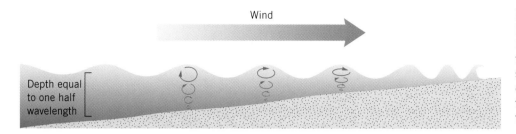

Wind

Depth equal
to one half
wavelength

Figure 19.3 The changes in a wave as it approaches shore. As the wave begins to feel bottom, friction retards its motion and changes the shape of its orbit. The wave slows down and its height increases. Close to shore, the wave becomes gravitationally unstable and collapses.

When a wave approaches a shoreline, another factor begins to influence its motion—the ocean bottom. As the water becomes shallower, contact between the wave and the ocean bottom increases. When the depth of the water is about half the wavelength, the wave begins to "feel bottom." Friction retards its motion, which leads to a series of transformations that intensify as the wave approaches shore. The incoming wave slows down drastically. Thus its length shortens and its height increases markedly, because the same amount of water is packed into a shrinking space. When the wave height reaches about one-seventh of the wavelength, the entire structure becomes gravitationally unstable, pitches forward, and collapses. Thus the energy contained in the wave is transferred to the shore. Because the depth at which a wave feels bottom depends upon the depth of the wave, a longer wave will break farther from the shore than a shorter wave will.

The shallow sea bottom also changes the direction of incoming waves that approach the coast at an oblique angle. The segment of the wave that encounters shallow water first slows down, while the other segment continues to approach shore at deep-water speed. Thus the wave pivots so that its approach is nearly at right angles to the shoreline. This bending effect of waves upon entering shallow water is called *refraction* (Figure 19.4). You may recall that seismic waves display similar refraction effects as they speed up and slow down through the layers of the Earth's interior.

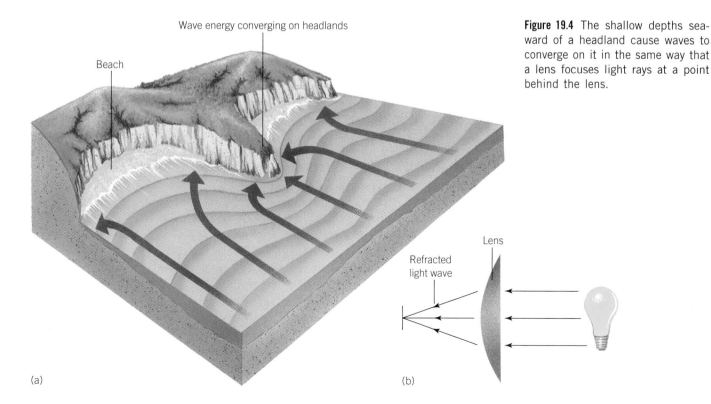

Wave energy converging on headlands

Beach

Figure 19.4 The shallow depths seaward of a headland cause waves to converge on it in the same way that a lens focuses light rays at a point behind the lens.

Lens

Refracted
light wave

(a)   (b)

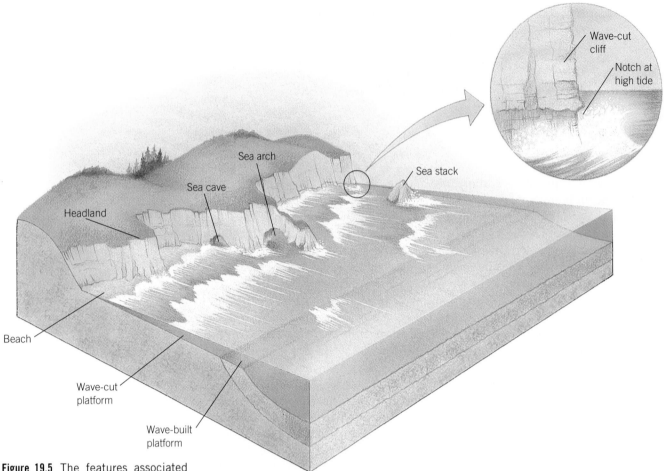

Figure 19.5 The features associated with the erosion of a rocky headland.

wave-cut cliff
A coastal cliff formed by wave erosion.

wave-cut platform
A more or less smooth sloping surface planed by wave erosion.

sea stack
An erosional remnant of the former headland.

sea arch
A bridgelike erosional remnant of coastal rocks.

Headland Erosion and Bay Deposition

Figure 19.4 shows incoming waves attacking an irregular stretch of coast. The unprotected headlands are prime candidates for wave attack, because the shallow sea bottom that surrounds them slopes off into the deeper bays on either side. The depth change causes the waves to converge and concentrate energy at the headlands in a manner similar to the way a lens concentrates light at its focal point.

As waves break against the headlands, they erode rocks in a number of ways: by direct impact, by forcing compressed air into cracks, and by abrasion as rock fragments are hurled back against the cliffs. Headland erosion is also aided by chemical weathering and mass wasting in this moist environment. Gradually, a deep cut or *notch* will be carved in the headland at the high-tide watermark (Figure 19.5). The rock overlying the notch will eventually lose support and collapse, resulting in a near-vertical **wave-cut cliff** facing the ocean. Retreat of the cliff produces a gently sloping ramp planed by wave erosion—a **wave-cut platform.** Isolated remnants of the original headland may remain on the platform as **sea stacks** and **sea arches.** In eroding the notch, the waves may also hollow out soft portions of the wave-cut cliff to produce sea caves.

What happens to the fragments eroded from the cliff? Gradually, abrasion causes them to become rounded and smaller. Eventually, many of them are transported to calmer, deeper waters seaward of the wave-cut platform. Transport is accomplished by the back-and-forth motion of the waves and by water from spent breakers, which carries sediment off the platform. Gradually, a depositional *wave-built platform* forms as a seaward extension of the wave-cut platform.

At the same time that refraction concentrates energy at the headlands, wave action is diminished in the bays, because refraction has spread the energy over a wider area.

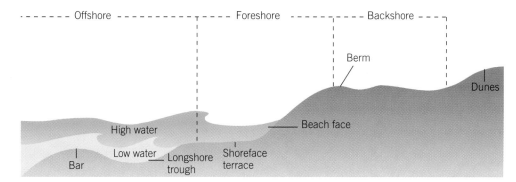

Figure 19.6 The components of a typical beach profile.

Therefore, sediment is able to accumulate in this calmer, protected environment, forming a **pocket beach.** The major sources of sediment are rock fragments eroded from the headlands and sand and silt swept into the bay by shallow currents close to shore. Given sufficient time, erosion of the headlands and deposition in the bays will lead to straightening of the coast—provided, of course, that the coast is of uniform composition and all the other factors that affect the coast remain constant. With straightening, wave energy is distributed evenly up and down the coast.

pocket beach
A small beach nestled between adjacent headlands.

Beach Formation and Shoreline Processes

Oceanographer Douglas L. Inman succinctly describes most beaches as "essentially long rivers of sand that are moved by waves and currents, and are derived from the material eroded from the coast and brought to the sea by streams." More formally, a **beach** consists of fragmented material (it need not be sand) that is subject to wave action at the interface of land and sea. It extends from the land exposed at low tide to the highest point that storm waves can reach or the point where there is a distinct break in landform—for example, a sea cliff, a line of dunes, or permanent vegetation.

beach
A sloping portion of the shore composed of sediments deposited and moved by waves, tides, and longshore currents.

The Beach Profile

The profile of a typical beach can be divided into three units of distinctly different form. Proceeding from the most inland portion of the beach to the ocean, these units are the backshore, the foreshore, and the offshore (Figure 19.6).

The nearly flat **backshore** starts at the base of a sea cliff, sand dunes colonized by vegetation, or any other distinct boundary that marks the outer limit of wave attack. It consists of one or two narrow terraces, called **berms,** that slope gently inland. The berm closest to the ocean ends in a low berm crest facing the ocean. The backshore is disturbed only occasionally by great storm waves. The inner portion, closest to the water, is where most people prefer to spread their blankets.

The **foreshore** slopes in a graceful, concave arc from the berm crest to the surf zone of frothing, breaking waves. The arc resembles the longitudinal profile of a stream—steep at the head and tapering gently toward and beneath the shallow waters where the waves churn. The steep exposed portion of the arc—the *beach face*—is where water from breaking waves is propelled upward. The gentler portion tapers into the flat *shoreface terrace,* the region of the surging surf zone. The foreshore is the active, working portion of the beach.

Closely connected to the beach is the shallow, permanently submerged region just offshore, where waves feel bottom, rise in height, and begin to break. Appropriately enough, it is called the **offshore.**

The profile of the beach is constantly adjusting to the incoming wave pattern, the controlling factor being the depths at which the waves break. Waves that arrive from

backshore
The upper zone of the beach, reached by waves only during storms and very high tides.

berm
A terracelike structure of sediments washed by waves up onto the foreshore.

foreshore
The sloping surface of the beach that falls within the tidal zone and regularly bears the brunt of breaking waves.

offshore
The submerged zone just seaward of the beach proper.

Figure 19.7 The seasonal cycle of a beach. (a) Short-wavelength storm waves wear back the beach face and steepen the profile. The eroded sediment is deposited offshore in deeper water. (b–c) In periods of calm, long-wavelength swells break offshore and move sediment back onto the beach face. Eventually the profile is restored.

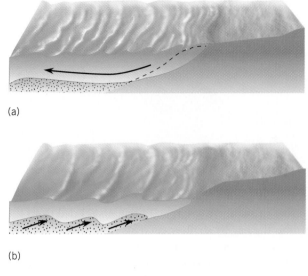

(a)

(b)

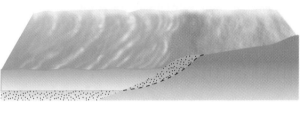

(c)

longshore bar
A shallow sand ridge created by wave action seaward of the shoreline.

nearby storms are short, high, and choppy; they break close to shore, wearing back the beach face, and leaving behind a narrow berm. During severe storms, they may even breach the dunes and beat against unprotected houses, highways, and sea walls. In extreme cases, the waves may strip away the entire beach. The obvious effect of these waves is to steepen the foreshore profile (Figure 19.7). Most of the eroded sediment is not lost, but is carried to the offshore beyond the short breakers to form sand bars. This supply of sediment will be used to reconstruct the beach during calmer weather.

During calmer weather, the beach receives the long, low swell of distant storms. The waves break far offshore and mold the bottom sediment into **longshore bars.** As the waves grow longer during very calm weather, they tend to break against the bars and wash sediment over onto the landward side, in much the same way that winds blow sand over onto the lee side of a dune. In this way, the bars migrate landward and are plastered to the narrow storm berm. The foreshore grows steadily, and a wide, calm-weather berm is created. Thus the beach profile is broadened and given a gentler slope.

Short, choppy storm waves are most common in winter, and long swell is most common in summer. This pattern imposes a seasonal cycle upon the beach. Such is the case along the west coast of the United States. The narrow winter berm is built outward during the spring and summer and is worn back to nearly the same place during the fall and following winter. East and Gulf coast beaches also display cyclical patterns that vary with the seasons. However, the Gulf and southern Atlantic coasts are plagued by late summer and early fall hurricanes. The New England and central Atlantic coasts are hit by violent storms called northeasters during late fall, winter, and early spring, as well as the occasional hurricane. The erosional cycles initiated by these events often mask the normal seasonal profile changes.

In bearing the brunt of breaking waves, giving ground during storms, and rebuilding during peaceful periods, beaches serve as energy sponges—flexible shock absorbers that protect the coast behind the beach from all but the fiercest attack.

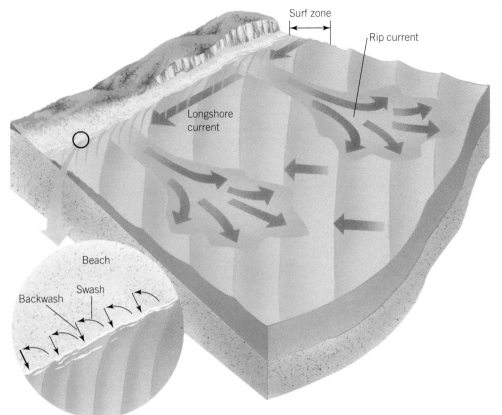

Figure 19.8 The nearshore circulation loop. Sediment moves parallel to the beach via swash and backwash, and via longshore currents.

The Nearshore Circulation Loop

A breaking wave drives water onto the beach as **swash,** which then is pulled back off the beach by gravity as **backwash** (Figure 19.8). If the wave strikes at an oblique angle, it also drives water down the beach as a **longshore current.** Backwash adds to and mixes with this current in the **surf zone.** At intervals along the beach, strong, narrow *rip currents* transfer the water pumped into the surf zone back out to sea. Beyond the surf zone, the rip current widens into a broad *rip head* that mixes with the calmer, deeper waters. The spacing of rip currents is determined by a complex set of circumstances. Basically, it depends upon beach and bottom contours, shore obstacles, and irregularities in the height and period of waves along the beach. In sum, the net effect of waves breaking against the shore at an oblique angle is to set up a nearshore circulation loop whose essential features include the breaking waves, longshore currents, rip currents, and rip heads.

Each portion of the circulation loop directly or indirectly contributes to the formation of the beach. The cycle of swash and backwash moves sediment landward, seaward, and also along the beach; it is primarily responsible for the construction of the beach profile. Longshore currents move sediments parallel to the shore and thus alter the beach in map view, creating a variety of features we will soon discuss. Triggered as they are by breaking waves, the two processes work in tandem.

The swash and backwash of breaking waves combine with the longshore current into a system of **longshore sediment transport** parallel to the beach. Note in Figure 19.8 that the paths of swash and backwash do not match. The former is driven obliquely up the beach in the direction of the breaking wave, but gravity moves the latter off the beach by the shortest possible route, which is directly downslope. A sand grain carried by a single swash and backwash cycle will thus trace an archlike, zigzag pattern called *beach drift.*

swash
The water from breaking waves that washes up onto the beach.

backwash
The return flow of breaking waves down the beach face.

longshore current
A shallow current parallel to the coast, caused by waves that approach the shore at an oblique angle.

surf zone
The zone of breaking waves.

longshore sediment transport
The movement of sediment parallel to the shore by longshore currents, swash, and backwash.

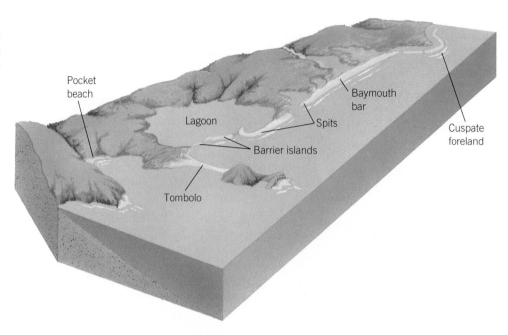

Beach Types

In the process of longshore sediment transport, beaches and related features are molded into a variety of forms, dependent upon the configuration of the coast, the resistance of rocks to erosion, wave direction and intensity, tidal action, organic activity, sediment supply, bottom topography, and sea-level changes. The following list describes some common beach forms (Figure 19.9).

- A pocket beach forms in the low-energy wave environment between bold head-lands; much of its sediment comes from headlands eroded in the manner previously described.
- An island may be connected to the mainland by a strip of beach called a **tombolo.** The island, often a sea stack, acts as a barrier to breaking waves, and deposition occurs on the protected side between island and mainland.
- Between the mainland and the **barrier island** that extends more or less parallel to it is a protected lagoon. Many barrier islands are more than mere beaches, having dunes, thick vegetation, and swampy marshes on the side facing the lagoon. Where sediment supply is great, the lagoon shallow, and waves and current erosion low, the lagoon may be filled in to form a low, marshy wetland between mainland and barrier island (see Figure 19.11).
- A **spit** is a curved, fingerlike projection of beach that extends into the sea, elongating the shoreline. The glacial deposits that compose Cape Cod, Massachusetts, have been shaped by longshore currents into the classic example of a complexly curved spit. It is also the world's longest spit.
- A **baymouth bar** extends partially or completely across the mouth of a bay. In the former case, it is a type of spit; in the latter case, it connects headlands and straightens an irregular coast.
- When spits, coastal beaches, or barrier islands join, a seaward projection called a **cuspate foreland** is formed. Cape Hatteras and Cape Lookout, North Carolina, are spectacular examples, as is Cape Canaveral, Florida.

The Sediment Budget

A beach grows, shrinks, and migrates along the coast as part of a much larger sediment circulation system. Figure 19.10 illustrates a sand circulation system typical of the Cal-

---

tombolo
A strip of beach connecting the mainland to an island or islands to one another.

barrier island
A striplike island that parallels the coast.

spit
A curving, fingerlike projection of beach.

baymouth bar
A strip of beach extending from a headland into the mouth of a bay.

cuspate foreland
A coastal landform composed of seaward-projecting beaches that meet at a point.

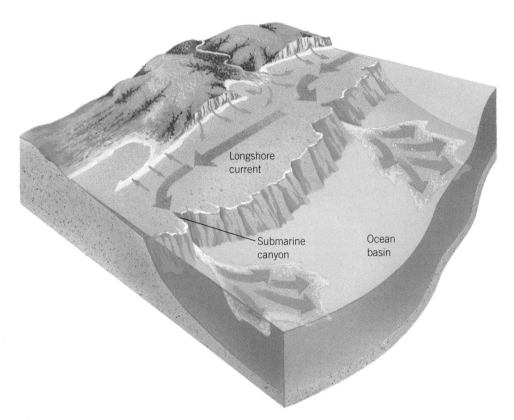

Figure 19.10 The sediment transport system off the California coast. Sediment is introduced to the system via streams and wave-cut cliffs. Longshore currents sweep the sediment into submarine canyons, where it is conducted via turbidity currents to the deep-sea floor.

ifornia coast. Sand enters the system via rivers and wave-cut cliffs and is temporarily stored on the beach. Longshore currents sweep the sediment to nearby submarine canyons, where bottom currents conduct it to deep-ocean basins. Note that the form of the beach is maintained *as the sediment passes through.* The beach grows when more sand arrives than waves, wind, and currents can carry away. The beach retreats when supply declines or when waves, winds, and currents intensify. In other words, a beach grows, shrinks, or remains static depending upon the balance between the amount of sand added to the system and the amount taken away. Its economy can be described in terms of a profit or loss budget.

Direct wave action tends to alternately build and erode the beach profile cyclically in accordance with the seasons. Over the long term, there is little net sediment loss or gain by this means. Longshore transport, more than any other factor, determines the long-term sediment balance of most beaches, although any of the factors on the beach may be dominant at a particular time or place.

This dynamic yet essentially stable arrangement is complicated by a rising sea level of about 15 centimeters per century on the east coast. The rise concentrates wave attack and tides at a higher level and drives the beaches inland. Barrier islands along the east coast illustrate the process (Figure 19.11). The rising sea forces waves to break higher on the foreshore and erode the berms. During great storms, the waves breach the dunes and spread sediments into the lagoons and bays as *overwash* fans behind the dunes. Storms cut through the entire island and link the lagoon to the open ocean via a *tidal inlet.* Eventually, the part of the barrier island that faced the lagoon becomes exposed to the open ocean as the rising sea forces the island to "roll over" toward the mainland. But there is no net sand loss.

All these local events—which may be either disastrous or beneficial to the people who depend on the shoreline for a living—do not change the overall stability of the system. The tidal deltas that form at the tidal inlets and the overwash fans that form behind the dunes are essentially sediment storehouses that become part of the active beach budget as the front of the beach is rolled over. In a matter of hours, hurricanes and northeasters may flush out the sediment that has accumulated in the bays, estuaries, and

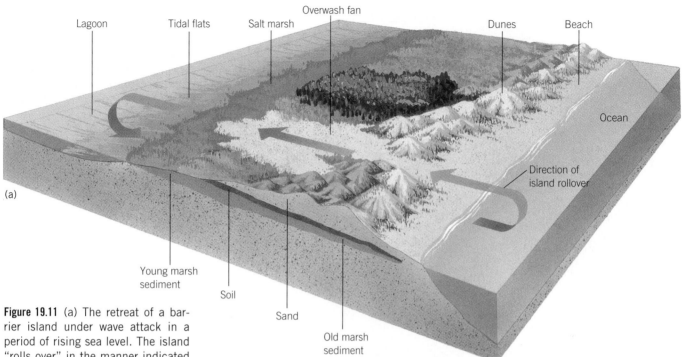

Lagoon   Tidal flats   Salt marsh   Overwash fan   Dunes   Beach

Ocean

Direction of island rollover

(a)

Young marsh sediment

Soil

Sand

Old marsh sediment

Figure 19.11 (a) The retreat of a barrier island under wave attack in a period of rising sea level. The island "rolls over" in the manner indicated by the arrows. (b) These trees on Tilghman Island, Maryland, are now exposed to the open sea, but they grew to maturity when this part of the island was on the protected side facing the lagoon.

(b)

lagoons over centuries. These sediments replace those that regularly leave the beach via bottom and rip currents. If the sediment supply remains constant, so will the essential form of the retreating, migrating beach—provided that waves and storms that mold the sediment attack from the same direction and with the same intensity.

 ## Tampering with the Sediment Supply

The fate of a beach is often determined by factors outside the circulation system—some natural, some human. Tectonic uplift of the land can lead to accelerated stream erosion and broadening of the beach (caused by an excess supply of sediments). Or sea level can rise for a variety of reasons and erode the shore; or natural barriers may develop that

Figure 19.12 The installation of groins caused the condition that exists along Cape May, New Jersey. Deposition increases on the upcurrent side of the groin, whereas erosion is accelerated on the downcurrent side.

impede longshore sediment transport. Or human activities far from the coast may have a profound effect. The Aswan Dam of Egypt, for example, catches the sediment meant for the Nile Delta and the beaches along the Mediterranean coast. With less sand available to replace wave and current erosion, the balance has shifted in favor of the Mediterranean Sea. The waves are wearing back the natural dunes along the shore at the rate of nearly a meter per year. In a similar fashion, nearly all the major rivers of the United States are dammed, harnessed, and regulated for multiple purposes, especially irrigation. Sediment is trapped inland, which leads to starvation of the beaches. Large percentages of United States beaches and shorelines are seriously eroding; the problem is particularly acute along our so-called fourth coast, the Great Lakes.

Tampering with Beaches

The rather abstract concept of the beach budget has concrete consequences for those who own beachfront property. They are concerned with beach preservation and employ a variety of hard-stabilization techniques to achieve this end. Hard stabilization uses immovable objects in the attempt to hold static what is essentially a fluid system.

For example, suppose resort community A wishes to prevent the longshore current from eroding the beach. The typical solution is to catch sediment by building a barrier, called a **groin,** out to sea at right angles to the beach. The beach indeed grows on the upstream side of the groin, but at the expense of trapping the sediment that would ordinarily be deposited on the downstream side of the groin. It also intensifies the currents that sweep around the groin. These two results lead to accelerated erosion downstream of the groin and the destruction of once-stable beaches and human-made facilities (Figure 19.12). The downstream erosion sets up an unsightly domino effect as property owners construct groins to protect *their* stretch of beach. If the entire community's beach is eventually protected—at great expense—the problem is merely transferred, in intensified form, to the neighboring beach community B, which is attacked by the same sediment-deprived longshore currents. It has been said that groins are like lawyers: nobody should get one, but once somebody does, we all need one.

However, environmental lawyer Katherine Stone has proposed another solution, the "sand rights" theory. According to traditional law, certain resources, like the air, belong

groin
A structure projecting perpendicular to the beach, designed to trap sediments and prevent beach erosion.

Apply and Decide 19.1

Coastal Development and Storms

There is scarcely a stretch of coast anywhere in the world that has not been buffeted by severe storms or hurricanes. What factors can protect the coastal communities located in their path, and what factors intensify the damage? Geologists Robert Thieler and David Bush of Duke University attempted to answer this question by investigating two highly developed coastal regions after each was struck by hurricanes of roughly equal severity. Hurricane Gilbert struck the Yucatán coast in September 1988, and Hugo struck the South Carolina coast in September of the following year.

Thieler and Bush identified a number of factors related to construction and planning decisions that affected the extent of the damage to the communities along these coasts. Some of the lessons the authors learned from these decisions after surveying the storm sites are listed below.

Lesson 1: Proper building construction protects homes. Although the eye of Hurricane Hugo passed almost directly over the beachfront house on the Isle of Palms shown in Figure 19A, it survived with minimal damage. Factors contributing to its survival include large and deeply sunk pilings; strong connections between the roof, walls, foundation, and pilings; and a generally low profile, with many rounded corners that lower wind resistance. Cinder block houses, such as the former vacation rental on Folly Beach, South Carolina, shown in Figure 19B, fared poorly in Hurricane Hugo because they are typically not reinforced against or elevated above storm surge and waves. Many houses along the South Carolina coast floated free during Hugo, often colliding with other houses a block or more from the beach (Figure 19C). Damage to inland houses would likely have been minimal had they not been hit by other houses.

Lesson 2: Lack of obstructions and wide beaches protect. The poststorm beach in South Carolina after Hugo is narrower on

Figure 19B

Figure 19C

stabilized shorelines than on unstabilized shorelines. On natural shorelines, some sand was moved inland over the dunes and later was returned to the beach. But obstructions such as sea walls, revetments, and large buildings prevented this movement, and most of the sand was washed offshore. Updrift of the Yukalpeten jetties on the north coast of the Yucatán, nearly 300 meters of beach has accreted. Buildings sited behind this wide beach suffered little damage, because storm waves expended their energy on the beach.

Lesson 3: Dunes protect. On Pawleys Island, South Carolina, the house on the left in Figure 19D was sited at low elevation; preexisting dunes were removed for its construction. The house on the right, however, was built on top of and behind the dune. The storm destroyed the house on the left; the one on the right sustained minimal damage.

Lesson 4: Rocky shorelines protect. Hotels at the northern end of the island of Cancún are built atop a ridge of resistant limestone (Figure 19E). The elevation and minimal shoreline retreat of this area protected the hotels from wave damage. Hotels built on the heavily eroded sandy shoreline farther south on the island, however, suffered severe damage (Figure 19F).

Figure 19A

Figure 19D

Figure 19F

Lesson 5: Proper street layouts protect. Roads that run perpendicular to the shore become funnels for overwash and storm surges. Overwash penetration and associated storm damage was an order of magnitude greater along shore-perpendicular streets in Surfside, South Carolina (Figure 19G). Overwash not only deposited large quantities of sand, but also propelled debris and even automobiles forcefully into buildings and utility poles, causing extensive damage. Roads should be set at diagonal angles to the beach.

Public Versus Private Responsibilities

A coastal beach is an inherently unstable strip of land. Buffeted by hurricanes and storms, stripped of sediment by longshore currents, retreating in the face of rising sea levels, its dimensions are in a constant state of flux. Like the levee systems installed to prevent stream flooding, engineering solutions designed to stabilize beaches are temporary at best, generally expensive, and often do more harm than good. Unlike those who settle on flood plains miles from the river, the people who build on beachfront property are usually fully aware of the dangers. Nevertheless, they insist on building there, hoping they will remain lucky and knowing that they can rebuild with

Figure 19G

help from the federal government if their homes are destroyed.

The following questions explore the somewhat murky boundary between public and private responsibility for the effects of natural disasters.

THOUGHT QUESTIONS

1. Imagine yourself a regional planner with authority over coastal development of the regions described above. How would you deal with the ongoing problems of poorly sited beachfront homes and hotels? What construction codes and other precautions would you institute to minimize future problems?
2. To what extent should taxpayers' money be used to protect and insure private beach developments? Is your answer the same for public beaches and recreational facilities? (Don't forget to include economic considerations, such as jobs, tax revenues, and tourism, in your decision making.)

Photos and accompanying information with permission from Robert Thieler and David Michael Bush, Duke University.

Figure 19E

Figure 19.13 The results of hard stabilization. Forty years ago, Seabright, New Jersey, was known for its wide, beautiful beach. Then the Army Corps of Engineers built the retaining wall to protect the apartment houses from wave attack during the storms, but the wall precipitated the destruction of the beach. Today it is the only thing that stands between the buildings and the sea.

to society at large. Stone argues that beach sand is such a resource, and that those who damage the beach by interrupting the natural flow of sand should be required to pay for the restoration of the beach. For example, community A should be required to pay the cost of restoring community B's beach; if that is not possible, the groin should be removed. The same principles would be applied to communities that construct dams that trap sediment intended for the beach.

The construction of retaining walls is another common hard-stabilization technique, designed to protect homes and shorefront facilities. They may indeed protect these structures for a time; but they do not prevent the erosion of sand from the beach, so that the beachfront property ends up minus its beach. The retaining wall itself is an artificial sea cliff subject to the energy of breaking waves. Eventually, it too will be undercut and eroded, leaving the shorefront structure exposed. The stretch of New Jersey coast shown in Figure 19.13 is an illustration of the sad results of relying on retaining walls to prevent beach erosion.

Mindful of the damages of hard stabilization, coastal engineers have developed soft-stabilization techniques that work in harmony with the natural processes that affect the beach. The simplest of them is to pump sand back onto the eroding beach. The source of the sand may be the far offshore, lagoons behind the beach, or sediment trapped by adjacent groins. Although this strategy is less harmful than hard stabilization, the pumped-in sand erodes more rapidly than the natural beach and in any case has little influence on the factors that cause the beach to erode. The high rate of replenishment and the huge sand volumes needed make this an expensive stopgap measure.

A number of geologists, environmentalists, and budget-conscious officials believe that hard and soft stabilizations are expensive and ultimately fail. The only rational solution, they believe, is to discourage settlement on beaches, and, where possible, relocate those who live there.

Organic Activity and Shorelines

Organic activity often plays a major role in shaping shorelines. In the absence of land-derived sediment, shell matter ground up by waves serves as the major component of beach sediment. Submerged coral reefs and similar structures act as sediment traps and

Figure 19.14 A mangrove coast, Queensland, Australia. The complex root systems of the trees act as sediment traps, expanding the coast.

barriers to waves and currents attacking from the open ocean. Frequently, they are the major architectural units of continental shelves. They exist in a belt between approximately 30° north and south of the equator and are especially prevalent on the western margins of the oceans, where they catch the warm currents. Corals grow upward from any firm base of support and are common on rocky shores and on the sloping flanks of volcanoes. Once in place, they are formidable barriers that protect the coast from wave attack. The Great Barrier Reef, 1600 kilometers in length, protects the coast of northeastern Australia.

Corals need not only a warm climate and a firm base of support but also clear, flowing water. Other organisms become the coast builders in muddier waters. Mangrove trees are prevalent along the beaches of shallow bays (Figure 19.14). Their deep, tangled root systems serve as traps that retain sediment and provide shelter and nutrients to other organisms. The beach is covered over with vegetation, allowing the mangroves to grow farther out into the bay. In this way, the shore is extended. The southern Florida coast displays both types of organic coasts. The Florida Keys are coral reefs deriving their nutrients from the warm Gulf Stream; behind the Keys, on the protected Gulf side, mangrove forests and swamps grow.

The Work of Tides

Wind-driven waves are the most active agents in building beaches and shaping coasts, as we have seen, and their work is greatly influenced by a number of factors. Chief among them are **tides**—the rhythmic rising and falling of the sea surface—which focus the point of wave attack and carry on a number of other important activities as well.

The energy of tides is supplied by the gravitational attraction between the Earth and the Moon and, to a lesser extent, between the Earth and the Sun. The attractions lead to the regular rise and fall of sea level observed along the shorelines of the oceans and large lakes of the world. The work of tides is cumulative and, with rare exceptions, underestimated, in that it is often masked by the effects of wind-driven waves.

As we all know, the Moon revolves around the Earth, but the system is not really that simple. Consider Figure 19.15. In this model, mutual gravitational attraction causes the Earth and the Moon to revolve about a common center of mass, which is *not* the center of the Earth (Figure 19.15a). Thus each point on the Earth is subject to two forces: the attractive gravitational force that the Moon exerts on the Earth at that point, and

tide
The rhythmic rise and fall of the sea surface caused by the unequal gravitational attraction of the Moon—and, to lesser degree, of the Sun.

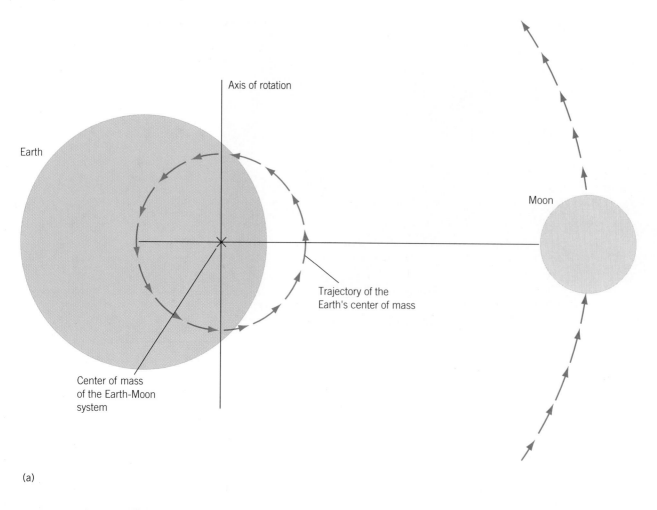

Axis of rotation

Earth

Moon

Trajectory of the
Earth's center of mass

Center of mass
of the Earth-Moon
system

(a)

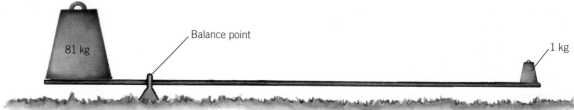

Balance point

81 kg

1 kg

(b)

Figure 19.15 The tide-producing mechanism. (a) Mutual gravitational attraction causes the Earth and the Moon to revolve around a common point, the center of mass of the Earth-Moon system. (b) As with the Earth-Moon system, the balance point of the board is closer to the more massive object. (c) Every point on the Earth is subject to the force of the Moon's gravitational attraction, and to the centrifugal force caused by the rotation of that point around the common center of mass. The net difference between the two forces generates a small tidal force, which, depending on the location of the point, is directed either toward point A or point B.

the outward-directed centrifugal force generated as the point revolves about the common center (Figure 19.15b). Only at the exact center of the Earth are these forces in perfect balance (Figure 19.15c). At all points on the side of the Earth facing the Moon, gravitational force exceeds centrifugal force. The difference between the two forces generates a small *tidal force*. This force drives water toward point A, the point closest to the Moon; the water at point A produces a tidal bulge (or high tide).

The force relations are reversed on the side of the Earth facing away from the Moon. There, centrifugal force exceeds gravitational force. Thus the direction of the tidal forces

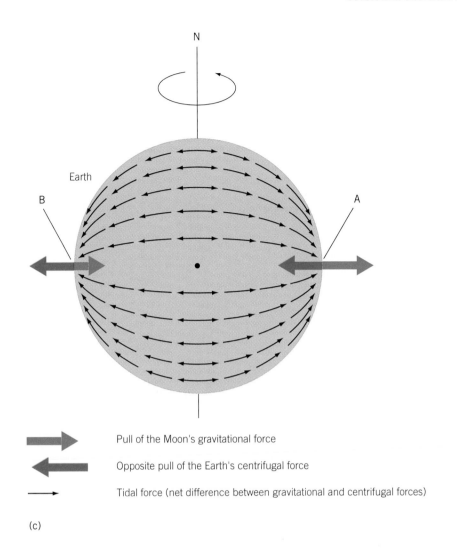

→ Pull of the Moon's gravitational force

← Opposite pull of the Earth's centrifugal force

→ Tidal force (net difference between gravitational and centrifugal forces)

(c)

is reversed, and a high tide is created at point B, opposite to point A. When A and B experience high tide, the region midway between the two is subject to low tides. Because the Earth completes one rotation about its axis every 24 hours, each point on Earth's surface *should* experience two high and two low tides per day. In other words, tides are waves. In our simplified model, high tides are the crests and low tides the troughs, the tidal wavelengths are equal to the diameter of the Earth, and the periods are on the order of 24 hours.

Of course, the Earth differs from the simplified water-covered model of Figure 19.15. It has continents, irregular coasts, and varying bottom contours that retard tidal motion. Because tides are very long waves, they feel bottom throughout the ocean. It is this friction that causes tides to lag behind the passage of the moon.

The interaction of these various factors produces three major tidal types, each common to a specific portion of the world's oceans:

1. *semidiurnal* (twice daily), having two highs and two lows per day, with a period of approximately 12 hours (the east coast of the United States)
2. *diurnal* (once daily), having one high and one low per day, with a period of 24 hours (the Gulf Coast)
3. *mixed,* having irregular heights and periods (the eastern Pacific Ocean)

Superimposed on these lunar tides is the gravitational effect of the Sun on the Earth, which exerts a tidal force of its own. **Neap tides,** having minimal tidal variation, occur twice each month when the Sun is at right angles to the Earth and Moon. **Spring tides,** having maximum tidal range, occur twice monthly when Earth, Moon, and Sun are

neap tide
A low-amplitude, twice-monthly tide.

spring tide
A high-amplitude, twice-monthly tide.

Figure 19.16 A satellite image of the drowned mid-Atlantic coast of the United States. Chesapeake Bay, an estuary, is the most prominent feature; as sea level rose following the last ice age, an intricate stream system of branching tributaries was invaded by the sea. Delaware Bay, to the north, is a similar, though less elaborate feature. Numerous beach forms, such as barrier islands and spits, are visible along the coast. The Susquehanna River, at the top left, cuts through resistant Appalachian ridges and enters into upper Chesapeake Bay.

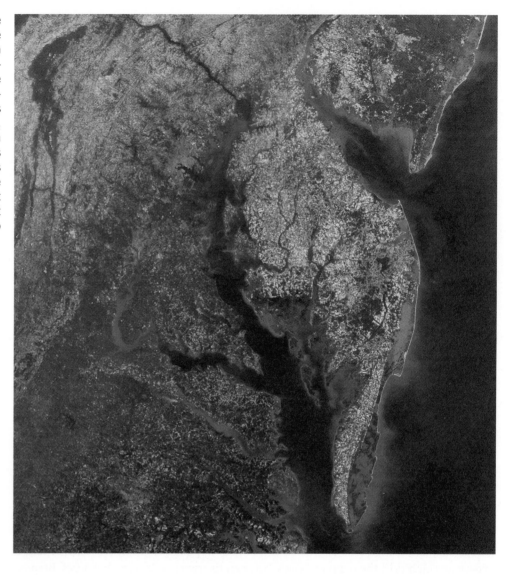

tidal current
The current of a rising or falling tide forced through a narrow constriction of coast.

tidal delta
A delta deposited by tidal currents on the seaward side of a lagoon.

aligned; in this position, the tidal forces produced by the Sun add to those generated by the Moon.

Because tides feel bottom, they behave like shallow-water waves. They are affected by water depths and by the configuration of the coast as they approach the shore. When a high tide approaches a bay that is constricted by headland, barrier beach, or spit, the water level of the open ocean stands above the water in the bay. Thus water is forced through the constriction under great pressure. Swift **tidal currents** are generated that can scour the bottom to great depth: 112 meters at the mouth of San Francisco Bay; 390 meters in the straits between the islands of Kyusho and Shikoku, Japan. The currents reverse during low tide. Therefore, sediment is periodically washed in and out of the bay. Depending upon the strength of incoming and outgoing tidal currents, the bottom of the bay will either be filled in or flushed. For this reason, communities that dispose of their waste by carelessly dumping it in the bay during low tide may, to their surprise, find it returning to their shores at high tide.

Tidal currents help shape barrier beaches, spits, and deltas. In some situations, they are more potent agents for shaping coasts and beaches than longshore currents. Tidal currents are especially effective in conjunction with great storm waves. When a barrier island (such as Fire Island, New York, or Cape Hatteras, North Carolina) is breached by high waves during a storm, tidal currents pour through the gap and transport sediment from the open ocean and foreshore of the beach into the quiet waters behind the island, producing a **tidal delta.** In time, longshore currents may seal the gap.

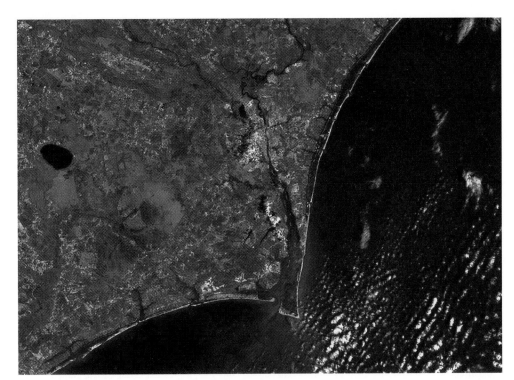

Figure 19.17 Cape Hatteras, North Carolina, is at the apex of a cuspate foreland molded by waves and long-shore currents.

Coasts

A *coast* is a strip of land bordering the ocean and directly subject to marine influences. The seaward border is the shoreline. The inland border is harder to define but is generally signaled by a marked change in topographic features. Nowhere else on Earth's surface do more geological processes and agents converge, which makes coasts difficult to classify. One way to study them is to divide them into two general types. *Primary coasts* owe their configuration mainly to nonmarine processes, such as land erosion or deposition, volcanic activity, or crustal movements. For example, Chesapeake Bay, on the eastern U.S. coast, was a river valley that was later submerged when the sea level rose following the retreat of the great North American ice sheet (Figure 19.16). *Secondary coasts* are molded primarily by marine influences, such as wave deposition or erosion, or the work of organisms, like coral reefs or mangrove trees. For example, just south of the Chesapeake Bay is Cape Hatteras, whose projecting shape and surrounding barrier beaches were formed by marine processes (Figure 19.17). Table 19.1 describes some further subdivisions within the two categories, primary and secondary. However, all coastlines are fundamentally controlled by two large-scale processes. The first, plate tectonics, provides the structural setting of coasts and shorelines. The second, the worldwide sea-level rise that accompanied the melting of the most recent ice sheet, has defined the line of land-sea interactions.

Tectonic Setting

Take a look at the familiar map in Figure 19.18(a) that shows the placement of North and South America with respect to tectonic plate boundaries. The western edge of each continent is adjacent to a plate boundary and has an *active margin*. The eastern edge of each continent is far from plate boundaries and, as a consequence, has a *passive margin*. These margins differ markedly from one another, and tectonic activity is the controlling factor (Figure 19.18b).

Most stretches of the west coast's border on a subduction zone, where a deep-sea trench or similar structure is found almost directly offshore. Other stretches border rift zones or transform boundaries. These plate boundaries are the sites of intense tectonic

Table 19.1 Classification of Coasts

Shaping Processes	Coast Types	Features	Examples
Primary coasts (nonmarine processes dominant)			
Erosion of the land surface and subsequent drowning by sea-level rise, sinking of land, or melting of ice caps	Drowned river-cut valleys	Relatively shallow estuaries	Ria coast of Spain; Chesapeake Bay (see Figure 19.16)
	Drowned glacial erosion coasts	Includes fjords and glacial troughs	Alaskan coast; Norwegian coast (see fjords, Chapter 17)
Retreat of shoreline caused by land-derived deposits	River deposition coasts	Deltaic coasts with distributary stream channels	Nile Delta; Mississippi Delta (see Chapter 15)
	Glacial deposition coasts	Partially submerged moraine; straightened by marine deposition and/or erosion	Long Island, New York
Volcanic activity	Volcanic coasts	Concave bays on the sides of volcanoes	Hanauma Bay, east of Honolulu
Tectonic movements	Fault coasts	Straight fault scarps	East coast of the Baja Peninsula on the Gulf of California; San Clemente Island, California (see Figure 19.19)
Secondary coasts (marine processes dominant)			
Wave action	Wave erosion coasts	Straightened by wave erosion, with wave-cut terraces	Seaward side of Cape Cod, Massachusetts; Monmouth, New Jersey
	Barrier coasts	Separated from mainland by lagoons or marshes	Southeastern coast of United States (see Figure 19.22)
	Cuspate forelands	Large projecting sandy beaches	Cape Hatteras (see Figure 19.17)
Organisms acting as barriers and sediment traps	Coral reef coasts	Built out by corals and algae	Atolls of the Pacific Ocean (see Chapter 2)
	Mangrove coasts	Sediments built up around tree roots to produce marshy land	Southwest Florida, between the Keys and the mainland
Nonspecific coasts			
Human activity	Altered coasts	Sea walls, dikes, harbors, reclaimed marshes, and so on	Reclaimed coasts of the Netherlands; portions of San Francisco Bay

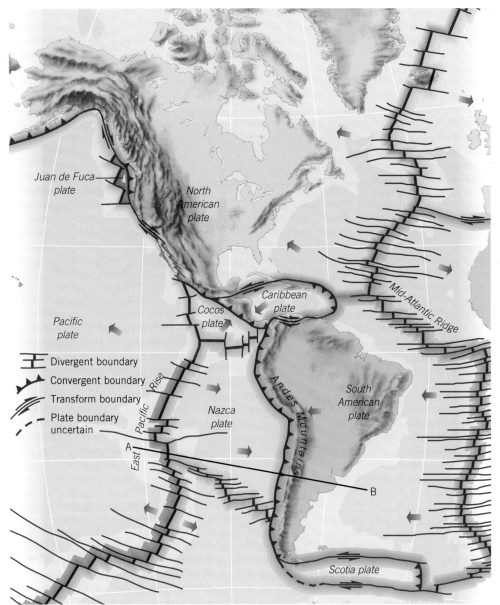

Divergent boundary

Convergent boundary

Transform boundary

Plate boundary uncertain

Juan de Fuca plate

North American plate

Pacific plate

Cocos plate

Caribbean plate

Mid-Atlantic Ridge

East Pacific Rise

Nazca plate

Andes Mountains

South American plate

Scotia plate

A

B

(a)

Figure 19.18 (a) The general tectonic setting of North and South America. (b) The narrow western coast of South America, adjacent to a subduction zone, is an active margin built on a tectonically unstable crust. The broad eastern coast of South America, far from plate boundaries, is a passive margin built on a tectonically stable crust.

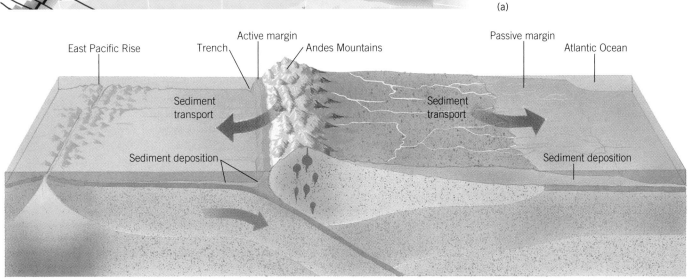

East Pacific Rise

Active margin

Trench

Andes Mountains

Passive margin

Atlantic Ocean

Sediment transport

Sediment transport

Sediment deposition

Sediment deposition

(b)

Figure 19.19 The elevated marine terraces and wave-cut cliffs of San Clemente Island, California.

submarine canyon
A steep canyon, resembling a V-shaped stream valley, that is cut into the continental shelf or slope.

activity and volcanism. As a consequence, the west coasts of North and South America are generally straight, mountainous, and extremely rugged. Few major streams enter directly into the ocean, and the west coast receives comparatively less sediment than the east coast. The continental margin is extremely narrow; and although there are broad beaches, they are of the wave-cut cliff and terrace variety. Some of these wave-cut cliffs and terraces have been uplifted; they ascend, steplike, high into the coast, which is striking evidence of recent uplift (Figure 19.19). This unstable tectonic environment does not favor the construction of a broad continental shelf. Instead, the sediment brought in by streams is funneled to deep trenches and abyssal fans via **submarine canyons** located close to shore.

The east coast runs along a passive continental margin located far from the Mid-Atlantic Ridge and other centers of tectonic activity. In this stable crustal environment, the margin was able to develop a broad continental shelf and slope. It is essentially a thick sediment wedge deposited over millions of years by streams emptying into the Atlantic Ocean. Except where ancient highlands intersect the shore, the coast is generally low and marked by a wide coastal plain, an extension of the continental shelf.

Sea-Level Changes

Because the Atlantic continental shelf is tectonically stable, it is a good place to trace the worldwide sea-level changes that have occurred since the last ice age. The emergence of the continental shelf happened recently, from a geological perspective, for the ice stood at its maximum a mere 18,000 years ago. During the ice ages, streams flowed across the shelf to the shelf break; like all streams, they cut channels and deposited sed-

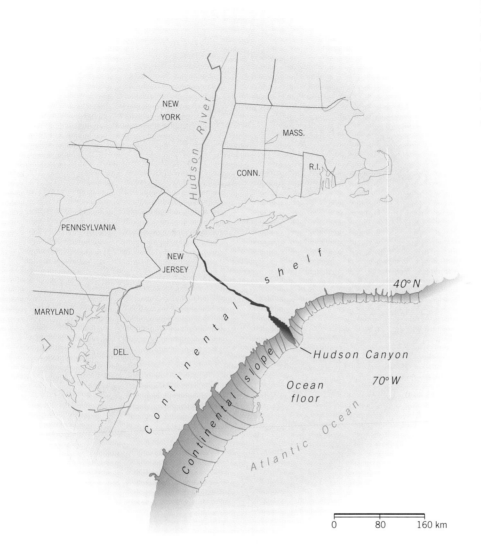

Figure 19.20 The channel of the Hudson River extends across a submerged continental shelf—proof that the Hudson flowed across the shelf during the last glacial age, when sea level was lower. The submarine Hudson Canyon was cut in the soft continental slope by thick turbidity currents.

iment in accordance with this lower base level. Many of these effects of the vast sea-level retreat are still visible on the shelves (Figure 19.20).

From analyses of sediments, fossils, and submerged shoreline features, we know that the seas have been advancing ever since. Figure 19.21 shows the sea-level rise in the past 18,000 years, since the melting of the last ice sheet. Note that the rate was rapid until 6000 years ago, when sea level reached approximately its present position, and has since leveled off. Sea level is currently rising about 4 millimeters per year off the Atlantic coast (about 16 inches per century). The effects of sea-level rise have been profound. For example, all of the low-lying coasts have been invaded by the sea. Many of today's islands were formerly inland hills. (In some cases, these hills were quite large—for example, the British Isles, which had been connected to mainland Europe.)

If you look at the East Coast of the United States on the map in Figure 19.22, you will notice that the coastline is highly irregular and indented. The landward fingers of the sea are **estuaries,** flooded river mouths where fresh- and saltwater mix and are controlled by tides. The Chesapeake Bay and the Hudson River are striking examples. In fact, the mouth of virtually every river in the world that emptied into the sea prior to this most recent sea-level rise is an estuary.

Sea-level rise is responsible for the numerous barrier islands that rim the East and Gulf coasts. Keep in mind that the slope of the shelf is so slight that a 1-meter sea-level

estuary
A mouth of a river invaded by the ocean, where fresh- and saltwater mix.

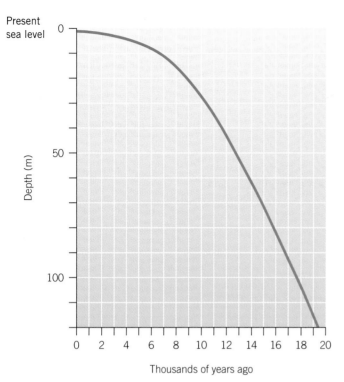

Present sea level

Depth (m)

Thousands of years ago

Figure 19.21 By studying the submerged continental shelf, geologists have found that 20,000 years ago, sea level was at least 120 meters lower than at present. A period of warming ensued, which caused the ice sheets to melt and sea level ro rise rapidly. Note that the rate has leveled off over the last few thousand years.

Figure 19.22 The East and Gulf Coasts of the United States. Rising sea level has created a complex coast of embayments, barrier islands, spits, cuspate forelands, and beaches.

rise causes the coastline to retreat by 1000 meters or more. The islands most probably originated as beaches on the emergent continental shelf during the ice age, when sea level was lower. The rising sea flooded the land behind the beaches, isolating them from the coast. The rising sea also elevated the point of wave attack on the beach faces, causing them to roll over in the manner discussed earlier. Thus the barrier islands have migrated toward the coast. Since migration and coastal retreat occur in tandem, the lagoons between them have been maintained. Occasionally, where a slightly higher coast has slowed retreat of the shoreline, the barrier islands have rolled up against the mainland and have closed out the lagoons and bays.

On the west coast, the mountains bordering Washington, Oregon, and Canada were covered by ice during the last glacial advance. Stream valleys were deepened by the ice, and streams flowed further west than at present. With the rise in sea level following melting of the ice, the valleys were flooded. Puget Sound is such a flooded valley. Thus the northwest coast is similar to the coasts of Maine and Nova Scotia, but far more rugged. As in the east coast, many small islands were cut off from the mainland by the sea-level rise.

Molded by so many factors—tectonic forces, waves, winds, currents, sediment supply, climate, tides, biological activity, sea-level changes, the works of human beings—coastal and shoreline processes are complex and vary with time. There is still much to learn about this subject. Perhaps the key concept to keep in mind is the dynamic nature of the coastal and shoreline environments. For this reason, short-term solutions to local problems can lead to disastrous mistakes on a wider scale. As with most choices involving conflicting interests, decisions are never clear-cut and often amount to a series of unsatisfactory trade-offs.

STUDY OUTLINE

I. WIND-DRIVEN WAVES

   A. Wind-driven **water waves** are vibrations by which energy is transferred from ocean to shore.

      1. Wave height is the vertical distance between crest (high point) and trough (low point); a wavelength is the distance between crests; wave period is the time it takes successive crests to pass a given point.

      2. At a storm center, waves of different lengths, periods, and heights interfere with one another and create a chaotic sea surface. Away from the storm center, the waves disperse; long waves outrun short, and the sea becomes calmer.

   B. As a wave approaches the shore, its movement is retarded by bottom friction. Close to shore, its length shortens and its height increases, but its period remains the same. Eventually, it collapses or breaks.

      1. A wave that approaches the shore at an oblique angle refracts; the segment closest to shore slows down first, and the entire wave pivots in a direction roughly parallel to the shoreline.

      2. Waves that approach an irregular coast erode the headlands and deposit sediment in the bays to form **pocket beaches.** Erosional features of the headlands include **wave-cut cliffs, wave-cut platforms, sea stacks,** and **sea arches;** the sediment from the sea cliffs is deposited behind the wave-cut platform as a wave-built platform. Erosion and deposition processes acting on an irregular coast lead to coastal straightening.

II. BEACH FORMATION AND SHORELINE PROCESSES

   A. Materials brought to the sea by streams and eroded from the coast are shaped by waves and currents into moving rivers of sand called **beaches.**

1. The beach profile is divided into three distinct units: **backshore, foreshore,** and **offshore.** The backshore consists of one or two narrow terraces or **berms** and the foreshore is the zone of breaking waves. The offshore contains submerged **longshore bars** and troughs that parallel the shoreline.

2. In calm weather, long waves that break offshore and sweep sediment onto the foreshore flatten and broaden the beach profile. During storms, short and choppy waves attack the foreshore and carry sediments to the offshore, steepening the beach profile.

B. A wave striking the beach at an oblique angle sets in motion a nearshore circulation loop consisting of **swash** and **backwash, longshore current,** rip current, and rip head. The circulation loop, which encompasses the **surf zone,** is instrumental in **longshore sediment transport.** Swash and backwash move sediment in the direction of the longshore current.

C. Longshore sediments mold beaches into a variety of forms that are also dependent on such factors as the resistance of rocks to erosion and wave direction and intensity. Some common beach forms include the pocket beach, the **tombolo,** the **barrier island,** the **spit,** the **baymouth bar,** and the **cuspate foreland.**

D. A typical beach is part of a larger sediment circulation system. Sand enters the system via wave-cut cliffs and rivers. Sand leaves the system via longshore and bottom currents.

1. Rising sea level forces barrier islands to roll over and migrate landward.

2. Changes in the amount of sediment entering the beach may be natural or human-induced (for example, dams trap sediments inland and starve the beach). As a consequence, many beaches and shorelines of the United States and around the world are seriously eroding.

3. A **groin** is a barrier built seaward to prevent longshore drift. It leads to accelerated erosion downcurrent of the groin. Retaining walls protect structures but intensify beach erosion. Many geologists consider beaches to be inherently fluid systems from which construction should be banned under normal circumstances.

4. A study of two storm sites found less damage to beach communities where buildings were constructed *behind* the dunes of wide, unobstructed beaches.

E. Organic activity, such as the growth of coral reefs or mangrove trees, plays a major role in shaping some shorelines.

F. **Tides** are the rhythmic rising and falling of the sea surface caused by the interplay of the gravitational pull of the Moon and centrifugal forces generated as the Earth revolves about the center of mass of the Earth-Moon system.

1. There are three major types of tides: semidiurnal, diurnal, and mixed.

2. The Sun also produces a tidal effect. Very high **spring tides** and very low **neap tides** each occur twice monthly.

3. A swift **tidal current** in conjunction with storm waves may breach a barrier island and deposit sediment as a **tidal delta** in the protected waters behind the island.

III. COASTS. A coast is a strip of land bordering the ocean.

A. Primary coasts are molded mainly by nonmarine processes: land erosion and deposition, volcanism, and tectonism. Secondary coasts are formed primarily by marine influences: wave deposition or erosion, coral reefs, and the like.

B. Proximity to tectonic plate boundaries provides the broad structural setting within which coastal processes occur.

1. The west coasts of North and South America are parts of tectonically active continental margins and are generally straight and narrow. Sediment brought to the narrow continental shelf is funneled by way of **submarine canyons** to the deep trenches or basins offshore.

2. The east coasts are parts of passive continental margins. Far from activity associated with plate boundaries, the margin consists of a broad continental shelf, slope, and rise. The coasts are low, intricately configured, and marked by coastal plains that are extensions of the continental shelves.

C. The sea-level rise that accompanied the melting of the last ice sheet caused invasion of the older river valleys, forming **estuaries,** or drowned river coasts and barrier islands, which were subsequently beaten back toward the mainland.

STUDY TERMS

backshore (p. 519)
backwash (p. 521)
barrier island (p. 522)
baymouth bar (p. 522)
beach (p. 519)
berm (p. 519)
cuspate foreland (p. 522)
estuary (p. 537)
foreshore (p. 519)
groin (p. 525)
longshore bar (p. 520)
longshore current (p. 521)
longshore sediment transport (p. 521)
neap tide (p. 531)
offshore (p. 519)

pocket beach (p. 519)
sea arch (p. 518)
sea stack (p. 518)
spit (p. 522)
spring tide (p. 531)
submarine canyon (p. 536)
surf zone (p. 521)
swash (p. 521)
tidal current (p. 532)
tidal delta (p. 532)
tide (p. 529)
tombolo (p. 522)
water wave (p. 516)
wave-cut cliff (p. 518)
wave-cut platform (p. 518)

CRITICAL THINKING QUESTIONS

1. Describe the sediment budget, of which the beach-nearshore system is a part. What natural changes can affect it? What human-induced activities can change it? What has been the effect of dam construction on beaches?
2. What is your opinion of the sand rights proposition mentioned in this chapter? Explain.
3. Should people be allowed to build permanent homes on beaches? Should they receive disaster aid following storms? Discuss both the pro and con positions.
4. Discuss the pros and cons of hard- and soft-stabilization techniques.
5. Describe some of the geological work of tides.
6. Describe how plate tectonics influences coastal development. What features are common to active margins? To passive margins?
7. Why are barrier islands more prevalent on the East and Gulf coasts than on the West Coast of the United States?

Mineral and Energy Resources

This chapter focuses on the origins of the **resources** extracted from the Earth. In the broadest sense, the term applies to the naturally occurring materials that may now or in the future be profitably extracted. Included in the definition of *resources* are metallic and nonmetallic **mineral deposits,** as well as **hydrocarbon deposits**—the accumulations of coal, petroleum (oil), and natural gas upon which modern civilization depends.

Geologists classify a resource according to the degree of certainty of its existence and on the economics of its extraction (Figure 20.1). A mineral deposit proven to exist is called an *identified resource;* it is also considered a **reserve** if it can be extracted at a profit under current economic and technological conditions. If not, it is a subeconomic resource. An *undiscovered resource* is a deposit that has been inferred to exist but has not been physically identified. For example, geologists may infer that a well-studied mining district may still contain deposits that have not yet been found. They may also infer from past experience or from geological theory that an unexplored region contains a mineral deposit or that a deposit exists in an unknown form. These, too, are undiscovered (speculative) resources.

In assessing the long-term supply of a resource, we must keep in mind the relationship between the rate at which it is used and the rate at which it is being replenished by natural processes. Consider, for example, hydroelectric power, which relies on the release of water amassed in dams at high elevation. The water comes from rain and snow and is an integral part of the hydrologic cycle. Because the major part of this cycle acts rapidly (in a matter of weeks), hydroelectric power is considered a **renewable resource**—one that can be regenerated in a time frame useful to human beings.

resource
Any Earth material of value to society that is known or believed to exist, and that may eventually become available for use.

mineral deposit
A naturally occurring accumulation of mineral matter of potential economic value.

hydrocarbon deposit
An accumulation of organic compounds in the solid, liquid, or gaseous state.

reserve
An identified resource that can be profitably extracted from the Earth at current market conditions and levels of technology.

renewable resource
A resource that is replenished in relatively brief time periods, so it can be used again by human beings.

◀ Different kinds of resources are extracted from this California field: petroleum from the subsurface, crops from the soil, and electricity from the wind. (Note the windmill at the far left.) The Sun, the energy resource ultimately responsible for all of them, hovers in the background.

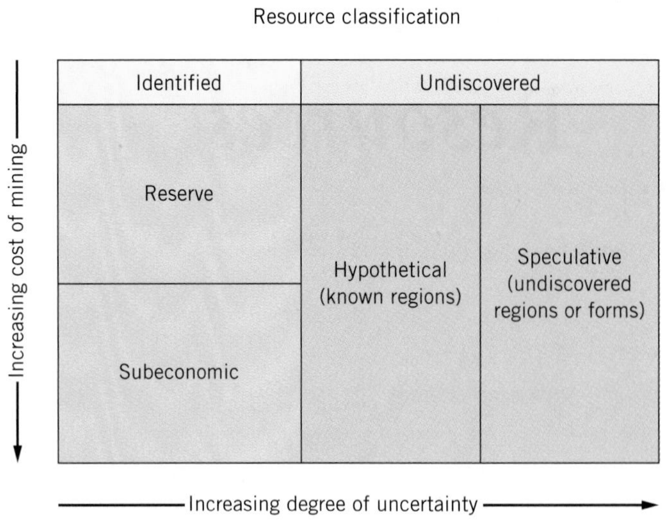

Figure 20.1 The U.S. Geological Survey classification of resources. The term *reserve* is limited to identified re-sources that can be extracted profitably at current costs and levels of technology.

nonrenewable resource
A resource that is replenished in time periods that are very long by human standards.

Well-cared-for soils and certain kinds of timber also fit this category. By contrast, petroleum is a **nonrenewable resource.** It takes millions of years to form, so that whatever oil we burn depletes—essentially forever—the amount available for future use. We concentrate on nonrenewable resources in this chapter.

Where are resources located? What are their origins? What makes them profitable to extract? Will we run out of them? What is the environmental impact of mining and refining them? We will discuss these questions in the course of this chapter.

The Origins of Mineral Resources

ore
The naturally occurring material from which an economically valuable mineral can be profitably extracted.

To be of economic value, a mineral resource must be suitably concentrated as well as readily extractable (Table 20.1). A mineral deposit that meets these requirements contains **ore.** The grade of an ore deposit refers to its concentration; low-grade ores have less intrinsic value than high-grade ores because extraction costs rise as concentrations

Table 20.1 Approximate Concentration of Ore Elements

Element	Approximate Concentration in Average Igneous Rocks (%)	Approximate Concentration in Ores (%)	Concentration Factor to Make Ore
Aluminum	8.0	35	4
Iron	5.0	50	10
Copper	0.007	0.5–5	70–700
Zinc	0.013	1.3–13	100–1000
Lead	0.0016	1.6–16	1000–10,000
Tin	0.004	0.01*–1	2.5–250
Silver	0.00001	0.05	5000
Gold	0.0000005	0.0000015*–0.01	2500
Uranium	0.0002	0.02	1000
Tungsten	0.003	0.5	170
Molybdenum	0.001	0.6	600

*Placer deposits.

Figure 20.2 Strip mining of coal, Navajo Indian Reservation, Window Rock, New Mexico. The coal seams are visible in the excavation on the right.

fall. Notice in Table 20.1 that necessary concentrations vary widely. Gold must be concentrated 2500 times its normal abundance in the crust to show a profit, whereas aluminum may be mined profitably at only 4 times its normal abundance.

Aluminum is the third-most abundant element of the crust. However, most of the world's aluminum is tightly bound up in the crystal lattices of silicate minerals, from which it is extremely difficult to separate. Only where natural processes have removed the silica does it become profitable to mine aluminum. This is why aluminum is extracted from the ore bauxite. You may recall from Chapter 13 that bauxite forms in humid, tropical climates where deep weathering has leached silica from the original minerals of certain igneous rocks, leaving behind the aluminum hydroxide or bauxite residue. Over geologic time, nature performs a separation task that for humans is prohibitively expensive and wasteful of energy.

The profitability of a given mineral deposit—whether it should be classified as a reserve or a subeconomic resource—is also extremely sensitive to a host of nongeological factors that affect price: the extraction technology, political considerations, environmental regulations, proximity to markets, competition from other sources, labor costs, and the world and local economies.

 Mineral Extraction and the Environment

The environmental toll of mineral extraction and the cleanup costs of mining operations are major factors affecting the overall profitability of a mineral deposit. Mining invariably produces undesirable side effects that can only be reduced, never eliminated. Because the valued substance, typically a metallic ore, usually constitutes a minute percentage of the matter present at the site, there is no way of getting at the ore mineral without disturbing the site (Figure 20.2). First, the surface vegetation and other habitats are destroyed and the land surface is scarred by road construction and heavy machinery, whether the mining operation is surface or subsurface. Excavation creates quantities of rock that are piled in (appropriately named) spoil heaps. As a general rule, the ore removed from the mine is processed in on-site or nearby mills. There, the ore minerals are physically separated from the undesirable substances or *gangue,* which is piled high as tailings. The metals in the ore minerals are then removed and purified in smelting plants.

If uncorrected, these activities leave ample opportunities for pollution. Removal of the natural surface accelerates erosion of the countryside, which leads to the clogging

Figure 20.3 Hazards generated by surface mining operations. Damage to the environment is unavoidable, but sound mining practices and land reclamation supervised by geologists may limit the harm.

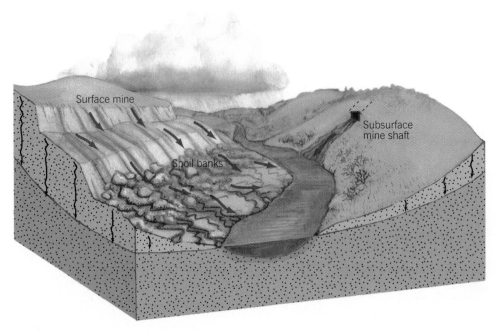

of streams with sediment. Spoil heaps and tailings are leached by groundwater that carries toxic chemicals to the water table (Figure 20.3). Smelting is also frequently a dirty activity, leading to the release of sulfur dioxide, particulates, and harmful metals into the air and soil. In fact, mining is but the first phase of a sequence that also involves the manufacture, use, and disposal of the resource. Each has its environmental impact. Today, the mineral industry is showing greater environmental awareness, and modern smelters and mines are being held to stricter standards.

Concentration Through Processes at Depth

Table 20.2 classifies the common mineral deposits by origin. On this basis, they fall neatly into two categories: those that form from processes occurring at depth, such as igneous activity and metamorphism, and those that form from processes occurring at or near the surface, such as weathering and sedimentation. As you can see, virtually every geologic process contributes directly or indirectly to one kind of mineral deposit or another.

Most ore-forming processes that occur at depth are closely associated with specific plate tectonic settings. The basic cycle works this way: metallic ores are concentrated in the oceanic crust along divergent plate boundaries—in mid-ocean rift valleys. They are a natural consequence of the creation of oceanic crust. The ores are then swept via plate movements to subduction zones. There, some of them are thrust physically onto the continents within accreted terranes; others are recycled beneath the crust, where they are remelted and concentrated through magmatic processes. Still other metals seem to be concentrated by other subduction processes alone. Thus the places to search for most mineral deposits are present and past plate boundaries.

Divergent Boundary Processes

Mid-ocean ridge hydrothermal activity of the kind that produces hydrothermal vents (chimneys or black smokers) is a very important ore-forming process. As discussed in previous chapters, the rift valley fractures are, in effect, an elaborate plumbing system, similar to the kind used in a large apartment house; the magma beneath the rift valley plays the role of basement boiler. Cold water enters the fractures and percolates deep into the crust, where it is superheated by the magma. The hot water then expands, rises through

Table 20.2 Classification of the Origins of Mineral and Hydrocarbon Deposits

Process	Origin	Mineral Examples
Processes at depth		
Magmatic	Deposits formed directly from the crystallization of magma into ingneous rock	Chromium and platinum deposits of the Bushveld Complex of South Africa
Hydrothermal	Deposits associated with circulating hot water (heated by magma) and gaseous solutions in the crust	Gold, zinc, lead, and copper ores
Metamorphic	Deposits formed in rocks transformed while in solid state by heat, pressure, and chemically active fluids; stones and asbestos	Certain iron ores, marble, serpentine
Near-surface processes		
Leaching	The insoluble residues of rocks whose soluble elements have been dissolved and carried away by groundwater residual deposits	Bauxite (aluminum ore) and pisolite (iron ore)
Groundwater deposition	Secondary enrichment deposits of the metallic elements leached from the surface and redeposited at or near the water table	Many valuable metallic ores, especially sulfide ores of copper and nickel
Mechanical concentration	Placer deposits of heavy-metal grains in streambeds, beaches, and so on; weathered and transported from a primary igneous source or "mother lode"	Gold, tin, diamond deposits
Evaporation	Precipitation of normally soluble elements from seawater or freshwater trapped in restricted basins where arid climates prevail (evaporites)	Salt, potash, gypsum, borates, carbonates, and important sedimentary iron ore deposits
Biochemical precipitation	Elements precipitated through the metabolic activities of organisms	Sedimentary iron ores
Hydrocarbon preservation	Organic deposits made up of decomposed remains of organisms preserved in subsurface rock and sediment	Coal, petroleum, oil shale, and natural gas

the fractures, and escapes to the ocean floor as metal-rich black smokers (Figure 20.4).

Chemical changes accompany this circulation. The cold, oxygen-rich seawater is initially able to hold only minute quantities of metal. As it is heated, it combines with silica in the crust to form low-grade metamorphic minerals such as chlorite, talc, and zeolites (see Chapter 9). These reactions release hydrogen ions into the seawater, converting it to an acid solution. The hot, highly acidic seawater then goes to work dissolving acid-sensitive metals in the crust: copper, cobalt, iron, manganese, zinc, and others. In this way, metals that exist in only trace percentages of the oceanic crust are now concentrated in the hot, upward-moving hydrothermal brines.

The environment of these brines changes as they rise in the crust. Cold, normal seawater may leak into the crust and mix with them, forcing precipitation of the metals in the fractures and voids of the crust. These metals generally combine with sulfur as they precipitate, forming *metal sulfide* ores [chalcopyrite, (Cu,Fe)S, is an example]. Those brines that vent to the surface precipitate metals in an oxygen-rich environment,

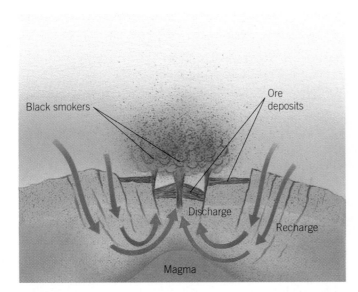

Figure 20.4 Hydrothermal circulation beneath mid-ocean rift valleys concentrates valuable elements in black smokers and veins of ore that line the fractures in the oceanic crust. The ores are then transferred to the continental margins through sea-floor spreading and subduction.

forming oxides, hydroxides, and silicate ore minerals. These minerals build the chimneys through which the brines escape. Often, the metals may precipitate directly around a small rock fragment, or shark's tooth, to form potato-shaped manganese nodules. This name is misleading, because manganese nodules also contain copper, iron, cobalt, nickel, and chromium (Figure 20.5). Billions of them are strewn throughout the sea floor. Their potential value is incalculable, and the nations of the world are busily working out treaties that will allow for their mining.

The stage of development of the ocean basin influences the nature of the deposits that accumulate in the rift valley. For example, the geologically young Red Sea basin is a narrow extension of new oceanic crust between the separating African and Arabian plates. In this structural setting, ocean circulation is restricted; in the arid climates that prevail there, seawater evaporates. This leads to the deposition of thick salts, such as halite and gypsum, on the rift valley floor. These salts are then dissolved by the hot water rising from the rift valley. The added salts turn the emerging hot-water brine into a dense, highly reactive solution that collects in pools within the deeper portion of the rift valley basin. These brines concentrate huge quantities of metals leached out of the crust. Eventually, the metals precipitate and settle in the muds that are deposited on the rift valley floor. The layers of metal-rich sediment reach a thickness of 10 kilometers in places—a resource of enormous potential, containing iron, zinc, copper, silver, and gold.

Concentration at Subduction Zones

We have seen that hydrothermal solutions deposit metals upon or within newly created oceanic crust at divergent boundaries. This crust is then swept conveyor belt style toward subduction zones. Though many of the metals are cycled back into the mantle within these zones, huge slices of metal-rich oceanic crust are sheared off the descending plates

Figure 20.5 A large field of metal-rich manganese nodules on the floor of the northeast Atlantic Ocean. (See Figure 8.27 on page 243 for a close-up view of a nodule.)

and shoved onto the bordering land-masses (see Chapter 12). These bodies of oceanic crust emplaced on land are called *ophiolites*. The island of Cyprus is an ophiolite formed from oceanic crust uplifted by the collision of the African and Eurasian plates. The name *Cyprus* is derived from *kyprus,* the ancient Greek word for copper—an appropriate name, considering the rich copper deposits of that island.

Subduction zones are also environments conducive to the development of hydrothermal circulation systems. The plates may partially melt as they descend into the mantle and release the metals that were concentrated in the upper oceanic crust by hydrothermal fluids at the mid-ocean rift valleys. At an ocean-continent plate boundary, subduction may also trigger melting of the overlying granitic crust. We know from the frequent earthquakes in the vicinity of subduction zones that the rocks overlying

Figure 20.6 Bands of platinum (light) and chromium (dark) exposed in the wall of a mine in the Precambrian Bushveld Complex of South Africa—the world's richest metallic mineral deposit. These layers were segregated during crystallization of the vast ultramafic intrusion.

the subducting plate are highly fractured. These fractures provide avenues for underground water to penetrate the crust, absorb heat from the magma, and rise to the surface. They also provide the means for seawater trapped within the descending plates to escape. These hot waters react with crustal rocks and rocks of the upper mantle to dissolve metals and reprecipitate them within the crust, either concentrated as thick veins or scattered as **disseminated deposits.**

disseminated deposit
An ore finely scattered within a host rock body.

Concentration Through Magmatic Processes

We have so far discussed magma in terms of its relatively limited roles of supplying the heat necessary to keep hydrothermal solutions circulating through the crust. But many valuable mineral deposits are formed through direct precipitation within the main body of the magma as it crystallizes to form igneous rock. Recall that a magma is an extremely complex silicate melt in which minerals crystallize over a broad range of temperatures. In many mafic magmas, metals such as nickel, chromium, tungsten, platinum, and cobalt crystallize at higher temperatures than most of the silicate minerals. Because they are denser than the melt, they settle to the bottom of the magma chamber, where they collect in thick, easily minable layers. This is the origin of many of the metals of the fabulous *Bushveld Complex* of South Africa, one of the richest deposits of rare, heavy metals on the Earth (Figure 20.6). The smaller but nevertheless important Stillwater Complex of Montana was formed in a similar manner. Each is of Precambrian age, and their exact origins are obscure. But association of the metals with mafic rocks indicates that they came from the mantle. Perhaps they crystallized at a rift zone or from mantle plume magmas that intruded the ancient crust. Valuable metal deposits are also concentrated in silica-rich magmas derived from the continental crust bordering subduction zones. The great Chilean copper deposits, for example, are closely associated with the andesite porphyries and granodiorites of the Andes.

Late-stage residual fluids of granitic magmas are an important source of mineral deposits. They are rich in silica, potassium, and rare elements that do not fit easily into the crystal structure of the major minerals composing the igneous rock mass. As the rock continues to cool and contract, the residual fluids are mobilized and pierce both

The 700,000 year-old Galeras volcano in the Colombian Andes spews half a kilogram of gold into the air each day and may be depositing 20 kilograms of gold a year into the rocks that line the crater.

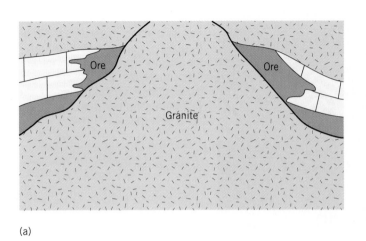

(a)

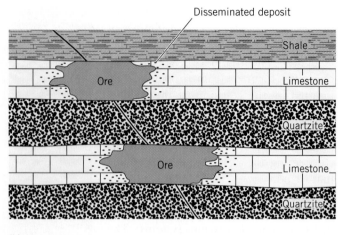

(b)

Figure 20.7 (a) An ore may form adjacent to an igneous intrusion in host rocks that are subjected to contact metamorphism. (b) An ore may also occur in disseminated form as fluids rise, escape from a magma, and react with host rocks. The ore is then deposited in minute fractures and faults. In both processes, some rocks, such as limestone, are more reactive than others.

the igneous rock and the surrounding host rock, crystallizing to a coarse-grained pegmatite. Some pegmatites contain deposits of gold, silver, uranium, lithium, bismuth, tungsten, tin, and antimony. In rare instances, protected cavities within the pegmatites contain beautiful gem minerals: varieties of beryl (emerald), topaz, garnet, and tourmaline.

An igneous intrusion may radically alter the surrounding country rock, producing a contact aureole or halo. The halo is the product of *contact metamorphism* in which fluids and gases are driven from the country rock. In some cases, valuable residues, such as magnetite iron ore, are left behind. Most contact-metamorphic deposits, however, are formed by the introduction of fluids as well as heat from the magma; they are products of *contact metasomatism* (see Chapter 9). Metal ions in the fluids react with and replace the ions present in the minerals of the host rock. Metasomatism is most effective when magmas intrude impure limestone, causing a recrystallization of calcite into a coarse-grained marble and the formation of many other minerals (Figure 20.7). Frequently, these minerals are of gem quality; the finest rubies and sapphires, found in the Magok region of Burma, are mined from aureoles and from nearby gravels weathered from the aureoles. Rocks subjected to contact metasomatism also yield iron, zinc, copper, and tungsten.

Concentration Through Surface Processes

Surface processes further concentrate the metals originally deposited in the crust by hydrothermal solutions and magmatic activity. Weathering, groundwater action, erosion and sedimentation, and organic activity all play significant roles (see Table 20.2).

Residual Ores

residual ore
A mineral deposit concentrated at the surface through the removal of soluble materials by weathering and groundwater leaching.

Often, rainwater seeping into the ground leaches soluble elements from exposed rocks, leaving behind insoluble **residual ores.** We have mentioned that the aluminum ore bauxite is formed this way—as the aluminum hydroxide weathering product of silicate igneous rocks in humid, tropical climates. Formed in similar fashion are extremely valuable iron ore deposits, such as central Australia's iron hydroxide *pisolite* ores (Figure 20.8).

The iron ores of the Lake Superior district and of western Australia were enriched through leaching (Figure 20.9). Apparently, they were originally precipitated in an iron-saturated Precambrian ocean as a by-product of the respiration of microscopic organisms that extracted energy from compounds in the water. The interesting point is that iron can be held in solution only in water that is chemically reduced—that is, lacking in free oxygen. Thus the oceans in which the iron was deposited must also have been

Figure 20.8 Pisolitic iron ore, a valuable residual deposit. Notice the iron-rich spherules.

Figure 20.9 Lake Superior banded iron deposit. The red bands are jasper, a form of chert. The darker bands are hematite iron ore.

chemically reduced; yet oxygen is plentiful in today's oceans, because they interact with the oxygen-rich atmosphere. This discrepancy has led some geologists to hypothesize that the Precambrian atmosphere was oxygen-deficient at the time the iron was first deposited.

Secondary Enrichment

The acids in groundwater also leach soluble metal ions from rocks exposed to the oxygen-rich surface environment. Left behind at the surface is a *leached zone* of highly insoluble iron oxide and other metallic residues. The leached zone is also called the *gossan* (German for "iron hat"), because it caps the mineral deposit. The leached metal ions are then transported to the chemically reduced zone of saturation, within which **secondary enrichment** occurs. In the absence of oxygen, the leached metal ions combine with sulfur, and precipitate as metal sulfide ores. Zinc, lead, iron, copper, gold, silver, and mercury accumulate in this way. So two distinct minable zones are created: one at the surface formed by leaching and oxidation, and another, commonly richer, formed by secondary enrichment (Figure 20.10). Outstanding examples of minable secondary enrichment deposits include the copper ores of Butte, Montana, and La Escondida, Chile (Figure 20.11).

secondary enrichment
The process by which metals are leached from the surface and precipitated below the water table.

Placer Deposits

Many native metals and valuable minerals are weathered from rock outcrops and are carried along by streams. They are physically resistant and do not dissolve readily, so

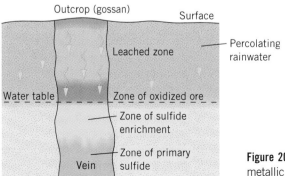

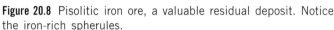

Figure 20.10 The various zones of metallic ores in secondary enrichment.

Figure 20.11 The vast Phelps-Dodge Morenci open pit copper mine, Clifton, Arizona. The operation exploits the various zones of enriched ore.

placer deposit
An accumulation of mechanically weathered mineral particles, transported and deposited by running water.

they tend to survive the rigors of transport. Because they are dense, they accumulate wherever the transporting water loses velocity (Figure 20.12): in the point bars of meandering streams, in stream channel pools, and in the upper berm faces of beaches, where backwash lacks the power to transport them back to the ocean. Mechanical concentrations like these are called **placer deposits.**

The gold sifted from stream gravels and sands at Sutter's Mill in northern California in 1848 probably constitutes this nation's most famous placer deposit. In sparking the Gold Rush, it also sparked a major population push to the West Coast. By panning the streams that flowed west from the Sierras, prospectors knew that the source from which the gold had weathered—the elusive "mother lode"—lay to the east, within the mountain range. Indeed, the mother lode was later found to be gold disseminated in scattered quartz veins within the granites of the western Sierras.

With rare exceptions, those who invested in outfitting the prospectors in clothing, food, and rigging profited more than the prospectors themselves. The same cycle seems to be repeating itself in the Amazon basin of Brazil, where vast numbers of poor prospectors have flocked in search of gold. Most will have little to show for much hard work that damages the rain forest. This damage goes beyond clearing patches of the forest. The mercury used to extract the gold is contaminating streams throughout the Amazon basin, and it has become a major environmental problem.

Placer deposits are commonly mined for rich accumulations of platinum, tungsten, and tin. The diamond placers deposited on South African beaches alerted prospectors to the mother lode found inland: the dark, igneous, cylindrical kimberlite pipes of South Africa. Submerged stream and beach placers of the Atlantic continental shelf are potentially valuable metal resources, provided that ways can be found to mine them economically without disturbing the marine environment.

Energy Resources

Coal, petroleum, and natural gas are derived from the partially decomposed remains of ancient organisms preserved within sedimentary rocks; that is why these hydrocarbon

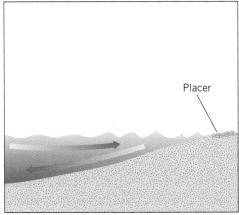

(a) High on beach berms

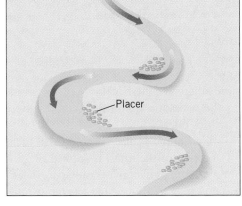

(b) In potholes

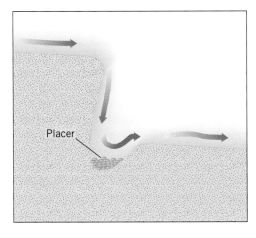

(c) Below waterfalls

(d) Inside meander loops

Figure 20.12 Four natural processes that enable dense placer deposits to accumulate through gravity separation.

resources are referred to as **fossil fuels.** They are residues of the carbon-oxygen cycle. At the heart of the cycle is photosynthesis, whereby plants, using sunlight, extract carbon from the carbon dioxide in the atmosphere and incorporate the carbon into their living tissues as organic (hydrocarbon) molecules. The plants also release oxygen to the atmosphere as a by-product of photosynthesis, and this oxygen is extracted from the atmosphere by the respiration of bacteria and animals. During respiration (and during the decay of organic wastes), the oxygen recombines with the carbon of the organisms to form carbon dioxide, which is then released back to the atmosphere, where it is used again in photosynthesis. This process of cycling carbon into and out of the atmosphere through photosynthesis and respiration/decay is remarkably efficient, but not perfect. Over millions of years, a tiny fraction of carbon in the form of organic matter is diverted and buried along with other sediments. Depending upon the prevailing geological conditions, this organic matter is converted either to coal or to oil and natural gas.

The economic value of coal and oil rests on two fundamental properties of carbon molecules. First, the carbon of oil and gas is in a chemically reduced state. In this state, the carbon atoms form bonds with other carbon atoms or with hydrogen atoms that require a high degree of energy to maintain. In the presence of oxygen, however, these bonds break down. The carbon reacts with oxygen to form carbon dioxide, which is held together by a less energetic bond. The excess energy is released as heat, causing combustion. Thus large accumulations of coal and oil represent huge supplies of chemical potential energy in concentrated, transportable form. Second, carbon is the most chemically versatile of the elements, capable of forming compounds of enormous size and complexity. Plastics, synthetic fibers, insecticides, vinyl—these carbon compounds

fossil fuel
Any hydrocarbon that can be used for fuel.

(a)   (b)   (c)

Figure 20.13 Stages in the burial and compaction of swamp plant matter leading to its transformation into coal.

and others are the backbone of the modern petrochemical industry. To produce them in the necessary quantities requires the enormous supply of carbon preserved in the Earth as oil.

The United States relies on fossil fuels for 87 percent of its energy consumption. Oil (44 percent of the total) is applied mainly to transportation, natural gas (21 percent) to industrial manufacturing and space-heating purposes, and coal (22 percent) to electrical power generation. Nuclear energy and water power comprise 11 of the remaining 13 percent. Per capita energy consumption in turn correlates strongly with living standards, for the energy is spent on goods, services, and trade. It is a sobering fact that the United States, the world's highest per capita energy consumer, now imports more oil than it produces.*

Coal

coal
A combustible carbonaceous rock formed from the compaction of altered plant remains.

Coal is a brown to black rocklike substance composed of the altered remains of ancient plants that thrived in freshwater and brackish swamps. The swamps were located in protected portions of coastal plains, estuaries, lagoons, deltas, and similar low-lying regions. Water circulation was often restricted in these environments, and the oxygen at the bottom was depleted, a necessary condition for the preservation of plant matter. The Great Dismal Swamp in Virginia is probably a modern equivalent of these ancient swamps.

The transformation of dead plant matter to coal involves bacterial action, which partially decomposes the mosses, ferns, trees, and other plants into a dark brown to black residue called *peat,* the precursor of most coal. Subsequent stages are triggered by burial under younger sediment (Figure 20.13). When the peat is compressed to about a tenth of its original thickness and then heated, the volatiles are driven off, increasing the carbon content. Gradually, the peat is converted to coal, which is ranked as *lignite, bituminous,* and *anthracite* on the basis of increasing carbon content.

More than mere burial is generally required to produce anthracite; it tends to occur in regions of moderately folded and compressed rock strata (such as eastern Pennsyl-

---

*Although *production* is the commonly used term, the more appropriate word is *extraction,* as nothing is actually being produced.

vania). Figure 20.14 gives the classification of the major kinds of coal as well as their per kilogram heat yields. Anthracite and some bituminous coals have the highest heat yields. Sulfur content is another factor affecting coal quality. Released to the atmosphere during combustion, sulfur combines with hydrogen and oxygen to form pollutants that cause or aggravate respiratory disorders; it is also a key ingredient of acid rain. Thus low-sulfur coal is preferred.

The swampy conditions that favor the formation of coal were prevalent over broad expanses of the east-central United States and central Europe during the Carboniferous period, some 300 to 350 million years ago, when the continents were near the equator. North America and Europe were either joined or close together during this period; the collision between the two continental blocks produced the Appalachian Mountains. Streams issued down from these mountains onto vast deltaic coastal plains spread over what is now eastern North America and Western Europe. These plains were subjected to shallow invasions and retreats of the sea. Thus swamps were repeatedly created and destroyed, which led to the formation of layer upon layer of coal between flood-plain sandstones and shales. At about the same time, similar conditions existed farther east, along the flanks of the Ural Mountains of Russia, producing thick coal deposits there.

The United States is fortunate. When North America and Europe separated some 90 million years ago, the broad area that later became the United States retained a huge share of coal. Millions of years later, during the Triassic period, this stock was augmented by enormous deposits of lignite and bituminous coal that blanket the western plains and the Rocky Mountains states. Today, most of the nation's coal is mined in these western states, in part because of the low sulfur content and because it is found in surface or near-surface locations where it is easily extracted by open-pit or strip-mining methods.

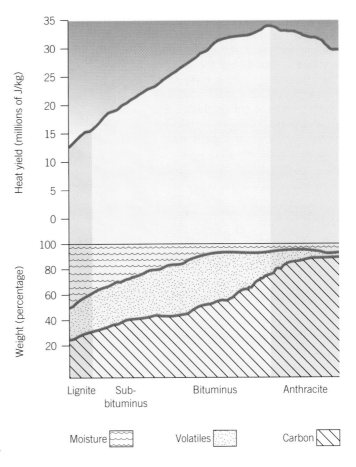

Figure 20.14 The classification of coal based on heat yield and carbon content. Coal with a high volatile content is likely to emit a higher percentage of pollutants, such as sulfur dioxide.

Petroleum

Stripped to its Greek root, **petroleum** means "rock oil." It is a complex tangle of hydrocarbon compounds found almost exclusively in sedimentary rocks. Petroleum geologists recognize three stages in the origin of the most profitable accumulations: generation, primary migration, and secondary migration.

petroleum
A naturally occurring liquid of complex hydrocarbon composition. (May include natural gas.)

Generation

Most known oil deposits were probably generated either in shallow basins within the continents, which at times had been invaded by the sea, or in depressions that lay offshore, within the continental shelves. Such modern-day basins as the Gulf of Mexico and the Orinoco River delta of South America are believed by petroleum geologists to be typical oil-forming environments. These basins are receptacles of three kinds of sediment: (1) detrital sands, clays, and silts washed in from the continents; (2) chemically and biochemically precipitated coral reefs, shell limestones, salts, gypsums, and cherts; and (3) the remains of organisms brought in by streams or that lived in the basins and collected on the sea bottom.

The organic matter is preserved in oxygen-poor portions of the basins, where decay is at a minimum, such as shallow lagoons or deltas closed off by barriers that restrict

Figure 20.15 The refining of crude oil. As pressures and temperatures are steadily increased, the products are distilled and separated according to their densities. Gaseous compounds rise while the heaviest liquids sink to the bottom. A similar though less exact process occurs during the conversion at depth of organic matter into oil and natural gas.

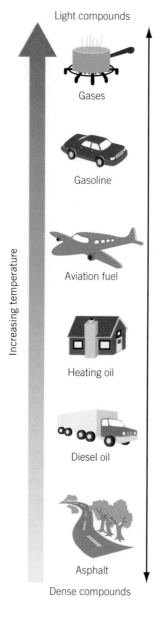

circulation. The oxygen in the water is soon depleted in this stagnant environment as organic remains accumulate in the muds. The organic matter is then slowly broken down into hydrocarbon compounds (petroleum) within these muds. Bacterial action plays an important role at first; but as the organic remains are covered by thick layers of younger sediments, the effects of compaction and the elevated temperatures that accompany burial become dominant.

What takes place is basically a natural and milder version of *cracking*—the process that oil refineries use to separate various hydrocarbons from one another. As the petroleum is heated under high pressure, the lighter compounds—those containing relatively few carbon atoms—are synthesized and rise (Figure 20.15). So as temperatures and pressures are increased, the denser residues—which contain a greater number of carbon atoms per molecule than the lighter compounds—remain behind. At least 500 meters of burial is required to produce oil of commercial value. Below about 10,000 meters, however, hydrocarbons begin to break down to pure carbon (graphite) and methane (natural gas).

Primary Migration

When first deposited, the typical organic muds within which oil is generated consist of about 75 percent water. Compacted by burial beneath younger sediments to depths of 2000 meters, that same mud, now converted to shale, holds about 10 percent water. The oil and natural gas squeezed out with the water are driven from their source sediment or rock to regions of low hydrostatic pressure, where they accumulate. This primary migration is generally upward, toward the surface, but the path is also controlled by the porosity and permeability of the sedimentary rocks. Eventually, the water, oil, and gas will become concentrated in rocks with large, continuous pore spaces. Well-sorted beach sandstones, fragmented limestone, and reef deposits made of the skeletal framework of ancient corals make good **reservoir rocks.** If impermeable barriers lie above and below such reservoirs, preventing escape, the water, oil, and gas will be trapped.

Lest anyone think migration and accumulation in a reservoir is rapid, remember that the oil and water must trickle through minute spaces in which surface tension and friction exert tremendous retarding forces. A drop of oil and water probably moves from 2 to 10 centimeters per year. At this rate, it takes millions of years for a large oil field to form, whereas the oil field itself may be depleted in a matter of decades—which is why we call oil a nonrenewable resource.

Secondary Migration: Oil Traps

Having become concentrated in reservoir rock during the primary migration stage, oil continues to migrate within the reservoir itself. This latter journey is called secondary migration. As in primary migration, the oil is driven from regions of high to low pressure, where it occupies the pore spaces between sediment grains, forming a *pool*. The location of the pool is controlled by an **oil trap,** a set of structural and/or stratigraphic

reservoir rock
Any porous and permeable rock that could contain oil or natural gas deposits.

oil trap
A structural or stratigraphic arrangement within a sedimentary basin that allows oil to accumulate and prevents its escape.

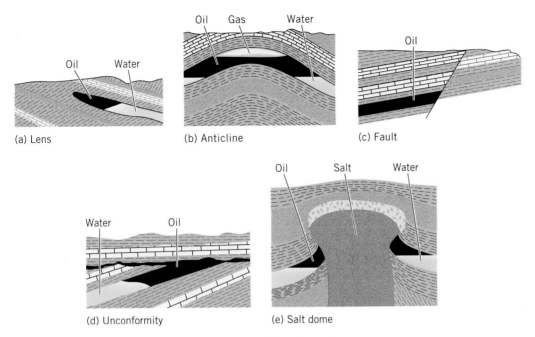

Figure 20.16 Some common occurrences of oil. Fluids migrate in a reservoir rock from a region of high to low pressure, where they are trapped and sealed by impermeable layers. The fluids within the traps are separated according to density into water, oil, and gas. Part (a) shows a stratigraphic trap, caused by a change in the permeability of the reservoir rock. Parts (b–e) show various structural traps. Of these, anticlines and salt domes are especially productive.

conditions that prevent further migration of the oil (Figure 20.16). In a *structural trap*, the location of the pool is controlled by folds, faults, and the dip of the rock strata. In a *stratigraphic trap*, location of the pool is controlled by changes in the porosity and permeability of the reservoir or neighboring rock, so the oil is no longer free to move through the strata. Often, a number of pools and traps occur in close proximity, related by similar structural and stratigraphic conditions, forming an **oil field.**

Figure 20.16(b) illustrates a typical structural trap in which the sequence of rock strata, including the reservoir rock, has been warped into an arch, or anticline. Because the apex of the arch is subjected to the least pressure, fluids are forced from the flanks to the axis of the anticline. The fluids are prevented from escaping by **cap rock,** impermeable strata that seal the oil-bearing reservoir rock. The fluids that accumulate in the reservoir separate according to density: natural gas (if present) on top, oil in the center, water below. A similar pattern exists in all oil traps.

Clearly, systematic oil and gas exploration requires detailed knowledge of the geology of the entire sedimentary basin under consideration. First, the petroleum geologist must know the various environments that existed within the basin at the time the sediments were deposited. These environments largely determine the location of source beds, reservoir rock, and stratigraphic traps, such as the gradual tapering of oil-bearing strata. Second, it helps to know the geometry and thickness of the sedimentary strata in the basin. The thicker the strata, the greater the probability of having organic matter in the sediment converted to oil and the greater the probability of trapping the oil. The geometry of the basin helps determine the pressure relations within the strata and the directions the oil will migrate. Third, the geologist should be familiar with the tectonic history of the basin; that is, the history of folding and faulting of the strata, as well as the periods of erosion followed by deposition of younger strata (unconformities). These events will determine the location of structural traps and important oil pools.

oil field
A geologically and spatially related feature containing two or more oil accumulations.

cap rock
An impermeable rock layer overlying oil-bearing reservoir rock.

Figure 20.17 The extent of the Green River formation, a vast oil shale deposit.

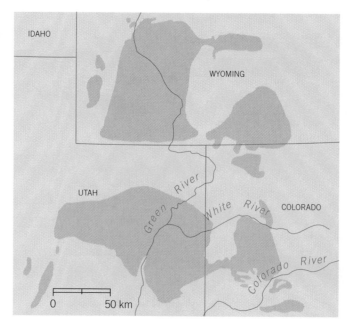

Oil Shales and Oil Sands

Nearly all of the world's oil of current commercial value is pumped as liquid from buried pools. Similar quantities exist at or near the Earth's surface but in forms that currently make them more costly to extract, both in financial and environmental terms.

oil shale
A fine-grained sedimentary rock from which oil or gas can be distilled.

Oil shales are fine-grained rocks consisting of clay minerals and silt-sized quartz grains. The spaces between grains are microscopic, making the rock highly impermeable. Trapped within these spaces is a solid, combustible hydrocarbon substance called *kerogen*. When heated, the kerogen breaks down and yields oil. We may think of oil shales as source rocks from which oil has not migrated.

Oil shale, like most hydrocarbon deposits, forms in stagnant lakes, in oxygen-deficient portions of shallow deltaic marshes and lagoons, or in the deeper oceans. These conditions preserve the accumulation of organic matter in the bottom muds. Typical are the oil shales of the Green River formation of Colorado, Utah, and Wyoming (Figure 20.17). Though variable in composition, they constitute the world's largest single oil shale accumulation, holding the equivalent of 1.8 trillion barrels of oil, or about twice the world's reserves of conventional oil. There are important limitations on the extraction of oil from the Green River shales, however. Chief among them is that the extraction process requires huge quantities of water in a region where it is already scarce and divided among competing groups: ranchers, industry, and urban communities. Extraction also requires considerable energy—far more than conventional oil. The other obstacle is what to do about the enormous piles of leftover shale and the inevitable environmental damage this operation would cause to the ecosystem. All these problems make the extraction costs prohibitive at the present time.

oil sand
Broadly defined as a porous sand deposit from which petroleum can be extracted.

Oil sands are sediments impregnated with viscous, asphaltlike crude oil. The oil probably seeped upward through porous reservoir rock to the surface, as in the famous La Brea tar pits of Los Angeles, California. Probably the world's best known tar sand deposits are located in Alberta, Canada. It appears that the tars, of Cretaceous age, were derived from oil trapped in a far more ancient Devonian reef. The reef was exposed by erosion, and much of the oil evaporated. The asphalt residue was then carried away and deposited with conventional sands.

Oil sands in general represent very large potential reserves. Although it is easier to extract the oil from them than from oil shales, the process is still difficult, requiring large amounts of energy and endangering fragile environments.

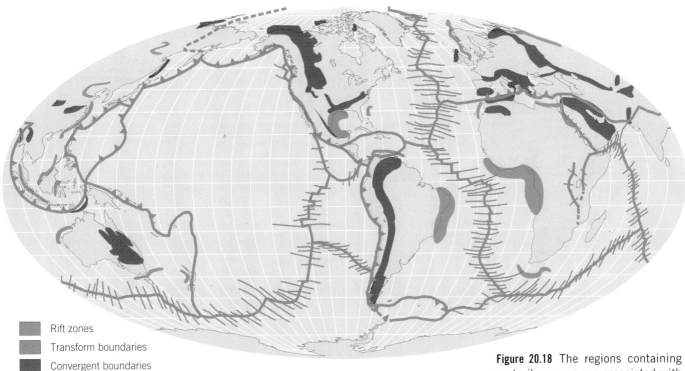

Rift zones
Transform boundaries
Convergent boundaries

Figure 20.18 The regions containing vast oil reserves are associated with present or past plate boundaries. (Adapted from David G. Howell, Kenneth J. Bird, and Donald L. Gautier, "Oil: Are We Running Out?" *Earth,* March 1993, pp. 26–33.)

Plate Tectonics and Petroleum

Earlier in this chapter, we described the close connection between plate tectonics and the concentration of valuable metals in the crust. We find a similar connection between plate tectonics and petroleum accumulations. Indeed, in each case, our knowledge of the occurrence of these deposits is incomplete without reference to the plate tectonic history of the region of interest.

As Figure 20.18 shows, the divergent, convergent, and horizontal sliding motions of lithospheric plates over the globe provide the structural setting for sediments to accumulate and for oil generation and preservation. The embryonic stage in the development of an ocean basin consists of a narrow system of rift valleys between separating segments of continental crust. Seawater circulation is restricted in the rift, which leads to deposition of thick salt deposits and other sediments. The restriction of seawater also leads to the accumulation of organic remains in stagnant portions of the rift, which are contained by faults and stratigraphic traps.

In later stages of ocean basin development, the creation of new oceanic crust and sea-floor spreading cause the continents to be widely separated. Because the salt layers rest on old oceanic crust moved aside by sea-floor spreading, they are found along the flanks of the continents, far from the rift valley where they were precipitated. Rivers flowing down from the continents to the sea deposit thick wedges of sediments along the continental margins, burying the ancient salt layers. The salt is less dense than the overlying sediment and easily deforms in response to pressure. As sediments accumulate above it, the entire basin compacts. The salt is mobilized into streamlined diapirs, similar in shape to the granitic magmas that rise in the crust. The diapirs rise and pierce the sediment as salt domes. The salt domes act as traps for petroleum derived from the organic-rich sediments of the original rift zone and from the sediments that accumulate on the continental margin. The oil deposits of the Gulf of Mexico and off the Nigerian coast formed in this manner as a result of the widening of the Atlantic Ocean.

Major oil fields are also found in those regions where plates collide—near subduction zones that border continents and the marginal seas between island arcs and continents. Oil-bearing sediments accumulate in the tectonically active basins of these

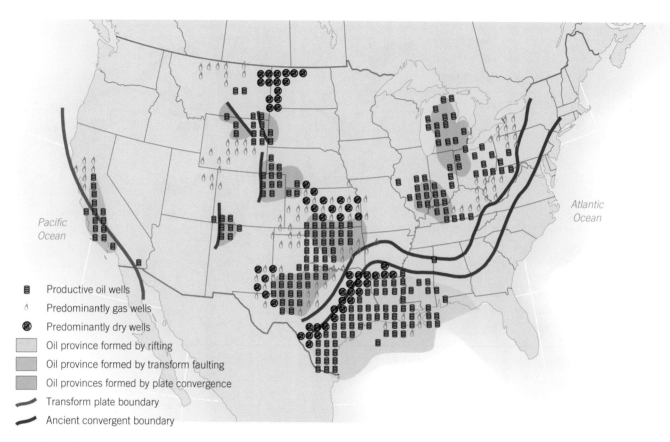

Productive oil wells
Predominantly gas wells
Predominantly dry wells
Oil province formed by rifting
Oil province formed by transform faulting
Oil provinces formed by plate convergence
Transform plate boundary
Ancient convergent boundary

Figure 20.19 Oil and gas provinces correlate closely with past and present plate tectonic patterns. (Adapted from David G. Howell et al.)

regions, where conditions favor the development of many structural and stratigraphic traps. Figure 20.19 shows that U.S. oil and gas fields correlate closely with past and present tectonic patterns. Notice, for example, that California's productive oil fields match up with the southern segment of the San Andreas fault.

 The Petroleum Supply

In view of the importance of economic resources, it is fair to ask whether we are in danger of running out of them. The answer to that question is different for each kind of raw material. Geological scarcity, current and projected rates of extraction, and technological advances must all be considered. We will discuss the future of one very important resource, petroleum. But the principles that we will introduce apply to most other resources.

Three factors should be kept in mind when we assess the supply of petroleum resources and reserves. The first is that only 30 percent of what is in the ground is currently recoverable. Much of the oil clings to rock particles and clay minerals by surface tension. Various recovery methods such as blasting, high-pressure water pumping, and the introduction of detergents are employed to work some of the bound oil loose, but they are of limited effect. The second factor is that there are many small oil fields, but relatively few of them contain enough oil to be extracted profitably. Most of the world's recoverable oil is concentrated in the oil fields of the Middle East. The third factor is that the United States contains only about 3 or 4 percent of the world's recoverable oil and about 6 percent of its natural gas. As mentioned previously, we import more oil than we produce.

Independent information on the abundance of remaining oil resources and reserves is often difficult to obtain. Much of it is held confidentially by oil companies that have sponsored explorations, and they are not about to release it to competitors; in some nations, information concerning oil reserves is a state secret. Also, there is much debate over the scientific and statistical methods of measuring resources and reserves. Therefore, it is not surprising that estimates of the abundance of recoverable oil vary widely. The estimate of three U.S. Geological Survey specialists is shown in Table 20.3. According to them, the world has about a 70-year supply of oil at current consumption rates—much less if we consider population growth and expanding economies, much more if stringent conservation measures are taken.

Until about 1800 and the advent of the Industrial Revolution, humans relied mainly on renewable energy resources: crops, timber, water, animal, and wind power—all of them extensions of solar energy. Emerging technologies required ever-more concentrated and portable energy, and hence there was a conversion to nonrenewable forms—first coal and, increasingly in the twentieth century, oil. Figure 20.20 shows that these fuels passed through a period of peak utilization as a percentage of total energy consumed and were then replaced by more abundant, less expensive, or more convenient fuels. The projections are that oil production will decline sometime early in the next century, to be replaced by natural gas, perhaps supplemented by oil shale. This will be an interim step, however, for natural gas faces the same basic arithmetic of depletion as oil production. Of course, if all else fails, civilization can revert back to coal as its primary energy source. Coal spurred the Industrial Revolution. It is still an exceedingly important fuel, and technologies exist that can convert coal to fluid form and make it more attractive for uses other than electrical power generation. There are still several centuries worth of it left in the ground. But many scientists warn us that heavier reliance on coal would be unwise. It would require strip mining on a gigantic scale; productive farmland and countryside would be destroyed. The sulfur released from coal would pollute groundwater, making surrounding regions uninhabitable. The burning of the fuels would intensify acid rain, as well as help induce global warming. Respiratory ailments would rise. After the fuel is spent, we would be left with disposing of massive quantities of ash. All these coal-induced problems exist today, but on a far smaller scale. Such drastic consequences need not occur, however. There have been significant technological advances in site reclamation, containment of contaminants, and air-quality control. Most mining companies in the United States and Canada have attempted to comply with strict state and federal regulations. What is required is a commitment to maintain high standards.

We should keep in mind that in coming decades, most of the adverse impact of coal and petroleum use is likely to come from developing nations. For example, as of 1991, the per capita carbon emissions to the atmosphere for the United States were nearly 20 tons a year, or ten times China's. But China has huge coal reserves. If its economy expands at close to its present rate over the next 25 years, China could well become the largest source of emissions. Clearly international treaties are needed to moderate the environmental impact of energy use.

There are a number of other energy sources on the horizon, some old and some new, waiting for their chance to replace or supplement fossil fuels. *Nuclear fission,* the release of energy by the splitting of uranium or related atoms, is capable of generating electrical power. Although the fuel is fairly abundant, it is not portable because of the dangerous radioactivity it emits. For this reason, it is doubtful that anyone would want to run a car with a nuclear reactor powering the engine. Also remaining to be resolved to

Table 20.3 Estimates of Current and Future World Oil Supply (in billions of barrels)

	United States	World
Production (through 1990)	55	650
Reserves (1991)	25	~950
Undiscovered recoverable oil	~40	~500
Anticipated reserve growth	~20	Unknown
Total recoverable oil	~140	~2100
Total oil left for the future	~85	~1550

Figure 20.20 This graph projects the usage of various energy resources into the twenty-first century. As the use of wood declined in the nineteenth century, it was replaced with coal, which later peaked, but is now in decline. Oil has replaced coal, but is expected to decline sometime early in the next century. An important question we must answer is: what will replace oil? (Adapted from David G. Howell et al.)

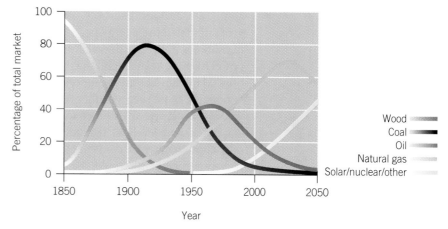

Figure 20.21 Nuclear fission in a breeder reactor relies on the conversion of uranium-238 into the highly fissionable plutonium-239 by neutron capture. Each step of plutonium fission generates energy. By this method, the nuclear industry can rely on the uranium-238 isotope for fuel rather than on the much rarer uranium-235 isotope.

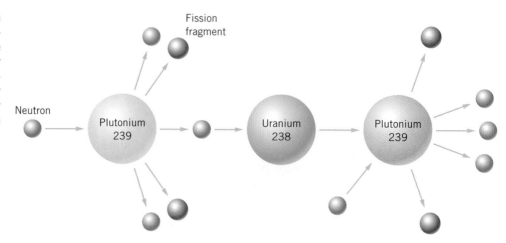

the satisfaction of experts and the public at large are important questions of plant safety and radioactive-waste disposal (see Chapter 16). In addition, for nuclear fuel to be viable in the long term, the nonfissionable uranium-238 isotope must be converted to highly fissionable plutonium in breeder reactors (Figure 20.21). Plutonium is the basic component of atomic bombs, and its production obviously raises national security as well as safety problems.

Alternative Energy Resources

Nuclear fusion, the conversion of hydrogen into helium, is potentially a very attractive power source; the Sun operates on the same principle. Tremendous energy is liberated in this manner, and hydrogen fuel is cheap—you can extract it from water. The problem is that the fusion reaction is difficult to control for more than a split second. At our present state of technology, it takes more energy to contain the reaction than the reaction itself yields. Physicists have been trying to control fusion since the early 1950s, and most experts predict that they will not succeed for at least another 30 years.

Watching volcanoes erupt, geysers spout, and black smokers vent suggests a logical question: Can we harness the Earth's **geothermal energy** to generate electrical power, warm homes and factories, and grow crops within enclosed, climate-regulated spaces? The answer is yes. Japan, Iceland, New Zealand, San Francisco, and Italy currently have geothermal systems in operation for these purposes. A practical geothermal system needs an accessible source of internal heat—a body of magma or hot rock—within 15 kilometers of the Earth's surface. Then engineers must be able to install a plumbing system to tap the heat (Figure 20.22). Regions of active volcanism (such as Hawaii) and tectonism (subduction zones, rift valleys, and hot spots) are the most promising to explore. In the long run, the energy yield of localized geothermal energy is probably comparable to the yield of an average nuclear power plant.

Solar energy may be the most promising alternative fuel on the near horizon. It is clean, safe, and inexhaustible. The sunlight that falls on the United States, for example, generates far more energy per year than we use by employing conventional fuels. Many new homes in the Southwest are partially solar-powered, and some are completely solar-powered. And solar energy has potential beyond space heating and hot-water generation. Many devices are being tested as steady sources of electrical and machine power. It remains to be seen whether solar power can generate the energy the world needs at prices competitive with other sources.

At current levels of technology, none of these alternative energy sources matches oil's combination of convenience, portability, accessibility, and concentrated energy. However, their status could change drastically as oil grows more expensive and the processing and delivery systems of the alternative energy sources grow more efficient. Costs may drop dramatically as demand increases and economies of scale are reached.

geothermal energy
Hot steam from the Earth's interior useful for generating power.

Figure 20.22 This geothermal energy field in northern California used to supply electricity to San Francisco. The plant tapped the stream generated by shallow hot rock or magma. The Earth's internal heat has great potential as an alternative energy resource.

STUDY OUTLINE

I. THE ORIGINS OF MINERAL RESOURCES

  A. Geologists are concerned with the origin and occurrence of mineral and energy **resources** that now or in the future may be profitably extracted from the Earth.

    1. A concentration of metallic or nonmetallic minerals at a specific location in the crust is a **mineral deposit.** Coal, oil, and natural gas are **hydrocarbon deposits.**

    2. A **reserve** is an identified resource that can be mined at a profit under current market conditions. Most resources are either currently not profitable or are as yet undiscovered.

    3. A **renewable resource,** such as running water, is regenerated within time spans useful to human beings. A **nonrenewable resource,** such as petroleum, is regenerated over longer time spans.

  B. A mineral deposit of economic value contains **ore.** To be profitable, an ore must be suitably concentrated and in readily extractable form. Extraction technology, political considerations, and market conditions also affect profitability.

    1. Mining is the first phase of a resource cycle that also includes manufacturing, use, and disposal. Each phase of the cycle has undesirable side effects.

    2. Mineral deposits are formed by processes occurring at depth and on the surface. Plate tectonics plays an active role in the former. Metallic sulfide ores are concentrated at mid-ocean ridge rift valleys by hydrothermal

solutions. They are swept up onto the continents via plate subduction. Hydrothermal processes in subduction zones also produce gold, copper, mercury, and other metals.

3. Magmatic processes involving mantle-derived mafic magmas produce heavy-metal concentrations such as chromium.

4. In subduction zones, late-stage granitic fluids form pegmatites that contain gem-quality minerals, uranium, gold, and silver. These fluids also produce fine-grained **disseminated deposits** of ore. Contact metamorphism and contact metasomatism are also important ore- and gem-forming processes.

C. Metals and other valuable minerals are concentrated in the crust through weathering, groundwater action, sedimentation, chemical precipitation, and organic activity.

1. **Residual ores** such as bauxite and pisolite form in humid tropical climates through weathering and groundwater leaching.

2. **Secondary enrichment** involves the leaching of soluble metals from surface rocks under oxidizing conditions and precipitation of the metals beneath the water table under reducing conditions.

3. **Placer deposits** are mechanically concentrated native metals and valuable minerals that are weathered from rock outcrops, transported by streams, waves, or currents, and deposited wherever the water loses velocity.

II. ENERGY RESOURCES

A. Petroleum, coal, and natural gas are known as **fossil fuels** because they are composed of the partially decomposed remains of once-living organisms. When oxidized, these organic compounds release heat. Fossil fuels account for 87 percent of U.S. energy consumption, a figure that correlates closely with the high U.S. living standard.

B. Coal forms by the burial of plant matter under reducing conditions in swampy environments that occur in coastal plains, lagoons, and deltas. Most of the coal found in the United States and Europe formed under swampy conditions caused by tectonic plate collisions during the Carboniferous period some 300 to 350 million years ago.

C. Petroleum is a fluid composed of hydrocarbon compounds found almost exclusively in sedimentary rocks. Profitable accumulations generally form in three stages.

1. During generation, the organic matter accumulates in an oxygen-deficient part of a sedimentary basin, where it is buried and compacted.

2. During primary migration, the oil is driven by pressure from source regions to regions of low pressure within the basin, where it is concentrated in a porous and permeable **reservoir rock.**

3. During secondary migration, the oil, water, and natural gas journey within the reservoir rock to a location where conditions allow it to accumulate and form a pool. The location of the pool is controlled by an **oil trap.** An impermeable **cap rock** prevents upward migration of the oil and seals the reservoir rock. A group of traps and pools constitutes an **oil field.**

4. **Oil shales** are fine-grained deposits of kerogen, a hydrocarbon compound that, when heated, breaks down to form oil. **Oil sands** are sediments impregnated with asphalt. Environmental problems are associated with extracting oil from shales and sands.

5. There is a strong correlation between plate tectonic setting and the occurrence of petroleum. Processes at convergent, divergent, and transform boundaries provide both the environmental settings conducive to the formation of oil and the structural traps that hold the oil pools.

6. Only 30 percent of the oil in the ground is currently recoverable; most of it is in the Middle East. The United States is the world's largest consumer of oil; we import more than we produce.

7. Coal remains an abundant source of power, but the environmental problems connected with its extraction and use are great.

D. Nuclear fission remains a viable energy source, but problems of safety, disposability, and national security also remain.

E. There are three possible future alternative energy sources. Nuclear fusion power is clean and plentiful, but physicists will probably not be able to harness it in the near future. **Geothermal energy** and solar power have yet to be tested on a large scale.

STUDY TERMS

cap rock (p. 557)
coal (p. 554)
disseminated deposit (p. 549)
fossil fuel (p. 553)
geothermal energy (p. 562)
hydrocarbon deposit (p. 543)
mineral deposit (p. 543)
nonrenewable resource (p. 544)
oil field (p. 557)
oil sand (p. 558)
oil shale (p. 558)

oil trap (p. 556)
ore (p. 544)
petroleum (p. 555)
placer deposit (p. 552)
renewable resource (p. 543)
reserve (p. 543)
reservoir rock (p. 556)
residual ore (p. 550)
resource (p. 543)
secondary enrichment (p. 551)

CRITICAL THINKING QUESTIONS

1. Using the discussion of resources and reserves as a guide, how would you classify your checking account? Your charge card? Your projected lifetime earnings?

2. Is it possible to run out of a renewable resource? Explain. Can a renewable resource be turned into a nonrenewable one?

3. Discuss in general terms the environmental problems engendered in each of the following phases of resource use: mining and smelting, product manufacturing, product use and disposal.

4. Explain how processes related to plate interactions can create environments conducive to the origin and accumulation of oil. Many of the oil deposits shown in Figure 20.18 located far from past or present plate boundaries may nevertheless have formed by plate interactions. Explain.

5. We have mentioned that our nation's high per capita energy consumption correlates with its high living standard. What other factors may account for the high energy consumption of the United States compared with, say, Western Europe's?

6. What steps may a nation take to prolong the lifetime of oil reserves? What steps has the United States taken? Is it necessarily a disadvantage to depend on foreign sources for fuel and mineral supplies? Explain.

Planetary Geology

The objective of this chapter is to explore the geological similarities and differences between the Earth and the other planets—in particular, the terrestrial planets, which are the Earth's closest neighbors and about which we have the best information. These planets are interesting in their own right, of course, but comparing them with the Earth offers a unique perspective from which to view the Earth's origin, structure, and functioning.

Scientists have traced connections between the Earth and other objects in space that go far beyond the Solar System. The very elements of our bodies and those that comprise the rocks, water, and air of the Earth were manufactured in the interior of stars that exploded long ago. These explosions seeded space with the elements from which the Sun and the planets evolved. The manufacture of the elements within stars is, in turn, related to a long chain of events that can be traced back to the origin of the universe itself. To explore these connections and to provide context, we will begin this chapter at the cosmic beginning. The first step is to get a sense of where the Earth fits into the scale of the universe.

In relation to the dimensions of the **universe,** the Earth, our home, is a tiny scrap of rock and iron, bound to an undistinguished star in the suburbs of a galaxy that is like billions of others. Not until recently in human history have we been able to leave the Earth. And although our travels have vastly improved our education, from the point of view of the astronomical distances covered, they haven't really amounted to much—nor will they ever.

To understand why, mentally shrink the Sun to the size of a medicine ball (about 1 meter in diameter) and shrink the other objects and distances of the Solar System

universe
The totality of space and time—past, present, and future.

◄ This huge volcano is on Venus, one of the terrestrial planets that are the Earth's closest neighbors. The geological similarities and differences between the Earth and the terrestrial planets are the main focus of this chapter.

567

Figure 21.1 NASA Hubble Space Telescope image of a remote cluster of galaxies. The light from these galaxies takes so long to reach the Earth that this image shows the universe when it was two-thirds its present age.

galaxy
An enormous collection of stars held together by gravitational attraction.

accordingly. On this scale, the Earth is the size of a pea, and the Moon, one-quarter that size, is located about 25 centimeters away. Both are about a football field (91 meters) from the medicine ball Sun. Baseball-sized Jupiter is the largest planet, and Pluto, the most distant, circles the Sun in an orbit of 3.86-kilometer radius. In this scale, the nearest star to the Sun is 26,828 kilometers away! Our Sun would not be visible if viewed from that star with the naked eye; it is too dim. The physical isolation of the Solar System defies the imagination, and yet the Solar System is part of something much larger.

The Sun is one of a hundred billion stars belonging to a disk-shaped **galaxy** called the Milky Way. It takes a light beam, whose speed is 3×10^5 kilometers per second, 100,000 years to cross the diameter of this island universe. Since it is a basic consequence of Einstein's theory of special relativity that no object can travel as fast as the speed of light, it would take a future space vehicle a longer time to cover that same distance. Yet the Milky Way is a rather ordinary star collection; there are probably 20 billion other galaxies in a variety of shapes and sizes. Now scale down the universe so that the Milky Way is the size of a penny. The billions of galaxies are then spread through space at an average of one every 3 meters. If it takes light 100,000 years to cross one of these pennies, we can see why our travels in space will remain limited, notwithstanding the imaginations of science fiction writers.

The Expanding Universe

Though the universe lacks well-defined boundaries, there is compelling evidence that it is expanding, swelling like dough in a gigantic oven, with the galaxies like raisins in the dough. The evidence for the expansion is provided by analysis of light signals from galaxies so distant they appear as mere points in the night sky (Figure 21.1). Typically,

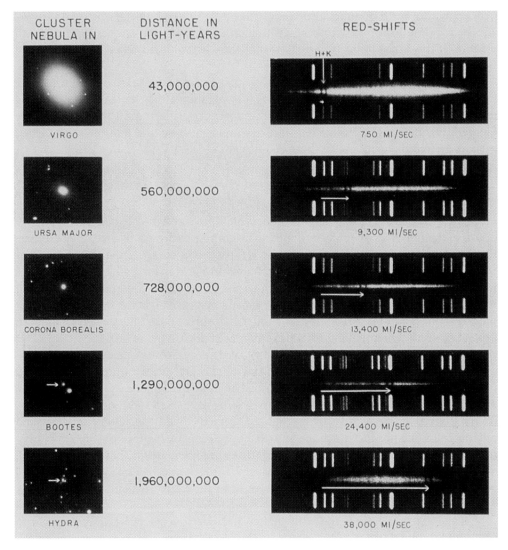

CLUSTER NEBULA IN	DISTANCE IN LIGHT-YEARS	RED-SHIFTS
VIRGO	43,000,000	750 MI/SEC
URSA MAJOR	560,000,000	9,300 MI/SEC
CORONA BOREALIS	728,000,000	13,400 MI/SEC
BOOTES	1,290,000,000	24,400 MI/SEC
HYDRA	1,960,000,000	38,000 MI/SEC

Figure 21.2 The relation between red shift and distance for nebulae (interstellar gas clouds) belonging to various galaxies. The left column shows visual images of the nebulae arranged in order of increasing distance. The middle column indicates the calculated distances from the Earth of the nebulae. (Note: One light-year equals about 9.6 trillion kilometers.) The right column shows the observed red shifts of each nebula. Arrows indicate shifts for calcium spectral lines H and K as compared with the solar reference spectrum. The more distant the galaxy, the greater the red shift. The velocities of the galaxies, which are proportional to the red shift, are shown below each image.

these signals are recorded electronically or on sensitive photographic plates. Often, the plates must remain exposed for days to collect enough light to record the images of the galaxies.

The light signal from a galaxy is separated into its component wavelengths by passing it through a prism in a device called a *spectroscope*. The result is a broad play of colors called a *spectrum*. The spectrum is produced by the glowing gases within stars and dust clouds of the galaxy. Each gaseous element emits a series of characteristic wavelengths, or spectral lines, that occupy specific positions within the spectrum and no other. Thus spectral lines are optical fingerprints that serve to identify the elements and the composition of the galaxy (Figure 21.2). When the spectra of distant galaxies are compared with the spectrum of our Sun, which is used as a reference, we discover an important fact: invariably, the spectral lines of the galaxies are displaced toward the long-wave end of the solar spectrum—the red end. This phenomenon is known as the *red shift.*

The red shift is strong evidence that the galaxies and the Earth are moving apart, because it is the expected effect of a wave source that recedes from the observer. The Earth receives less light energy per second from a galaxy that is receding from it, which shows up as a shift of wavelengths toward the less energetic part of the reference spectrum—that is, the red. Had the galaxy been approaching the Earth, its spectrum would have shifted toward the violet (short-wavelength) end. Because there is no reason to believe that the Earth occupies a special vantage point in the universe, the galaxies must also be receding from one another in a fashion similar to what we here on the Earth observe. If this assumption is correct, then we are left with no other conclusion than that the universe is expanding.

The magnitude of a red shift is directly proportional to the velocity with which the source is receding from the observer. Thus it is significant that the galaxies displaying the greatest red shifts are also found by other means to be farthest from the Earth (see Figure 21.2). When we translate red shifts into velocities, we find that the most distant galaxies are receding at close to the speed of light.

The Big Bang Theory

big bang theory
A model of the evolution of the universe that postulates its origin from a hot, dense mass that expanded rapidly and cooled.

From the evidence of expansion, astronomers have proposed the **big bang theory** of the origin of the universe. By projecting the expansion rate backward in time, they estimate that 15 to 20 billion years ago, the universe was packed into a space the size of the Solar System. They have calculated that temperatures inside this incredible "primeval egg" were on the order of trillions of degrees Celsius. Under these conditions, nothing but energy in the form of intense electromagnetic radiation existed. For reasons still debated, the egg burst and splattered the radiation in all directions. The universe has been expanding from this initial big bang explosion ever since.

With expansion came cooling. At billions of degrees Celsius, subatomic particles—electrons, protons, neutrons—condensed out of the fireball like droplets from a fog (energy being converted to matter, as described by Einstein's famous equation, $E = mc^2$). At 100 million degrees Celsius, tiny negative electrons hooked up in orbit around positively charged protons to form the lightest element, hydrogen. Other linkages and nuclear collisions produced helium, in which a nucleus of two protons and two neutrons is surrounded in orbit by two electrons. Astronomers tell us that by this time, all of 3 minutes had elapsed since the big bang. Hydrogen and helium comprise over 90 percent of the matter in the universe, and most of these two elements that exist today were created during the big bang.

As the vast expanding stream of matter and electromagnetic waves continued to cool, random disturbances—perhaps local electrical storms or heat flashes—upset the homogeneity of the universe. Matter gradually became separated into enormous clouds of hydrogen, helium, and subatomic dust, which by this time had cooled to icelike particles. These clouds were the primitive galaxies, wandering randomly through space. Throughout the billions of years since the big bang, gravitational attraction within the cold galactic clouds has led to the formation of stars, huge nuclear pressure cookers, where the hydrogen and helium have fused to form heavier elements, with light and heat given off as by-products. Near the stars, concentrations of matter too small to sustain nuclear reactions formed **planets** and other satellites.

planet
A large body that revolves around a star and is incapable of generating light internally.

Although the big bang explosion occurred 15–20 billion years ago, scientists can still detect the remaining noise. By training their radio telescopes in all directions, they find that space is permeated by a faint but unmistakable background signal, a sort of static that corresponds to a temperature of 3 kelvin (-270 °C). The existence of just such a signal, the remnant of the initial fireball that expanded and cooled, had been predicted by advocates of the big bang theory long before the discovery of the signal.

The Solar System

The Sun has been referred to as an undistinguished star in the sense that its vital statistics—mass, size, surface temperature, light output—are about average. There are probably billions of stars in the Milky Way like it, although the Sun is somewhat unusual because, as far as we know, it occurs alone rather than in a pair or a cluster, as do most stars. Average or not, by human standards, the Sun is a formidable piece of work. It

Figure 21.3 A cross section of the interior of the Sun and its various surface features.

shines with the power of a 380-million-billion-billion-watt bulb. This light is radiated in all directions. Fortunately, the Earth is far enough away from the Sun, and so tiny in comparison to it, that only a minute percentage of the Sun's light reaches the Earth.

The Sun came into being when an immense cloud of cold dust swirling through space started to fall in on itself. What triggered the collapse of this stellar nebula is uncertain; shock waves generated by exploding stars is a favored hypothesis. Whatever the cause, once started, mutual gravitational attraction pulled the dust relentlessly inward. And as the dust (composed of about 80 percent hydrogen, 20 percent helium, and traces of heavier elements) sped inward, it gained velocity. In the center of the cloud, hydrogen atoms collided with enough energy to engage in nuclear fusion reactions, and the Sun was born. Specifically, hydrogen was converted to helium in a series of elaborate steps. These reactions in turn liberated enough energy in the form of heat, light, and other electromagnetic waves to counterbalance the relentless gravitational pressure and maintain the Sun at its present size.

That is where the situation stands today: nuclear reactions in the Sun's core exactly balance gravitational pressure. To sustain this balance, 657 million tons of hydrogen are converted to 652 million tons of helium each second: the difference, 5 million tons of matter per second, goes off as energy. The Sun has been burning this hydrogen fuel for at least 4.6 billion years and will continue for at least another 4 billion years or so—it is a middle-aged star.

Notice in Figure 21.3 that the Sun's interior is divided into well-defined zones. By the time the energy goes from the core region, where nuclear reactions occur, to the surface, temperatures lower drastically. The surface of the Sun is a mere 6000 kelvin (5727 °C). It is never at peace, as indicated by the unceasing display of prominences, sunspots, coronas, and streams of charged particles (solar wind) ejected into space.

As it turns out, the Earth is locked into a star—the Sun—whose relatively low mass makes it incapable of carrying out fusion reactions beyond the conversion of hydrogen into helium. More massive stars are capable of using helium and heavier elements to fuel nuclear reactions in their centers. Indeed, this is how the elements up to iron on the periodic table of the elements are created—by the sequential fusion of helium and ever more massive elements within massive stars. But each time a more massive element is used as a fuel, more energy is required to make the reaction work, so that the net energy yield of the reaction is less. There comes a crossover point—reached at about the element iron—beyond which the energy liberated by the nuclear reaction is insufficient to prevent the gravitational collapse of the star. A split second of unimaginable violence follows, during which the elements heavier than iron are fused, and the interior of the star is spewed into space in a *supernova* explosion. It is among the most spectacular events in the universe.

About 0.2 percent of the Sun consists of elements heavier than hydrogen and helium. Because the Sun lacks the mass to create these elements, they must have formed in the interiors of older, more massive stars. The Sun, therefore, is a recycled star, its original dust cloud a cosmic recipe of countless supernova explosions, the contributions of anonymous stars long buried in dark space. The other bodies of the Solar System are also built from these stellar contributions. The same is true of the carbon, nitrogen, sulfur, iron, magnesium, and sodium of the human body, all of which the Sun lacks the power to create. This means that there are definite if not easily traceable connections within each individual to objects very far away and to events that predate the birth of the Sun.

The Planets

In terms of mass and size, the Sun dominates all the other bodies of the Solar System. For example, 342 Earths could be strung like beads around the circumference of the Sun. The Sun contains 99.86 percent of the Solar System's mass, and it is this mass concentration that binds the system together by means of gravitational attraction. This attraction keeps the components in revolution about the Sun. The remaining 0.14 percent of the mass of the Solar System is divided unevenly among the following bodies:

terrestrial planets
The Earth and the planets similar in size and composition to the Earth: Mercury, Venus, and Mars.

asteroid
One of several thousand small planetlike bodies of rock and iron that revolve around the Sun.

1. The relatively small, dense, rocky **terrestrial planets** closest to the Sun (Mercury, Venus, Earth, Mars); their natural satellites (or moons), and a belt of **asteroids,** which are small, planetlike bodies in orbit beyond Mars.
2. The large, low-density Jovian planets farthest from the Sun (Jupiter, Saturn, Neptune, Uranus, and their satellites), whose outer layers are composed mostly of frozen or liquid hydrogen, helium, ammonia, and methane.
3. Pluto, the farthest planet, of which little is known. It is a relatively light body having a density of about 2 g/cm^3 and a moon half as large as itself.
4. *Comets.* The Dutch astronomer, Jan Oort, showed in 1950 that comets form a diffuse spherical cloud located in the far reaches of the Solar System, about 40,000 times the distance of the Earth to the Sun; it is now called the Oort cloud. Like moths circling a flame, some comets periodically swoop toward the Sun, only to retreat again to the dark outer fringes. Halley's comet was intensively studied via space probes during its 1986 visit to our part of the Solar System. It turned out to be a disintegrating mass of rock and ice, with a nucleus about the size of the island of Manhattan and trailing matter in a long tail.

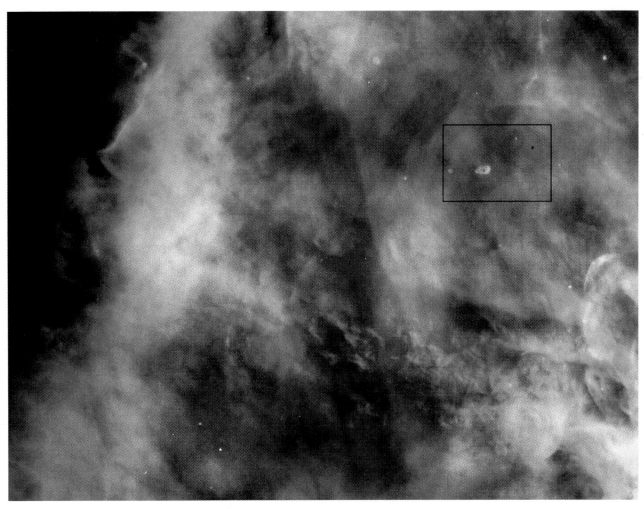

Origin of the Planets

There are regions in space today where astronomers can see what appear to be new stars forming (Figure 21.4). Surrounding the young stars are glowing clouds of gases and dust particles, flattened into wide orbiting disks, that are maintained by the stars' strong gravitational attraction. Mixed in with the overwhelming abundance of hydrogen and helium of the cloud are minute percentages of the same elements that compose the planets of our Solar System. There is every reason to believe that a similar disk of gaseous matter surrounded our embryo Sun in the past. Nearly all the planets and their satellites orbit the Sun in what is very close to a common plane—a feature we would expect if they condensed from the same orbiting disk.

Planetary geologists believe that solid particles condensed out of the gaseous disk surrounding the primitive Sun in a sequence that was temperature-controlled. Closest to the Sun came the light metals, such as aluminum, titanium, calcium, and their oxides (Figure 21.5). Somewhat farther from the Sun followed iron, nickel, and silicate minerals; included among the latter were the hydrous silicates, those that hold chemically bound water within them. These metals and silicates constituted a tiny percentage of the dust cloud, but the close-in terrestrial planets and asteroids are composed of them. At still greater distances, with temperatures approaching 200 kelvin (-73 °C), substances we normally think of as gaseous were converted to ice, and they formed the distant Jovian planets. These planets consist mainly of layered hydrogen, ammonia, and methane ice wrapped around tiny rock-iron cores. Table 21.1 lists some of the basic properties of the planets.

Figure 21.4 A stellar nebula from which primitive solar systems condense. The disklike structure in the rectangle at the upper right is a gaseous ring encircling a sun. Spectral analysis helps determine composition: Blue codes the presence of oxygen; red, the presence of nitrogen.

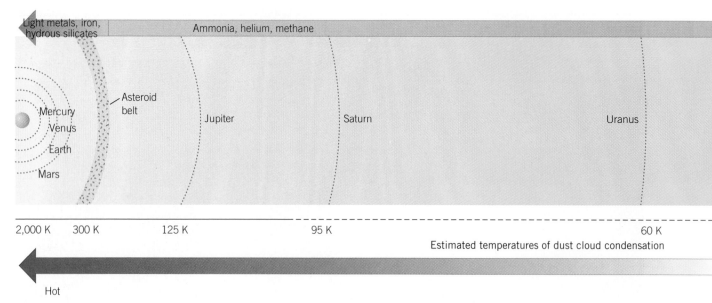

Light metals, iron, hydrous silicates

Ammonia, helium, methane

Mercury
Venus
Earth
Mars
Asteroid belt
Jupiter
Saturn
Uranus

2,000 K   300 K   125 K   95 K   60 K

Estimated temperatures of dust cloud condensation

Hot

Figure 21.5 The orbits of the planets with respect to the Sun. The arrow above shows the compositional variation of the planets; the arrow below shows the temperatures at which the primitive dust cloud condensed to form the planets. The terrestrial planets were formed of metals and silicates that condensed close to the sun at high temperatures, while the Jovian planets are made of ammonia, helium, and methane that condensed far from the Sun, where temperatures were low.

Support for the condensation hypothesis comes from samples of the Solar System in primitive condition in the form of meteorites, hunks of rock and iron that have fallen from space. Mostly derived from the asteroid belt, they are relics left over from the birth of the Sun and the planets (see Figure 21.5). Their age, determined from the analysis of the decay of radioactive elements found within them, shows that they were formed 4.6 billion years ago—the probable age of the Solar System. The structure of the most common type, the *chondrites*, is an aggregate of condensed, pea-sized mineral matter—the original hot rain of the cooling dust cloud. In addition, the chemical composition of an important subclass, the carbonaceous chondrites, closely matches the composition of the Sun minus its gaseous hydrogen and helium. This analysis strongly suggests that the chondrites and the Sun have a common origin; by extension, so do the Sun and the other rocky bodies of the Solar System.

The largest asteroid is less than one-third the diameter of the smallest planet, which leaves open the question of how the planets evolved from the droplets of the original dust cloud into the large bodies they are today (Figure 21.6). One popular hypothesis favors the collection through gravitational attraction of condensed matter into small asteroid-sized bodies called *planetesimals* that collided, fragmented, and re-formed over

Table 21.1   Properties of the Planets

Planet	Distance from Sun (millions of km)	Diameter (equatorial km)	Mass Relative to Earth	Density (g/cm^3)	Mean Surface Temperature (°C)
Mercury	58	4880	0.056	5.42	350 day/−170 night
Venus	108	12,100	0.815	5.25	475
Earth	150	12,760	1.000	5.52	22
Mars	228	6800	0.107	3.94	−23
Jupiter	778	142,800	317.9	1.31	−150
Saturn	1425	120,660	95.15	0.69	−180
Uranus	2870	50,800	14.54	1.19	−210
Neptune	4490	48,600	17.23	1.66	−220
Pluto	5900	3000	0.002	2.1	−230

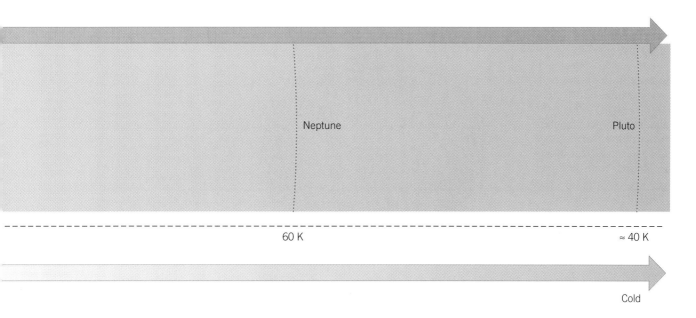

a period of a few million years. Some gradually acquired enough mass to sweep up the smaller bodies in the surrounding space like gravitational vacuum cleaners. They became the primitive planets.

We do not know the precise steps whereby the primitive terrestrial planets were transformed into the bodies we see today. But we are fairly certain that they all went through an early process of remelting and core formation. The necessary heat was supplied by meteorite impacts, which, in this early period of the Solar System, were far more numerous than they are today; by the compaction of matter in the planet; and by the decay of abundant radioisotopes. The heat derived from these diverse sources melted the iron scattered within the interior of the planets. Being denser than the silicates of the planets, the molten iron sank, forming dense planetary cores. The sinking process liberated still more heat, which led to the total or partial melting of the silicate minerals of the planets. Being lighter than the core, they rose to form planetary mantles and lithospheres. Figure 21.7 shows relative percentages of core in each planet. It also compares planetary sizes. Notice that Mercury is a planet characterized by an extremely large iron core, which gives it a strong magnetic field.

Comparative Geology of the Terrestrial Planets

Although the terrestrial planets and the Moon are of roughly similar composition and origin, they differ markedly from the Earth and from one another (Table 21.2). Each has followed a unique path of planetary evolution, a path that can be read from its surface. But there are only four types of processes that appreciably affect planetary surfaces: (1) those that result from the collision of foreign bodies with the planet's surface, mainly **impact cratering;** (2) those related to the dissipation of internal heat through volcanism; (3) those related to uplift, depression, folding, and faulting of the crust—tectonism; and (4) those related to interaction of the atmosphere with the planetary surface, mainly the erosion, transport, and deposition of material.

Some important properties of the planets are apparent if viewed from this perspective. For example, large impact craters are rare on the Earth's surface but are common on the Moon, and it is easy to see the reason for the difference. The Earth has an efficient erosion system and an equally efficient system of recycling the crust through plate tectonics—both of which rapidly destroy the craters. So the presence of abundant

impact cratering
The crustal deformation caused by a meteor collision with a planetary surface.

Figure 21.6 The condensation hypothesis. (a) The primitive solar nebula. (b) The nebula flattens into a rotating disk; most matter streams in toward the center to form the primitive Sun; some matter remains outside. (c) The disk cools, forming tiny particles that grow into planetesimals. (d) A few of the planetesimals grow large enough to attract others, as well as debris from the surrounding space. Eventually they grow into the primitive planets.

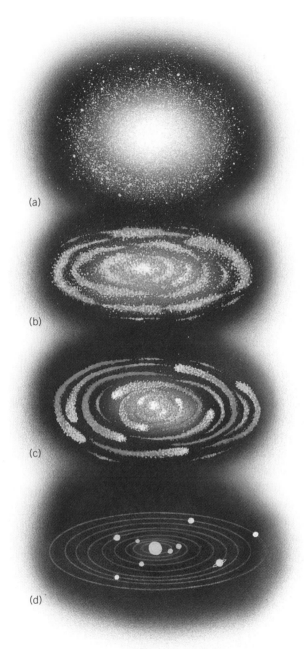

(a)

(b)

(c)

(d)

Table 21.2 Properties of the Terrestrial Planets

	Earth	Moon	Mars	Mercury	Venus
Revolution (Earth days)	365.26	27.32	687	88	224.7
Rotation (Earth days)	0.9983	27.32	1.026	59	243 *
Atmosphere (major components)	Nitrogen, oxygen	None	Carbon dioxide	None	Carbon dioxide
Atmosphere (minor components)	Carbon dioxide	None	Inert gases, nitrogen	None	Inert gases; hydrochloric, hydrofluoric, and sulfuric acids
Atmospheric pressure (millibars)	1000	0	6	0	90,000
Surface gravity (Earth = 1)	1	0.16	0.38	0.37	0.88

*In opposite direction to Earth's.

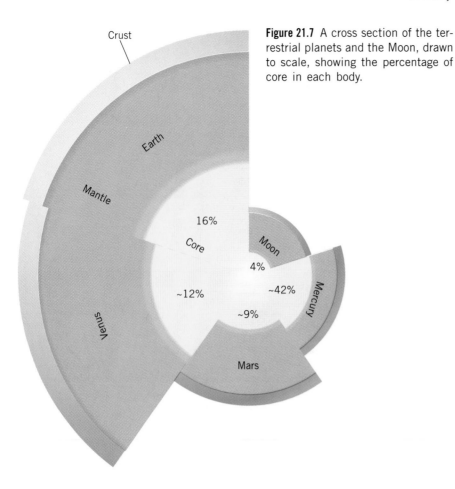

Figure 21.7 A cross section of the terrestrial planets and the Moon, drawn to scale, showing the percentage of core in each body.

large craters is a sign of a planetary surface of great age. Of equal importance, a planetary surface nearly everywhere covered by craters is the sign of a "dead planet," one lacking both the heat necessary to sustain volcanic and tectonic activity and the atmosphere to sustain erosion. Mercury and the Moon fit this description.

Why has the Earth been able to maintain an abundant supply of internal heat and an atmosphere for 4.6 billion years, whereas the Moon has been unable to do so? Geologists believe that the answer lies in the masses of the respective bodies. The Moon lacks the mass to retain an atmosphere through gravitational attraction. Also, because the Moon is a sphere of small volume in comparison to the Earth, it has a higher percentage of exposed surface through which heat can escape. The Moon lacked the mass and the volume to prevent the escape of its internal heat; after a relatively brief period of activity, it became dead.

Geology of the Moon

When Galileo first studied the Moon through the telescope he had invented, he was able to identify two types of terrain. One—smooth, dark, and level—reminded him of the ocean at night, so he named those regions **maria** (Latin for "seas") (Figure 21.8). The maria seemed to lap onto the shores of lighter-colored rock of higher elevation, so he named these regions *terrae,* or land. In a way, he was correct, for the maria are frozen seas, flood basalts that poured out onto the lunar surface and, like any liquid, covered all in their path. The terrae, or **lunar highlands,** on the other hand, represent the older crust, that portion of the surface that stood above the maria. If we superimpose the features of meteor impact cratering and cosmic ray bombardment on the maria and highlands, we have in place the major aspects of the lunar surface geology.

maria (singular, **mare**)
Dark-colored basaltic lowland regions of the Moon.

lunar highland
Highly cratered elevated region of the Moon.

Figure 21.8 A false color composite photograph of the Moon taken by the Galileo spacecraft on December 7, 1992. The lunar highlands are in red, whereas the mare lava flows are in shades of blue and orange. Notice the ejecta radiating from the impact craters and the fresh appearance of the 85-kilometer-wide Tycho crater at the bottom. The freshness, plus the fact that the ejecta cuts across most of the other lunar features, establishes that Tycho is a relatively young crater.

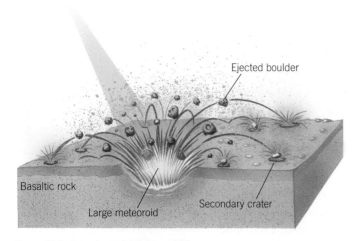

Figure 21.9 A meteoroid that collides with the lunar surface disintegrates upon impact. Ejecta from the meteoroid and excavated crater materials are thrown out in all directions. Large fragments striking the surface form secondary craters.

When a large meteoroid crashes into the lunar surface, it accomplishes many things in the course of blasting a deep crater. Upon impact, it tosses out pulverized rock and other debris, or *ejecta* (Figure 21.9). Most of it collects around the crater rim, but some of it trails off over great distances in star-shaped rays. The impact also shatters and partially melts the rocks at considerable depth beneath the crater floor. Some of the angular fragments then fuse into a distinctive impact breccia. If the meteoroid is of large size and strikes with great force, a huge impact basin is formed, complete with circular ring fractures as well as considerable readjustment by slumping and settling of the surrounding surface (Figure 21.10a). In the past, some of these deep fractures tapped extensive reservoirs of semimolten rock below the lunar crust. The rock then melted and rose up through the fractures to fill the basins and craters, forming the maria (Figure 21.10b).

The lunar highlands crust is primarily composed of anorthosite, an igneous rock with an extremely high content of calcium-rich plagioclase and similar to rocks found in the Adirondack Mountains of New York. Having been exposed to repeated impacts, it is almost completely brecciated, amounting to about 85 percent of the lunar surface. Even small meteoroid fragments that would burn up in the Earth's atmosphere (the shooting stars of the night sky) have left their mark on the lunar surface. The result is that the original hard rock surface of the Moon has been reduced to a fine, powdery lunar regolith. Because of repeated impacts, large and microscopic, the surface has been churned and rechurned for billions of years, as if by a gardener's rake.

Dating Lunar Events

Impact craters have proved extremely useful in the relative dating of the lunar surface. In essence, the older a given portion of the lunar surface, the more impacts per square kilometer it has absorbed over the course of geologic time. Thus *crater counts*—the number of craters per square kilometer—are indicators of the relative age of various portions of the lunar surface. Observe the lunar highlands, which, because they pass under the maria lavas, we take to be the remnants of the original lunar crust. They are regions having an extremely high crater count. On the other hand, the maria lavas that cover portions of the highland rocks have fewer craters. The youngest lavas that cover the older maria are quite clean, except for occasional sharply defined craters with their star-shaped pattern of ejecta that cut across the older features of the lunar surface.

When the lunar astronauts came home to the Earth, they brought with them samples of highland rocks and various maria basalts. By dating these samples, lunar geologists were able determine the ages of the surfaces that received the meteor impacts. In this manner, a specific crater count was associated with a specific radiometric age. They were thus able to use crater counts to date the major events of lunar history. Notice in Figure 21.11(a) that most of the cratering took place about 4 to 3.8 billion years ago and then declined drastically to its present low level. We may assume that during that time span, meteors or planetesimals were far more numerous than they are today. The early cratering probably represented the final stage of the Moon's accumulation into a planetary body, and it involved the highland rocks that formed the young lunar crust (Figure 21.11b). Overlapping this period of heavy bombardment was a declining period of lava outpourings that formed the maria basalts. The Moon has remained essentially unchanged for about 3 billion years, except for the effect of an occasional large impact. These meteors, probably derived from the asteroid belt, are the leftover remnants of the original matter that formed the terrestrial planets.

Recall from Chapter 10 that fossils are associated with a specific interval of geologic time and that identical fossils collected from strata on different continents establish that the strata are the same age. Crater counts are planetary "fossils"; they are used in the same way to determine the ages of the planetary surfaces of Mercury, Mars, and Venus. All that is needed to roughly date an area of a planet's surface is to match its crater count to the Moon's.

The Moon and the Earth constitute two extremes of planetary activity (Figure 21.12). At one extreme, we have the Moon, smallest of the terrestrial bodies if we exclude the asteroids. It lacks the mass to retain sufficient internal heat or an atmosphere. At the other extreme, we have the Earth, largest of the terrestrial bodies. Its great mass retains significant internal heat and exerts enough gravitational force to retain an atmosphere. Between these two extremes lie Mars and Venus. Mars is intermediate in mass and size between the Earth and the Moon and is more or less intermediate in geologic activity. It displays both a primitive crust and a high level of past volcanism, tectonism, and atmospheric alteration. Venus, nearly the same size as the Earth, has proved to be a very active planet, though its tectonic and volcanic style differs markedly from the

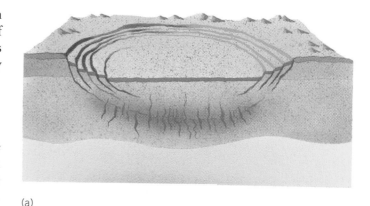

(a)

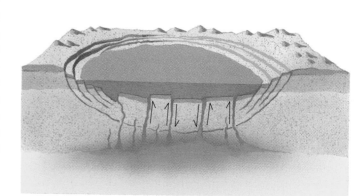

(b)

Figure 21.10 Large meteor impacts and the mare basalts. (a) The meteor excavates a deep crater and causes a series of ring fractures in the lunar crust. (b) The fractures tap the magma beneath the lunar crust, causing the upwelling lava to flood the crater and form mare basalts.

Apply and Decide 21.1

Meteoroid Impact Hazards

Although impact craters are rare on the Earth in comparison with the Moon and the other terrestrial planets, we have no reason to believe that large meteoroids strike the Earth less frequently than they strike other bodies. Instead, many craters have been obliterated by the Earth's active erosion and crustal-recycling processes. As discussed in Aside 11.1, impact cratering is now believed to have had a profound influence on the course of geologic history. Figure 21A shows the locations of most of the 150 impact craters that have been discovered on the continental crust.

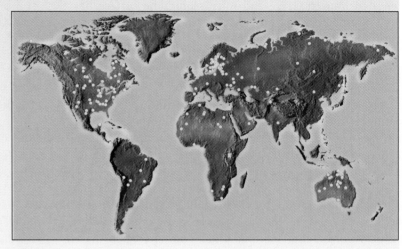

Figure 21A Impact Craters

The Origins of Meteorites

The early meteorites that crashed into the primitive Earth and the lunar highlands may have been derived from nearby sources in the Solar System. More recent meteorites, however, are most likely derived from the asteroid belt. Orbiting the Sun within the belt are more than 1 million asteroids with diameters exceeding 1 kilometer, 1000 with diameters exceeding 30 kilometers, and 200 with diameters exceeding 100 kilometers. Most astronomers suspect that Jupiter's powerful gravitational pull has prevented these millions of fragments from coalescing into a single planet. A *meteoroid* is formed when a collision within the asteroid belt sends a fragment of one of these bodies on an eccentric path. Occasionally the path of a meteoroid intercepts the Earth's gravitational field. As the meteoroid enters the Earth's atmosphere, friction sets it afire; the light we see streaking across the night sky is called a *meteor* or "shooting star." The solid rocky or metallic remains of meteoroids are called *meteorites.*

Comets that circle the Sun in highly eliptical orbits represent another plausible source of meteorites. Like cosmic slingshots, they accelerate as their orbits swing close around the Sun. Studies of Halley's comet, which passed relatively close to the Earth in 1986, revealed that this body has a nucleus composed of rock and iron fragments held together by a disintegrating mass of ice. Comets that pass close to the Sun frequently break up and leave a trail of debris behind in the inner Solar System. Clusters of these meteoroids eventually intercept the Earth's atmosphere and send showers of meteors toward the surface.

The Frequency of Impact Cratering

The importance of impact cratering as a geologic process depends upon the severity of the impacts and the frequency of their occurrence. Once again, the Moon can serve as a guide. Because the Moon lacks the ability to erode its surface or to re-

cycle its crust through plate tectonics, the entire history of impacts remains on its surface for us to observe.

Early Impacts

As mentioned earlier in the chapter, the most intense period of lunar bombardment occurred approximately 4.1 billion years ago and affected the most primitive crust, the lunar highlands. If we assume that the Earth's impacts occurred around the same time, then it is clear why impact craters are relatively rare on the Earth. Rocks older than 3.8 billion years have yet to be found on Earth because the planet's ancient crust has been demolished by erosion and tectonic processes—and perhaps by further meteoroid impacts as well.

Impact cratering must have made the Earth an extremely violent place during the first few hundred million years of its existence. According to one hypothesis based on computer simulations of asteroid trajectories, a glancing blow from a huge meteoroid during one cataclysmic event sent splinters of the Earth into space to form the Moon! The collision knocked the Earth's rotational axis off-tilt, which is why it is inclined about 23.5 degrees with respect to the plane of the Earth's orbit about the Sun. The meteorite vaporized on impact as it blasted its way into the Earth's mantle. The ejecta released from the mantle and the meteorite was tossed into orbit, where it coalesced into the lunar nucleus. Subsequent impacts added to the Moon's volume.

On a more constructive note, many planetary geologists believe that comets, which were vastly more numerous 4 billion years ago, may have delivered water to the Earth and contributed, along with volcanic activity, to the creation of the oceans. Wherever liquid water is found, the possibility of life exists. Indeed, some biologists are convinced that the water-bearing comets may have seeded the Earth with primitive microbes or DNA.

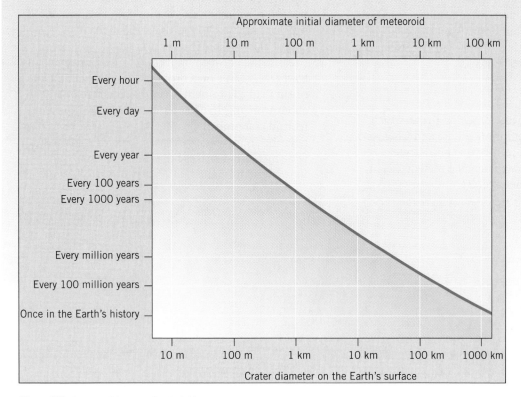

Figure 21B Graph of Impact Probability

(Figure 21B). Thus the likelihood that a dangerous meteoroid will strike the Earth during your lifetime—or the lifetime of your grandchildren—is remote. Note, however, that nothing in theory would exclude a large meteoroid from wiping out the human race in the future, as many believe the dinosaurs were wiped out 65 million years ago. Given sufficient time, the unlikely will eventually occur.

Even relatively small impacts can be extremely dangerous, and their probabilities are higher. Meteoroids measuring between 100 and 1000 meters in diameter impact every 1000 to 200,000 years. If one of roughly 500-meters diameter were to strike New York City, it would probably blast a crater 5 kilometers wide, trigger a magnitude 15 or greater earthquake, destroy 4000 square kilometers of property, and kill 25 million people.

If this seems like an implausible scenario, consider that in 1989 an asteroid of 500 meters diameter passed a point in space that the Earth had occupied only six hours earlier. There have been other narrowly averted catastrophes. In 1908, for example, a comet or meteoroid fragment exploded over Tunguska, Siberia, flattening 1000 square kilometers of forest—an area about the size of Washington, D.C. Had the fragment arrived a few hours later, it would have hit west of Tunguska along the same latitude, possibly obliterating the Russian city of St. Petersburg or Oslo, the capital of Norway.

Recently a number of astronomers have begun to survey the Solar System for potential hazards to Earth. To date they have identified an estimated 2 or 3 percent of what's out there. A current debate swirls around the amount of public and private funds that should be diverted to searching for potential meteoroid hazards, both inside and outside the Solar System.

As with any perceived threat, the possibility of a catastrophic impact may generate varying levels of response, depending upon the ratio of cost to risk. Prior to 1994, when the comet Shoemaker-Levy 9 crashed into Jupiter, the prevailing wisdom among NASA and other scientists was that the odds against a major impact event in our lifetime dictate that we do nothing. Since the Shoemaker-Levy 9 event, how-

Later Impacts

As noted earlier, large impacts declined rapidly after the early lunar highlands bombardment, thus allowing the 3.2-billion-year-old maria lava plains that cover portions of the highlands to remain relatively unscarred. *Relatively unscarred* does not mean untouched, however. In fact, the maria plains have received 29 major impacts, leaving behind craters with diameters greater than 25 kilometers. If we were to scale the surface area of the maria up to match the Earth's, it would have received 80 times this number of large impacts, producing approximately 2400 craters! Assuming a rough equivalence with the lunar maria, we should expect the Earth to have received the same number of large impacts over the last 3.2 billion years. By analyzing the frequency of lunar impacts over time, and by observing the paths of asteroids and comets in the Solar System, it is possible to estimate how often a meteorite of a given size will strike the Earth.

The one that formed the Barringer Crater of northern Arizona struck the Earth a mere 50,000 years ago (see Aside 11.1 The Geology of Meteorite Craters).

How Safe Are We?

Because larger asteroids and comet fragments are rarer than smaller ones, there is an inverse relationship between the size of the body and the probability that it will hit our planet

ever, a number of scientists have suggested the formation of a limited space survey program to locate potential hazards so that we may better evaluate the risks and institute an early warning system. A few astronomers and military experts have proposed an extensive (and expensive) program that would apply "star wars" nuclear weapon technology to intercepting and destroying potential threats.

THOUGHT QUESTIONS

1. If it were brought to the electorate for a vote, which of the above options would you choose and why? Assuming finite resources, would you concentrate research on higher-frequency smaller meteoroids or lower-frequency large meteoroids?

2. How may the need to preserve their jobs and receive funding for their research projects color scientists' recommendations concerning these options? Cite two hypothetical or real examples. If a degree of self-interest is involved, does that necessarily negate a scientist's recommendations.

3. Because the Earth contains far more ocean area than land area, a large meteoroid is more likely to hit an ocean than a continent. Using what you learned about tsunamis in Chapter 3, Earthquakes, and meteoroid cratering in Chapter 11, Rock Deformation, describe an imaginary event in which a 1-kilometer-wide meteoroid strikes the Pacific Ocean just east of the Hawaiian Islands.

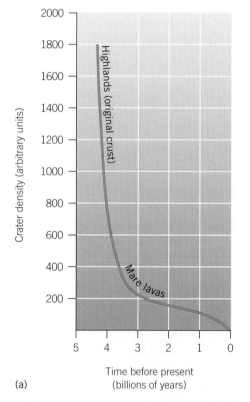

(a)

Figure 21.11 Part (a) shows the relationship of crater densities to various ages of the lunar surface. The craters are numerous in the older lunar highlands, but relatively scarce in the younger mare basalts. (Courtesy of Don E. Williams, "Moon," in Michael H. Carr, ed. *The Geology of the Terrestrial Planets,* Washington, D.C.: NASA, 1984, p. 194.) Parts (b–d) illustrate how the lunar surface has changed over time.

(b)

(c)

(d)

Figure 21.12 A model of planetary evolution. The level of activity that a planet is able to maintain over time depends upon its ability to generate and maintain internal heat. The Moon lacks the size to maintain internal heat; its dynamism ceased about 3 billion years ago. The Earth and Venus, however, are large enough to remain internally dynamic planets. (Data from J. A. Wood, *The Solar System,* Englewood Cliffs: Prentice-Hall, 1979.)

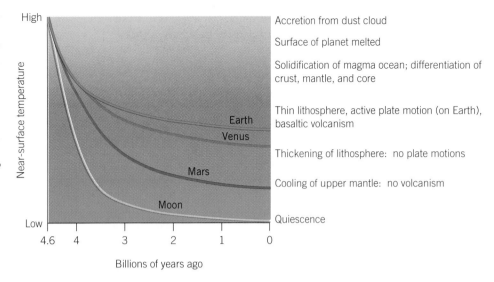

Accretion from dust cloud

Surface of planet melted

Solidification of magma ocean; differentiation of crust, mantle, and core

Thin lithosphere, active plate motion (on Earth), basaltic volcanism

Thickening of lithosphere: no plate motions

Cooling of upper mantle: no volcanism

Quiescence

Earth's. We also see that factors other than mass and size play a significant role in planetary evolution. Probably the most important of these factors is proximity to the Sun, which controls to a great extent the composition of the atmospheres and the likelihood of life on the planetary surface.

Geology of Mars

Two-thirds of the Martian surface, mainly in the southern hemisphere, resembles the heavily cratered lunar highlands. There are also cratered plains on Mars that resemble the lunar maria. Nevertheless, Mars is a very different planetary body from the Moon. It has volcanoes, huge rift valleys, and unmistakable signs that erosion and deposition have altered the surface of the planet.

The enormous Martian volcanoes are among the most impressive planetary features of the Solar System. Alba Patera, for example, is 1500 kilometers in diameter. Olympus Mons is 550 kilometers wide, and its summit rises 25 kilometers above the surrounding plains (Figure 21.13). Both are shield volcanoes like Mauna Loa, the Earth's largest volcano, but Mauna Loa rises a paltry 9 kilometers above the Pacific Ocean floor and is a mere 120 kilometers wide. Most of the Martian volcanoes are arrayed in two broad bands that cross the equator of the planet. One of them is associated with the 10-kilometer-high Tharsis Bulge that rises like an enormous cracked blister above the general level of the planet's surface. Crater counts of Martian lava flows indicate that the planet has remained volcanically active throughout most of its history, though at a declining rate. While volcanism on Mars has probably ceased, we cannot rule out the possibility that some of the volcanoes might still be active.

The Martian surface also bears the scars of large-scale tectonism. The Valles Marineris, 600 kilometers wide and 7 kilometers deep, runs for 4500 kilometers. This is one-quarter the circumference of the planet, amounting to the width of the United States (Figure 21.14). The Valles closely resembles the East African Rift Valley in form, though it is longer, wider, and deeper. Extensive canyon systems branch from it.

Figure 21.13 Mars's Olympus Mons, the largest volcano in the Solar System. The Hawaiian Islands, the Earth's largest volcanoes, are superimposed for scale.

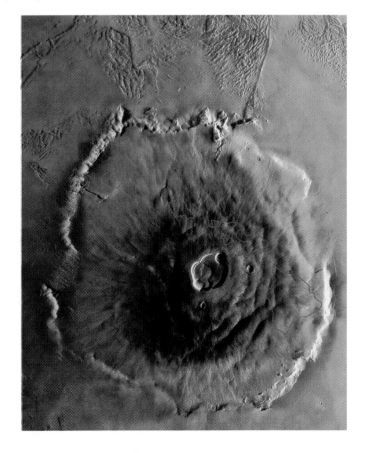

Figure 21.14 A mosaic image of Mars, taken by the Viking spacecraft. The Valles Marineris slashes across the surface for about 5000 kilometers, or the length of the entire United States. The Tharsis Bulge, within which a series of large shield volcanoes are located, is at the far left.

However, the evidence of relatively recent tectonic activity is absent, judging from the cratered terrain surrounding the Valles.

Mars possesses a very thin carbon dioxide atmosphere, and the evidence of this atmosphere's interaction with the surface of the planet is striking. The *Viking 2* landing gained evidence that iron on Mars's surface has weathered to a deep red, which gives the Red Planet its distinctive color. Large ice caps cover both poles; mixtures of frozen carbon dioxide and ice, they wax and wane with the Martian seasons (Figure 21.15). The winds across the cold Martian surface are harsh, and you can see extensive dune fields and plains of layered windblown sediments.

Most intriguing are the branching stream channels of Mars (Figure 21.16). It is difficult to imagine any agent other than running water that could be responsible for them. Where did the water come from, and where did it go? Plainly, the Martian surface is too cold to support water in the liquid state. Probably the water came from ice within the Martian soil that periodically melted in the past. (Today, there is abundant evidence of ice beneath the surface.) Or the planet could simply have been warmer early in its history and subsequently cooled. In either event, the erosive effects of running water on Mars did not approach the importance they have on the Earth. We can see, for example, that stream erosion has been unable to wear away ancient volcanoes and craters.

Although the Earth and Mars share many processes in common, Mars lacks the recycling mechanisms of plate tectonics and the hydrologic cycle. Nothing like our mid-ocean ridge system exists. Nor do subduction zones, fold-mountain belts, or island arc volcanoes exist. The entire Martian surface may be thought of as consisting of a single, very thick lithospheric plate. The Martian volcanoes owe their great size to two factors.

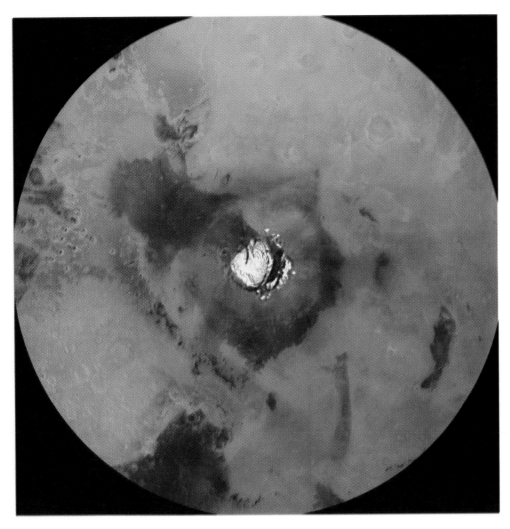

Figure 21.15 A polar view of Mars showing its ice cap of water, carbon dioxide, and dust.

First, the plate has remained stationary for billions of years; second, it has been rigid enough to support the weight of the volcanoes without buckling. The surface of Mars is a vast frigid desert, having an extremely ineffective erosion system. As a consequence, its features remain largely as they were when formed—huge volcanoes, impact craters, deep canyons, and all.

In sum, we can describe Mars as a formerly active planet that seems to have lacked the internal heat to recycle its surface. Therefore, the entire history of Mars remains written on its crust. The Earth, however, is subject to the processes that Mars lacks. As a consequence, the early history of our planet has been largely destroyed.

Geology of Venus

Planetary geologist Robert Grimm uses this analogy to depict the fascination of Venus: Suppose we make two pies from the same recipe and put them in the oven simultaneously. An hour later we remove them from the oven, only to find that they are markedly different. What happened? In other words, how could the Earth and Venus—neigh-

Figure 21.16 These branching channels seem identical to stream channels on the Earth, and virtually offer proof that in the past, running water operated on the Martian surface.

Figure 21.17 A radar image of Venus, taken by the *Magellan* spacecraft in 1992. Aphrodite Terra, the broad, light band running across the center of the image, can be divided into three parts: (1) Alta Regio (the bright region at center right), a complex dome of shield volcanoes with lava flows radiating from the central region; (2) the highland plateau of Thetis Regio (the bright region to the left), which was created by crustal compression; and (3) the Chasmata (between Thetis and Alta Regio), which are prominent canyons. The dark patches scattered over the surface of the planet are impact craters.

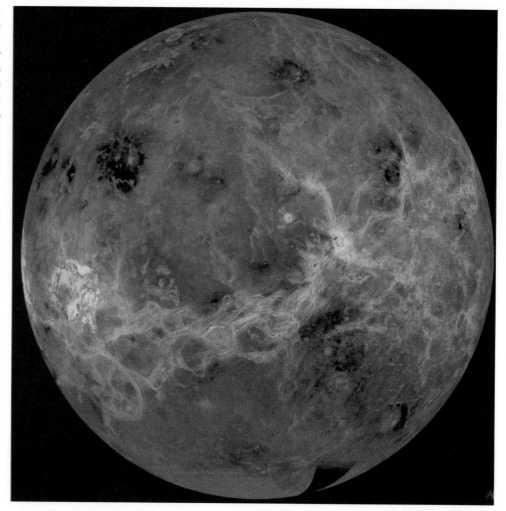

boring planets of about the same size, age, and composition—have evolved so differently?

The most striking feature of Venus is its atmosphere. It is a witch's brew of carbon dioxide and sulfuric acid clouds so dense that it exerts a pressure 90 times greater than the Earth's atmosphere. This incredible steam bath of acid keeps the surface temperature of the planet at about 500 °C, hot enough to melt lead. On Venus, the heat-trapping capacity of the atmosphere is so extreme that it is called a runaway greenhouse effect. Needless to say, Venus has no rivers or oceans, the result of the planet's nearness to the Sun, which prevents the formation and cycling of liquid water.

The surface of Venus is not visible through its thick clouds. In 1989, however, the orbiting satellite *Magellan* invaded the planet's privacy by sending out a wide swath of radar beams that penetrated its atmosphere and scanned the surface. The reflection, absorption, and scatter patterns of the beams enabled scientists to image about 90 percent of the planet. They discovered a surprising planetary topography (Figure 21.17).

Although broad upland surfaces are evident on Venus, they do not resemble the Earth's continents, which are composed of granite and whose elevation is the expression of the difference in composition between them and the basaltic oceanic crust. Instead, the entire crust of Venus appears to be basaltic. The planet's topography seems to have been created by rising mantle plumes that arched the crust and formed volcanic hot spots, and by sinking plumes that depressed the crust (Figure 21.18). Virtually the entire surface of the planet consists of volcanic rocks and their windblown weathering products. It is a volcanologist's paradise—provided that he or she can stand the humidity. More than 1660 volcanic landforms have been identified, including approximately 550 shield volcanoes that are typically associated with regions of mantle upwelling.

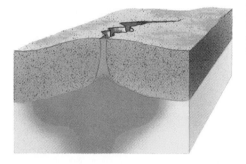

Figure 21.18 Rising mantle plumes, similar to those on Earth, probably caused the Alta Regio topography of Venus.

These features, which are spread more or less evenly across the planet's surface, represent the consequence of mantle convection. Fold-mountain belts and rift-type faults are common but largely confined to regions of rising and sinking plumes (Figure 21.19). Indeed, evidence of Earthlike, large-scale tectonic plates is lacking on Venus. We do not see many features that remind us of mid-ocean ridges, subduction zones, island arcs, or large-scale transform faults.

The observation that tectonic and igneous activity on Venus resembles the Earth's hot-spot mode rather than its plate tectonics is interesting but perhaps unremarkable. What *is* remarkable is the fact that impact crater counts reveal the entire surface of Venus to be approximately 500 million years old! The craters are pristine—untouched since the day they were formed (Figure 21.20). No lava has flowed into them, nor have their dimensions been offset by faults or distorted by folds. Apparently the entire surface of Venus was repaved with lava some 500 million years ago, obliterating 4.1 billion years of prior history. Venus has been tectonically and volcanically quiescent ever since.

If Venus worked the same way as the Earth, we would find surface features representing a broad age spectrum, from the most ancient to the most recent. Instead, every rock on the Venutian crust appears to be roughly the same age as every other rock. How did this happen? Did Venus erupt cataclysmically 500 million years ago, and then "die" like the Moon? Or does Venus behave cyclically, storing internal heat for 500 million years until magma boils to the surface and the planet remelts? The

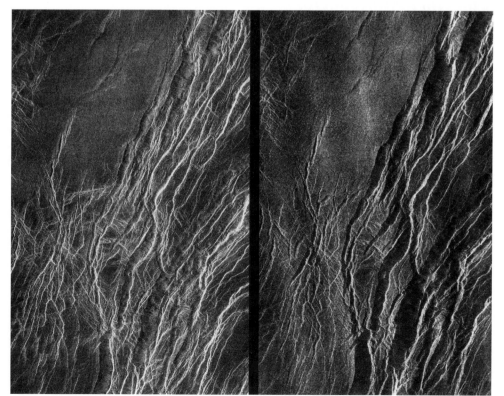

Figure 21.19 A surface detail of Venus. The long lines in this stereo (double) image are linear tension fractures. The topography resembles the horst and graben structures of the Basin and Range Province of the western United States.

Figure 21.20 Oblique view of Maat Mons of Alta Regio, Venus. It stands 6 kilometers above the surrounding plains, and is several hundred kilometers wide. A 25-kilometer-wide impact crater is evident in the central part of the image.

surface of Venus certainly appears to contradict the uniformitarian model, which postulates that geological activity is generally continuous throughout the history of a planet. We must, however, collect and analyze a good deal more data on the planet's internal structure, composition, and heat flow before we can theorize confidently about how it works.

Planetary Atmospheres

Judging from their densities and other evidence, Venus, Earth, and Mars are roughly similar in composition and origin. So it is probable that their early atmospheres were derived from roughly similar gases by a roughly similar process—volcanic eruptions during the initial remelting stage of the planets. The atmospheres of the three planets are now quite different, however, reflecting the planets' different evolutionary paths. Although the atmospheres of Venus and Mars are roughly 95 percent carbon dioxide, the first is a suffocating blanket with pressure 90 times greater than the Earth's; the other is so thin that it exerts a pressure equal to 1 percent of the Earth's (Figure 21.21). In contrast, the Earth's atmosphere contains 78 percent nitrogen, 21 percent oxygen, and a mere 0.035 percent carbon dioxide.

Putting the details of planetary atmospheres aside, it is fair to say that advanced life as we know it owes its existence to the Earth's fortunate orbit with respect to the Sun (Figure 21.22). If the Earth were 5 percent closer to the Sun, a Venus-like greenhouse effect would produce surface temperatures close to 450 °C. If, on the other hand, the Earth were a mere 1 percent farther from the Sun, vast glaciation would produce a barren Martian-like crust.

As it is, the Earth has been capable of maintaining water in the liquid state for at least 4 billion years. The condensation of water vapor and the resulting formation of oceans set off a chain of events that scrubbed the atmosphere nearly clean of carbon dioxide and produced the nitrogen- and oxygen-rich atmosphere of today. First, the ocean was able to absorb 50 times more carbon dioxide than is found in the present atmosphere. Second, the ready water supply enabled the early evolution of algae and

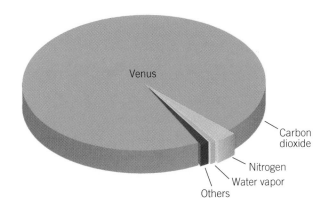

Venus

Carbon
dioxide

Nitrogen
Water vapor
Others

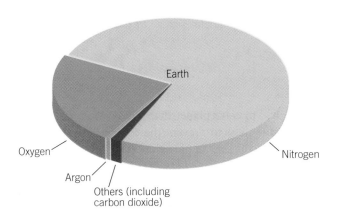

Earth

Oxygen

Argon

Others (including
carbon dioxide)

Nitrogen

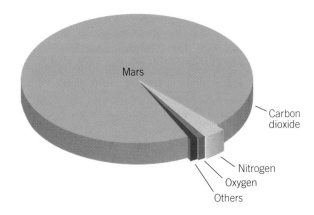

Mars

Carbon
dioxide

Nitrogen
Oxygen
Others

Figure 21.21 Although the atmospheres of Venus and Mars are similar in composition, Venus's atmosphere is very thick, whereas Mars's is extremely thin. The Earth's store of carbon dioxide is located mainly in its biosphere, rocks, and oceans, rather than in its atmosphere.

plant life, which extracted additional carbon from the atmosphere and released oxygen through photosynthesis. Third, as living things proliferated, large quantities of carbon, derived from dead organisms, accumulated over hundreds of millions of years in rocks as shell matter, coal, and oil. Fourth, huge quantities of atmospheric carbon dioxide were incorporated into rocks and soils through weathering reactions. Thus rocks and soils came to contain overwhelmingly the largest fraction of the Earth's carbon— some 99.6 percent. These carbon-extracting processes continue to this day. If the carbon held in these various reservoirs were returned to the atmosphere, the probable result would be a carbon dioxide-induced greenhouse effect similar to the one observed on Venus.

The Martian atmosphere also contrasts interestingly with the Earth's. First, remem-

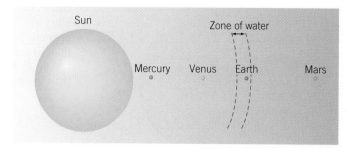

Figure 21.22 The Earth orbits the Sun within a narrow zone in which water can exist in the liquid state, thus making life possible. If the Earth were located only slightly closer to the Sun, the water would vaporize, leading to the runaway greenhouse effect of Venus. If Earth were a fraction farther from the Sun, the water would freeze, causing conditions similar to those on Mars.

ber that Mars is a much less massive planet than the Earth and exerts less gravitational attraction. It simply lacks the mass to prevent the escape of nitrogen—a key element of life—from its atmosphere. Second, remember that the Martian ice caps are largely composed of carbon dioxide, and much carbon dioxide is bound up in the Martian soil. In other words, most of it is someplace other than in the atmosphere, just like it is on the Earth.

But on the Earth, carbon dioxide is constantly recycled through volcanism at subduction zones. The released carbon dioxide in turn traps solar energy in the form of infrared waves rising from the Earth's surface—exerting a strong greenhouse effect and keeping water liquid. Volcanic activity on Mars, however, probably ended millions of years ago. Therefore, not much of the carbon dioxide held in the soil through weathering or in the polar ice caps has been recycled to the atmosphere, and the Martian atmosphere is but a thin carbon dioxide blanket. This thin blanket leads to a lessened greenhouse effect and a cold, dry surface. The lack of liquid water in turn lessens the probability of life on Mars, although the possibility that relatively simple life-forms exist there today or existed sometime in the Martian past has not been ruled out. For example, there are some forms of bacteria on the Earth that are capable of surviving under the extremely cold temperatures prevalent on Mars. Also, local pockets of warmth may exist on or beneath the Martian surface where water remains liquid and life can thrive. Some evidence suggests that periodic shifts in Mars's orbit and axis of rotation may warm the planet and melt its ice caps in a sort of Martian Milankovitch effect. Finally, we cannot rule out the possibility that life has adapted to the harsh Martian environment in ways we have not observed on the Earth. As with all things geological, we must keep an open mind about Martian life until scientists have been able to systematically explore this most congenial of our neighboring planets.

In sum, the differences between the atmospheres of the three neighbors in space—and the conditions on their surfaces—are controlled by many factors: proximity to the Sun, planetary composition, volcanic activity, recycling capacity, and planetary mass. All these factors can be traced back to events that occurred during the origin of the Solar System and, before that, the universe.

We close this book by commenting on what a remarkable planet the Earth is. Not only is it a geologically dynamic planet, but dynamic life has evolved on its surface. From this long history of life has evolved an organism capable of contemplating its relationship to this planet and the universe as a whole.

STUDY OUTLINE

The Sun, thousands of times larger than the Earth, is one of the hundred billion stars scattered hundreds of billions of kilometers apart in the Milky Way **galaxy.** The Milky Way is one of the billions of galaxies in the **universe.**

I. **THE EXPANDING UNIVERSE.** Evidence of the expanding universe has led to the **big bang theory,** which states that the universe began as a high-temperature ball of energy that exploded 15 to 20 billion years ago. As the ball expanded, it cooled, and energy was converted to matter. Over the billions of years since, gravitational attraction has pulled the particles together to form isolated galaxies, within which stars and **planets** evolved.

II. **THE SOLAR SYSTEM**
   A. The Sun formed at the center of the Solar System as a cloud of hydrogen and helium that collapsed inward.

   B. The Solar System also consists of the dense **terrestrial planets** (Mercury, Venus, Earth, Mars), their moons, and a belt of **asteroids;** the light Jovian planets (Jupiter, Saturn, Neptune, Uranus) and their moons; Pluto, the farthest planet from the Sun; and comets, small masses of rock and ice that occupy the Oort cloud at the farthest distance from the Sun.

   C. The planets began as clouds of gases and dust particles. Light metals, iron, and hydrous silicate minerals condensed closer to the Sun to form the terrestrial planets; hydrogen, ammonia, water vapor, and methane gases condensed farther out to form the Jovian planets.
      1. Gravitational attraction further condensed matter into small asteroid-sized bodies called planetesimals, some of which gradually acquired enough mass to continue growing to the size of a planet.
      2. Through heat supplied by meteor impact and the decay of abundant radioisotopes, all the planets went through core formation and remelting early in their histories.

III. **COMPARATIVE GEOLOGY OF THE TERRESTRIAL PLANETS**
   A. Four processes affect planetary surfaces: **impact cratering,** volcanism, tectonism, and interaction with an atmosphere. Sufficient mass enables a planetary body to maintain internal heat and retain an atmosphere.

   B. Because of its low mass, the Moon lacks internal heat and an atmosphere.
      1. There are two major terrains on the Moon:
         a. the lunar **maria,** composed of flood basalts that rose up through the fractures formed by past impact cratering, and
         b. the **lunar highlands,** older regions of crust marked by numerous impact craters.
      2. Crater counts are indicators of the relative age of various portions of the lunar surface.

   C. Mars is intermediate in mass and size between the Earth and the Moon. Most of the Martian surface resembles the lunar crust, but its history includes periods of volcanism, tectonism, and atmospheric alteration.
      1. The Martian surface is punctuated by huge volcanoes, such as the Olympus Mons. Cratered lava flows indicate active volcanism through most of

the planet's history. Features such as the Valles Marineris indicate large-scale rifting, but there is no evidence of plate tectonics.

2. The presence of stream channels indicates that water was present on the surface of the planet in the past. Today, the water is bound up in the Martian ice caps and soil.

D. The surface of Venus consists of volcanoes and lava plateaus associated with rising and sinking mantle plumes; compressional and rift structures closely associated with these plumes; and impact craters.

1. Evidence of the existence of large-scale tectonic plates is lacking.
2. Crater counts suggest that the entire surface of Venus is 500 million years old and has been quiescent since that time.

E. Planetary atmospheres

1. The atmosphere of Venus is essentially a thick carbon dioxide blanket that creates a runaway greenhouse effect and a 500 °C planetary surface. The atmosphere of Mars is a thin tissue of carbon dioxide. The atmosphere of the Earth contains little carbon dioxide but is rich in nitrogen and oxygen.
2. The Earth's precise distance from the Sun allows liquid water to form; hence, we have a hydrologic cycle, the development of photosynthetic life, and the extraction of carbon dioxide from the atmosphere. Venus is too close to the Sun and Mars is too far from the Sun to allow for the existence of liquid water.
3. It is unlikely that the harsh Martian environment could support advanced life. The possibility remains, however, that simple life-forms presently survive there or existed there in the past.

STUDY TERMS

asteroid (p. 572)
big bang theory (p. 570)
galaxy (p. 568)
impact cratering (p. 575)
lunar highland (p. 577)

maria (p. 577)
planet (p. 570)
terrestrial planets (p. 572)
universe (p. 567)

CRITICAL THINKING QUESTIONS

1. If the stars and planets formed in the manner described in this chapter, what is the likelihood of the existence of life elsewhere in the universe?
2. Venus is about the same size and density as the Earth, so it probably has a metallic core. Yet it lacks a magnetic field. Why? (Hint: Compare the planets' rotation rates in Table 21.1.)
3. Are there weathering and erosion on the surface of Venus? If so, what forms might they take?
4. It is said that Venus cannot get carbon dioxide *out* of its atmosphere and that Mars cannot get carbon dioxide *into* its atmosphere. Explain. What other planet in the Solar System has mobile carbon dioxide? What are the consequences of this mobility?
5. Aside from the Earth, the Moon is the only planetary body that has been dated through radiometric means. How are we able to date the planetary surfaces of Mercury, Venus, and Mars? What assumptions underlie these dates?

6. Compare the tectonic styles of Venus, Mars, and the Earth.
7. The topography of the Moon is complex, but its history is relatively simple. Explain.

Conversion Table for Metric and English Units

Length

1 kilometer (km)	= 0.6214 miles
	= 328 yards
	= 1000 meters
1 meter (m)	= 1.0936 yards
	= 3.2808 feet
	= 100 centimeters
1 centimeter (cm)	= 0.3937 inches
1 mile (mi)	= 1.6093 kilometers
	= 1760 yards
1 yard (yd)	= 0.9144 meters
	= 3 feet
1 foot (ft)	= 0.3048 meters
	= 12 inches
1 inch (in.)	= 2.54 centimeters

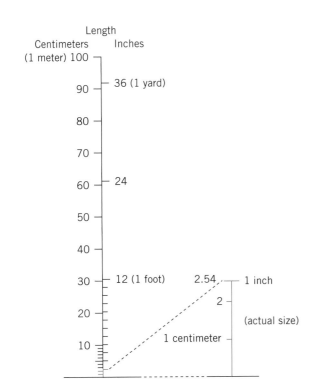

Area

1 square kilometer (km^2)	= 0.386 square miles
	= 100 hectares
1 hectare	= 2.471 acres
	= 10,000 square meters
1 square meter (m^2)	= 1.20 square yards
	= 10.76 square feet
	= 10,000 square centimeters
1 square centimeter (cm^2)	= 0.15 square inches
1 square mile (mi^2)	= 2.59 square kilometers
	= 640 acres
1 acre	= 4033 square meters
	= 4840 square yards
1 square yard (yd^2)	= 0.833 square meters
	= 9 square feet
1 square foot (ft^2)	= 0.093 square meters
	= 144 square inches
1 square inch ($in.^2$)	= 6.45 square centimeters

Volume

1 cubic kilometer (km^3)	= 0.24 cubic miles
	= 1,000,000,000 cubic meters
1 cubic meter (m^3)	= 1.31 cubic yards
	= 35.31 cubic feet
	= 1000 liters
1 liter (L)	= 0.26 gallons
	= 1000 cubic centimeters
1 cubic centimeter (cm^3)	= 0.06 cubic inches
1 cubic mile (mi^3)	= 4.17 cubic kilometers
	= 5,460,000,000 cubic yards
1 cubic yard (yd^3)	= 0.76 cubic meters
	= 27 cubic feet
1 cubic foot (ft^3)	= 0.028 cubic meters
	= 1728 cubic inches
1 cubic inch ($in.^3$)	= 16.67 cubic centimeters
1 gallon (gal)	= 3.785 liters

Mass

1 metric ton (t)	= 1.1 tons
	= 1000 kilograms
1 kilogram (kg)	= 2.205 pounds
	= 1000 grams
1 gram (g)	= 0.035 ounces
1 ton	= 0.91 metric tons
	= 2000 pounds
1 pound (lb)	= 0.45 kilograms
	= 16 ounces
1 ounce (oz)	= 28.35 grams

Temperature

Kelvin (K)	= °C + 273
°C	= $\frac{5}{9}$ (°F − 32)
°F	= $\frac{5}{9}$ °C + 32

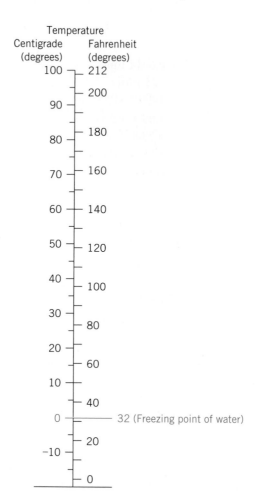

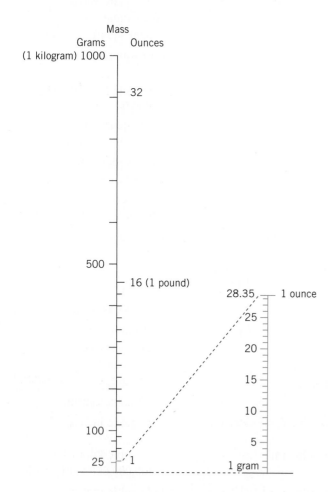

APPENDIX B

Periodic Table of the Elements

Inert elements → VIIIA

Metals ↔ Nonmetals

Transition metals

Atomic number — 1
Element symbol — H
Atomic mass — 1.008

IA	IIA	IIIB	IVB	VB	VIB	VIIB	VIIIB			IB	IIB	IIIA	IVA	VA	VIA	VIIA	VIIIA
1 **H** 1.008																	2 **He** 4.003
3 **Li** 6.941	4 **Be** 9.012											5 **B** 10.81	6 **C** 12.01	7 **N** 14.01	8 **O** 16.00	9 **F** 19.00	10 **Ne** 20.18
11 **Na** 22.99	12 **Mg** 24.31											13 **Al** 26.98	14 **Si** 28.09	15 **P** 30.97	16 **S** 32.06	17 **Cl** 35.45	18 **Ar** 39.95
19 **K** 39.10	20 **Ca** 40.08	21 **Sc** 44.96	22 **Ti** 47.88	23 **V** 50.94	24 **Cr** 52.00	25 **Mn** 54.94	26 **Fe** 55.85	27 **Co** 58.93	28 **Ni** 58.69	29 **Cu** 63.55	30 **Zn** 65.38	31 **Ga** 69.72	32 **Ge** 72.59	33 **As** 74.92	34 **Se** 78.96	35 **Br** 79.90	36 **Kr** 83.80
37 **Rb** 85.47	38 **Sr** 87.62	39 **Y** 88.91	40 **Zr** 91.22	41 **Nb** 92.91	42 **Mo** 95.94	43 **Tc** (98)	44 **Ru** 101.1	45 **Rh** 102.9	46 **Pd** 106.4	47 **Ag** 107.9	48 **Cd** 112.4	49 **In** 114.8	50 **Sn** 118.7	51 **Sb** 121.8	52 **Te** 127.6	53 **I** 126.9	54 **Xe** 131.3
55 **Cs** 132.9	56 **Ba** 137.3	57 **La*** 138.9	72 **Hf** 178.5	73 **Ta** 180.9	74 **W** 183.9	75 **Re** 186.2	76 **Os** 190.2	77 **Ir** 192.2	78 **Pt** 195.1	79 **Au** 197.0	80 **Hg** 200.6	81 **Tl** 204.4	82 **Pb** 207.2	83 **Bi** 209.0	84 **Po** (209)	85 **At** (210)	86 **Rn** (222)
87 **Fr** (223)	88 **Ra** 226	89 **Ac†** (227)	104 **Unq**	105 **Unp**	106 **Unh**	107 **Uns**	108 **Uno**	109 **Une**									

*Lanthanides

58 **Ce** 140.1	59 **Pr** 140.9	60 **Nd** 144.2	61 **Pm** (145)	62 **Sm** 150.4	63 **Eu** 152.0	64 **Gd** 157.3	65 **Tb** 158.9	66 **Dy** 162.5	67 **Ho** 164.9	68 **Er** 167.3	69 **Tm** 168.9	70 **Yb** 173.0	71 **Lu** 175.0

†Actinides

90 **Th** 232.0	91 **Pa** (231)	92 **U** 238.0	93 **Np** (237)	94 **Pu** (244)	95 **Am** (243)	96 **Cm** (247)	97 **Bk** (247)	98 **Cf** (251)	99 **Es** (252)	100 **Fm** (257)	101 **Md** (258)	102 **No** (259)	103 **Lr** (260)

Manufactured

Table of Atomic Masses*

Element	Symbol	Atomic Number	Atomic Mass
Actinium	Ac	89	(227)†
Aluminum	Al	13	26.98
Americium	Am	95	(243)
Antimony	Sb	51	121.8
Argon	Ar	18	39.95
Arsenic	As	33	74.92
Astatine	At	85	(210)
Barium	Ba	56	137.3
Berkelium	Bk	97	(247)
Beryllium	Be	4	9.012
Bismuth	Bi	83	209.0
Boron	B	5	10.81
Bromine	Br	35	79.90
Cadmium	Cd	48	112.4
Calcium	Ca	20	40.08
Californium	Cf	98	(251)
Carbon	C	6	12.01
Cerium	Ce	58	140.1
Cesium	Cs	55	132.9
Chlorine	Cl	17	35.45
Chromium	Cr	24	52.00
Cobalt	Co	27	58.93
Copper	Cu	29	63.55
Curium	Cm	96	(247)
Dysprosium	Dy	66	162.5
Einsteinium	Es	99	(252)
Erbium	Er	68	167.3
Europium	Eu	63	152.0
Fermium	Fm	100	(257)
Fluorine	F	9	19.00
Francium	Fr	87	(223)
Gadolinium	Gd	64	157.3
Gallium	Ga	31	69.72
Germanium	Ge	32	72.59
Gold	Au	79	197.0
Hafnium	Hf	72	178.5
Helium	He	2	4.003
Holmium	Ho	67	164.9
Hydrogen	H	1	1.008
Indium	In	49	114.8
Iodine	I	53	126.9
Iridium	Ir	77	192.2
Iron	Fe	26	55.85
Krypton	Kr	36	83.80
Lanthanum	La	57	138.9
Lawrencium	Lr	103	(260)
Lead	Pb	82	207.2
Lithium	Li	3	6.941
Lutetium	Lu	71	175.0
Magnesium	Mg	12	24.31
Manganese	Mn	25	54.94
Mendelevium	Md	101	(258)
Mercury	Hg	80	200.6
Molybdenum	Mo	42	95.94
Neodymium	Nd	60	144.2
Neon	Ne	10	20.18
Neptunium	Np	93	(237)
Nickel	Ni	28	58.69
Niobium	Nb	41	92.91
Nitrogen	N	7	14.01
Nobelium	No	102	(259)
Osmium	Os	76	190.2
Oxygen	O	8	16.00
Palladium	Pd	46	106.4
Phosphorus	P	15	30.97
Platinum	Pt	78	195.1
Plutonium	Pu	94	(244)
Polonium	Po	84	(209)
Potassium	K	19	39.10
Praseodymium	Pr	59	140.9
Promethium	Pm	61	(145)
Protactinium	Pa	91	(231)
Radium	Ra	88	226
Radon	Rn	86	(222)
Rhenium	Re	75	186.2
Rhodium	Rh	45	102.9
Rubidium	Rb	37	85.47
Ruthenium	Ru	44	101.1
Samarium	Sm	62	150.4
Scandium	Sc	21	44.96
Selenium	Se	34	78.96
Silicon	Si	14	28.09
Silver	Ag	47	107.9
Sodium	Na	11	22.99
Strontium	Sr	38	87.62
Sulfur	S	16	32.06
Tantalum	Ta	73	180.9
Technetium	Tc	43	(98)
Tellurium	Te	52	127.6
Terbium	Tb	65	158.9
Thallium	Tl	81	204.4
Thorium	Th	90	232.0
Thulium	Tm	69	168.9
Tin	Sn	50	118.7
Titanium	Ti	22	47.88
Tungsten	W	74	183.9
Uranium	U	92	238.0
Vanadium	V	23	50.94
Xenon	Xe	54	131.3
Ytterbium	Yb	70	173.0
Yttrium	Y	39	88.91
Zinc	Zn	30	65.38
Zirconium	Zr	40	91.22

*The values given here are to four significant figures.
†A value given in parentheses denotes the mass of the longest-lived isotope.

APPENDIX B

APPENDIX C

Mineral Identification Table

Mineral	Composition	Cleavage/Fracture	Hardness	Color/Streak	Miscellaneous Properties
Actinolite	Complex Ca, Mg, Fe hydrous silicate	Prismatic (at 56° and 124°)/subconchoidal, uneven	5–6	White to light green/colorless	Slender glassy crystals; vitreous to pearly luster
Amphiboles (see actinolite, tremolite)					
Andalusite	$AlO(SiO_4)$	One direction, distinct	7.5	Flesh, reddish brown, olive green/colorless	Blunt, nearly square prisms; vitreous luster
Anhydrite	$CaSO_4$	Three directions (at 90°)/uneven, sometimes splintery	3–3.5	Colorless to bluish or violet/grayish white	Brittle, somewhat greasy; vitreous to pearly luster
Apatite	$Ca_3(F \cdot Cl \cdot OH)(PO_4)_3$	One direction, imperfect/conchoidal and uneven	5	Green or brown/white	Prominent pyramidal crystals; vitreous to subresinous luster
Augite	$Ca(Mg,Fe,Al)(Si, Al)_2O_6$	Prismatic (at 87° and 93°)/uneven to conchoidal	5–6	Dark green to black/greenish gray	Squarish cross section; vitreous luster
Azurite	$Cu_3(CO_3)_2(OH)_2$	One direction, perfect/conchoidal	3.5–4	Intense azure/blue	Effervesces in HCl; vitreous luster
Barite	$BaSO_4$	Two directions, perfect/uneven	3–3.5	Colorless, white, blue, yellow, red/white	Tabular crystals, also globular; vitreous luster; 4.5 specific gravity
Bauxite	$Al_2O_3 \cdot 2H_2O$	Uneven	1–3	White, gray, yellow, red/colorless	Pisolitic, concretionary grains; dull to earthy luster
Beryl	$Be_2Al_2(Si_6O_{18})$	Poor/conchoidal to uneven	7.5–8	Bluish green or yellow/white	Hexagonal crystal form; vitreous luster
Biotite	$K(Mg,Fe)_3(AlSi_3O_{10})(OH)_2$	One direction parallel to base/none	2.5–3	Dark green, brown, or black/colorless	Thin, flexible sheets; splendent luster
Calcite	$CaCO_3$	Rhombohedral, perfect/conchoidal	3 on cleavage; 2.5 on base	White, colorless/white or grayish	Effervesces in HCl; vitreous to earthy luster
Chalcopyrite	$CuFeS_2$	Sometimes distinct/uneven	3.5–4	Brass yellow, often tarnished/greenish black	Brittle; metallic luster; 4.1–4.3 specific gravity
Chlorite	$(Mg,Fe)_5(Al,Fe)_2Si_3O_{10}(OH)_8$	One direction	2–2.5	Light to dark green, yellowish/white or colorless	Flaky

Mineral	Composition	Cleavage/Fracture	Hardness	Color/Streak	Miscellaneous Properties
Cinnabar	HgS	One direction, perfect	2.5	Vermilion to brownish red/scarlet	Slightly sectile; adamantine to dull earthy luster; 8.1 specific gravity
Copper	Cu	None/hackly	2.5–3	Copper red on fresh surface/red	Highly ductile and malleable; metallic luster
Corundum	Al_2O_3	None	9	Brown, pink, blue, or yellow/colorless	Rhombohedral parting; adamantine to vitreous luster
Diamond	C	Octahedral, perfect/conchoidal	10	White or colorless to pale yellow/white	Very high refractive index; adamantine to greasy luster
Diopside	$CaMg(Si_2O_6)$	Prismatic (at 87° and 93°), imperfect/uneven to conchoidal	5–6	White to light green/grayish green	Squarish cross section; vitreous luster
Dolomite	$CaMg(CO_3)_2$	Rhombohedral, perfect/subconchoidal	3.5–4	Pink, flesh, colorless, or white/colorless	Does not effervesce in cold HCl; vitreous luster
Epidote	Complex Ca, Al, Fe hydrous silicate	Perfect, parallel to base/uneven	6–7	Yellowish green to blackish green/colorless	Peculiar yellowish green color; vitreous luster
Feldspar group (see plagioclase feldspars, potassium feldspars)					
Fluorite	CaF_2	Octahedral, perfect/flat conchoidal	4	Green, yellow, or purple/white	Cubic crystals; vitreous luster
Galena	PbS	Cubic/flat subconchoidal	2.5	Lead gray/lead gray	Brittle; metallic luster; 7.4–7.6 specific gravity
Garnet group	General formula $A_3B_2(SiO_4)_3$	None/conchoidal to uneven	6.5–7	Red or brown/white	12-sided crystals; vitreous to resinous luster; 3.5–4.3 specific gravity
Goethite (limonite)	$FeO(OH)nH_2O$	None	5.5–6	Dark brown to black/yellowish brown	Globular form; silky, often submetallic, earthy luster
Gold	Au	None/hackly	2.5–3	Various shades of yellow/golden yellow	Very malleable and ductile; metallic luster; 19.3 specific gravity when pure
Graphite	C	Basal, perfect	1–2	Black to steel gray/black	Greasy feel, foliated; metallic or dull earthy luster
Gypsum	$CaSO_4 \cdot 2H_2O$	Four directions; three unequal/fibrous in one direction; conchoidal	2	Colorless, white, gray/white	Sometimes in fibrous masses; vitreous, pearly, silky luster

Mineral	Composition	Cleavage/Fracture	Hardness	Color/Streak	Miscellaneous Properties
Halite	$NaCl$	Cubic, perfect/conchoidal	2.5	Colorless or white/colorless	Salty taste; transparent to translucent
Hematite	Fe_2O_3	None	5.5–6	Reddish brown to black/light to dark red	Breaks with almost cubic angles; metallic to dull earthy luster; 5.3 specific gravity
Hornblende	Complex Ca, Mg, Fe, Al hydrous silicate	Prismatic (at 56° and 124°) perfect/subconchoidal, uneven	5–6	Dark green to black/colorless	Long prisms; vitreous or silky luster
Kaolinite	$Al_2(Si_2O_5)OH)_4$	Coarse masses, direction seldom distinct/none	2–2.5	White with colored impurities/colorless	Earthy, soapy feeling; dull earthy luster
Magnetite	Fe_3O_4	Not distinct/subconchoidal to uneven	6	Iron black/black	Strongly magnetic, octahedral parting; metallic and splendent luster; 5.2 specific gravity
Malachite	$Cu_2CO_3(OH)_2$	One direction, perfect/subconchoidal to uneven	3.5–4	Bright green/pale green	Effervesces in HCl; adamantine to vitreous luster
Micas (see muscovite, biotite)					
Muscovite	$KAl_2(AlSi_3O_{10})(OH)_2$	One direction parallel to base/none	2–2.5	Colorless, yellow, brown, or green/colorless	Thin, flexible sheets; vitreous, silky, or pearly luster
Olivine group	$(Mg,Fe)_2(SiO_4)$	Poor/conchoidal	6.5–7	Pale yellow green to brownish black/white or gray	Granular texture; vitreous luster; 3.8–4.4 specific gravity
Plagioclase feldspars					
Albite	$(NaAlSi_3O_8)$ Varies between above 100% sodium aluminum silicate and below 100% calcium silicate	Two directions (at 93°), good/uneven to conchoidal	6	Colorless, white, gray, greenish, or yellow/colorless	Striations on cleavage; vitreous to pearly luster
Anorthite	$(CaAl_2Si_2O_8)$	Two directions (at 94°), good/conchoidal to uneven	6	Colorless, white, gray, or reddish/colorless	Striations on cleavage; vitreous to pearly luster
Potassium feldspars (aluminosilicates)					
Microcline	$K(AlSi_3O_8)$	Three directions (at 89°), good/uneven	6	White, yellow, pink, or green/colorless	Twinned crystals (reversed pairs); vitreous luster
Orthoclase	$K(AlSi_3O_8)$	Two directions (at 90°), good/conchoidal to uneven	6	Colorless, white, gray, flesh, or red/colorless	Twinned crystals (reversed pairs); vitreous luster

Mineral	Composition	Cleavage/Fracture	Hardness	Color/Streak	Miscellaneous Properties
Pyrite	FeS_2	Cubic, poor/conchoidal to uneven	6–6.5	Pale brass yellow/greenish or brownish black	Brittle, cubic crystals; metallic, splendent luster; 5.0 specific gravity
Pyroxene (see augite, wollastonite)					
Quartz	SiO_2	None/conchoidal	7	Colorless or white, with colored impurities/colorless	Crystals end in prisms; vitreous, greasy, splendent luster
Serpentine	$Mg_3(Si_2O_5)(OH)_4$	Sometimes distinct/conchoidal or splintery	2–5	Mixed shades of green/white	Fibrous or platy variety; greasy, silky, and waxy luster
Sillimanite	Al_2SiO_5	One direction	6–7	White to gray/white	Long, slender crystals or fibrous; vitreous luster
Sphalerite	ZnS	Six directions, perfect/conchoidal	3.5–4	Yellow brown, dark brown, or black/white or yellow to brown	Brittle; resinous to adamantine luster; 3.9–4.1 specific gravity
Sulfur	S	None/irregular	1.5–2.5	Bright yellow/colorless	Insoluble; resinous to dull luster
Talc	$Mg_3(Si_4O_{10})(OH)_2$	One direction parallel to base/none	1	Light green, silvery white/white	Soapy feeling; pearly, greasy luster
Topaz	$Al_2(SiO_4)(F \cdot OH)_2$	Perfect, parallel to base/subconchoidal to uneven	8	Colorless to yellow/colorless	Prismatic crystals; vitreous luster
Tourmaline	Complex silicate of B and Al	Poor/subconchoidal to uneven	7–7.5	Commonly black/colorless	Crystals triangular in cross section; vitreous to resinous luster
Tremolite	Complex Ca, Mg hydrous silicate	Prismatic (at 56° and 124°) perfect/splintery surface	5–6	White to light green/colorless	Radiating, bladed columns; vitreous or silky luster
Wollastonite	$Ca(SiO_3)$	Two directions (at 84° and 96°), perfect/uneven	5–5.5	Colorless, white, or gray/white	Sometimes fibrous; vitreous or silky luster
Zircon	$Zr(SiO_4)$	Poor/conchoidal	7.5	Brown, colorless, or gray/colorless	Terminated prisms; adamantine luster; 4.7 specific gravity

Basic Guide to Geologic Maps*

A geologic map uses a combination of colors, lines, and symbols to depict the composition, structure, and age relationships of the rock units within a given portion of the Earth's crust. It shows the Earth as it would appear from the air with all of the loose materials stripped away to reveal the bedrock below. In some regions—such as the American West—this bedrock is already extensively exposed and we are able to trace the various rock units for many kilometers. In most other regions, however, rock exposures are widely scattered, and the map must be constructed from observing cliff faces, ledges, stream banks, road cuts, and other excavations.

Map 1 (Figure D.1) illustrates the first phase in the construction of a geologic map. Each of the symbols (circles, dots, dashed lines, wavy lines, etc.) represents an outcrop that a team of geologists has examined. Each symbol, plotted in its correct relationship to the stream, house, roads, and other outcrops, represents a different kind of rock. Dots, for example, are often used to represent sandstone; circles, conglomerate; straight lines, shale; and wavy lines, schist or gneiss. The symbols depict the basic rock units, or *formations*, of the region.

The flat T-shaped symbols with numbers like 3°, 10°, 25°, and 40° are called *strike-and-dip* symbols and give the amount and direction of slope of the layers in the various outcrops. If such a symbol were used to describe a sloping roof, the long line (the *strike* line) would parallel the direction of the ridge pole, and the short line (the *dip* line) would point directly down the slope of the roof in the direction that water would drain. The number gives the inclination (or dip) of the roof in degrees measured below a horizontal plane. For a flat roof, the dip would be 0°, and for a vertical wall, 90°.

The strike-and-dip symbols tell whether the layers at each place go down into the Earth at a steep or gentle angle and in what direction. The geologists can predict the subsurface geology by the strike-and-dip symbols and the distribution pattern of the rock formations, even though no cliffs or other steep cuts show the vertical dimensions and depths directly. The other numbers (82 to 96) near some of the outcrops refer to entries in the geologists' field notebooks in which they record numerous additional observations on the rocks that cannot be conveniently represented by map symbols. For example, the geologists might notice that the rocks along a line running north-south, just west of the house shown in Map 1 appear more fractured than normal, suggesting that these outcrops lie near an important break or *fault* in the rocks. They may also note interruptions in the characteristic fold patterns of these formations; formations of different ages abut against one another.

Their next step is to construct Map 2 (Figure D.2) by sketching in the inferred boundaries or contacts between the various formations. By drawing cross sections in different directions across the area of Map 2, we can see how it shows the subsurface geology (Figure D.3). We can predict what might be seen in the walls of imaginary trenches 120 or 150 meters deep if they were dug along the lines *A–A'* and *B–B'*. The top line of each cross section shown here represents the surface of the ground. Proceeding from *A* toward *A'* along this top line, the map shows schist, symbolized by wavy lines. About 75 meters from *A*, the schist gives way to sandstone, symbolized by a stipple pattern. The strike-and-dip symbol just north of the stream, Map 2, indicates that this sandstone layer is inclined toward the east at an angle of 45°. It is known by scaling from the map (see bar scale, Map 2) how far it is from *A* to the top and bottom of

*Adapted from Walter S. White, *Geologic Maps: Portrait of the Earth*, U.S. Geological Survey pamphlet, U.S. Government Printing Office, 0-333-138: QL 3, 1992.

Figure D.1

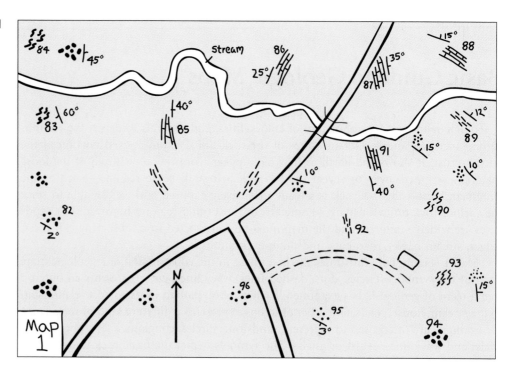

Figure D.2

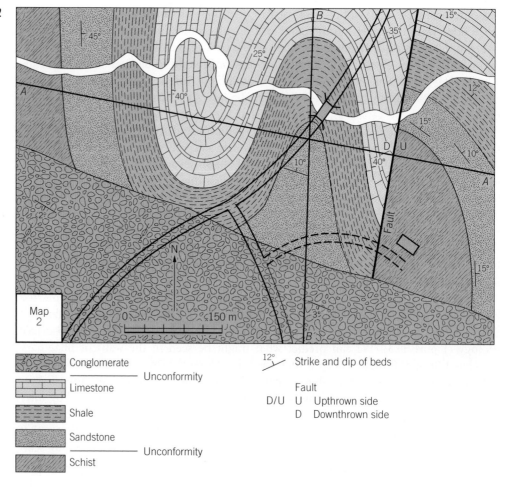

Conglomerate

———— Unconformity

Limestone

Shale

Sandstone

———— Unconformity

Schist

Strike and dip of beds

Fault
D/U   U   Upthrown side
      D   Downthrown side

Figure D.3

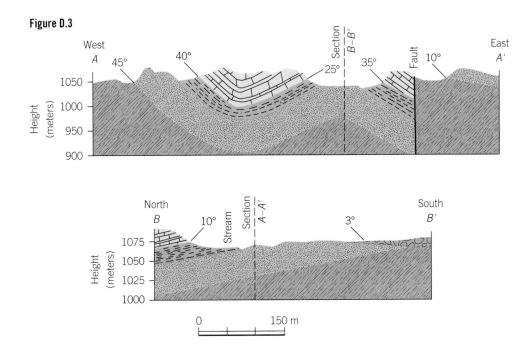

the sandstone layer, and it is known that the sandstone extends downward into the Earth toward the east at an angle of 45°; therefore we can tentatively draw it in with this inclination for at least a few hundred feet below the surface in the cross section. Proceeding eastward in the same way, we can fill out the near-surface part of the section all the way from A to A′. (The heavy line labeled "fault" on the map represents a break in the Earth's crust. The strike-and-dip symbols do not show the fault's attitude, but let us assume that it is vertical.) If the boundary lines of the various rock units are extended downward parallel to their dips, and the patterns are filled in, we can see how rock formations in the western two-thirds of the section probably join below the surface in a trough-shaped structure, or a syncline. Familiarity with similar folded structures elsewhere leads the geologists to round off the bottom as is shown in the cross-sectional drawing.

Eastward, the east side of the *syncline* is the west side of an upwarp, or *anticline,* and everything down to the top of the sandstone can be drawn without any serious stretch of imagination. From the cross section alone, the thickness of the sandstone is not known. However, if the geologists know the thickness of the sandstone in other places where it comes to the surface within and outside the map area, they may be able to predict its thickness in the anticline and complete this part of the cross section with some confidence. Otherwise they might use the thickness of the sandstone layer found in the western part of the section and dash the base of the section to show uncertainty. The upper surface of the schist is the floor on which the sandstone was deposited. The attitude of the schist layers suggested by wavy lines need not conform to the dip of the layers in the sandstone and overlying rocks.

The fault is a break or dislocation in the rock layers, and each rock layer ends against it on the west side and starts anew between the fault and point A′. If the layers almost join across the fault, the displacement is small; if the fit is poor, the displacement is large. Estimates of the actual vertical displacement can be made by measuring how far the rocks on one side of the fault must be shifted to bring the base of the sandstone to the same level. Cross sections can be drawn in other directions and places to determine what lies below the surface in all parts of the area. In section B–B′, drawn nearly north–south, the dip of 10° measured in sandstone gives the same thickness for the shale that was obtained from the construction of section A–A′. Such a check for internal consistency between geologic maps and sections makes both the map and the cross section more credible.

The conglomerate formation (open circle pattern) in the southern part of the area is shown as a thin sheet inclined gently south, as the 3° dip of the bedding suggests. Map 2 shows a discontinuity between the conglomerate and the rocks to the north—the northward limit of the conglomerate is a straight line that cuts across other formation boundaries. This suggests a fault, but the map does not indicate that a fault is necessarily present. If the conglomerate included fossils younger than the other rocks, or if it contained pebbles of rock units to the north, it would be possible to interpret the conglomerate as a broad gravel bank that lies on top of the other rocks; under these circumstances, no fault need be present. This latter interpretation is shown in cross section *B–B′*. The existence of these two possible interpretations shows how a geologic map may correctly represent the distribution of rocks at the surface and still be ambiguous.

We can reconstruct the geologic history of the area encompassed by the map by applying some basic geological concepts to the cross sections. For example, from the principle of superposition, we know that the sandstone in section *A–A′* was originally laid down as a sheet on a relatively flat surface that cuts across the layering in the schist, marking a great time-break, or unconformity. Still earlier, the schist must have been deformed, and erosion then carved a level surface on this rock before the sandstone was deposited. Layers of shale were deposited on the sandstone and this shale was overlain, in turn, by limestone. Detailed study of these formations would probably reveal whether they were laid down on the sea floor, in lakes, or on land, because each kind of rock has physical or chemical characteristics that reflect its environment of deposition. Fossils in the rock then provide information about both the environment and the time of deposition.

After the limestone was deposited, the area was crumpled into folds and later faulted. Again the area was subjected to erosion, and a relatively flat surface was cut across all the rocks. The conglomerate was deposited on this surface, which marks the second unconformity of the region. The area was then tilted slightly toward the south to give the conglomerate its present low dip in that direction. The latest event that produced the land surface was the erosion that is still going on.

In this brief account of how geologists make and use maps to organize field observations for study and analysis, we have described what they do as a series of steps taken in a regular sequence, with each completed before the next is begun. Actually, because a geologic map represents an interpretation, it is used as a diagram to test hypotheses. It may be drawn and redrawn many times before a version satisfies all pertinent facts. If more than one version survives these tests, additional field or laboratory observations may be needed to select the best depiction.

Glossary

aa lava A lava flow with a rough, jagged surface. (p. 199)

ablation The loss, or wastage, of ice from a glacier through melting, evaporation, and calving. (p. 459)

abrasion The mechanical wearing or grinding of rock surfaces by friction and impact of rock particles transported by wind, ice, waves, running water, or gravity. (p. 462)

absolute age The age of object or event in years as determined by radiometric dating. (p. 30)

absolute time The ages of objects or events as measured in time units such as years; determined by radiometric-dating techniques. (p. 271)

abyssal plain A flat, level area of the ocean floor that begins at the foot of the continental rise. (p. 64)

accreted terrane A body of rock that is foreign to its surroundings and is bordered on all sides by faults. (p. 331)

accretionary wedge A large mass of sediment and lava scraped off a descending plate that accumulates on the margin of a continent or island arc bordering a subduction zone. (p. 75, 330)

accumulation The mass of ice gained by a glacier because of snowfall. (p. 459)

active continental margin A boundary between an ocean and a continent marked by plate interaction and thus by frequent volcanic or seismic activity. (p. 78)

aftershocks A series of lower-magnitude earthquakes that may directly follow a higher-magnitude earthquake. (p. 102)

aggradation The process of building up a surface by deposition. (p. 405)

A horizon The uppermost zone of the soil horizon. (p. 357)

alluvial fan A gently sloping, fan-shaped mass of sediment deposited by a stream where it issues from a narrow canyon onto a plain or valley floor. (p. 403)

alpine glacier A glacier in mountainous terrain that flows downslope in a valley previously made by a stream. (p. 456)

amphibole A group of double-chain ferromagnesian silicates whose cleavage planes intersect at approximately 120° and 60°. (p. 158)

andesite A dark, fine-grained, extrusive igneous rock composed of intermediate plagioclase and a ferromagnesian mineral. (p. 172)

angle of repose The maximum slope, or angle, at which a pile of loose material remains stable. (p. 371)

angular unconformity A surface formed by the deposition of sediments on the eroded, upturned edges of older, tilted strata. (p. 273)

anticline A fold with a core of stratigraphically older rocks; usually its convex side is upward. (p. 302)

aphanitic An igneous rock texture in which the mineral components are too small to be identified by the unaided eye. (p. 170)

aquifer A permeable body of subsurface sediment or fractured rock that conducts water. (p. 427)

arête A divide between adjacent valleys, narrowed to a knife-edged ridge by the backward erosion of the valley walls. (p. 463)

artesian well A well under sufficient hydrostatic pressure to force the water to rise above the top of the aquifer. (p. 427)

ash Pyroclastic material less than 2 millimeters in diameter. (p. 201)

asteroid One of several thousand small planetlike bodies of rock and iron that revolve around the Sun. (p. 572)

asthenosphere A semimolten zone of the mantle just below the lithosphere whose depth extends from about 100 to 200 kilometers and is characterized by diminished seismic wave velocities. (p. 69, 129)

atom The smallest unit of an element that still retains the properties of the element; a dense, positively charged nucleus surrounded by negatively charged electrons. (p. 141)

atomic mass The average mass of the atoms of an element; to a close approximation, the number of protons plus the number of neutrons in an atom of the element. (p. 144)

atomic number The number of protons in the nucleus of an atom. (p. 143)

aureole A halolike region of contact metamorphism surrounding an igneous intrusion. (p. 256)

axial plane An imaginary planar surface that divides a fold symmetrically. (p. 302)

backshore The upper zone of the beach, reached by waves only during storms and very high tides. (p. 519)

backwash The return flow of breaking waves down the beach face. (p. 521)

bajada Alluvial fans that coalesce to form a continuous apron along the base of mountains. (p. 498)

barrier island A striplike island that parallels the coast. (p. 522)

basal slip Movement in which a glacier decouples from the valley floor and slides downslope. (p. 459)

basalt A dark, fine-grained, mafic, extrusive igneous rock composed largely of plagioclase feldspar, pyroxene, and olivine. (p. 172)

base level The limiting level below which a stream cannot erode its bed. (p. 396)

basin A synclinal circular structure with rocks dipping gently toward the center. (p. 305)

basin and range A topography or landscape characterized by a series of tilted fault blocks forming longitudinal, asymmetric ridges or mountains with broad, intervening basins. (p. 314)

batholith An igneous intrusion with a large mass, a surface area greater than 100 km², and no known floor. (p. 168)

baymouth bar A strip of beach extending from a headland into the mouth of a bay. (p. 522)

beach A sloping portion of the shore composed of sediments deposited and moved by waves, tides, and longshore currents. (p. 519)

bedding The stratification or layering of sedimentary rock. (p. 232)

bedding plane The plane that marks the boundary of each layer of stratified rock. (p. 232)

bed load Large or dense particles that are transported by streams on or immediately above the stream bed or by winds on or immediately above the ground. (p. 398)

beds (strata) Visually distinguishable layers of sedimentary rock. (p. 232)

belt of soil moisture The thin layer of moisture just beneath the land surface; the uppermost subdivision of the zone of aeration. (p. 425)

berm A terracelike structure of sediments washed by waves up onto the foreshore. (p. 519)

B horizon The zone of the soil profile consisting of enriched clays and precipitates leached from the A horizon. (p. 358)

big bang theory A model of the evolution of the universe that postulates its origin from a hot dense mass that expanded rapidly and cooled. (p. 570)

biochemical sediment A sediment precipitated directly or indirectly by the activities of organisms. (p. 220)

block A pyroclastic fragment larger than 64 millimeters in diameter that is ejected in a solid state. (p. 201)

body wave A seismic wave that travels through the Earth's interior. (p. 94)

bomb A clump of partially molten lava, ejected from a volcano, whose shape is streamlined in flight; larger than 64 millimeters in diameter. (p. 201)

Bowen's reaction series A proposed sequence of mineral crystallization from basaltic magma, based on experimental evidence. (p. 175)

braided stream A stream that divides into branching and intertwining subchannels separated by islands or sandbars. (p. 403)

breccia A detrital rock consisting of angular pebble-size or larger rock fragments commonly set in a matrix of silt or sand. (p. 223)

brittle response The fracturing of a rock in response to stress with little or no permanent deformation prior to its rupture. (p. 299)

burial metamorphism Metamorphism that results in response to the pressure exerted by the weight of the overlying rock. (p. 250)

calcification The encrusting of a soil with alkaline compounds, particularly calcium oxides. (p. 360)

caldera A volcanic crater larger than 1 kilometer in diameter, usually formed by explosion or collapse. (p. 195)

caliche Desert sand or soil cemented with porous calcium carbonate. (p. 360)

calving The process through which blocks of ice break off from ice shelves and float out to sea as icebergs. (p. 460)

capacity The transporting ability of a stream as measured by the quantity of sediment that the stream carries in a given time interval. (p. 400)

capillary fringe The thin belt just above the water table within which moisture is drawn up by surface tension and partly fills the voids in sediments and rocks. (p. 425)

cap rock An impermeable rock layer overlying oil-bearing reservoir rock. (p. 557)

catastrophism The doctrine that a series of sudden, violent, and short-lived worldwide events are responsible for the state of the Earth's crust and for the variety of life-forms that live on it. (p. 26)

cave A natural cavity beneath the surface of the Earth, usually formed by the dissolution of limestone by groundwater; large caves are also called caverns. (p. 436)

cementation The process by which precipitates bind together the grains of a sediment, converting it into rock. (p. 222)

Cenozoic era The most recent of the four eras of geologic time, beginning at the end of the Mesozoic era (66 million years ago) and extending to the present. (p. 282)

chemical bond The forces exerted between atoms that hold the atoms together. (p. 147)

chemical sediment A sediment composed of ions precipitated directly from water. (p. 220)

chemical weathering The surface process that decomposes rocks and minerals through chemical reactions. (p. 347)

chert A hard sedimentary rock formed from the lithification of biochemical silica. (p. 232)

C horizon The layer of weathered bedrock at the base of the soil horizon. (p. 358)

cinder Glassy, vesicular airborne fragment from 4 to 32 millimeters in diameter. (p. 201)

cinder cone A steep-sided volcano formed by the accumulation of ash, cinders, and other debris close to the vent. (p. 204)

cirque A deep, curving, steep-walled depression scooped out of the bedrock at the head of a valley glacier. (p. 462)

clastic texture Texture of a rock composed mainly of fragments of other rocks and minerals; most commonly used to describe detrital rocks. (p. 223)

cleavage The tendency of a mineral to break along parallel planes of weak bonding. (p. 150)

coal A combustible carbonaceous rock formed from the compaction of altered plant remains. (p. 232, 554)

columnar joints Cracks that form as the result of contraction during the cooling of lava flows, dividing the lava into columns. (p. 199)

compaction Reduction in volume of sediments resulting from the weight of newly deposited sediments above. (p. 221)

competence The transporting ability of a stream, as measured by the largest particles that it can carry. (p. 400)

compound A substance formed by the chemical combination of two or more elements in definite proportions and commonly having properties different from those of its constituent elements. (p. 147)

compressive stress The stress generated by forces directed toward one another on opposite sides of a real or imaginary plane. (p. 249, 298)

concordant Pertaining to igneous rock bodies that intruded and solidified parallel to the layers of country rock. (p. 168)

cone of depression A concentric indentation in the water table that develops around a pumping well. (p. 430)

confining bed An impermeable layer adjacent to an aquifer that prevents escape of groundwater from the aquifer. (p. 427)

conglomerate A detrital rock consisting of rounded pebble-size or larger rock fragments commonly set in a matrix of silt or sand. (p. 223)

contact metamorphism The transformation of rocks caused by heat escaping from an igneous intrusion. (p. 250)

continental drift A hypothesis suggesting that the continents move over the Earth's surface. (p. 39)

continental glacier See *ice sheet.*

continental rise That part of the continental margin that extends gently downward from the continental slope to the abyssal plain. (p. 75)

continental shelf A very gently sloping surface that extends from the shoreline to the continental slope. (p. 75)

continental slope A more steeply sloping surface that extends from the continental shelf to the continental rise or oceanic trench. (p. 75)

convection A circular heat transfer mechanism in which material heated from below rises because it is less dense than the cooler material above it. The warm material then surrenders heat, becomes denser, and sinks down to the bottom, where it is reheated. (p. 132)

convection current The flow resulting from the rise of warmer, less dense material and the descent of cooler, denser material. (p. 83)

convergent boundary The border between two plates that are colliding with one another. Crustal material is subducted back into the mantle at these boundaries. (p. 69)

Coriolis effect The apparent deflection of moving objects, such as winds, viewed from a rotating Earth; deflection is to the right of the initial path in the Northern Hemisphere and to the left in the Southern Hemisphere. (p. 490)

correlation The establishing of equivalence, either in age or rock type, of separated rock units. (p. 277)

covalent bond A type of chemical bond in which electrons are shared by atoms. (p. 147)

crater A circular depression; a volcanic crater contains the vent or vents of the volcano. (p. 195)

craton The stable core of the continental crust that includes basement rock, shield, and platform. (p. 324)

creep The imperceptibly slow downslope movement of soil and rock particles, mainly occurring in cold climates where water alternately freezes and thaws. (p. 382)

crevasse A fracture in the rigid outer layer of a glacier caused by glacier movement. (p. 457)

cross-bedding Thin strata laid down by currents of wind or water at an oblique angle to the main bed. (p. 233)

cross-cutting relationships The principle that an intrusion or fault is younger than the rock that it cuts. (p. 275)

crust The outermost layer of the Earth, about 2 to 50 kilometers thick, consisting of continental and oceanic components of distinctly different density and composition. The crust represents less than 0.1 percent of the Earth's total volume. (p. 37, 126)

crystal A solid element or compound whose atoms display a definite, orderly atomic arrangement repeated throughout the solid. (p. 149)

crystal face The smooth surface of an unbroken crystal. (p. 149)

cuspate foreland A coastal landform composed of seaward-projecting beaches that meet at a point. (p. 522)

cutbank A steep slope caused by erosion of the outer bank of a meander. (p. 404)

data Items of factual or statistical information. (p. 12)

daughter product An element formed from the decay of a parent isotope. (p. 284)

debris flow Downslope movement of a viscous mass of rock and soil particles, more than half of which are larger than a sand grain. (p. 381)

deep-sea trench A deep, narrow, elongated depression in the sea floor. (p. 55)

deflation The removal of loose materials by wind, which often results in the lowering of the land surface. (p. 500)

delta A triangle-shaped landform built up over time where a stream enters a calmer body of water and deposits its sediment load. (p. 410)

dendritic drainage A stream drainage pattern resembling a branching tree that is common to regions underlain by materials of uniform composition or horizontal layers. (p. 416)

density The mass per unit volume of a substance. (p. 121)

deposition The gravitational settling of rock-forming materials out of such natural agents as water, wind, or ice. (p. 32, 221)

desert A region with an average annual precipitation of less than 25 centimeters, too dry to support more than sparse vegetation. (Climatologists define deserts more complexly on the basis of temperature-precipitation-evaporation ratios.) (p. 488)

desertification Declining productivity of arable land due to mismanagement, leading to a state resembling a desert. (p. 510)

desert pavement The remaining pebble- to cobble-size rock fragments covering a desert surface after removal of lighter fragments by wind. (p. 500)

desert varnish A dark, shiny, enamel-like film of iron oxide and manganese that coats the surface of rocks in desert regions. (p. 492)

detrital sediment Fragments derived from the weathering of rocks, transported by water, wind, or ice, and deposited in loose layers on the Earth's surface. (p. 220)

dike A tabular, discordant igneous intrusion. (p. 168)

dip The angle formed by the intersection of a bedding or fault plane and the horizontal plane; measured in a vertical plane perpendicular to strike. (p. 301)

dip-slip fault A fault in which the movement is parallel to the dip of the fault plane. (p. 310)

direct solution The dissolving of rock or mineral materials. (p. 351)

discharge The volume of water moving past a particular point in a stream over a given time interval; the product of the cross-sectional area of the channel and the stream velocity. (p. 396)

discharge area The area—primarily stream channels—that receives groundwater from the zone of saturation and conducts it away. (p. 429)

disconformity An unconformity between parallel sedimentary layers. (p. 273)

discordant Pertaining to igneous rock bodies that cut across the layers of country rock. (p. 168)

disseminated deposit An ore finely scattered within a host rock body. (p. 549)

dissolved load The portion of a stream load carried in solution. (p. 399)

distributary One of a system of small channels that carry water and sediment from the mainstream channel and spread them over the delta surface. (p. 410)

divergent boundary The border between two plates that are moving away from one another as new crust is formed. (p. 69)

dome An anticlinal circular structure with rocks dipping gently away from the center. (p. 305)

drainage basin The total area that contributes water to a stream system. (p. 394)

drift Glacial deposit of rock fragments. (p. 463)

drumlin A rounded, low, and elongated hill of compacted till, with its blunter end pointing upstream and its tapering end pointing in the direction of ice flow. (p. 466)

ductile response The permanent deformation of a body, without fracture, in the shape of a solid. (p. 299)

dune A mound or hill of windblown sand. (p. 501)

earthflow Downslope movement of a loose mass consisting mainly of rock fragments and soil in a semifluid state. (p. 379)

earthquake Vibrations within the Earth set in motion by the sudden release of accumulated strain energy. (p. 89)

effluent stream A stream that intersects the water table and gains water from the zone of saturation. (p. 427)

elastic limit The maximum amount of stress that a material can withstand before it deforms permanently. (p. 299)

elastic-rebound theory The concept that earthquakes are generated by the sudden slippage of rocks on either side of a fault plane. In the process, the rocks release gradually accumulated strain energy and are returned to an unstrained condition. (p. 91)

elastic response The deformation of a body in proportion to the applied stress and its recovery once the stress is removed. (p. 299)

electron A tiny particle with a negative charge—equal to a proton's positive charge—that orbits the nucleus of the atom. (p. 141)

electron shell A region surrounding the nucleus occupied by electrons having approximately the same energy. (p. 144)

element A substance made up entirely of atoms of the same atomic number that cannot be decomposed into a simpler substance by ordinary chemical or physical means. (p. 37, 142)

epicenter The point on the Earth's surface that is directly above the focus of an earthquake. (p. 92)

erosion The wearing down of rocks or soil, by weathering, mass wasting, running water, wind, and ice. (p. 32)

esker A snakelike ridge of roughly stratified gravel and sand, left behind when glacial ice melted. (p. 467)

estuary A mouth of a river invaded by the ocean, where fresh- and saltwater mix. (p. 537)

evaporite Deposit from the evaporation of aqueous solutions. The most common example is rock salt. (p. 231)

exfoliation The flaking or stripping of a rock body in concentric layers. (p. 348)

extrusive Igneous rock formed from lava flows or pyroclastic materials that were spewed to the Earth's surface. (p. 170)

facies The characteristics and appearance of a sedimentary rock unit whose conditions of origin differentiate it from neighboring units. (p. 226)

fall The free downward movement of detached rock fragments through the air from a cliff or steep slope. (p. 372)

fault A fracture in bedrock along which rocks on one side have moved relative to the other side. (p. 54, 299)

fault breccia A metamorphic rock consisting of angular fragments that are the result of the grinding and shattering action that occurs along active fault zones. (p. 254)

fault scarp A cliff created by the movement along a fault. (p. 311)

faunal succession The principle that fossil organisms succeed one another in definite and recognizable order, so that rocks containing identical fossils are identical in age. (p. 29, 279)

feldspar A common group of aluminum silicate rock-forming minerals that contain potassium, sodium, or calcium and display right-angle cleavage. (p. 156)

ferromagnesian Category of silicate minerals rich in iron and magnesium. (p. 156)

firn The pelletlike form assumed by snow in its transition to glacial ice. (p. 455)

fissure eruption A volcanic eruption through a long fracture rather than a central vent. (p. 205)

fjord A deep, narrow arm of the sea, formed when a glacial trough that extended to the sea became submerged after melting of the ice. (p. 463)

flood basalt Highly fluid basaltic lava produced during a fissure eruption. (p. 205)

flood plain Layers of sediment beds deposited across a valley as a stream periodically overflows its banks and meanders laterally. (p. 405)

flow Downslope movement of loose rock and soil as a viscous fluid mass. (p. 379)

focus The initial point of rupture within the Earth from which seismic waves are generated in an earthquake. (p. 92)

fold Permanent wavelike deformation in layered rock or sediment. (p. 299)

fold axis A line formed by the intersection of the axial plane and a layer or bed surface. (p. 302)

foliation The arrangement of a rock in parallel planes or layers; in metamorphic rocks, caused by parallel alignment of the minerals. (p. 249)

footwall The rock mass beneath an inclined vein or fault. (p. 310)

foreshocks A series of lower-magnitude earthquakes that may directly precede a higher-magnitude earthquake. (p. 102)

foreshore The sloping surface of the beach that falls within the tidal zone and regularly bears the brunt of breaking waves. (p. 519)

formation A distinctive mappable rock unit. (p. 272)

fossil The remains, trace, or imprint of a plant or animal preserved in rock. (p. 28, 234)

fossil fuel Any hydrocarbon that can be used for fuel. (p. 553)

fracture The manner in which minerals break other than along planes of cleavage. (p. 152)

fracture zone A region of closely spaced cracks or faults in rocks. Also, a prominent crack or fault that runs perpendicular to, and offsets, the mid-ocean ridge. (p. 55)

frost wedging A process by which water seeps into rock joints, freezes, and wedges the rock apart. (p. 348)

fumarole A vent or ground opening that spews volcanic fumes or vapors. (p. 202, 442)

galaxy An enormous collection of stars held together by gravitational attraction. (p. 568)

geologic map A visual display of the rock types, distribution, ages, and structural features of a given portion of the Earth's surface. (p. 15)

geologic time scale A chronological ordering of geologic events and time units listed in sequence from the oldest on the bottom to the youngest on the top. (p. 29)

geologist One who investigates the materials, processes, products, and history of the Earth. Geologists conduct basic scientific research in order to increase our understanding of the Earth, and they apply their knowledge to improve our lives in many ways. (p. 3)

geology The study of the materials, processes, products, and history of the Earth. (pp. 3, 25)

geothermal energy Hot steam from the Earth's interior useful for generating power. (p. 562)

geothermal gradient The rate of increase of the temperature in the Earth with depth; it varies from region to region. The geothermal gradient averages 25 °C per kilometer of depth in the crust. (p. 131)

geyser A periodic eruption of hot water and steam, caused by the heating of pressurized groundwater in a network of underground channels in contact with hot rock or magma. (p. 441)

glacial age An interval of geologic time in which glacial ice sheets advance. (p. 471)

glacial erratic A large rock fragment or boulder, carried by glacial ice away from its place of origin and usually deposited on bedrock of a different type. (p. 466)

glacial striations Parallel grooves and scratches that were gouged in the bedrock by rock fragments attached to the bottom of the ice as the glacier slid over it. (p. 462)

glacial terminus The extremity or outer margin of a glacier; the glacial front. (p. 457)

glacial trough A U-shaped valley carved by a mountain glacier in what was previously a V-shaped stream valley. (p. 463)

glacier A large ice mass formed on land by the compaction and recrystallization of snow that survives from year to year and shows evidence of present or past flow. (p. 455)

glassy The texture of an igneous rock having a high content of glass. (p. 170)

gneiss A foliated metamorphic rock in which bands of granular minerals alternate with bands of flaky minerals. (p. 253)

graben An elongated, depressed block bounded by normal faults on its long sides. (p. 313)

gradation The balance between erosion and deposition that maintains a general slope of equilibrium trending toward sea level. (p. 347)

graded bedding Stratification in which particle size changes from coarse to fine from the bottom to the top of each bed. (p. 232)

graded stream A stream whose slope and channel are so adjusted that it has just enough energy to transport its load; no excess energy is present to erode the channel, nor is there a deficiency of energy that leads to deposition of sediment in the channel. (p. 400)

gradient The steepness or slope over a specific length of a stream channel. (p. 396)

granite A coarse-grained, intrusive igneous rock containing quartz and feldspar (primarily potassium feldspar). (p. 173)

greenhouse effect The warming of the Earth's atmosphere through the presence of atmospheric gases that absorb and re-radiate the heat rising from the surface. (p. 212)

groin A structure projecting perpendicular to the beach, designed to trap sediments and prevent beach erosion. (p. 525)

ground moraine A rough blanket of till that accumulated under a glacier and was exposed as the terminus receded toward the head of the glacier. (p. 465)

groundwater Underground water within the zone of saturation, below the water table. (p. 424)

guyot A sea-floor volcano that at one time rose above sea level, where its top was planed by wave erosion. (p. 65)

half-life The time required for half of a given parent isotope to decay to its daughter product. (p. 285)

hanging valley A glacial valley, formed in what was previously a tributary stream valley that was cut off and left stranded at a higher elevation from the main glacial valley. (p. 463)

hanging wall The rock mass overlying an inclined vein or fault. (p. 310)

hardness The resistance of the surface of a mineral to scratching. (p. 152)

headward erosion The lengthening of a stream or gully by erosion at the head of the stream valley. (p. 416)

horn A steeply carved rock peak, formed by the walls of three or more cirques. (p. 463)

hornfels Fine-grained rock formed by contact metamorphism. (p. 255)

horst An uplifted block bounded by normal faults on its long sides. (p. 313)

hot spot An area of volcanic activity produced by a plume of magma rising from the mantle. (p. 68)

hot spring A spring whose waters have been heated above body temperature (36.7 °C) by hot rock or magma. (p. 442)

hydraulic gradient The slope of the water table, measured by the difference in elevation of any two points along the water table divided by the horizontal distance between them. (p. 428)

hydrocarbon deposit An accumulation of organic compounds in the solid, liquid, or gaseous state. (p. 543)

hydrologic cycle The continuous circulation pattern of the world's water, from ocean to atmosphere to Earth's land surface to ocean again. (p. 31)

hydrolysis The reaction between minerals (especially silicates) and water. (p. 351)

hydrothermal metamorphism The transformation of rock through the action of high-temperature solutions. (p. 251)

hypothesis A tentative explanation of a phenomenon or process that is tested for validity by repeated observation or experimentation. (p. 4)

iceberg A block of glacial ice floating on a body of water, with 80 percent or more of its volume below the surface. (p. 460)

ice cap A glacier less than 50,000 square kilometers in area; either dome-shaped, blanketing the summit of a mountain mass, or plate-shaped, covering a flat landmass, such as an Arctic island. (p. 456)

ice sheet A glacier more than 50,000 square kilometers in area that spreads in all directions unconfined by underlying topography. (p. 456)

ice shelf A sheet of thick ice terminating in steep cliffs. One end remains attached to the land; the other end extends out into the sea, where it floats. (p. 459)

igneous rock Rock that has cooled from the molten, or magmatic, state. (p. 33)

impact cratering The crustal deformation caused by a meteoroid collision with a planetary surface. (p. 575)

incised meander A deep, narrow winding stream valley caused by the combination of regional uplift and stream downcutting. (p. 414)

index fossil A fossil used to accurately establish the relative age of the stratum within which it is found. (p. 280)

index mineral A mineral that characterizes a given intensity of metamorphism, having developed under a specific range of temperature and pressure conditions. (p. 262)

influent stream A stream that lies above the water table and loses water to the zone of saturation. (p. 428)

inner core The solid spherical center of the Earth, extending from the outer core to the Earth's crust. (p. 38, 125)

inselberg An isolated knob or hill, remnant of a heavily eroded mountain in a desert region; from the German meaning "island mountain." (p. 499)

interglacial age An interval of geologic time in which glacial ice sheets retreat. (p. 471)

intermittent stream A stream whose channel conducts water during periods of high rainfall and is dry during intervening periods. (p. 493)

intraplate earthquake An earthquake whose epicenter is far from a lithospheric plate boundary. (p. 105)

intrusive An igneous rock derived from magma that solidified within the mantle or crust. (p. 170)

ion An atom with either a positive or a negative charge caused by the loss or gain of electrons. (p. 147)

ionic bond A chemical bond that holds two oppositely charged ions together through electron transfer. (p. 147)

island arc A curved chain of islands with its convex side to the ocean, emerging from the deep-sea floor. Generally located close to a continent. (p. 58)

isograd A line on a map connecting points of equal metamorphic intensity; usually indicated by the first appearance of a given index mineral. (p. 262)

isostasy The condition of balance, or equilibrium, in which the crust floats on the mantle. (p. 37)

isotope A variety of an element that differs from other varieties of the same element in the number of neutrons it contains and thus in its atomic mass. (p. 144)

joint A fracture in a rock, without noticeable movement along the plane of fracture. (p. 299)

joint set A group of joints, generally parallel. (p. 309)

kame A mound or short ridge of stratified sand and gravel deposited by water streaming under or trapped within glacial ice; from the Scottish word meaning "steep-sided monument." (p. 468)

karst topography Topography formed by the intensive dissolution of underlying limestone bedrock, featuring sinkholes, intricate cave networks, and diversion of surface drainage underground. (p. 439)

kettle A depression in a moraine or outwash plain formed when a large block of ice was isolated from the retreating glacier and buried in the drift, melting afterward. (p. 468)

laccolith A mushroom-shaped, concordant igneous intrusion that has domed the overlying crustal rocks. (p. 168)

lahar A mudflow of volcanic material. (p. 201)

laminar flow A type of flow in which water moves in straight paths parallel to the channel. (p. 398)

landslide The rapid downslope movement along shear planes of a mass of rock fragments and soil. (p. 374)

lapilli Pyroclastic particles that range in size from 2 to 64 millimeters in diameter. (p. 201)

lateral moraine A long, narrow mound of till that lies perpendicular to the glacier front, formed by plucking and rockfalls along the valley walls. (p. 466)

laterite A highly weathered red soil typical of tropical and semitropical climates, enriched in iron and aluminum oxides. (p. 359)

lava Magma that flows out onto the surface of the Earth; also refers to the rock body formed after the magma cools. (p. 195)

lava dome A convex structure of solidified lava, extruded from the vent of a volcanic crater. (p. 200)

lava plateau An elevated, flat-topped region composed of a thick succession of horizontal lava flows. (p. 205)

leaching The transfer in solution of organic matter and other elements from upper to lower soil levels. (p. 356)

limestone The consolidated product of calcareous sand, limy mud, and/or crushed shells. (p. 228)

liquefaction The transformation of saturated sediment or soil to liquid when ground shaking causes the particles to lose contact. (p. 112)

lithification The process by which sediment is converted to sedimentary rock through compaction, cementation, or crystallization. (p. 34, 221)

lithosphere A rigid zone of the Earth that includes the crust and a sliver of upper mantle and that rests directly on the asthenosphere. (p. 69, 127)

lithostatic stress The uniform stress in the Earth's crust, caused by the weight of the overlying rocks. (p. 249)

local base level A transitory, local feature, such as a lake, dam, or resistant rock layer, below which a stream cannot erode its channel. (p. 396)

loess A thick deposit of fine windblown dust consisting of unstratified silt-size calcium carbonate and bits of clay, which has the ability to maintain nearly vertical walls despite its weak cohesion. (p. 501)

longitudinal dune A long ridge of sand standing parallel to the prevailing wind. (p. 503)

long profile A cross section of a stream channel showing the gradient from the source to the mouth of the stream. (p. 396)

longshore bar A shallow sand ridge created by wave action seaward of the shoreline. (p. 520)

longshore current A shallow current parallel to the coast, caused by waves that approach the shore at an oblique angle. (p. 521)

longshore sediment transport The movement of sediment parallel to the shore by longshore currents, swash, and backwash. (p. 521)

lower mantle The part of the mantle that extends from 670 to 2900 kilometers in depth. (p. 130)

lunar highland Highly cratered elevated region of the Moon. (p. 577)

luster The quality and intensity of light reflected from the surface of a mineral. (p. 153)

magma Molten (hot-liquid) rock material, generated within the Earth, that forms igneous rocks when solidified. (p. 33, 167)

magnetic field The region surrounding a magnetized body that is subject to magnetic force. (p. 122)

magnetic reversal A switch in the direction of the Earth's magnetic field such that a compass needle that today points north would, at the time of reversal, have pointed south. (p. 61)

magnitude A measure of earthquake strength as interpreted from the maximum wave amplitude recorded by a seismograph. (p. 97)

mantle The layer of the Earth located between the crust and the outer core whose depth extends from about 20 to 2900 kilometers. (p. 37, 125)

mantle plume A pipe-shaped mass of heat-softened light rock that rises from the mantle toward the crust. (p. 133)

marble A metamorphic rock consisting mainly of recrystallized calcite and/or dolomite. (p. 255)

maria (singular *mare*) Dark-colored basaltic lowland regions of the Moon. (p. 577)

mass wasting The downslope movement of rocks and soil caused by gravity alone, without the aid of a transport medium such as a stream, glacier, or lava flow. (p. 32, 369)

matrix The fine-grained material surrounding larger grains in a rock. (p. 222)

meander *As a verb:* to flow in a winding, sinuous course. *As a noun:* a loop created as a stream winds back and forth across a valley. (p. 403)

meander cutoff A new channel created when a stream takes the shorter route across the neck of a meander. (p. 405)

mechanical weathering The combination of physical processes that disintegrates a rock without chemical change. (p. 347)

medial moraine The joining of lateral moraines where two valley glaciers merge. (p. 466)

mélange A chaotic assemblage of accretionary wedge sediments and oceanic crust thrown up onto land at a subduction zone. (p. 330)

Mercalli intensity scale A 12-point scale that measures earthquake severity in terms of the damage inflicted. (p. 100)

Mesozoic era The era preceding the Cenozoic, extending from about 245 million years to 66 million years ago. (p. 282)

metallic bond A chemical bond created in electron-donating elements through the merging of electron shells. (p. 148)

metamorphic facies Mineral assemblages that have formed within a well-defined set of pressure-temperature conditions. (p. 262)

metamorphic rock A rock that has been altered from its original state through metamorphism. (p. 35)

metamorphic zone An area of equal metamorphic intensity between isograds in which the mineral content of rocks remains constant. (p. 262)

metamorphism The structural and mineralogical changes that occur in solid rock through the action of heat, pressure, and chemically active fluids. (p. 35)

metasomatism A replacement process whereby the elements of a rock are exchanged with those of a magmatic fluid. (p. 251)

mid-ocean ridge A seismically and volcanically active mountain range, marked by a central rift valley, that extends continuously through the major ocean basins. (p. 54)

migmatite From the Greek *migma,* meaning "mixture"; a complex mixture of metamorphic and igneous rock components (usually granite). (p. 253)

mineral A naturally occurring and inorganic solid with a definite chemical composition and orderly internal atomic arrangement. (p. 141)

mineral deposit A naturally occurring accumulation of metallic or nonmetallic minerals having potential economic value. (p. 543)

model A hypothesis expressed as a visual or statistical simulation, or as a description by analogy of phenomena or processes that are difficult to observe and describe directly. (p. 11)

Mohorovičić discontinuity (Moho) The surface that defines the boundary between the Earth's crust and mantle. (p. 126)

moment of inertia A measure of distribution of mass within an object that determines the ease with which it rotates. (p. 122)

monocline A sudden steepening in an otherwise gently dipping strata. (p. 305)

moraine A landform composed of till left behind by a retreating glacier. (p. 465)

mud crack Polygonal crack formed by the drying and shrinking of mud, silt, or clay. (p. 234)

mudflow Downslope movement of a fluidized mass of clay and other fine-grained materials. (p. 379)

multiple working hypotheses An approach to geologic research in which several possible explanations of a phenomenon are developed and evaluated simultaneously and impartially. (p. 11)

natural levee A ridge of sand and silt that parallels the stream channel and that is deposited over time, when the stream overflows its banks during floods. (p. 405)

neap tide A low-amplitude, twice-monthly ocean tide. (p. 531)

neutron A particle in an atomic nucleus with a mass virtually equal to the proton's but with no electric charge. (p. 141)

nonconformity An unconformity with stratified rocks above and igneous or metamorphic rocks below. (p. 273)

nonrenewable resource A resource that is replenished in time periods that are very long by human standards. (p. 544)

normal fault A fault in which the hanging wall has been moved downward relative to the footwall. (p. 310)

nucleus The dense center of an atom composed of protons and neutrons. Nearly all of the mass of an atom is concentrated in the nucleus. (p. 143)

nuée ardente A turbulent, ground-hugging, gaseous cloud erupted from a volcano. (p. 201)

obsidian A volcanic glass, either black or dark-colored, usually of rhyolite composition, and characterized by conchoidal fracture. (p. 199)

offshore The submerged zone just seaward of the beach proper. (p. 519)

O horizon The top layer of soil consisting of decayed plant matter. (p. 357)

oil field A geologically and spatially related feature containing two or more oil accumulations. (p. 557)

oil sand Broadly defined as a porous sand deposit from which petroleum can be extracted; also refers to certain sedimentary deposits impregnated with tarlike petroleum residue. (p. 558)

oil shale A fine-grained sedimentary rock from which oil or gas can be distilled. (p. 558)

oil trap A structural or stratigraphic arrangement within a sedimentary basin, consisting of permeable reservoir rock overlain by impermeable cap rock, which allows oil to accumulate and prevents its escape. (p. 556)

olivine A ferromagnesian silicate common in mafic and ultramafic rocks having an internal structure of isolated tetrahedra. (p. 158)

ore The naturally occurring material from which an economically valuable mineral can be profitably extracted. (p. 544)

original horizontality The principle that sediments are deposited in horizontal layers, parallel to the Earth's surface. (p. 272)

orogeny The process of mountain building. (p. 325)

outer core The upper layer of the Earth's core whose depth extends from about 2900 to 5140 kilometers. It is presumed to be molten. (p. 37, 125)

outwash plain A broad, gently sloping sheet of stratified sand and gravel sediment deposited by meltwater streaming out of the front of a glacier. (p. 467)

overturned fold A fold in which one limb has rotated past the perpendicular, so that both limbs dip in the same direction. (p. 302)

oxbow lake A crescent-shaped abandoned meander channel isolated from the stream channel by a meander cutoff and sedimentation. (p. 405)

oxidation The process by which an element combines with oxygen. (p. 352)

pahoehoe lava A lava flow with a smooth, ropy surface. (p. 199)

paleomagnetism The study of the Earth's past magnetism as recorded in rocks at the time of their formation. (p. 45)

Paleozoic era The second of the geologic time eras, extending from the end of the Precambrian (about 570 million years ago) to the beginning of the Mesozoic (about 245 million years ago). (p. 282)

Pangaea The name of the supercontinent that existed between 200 and 300 million years ago, which has since fragmented into the present continents. (p. 39)

parent isotope An isotope undergoing radioactive decay. (p. 284)

passive continental margin A margin between an ocean and a continent that does not include a plate boundary. It is also free of volcanic or seismic activity. (p. 75)

paternoster lake A small lake in a depression carved in a glacial valley floor. (p. 463)

pediment A broad erosional surface formed by running water that slopes gently away from a receding mountain front; the bedrock may be exposed or thinly covered with debris. (p. 499)

perched water table The upper limit of a local groundwater body stranded above the regional water table by an impermeable layer of rock or sediment. (p. 427)

peridotite A dark, coarse-grained, intrusive igneous rock composed mainly of olivine, small amounts of pyroxene, and little or no plagioclase. (p. 172)

periodic table of the elements A chart of the elements arranged in order of increasing atomic number and according to similarities of electron structure. As a consequence, the elements are divided into groups having similar chemical properties. (p. 145)

permafrost A condition of permanently frozen soil or subsoil occurring in cold climates. (p. 383)

permeability The capacity of a material to conduct a fluid; measured by the volume of fluid that will move through a cross-sectional area in a given period of time, subject to a given pressure difference. (p. 426)

petroleum A naturally occurring liquid of complex hydrocarbon composition. Depending on context, the term may include natural gas. (p. 555)

phaneritic An igneous rock texture in which the mineral components are visible to the unaided eye. (p. 170)

Phanerozoic eon That part of geologic time represented by rocks in which the evidence of life is abundant. (p. 282)

phenocryst A relatively large crystal embedded in a finer-grained matrix. (p. 199)

phreatic eruption A volcanic eruption of steam, mud, or ash initiated by the contact of water with magma or hot rock. (p. 198)

piedmont glacier A glacier formed by the joining of two or more mountain glaciers as they flow out onto the lowlands at the base of a mountain range. (p. 456)

pillow lava A term applied to lavas of ovoid or pillow shape. (p. 198)

placer deposit An accumulation of mineral particles, derived from the mechanical weathering of a mother lode or source rock, that is transported and deposited by running water. (p. 552)

plagioclase A group of feldspars in which calcium and sodium substitute for one another in solid solution. (p. 162)

planet A large body that revolves around a star and is incapable of generating light internally. (p. 570)

plate A rigid segment of the lithosphere that moves as a unit over the asthenosphere. (p. 68)

plate tectonics The theory that proposes that the lithosphere is divided into plates that interact with one another at their boundaries, producing tectonic activity. (p. 68)

platform That part of the craton consisting of essentially horizontal sedimentary strata overlying the older basement rocks. (p. 324)

playa A flat area at the bottom of a desert basin that becomes an intermittent (playa) lake in the wet season. (p. 498)

Pleistocene epoch An interval characterized by widespread glacial advances and retreats that began 2 million years ago and ended about 11,000 years ago. (p. 471)

plucking An erosion process in which meltwater trickles into rock fractures, refreezes, and expands, breaking off rock fragments. (p. 461)

plunging fold A fold whose axis is inclined rather than horizontal. (p. 304)

pluton An intrusive rock body. (p. 168)

pluvial lake A body of water in a nonglaciated region formed by abundant rainfall due to the cooler climate caused by the growth of ice sheets. (p. 470)

pocket beach A small beach nestled between adjacent headlands. (p. 519)

podzolization The process of nutrient leaching of the A horizon that produces the ashy gray podzol soils of coniferous forests in cool, moist climates. (p. 361)

point bar A low, crescent-shaped deposit of sand and gravel developed on the inner bank of a meander where the water velocity is low. (p. 404)

polymorph A mineral having a different crystal structure from minerals of the same composition. (p. 149)

porosity The ratio of open space to total volume of a subsurface material. (p. 425)

porphyritic An aphanitic igneous-rock texture in which large crystals are embedded within an aphanitic or glassy matrix. (p. 170)

pothole A smooth, bowl-shaped depression in the bed of a stream caused by abrasion when turbulent currents circulate stones or coarse sediment. (p. 398)

Precambrian All geologic time prior to the beginning of the Paleozoic era (about 570 million years ago). (p. 282)

primary (P) wave A seismic wave that propagates through the Earth as a series of compressions and expansions. (p. 94)

principle of cross-cutting relationships See *cross-cutting relationships.*

principle of faunal succession See *faunal succession.*

principle of original horizontality See *original horizontality.*

principle of superposition See *superposition.*

proton A positively charged particle in an atomic nucleus. (p. 141)

pumice A very light, cellular, glassy rock; often floats in water and is commonly used as an abrasive. (p. 199)

pyroclastic Pertaining to rock material formed by an explosive ejection from a volcanic vent. (p. 171)

pyroclastic rock A rock of any size formed from the cementation or welding of volcanic fragments. (p. 201)

pyroxene A ferromagnesian silicate common in mafic and ultramafic rocks having single-chain structure and right-angle cleavage. (p. 158)

quartz A mineral composed exclusively of silicon dioxide, whose tetrahedra are linked in a framework structure. (p. 156)

quartzite A metamorphic rock composed mainly of quartz, formed by the recrystallization of sandstone. (p. 255)

radial drainage A stream pattern that radiates like the spokes of a wheel from the summit of a volcano. (p. 416)

radioactive decay The disintegration of certain isotopes by the emission of subatomic particles. (p. 144, 283)

rain shadow A dry region on the leeward side of a mountain range where rainfall is significantly less than on the windward side; the region is said to be "in the rain shadow" of the mountain range. (p. 490)

recessional moraine A lateral or end moraine accumulated during a pause in a glacier's retreat. (p. 466)

recharge area The area—mainly the broad upland between stream valleys—that receives precipitation and adds water to the zone of saturation. (p. 429)

recrystallization The formation of new crystalline mineral grains in a rock. (p. 222)

rectangular drainage A drainage pattern commonly found in homogeneous igneous and metamorphic rocks in which tributary streams display right-angled bends that follow joints and faults. (p. 417)

recumbent fold An overturned fold whose axial plane is horizontal. (p. 302)

reflection The return of a wave to its original medium upon striking the boundary of another medium. (p. 123)

refraction The change of direction that occurs when a wave passes from one medium to another; also, the bending of water waves upon entering shallow water. (p. 123, 503)

regional metamorphism Metamorphism of an extensive area of the crust, generally associated with intensive compression and mountain building. (p. 250)

regolith All the fragmented and unconsolidated material overlying bedrock. (p. 356)

relative age The age of an object or event expressed relative to the age of other objects or events but not in relation to time units such as years. (p. 28)

relative time The chronological ordering of events without reference to their age in years. (p. 271)

renewable resource A resource that is replenished in relatively brief time periods, so it can be used again by humans. (p. 543)

reserve An identified resource that can be profitably extracted from the Earth at current market conditions and levels of technology. (p. 543)

reservoir rock Any porous and permeable rock, such as sandstone, that might contain oil or natural gas deposits. (p. 556)

residual ore A mineral deposit containing metals or other valuable elements that is concentrated at the surface through the removal of soluble materials by weathering and groundwater leaching. (p. 550)

resource Any Earth material of value to society that is known to exist, is inferred to exist, or may eventually become available. (p. 543)

reverse fault A fault in which the hanging wall has been raised relative to the footwall. (p. 310)

Richter magnitude scale A logarithmic scale of earthquake magnitudes based on the maximum-amplitude wave recorded by a standard seismograph and corrected for distance of the seismograph from the epicenter. (p. 97)

ripple mark A corrugated form displayed in sedimentary rocks caused by currents of air or water that moved over the sediments prior to lithification. (p. 234)

roche moutonnée French term for a glacially sculpted knob of rock, gentle and smooth on its upstream side; steep and rough on its downstream side. (p. 462)

rock cycle A model that describes the formation, breakdown, and reformation of a rock as a result of sedimentary, igneous, and metamorphic processes. (p. 36)

rockslide The rapid downslope movement along shear planes of a mass consisting mostly of large chunks of bedrock. (p. 374)

saltation Transport by streams in which bed load particles are moved in short skips and bounces. (p. 400)

saltwater encroachment The displacement in the zone of saturation of freshwater by saltwater. (p. 433)

sandstone A detrital rock consisting primarily of sand held together by a cementing agent. (p. 223)

schist A foliated, coarse-textured metamorphic rock; most of the minerals display a pronounced parallelism. (p. 253)

scientific method A process of investigation in which a problem is identified, data are collected and analyzed, and a hypothesis is formulated and tested. (p. 11)

sea arch A bridgelike erosional remnant of coastal rocks. (p. 518)

sea-floor spreading The theory that new oceanic crust is created at the mid-ocean ridges, spreads laterally, and descends back into the mantle at the deep-sea trenches. (p. 60)

seamount A sea-floor volcano that has never risen above sea level. (p. 65)

sea stack An erosional remnant of a former headland. (p. 518)

secondary enrichment The process by which acidic water leaches metals downward from the surface and precipitates them below the water table to form a metallic mineral deposit, usually composed of sulfide minerals. (p. 551)

secondary (S) wave A seismic wave that causes the components of a rock to vibrate perpendicularly to the direction of wave propagation. Also called a shear wave. (p. 94)

sediment Particles that are mechanically transported by water, wind, or ice, or chemically precipitated from solution, or secreted by organisms, and deposited in loose layers on the Earth's surface. (p. 26, 219)

sedimentary rock A layered rock formed from the consolidation of sediment. (p. 26)

seismic belt A long, narrow zone of earthquake activity associated with lithospheric plate boundaries. (p. 68)

seismic gap A seismically inactive segment of a fault within which strain is accumulating; these gaps are bracketed by the epicenters of relatively recent great earthquakes. (p. 107)

seismic tomography A computer-imaging technique based upon the analysis of the variations in velocities that occur along the path of seismic waves. (p. 132)

seismic wave A wave generated within the Earth by sudden fault slippage or an explosion. (p. 92)

seismogram A record of seismic waves detected by a seismograph. (p. 95)

seismograph A device that detects and records seismic waves. (p. 95)

settling velocity The velocity required for a suspended particle of a given size to sink to the bottom of the stream channel. If the stream velocity is greater than this value, the particle will remain in suspension; if less, it will sink. (p. 400)

shadow zone A region between 105° and 142° from the epicenter of an earthquake where no direct P and S waves reach the Earth's surface, owing to the properties of the outer core. (p. 125)

shale A fine-textured detrital rock with a layered structure resulting from compaction of clay, mud, or silt. (p. 223)

shear metamorphism The transformation of rocks within the shear zone associated with active fault movement; mainly involves grinding, pulverizing, and recrystallization of the rocks. (p. 250)

shear strength The internal resistance of a body to shear stress, resulting from particle friction and cohesion, and moisture surface tension in pore spaces. (p. 370)

shear stress Stress (force per unit area) that acts parallel to a (fault) plane and tends to cause the rocks on either side of the plane to slide by one another. (p. 91, 298)

sheeting A type of jointing parallel to the rock surface caused by pressure release. Similar to exfoliation. (p. 348)

shield A large region of exposed metamorphic and igneous basement rocks, generally having a gently convex surface and surrounded by a sediment-covered platform. (p. 324)

shield volcano A broad, low-profile volcanic cone, commonly composed of basaltic flows. (p. 203)

shock metamorphism Changes in rock and minerals caused by shock waves from high-velocity impacts, mainly from meteorites. (p. 251)

silicate A compound containing silicon and oxygen ions arranged as negatively charged ions. (p. 155)

silicon-oxygen tetrahedron An arrangement in which four oxygen ions surround one silicon ion, forming a four-sided structure of negative charge. (p. 158)

sill A tabular, concordant igneous intrusion. (p. 168)

sinkhole A circular surface depression that occurs when the roof of an underground cave collapses or dissolves. (p. 438)

slate A fine-grained, foliated metamorphic rock; mostly formed from the transformation of shale. (p. 253)

slide The rapid downslope movement along shear planes of more or less consolidated rock or fragmented materials. (p. 374)

slip face The steeper downwind (or leeward) side of a dune. (p. 502)

slope wash Unchanneled water that moves downhill; also, the rock and soil that is transported by the water. (p. 415)

slump Downslope movement of rock or loose debris along a concave plane. (p. 379)

snowline Altitude above which the snow is permanent. (p. 455)

soil Unconsolidated material capable of supporting vegetation. (p. 356)

soil horizon A soil layer with physical and chemical properties that differ from those of adjacent soil layers. (p. 357)

soil profile A vertical cross section of soil that displays all soil horizons. (p. 357)

solid solution Variations in the composition of a mineral caused by the substitution of metal ions within the crystal structure. (p. 160)

solifluction The slow downslope movement of waterlogged soil caused by repeated freezing and thawing in cold climates. (p. 383)

sorting The process by which the agents of transportation (principally, running water) separate sediments according to shape, size, or density. (p. 223)

specific gravity The density of a substance compared with the density of water. (p. 153)

spheroidal weathering The peeling of small rock bodies into onionlike layers. (p. 349)

spit A curving, fingerlike projection of beach. (p. 522)

spreading center The segment of the mid-ocean ridge that is the site of active rifting and sea-floor spreading. (p. 72)

spring A place where the land surface intersects the water table and groundwater seeps or flows naturally out of the ground. (p. 428)

spring tide A high-amplitude, twice-monthly ocean tide. (p. 531)

stalactite A calcite deposit that projects from a cave roof, precipitated from groundwater supersaturated with calcium carbonate. (p. 437)

stalagmite A cone-shaped calcite deposit, growing up from a cave floor, precipitated from groundwater supersaturated with calcium carbonate. (p. 437)

strain The result of stress applied to a body, causing the deformation of its shape and/or a change of volume. (p. 298)

strata (beds) Visually distinguishable layers of sedimentary rock. (p. 26, 226)

stratified drift Deposits left by streams or pools of glacial meltwater that are sorted and layered according to size. (p. 465)

stratovolcano (composite cone) A volcanic cone consisting of alternating layers of pyroclastic deposits and lava. (p. 206)

streak The color of the powder of a mineral when scratched on a porcelain plate. (p. 153)

stream Flowing water within a channel of any size. (p. 393)

stream divide The boundary that separates adjacent stream valleys. (p. 415)

stream piracy The diversion of a stream into another stream that has a steeper gradient. (p. 417)

stream terrace An elevated shelflike surface upon which the stream formerly flowed; typically, an ancient flood-plain remnant abandoned as the stream cut downward and established a course at a lower level. (p. 414)

stress The force applied to a plane divided by the area of the plane. (p. 298)

strike The direction of the line formed by the intersection of a horizontal plane with a bedding or fault plane. (p. 301)

strike-slip fault A fault in which the movement is parallel to the strike of the fault plane. (p. 310)

subduction The downward plunging of one lithospheric plate under another. (p. 75)

subduction zone The region where one lithospheric plate thrusts downward under another. (p. 75)

submarine canyon A steep canyon, resembling a V-shaped stream valley, that is cut into the continental shelf or slope. (p. 536)

subsidence The sinking of an area of the Earth's surface upon compaction or dissolution of subsurface materials, often caused by groundwater extraction. (p. 432)

superposition The principle that in a sequence of sedimentary strata, the oldest layer is located on the bottom and followed in turn by successively younger layers, on up to the top of the sequence. (p. 27, 272)

surface wave A seismic wave that travels along the Earth's surface and affects the interior to depths that are dependent upon its wavelength. (p. 94)

surf zone The zone of breaking waves. (p. 521)

surge A brief period of rapid glacial flow. (p. 459)

suspended load The fine particles carried in the mass of a stream and kept aloft by turbulence. (p. 399)

suture zone The narrow region that marks the juncture of two colliding blocks of continental crust. (p. 335)

swash The water from breaking waves that washes up onto the beach. (p. 521)

syncline A fold with a core of stratigraphically younger rocks; usually its concave side is upward. (p. 302)

talus Coarse angular rock fragments that collect at the base of the cliffs from which they were dislodged. (p. 373)

tarn A lake in a cirque. (p. 463)

tensile stress The stress generated by forces directed away from one another on opposite sides of a real or imaginary plane. (p. 298)

tephra A general term that refers to all airborne pyroclastic debris. (p. 201)

terminal moraine A thick pile of till deposited at the line of maximum glacial advance. (p. 465)

terrestrial planets The Earth and the planets similar in size and composition to the Earth: Mercury, Venus, and Mars. (p. 572)

theory A widely accepted explanation for a group of known facts. A theory is a hypothesis that has been elevated to a high level of confidence by repeated confirmation through testing and experimentation. (p. 13)

thrust fault A low-angle reverse fault that generally dips at about 15° or less. (p. 310)

tidal current The current of a rising or falling tide forced through a narrow constriction of coast. (p. 532)

tidal delta A delta deposited by tidal currents on the seaward side of a lagoon. (p. 532)

tide The rhythmic rise and fall of the sea surface caused by the unequal gravitational attraction of the Moon—and, to a lesser degree, of the Sun. (p. 529)

till Unsorted, unlayered drift deposited directly by glacial ice. (p. 465)

tombolo A strip of beach connecting the mainland to an island or islands to one another. (p. 522)

transform boundary The border between two plates that are sliding by one another horizontally without either creating or destroying oceanic crust. It connects offset ridge segments, offset trenches, or ridge segments to trenches. (p. 69)

transform fault Type of fault where the plates slide by one another horizontally without either creating or destroying oceanic crust. (p. 81)

transition zone The upper mantle region, between 400 and 700 kilometers in depth, characterized by a series of steplike increases in seismic wave velocities and the conversion of minerals to denser forms. (p. 130)

transverse dune A long ridge of sand standing at right angles to the prevailing wind. (p. 503)

trellis drainage A drainage pattern in which the main stream cuts across the regional structure (typically folded or tilted strata), and tributaries follow parallel belts of weak strata that lie perpendicular to the main stream. (p. 416)

tributary stream A stream that flows into a larger stream. (p. 394)

tsunami The Japanese word for a large, often deadly sea wave set in motion by an undersea earthquake. (p. 110)

tuff A rock of consolidated volcanic ash. (p. 201)

turbidite The deposited sediment of a turbidity current; characterized by graded bedding and light sorting. (p. 227)

turbulent flow A type of flow in which irregular, chaotic motions are superimposed on the general current direction. (p. 398)

ultimate base level The lowest possible level to which a stream can erode its channel; with rare exceptions, sea level. (p. 396)

unconformity An erosion surface bounded by rocks of markedly different age and signifying a break in the geologic record. (p. 273)

uniformitarianism The principle that is based on the concept that past geological events can be explained by forces occurring today. "The present is the key to the past." (p. 26)

universe The totality of space and time—past, present, and future. (p. 567)

upper mantle The outermost part of the mantle whose depth extends from about 20 to 670 kilometers. (p. 126)

valley glacier See *alpine glacier.*

van der Waals bond A weak bond between neighboring compounds or between atomic layers. (p. 148)

varve Two layers of sediment deposited in one year's time in a glacial lake: one layer of light sand and silt deposited in summer and one layer of dark clay deposited in winter. (p. 471)

vent A conduit through which magma rises to the surface. (p. 194)

ventifact A rock that has been polished and faceted by the action of wind-blown sand. (p. 501)

volcanic breccia A pyroclastic rock composed of angular fragments that are larger than 64 millimeters in diameter. (p. 201)

volcanic cone An accumulation of lava and/or pyroclastics around a volcanic vent. (p. 193)

volcano A vent in the surface of the Earth through which magma, gases, and rock fragments erupt; also the term for the landform that develops around the vent. (p. 194)

water table The surface marking the upper limit of the zone of saturation. (p. 424)

water wave A repetitive circular or elliptical motion in an ocean or lake, causing the water surface to rise and fall. (p. 516)

wave-cut cliff A coastal cliff formed by wave erosion. (p. 518)

wave-cut platform A more or less smooth sloping surface planed by wave erosion. (p. 518)

weathering The physical and chemical alteration of rocks exposed to the atmospheric influences on the Earth's surface. (p. 32, 221)

whole-mantle convection The theory that the entire mantle circulates and mixes in the course of bringing heat from the outer core to the Earth's surface. (p. 133)

zone of aeration The subsurface zone between the water table and the surface; retains little moisture except within the belt of soil moisture and the capillary fringe. (p. 424)

zone of fracture The rigid outer layer of a glacier where the ice fractures as a result of stress caused by glacial movement. (p. 456)

zone of plastic flow The inner mass of a glacier where the ice moves without fracturing. (p. 457)

zone of saturation A subsurface zone where groundwater accumulates and completely fills the voids in sediments and rocks. (p. 424)

Selected Readings

Introduction

Carpi, J. 1993. Research vacations. *Earth* 2, no. 3 (May): 50–53.

Hsü, K. J. 1983. *The Mediterranean was a desert: A voyage of the Glomar Challenger.* Princeton, N.J.: Princeton University Press.

Morrison, P., and P. Morrison. 1987. *The ring of truth: An inquiry into how we know what we know.* New York: Random House.

Stanley, D. J. 1990. Med desert theory is drying up. *Oceanus* 33 (Spring): 14–23.

Chapter 1

Adams, F. D. 1954. *The birth and development of the geological sciences.* New York: Dover Press (originally published in 1938).

Badash, L. 1989. The age-of-the-Earth debate. *Scientific American* 261 (August): 90–96.

Berggren, W. A., and V. A. Couvering, eds. 1984. *Catastrophes and Earth history: The new uniformitarianism.* Princeton, N.J.: Princeton University Press.

Brice, W. R. 1982. Bishop Usher, John Lightfoot and the age of creation. *Journal of Geological Education* 30:18–24.

Dietz, R. S., and J. C. Holden. 1970. The break-up of Pangaea. *Scientific American* 222 (April): 30–41.

Gould, S. J. 1965. Is uniformitarianism necessary? *American Journal of Science* 263:223–228.

Hurley, P. M. 1968. The confirmation of continental drift. *Scientific American* 218 (April): 52–64.

Marvin, U. 1973. *Continental drift: The evolution of a concept.* Washington, D.C.: Smithsonian Institution Press.

Sullivan, W. 1974. *Continents in motion.* New York: McGraw-Hill.

Wegener, A. 1966. *The origin of the continents and oceans.* New York: Dover Press (originally published in German, 1915).

Wilson, J. T. 1963. Continental drift. *Scientific American* 208 (April): 86–100.

Chapter 2

Ballard, R. D. 1983. *Exploring our living planet.* Washington, D.C.: National Geographic Society.

Bonatti, E. 1987. The rifting of continents. *Scientific American* 256 (March): 74–81.

Burke, K. C., and J. T. Wilson. 1976. Hotspots on the Earth's surface. *Scientific American* 235 (August): 46–57.

Cox, A., and R. B. Hart. 1986. *Plate tectonics: How it works.* Palo Alto: Blackwell.

Hallan, A. 1973. *A revolution in the earth sciences: From continental drift to plate tectonics.* New York: Oxford.

Moores, E., ed. 1990. *Plate tectonics: Readings from Scientific American.* New York: W. H. Freeman.

Vine, F. J. 1966. Spreading of the ocean floor: New evidence. *Science* 154:1405–1415.

Vink, G. E., et al. 1985. The Earth's hot spots. *Scientific American* 252 (April): 50–57.

Chapter 3

Allen, C. R. (chairman). 1976. *Predicting earthquakes: A scientific and technical evaluation with implications for society.* Washington, D.C.: National Academy of Sciences.

Atwater, T. 1970. Implications of plate tectonics for the Cenozoic evolution of western North America. *Geological Society of America Bulletin* 81, no. 12:3513–3535.

Bolt, B. A. 1993. *Earthquakes,* 2d ed. New York: W. H. Freeman.

Davidson, K. 1994. Predicting earthquakes. *Earth* 3, no. 3: 56–63.

Fischman, J. 1992. Falling into the gap. *Discover* (October): 57–63.

Isacks, B., J. Oliver, and L. R. Sykes. 1968. Seismology and the new global tectonics. *Journal of Geophysical Research* 73, no. 18:5855–5899.

Johnston, A. 1992. New Madrid: The rift, the river and the earthquake. *Earth* 1, no. 1:34–43.

Monastersky, R. 1993. Lessons from Landers. *Earth* 2, no. 3: 41–47.

———. 1994. Los Angeles quake: A taste of the future? *Science News* 145:53.

Soren, D. 1988. The day the world ended for Kourion: Reconstructing an ancient earthquake. *National Geographic* (July): 30–53.

USGS Staff. 1964. *Alaska's Good Friday earthquake, March 27, 1964.* U.S. Geological Survey Circular 491.

———. 1990. The Loma Prieta California earthquake: An anticipated event. *Science* 247:286–293.

Chapter 4

Anderson, D. L., et al. 1992. Plate tectonics and hot spots: The third dimension. *Science* 256 (June): 1645–1651.

Clark, S. P. 1970. *Structure of the Earth.* Englewood Cliffs, N.J.: Prentice Hall.

Dawson, J. 1993. Cat scanning the Earth. *Earth* 2 (May): 37–41.

Elsasser, W. M. 1958. The Earth as a dynamo. *Scientific American* 199 (May): 50–62.

Jeanloz, R., and T. Lay. 1993. The core-mantle boundary. *Scientific American* 268 (May): 48.

McKenzie, D. P. 1983. The Earth's mantle. *Scientific American* offprint 249 (3):50–62.

Murphy, J. B., and R. D. Nance. 1992. Mountain belts and the supercontinent cycle. *Scientific American* 266 (April): 84–91.

Ringwood, A. E. 1983. The Earth's core: Its composition, formation, and bearing on the origin of the Earth. *Proceedings of the Royal Society.* London. 395:1–46.

Vidale, J. E., and T. Lay. 1993. Phase boundaries and mantle convection. *Science* 261 (September): 1401.

Vogel, S. 1994. The big flush. *Earth* 3, no. 2:39–43.

Wyllie, P. 1976. *The way the Earth works.* New York: John Wiley and Sons.

Chapter 5

Berry, L. G., B. Mason, and R. V. Dietrich. 1983. *Mineralogy,* 2d ed. New York: W. H. Freeman.

Chesterman, C. W. 1979. *The Audubon Society field guide to North American rocks and minerals.* New York: Alfred A. Knopf.

Dana, J. D. 1985. *Manual of mineralogy,* 20th ed. (revised by C. S. Hurbut, Jr., and C. Klein). New York: John Wiley and Sons.

Ernst, W. G. 1969. *Earth materials.* Englewood Cliffs, N.J.: Prentice Hall.

Sorrell, C. A., and G. F. Sandstrom. 1973. *Rocks and minerals: A guide to field identification.* New York: Golden Press.

Chapter 6

Barker, D. S. 1983. *Igneous rocks.* Englewood Cliffs, N.J.: Prentice Hall.

Bonatti, E. 1994. The Earth's mantle below the oceans. *Scientific American* 270, no. 3:44–52.

Bowen, N. L. 1928. *The evolution of igneous rocks.* Princeton, N.J.: Princeton University Press (reprinted by Dover Press).

Holden, A., and P. Singer. 1960. *Crystals and crystal growing.* New York: Doubleday.

MacKenzie, W. S., C. H. Donaldson, and C. Guilford. 1982. *Atlas of igneous rocks and their textures.* New York: Halstead Press.

Chapter 7

Blong, R. J. 1984. *Volcanic hazards.* Sydney: Academic Press.

Coffin, M. E., and O. Eldholm. 1993. Large igneous provinces. *Scientific American* 269, no. 4 (October): 42–49.

Decker, R., and B. Decker. 1989. *Volcanoes.* New York: W. H. Freeman.

Earthquakes and volcanoes (a bimonthly publication). U.S. Geological Survey.

MacDonald, G. A. 1972. *Volcanoes.* Englewood Cliffs, N.J.: Prentice Hall.

Simkin, T., et. al. 1981. *Volcanoes of the world.* Stroudsburg, Pa.: Hutchinson Ross.

Tilling, R. I. 1985. *Eruptions of Mount Saint Helens: Past, present, and future.* Washington, D.C.: U.S. Geological Survey.

Tilling, R. I., C. Heliker, and T. L. Wright. 1987. *Eruptions of Hawaiian volcanoes.* Washington, D.C.: U.S. Geological Survey.

Williams, H., and A. R. McBirney. 1979. *Vulcanology.* San Francisco: Freeman Cooper.

Chapter 8

Blatt, H., G. V. Middleton, and R. C. Murray. 1992. *Origin of sedimentary rocks,* 3d ed. Englewood Cliffs, N.J.: Prentice Hall.

Pettijohn, F. J. 1975. *Sedimentary rocks,* 3d ed. New York: Harper and Row.

Prothero, D. R. 1990. *Interpreting the stratigraphic record.* New York: W. H. Freeman.

Chapter 9

Barth, T. F. 1978. Metamorphic rocks, metamorphism, and metasomatism. In *McGraw-Hill encyclopedia of the Earth sciences,* 5th ed., 462–475. New York: McGraw-Hill.

Best, M. G. 1982. *Igneous and metamorphic petrology.* San Francisco: W. H. Freeman.

Miyashiro, A. 1973. *Metamorphism and metamorphic belts.* New York: John Wiley and Sons.

Philpotts, A. R. 1990. *Principles of igneous and metamorphic petrology.* Englewood Cliffs, N.J.: Prentice Hall.

Chapter 10

Alvarez, W., and F. Asaro. 1990. What caused the mass extinction? An asteroid impact. *Scientific American* 262, no. 10:78–84.

Brush, S. G. 1982. Finding the age of the Earth by physics or faith?! *Journal of Geologic Education* 30:34–58.

Courtillot, V. E. 1990. What caused the mass extinction? A volcanic eruption. *Scientific American* 262, no. 10:85–92.

Dalrymple, G. B. 1991. *The age of the Earth.* Palo Alto: Stanford University Press.

Eicher, D. C. 1976. *Geologic time,* 2d ed. Englewood Cliffs, N.J.: Prentice Hall.

Gould, S. J. 1987. *Time's arrow, time's cycle: Myth and metaphor in the discovery of geological time.* Cambridge, Mass.: Harvard University Press.

Schoch, R. M. 1989. *Stratigraphy: Principles and methods.* New York: Van Nostrand Reinhold.

Chapter 11

Billings, M. P. 1973. *Structural geology,* 3d ed. Englewood Cliffs, N.J.: Prentice Hall.

Chew, B. 1993. Anatomy of a mountain range. *Earth* 2, no. 1 (January): 36–43.

Davis, G. H. 1984. *Structural geology of rocks and regions.* New York: John Wiley and Sons.

Oberlander, T. M. 1985. Origin of drainage transverse to structures in orogens. In *Tectonic geomorphology,* M. Morisawa and J. T. Hack, eds., 55–182. London: Allyn and Irwin.

Chapter 12

Cook, F. A., L. D. Brown, and J. C. Oliver. 1980. The southern Appalachians and the growth of continents. *Scientific American* 243, no. 4:156–168.

Davidow, B. 1993. The high one (Denali National Park, Alaska, how it was stitched together). *Earth* 2, no. 3:44–51.

Dewey, J. F., and J. M. Bind. 1970. Mountain belts and the new global tectonics. *Journal of Geophysical Research* 75: 2625–2647.

Hoffman, P. 1988. United plates of America, the birth of a craton: Early Proterozoic assembly and growth of Laurentia. *Annual Review of Earth and Planetary Sciences* 16:543–604.

Howell, D. C. Terranes. 1985. *Scientific American* 253, no. 5:116–126.

Jones, D. L., et al. 1982. The growth of western North America. *Scientific American* 247, no. 5:70–128.

Molnar, P., and P. Jupponier. 1977. The collision between India and Eurasia. *Scientific American* 236, no. 4:30–41.

Chapter 13

Birkeland, P. W. 1984. *Soils and geomorphology.* New York: Oxford University Press.

Brown, L. R., and E. C. Wolf. 1984. Soil erosion: Quiet crisis in the world economy. In *Worldwatch Paper* 60: Worldwatch Institute.

Goldich, S. S. 1938. A study of rock weathering. *Journal of Geology* 46:17–58.

Hunt, C. B. 1972. *Geology of soils.* San Francisco: W. H. Freeman.

Repetto, R. 1990. Deforestation of the tropics. *Scientific American* 262, no. 4:36–42.

Waters, T. 1993. Roof of the world: How the Himalayas change our climate. *Earth* 2, no.4:26–35.

Chapter 14

Kiersch, G. A. 1965. The Vaiont Reservoir disaster. In *Mineral Information Service* 18, no. 7:129–138. California Division of Mines and Geology.

Schuster, R. L., ed. Landslides: Analysis and control. In *Transportation Board Special Report* 176:11–33. Washington D.C.: National Academy of Sciences.

Small, R. J., and M. J. Clark. 1982. *Slopes and weathering.* Cambridge, U.K.: Cambridge University Press.

Voight, B., ed. 1978. *Rockslides and avalanches: Part I, natural phenomena.* New York: Elsevier.

Williams, P. J. 1979. *Pipelines and permafrost: Physical geography and development of the circumpolar north.* New York: Longman.

Chapter 15

Holloway, M. 1994. Nurturing nature. *Scientific American* 270, no. 4:98–108.

Leopold, L. B., M. G. Wolman, and J. P. Miller. 1964. *Fluvial processes in geomorphology.* San Francisco: W. H. Freeman.

Mackin, J. H. 1948. Concept of the graded river. *Bulletin of the Geological Society of America* 59:463–512.

McPhee, J. 1989. *The control of nature.* New York: Farrar Straus Giroux.

Morgan, J. P. 1970. Deltas: A resource. *Journal of Geological Education* 18, no. 3:107–117.

Morisawa, M. 1986. *Rivers: Form and process.* New York: Longman.

Chapter 16

Armstead, C. H. 1978. *Geothermal energy.* New York: John Wiley and Sons.

Environmental Protection Agency. 1990. *Citizen's guide to groundwater protection.* Washington, D.C.: Environmental Protection Agency.

Fetter, C. W. 1988. *Applied hydrology,* 2d. ed. Columbus, Ohio: Merrill.

Grossman, D., and S. Shulman. 1994. Verdict at Yucca Mountain. *Earth* 3, no. 2:54–63.

Heath, R. C. 1983. Basic groundwater hydrology. In *Water Supply Paper* 2220. Washington, D.C.: U.S. Geological Survey.

Sweeting, M. 1972. *Karst landforms.* New York: Columbia University Press.

Chapter 17

Flint, R. F. 1971. *Glacial and Pleistocene geology.* New York: John Wiley and Sons.

Hays, J. D., J. Imbrie, and N. J. Shacketon. 1976. Variations in the Earth's orbit: Pacemaker of the ice ages. *Science* 194:1121–1132.

Imbrie, J., and K. P. Imbrie. 1986. *Ice ages: Solving the mystery.* Cambridge, Mass.: Harvard University Press.

National Academy of Sciences. 1989. *Ozone depletion, greenhouse gases, and climate change.* Washington, D.C.: National Academy Press.

———. 1992. *Global environmental change.* Washington, D.C.: National Academy Press.

Chapter 18

Bagnold, R. A. 1941. *The physics of windblown sand and desert dunes.* London: Methune and Company.

Byers, H. R. 1974. *General meteorology.* New York: McGraw-Hill.

El-Ashry, M., and D. C. Gibbons, eds. 1988. *Water and arid lands of the western United States.* Washington, D.C.: World Resources Institute.

Ellis, W. 1987. Africa's stricken Sahel. *National Geographic* 172:140–179.

Kotyakou, V. M. 1991. The Aral Sea: A critical environmental zone. *Environment* 33, no. 1:4–9, 36–39.

Mabbutt, J. A. 1977. *Desert landforms.* Cambridge, Mass.: MIT University Press.

Walker, A. S. 1992. *Deserts: Geology and resources.* Washington, D.C.: U.S. Geological Survey.

Chapter 19

Bird, E. F. 1984. *Coasts: An introduction to coastal geomorphology.* Oxford, U.K.: Blackwell.

Flanagan, R. 1993. Beaches on the brink. *Earth* 2, no. 6:24–33.

Inman, D. L. 1988. Nearshore sedimentary processes. In *McGraw-Hill encyclopedia of the geological sciences,* 2d. ed., 407–414. New York: McGraw-Hill.

Shepard, F. P. 1971. *Our changing shorelines.* New York: McGraw-Hill.

Chapter 20

Bartlett, A. 1980. The forgotten fundamentals of the energy crisis. *Journal of Geological Education* 28:4–35.

Craig, J. R., D. J. Vaughn, and B. J. Skinner. 1988. *Resources of the earth.* Englewood Cliffs, N.J.: Prentice Hall.

Harben, P. 1992. Strategic minerals. *Earth* 1, no. 4:36–45.

Howell, D. G., K. T. Bird, and D. L. Gautier. 1993. Oil: Are we running out? *Earth* 2, no. 2:26–33.

Hunt, J. M. 1979. *Petroleum geochemistry and geology.* San Francisco: W. H. Freeman.

Skinner, B. J. 1986. *Earth resources,* 3d. ed. Englewood Cliffs, N.J.: Prentice Hall.

Weiner, J. 1990. *The next one hundred years: Shaping the future of our living earth.* New York: Bantam.

Chapter 21

Carr, M. H., ed. *The geology of the terrestrial planets.* Washington, D.C.: National Aeronautics and Space Administration.

Magellan at Venus. 1992. *Journal of Geophysical Research* 97, no. E8 and E10.

The planets 1975–1983. *Reprints from Scientific American.* San Francisco: W. H. Freeman.

Wood, J. 1978. *The solar system.* Englewood Cliffs, N.J.: Prentice Hall.

Acknowledgments

pp. ii–iii, Elizabeth Morales

Introduction

p. 2, N. Alexanian/Stock, Boston; **p. 4** (Fig. I.1), GEOPIC®, Earth Satellite Corporation; **p. 6** (Fig. I.4a), Elizabeth Morales; (fig. I.4b), Courtesy of Texas A & M; **p. 7** (Fig. I.5a–b), Courtesy of Professor Kenneth J. Hsü; (Fig. I.5c), BPS; **p. 8** (Fig. I.6), Elizabeth Morales; **p. 12** (Fig. I.9), Odyssey/Frerck/Chicago; (Fig. I.10), Graph from *Drifting Continents and Colliding Paradigms*, p. 46, Copyright ©1990 by Indiana University Press. Reprinted by permission of the publisher; **p. 14** (Fig. I.11a), Krafft/Explorer/Photo Researchers Inc.; (Fig. I.11b), Stephen Trimble; (Fig. I.11c), Terraphotographics/BPS; **p. 19,** Spenser Grant/Photo Researchers; **p. 21** (Fig. I.13), Liaison International; (Fig. I.14), Terraphotographics/BPS

Chapter 1

p. 24, Stephen Trimble; **p. 26** (Fig. 1.1), Landform Slides; **p. 27** (Fig. 1.2), Elizabeth Morales; **p. 28,** Gamma-Liaison; **p. 29** (Fig. 1.3), Sinclair Stammers/Science Photo Library/Photo Researchers Inc.; **p. 32** (Fig. 1.5) Elizabeth Morales; **p. 33** (Fig. 1.6a), Robert Isaacs/Photo Researchers Inc.; (Fig. 1.6b), GEOPIC®, Earth Satellite Corporation; **p. 34** (Fig. 1.7a), David R. Frazier/Photo Researchers Inc.; (Fig. 1.7b), Richard Gross; **p. 35** (Fig. 1.8a), William Ferguson; (Fig. 1.8b), Richard Gross; (Fig. 1.9a), John Shelton; (Fig. 1.9b–d), E. R. Degginger; **p. 36** (Fig. 1.10), Elizabeth Morales; **p. 38** (Fig. 1.11), Elizabeth Morales; **pp. 40–41** (Fig. 1.13), Elizabeth Morales; **p. 42** (Fig. 1.14), Elizabeth Morales; **p. 44** (Fig. 1.15), Elizabeth Morales; **p. 45** (Fig. 1.16), Illustration by Juan Barbera, *New York Times*, July 19, 1988; (Fig. 1.17), Elizabeth Morales; **p. 46** (Fig. 1.19a–b), Photri; **p. 47** (Fig. 1.21), Elizabeth Morales; **p. 48** (Fig. 1.22), Elizabeth Morales

Chapter 2

p. 52, NOAA map; **p. 54** (Fig. 2.1), Courtesy of Marie Tharp; **pp. 56–57** (Fig. 2.3), Courtesy of Marie Tharp; **p. 58** (Fig. 2.4), Courtesy of Woods Hole Oceanographic Institute; (Fig. 2.5), Peter Ryan/Science Photo Library/Photo Researchers Inc.; **p. 60** (Fig. 2.7), Elizabeth Morales; (Fig. 2.8), Elizabeth Morales; **p. 65** (Fig. 2.13), Elizabeth Morales; **p. 66** (Fig. 2A), Douglas Faulkner/Photo Researchers Inc.; (Fig. 2B), Elizabeth Morales; **p. 67** (Fig. 2.15a), Courtesy of Marie Tharp; (Fig. 2.15b), Elizabeth Morales; **p. 72** (Fig. 2.19), Elizabeth Morales; **p. 73** (Fig. 2.20a–b), Elizabeth Morales; (Fig. 2.20c), Courtesy of Marie Tharp; **p. 76** (Fig. 2.22), Elizabeth Morales; **p. 77** (Fig. 2.23), Elizabeth Morales; **p. 78** (Fig. 2.24), Elizabeth Morales; **p. 79** (Fig. 2.25), Elizabeth Morales

Chapter 3

p. 88, Haruyoshi Yamaguchi/SYGMA; **p. 91** (Fig. 3.1), USGS Photo Library; **p. 92** (Fig. 3.2a), G. K. Gilbert/USGS; (Fig. 3.2b), R. E. Wallace/USGS; (Fig. 3.2c), USGS/Science Photo Library/Photo Researchers Inc.; **p. 93** (Fig. 3.3), Elizabeth Morales; (Fig. 3.4), Elizabeth Morales; **p. 94** (Fig. 3.5), Elizabeth Morales; **p. 95** (Fig. 3.7b), Cindy Charles/Gamma-Liaison; **p. 98** (Fig. 3.12), Illustration from *Earthquakes* by Bruce Bolt, p. 105. Copyright ©1978 by W. H. Freeman and Company. Used with permission; **p. 102** (Fig. 3.15), Elizabeth Morales; **p. 104** (Fig. 3.17), Georg Gerster/COMSTOCK; **p. 109** (Fig. 3.24), Phil Green/Science Photo Library/Photo Researchers Inc.; **p. 111,** Elizabeth Morales; **p. 112** (Fig. 3.26), Wide World Photos; (Fig. 3.27), USGS; **p. 113** (Fig. 3.28), Figure from *Earthquakes* by Bruce Bolt, p. 103. Copyright © 1978 by W. H. Freeman and Company. Used with permission; **p. 116** (Fig. 3.29), Elizabeth Morales; **p. 117** (Fig. 3.30), David Kennerly/Gamma-Liaison

Chapter 4

p. 120, Courtesy of Paul Morin; **p. 123** (Fig. 4.3), E. R. Degginger; **p. 129** (Fig. 4.10), Elizabeth Morales; **p. 130** (Fig. 4.11), Landform Slides; **p. 132** (Fig. 4.13a–b), John H. Woodhouse/Oxford University

Chapter 5

p. 140, Erica and Harold van Pelt; **p. 142** (Fig. 5.1a), E. R. Degginger; (Fig. 5.1b), Breck Kent/Smithsonian; **p. 150** (Fig. 5.11), (galena) Breck Kent, (zircon) American Museum of Natural History K13794, (quartz) E. R. Degginger, (olivine) M. Claye Jacana/Photo Researchers Inc., (gypsum) Photo Researchers Inc., (kyanite) Breck Kent; **p. 152** (Fig. 5.13), E. R. Degginger; **p. 153** (Fig. 5.14), Paul Silverman/Photo Researchers Inc.; (Fig. 5.15), Breck Kent; **p. 154** (Fig. 5.16a), Breck Kent; (Fig. 5.16b), H. Rudolf Becker/Phototake; (Fig. 5.17), Breck Kent; **p. 157** (Fig. 5B), Dr. George E. Harlow. Courtesy of the Museum of Natural History

Chapter 6

p. 166, Alfred Pasieka/Science Photo Library/Photo Researchers Inc.; **p. 168** (Fig. 6.1), Elizabeth Morales; **p. 169** (Fig. 6.2), Landform Slides; (Fig. 6.3), Tom Bean; **p. 170** (Fig. 6.5), E. R. Degginger; (Fig. 6.6), E. R. Degginger; **p. 171** (Fig. 6.7a), John Shelton; (Fig. 6.7b), Bruce Iverson; (Fig. 6.8), E.R. Degginger; (Fig. 6.9), Landform Slides; **p. 172** (Fig. 6.10), E. R. Degginger; (Fig. 6.11), E. R. Degginger; **p. 173** (Fig. 6.12), E. R. Degginger; (Fig. 6.13), E. R. Degginger; **p. 174** (Fig. 6.15a–b), Landform Slides; **p. 175** (Fig. 6.16), Landform Slides; (Fig. 6.17), Alfred Pasieka/Science Photo Library/Photo Researchers Inc.; **p. 176** (Fig. 6A), John Buitenkant/Photo Researchers Inc.; **p. 179** (Fig. 6.20), BPS; **pp. 180–181** (Fig. 6.21), Elizabeth Morales; **p. 182** (Fig. 6.22), Elizabeth Morales; **p. 183** (Fig. 6.23), Peter Ryan/Scripps/Science Photo Library/Photo Researchers Inc.; **p. 185** (Fig. 6.26), William Ferguson; **p. 189** (Fig. 6.27), Landform Slides

Chapter 7

p. 192, Soames Summerhays/Photo Researchers Inc.; **p. 194** (Fig. 7.1), F. Gohier/Photo Researchers Inc.; **p. 195** (Fig. 7.2), Elizabeth Morales; **p. 198** (Fig. 7.6), William Ferguson; (Fig. 7.7), Mats Wibe Lund; **p. 199** (Fig. 7.8a–c), E. R. Degginger; **p. 200** (Fig. 7.9), Tom Till; (Fig. 7.10), John Shelton; **p. 201** (Fig. 7.11), David Weintraub/Photo Researchers Inc.; **p. 202** (Fig. 7.12a), Breck Kent; (Fig. 7.21b), E. R. Degginger; (Fig. 7.13), USGS; **p. 203** (Fig. 7.14), Elizabeth Morales; **p. 204** (Fig. 7.15), Copyright © 1986 Dynamic Graphics, Inc. Used with permission; (Fig. 7.16), John Shelton; **p. 205** (Fig. 7.17), Michael Freeman/Bruce Coleman Ltd.; (Fig. 7.18), USGS; **p. 207** (Fig. 7.21), Elizabeth Morales; **p. 208** (Fig. 7.22), E. R. Degginger; **p. 209** (Fig. 7.24), USGS; (Fig. 7.25), USGS; **p. 210** (Fig. 7.26), USGS; **p. 213** (Fig. 7.28), Courtesy of the National Oceanic and Atmospheric Administration; **p. 214** (Fig. 7.29), Pat and Tom Leesson/Photo Researchers Inc.

Chapter 8

p. 218, Tom Bean; **p. 220** (Fig. 8.1), Elizabeth Morales; **p. 224** (Fig. 8.3a), K.H. Switak/Photo Researchers Inc.; (Fig. 8.3b), Landform Slides; **p. 225** (Fig. 8.5), William Ferguson; (Fig. 8.6), Bruno Maso/Photo Researchers Inc.; **p. 226** (Fig. 8.7), Landform Slides; (Fig. 8.8), William Ferguson; **p. 227** (Fig. 8.10), Landform Slides; **p. 228** (Fig. 8.11), Tom Bean; **p. 229** (Fig. 8.12), Topham/The Image Works; (Fig. 8.13), William Ferguson; **p. 230** (Fig. 8.14), Elizabeth Morales; **p. 231** (Fig. 8.15), John Shelton; (Fig. 8.16a), Landform Slides; (Fig. 8.16b), Ray Simons/Photo Researchers Inc.;

p. 232 (Fig. 8.17), William Ferguson; **p. 233** (Fig. 8.19), W. R. Hansen/USGS; **p. 234** (Fig. 8.20), Tom Bean; (Fig. 8.21), Tom Bean; (Fig. 8.22), Stephen Trimble; **p. 235** (Fig. 8.23a), Paul Zohl/Photo Researchers Inc.; (Fig. 8.23b), Tom Bean; (Fig. 8.23c), E. R. Degginger; (Fig. 8.23d), William Ferguson; (Fig. 8.23e), Tom Bean; (Fig. 8.23f), Landform Slides; **p. 236**, Jean Gerard Sidamor/Photo Researchers; **p. 238** (Fig. 8.24), Thomas Taylor/Photo Researchers Inc.; **p. 239**, Tom Bean; **p. 241** (Fig. 8.25a–b), E. R. Degginger; **p. 243** (Fig. 8.27), Tom Pantages

Chapter 9

p. 246, Massimo Borchi/Bruce Coleman Ltd.; (inset), Vandystadt/Photo Researchers Inc.; **p. 250** (Fig. 9.3), John Shelton; **p. 254** (Fig. 9.5), E. R. Degginger; (Fig. 9.6), E. R. Degginger; (Fig. 9.7a), Breck Kent; (Fig. 9.7b), Landform Slides; **p. 255** (Fig. 9.8), William Ferguson; (Fig. 9.9), Tom Bean; **p. 256** (Fig. 9.10a–d), John Shelton; (Fig. 9.11), A. W. Ambler/Photo Researchers Inc.; **p. 257** (Fig. 9.12a), E. R. Degginger; (Fig. 9.12b), Bruce Iverson; **p. 258** (Fig. 9.13), Three photomicrographs adapted from Myron G. Best, *Igneous and Metamorphic Petrology,* p. 416. Copyright © 1982 by W. H. Freeman and Company. Reprinted with permission; **p. 259** (Fig. 9.14a–b), Landform Slides; **p. 260** (Fig. 9.16), John Shelton; (Fig. 9.17), Landform Slides; **p. 261** (Fig. 9.18a), E. R. Degginger; (Fig. 9.18b), Landform Slides; (Fig. 9.19), Stephen Trimble; **p. 262** (Fig. 9.20), John Shelton; **p. 264** (Fig. 9.22), Elizabeth Morales; **p. 265** (Fig. 9.23), Elizabeth Morales; (Fig. 9.24), Courtesy of the Woods Hole Oceanographic Institute; **p. 266** (Fig. 9.25), Elizabeth Morales

Chapter 10

p. 270, Tom Bean; **p. 273** (Fig. 10.1a), Landform Slides; (Fig. 10.1b), John Shelton; (Fig. 10.1c), USGS; **p. 274** (Fig. 10.3), John Shelton; **pp. 275–276** (Fig. 10.4), Adapted, with the author's permission, from *Geology Illustrated* by John S. Shelton, originally published by W. H. Freeman and Co. (1966). Redrawn by Elizabeth Morales; **p. 283** (Fig. 10.10), J. Amos/Photo Researchers Inc.; **p. 287** (Fig. 10.4), Elizabeth Morales; **p. 290** (Fig. 10A), Courtesy of W. C. Elliott; **p. 291** (Fig. 10B), Courtesy of Lunar and Planetary Institute; (Fig. 10C), JPL/NASA

Chapter 11

p. 296, John Shelton; **p. 300** (Fig. 11.4), Landform Slides; **p. 303** (Fig. 11.9), USGS; **p. 304** (Fig. 11.10), Landform Slides; (Fig. 11.11a), Aerofilms; **p. 305** (Fig. 11.12), Elizabeth Morales; **p. 306** (Fig. 11.13), USGS; **p. 307** (Fig. 11.14a), Landform Slides; (Fig. 11.14b), Elizabeth Morales; (Fig. 11.15a), H. R. Joesting/USGS; (Fig. 11.15b), Elizabeth Morales; **p. 310** (Fig. 11.18a), Jules Cowan/Bruce Coleman Ltd.; (Fig. 11.18b), Elizabeth Morales; **p. 312** (Fig. 11.21), Elizabeth Morales; **p. 313** (Fig. 11.23), NASA; (Fig. 11.24), Lohman/USGS; **p. 314** (Fig. 11.26), Elizabeth Morales; **p. 315** (Fig. 11.27), John Shelton; (Fig. 11.28), Elizabeth Morales; **p. 316**, John Shelton; **p. 319** (Fig. 11.29a), Breck Kent; (Fig. 11.29b), Elizabeth Morales

Chapter 12

p. 322, William Thompson; **p. 325** (Fig. 12.2), USGS; (Fig. 12.3), Terraphotographics/BPS; (Fig. 12.4), Elizabeth Morales; **p. 326** (Fig. 12.5), USGS; **p. 327** (Fig. 12.6), Elizabeth Morales; **p. 328** (Fig. 12.7), Elizabeth Morales; **p. 330** (Fig. 12.9), Elizabeth Morales; **pp. 332–333** (Fig. 12.10), Adapted from David G. Howell, "Terranes" from *Scientific American,* November 1985, pp. 122–125. Copyright © 1985 by Scientific American, Inc. All rights reserved; **p. 336** (Fig. 12.14), Elizabeth Morales; **p. 338** (Fig. 12.16), Seismic profile of the Southern Appalachians reprinted by permission of Frederick Cook, University of Calgary, Canada; **p. 339** (Fig. 12.17), Elizabeth Morales

Chapter 13

p. 344, Owen Franken/Stock, Boston; **p. 346** (Fig. 13.1), Elizabeth Morales;

(quote), *Love Is Here To Stay,* Music and lyrics by George Gershwin and Ira Gershwin, © 1938 (Renewed 1965) George Gershwin Music and Ira Gershwin Music. All Rights administered by WB Music Corp. All Rights Reserved. Used by Permission. Warner Bros. Publication, Inc., Miami, FL 33014; **p. 348** (Fig. 13.3), Landform Slides; **p. 349** (Fig. 13.4), Tom Bean; **p. 350** (Fig. 13.5), Landform Slides; (Fig. 13.6), Gregory Dimigian/Photo Researchers Inc.; **p. 354** (Fig. 13.9), Illustration from Robert A. Muller and Theodore M. Oberlander, *Physical Geography Today,* Third Edition, pp. 190–191. Copyright © 1984. Reprinted by permission of McGraw-Hill, Inc.; (Fig. 13.10a–b), Illustrations from Robert A. Muller and Theodore M. Oberlander, *Physical Geography Today,* Third Edition, pp. 441 and 445. Copyright © 1984. Reprinted by permission of McGraw-Hill, Inc.; **p. 355** (Fig. 13.11), Courtesy of Owen Beattie/University of Alberta; **p. 356** (Fig. 13.12a–b), Courtesy of Charles Adler; **p. 357** (Fig. 13.13), Tom Bean; **p. 359** (Fig. 13.15), USDA; **p. 360** (Fig. 13.16), George Whiteley/Photo Researchers Inc.; (Fig. 13.17), John Shelton; **p. 361** (Fig. 13.18), USDA; **p. 362** (Fig. 13A), Adrian Davies/Bruce Coleman Ltd.; **p. 363** (Fig. 13B), Frederica Georgia/Photo Researchers Inc.; **p. 364** (Fig. 13.19), Bill Bachman/Photo Researchers Inc.; **p. 365** (Fig. 13.21), John Shelton

Chapter 14

p. 368, John Shelton; **p. 370** (Fig. 14.1), Jules Cowan/Bruce Coleman, Ltd.; **p. 373** (Fig. 14.4), J. Serreo/Photo Researchers Inc.; (Fig. 14.5), John Shelton; **p. 375** (Fig. 14.7a), USGS; **p. 376** (Fig. 14.8a), Geological Survey of Canada; **p. 377** (Fig. 14.9a), John Shelton; **p. 378** (Fig. 14.10a), John Shelton; (Fig. 14.11a), John Shelton; **p. 380** (Fig. 14.12d), USGS; **p. 381** (Fig. 14.13), Bouvet/Hires/Duclos/Gamma-Liaison; (Fig. 14.14), USGS; **p. 382** (Fig. 14.15), Elizabeth Morales; **p. 383** (Fig. 14.16), Landform Slides; **p. 384** (Fig. 14A), USGS; **p. 385** (Fig. 14B), Elizabeth Morales; **p. 386** (Fig. 14.17), Fletcher & Baylis/Photo Researchers Inc.; **p. 388** (Fig. 14.19), Elizabeth Morales; **p. 389** (Fig. 14.20), Gary Williams/Gamma-Liaison

Chapter 15

p. 392, Jim Steinberg/Photo Researchers Inc.; **p. 394** (Fig. 15.1), Elizabeth Morales; **p. 399** (Fig. 15.9), Landform Slides; **p. 402** (Fig. 15.12a), Norman Tomalin/Bruce Coleman Ltd.; (Fig. 15.12b), Bruce Coleman Ltd.; **p. 403** (Fig. 15.14), Tom Bean; **p. 404** (Fig. 15.15), John Shelton; **p. 405** (Fig. 15.16), Elizabeth Morales; **p. 406** (Fig. 15.17), Elizabeth Morales; **p. 407** (Fig. 15.18), Jeff Christensen/Gamma-Liaison; **p. 409** (Fig. 15.20), Cameron Davidson/COMSTOCK; **p. 410** (Fig. 15.21), Earth Satellite Corporation/Science Photo Library/Photo Researchers Inc.; **p. 413**, NASA photo; **p. 415** (Fig. 15.24), Landform Slides; (Fig. 15.25), Tom Bean

Chapter 16

p. 422, Tom Till; **p. 424** (Fig. 16.1), Elizabeth Morales; **p. 426** (Fig. 16.3), Elizabeth Morales; **p. 427** (Fig. 16.4), Elizabeth Morales; **p. 428** (Fig. 16.5), Elizabeth Morales; **p. 433** (Fig. 16.11), USGS; **p. 434** (Fig. 16.12), Elizabeth Morales; **p. 436** (Fig. 16.14), Elizabeth Morales; **p. 437** (Fig. 16.15), Kevin Downey; **p. 438** (Fig. 16.16), Kevin Downey; **p. 439** (Fig. 16.17), Carolina Biological Supply/Phototake; (Fig. 16.18), AP/Wide World Photos; **p. 440** (Fig. 16.19), Landform Slides; **p. 441** (Fig. 16.20), William Ferguson; **p. 442** (Fig. 16.21), Biological Photo Service; **pp. 444–445** (Fig. 16.23), Elizabeth Morales; **p. 448** (Fig. 16.24), Elizabeth Morales

Chapter 17

p. 452, Tom Bean; **p. 454** (Fig. 17.1), NRSC Ltd./Science Photo Library/Photo Researchers Inc.; **p. 456** (Fig. 17.4), William Ferguson; **p. 457** (Fig. 17.5), Tom Bean; **p. 458** (Fig. 17.6), Elizabeth Morales; **p. 460** (Fig. 17.7), Elizabeth Morales; **p. 461** (Fig. 17.8), Tom Bean; (Fig. 17.9), Paul Hanny/Gamma-Liaison; **p. 462** (Fig. 17.10), Tom Bean; **p. 463** (Fig. 17.11), Landform Slides; **p. 464** (Fig. 17.12), Elizabeth Morales; **p. 465** (Fig. 17.13), Elizabeth Morales; **p. 466** (Fig. 17.14), Tom Till; **p. 467** (Fig. 17.15), Tom Bean; **p. 468** (Fig. 17.16), Elizabeth Morales; (Fig. 17.17), Tom Bean; **p. 469** (Fig. 17.18), Terraphotographics/BPS; **p. 470** (Fig. 17.19), Photri;

p. 471 (Fig. 17.20), John Shelton; **p. 473** (Fig. 17.22, center), John Beck/Texas A & M; (Fig. 17.22, right top and bottom), © 1994 Dee Breger; **p. 475** (Fig. 17.24), Elizabeth Morales; **p. 479** (Fig. 17.28), Lucy B. Williams; **p. 481**, Tom Stack and Associates

Chapter 18

p. 486, Jim Steinberg/Photo Researchers Inc.; **p. 490** (Fig. 18.3), Elizabeth Morales; **p. 492** (Fig. 18.5), Jim Brandenburg/Minden Pictures; **p. 493** (Fig. 18.6), Stephen Trimble; **p. 494** (Fig. 18.7), William Jahoda/Photo Researchers Inc.; **p. 496** (Fig. 18.10), Phil Degginger; **p. 497** (Fig. 18.11), Tom Till; **p. 498** (Fig. 18.12), John Shelton; **p. 500** (Fig. 18.14), John Eastcott & Yva Momatiuk/The Image Works; **p. 501** (Fig. 18.15), John Shelton; **p. 502** (Fig. 18.16), Christopher Liu/ChinaStock; **p. 503** (Fig. 18.17), E. R. Degginger; **p. 505** (Fig. 18.19), Landform Slides; **p. 508** (Fig. 18.22), Gilles Saussier/Gamma-Liaison; **p. 510** (Fig. 18.25), Reza/Gamma-Liaison

Chapter 19

p. 514, Randy Taylor/Gamma-Liaison; **p. 517** (Fig. 19.4), Elizabeth Morales; **p. 518** (Fig. 19.5), Elizabeth Morales; **p. 521** (Fig. 19.8), Elizabeth Morales; **p. 522** (Fig. 19.9), Elizabeth Morales; **p. 523** (Fig. 19.10), Elizabeth Morales; **p. 524** (Fig. 19.11a), Elizabeth Morales; (Fig. 19.11b), M. R. Warren/Photo Researchers Inc.; **p. 525** (Fig. 19.12), John Shelton; **pp. 526–527** (Figs. 19A–G), Courtesy of Robert Thieler and David Bush; **p. 528** (Fig. 19.13), Phil Degginger; **p. 529** (Fig. 19.14), Franklin Viola/COMSTOCK; **p. 532** (Fig. 19.16), Earth Satellite Corporation; **p. 533** (Fig. 19.17), Earth Satellite Corporation; **p. 535** (Fig. 19.18b), Elizabeth Morales; **p. 536** (Fig. 19.19), John Shelton

Chapter 20

p. 542, Dale O'Dell/The Stock Market; **p. 543** (Fig. 20.2), Kenneth Murray/Photo Researchers Inc.; **p. 548** (Fig. 20.4), Elizabeth Morales; (Fig. 20.5), Institute of Oceanographic Sciences/NERC/Photo Researchers Inc.; **p. 549** (Fig. 20.6), JB Pictures; **p. 551** (Fig. 20.8), William Ferguson; (Fig. 20.9), William Ferguson; **p. 552** (Fig. 20.11), David Ball/The Stock Market; **p. 554** (Fig. 20.13), Elizabeth Morales; **p. 559** (Fig. 20.18), Adapted from David G. Howell, Kenneth J. Bird, and Donald L. Gautier, "Oil: Are We Running Out?" *Earth*, March 1993, pp. 26–33. Reprinted with permission; **p. 560** (Fig. 20.19), Adapted from David G. Howell, Kenneth J. Bird, and Donald L. Gautier, "Oil: Are We Running Out?" *Earth*, March 1993, pp. 26–33. Reprinted with permission; **p. 561** (Fig. 20.20), Adapted from David G. Howell, Kenneth J. Bird, and Donald L. Gautier, "Oil: Are We Running Out?" *Earth*, March 1993, pp. 26–33. Reprinted with permission; **p. 563** (Fig. 20.22), BPS

Chapter 21

p. 566, Courtesy of NASA; **p. 568** (Fig. 21.1), Courtesy of NASA; **p. 569** (Fig. 21.2), Courtesy of Palomar Observatory/CalTech; **p. 573** (Fig. 21.4), Courtesy of NASA; **p. 576** (Fig. 21.6), Elizabeth Morales; **p. 578** (Fig. 21.8), Courtesy of NASA; (Fig. 21.9), Elizabeth Morales; **p. 579** (Fig. 21.10), Elizabeth Morales; **p. 580**, Richard Grieve, Geological Survey of Canada in Ottawa, as seen in *Earth* magazine, August 1996, p. 29; **p. 582** (Fig. 21.11b–d), Paintings by D. Davis and D. Wilhelm/USGS; **p. 583** (Fig. 21.13), Courtesy of NASA; **p. 584** (Fig. 21.14), Courtesy of NASA; **p. 585** (Fig. 21.15), Courtesy of NASA; **p. 586** (Fig. 21.16), Courtesy of NASA; **p. 587** (Fig. 21.17), Courtesy of NASA; (Fig. 21.18), Elizabeth Morales; **p. 588** (Fig. 21.19), Courtesy of NASA; (Fig. 21.20), Courtesy of NASA

Inside back cover

Adapted from David G. Howell, "Terranes," from *Scientific American*, November 1985, pp. 122–125. Copyright © by Scientific American, Inc. All rights reserved.

Index

Chasmata, *587*
Chemical bonds, 147
Chemical limestones, 230–231, *231*
Chemical sedimentary rocks, 222, 227–232, *229–232*
Chemical sediments, 220–221
Chemical weathering, 221, 347, 350–352, *351*
  and dissolved loads, 399
  and headland erosion, 530
Chemically active fluids, metamorphism from, 249–250
Chert, 232, *232*
Chesapeake Bay, *532*, 533, 537
Chief Mountain, 314, *319*
Chilled rock zones, 176–177, *177*
Chimneys, 181–182, 495, *496*
Chlorofluorocarbons (CFCs), 479
Chondrites, 574
Chromium, *549*
Cinder cones, 204–205, *204–205*
Cinders, 201
Circulation loop, nearshore, 521, *521*
Cirques in glaciers, *458*, 462–463, *464*
Civilizations, decline of, 236–237
Classification
  of igneous rocks, 171–174
  of metamorphic rocks, 253–255, *254–256*
  of minerals, 157–158
  of sedimentary rocks, *222–223*, 222–232
Clastic texture, 223
Clays
  in deep-ocean sediments, 240–241
  in detrital rocks, 224
  in hydrolysis, 352
  porosity and permeability of, 426–427
  water effects on, 372
Cleavage
  of minerals, 150–151, *151*, 160
  slaty, 253, 260, *260*
Cliff-dwelling Indians, 348, *349*
Cliffs, wave-cut, 518
Clifton, Arizona, copper mining at, *552*
Climate, 472–473
  atmospheric variation in, 477–478, *478*
  Earth orbit effects on, 475–477, *476–477*
  future changes in, 478–482
  plate tectonic effects on, 473–474, *474–475*
  and soil formation, 358–361
  volcanic activity effects on, 212–213, *213*
  and weathering, 353–356, *354–357*
Clouds on Venus, 586
Coal, 232
  energy percentage derived from, 554
  environmental effects of, 561
  formation of, 554–555, *554–555*
  reliance on, *561*
Coarse detrital rocks, 224–227, *224–227*
Coarse sediments, 223
Coasts. *See also* Beaches
  nearshore circulation loop of, 521, *521*
  organic activity at, 528–529, *529*

and sea-level changes, 536–539, *537–538*
  tectonic settings for, *533*, 533–536, *535–536*
  tides at, 529–532, *530–532*
  wind-driven waves at, 515–519, *516–519*
Cocos plate, 78, 103, *103*
Cold accretion hypothesis, 136–137
Color
  of metamorphic rocks, 254
  of minerals, 152–153, *153*
  of sedimentary rocks, 239
Colorado Plateau, *274*, 275, 306, 495–497, *495–497*
Colorado River
  suspended load of, 399, 400
  water diverted from, 506
Columnar joints, 199, *200*
Comets, 572
Compaction
  in glacier formation, 455
  in lithification, 221–222
Competence of streams, 400
Composite cones, 206
Compounds, 147
Compressive stresses, 92, *93*, 249, *249*, 298, *298*, 309
Computer models, 15–17
Concave stream profiles, 396, *396*
Conchoidal fracture, 152
Concordant plutons, 168
Concretions, 238
Condensation hypothesis of solar system, 573–574, *576*
Cones of depression in wells, *430*, 430–431
Confined aquifers, 427, *427*, 431–435, *432, 435*, 505–506
Confining beds, 427
Conglomerates, 223, 225
Constancy of interfacial angles in minerals, 149
Contact metamorphic rocks, 256–258, *258–259*
Contact metamorphism, 250, 550, *550*
Contact metasomatism, 550
Continent-continent convergence, 78–81, *79–80*
Continental crust, 37, 323, *324*
  cratons in, 324, *324*
  growth and evolution of, 340–341
  minerals in, 156–157
  mountain belts in, 325–326, *326*
Continental drift, 39–41, *40–41*
  paleomagnetic evidence of, 45–48, *46–49*
  physical evidence of, 41–45, *42, 44–45*
  and plate tectonics, 48–49. *See also* Plate tectonics
  scientific papers on, *12*, 13
Continental glaciers, 456
Continental margins, 326–328, *327–328*
  and coasts, 536
  sedimentary environments in, 238–240
Continental rise, 75

Continental shelf, 75, *455*
Continental slope, 75
Continuous reaction series, 175
Continuous seismic profilers, 5–7, *6*
Convection currents, 83, *83*
Convection models
  of core, *135*, 135–136
  of mantle, 132–134, *133*
Convergent boundaries, 69. *See also* Subduction zones
  earthquakes at, 101
  processes of, *72*, 75–81, *77–80*, 84
Coquinas, 228, *229*
Coral reefs, 229–230, *230–231*
  and beaches, 528–529
  and global warming, 479
Core, 37–38, 130
  convection models of, *135*, 135–136
  of craters, 317
  detection of, 124–126, *126*
  of planets, 575, *577*
  of Sun, 571, *571*
  temperatures in, 131
Core samples, 6, 7, 453
Coriolis effect, 490–491
Correlation of rock units, 277–283, *278–279*
  and faunal succession, 279–280
  index fossils and overlapping ranges in, 280–281, *281*
Cortés, Hernando, 236
Cost versus benefit, 114
Covalent bonds, 147–148, *148*
Cox, Alan, 61
Cracking process, 556, *556*
Crandell, Dwight R., 211
Crater Lake, 193–194, *194*
Crater rim, 317
Cratering, 575, *577*
Craters
  on Mars, 584
  meteorite, *316–317*, 316–318
  on Moon, *578*, 578–579
  for relative dating, 579
  on Venus, 587
  volcanic, 195, 212
Cratons, 324, *324*
Creep, 368, 382–384, *383*
Crests of waves, 515–516, *516*
Cretaceous period and K-T boundary, 290–291, *290–291*
Crevasses in glaciers, 457, *458*
Cross-bedding, 233–234, *234*
Cross-cutting relationships, *274*, 274–275, 288, *288*
Crust, 37. *See also* Continental crust; Oceanic crust
  detection of, 126, *126*
  lithosphere in, 69
  minerals in, 155–158
  of terrestrial planets, *577*
Crustal uplift and climate, 474, *475*